Compendium of Plant Genomes

Series Editor

Chittaranjan Kole, President, International Climate Resilient Crop Genomics Consortium (ICRCGC), President, International Phytomedomics & Nutriomics Consortium (IPNC) and President, Genome India International (GII), New Delhi, India

Whole-genome sequencing is at the cutting edge of life sciences in the new millennium. Since the first genome sequencing of the model plant *Arabidopsis thaliana* in 2000, whole genomes of about 100 plant species have been sequenced and genome sequences of several other plants are in the pipeline. Research publications on these genome initiatives are scattered on dedicated web sites and in journals with all too brief descriptions. The individual volumes elucidate the background history of the national and international genome initiatives; public and private partners involved; strategies and genomic resources and tools utilized; enumeration on the sequences and their assembly; repetitive sequences; gene annotation and genome duplication. In addition, synteny with other sequences, comparison of gene families and most importantly potential of the genome sequence information for gene pool characterization and genetic improvement of crop plants are described.

More information about this series at https://link.springer.com/bookseries/11805

Chittaranjan Kole · Trilochan Mohapatra
Editors

The *Brassica juncea* Genome

Springer

Editors
Chittaranjan Kole
President, International Climate Resilient Crop Genomics Consortium (ICRCGC)
President, International Phytomedomics & Nutriomics Consortium (IPNC)
and President, Genome India International (GII)
New Delhi, India

Trilochan Mohapatra
Secretary, Department of Agricultural Research and Education (DARE), Government of India; and Director General, Indian Council of Agricultural Research (ICAR)
New Delhi, India

ISSN 2199-4781 ISSN 2199-479X (electronic)
Compendium of Plant Genomes
ISBN 978-3-030-91506-3 ISBN 978-3-030-91507-0 (eBook)
https://doi.org/10.1007/978-3-030-91507-0

This Springer imprint is published by the registered company Springer Nature Switzerland AG
The registered company address is: Gewerbestrasse 11, 6330 Cham, Switzerland

This book series is dedicated to my wife Phullara and our children Sourav and Devleena

Chittaranjan Kole

Preface to the Series

Genome sequencing has emerged as the leading discipline in the plant sciences coinciding with the start of the new century. For much of the twentieth century, plant geneticists were only successful in delineating putative chromosomal location, function, and changes in genes indirectly through the use of a number of "markers" physically linked to them. These included visible or morphological, cytological, protein, and molecular or DNA markers. Among them, the first DNA marker, the RFLPs, introduced a revolutionary change in plant genetics and breeding in the mid-1980s, mainly because of their infinite number and thus potential to cover maximum chromosomal regions, phenotypic neutrality, absence of epistasis, and codominant nature. An array of other hybridization-based markers, PCR-based markers, and markers based on both facilitated construction of genetic linkage maps, mapping of genes controlling simply inherited traits, and even gene clusters (QTLs) controlling polygenic traits in a large number of model and crop plants. During this period, a number of new mapping populations beyond F_2 were utilized and a number of computer programs were developed for map construction, mapping of genes, and for mapping of polygenic clusters or QTLs. Molecular markers were also used in the studies of evolution and phylogenetic relationship, genetic diversity, DNA fingerprinting, and map-based cloning. Markers tightly linked to the genes were used in crop improvement employing the so-called marker-assisted selection. These strategies of molecular genetic mapping and molecular breeding made a spectacular impact during the last one and a half decades of the twentieth century. But still they remained "indirect" approaches for elucidation and utilization of plant genomes since much of the chromosomes remained unknown and the complete chemical depiction of them was yet to be unraveled.

Physical mapping of genomes was the obvious consequence that facilitated the development of the "genomic resources" including BAC and YAC libraries to develop physical maps in some plant genomes. Subsequently, integrated genetic–physical maps were also developed in many plants. This led to the concept of structural genomics. Later on, emphasis was laid on EST and transcriptome analysis to decipher the function of the active gene sequences leading to another concept defined as functional genomics. The advent of techniques of bacteriophage gene and DNA sequencing in the 1970s was extended to facilitate sequencing of these genomic resources in the last decade of the twentieth century.

As expected, sequencing of chromosomal regions would have led to too much data to store, characterize, and utilize with the-then available computer software could handle. But the development of information technology made the life of biologists easier by leading to a swift and sweet marriage of biology and informatics, and a new subject was born—bioinformatics.

Thus, the evolution of the concepts, strategies, and tools of sequencing and bioinformatics reinforced the subject of genomics—structural and functional. Today, genome sequencing has traveled much beyond biology and involves biophysics, biochemistry, and bioinformatics!

Thanks to the efforts of both public and private agencies, genome sequencing strategies are evolving very fast, leading to cheaper, quicker, and automated techniques right from clone-by-clone and whole-genome shotgun approaches to a succession of second-generation sequencing methods. The development of software of different generations facilitated this genome sequencing. At the same time, newer concepts and strategies were emerging to handle sequencing of the complex genomes, particularly the polyploids.

It became a reality to chemically—and so directly—define plant genomes, popularly called whole-genome sequencing or simply genome sequencing.

The history of plant genome sequencing will always cite the sequencing of the genome of the model plant Arabidopsis thaliana in 2000 that was followed by sequencing the genome of the crop and model plant rice in 2002. Since then, the number of sequenced genomes of higher plants has been increasing exponentially, mainly due to the development of cheaper and quicker genomic techniques and, most importantly, the development of collaborative platforms such as national and international consortia involving partners from public and/or private agencies.

As I write this preface for the first volume of the new series "Compendium of Plant Genomes," a net search tells me that complete or nearly complete whole-genome sequencing of 45 crop plants, eight crop and model plants, eight model plants, 15 crop progenitors and relatives, and three basal plants is accomplished, the majority of which are in the public domain. This means that we nowadays know many of our model and crop plants chemically, i.e., directly, and we may depict them and utilize them precisely better than ever. Genome sequencing has covered all groups of crop plants. Hence, information on the precise depiction of plant genomes and the scope of their utilization are growing rapidly every day. However, the information is scattered in research articles and review papers in journals and dedicated Web pages of the consortia and databases. There is no compilation of plant genomes and the opportunity of using the information in sequence-assisted breeding or further genomic studies. This is the underlying rationale for starting this book series, with each volume dedicated to a particular plant.

Plant genome science has emerged as an important subject in academia, and the present compendium of plant genomes will be highly useful to both students and teaching faculties. Most importantly, research scientists involved in genomics research will have access to systematic deliberations on the plant genomes of their interest. Elucidation of plant genomes is of interest not only for the geneticists and breeders, but also for practitioners of an array of plant science disciplines, such as taxonomy, evolution, cytology, physiology, pathology, entomology, nematology, crop production, biochemistry, and obviously bioinformatics. It must be mentioned that information regarding each plant genome is ever-growing. The contents of the volumes of this compendium are, therefore, focusing on the basic aspects of the genomes and their utility. They include information on the academic and/or economic importance of the plants, description of their genomes from a molecular genetic and cytogenetic point of view, and the genomic resources developed. Detailed deliberations focus on the background history of the national and international genome initiatives, public and private partners involved, strategies and genomic resources and tools utilized, enumeration on the sequences and their assembly, repetitive sequences, gene annotation, and genome duplication. In addition, synteny with other sequences, comparison of gene families, and, most importantly, the potential of the genome sequence information for gene pool characterization through genotyping by sequencing (GBS) and genetic improvement of crop plants have been described. As expected, there is a lot of variation of these topics in the volumes based on the information available on the crop, model, or reference plants.

I must confess that as the series editor, it has been a daunting task for me to work on such a huge and broad knowledge base that spans so many diverse plant species. However, pioneering scientists with lifetime experience and expertise on the particular crops did excellent jobs editing the respective volumes. I myself have been a small science worker on plant genomes since the mid-1980s and that provided me the opportunity to personally know several stalwarts of plant genomics from all over the globe. Most, if not all, of the volume editors are my longtime friends and colleagues. It has been highly comfortable and enriching for me to work with them on this book series. To be honest, while working on this series I have been and will remain a student first, a science worker second, and a series editor last. And I must express my gratitude to the volume editors and the chapter authors for providing me the opportunity to work with them on this compendium.

I also wish to mention here my thanks and gratitude to the Springer staff, particularly Dr. Christina Eckey and Dr. Jutta Lindenborn for the earlier set of volumes and presently Ing. Zuzana Bernhart for all their timely help and support.

I always had to set aside additional hours to edit books beside my professional and personal commitments—hours I could and should have given to my wife, Phullara, and our kids, Sourav and Devleena. I must mention that they not only allowed me the freedom to take away those hours from them but also offered their support in the editing job itself. I am really not sure whether my dedication of this compendium to them will suffice to do justice to their sacrifices for the interest of science and the science community.

New Delhi, India Chittaranjan Kole

Preface

Brassica juncea, popularly known as Indian mustard, Chinese mustard, brown mustard, and oriental mustard, is a multipurpose crop primarily used as oilseed and condiment and also as a minor vegetable. Besides oil, the oil cake remaining after oil extraction is used as animal feed. However, *Brassica juncea* is predominantly cultivated as an oilseed crop in the Asian countries including India, China, Vietnam, Pakistan, Bangladesh, and Nepal. It is also cultivated in some pockets in USA, Canada, Japan, and Australia. It is also gaining importance nowadays because of its rich nutraceutical profile. It is a unique oilseed crop that adapts well to dryland cultivation and can grow under low moisture regimes and is therefore becoming a preferred crop even in the non-traditional regions.

All the species of the Brassicaceae family, specifically the Brassica species of the U's triangle, have been immensely benefited from the pioneering research outcomes of their allied species *Arabidopsis thaliana*—the leading model plant. This is especially true for the fields of molecular genetics and genomics research. *Brassica juncea* always lagged behind, in this respect, its close counterpart Brassica crop species including *Brassica napus*, *Brassica rapa*, and *Brassica oleracea* because of the complexity as a tetraploid crop and lack of enough genetic and genomic resources. However, stupendous researches have been accomplished in this crop species in all the fields of genetics, genomics and breeding. Sequencing of the genome of *Arabidopsis thaliana* in 2000 flagged off the beginning of genome sequencing in model and crop plants starting with rice in 2002. Along with the genome sequences of other crops, the sequence of a vegetable type *B. juncea* (2016) became available following its allied *Brassica* crop species, *B. rapa* (2011), *B. oleracea* (2014), and *B. napus* (2014) although we had to wait until 2021 to have reference-standard genome sequences of the oilseed *Brassica juncea*, brown and yellow mustard. Later on, re-sequencing of 480 global accessions shed light on the origin and domestication of *Brassica juncea*. Thus, the molecular and genomics tools have facilitated a comprehensive elucidation of the *B. juncea* genome and also made the publication of this stand-alone book highly relevant.

This book entitled, "The *Brassica juncea* Genome," includes 25 chapters. The first five chapters are dedicated to the basic aspects of the crop plant including its economic importance; botanical descriptions; enumeration of the nutritional and nutraceutical compounds present in its oil and oil cake; available germplasm; and assessment of genetic diversity and population structure analysis. The following four chapters present the conventional concepts and techniques of genetics and genetic improvement such as classical genetics and traditional breeding; fundamentals and application of heteroploidy; introgression breeding; and in vitro culture for micropropagation and other somatic techniques for genetic studies and crop improvement. The next two chapters delineate the strategies and achievements of genetic engineering and molecular linkage mapping in this crop species. Two chapters are devoted to comprehensive discussions on whole-genome sequencing in *Brassica juncea* including background history, different strategies and tools of sequencing employed, enumeration of the sequences and applications of the outcome in basic genetics, and breeding for different agronomic traits including biotic and abiotic stresses. These are followed by a chapter on the application of re-sequencing for elucidation of origin and diversity; large-scale genome analysis for a physical depiction of the genome and its chromosomal reorganization and repetitive DNA and their implications; and the chloroplast genome sequence. The next three chapters contain reviews on functional genomics research including transcriptomics, metabolomics, and proteomics and their impact on highlighting the roles of genes and their products. These are followed by a chapter on evolutionary genomics which represent the comparative genomics exploited to understand the evolutionary history and detection of conserved genomic regions in *B. juncea* and a chapter on the role of regulatory genes in the development and adaptation of the crop. The next three chapters deal with the advanced molecular and genomic strategies and tools in crop improvement specifically focusing on yield, oil and oil cake quality, and resistance to biotic and abiotic stresses. The last chapter deliberates on the potential and prospects of gene editing in the Brassica crops.

These chapters are contributed by 105 eminent scientists from eight countries including Canada, China, Germany, India, Italy, Malaysia, UK and USA. We remain thankful for their scientific contributions and sincere cooperation.

New Delhi, India

Chittaranjan Kole
Trilochan Mohapatra

Contents

Contributors

Sagar S. Arya The Energy and Resources Institute, TERI-Deakin Nano Biotechnology Centre, Gurugram, Haryana, India

Manju Bala Food Grains and Oilseed Processing Division, ICAR-Central Institute of Post-Harvest Engineering and Technology, Ludhiana, Punjab, India

Raju Balaji Department of Genetic Engineering, SRM Institute of Science and Technology, Chennai, India

S. S. Banga Oilseeds Section, Department of Plant Breeding and Genetics, Punjab Agricultural University, Ludhiana, Punjab, India

Surinder S. Banga Department of Plant Breeding and Genetics, Punjab Agricultural University, Ludhiana, Punjab, India

Helen Behn Plant Breeding Department, University of Bonn, Bonn, Germany

P. Bharadwaj Molecular Genetics Laboratory, Department of Botany, Central University of Punjab, VPO Ghudda, Bathinda, Punjab, India

Naveen C. Bisht National Institute of Plant Genome Research, New Delhi, India

Manosh Kumar Biswas Department of Genetics and Genome Biology, University of Leicester, Leicester, UK

Ferdinando Branca Department of Agriculture, Food and Environment (Di3A), University of Catania, Catania, Italy

Prerna Chaudhary School of Life Sciences, Jawaharlal Nehru University, New Delhi, India

Hao Chen College of Agronomy, Hunan Agricultural University, Changsha, China

Sandip Das Department of Botany, University of Delhi, New Delhi, India

Renu Deswal Department of Botany, University of Delhi, New Delhi, India

Shevin Rizal Feroz Department of Biological Sciences and Biotechnology, Faculty of Science and Technology, Universiti Kebangsaan Malaysia, Bangi, Selangor, Malaysia

Mehak Gupta Department of Plant Breeding and Genetics, Punjab Agricultural University, Ludhiana, Punjab, India

N. C. Gupta ICAR National Institute of Plant Biotechnology, New Delhi, India

S. Hamsa Department of Genetics, University of Delhi, New Delhi, India

J. S. Pat Heslop-Harrison Department of Genetics and Genome Biology, University of Leicester, Leicester, UK

Lei Kang College of Agronomy, Hunan Agricultural University, Changsha, China

Anamika Kashyap ICAR National Institute of Plant Biotechnology, New Delhi, India

Gurpreet Kaur Oilseeds Section, Department of Plant Breeding and Genetics, Punjab Agricultural University, Ludhiana, Punjab, India

Jagreet Kaur Department of Genetics, University of Delhi, New Delhi, India

Jasmeet Kaur Oilseeds Section, Department of Plant Breeding and Genetics, Punjab Agricultural University, Ludhiana, Punjab, India

Kawalpreet Kaur Department of Agricultural, Food and Nutritional Science, University of Alberta, Edmonton, AB, Canada

Rimaljeet Kaur Department of Biochemistry, College of Basic Sciences & Humanities, Punjab Agricultural University, Ludhiana, India

Snehdeep Kaur Department of Plant Breeding and Genetics, Punjab Agricultural University, Ludhiana, Punjab, India

Arun Vincent Kisku School of Life Sciences, Jawaharlal Nehru University, New Delhi, India

Arun Kumar ICAR-Directorate of Rapeseed-Mustard Research, Bharatpur, Rajasthan, India

Deepak Kumar Department of Botany, Institute of Science, Banaras Hindu University, Varanasi, India

Vinod Kumar ICAR-Directorate of Rapeseed-Mustard Research, Bharatpur, Rajasthan, India

Divyani D. Kumari The Energy and Resources Institute, TERI-Deakin Nano Biotechnology Centre, Gurugram, Haryana, India

Sapna Langyan ICAR-National Bureau of Plant Genetic Resources, New Delhi, India

Sangram K. Lenka The Energy and Resources Institute, TERI-Deakin Nano Biotechnology Centre, Gurugram, Haryana, India

Zhongsong Liu College of Agronomy, Hunan Agricultural University, Changsha, China

Bhanu Malhotra National Institute of Plant Genome Research, New Delhi, India

Annaliese S. Mason Plant Breeding Department, University of Bonn, Bonn, Germany

H. S. Meena ICAR-Directorate of Rapeseed-Mustard Research, Bharatpur, India

Purushothaman Natarajan Department of Biology, West Virginia State University, Institute, WV, USA;
Department of Genetic Engineering, SRM Institute of Science and Technology, Chennai, India

Nishakant Pandey School of Life Sciences, Jawaharlal Nehru University, New Delhi, India

Pratibha Pant Department of Genetics, University of Delhi, New Delhi, India

Kumar Paritosh Centre for Genetic Manipulation of Crop Plants, University of Delhi South Campus, New Delhi, India

Deepak Pental Centre for Genetic Manipulation of Crop Plants, University of Delhi South Campus, New Delhi, India

Sampath Perumal Global Institute for Food Security (GIFS), University of Saskatchewan, Saskatoon, SK, Canada

Paomipem Phazang School of Life Sciences, Jawaharlal Nehru University, New Delhi, India

Neelam Prabha Negi University Institute of Biotechnology, Chandigarh University, Mohali, India

Sudheesh K. Prabhudas Nutrition, Genetics and Biotechnology Division, ICAR—Central Institute of Brackishwater Aquaculture, Chennai, India

Subhash Chandra Prasad School of Life Sciences, Jawaharlal Nehru University, New Delhi, India

Priyamedha ICAR-Directorate of Rapeseed-Mustard Research, Bharatpur, Rajasthan, India

Lunwen Qian College of Agronomy, Hunan Agricultural University, Changsha, China

A. N. Rai Ex-Head, NABTD, Nuclear Agriculture & Biotechnology Division, Bhabha Atomic Research Centre, Mumbai, India

P. K. Rai ICAR-Directorate of Rapeseed-Mustard Research, Bharatpur, Rajasthan, India

Prajjwal Rai Institute of Agricultural Sciences, Banaras Hindu University, Varanasi, Uttar Pradesh, India

Ravi Rajwanshi Discipline of Life Sciences, School of Sciences, Indira Gandhi National Open University, New Delhi, India

Nirala Ramchiary School of Life Sciences, Jawaharlal Nehru University, New Delhi, India

Thakku R. Ramkumar Department of Microbiology and Cell Sciences, IFAS, University of Florida, Gainesville, FL, USA

Kusum Rana Department of Biotechnology, Punjab University, Chandigarh, India

M. Rao ICAR National Institute of Plant Biotechnology, New Delhi, India

Indu Rialch Department of Plant Breeding and Genetics, Punjab Agricultural University, Ludhiana, India

Giulio Flavio Rizzo Department of Agriculture, Food and Environment (Di3A), University of Catania, Catania, Italy

Suchandra Deb Roy School of Life Sciences, Jawaharlal Nehru University, New Delhi, India

Anjana Rustagi Department of Botany, Gargi College, New Delhi, India

Navinder Saini Division of Genetics, Indian Agricultural Research Institue, New Delhi, India

Neera Bhalla Sarin School of Life Sciences, Jawaharlal Nehru University, New Delhi, India

Mukesh Saxena School of Life Sciences, Jawaharlal Nehru University, New Delhi, India

Trude Schwarzacher Department of Genetics and Genome Biology, University of Leicester, Leicester, UK

Ankita Sehrawat Maitreyi College, University of Delhi, New Delhi, India

Anju Sharma Department of Biochemistry, College of Basic Sciences & Humanities, Punjab Agricultural University, Ludhiana, India

Anubhuti Sharma ICAR-Directorate of Rapeseed-Mustard Research, Bharatpur, Rajasthan, India

Hariom Kumar Sharma ICAR-Directorate of Rapeseed-Mustard Research, Bharatpur, India

Pankaj Sharma ICAR-Directorate of Rapeseed-Mustard Research, Bharatpur, India

Sanjula Sharma Oilseeds Section, Department of Plant Breeding and Genetics, Punjab Agricultural University, Ludhiana, Punjab, India

Shashi Shekhar School of Life Sciences, Jawaharlal Nehru University, New Delhi, India

Anandita Singh Department of Biotechnology, Institutional Area, TERI School of Advanced Studies, Vasant Kunj, New Delhi, India

K. H. Singh ICAR-Directorate of Rapeseed-Mustard Research, Bharapur, Rajasthan, India

V. V. Singh ICAR-Directorate of Rapeseed-Mustard Research, Bharatpur, Rajasthan, India

Simarjeet Kaur Sra Department of Plant Breeding and Genetics, Punjab Agricultural University, Ludhiana, Punjab, India

P. Suprasanna Ex-Head, NABTD, Nuclear Agriculture & Biotechnology Division, Bhabha Atomic Research Centre, Mumbai, India

Saad Tayyab Biomolecular Research Group, Biochemistry Programme, Institute of Biological Sciences, Faculty of Science, Universiti Malaya, Kuala Lumpur, Malaysia

Sapna Thakur Molecular Genetics Laboratory, Department of Botany, Central University of Punjab, VPO Ghudda, Bathinda, Punjab, India

Simone Treccarichi Department of Agriculture, Food and Environment (Di3A), University of Catania, Catania, Italy

Mohan Varghese National Institute of Plant Genome Research, New Delhi, India

Sujata Vasudev Division of Genetics, Indian Agricultural Research Institue, New Delhi, India

Harinder Vishwakarma Division of Germplasm Evaluation, ICAR-National Bureau of Plant Genetic Resources, New Delhi, India

Prashant Yadav ICAR-Directorate of Rapeseed-Mustard Research, Bharatpur, Rajasthan, India

D. K. Yadava ADG (Seeds), Indian Council of Agricultural Research, New Delhi, India

Pranjal Yadava Division of Plant Physiology, Indian Agricultural Research Institute, New Delhi, India

Satish K. Yadava Centre for Genetic Manipulation of Crop Plants, University of Delhi South Campus, New Delhi, India

Rashmi Yadav Division of Germplasm Evaluation, ICAR-National Bureau of Plant Genetic Resources, New Delhi, India

Sangita Yadav Division of Seed Science and Technology, Indian Agricultural Research Institue, New Delhi, India

Liu Yang College of Agronomy, Hunan Agricultural University, Changsha, China

Yashpal Division of Genetics, Indian Agricultural Research Institue, New Delhi, India

Mohd Aslam Yusuf School of Life Sciences, Jawaharlal Nehru University, New Delhi, India

Abbreviations

2,4-D	2,4-Dichlorophenoxyacetic acid
2D DIGE	Two-dimensional difference gel electrophoresis
2-DE	Two-dimensional gel electrophoresis
3T3-L1	Mouse 3T3 cell
4MI3G	4-Hydroxyindol-3-ylmethyl glucosinolate
6-BA	6-Benzyladenine
6-BAP	6-Benzylaminopurine
A-ARR	Type A Arabidopsis response regulator
ABA	Abscisic acid
ABC	ATP-binding cassette
ABC	ATP-binding cassette transporter
Ac2(t)	*Albugo candida resistance locus against race 2*
Ac2V1	*Albugo candida resistance locus against race 2V1*
ACA13	*Autoinhibited Ca2+-ATPases 13*
AcB1	*Albugo candida race Bharatpur 1*
AcB1-A4.1	*Albugo candida race Bharatpur 1 resistance locus LGA04*
AcB1-A5.1	*Albugo candida race Bharatpur 1 resistance locus LGA05*
ACCase	Acetyl-CoA carboxylase
ACP	Acyl carrier protein
Acr/Ac2a1	*Albugo candida resistance/race 2a1*
ACS	Aminocyclopropane-1-carboxylate synthase
ADS1	Acyl-CoA-Δ9 desaturases/acyl-lipid-Δ9 desaturase
AFB	Auxin signaling F box protein
AFDD	Allele frequency diversity divergence
AFLP	Amplified fragment length polymorphism
AGL24	*AGAMOUS LIKE 24*
AICRP-RM	All India Coordinated Research Project on Rapeseed Mustard
AIL	*AINTEGUMENTA-Like*
AITC	Allylisothiocyanate
Ala	Alanine
ALaT	*ALANINE AMINOTRANSFERASE*
AM	Association mapping
AMOVA	Analysis of molecular variance
AMT	Ammonium transporter
ANR/BAN	*ANTHOCYANIDIN REDUCTASE/BANYULS*

ANS	*ANTHOCYANIDIN SYNTHASE*
ANT	*AINTEGUMENTA*
AOP/GSL-ALK	2-Oxoglutarate dependent dioxygenase
AOP2	Alkenylhydroxalkyl-producing 2
AOX	Alternative oxidase
AP1	*APETALA 1*
AP2	*APETALA 2*
APS/ATPS	ATP sulfurylase
APSR	APS reductase
APX	Ascorbate peroxidase
ARD/ARD′	Acireductone dioxygenase
ARDRA	Amplified rDNA restriction analysis
ARF	*AUXIN RESPONSE FACTOR*
ARF	Auxin response factor
ARF5	Auxin response factor 5
Arg	Arginine
ARGOS	*AUXIN-REGULATED GENE INVOLVED IN ORGAN SIZE*
ARID	AT-rich interaction domain
ARP	*AUXIN-REPRESSED PROTEIN 1*
AS	Alternative splicing
As	Arsenic
ASAL	*Allium sativum* leaf lectin
Asn	Asparagine
ASN2–3	*ASPARAGINE SYNTHETASE 2-3*
ASSR	Anchored simple sequence repeat
At	*Arabidopsis thaliana*
AT	Associative transcriptomics
ATM3	*ABC transporter of the mitochondrion 3*
atpF	Adenosine triphosphate synthase subunit F
ATR1/MYB34	Altered tryptophan regulation 1
ATR4	*ALTERED TRYPTOPHAN REGULATION 4*
AUX/IAA	Auxin/indole-3-acetic acid
BA	Benzyladenine
BAC	Bacterial artificial chromosome
BAM	*BARELY ANY MERISTEM*
BAP	Benzylaminopurine
BAR	Glucobarbarin
B-ARR	Type B Arabidopsis response regulators
BAT5	Bile acid transporter 5
BC	Backcross
BCAT4	Branched-chain amino acid aminotransferase 4
BGD	Brassica Genomics Database
bHLH	Basic Helix-Loop-Helix
BIN2	*BRASSINOSTEROID-INSENSITIVE 2*
BLAST	Basic Local Alignment Search Tool
BnaPh1	*B. napus* pairing homoeologous 1
BnPIR	*Brassica napus* Pangenome Information Resource

BR	Brassinosteroid
BRAD	Brassica Database
BRI1	*BRASSINOSTEROID INSENSITIVE 1*
BSA	Bulked segregant analysis
BSA-RNAseq	BSA-RNA sequencing
BSR-Seq	Bulked segregant RNA-seq
BUSCO	Benchmarking universal single-copy ortholog
bZIP	Basic leucine zipper
bZIP	Basic leucine zipper domain
C4H	Cinnamate-4-Hydroxylase
CAGR	Compound annual growth rate
CALM	Calmodulin
CaM7	Calmodulin7
CAMTA	Calmodulin-binding transcription activator
CaMV	*Cauliflower mosaic virus*
CaMV 35S	*Cauliflower mosaic virus* promoter 35S
CAP	Cationic antimicrobial peptide
CAPS	Cleaved amplified polymorphic sequence
Cas9	CRISPR-associated protein 9
CAT	Catalase
CAT3	*CATALASE 3*
CATD	CENP-A centromere targeting domain
CAX1	Calcium exchanger 1
CBF	C-repeat binding factor
CC	Coiled-coil
Cd	Cadmium
CDC48A4	Cell division control protein 48 homolog A
CDF1	*CYCLING DOF FACTOR 1*
cDNA	Complementary DNA
CDPK	Ca-dependent protein kinase
CE	Capillary electrophoresis
CE/LIF	Capillary electrophoresis/laser-induced fluorescence
CE/MS	Capillary electrophoresis/mass spectrometry
CE/UV	Capillary electrophoresis/ultraviolet
CEGMA	Core eukaryotic genes mapping approach
CENH3	Centromere-specific histone 3
CET	Cation-Efflux Transporter
CG	Candidate gene
CG-AM	Candidate gene-based association mapping
CIS	Commonwealth of Independent States
CLV1	*CLAVATA 1*
cM	Centi-Morgan
CMS	Cytoplasmic male sterile
CMS	Cytoplasmic male sterility
CNL	Coiled-coil-NLR
CNV	Copy number variation
CNV	Copy number variant
CO	*CONSTANS*

CO-like	*CONSTANS-Like*
COR	*Cold Regulated* gene
COR14	*Cold Regulated 14*gene
COX1	*Cytochrome c oxidase I*
COX-2	*Cyclooxygenase-2*
CP	Chloroplast
cp	Chloroplast
CP	Cysteine protease
cpDNA	Chloroplast DNA
cpSL	Chloroplastic selenocysteine lyase
CPT	CDP-choline:diacylglycerol cholinephosphotransferase
CPU	Central processing unit
CRE1	*Cytokinin response 1*
CRISPR	Clustered regularly interspaced short palindromic repeat
CRoPS	Complexity reduction of polymorphic sequence
crRNA	CRISPR-RNA
CRT	CRISPR Recognition Tool
CTR1	*Copper transporter 1*
cv	Cultivar
CYP78A9	Cytochrome P450 78A9
CYP79A2	CYTOCHROME P450 79A2
Cys	Cysteine
d6D	Delta-6 desaturase
DAG	Diacylglycerol
DAG Kinase	Diacylglycerol kinase
DArT	Diversity array technology
dCAPS	Derived cleaved amplified polymorphic sequence
ddRAD-seq	Double digest restriction associated DNA sequencing
DE	Differential expression
DEFL	*Defensin-like*
DEG	Differentially expressed gene
DFFS	Diversity fixed foundation set
DFR	*DIHYDROFLAVONOL REDUCTASE*
DFR	Dihydroflavonol-4-reductase
DGAT	Diacylglycerol acyltransferase
DH	Doubled haploid
DHA	Docosahexanoic acid
DHAR	Dehydroascorbate reductase
DHN 2/3	*DEHYDRIN 2/3*
DM	Dry matter
DOF6.8	*DNA binding with one finger 6.8*
DPPH	2, 2-Diphenyl-1-picrylhydrazyl
DRE	Drought responsive element
DREB	Dehydration responsive element binding
DRM1	*DORMANCY-ASSOCIATED PROTEIN 1*
DRMR	Directorate of Rapeseed-Mustard Research
DSB	Double stranded break
DUF26	*DOMAIN OF UNKNOWN FUNCTION 26*

DW	Dry weight
e2f-dp	*E2 Transcription Factor-Dimerization Partner*
EA	Erucic acid
*Ea*DAcT	*Euonymusalatus* diacylglycerol acetyltransferase
EBI	European Bioinformatics Institute
EBV	Estimated breeding value
EF	Edema factor
EFA	Essential fatty acid
EFSA	European Food Safety Authority
eIF2Bβ	*Eukaryotic translation initiation factor 2B-beta*
EIN2/3	*ETHYLENE INSENSITIVE 2/3*
EIT	Electron impedance tomography
EMS	Ethylmethane sulphonate
EMSA	Electrophoresis mobility shift assay
ENU	Ethylnitroso urea
EPG	Electron penetration graph
ER	Endoplasmic reticulum
ERD	*EARLY RESPONSIVE TO DEHYDRATION*
ERF	*ETHYLENE response factors*
ESI	Electrospray ionization
EST	Expressed sequence tag
ET	Ethylene
EtBr	Ethidium bromide
ETC	Electron transport chain
ETI	Effector-triggered immunity
ETR	*ETHYLENE RESPONSE 1*
EXPA16	Expansin A16
EXPB1	Expansin B1
EβF	E-β-farnesene
F_1	First filial generation
F_2	Second filial generation
F_3	Third filial generation
F3H	*FLAVANONE-3-HYDROXYLASE*
F_5	Fifth filial generation
F_6	Sixth filial generation
F_7	Seventh filial generation
FAD	*FATTY ACID DESATURASE*
FAD	Fatty acid desaturase
FAD 4	Δ4 fatty acid desaturase
FAE	Fatty acid elongase
FAE1	*Fatty acid elongase 1*
FAF2	*FANTASTIC FOUR 2*
FAR1	*FAR RED IMPAIRED RESPONSE1*
FAS	Fatty acid synthase
FatB	FatB thioesterase
FAX1	Fatty acid export1
FCA	*FLOWERING CONTROL LOCUS A*

Fd-GOGAT	*Ferredoxin-dependent glutamate synthase*
FHA	*FORKHEAD-ASSOCIATED DOMAIN*
FISH	Fluorescence *in situ* hybridization
FISH	Fluorescent *in situ* hybridization
FLC	*Flowering Locus C*
FLC	*FLOWERING LOCUS C*
FLD	*FLOWERING LOCUS D*
FLK	*FLOWERING LOCUS K HOMOLOGY DOMAIN*
FMO	Flavin-monooxygenase
FPA	*Flowering time control protein*
FRAP	Ferric reducing antioxidant power
FRI	*FRIGIDA*
FSD2	*Fe SUPEROXIDE DISMUTASE 2*
FT	*FLOWERING LOCUS T*
FTMS	Fourier transform ion cyclotron mass spectrometry
FUL	*FRUITFULL*
G X E	Genotype x environment
G2-like	*Golden2-like*
GA	Gibberellin
GAI	*Gibberellininsensitive*
GAPDH	Glyceraldehyde-3-phosphate dehydrogenase
GAS1	*GA signal-transduction*
GBF	(bZIP) G-box binding factor
GBS	Genotyping-by-sequencing
GC	Gas chromatography
GC	Guanine cytosine
GC/MS	Gas chromatography/mass spectrometry
GCA	General combining ability
GDH2	*Glutamate dehydrogenase*
GDP	Gross domestic product
GE	Genetic engineering
GE	Genome editing
GeBP	*GLABROUS1 (GL1) enhancer-binding protein*
GFP	Green fluorescent protein
GH3 family	*GRETCHEN HAGEN 3* family
GI	*GIGANTEA*
GID1	*GIBBERELLIN-INSENSITIVE DWARF1*
GISH	Genomic *in situ* hybridization
GLA	Gamma linolenic acid
Gln	Glutamine
GLP13	*Germin-like proteins 13*
GLS	Glucosinolate
Glu	Glutamic acid
Gly	Glycine
Gly	Glyoxalase
GM	Genetically modified
GMS	Genetic male sterility

GNAT	*Gcn5-related N-acetyltransferases*
GNC	*GATA, NITRATE-INDUCIBLE, CARBON-METABOLISM INVOLVED*
GP	Genepool
GPOX	Glutathione peroxidase
GRF2	Growth-regulating factor 2
GRF7	Growth-regulating factor 7
GS	Genomic selection
GS	Glutathione synthetase
GS3A	Glutamine synthase
GSH	Glutathione
GSH	Glutathione reduced
GSL	Glucosinolate
GST	Glutathione-S-transferase
GTR	General time reversible
GUS	β-Glucuronidase
GWAS	Genome-wide association study/studies
H1299	Human non-small cell lung carcinoma cell line
HAD	*HISTONE DEACETYLASE*
HAST	High affinity sulphate transporter
HB-other	Homeobox-other
HDL	High-density lipoprotein
HDR	Homology directed repair
HEA	High erucic acid
HFD	Histone fold domain
HGAP	Hierarchical genome assembly process
Hi-C	Chromatin Conformation Capture
HIG	Hydroxyindole GLS
His	Histidine
Hi-TOM	High-throughput tracking of mutation
HL60	Human promyelocytic leukemia cell
HMA2/3/4	*HEAVY METAL ATPase 2/3/4*
HNRTs	Homoeologous non-reciprocal translocation
HO1	Hemeoxygenase 1
HOA	High oleic acid
HPLC	High-performance liquid chromatography
HPLC/MS	High-performance liquid chromatography/mass spectrometry
HR	Hypersensitive response
HRT	Homoeologous reciprocal translocation
HSF	Heat shock factor
Hv-TLP8	*Hordeum vulgare* (barley) thaumatin-like protein 8
I3C	Indole-3-carbinol
IAA	Indole acetic acid
IAA33	Indole-3-acetic acid inducible 33
ICAR	Indian Council of Agricultural Research
ICE	Inducer of CBF expression
IFN	Interferon

Ig	Immunoglobulin
IGMT	Indole GLSs methyltransferase
IHP	Inter-homeolog polymorphism
IKU1	*Leucine-rich repeat (LRR) KINASE gene 1*
IL	Introgression lines
IL	Interleukin
IL-1β	Interleukin-1 beta
IL-6	Interleukin-6
Ile	Isoleucine
ILP	Intron length polymorphism
InDel	Insertion/deletion
IP	Intron polymorphism
IPT	Adenylated i-methylallyl transferase
IR	Infrared spectroscopy
IR	Inverted repeat
IRAP	Inter-retrotransposon amplified polymorphism
IRT1	*Iron-Regulated Transporter 1*
ISH	*In situ* hybridization
ISSR	Inter-simple sequence repeat
iTRAQ	Isobaric tag for relative and absolute quantitation
ITS-1	Internal transcribed spacer 1
ITS-2	Internal transcribed spacer 2
JA	Jasmonic acid
JAZ	Jasmonate- ZIM domain
JLA	Junction of LSC and IRs
JLB	Junction of LSC and IRb
JMT	Jasmonic acid carboxyl-methyl transferase
JSA	Junction of SSC and IRa
JSB	Junction of SSC and IRb
KASP	Kompetitive allele specific PCR
KCS	β-Ketoacyl-CoA synthase
KCS6/KCS5	3-Ketoacyl-CoA synthase 6/5
LAST	*LOW-affinity sulfate transporter*
LBD	*LATERAL organ boundaries (LOB) domain*
LC	Liquid chromatography
LC/MS	Liquid chromatography/mass spectrometry
LD	Linkage disequilibrium
LD	*Luminidependens*
LDL	Low-density lipoprotein
LDOX	*LEUCOANTHO-CYANIDIN DIOXYGENASE*
LEA	Low erucic acid
LEA	Late embryogenesis abundant
Leu	Leucine
LF	Least fractionated
LFY	*LEAFY*
LG	Linkage group
LIF	Laser-induced fluorescence
lncRNA	Long non-coding RNA

LOA	Low oleic acid
LOD	Logarithm of odds
LPA	Lysophosphatidic acid
LPAAT	Lysophosphatidic acid acyltransferase
LPCAT	Lysophosphatidylcholine acyltransferase
LRR	Leucine-rich repeat
LRR-RLK	Leucine-rich repeat receptor-like kinase
LRR-RLP	Leucine-rich repeat receptor-like protein
LSC	Large single copy
LSU	Large subunit
LTR	Long terminal repeat
Luc	*Luciferase* gene
Lys	Lysine
LysM	Lysin motif
M_3	Mutated generation 3
MAALs	Monosomic alien addition lines
MAB	Marker-assisted breeding
MAF	Minor allele frequency
MAGIC	Multi-parent advanced generation intercross
MAM	Methylthioalkylmalate synthase
MAN7	*Endo-beta-mannase 7*
MAPK	Mitogen-activated protein kinase
MAPKKK	Mitogen activated kinase kinase kinase
MAS	Marker-assisted selection
MAT	Morphological analysis tool
MATE	Multidrug and toxin extrusion protein 1
MCF7	Michigan Cancer Foundation-7
MCU	*Mitochondrial calcium uniporter*
MD	Microspore-derived
MDA	Malondialdehyde
MEGAX	Molecular Evolutionary Genetics Analysis version-X
MeJ	Methyl jasmonate
MeJA	Methyl jasmonate
Met	Methionine
MF1	Medium fractionated
MF2	Most fractionated
mGWAS	Metabolic- GWAS
miRNA	MicroRNA
MKS	MAPK signalling
MN	Mega nucleases
MRM	Multiple reaction monitoring
MS	Mass spectrometry
MS	Murashige and Skoog
MSH4	MutS protein homolog 4
MT	Mitochondrial
MTA	Marker-trait association
MTA	Material transfer agreement
mtDNA	Mitochondrial DNA

mTERF	*Mitochondrial Transcription tERmination Factor (mTERF)*
MTL	Matrilineal
MTP11	*Metal tolerance protein 11*
MTPC2	*Metal tolerance protein c2.*
MUFA	Mono-unsaturated fatty acid
MUSCLE	Multiple sequence comparison by log-expectation
Mya	Million years ago
MYR	*MYROSINASE*
NA	Nicotinamide
NAA	Naphthalene acetic acid
NAD(P)H	Nicotinamide adenine dinucleotide phosphate
NB	Nucleotide binding
NBS	Nucleotide binding site
NCBI	National Center for Biotechnology Information
ncRNA	Noncoding RNA
ndhA	Nicotinamide adenine dinucleotide hydride-dehydrogenase subunit A
NF-kB	Nuclear factor kappa-light chain enhancer of activated B cell
NGS	Next-generation sequencing
NHEJ	Non-homologous end joining
NILs	Near-isogenic lines
NiR	*Nitrite reductase*
NIRS	Near-infrared spectroscopy
NLR	Nucleotide-binding site-leucine-rich repeat
NMR	Nuclear magnetic resonance
NPA	N-(1-naphthyl)-thalamic acid
NPR	Non-expressor of pathogenesis-related protein
NR	*Nitrate reductase*
NRAMP	Natural resistance-associated macrophage protein
NRT	*Nitrate transporters*
NST2	*NAC SECONDARY WALL THICKENING PROMOTING FACTOR 2*
nt	Nucleotide
NUE	Nitrogen use efficiency
OA	Oleic acid
OD	Optical density
OG	*Orphan gene*
ONT	Oxford Nanopore Technologies
OPR1	*Oxophytodienoate reductase 1*
OPR3	*Oxyphytodienoate reductase 3*
ORF	Open reading frame
ORG3	*OBP3 (OBF-binding protein 3)-responsive gene*
OXI1	*Oxidative Signal Inducible1*
PA	Phosphatidic acid
PA	Protective antigen
PAC	P1-derived artificial chromosome
PacBio	Pacific Biosciences

PAGE	Polyacrylamide gel electrophoresis
PAM	Protospacer-adjacent motif
PAMP	Pathogen-associated molecular pattern
PAP	Phosphatidic acid phosphohydrolase
PAVs	Presence and absence variations
PAVs	Presence/absence variants
PC	Phosphatidylcholine
PC	Phytochelatin
PCIB	p-Chlorophenoxyisobutyric acid
PCR	Polymerase chain reaction
PCR 2/6	*Plant cadmium resistance protein 2/6*
PCS1	*Phytochelatin synthase 1*
PDA	Photodiode array
PDAT	Phospholipid:diacylglycerol acyltransferase
PDR	Pleiotropic drug resistance
PGP	Plant genomics and phenomics
PGR	Plant growth regulator
PGRC	Plant Germplasm Registration Committee
Ph1	*Pairing homoeologous1*
Phe	Phenylalanine
PHT	Phosphate transporters
PHYB	*PHYTOCHROME B*
PIC	Polymorphism information content
PIF4	*Phytochrome-interacting bHLH factor*
PIP1	*Plasma membrane intrinsic protein 1*
PK	*Protein kinase*
PLA	*Patatin-related phospholipase a*
PLATZ	*Plant AT-rich sequence and zinc-binding proteins*
PLCD	*Phospholipase C Delta 1*
PMC	Pollen mother cell
POD	Peroxidase
PP2	Phloem protein 2
PP2A-B	Protein phosphatase 2A regulatory subunit B
PP2C	*Protein phosphatase 2c*
PPO	Polyphenol oxidase
PR	Pathogenesis-related
PR 1/5	*Pathogenesis-related protein 1/5*
PrBn	Pairing regulator in *B. napus*
PRC 2	*Polycomb repressive complex 2*
PRGdb	Pathogen Receptor Genes Database
PRO	Progoitrin
Pro	Proline
PTI	PAMP-triggered immunity
PTM	Post-translational modification
PUFA	Polyunsaturated fatty acid
PYL	*Pyrabactin resistance (PYR)/PYR1-like*
PYL	Pyrabactin resistance 1 (PYR) like
qRT-PCR	Quantitative real-time PCR

QTL	Quantitative trait locus
QTLs	Quantitative trait loci
QTOF	Quadrupole time-of-flight
R gene	Resistance gene
RAB18	*Responsive to ABA 18*
RAD	Restriction-site associated DNA
RAD Seq	Restriction-site associated DNA sequencing
RADseq	Restriction associated DNA sequencing
RAMP 4	Ribosome-associated membrane protein 4
RAPD	Random amplified polymorphic DNA
RBOH	*Respiratory burst oxidase homolog*
RBOH	Respiratory burst oxidase homolog
RBR	Restituted *Brassica rapa*
RCAR	*Regulatory components of ABA receptor*
RCE	*RUB* (related to ubiquitin 2) conjugating enzyme
RCO	*Reduce complexity*
RCSU	Relative synonymous codon usage
REF6	*Relative of Early Flowering 6*
Ren – Seq	Resistance gene enrichment sequencing
retr03	*R*ecessive TuMV resistance 03
Rf gene	*Fertility restorer* gene
RFLP	Restriction fragment length polymorphism
RGA	*Repressor of ga1-3*
RGA	R gene analog
RGA1	*Repressor of ga1*
RILs	Recombinant inbred lines
RIP	Ribosome inactivating protein
RLK	Receptor like kinase
RLP	Receptor-like protein
RNAi	RNA interference
RNA-seq	RNA-sequencing
RNP	Ribonucleoprotein
ROS	Reactive oxygen species
RPD3	*Reduced Potassium Dependency 3*
RPKM	R^2 (Pearson) correlation of gene expression
rRNA	Ribosomal RNA
RRS	Reduced representation sequencing
RTPP	Real time plant phenotyping
RuBisCO	Ribulose-1,5-bisphosphate carboxylase/oxygenase
SA	Salicylic acid
SA	Sinapic acid
SAD	Stearoyl-ACP desaturase
SAGE	Serial analysis of gene expression
SAR	Systemic acquired resistance
SAUR	*Small Auxin Up-Regulated RNA*
SB	Speed breeding
SBP	*Squamosa promoter binding protein*
SCA	Specific combing ability

SCAR	Sequence-characterized amplified region
SCR	*SCARECROW*
scRNA-seq	Single-cell RNA sequencing
SCT	*Sinapoylglucose:cholinesinapoyl transferase*
SDS	Sodium dedocyl sulfate
SE	Somatic embryogenesis
SEM	Scanning electron microscopy
Ser	Serine
SFA	Saturated fatty acid
SgRNA	Single guide RNA
SGT	*UDP-glucose:sinapate glucosyl transferase*
SHAM	Salicylhydroxamic acid
SHP1/2	*Shatterproof 1/2*
SK	Shikimate kinase
SLR1	*S-locus-related gene 1*
SLR2	*S-locus-related gene 2*
SMRT	Single-molecule real-time
SMT	*Selenocysteine methyltransferase*
SNP	Single nucleotide polymorphism
SnRKs1	*SNF1 (sucrose non-fermenting 1)-related protein kinases*
SOC1	*Suppressor of overexpression of CONSTANS*
SOD	Superoxide dismutase
SOPA	The Soybean Processors Association of India
SOS	Salt overlay sensitive
SOT16	*SULFOTRANSFERASE*
SOT17	Sulfotransferase 17
SP	Sinapine
SpCas9	*Streptococcus pyogenes* CRISPR-associated protein 9
SR	Stem rot
SRAP	Sequence-related amplified polymorphism
SRPE	Single rapeseed seed protein extraction
SRR1	*Sensitivity to red light reduced 1*
SSC	Short single copy
SSH	Suppression subtractive hybridization
SSLP	Simple sequence length polymorphism
SSN	Site-specific nuclease
SSP	Seed storage protein
SSR	Simple sequence repeat
SSS	Sub-stoichiometric shifting
SSU	Small subunit
STK	Serine threonine kinase
STMS	Sequence-tagged microsatellite site
STP4	Sugar transporter protein
STR	Short tandem repeat
STS	Sequence-tagged site
SUC2	*SUCROSE SYNTHASE2*
SULTR2;1	*SULPHATE TRANSPORTER 2;1*
SUR1	*SUPERROOT1*

SuS	Sucrose synthase
TAG	Triacylglycerol
TALEN	Transcription activator-like effector nuclease
TCA	Trichloroacetic acid
TDI	Tolerable daily intake
TDZ	Thidiazuron
TE	Transposable element
TEM	Transmission electron microscopy
TF	Transcription factor
TFL1 (Sdt1)	*TERMINAL FLOWER 1*
TGA 3	*TGACG motif-binding (TGA1a) related 3*
TGG 1	*Myrosinase 1/β-thioglucosideglucohydrolase 1*
Thr	Threonine
TILLING	Targeting induced local lesions in genomes
TIR	Toll interleukin-1 receptor
TIR1	Transport inhibitor response-1
TLC	Thin layer chromatography
TLR	Toll-like receptor
TM	Transmembrane
TNF	Tumor necrotic factor
TNF-α	Tumor necrosis factor-alpha
TNL	*Toll/interleukin-1 receptor-NLR*
TR	Transcription regulator
tra crRNA	*Trans*-activating CRISPR RNA
TRAF	*Tumor necrosis factor receptor-associated factor*
tRNA	Transfer RNA
*trn*L-*trn*F	*trn*L-*trn*F region of the chloroplast genome
TRO	Glucotropaeolin
Trp	Tryptophan
TSW	Thousand seed weight
TT 1	*TRANSPARENT TESTA 1* (C2H2 zinc-finger protein)
TT 2	*TRANSPARENT TESTA 2* (R2R3-MYB)
TT 8	*TRANSPARENT TESTA 8* (bHLH protein)
tTCLs	Transverse thin cell layers
TTG 1	*TRANSPARENT TESTA GLABRA1* (WD40 regulatory protein)
TuMV	*Turnip mosaic virus*
Tyr	Tyrosine
UDPG	UDP-glycosyltransferase
UFA	Unsaturated fatty acid
UGM	Unigene-derived microsatellite
UGT	*UDP-glucose:thiohydroximate S-glucosyltransferase*
UGT74	UDP-glucosyltransferase 74C1
UHPLC	Ultra HPLC
UM-UC-3	University of Michigan-Urothelial Carcinoma-3
UPM	Universidad Polite´cnica of Madrid
USDA	United States Department of Agriculture
USFA	Unsaturated fatty acid

UV	Ultraviolet
Val	Valine
var	Variety
VDAC 2	*Voltage-dependent anion channel 2*
VIN 3	*VERNALIZATION INSENSITIVE3 (VIN3)*
VIN3	*Vernalization insensitive 3*
VLCFA	Very long chain fatty acid
VLCMFA	Very long chain monounsaturated fatty acid
VLCUFA	Very long chain unsaturated fatty acid
VNTR	Variable number of tandem repeat
VPE	Vacuolar processing enzyme
VRN1/VRN 2	*VERNALIZATION 1/VERNALIZATION 2*
WGA	Wheat-germ agglutinin
WGRS	Whole-genome resequencing
WGS	Whole-genome sequencing
WGT	Whole-genome triplication
WRKY11	WRKY transcription factor 11
WRKY70	WRKY transcription factor 70
WRR	White rust resistance
WUE	Water use efficiency
XP- GWAS	Extreme phenotype -GWAS
XP-CLR	Cross-population composite likelihood ratio test
XTH9	Xyloglucan endotransglucosylase/hydrolase 9
YAC	Yeast artificial chromosome
YSL	Yellow stripe-like proteins
Zem	Zero erucic mustard
ZFN	Zinc-finger nuclease
ZFP	Zinc-finger protein
ZIM	Zinc-finger inflorescence meristem
γ-ECS	*γ-Glutamylcysteine synthetase*
γ-rays	Gamma rays
γ-TMT	γ-Tocopherol methyltransferase

Brassica juncea: A Crop for Food and Health

1

P. K. Rai, Prashant Yadav, Arun Kumar, Anubhuti Sharma, Vinod Kumar, and Prajjwal Rai

Abstract

Brassica juncea, L. (Czern. and Coss.), commonly known as mustard, is a member of family Brassicaceae and an important edible oilseed crop in the world, mainly cultivated in India, Pakistan, Bangladesh, China, and in some pockets in the USA, Canada, Japan, and Australia. Along with *B. napus* and *B. rapa*, *B. juncea* is the primary source of canola oil due to its high oil content (38–42%). Traditionally, mustard oil is the major source of cooking oil in the Indian subcontinent. Besides its oil, the seed meal of *B. juncea* is also used as animal feed and fertilizers. *B. juncea* is an amphidiploid ($2n = 36$, AABB) species evolved from the natural interspecific hybridization between *B. rapa* ($2n = 20$, AA) and *B. nigra* ($2n = 16$, BB) followed by genome duplication. Comparatively, the oil of *B. juncea* has balanced levels of saturated and unsaturated fatty acids and omega-6 and omega-3 polyunsaturated fatty acids (PUFAs). The oil of *B. juncea* is also known for its nutraceutical properties due to the presence of bioactive compounds viz. glycosides, flavonoids, phenolic compounds, sterols, and triterpene alcohols. Its oil has high percentage of erucic acid (40–57% of total fatty acids) which is not considered good for health, and therefore varieties with low levels of erucic acid have been developed. Although it is considered a hardy crop, *B. juncea* production is adversely affected by various biotic (stem rot, white rust, Alternaria blight, aphid) and abiotic stresses (heat, drought, frost, salinity).

1.1 Introduction

The Genus *Brassica* comprises a diversified range of oilseed crops cultivated worldwide for their importance for human nutrition and health. Brassica crops range from invasive weeds to leaf and root vegetables to condiment crops and are also used for fodder, potential biofuel, preparation, and seasoning of food, besides being an important oilseed crop (Raymer 2002; Rahman et al. 2018). The important species of oilseed Brassicas that are extensively cultivated are *Brassica juncea*, *Brassica rapa* (syn. *Brassica campestris*), and *Brassica napus*. Of these, *B. juncea* is predominantly a cultivated species in Asia including India, China, Pakistan, Bangladesh, and Nepal, while *B. napus* and *B. rapa* are

P. K. Rai (✉) · P. Yadav · A. Kumar · A. Sharma · V. Kumar
ICAR-Directorate of Rapeseed-Mustard Research, Sewar, Bharatpur, Rajasthan 321303, India
e-mail: Prashant.Yadav@icar.gov.in

P. Rai
Institute of Agricultural Sciences, Banaras Hindu University, Varanasi, Uttar Pradesh 221005, India

C. Kole and T. Mohapatra (eds.), *The Brassica juncea Genome*, Compendium of Plant Genomes,
https://doi.org/10.1007/978-3-030-91507-0_1

largely cultivated in Sweden, Germany, France, Canada, and Australia. Oilseed Brassicas have a natural potential to survive and grow at relatively low temperatures; they are one of the few edible oil sources that can be successfully cultivated in cool temperate regions. This distinctive characteristic makes them well adapted to cultivation at higher altitudes and as winter crops in the subtropics. In temperate regions, *B. napus* and *B. rapa* predominate, while in the subtropics of Asian regions, *B. juncea* is the major oilseed crop (OECD 2016).

India ranks third after Canada and China sharing about 11.0% of the global rapeseed-mustard production (72.37 mt) and shares 16.72% and 12.79% in terms of area and production, respectively, of oilseeds in India during 2018–19. Rapeseed-mustard production trends all over the world showed substantial growth in production trends of 58.2 mt from 31.2 mha area with 1860 kg ha^{-1} productivity during 2008–09 which enhanced to 72.4 mt from 36.6 mha with 1980 kg ha^{-1} during 2018–19 (Fig. 1.1). More than 90% of the world's rapeseed-mustard produced is in Asia, EU, and N. America, i.e., 31.83%, 35.44%, and 26.83%, respectively during 2010–11 to 2018–19. China alone produced 62.44% of the crop in Asia, Canada produced 93.83% of N. America and France produced 20.88% of the crop in Europe. The majority of the countries grow rapeseed (*B. napus*), however the largest area, around 80%, under Indian mustard (*B. juncea*) is in India (George and Loeser 2021).

1.2 Oilseed Brassicas

Among the *Brassica* species, *B. nigra* ($2n = 16$, BB), *B. oleracea* ($2n = 18$, CC), and *B. rapa* ($2n = 20$, AA) are the primary diploid species, while *B. carinata* ($2n = 34$, BBCC), *B. juncea* ($2n = 36$, AABB) and *B. napus* ($2n = 38$, AACC) are the amphidiploids (Nagaharu 1935). The amphidiploids have arisen through interspecific hybridization between primary diploid species in nature. The amphidiploid species *B. juncea* (AABB) comprises the genomes of both *B. rapa* (AA), and *B. nigra* (BB). Five related species cultivated worldwide as a source of edible seed oil are *B. juncea*, *B. napus*, *B. carinata*, *B. rapa*, *B. nigra* and *Eruca sativa*.

1.2.1 Classification and Nomenclature

The family Brassicaceae (Syn. Cruciferae) mainly includes 338 genera and 3709 species (Warwick et al. 2006). The Genus *Brassica* includes three distinct species, such as *B.* rapa (syn. *B. campestris*) (yellow sarson, brown sarson, and toria), *B. juncea* (Indian mustard, Chinese mustard, brown mustard), and *B. nigra*. The Indian *Brassica* belongs to two genera *Brassica* and *Eruca*. Leaf shape polymorphism in *B. juncea* also led to its classification under various specific and varietal names. Kumazawa and Akiya in the nineteenth century, used variations in stem characteristics to differentiate *B. juncea* in Taiwan, as reviewed by Dixon (2007). Such as cultivars like Ta-sin-t'sai had compact stems while cultivars like Ta-sin-t'sai had enlarged stems. Indian mustard types were classified on the basis of oil content 3-butelyl isothiocyanate, which is absent from Chinese mustard types. Nomenclature of *B. juncea* has been reviewed by several researchers like Bailey (1922), Vaughan and Gordon (1973), and Prakash and Hinata (1980). Based upon the views of different workers, the group developed by Vaughan et al. (1963) on the basis of the variation in volatile oil and seed protein in *B. juncea* appears to be sensible. The well-defined different seven groups on the basis of geographical and botanical features described by Dixon (2007) are given in Table 1.1.

1.2.2 *Brassica juncea* as a Crop

The word *Brassica* originated from Bresic of Bresych, which in Celtic means cabbage and a contraction of praesecare (to cut off early) since the leaves were traditionally removed for green fodder (Hegi 1919). The ability of Brassica

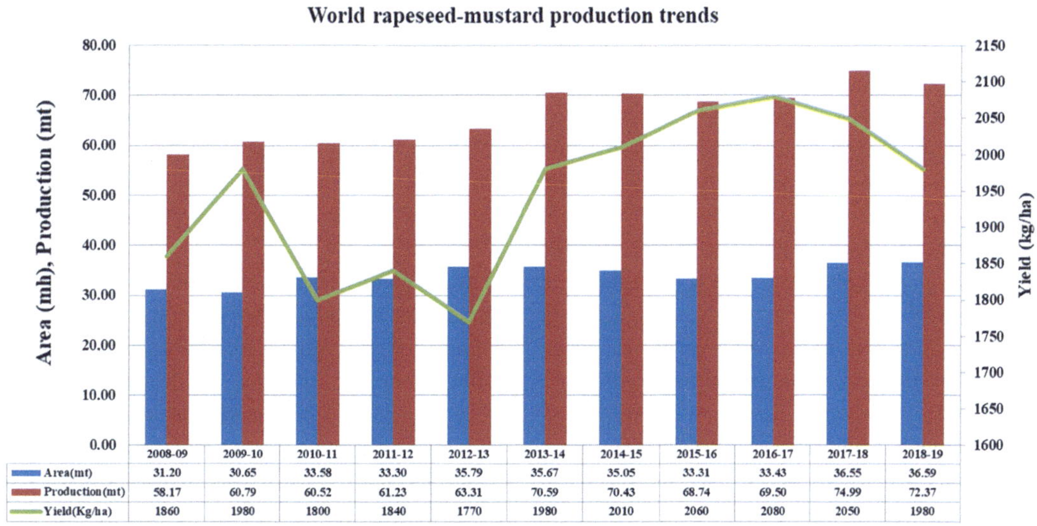

Fig. 1.1 Global rapeseed-mustard production (2008–09 to 2018–19)

Table 1.1 Subspecies of *B. juncea*, its morphology, and distribution in the world

S. No.	Brassica subspecies	Morphological characteristics	Geographical distribution
1	*B. juncea* var. *integrifolia*	Entire of lobed basal and succulent leaves	Grown in Southern and middle China, Japan, and South-Eastern Asia
2	*B. juncea* var. *japonica, foliosa*	Dissected basal, pubescent, dark green, and glabrous leaves	Grown in middle and Northern China, and Japan
3	*B. juncea* var. *crispifolia* group Azamina	Dissected crisp leaves	Grown for salads and as ornamentals in the USA
4	*B. juncea* Coss group Hakarashina	Pinnate leaves	Grown in India, central Asia, and Europe
5	*B. junceavar. napiformis* group Nekarashina	Enlarged roots	Grown in Mongolia, Manchuria, Northern China, and absent in Japan
6	*B. juncea* var. *bulbifolia* Mas, group Ta hsintsai	Succulent stems and elongated internodes, leaves are not edible	Grown China
7	*B. junceavar. rugosa* (Roxb)	Leaves with wide flat entire midribs and succulent	Grown in Southern China

plants to germinate and thrive at low temperatures has made Brassica oilseeds one of the few edible oil crops that can be cultivated in the temperate agricultural zones of the world, at high elevations and, as winter crops, under relatively cool growing conditions.

B. juncea is commonly known as oriental or brown mustard, Chinese mustard, vegetable mustard, or Indian mustard. The brown-seeded cultivars are referred to as brown mustard, while the yellow-seeded cultivars are called yellow or oriental mustard. *B. juncea* is considered to be one of the earliest domesticated plants referred to as condiments in early times. This crop is well adapted to drier conditions and matures relatively quickly. *B. juncea* crops are grown widely as an edible oilseed crop in South Asian countries including India, Pakistan, Bangladesh, Sri Lanka. In China, mustard has been cultivated for 6000–7000 years, where variations of the cultivated forms are found (Anonymous 1963). It is also cultivated in Northern parts of Africa and Central Asia which includes Southern and South-eastern provinces of former Soviet Union. *B. juncea* is grown in the European countries and in some considerable parts of North America (Dixon 2007). It has been mainly grown for condiment purposes in western Canada but has considerable potential as an edible oilseed crop (Woods et al. 1991), notably in Australia and New Zealand. In India, *B. juncea* is locally known as *rai*, *raya*, *rayeda,* or *laha* and is one of the most important edible oilseed crops. It has been an important component of the Indian cropping system since the Indus Valley civilization from 2300 to 1750 BC and where it was mentioned as "Rajika" (Prakash and Hinata 1980).

Middle East seems to be the place of origin of *B. juncea* because wild forms of *B. rapa* and *B. nigra* occur together (Olsson 1960; Shigesaburo 1967; Prakash and Hinata 1980). The regions of South-Western China and North-Western India are two secondary centers where there is enormous diversity (Vaughan et al. 1963; Vaughan and Gordon 1973). Restriction fragment length polymorphism (RFLP) studies support two centers of origin, Middle East and China (Song et al. 1988). Spect and Diederichsen (2001) support the view that the primary center of origin lies somewhere between China and Eastern Europe, where the progenitor species are sympatric. Major diversity of *B. juncea* occurs in western and central India. Oilseed Brassicas seem to be evolved from European-Mediterranean, with a secondary center of origin in Asia (Downey 1983).

1.3 Nutritional Significance of *Brassica juncea*

Brassica juncea is commonly known for nutraceutical properties of its bioactive compounds including glycosides, flavonoids, phenolic compounds, sterols, and triterpene alcohols, proteins, and carbohydrates (Sharma and Rai 2018). The importance of these secondary metabolites is their therapeutically interesting pharmacological properties. Its seeds and greens have been now often considered to be promising constituents and therefore often called a healthy Brassica. The preclinical researches conducted on mustard strongly suggest that it can be used as a potential source of affordable nutraceuticals, therapeutic drugs to prevent or cure diverse type of diseases.

Many elements of the human diet which were overlooked for a very long period are gaining prominence in curing diseases as well as providing nutrients. Over the last decade, diseases-preventing foods and ingredients (nutraceuticals) have been clinically proven beneficial to human health (Ketkar et al. 2014; Chakraborty and Bhattacharjee 2018). Today, the most beneficial nutraceutical compounds have antioxidant vitamins, dietary fibers, oligosaccharides, bioactive peptides, polyphenols, glucosinolates, and other products (Mishra et al. 2012; Sharma et al. 2016).

The medicinal properties of Indian mustard are due to presence of a variety of phytochemicals imparting the effect. Different parts of the mustard plant are used as medicinal and beauty products across the globe (Table 1.2).

Essential fatty acids (EFAs) present in mustard oil are healthy for many body systems viz. cardiovascular, reproductive, immune, and nervous systems. The essential fatty acids are exclusively synthesized by the plants and therefore must be supplied through the diet. EFAs are a necessary component of the cell membranes, enabling the cells to perform nutrition and excretion processes. EFAs are also responsible for the production of prostaglandins. The prostaglandins regulate important body functions viz. heart rate, blood pressure, fertility, and blood clotting. EFAs also play a crucial role in immunity by regulating inflammation and encouraging the human body to fight infection. There is a huge variability of fatty acid content observed among the members of the Brassicaceae family. Fatty acid composition regulates the functional and nutritional values of edible seed oils, varying considerably depending on the plant species. The oil obtained from the Brassica species consists of

Table 1.2 Health benefits of different parts of mustard

Plant parts	Uses and benefits
Leaves	cure scurvy, cure stomachic, relieve headache, muscular and skeletal pains
Seed	Increase blood flow in the skin after application, prevent cold, improve liver health, seed paste improve skin health
Powdered seed	Antibacterial activity, internally used for hiccup, augments the appetite, relieves the phlegm in cough
Oil	Aphrodisiac, lubricant, hair oil, preservative, counterirritant, emetics in drunkenness and in poisoning, skin eruptions and ulcers, colic, externally applied for arthritis, antiseptic and anti-inflammatory, seed oil, with salt is an effective gargle in dental infections and pyorrhea
Seed residue	Cattle feed, fertilizer
Other parts	Diuretic, for arthritis, foot ache, lumbago, and rheumatism its decoction is useful in amenorrhea
Roots	A galactagogue in Africa
Dried leaf and flower	A body odor repellent to mosquitoes, dengue fever
Total plant	In Java, the plant is used as an antisyphilitic emmenagogue, bronchitis, anorexia, dyspepsia, tumors, worm infestations, and splenic disorders

both saturated and unsaturated fatty acids. The saturated fatty acids (SFAs) present in the mustard oil have all the carbon valency saturated through a single bond except terminal carboxyl group and they generally have even C number, e.g., palmitic acid (C16:0). The unsaturated fatty acids (USFAs) contain one or more double bonds with a terminal carboxyl group. USFAs are either monounsaturated (MUPAs), i.e., oleic acid (C18:1) and erucic acid (C22:1) or polyunsaturated fatty acids (PUFAs) such as omega-6-linoleic acid (C18:2) and omega-3- alpha-linolenic acid (C18:3). A human requires both SFAs as well as USFAs at an optimal level for healthy life. The high level of both fatty acids is dangerous for human health. A higher amount of SFAs and Un SFAs lead to potential application in industrial purposes (Swati et al. 2015).

Mustard greens are high in vitamins and minerals (Table 1.3). The leaves of Indian mustard contain many important bioactive compounds 'polyphenols' which are considered good for human health. Polyphenols have antioxidant activities which scavenge the cancer-causing reactive oxygen species (ROS) at the sub-cellular levels (Morales-López et al. 2017). These polyphenols range from low molecular weight, simple aromatic cyclic compounds to complex tannins and derivatized polyphenols. Polyphenols are divided into two groups, flavonoids and non-flavonoids on the basis of number and the arrangements of the carbon atoms. Flavonoids consist of flavonols, flavones, flavan-3-ols, anthocyanidins, flavanones, isoflavones, and others whereas, non-flavonoids include phenolic acids, hydroxycinnamates, stilbenes, etc. The most common and heterogeneous groups of polyphenols present in mustard oil are flavonoids and hydroxycinnamic acids (Kapusta-Duch et al. 2012). Brassica vegetables also contain anthocyanins which cause pigmentation in red cabbage and broccoli sprouts (Aires 2015).

1.3.1 Major Antioxidant Compounds in Indian Mustard

Glucosinolates are an important secondary metabolite found in Brassica seeds and vegetables (Kristal and Lampe 2002; Ehlers et al., 2015). The basic structure of glucosinolates comprises β-D-thioglucose group, sulphonated oxime (–C=NOH) group, and variable side chain derived from amino acids (Fig. 1.2).

Only a few types of glucosinolates are biologically active (Sanlier and Guler 2018). Almost

Table 1.3 Chemical composition of *B. juncea* (per 100 g)

Principle	% of RDA	Vitamins	% of RDA
Energy	25	Folates	40
Carbohydrates	21	Niacin	30
Protein	46	Pantothenic acid	16
Total fat	121	Pyridoxine	31
Cholesterol	0	Riboflavin	20
Dietary fiber	32	Thiamin	67
Electrolytes		Vitamin A	1
Sodium	1	Vitamin C	12
Potassium	16	Vitamin E	132
Phyto-nutrients		Vitamin K	4
Carotene-ß	0.1	Minerals	–
Crypto-xanthin-ß	0.05	Iron	115
Lutein-zeaxanthin	0.9	Magnesium	92
Minerals		Manganese	106
Calcium	27	Selenium	378
Copper	71	Zinc	55

Source Agriculture marketing information network and USDA network database (2004)

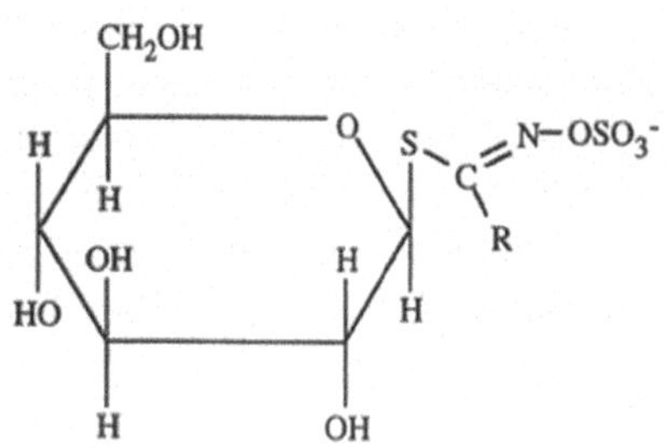

Fig. 1.2 Basic structure of glucosinolate (Sinigrin). *Source* Awasthi and Saraswathi (2016)

all cruciferous plants have enzyme β-thioglucosidase often named myrosinase which hydrolyzes glucosinolates into several potentially toxic compounds depending upon the reaction conditions and the presence of specifier proteins (Bones and Rossiter 2006; Kong et al. 2012). It has been mentioned by researchers that these glucosinolate degradation products are major contributors to the distinct taste and flavor of cruciferous plants and they also contribute towards an able defense mechanism against pathogens and herbivores (Wittstock and Burow 2010; Razis et al. 2011).

When the tissue is disrupted, the thioglucosidic bond breaks down, glucose and thiohydrosimate-O-sulphonate (unstable aglycone) are formed by β-thioglucosidase (Ehlers et al. 2015; Sharma et al. 2016).

Isothiocyanates have different modes of action, e.g., it suppresses the activation of carcinogenic compounds or they detoxify the poisonous effects of these carcinogenic compounds that have been activated. Isothiocyanates can also speed up their removal from the body. The major isothiocyanates with the strongest anticancer effects are phenylethyl isothiocyanate, benzyl isothiocyanate, and 3-phenylpropyl isothiocyanate. Xu and Thornalley (2001) explained that isothiocyanates can also conjugate rapidly with cells of glutathione in leukemia cells and further induce cellular oxidative stress and reduce the level of glutathione. However, in this medical era, drug-resistant cancer cells can be treated with isothiocyanate by a process of oxidative stress (Trachootham et al. 2008). A large number of isothiocyantes (for example

BITC, PEITC, and SFN, etc.) are able to induce cell cycle arrest in cultured cells (Lawson et al. 2015).

Sulphoraphane (1-isothiocyanat-(4R)-(methylsulfinyl) butane) is a naturally occurring isothiocyanate compound found in mustard plants. It is an important bioactive compound with antioxidant and antitumor properties. As per the reports from National Cancer Institute, US, sulforaphone is the most important chemopreventive molecule out of more than 40 agents including glucosinolate hydrolysis products, e.g., phenethyl isothiocyanate and indole-3-carbinol, etc. (Kelloff et al. 2000).

Phytochemicals present in *B. juncea* are beneficial in many ways. It reduces antioxidative stress and induces the activity of enzymes responsible for detoxification processes. It also helps in the proper functioning of the immune system. Phytochemicals inhibit the growth of cancerous cells and prevent cancerous mutations (Nawaz et al. 2018). Sulphoraphane protects human body from the harmful effects of reactive oxygen species (ROS). ROS cause tissue and membrane damage, DNA breaks, modulation of gene expression, base modification, and glucosinolate degradation products (e.g., isothiocyanates) also catalyze the incorporation of glutathione to form a substrate for glutathione S-transferase (GST), a family of phase II enzymes, involved in the detoxification of carcinogens, environmental toxins, and oxidative stress products (Kim and Park 2009; Sharma et al. 2016).

Bioactive substances present in mustard oil have the ability to reduce inflammatory responses. It activates the detoxification enzymes which remove free radicals and thus boost immunity (Fimognari et al. 2012). Because of its antibacterial and anti-inflammatory properties, cabbage, one of the Brassica vegetables, is widely used in traditional medicine to relieve symptoms associated with gastrointestinal disorders and also in the treatment of small incisions, wounds, and mastitis (Kapusta-Duch et al. 2012).

In modern era, novel therapeutic approaches are searched to control blood sugar which otherwise causes the increase of mortality and morbidity (Jiménez-Osorio et al. 2015; Xu et al. 2016). Functional foods having nutraceutical properties might be recent treatment modalities for type 2 *Diabetes mellitus* and preventing its long-term complications (Bahadoran et al. 2013). It is thought that oxidative stress is the major concern for diabetic implications, e.g., disorders in glucose and lipid metabolisms. However, the higher level of antioxidant concentrations in it is reported to be effective in reducing diabetic complications (Bahadoran et al. 2011; Xu et al. 2016).

Oilseeds, with diverse roles in nutrition, have significant contributions to the food security mission across the globe. It contributes in both ways, in terms of quality and quantity as well. To lower down the malnutrition in society, edible oils play a major role. These oils are a rich source of unsaturated fatty acids, essential fatty acids

Table 1.4 Indian mustard varieties developed in India with low erucic acid and glucosinolate

Quality trait	Variety
Single low (erucic acid < 2%)	Pusa Karishma (LES 39),
	Pusa Mustard-21 (LES 1-27),
	Pusa Mustard-24 (LET-18),
	RLC-1(ELM-079),
	Pusa Mustard 21 (NPJ 112),
	RLC 2,
	Pusa Mustard 29 (LET-36),
	Pusa Mustard 30 (LES-43),
	Pusa Mustard 32 (LES 54)
Double low (erucic acid < 2% & glucosinolate <30 μmole/g defatted seed meal)	RLC-3, PDZ-1

(linolenic and alpha-linolenic), vitamin E, and contain no dietary cholesterol. Indian mustard oil contains high amount of erucic acid (40–57% of total fatty acids) which is not considered good for health, and therefore emphasis was given to developing varieties with erucic acid < 2% (Table 1.4) (Sharma and Rai 2018).

In this concern, the first low erucic acid variety Pusa Karishma of *B. juncea* and first double low variety GSC 5 of *B. napus* were released in 2004 and 2005, respectively. Subsequently, eight 0-quality (single zero) varieties (low in erucic acid) and eight 00-quality (double zero) varieties (low in erucic and low glucosinolate, Canola quality) have been released in *B. juncea*. These are initial steps to lower down the malnutrition. The list of mustard germplasm registered by PGRC, New Delhi is presented in Table 1.5. A lot of efforts to tag the traits for low erucic acid and low glucosinolate content in Indian mustard with high yield potential are underway. The mustard oil may also be used for many high-value products (low erucic acid or low glucosinolate or both, rich in healthy mineral elements, and medicaments). It is also used to prepare Margarine which contains unsaturated "good" fats. Elements such as Ca, Fe, Mg, Mn, K, Zn are found in rich quantities in the leaves of oilseed Brassica. The oilseed Brassica seed meal contains abundant protein and other phytic and phenolic compounds. Other purposes include polymers, lubricants, plastic, cosmetics, pharmaceuticals, bio fumigants, and biofuel (Maina et al. 2020).

Some common examples of value addition to enhance nutritional security (i) Erucamide, made from erucic acid is a very important processing aid for olefin polymers namely polyethylenes and polypropylene. (ii) Mustard and castor oils are widely used as lubricating oil. The popular castrol is a special blend of lubricating oil containing castor oil.

1.4 Biotic and Abiotic Stresses

1.4.1 Biotic Stresses

Brassica crop is susceptible to various types of biotic stresses. The major diseases of Indian mustard are Alternaria blight (*Alternaria brassicae* and *A. brassicicola*), white rust (*Albugo candida*), downy mildew (*Peronospora brassicae*), stem rot (*Sclerotinia sclerotiorum*), and

Table 1.5 Germplasm of mustard registered for quality traits

S. No.	Crop	Doner identity	INGR No.	Trait
1	*B. juncea*	SWARNA {TERI(OE) M21}	98001	Zero erucic acid, yellow seeded, early maturing (117 days)
2	*B. juncea*	Heera	3033	Low glucosinolate content 16.96 micro moles/g of seed) and low erucic acid in oil (0.12%)
3	*B. juncea*	NUDH-YJ-5	3034	Low glucosinolate content (9.3 micro moles/g of seed and low erucic acid in oil (0.14%)
4	*B. juncea*	TERI- GZ-05	4078	Yellow seeded, double low quality and high oleic and linoeic acid
5	*B. juncea*	PRQ-2005-1	6013	Low erucic acid (1.08%) in oil and yellow seed coat
6	*B. juncea*	NDUH-YJ-6	13015	Low glucosinolate in seed meal, resistance to white rust, yellow seed coat color
7	*B. juncea*	DRMR 1-5	19020	Double low (<2% erucic acid in oil & <30 μmoles glucosinolate/g of defatted seed meal). White rust resistant. Yellow seed coat
8	*B. juncea*	DRMRQ1-16-27	21062	Double low, high antioxidant (phenol, tocopherol), low antinutritional factors (phytic acid)

powdery mildew (*Erysiphae cruciferarum*). Aphid (*Lipaphis erysimi*), mustard sawfly (*Athalia proxima*), and painted bug (*Bagrada hilaris*) are the major insect pests. These diseases/insect pests cause substantial yield reduction.

1.4.1.1 Alternaria Blight

In case of Alternaria blight, the total destruction of the crop is rarely observed, however under suitable disease conditions the yield losses can reach up to 47% (Kolte and Awasthi 1987). Besides the reduction in the yield due to disease infestation, the seed quality attributes such as size, viability, etc. are also compromised. Alternaria blight has variable severity on *B. juncea* which differs with seasons, regions, and also between the different genotypes grown in the same region. The cultivated species, *B. carinata* and *B. napus* are reported to be tolerant, followed by *B. nigra*, *B. juncea,* and *B. campestris* (Chauhan et al. 2011). *B. hirta* is reported to be resistant to Alternaria blight (Sharma and Singh 1992). Two strains of *B. juncea* RH 8114 and RH 8114, moderately resistant to this disease, were developed under the All India Coordinated Research Project on Rapeseed-Mustard program. These strains gave appreciably higher seed yields than the local variety. The variety RH 8113 has been released for general cultivation. Some cultivars of *B. juncea* viz. RC781, PHR 2, EC 399,301, PAB 9534, Divya, and JMM 915 have been identified as tolerant on the basis of multi-location testing under All India Coordinated Research Project on Rapeseed-Mustard (Chauhan et al. 2011).

1.4.1.2 White Rust

White rust caused by *Albugo candida* (Pers. Ex Fr.) Kuntz. is an obligate pathogen of all Brassica crops. Plants of 241 species in 63 genera of Brassicaceae family have been reported to be infected by *A. candida* (Biga 1955) worldwide. Two races, AC12 from *B. juncea* and AC13 from *B. rapa* var. toria have been reported. *B. napus* and *B. carinata* have been observed as resistant to the pathogen. White rust can result in yield losses up to 47 percent (Kolte and Awasthi 1987). With each percent of disease severity and stag-head formation, there is a loss in the yield of about 82 kg/ha and 22 kg/ha, respectively (Meena et al. 2012). A number of resistant varieties of *B. juncea* were identified against the pathogen and these include JMMWR 93-7, JMMWR 93-19, JMMWR 93-38 and NRCDR 515, DRMR-2019, DRMR-2035, DRMRMJB-35, DRMR 1-5, DRMRIJ-512, and DRMR 12-48 are registered genetic stock. The varieties Jawahar Mustard-1, Jawahar Mustard-2, and Basanti are resistant to white rust disease.

1.4.1.3 Stem Rot

Stem rot of *B. juncea*, is caused by a polyphagus fungus *Sclerotinia sclerotiorum* (Lib.) de Bary. The *S. sclerotiorum* has recently become one of the most important pathogens in India. Sclerotinia rot causes serious damage to the *B. juncea* crop leading to substantial yield losses reported worldwide. The earliest record of its occurrence on rapeseed and mustard was made from India (Shaw and Ajrekar 1915). The *S. sclerotiorum* has a wide host range, known to infect about 408 plant species (Boland and Hall 1994) with no proven source of resistance. High (up to 66%) disease incidence and severe yield losses (up to 39.9%) lead to discouragement of growers of the crop (Chattopadhyay et al. 2003).

1.4.1.4 Powdery Mildew

The powdery mildew disease of *B. juncea* is caused by the fungus *Erysiphe cruciferarum* (Sharma 1979). This disease occurred on rapeseed and mustard around the world. In certain states of India such as Gujarat, Haryana, and Rajasthan, the disease has been found to occur quite severely, resulting in considerable loss in yield.

1.4.1.5 Insect Pests

About 50 insect pests are known to attack Brassicas. Among these, the mustard aphid (*Lipaphis erysimi*) has emerged to be a key pest. Sawfly (*Athalia lugens proxima*), painted bug (*Bagrada hilaris*), pea leaf miner (*Chromatomyia horticola*) and Bihar hairy caterpillars (*Spilosoma obliqua* (Walker)) are other serious pests.

However, *L. Erysimi* (mustard aphid) continues to be a key pest on *B. juncea* crop in India. Mustard aphid can cause up to 9–96% yield reduction and up to 10% losses in oil content of the produce (Abraham and Bhatia 1994). There are several parameters developed for the assessment of aphid resistant sources viz. aphid infestation index, aphid fecundity, seedling survival, estimated aphid injury, and evaluation of genotypes in the field. Few strains of *B. juncea*, viz. T 6342, Glossy B 85, RH 7846, RH 7847, RH 7848, and CSR 1017 developed through hybridization, have been reported to be high yielding and aphid tolerant.

1.4.2 Abiotic Stresses

Abiotic stresses such as drought, high temperature, frost, and salinity are the major concerns to Indian mustard production and play important role in determining the productivity of this crop.

1.4.2.1 Drought

Indian mustard is often grown as a marginalized crop in rainfed conditions. About 25% of the total Indian mustard area is rainfed and it is grown on conserved moisture received during rainy seasons. *B. juncea,* owing to its low water requirement, fits well in the rainfed cropping system (Chauhan et al. 2011). Depending on planting time and winter rains, the crop is exposed to water stress at one or more phonological stages. In Indian mustard, those genotypes with thicker leaves had greater water-use efficiency (Singh et al. 2003). The adverse effects of drought during seed development may be alleviated by remobilization of assimilates to the sink and thereby increasing the harvest index. The genotypes which could extract moisture from deep soil profiles during the reproductive phase would yield higher. Osmotic adjustment and transpirational cooling have been reported to be controlled by single gene (Chaudhary et al. 1989; Singh et al. 1996). It is suggested that recurrent selection should be practiced for incorporation of drought tolerance. Several high-yielding varieties in *B. juncea*, e.g., Vardan, RH 781, RH 819, Pusa Barani, RB 50, NPJ 93, NPJ 112, RH 406, RH 725, and RGN 229 have been released for cultivation for drought-prone areas of India.

1.4.2.2 High Temperature (Heat)

High temperature (heat) is one of the most important stresses other than drought. It has serious impediments to the growth and development of the crop. In India, prevalence of high temperature (heat) stress at terminal stage is very important, as crop season of mustard gets delayed in Eastern and Northeastern regions and certain areas in Central and Eastern province of Uttar Pradesh due to sowing after rice and mixed/intercropping mustard with wheat culminating into serious fall in yield. For improvement in the yield of mustard in late sown conditions, genetic up-scaling of thermo-tolerance is very important. On the other hand, high temperature at the time of crop establishment is a very important factor in countries like India, and situation even gets worsened when there is delay in monsoon causing substantial loss in productivity. In Indian mustard, some potential donors like RH 0116, RH 8814, SKM 9928, SKM 0149, NRCDR 02, and BPR 543-2 developed having thermo-tolerance. Registered germplasm lines BPR 541-4 and BPR 543-2, BPR549-9, BPR 540-6 for high thermo-tolerance at terminal and juvenile stages are also available. Varieties of Indian mustard-like RGN 13, Urvashi, NRCDR 02, PM 25, Pant Rai 19, PM 27, Pusa Vijay, RH 0119, RH 406, RGN-229, RGN-236, RGN-298, and PM 22 for thermo-tolerance during crop establishment stage are under cultivation.

1.4.2.3 Frost

Freezing injury is a serious threat to the mustard crop. In Northern parts of India, frost is quite common in the month of January which coincides with the blooming and pod formation period. Yield losses up to 70% have been reported due to a single exposure to freezing temperature. Freezing injury reduces the rate of photosynthesis but slightly enhances the transpiration rate. Evaluation of frost tolerance can be done by counting the number of killed/undeveloped seeds having low-temperature stress. Promising donors being used in the breeding program of *B. juncea* are RH 8814, RH 8602, RH 8824, and RH 781. The cultivars RH 781, RH 819, RGN 48, and Swarnjyoti are known for frost tolerance.

1.4.2.4 Salinity

Soil salinity and irrigation with brackish water affect the large areas under mustard cultivation in India especially in the states of Rajasthan, Haryana, Punjab, and Uttar Pradesh. The yield reduction due to salinity/sodicity has been reported up to 33%. The salinity stress is also known to reduce oil content and alter the fatty acid composition. The tolerance to salinity is characterized on the basis of percent reduction in seedling emergence and dry weight. A large number of improved germplasm lines are available viz. CS 416, RH 8816, RH 8701, RK 9807, and RK 9703 which are used as donors under varietal development program. Breeding efforts have resulted in the release of salinity tolerant varieties viz. CS 52, CS 54, CS 56, CS 58 and CS 60.

References

Abraham V, Bhatia CR (1994) Testing for tolerance to aphids in Indian mustard, *Brassica juncea* (L.) Czern and Coss. Plant Breed 112:260–263

Aires A (2015) Brassica composition and food processing A2- Preedy, Victor processing and impact on active components in food, vol 3. Academic Press, pp 17–25

Anonymous (1963) Xian Banpo country, Special issue of archaeology. Archaeology Press, Beijing, China

Awasthi S, Saraswathi NT (2016) Sinigrin, a major glucosinolate from cruciferous vegetables restrains non-enzymatic glycation of albumin. Int J Biol Macromol 83:410-415. https://doi.org/10.1016/j.ijbiomac.2015.11.019

Bahadoran Z, Mirmiran P, Azizi F (2013) Potential efficacy of broccoli sprouts as a unique supplement for management of type 2 diabetes and its complications. J Med Food 16:375–382

Bahadoran Z, Mirmiran P, Hosseinpanah F, Hedayati M, Hosseinpour-Niazi S, Azizi F (2011) Broccoli sprouts reduce oxidative stress in type 2 diabetes: a randomized double-blind clinical trial. Eur J Clin Nutr 65:972–977

Bailey L (1922) The cultivated Brassicas I. Genet Herb 1:53–108

Biga MLB (1955) Review of the species of the genus *Albugo* based on the morphology of the conidia. Sydowia 9:339–358

Boland GJ, Hall R (1994) Index of plant hosts of *Sclerotinia sclerotiorum*. Can J Plant Pathol 16:93–108

Bones AM, Rossiter JT (2006) The enzymic and chemically induced decomposition of glucosinolates. Phytochemistry 67:1053–1067

Chakraborty S, Bhattacharjee P (2018) Design of lemon–mustard nutraceutical beverages based on synergism among antioxidants and in vitro antioxidative, hypoglycaemic and hypocholesterolemic activities: characterization and shelf life studies. J Food Meas Charact 12:2110–2120

Chattopadhyay C, Meena PD, Sastry RK, Meena RL (2003) Relationship among pathological and agronomic attributes for soilborne diseases of three oilseed crops. Indian J Plant Protec 31:127–128

Chaudhary BD, Singh GP, Singh P, Kumar A (1989) Inheritance studies of plant water relations in *Brassica juncea*. Biol Planta 31:202–206

Chauhan JS, Singh KH, Singh VV, Kumar S (2011) Hundred years of rapeseed-mustard breeding in India: accomplishments and future strategies. Indian J Agric Sci 81:1093–1109

Dixon GR (2007) Origins and diversity of Brassica and its relatives. In: Dixon GR (ed) Vegetable Brassicas and related crucifers. CABI, Wallingford UK, pp 1–33

Downey RK (1983) The origin and description of the Brassica oilseed crops. In: Karmar JKG, Sauer FDR, Pigden WJ (eds) High low Erucic acid rapeseed oils: production, usages chemistry and toxicological evalution. Academic Press, New York, pp 1–20

Ehlers A, Florian S, Schumacher F, Meinl W, Lenze D, Hummel M, Heise T, Seidel A, Glatt H, Lampen A (2015) The glucosinolate metabolite 1-methoxy-3-indolylmethyl alcohol induces a gene expression profile in mouse liver similar to the expression signature caused by known genotoxic hepatocarcinogens. Mol Nutr Food Res 59:685–697

Fimognari C, Turrini E, Ferruzzi L, Lenzi M, Hrelia P (2012) Natural isothiocyanates: genotoxic potential versus chemoprevention. Mutat Res Mutat Res 750:107–131

George B, Loeser E (2021) Oilseeds: world markets and trade. United States Department of Agriculture, Washington DC

Hegi G (1919) Illustrierte Flora von Mittel-Europa. 4 Munchen

Jiménez-Osorio AS, Gonzalez-Reyes S, Pedraza-Chaverri J (2015) Natural Nrf2 activators in diabetes. Clin Chim Acta 448:182–192

Kapusta-Duch J, Kopec A, Piatkowska E, Borczak B, Leszczynska T (2012) The beneficial effects of Brassica vegetables on human health. Rocz Państw Zakl Hig 63:389–395

Kelloff GJ, Crowell JA, Steele VE, Lubet RA, Malone WA, Boone CW, Kopelovich L, Hawk ET,

Lieberman R, Lawrence JA (2000) Progress in cancer chemoprevention: development of diet-derived chemopreventive agents. J Nutr 130:467S-471S

Ketkar SS, Rathore AS, Lohidasan S, Rao L, Paradkar AR, Mahadik KR (2014) Investigation of the nutraceutical potential of monofloral Indian mustard bee pollen. J Integr Med 12:379–389

Kim MK, Park JHY (2009) Cruciferous vegetable intake and the risk of human cancer: epidemiological evidence: conference on 'multidisciplinary approaches to nutritional problems' symposium on 'nutrition and health.' Proc Nutr Soc 68:103–110

Kolte SJ, Awasthi RP (1987) Assessment of yield losses due to Alternaria blight in rapeseed and mustard. Indian Phytopathol 40:209–211

Kong XY, Kissen R, Bones A (2012) Characterization of recombinant nitrile-specifier proteins (NSPs) of *Arabidopsis thaliana*: dependency on Fe (II) ions and the effect of glucosinolate substrate and reaction conditions. Phytochemistry 84:7–17

Kristal AR, Lampe JW (2002) Brassica vegetables and prostate cancer risk: a review of the epidemiological evidence. Nutr Cancer 42:1–9

Lawson AP, Long MJC, Coffey RT, Qian Y, Weerapana E, El Oualid F, Hedstrom L (2015) Naturally occurring isothiocyanates exert anticancer effects by inhibiting deubiquitinating enzymes. Cancer Res 75:5130–5142

Maina S, Misinzo G, Bakari G, Kim HY (2020) Human, animal and plant health benefits of glucosinolates and strategies for enhanced bioactivity: a systematic review. Molecules 25:3682

Meena PD, Rani A, Meena R, Sharma P, Gupta R, Chowdappa P (2012) Aggressiveness, diversity and distribution of *Alternaria brassicae* isolates infecting oilseed Brassica in India. Afr J Microbiol Res 6:5249–5258

Mishra A, Pragyandip D, Murthy PN, Siddique HH, Poonam K (2012) A classical review on Rajika (*Brassica juncea*). Res Rev J Biol Sci 1:8–23

Morales-López J, Centeno-Álvarez M, Nieto-Camacho A, López MG, Pérez-Hernández E, Pérez-Hernández N, Fernández-Martínez E (2017) Evaluation of antioxidant and hepatoprotective effects of white cabbage essential oil. Pharm Biol 55:233–241

Nagaharu U, Nagaharu N (1935) Genome analysis in Brassica with special reference to the experimental formation of *B. napus* and peculiar mode of fertilization. Jpn J Bot 7:389–452

Nawaz H, Shad MA, Muzaffar S (2018) Phytochemical composition and antioxidant potential of Brassica. In: Brassica germplasm—characterization. Breed Util, pp 7–26. https://www.intechopen.com/chapters/61233

OECD (2016) Brassica crops (Brassica species), safety assessment of transgenic organisms in the environment, Vol 5, OECD Consensus Documents. OECD Publishing, Paris, pp 151–291

Olsson G (1960) Species crosses within the genus Brassica I. Artificial *Brassica juncea* Coss. Hereditas 46:171–223

Prakash S, Hinata K (1980) Taxonomy, cytogenetics and origin of crop Brassicas, a review. Opera Bot, 1–57

Rahman M, Khatun A, Liu L, Barkla BJ (2018) Brassicaceae mustards: traditional and agronomic uses in Australia and New Zealand. Molecules 23:231

Raymer PL (2002) Canola: an emerging oilseed crop. Trends New Crop New Uses 1:122–126

Razis AFA, Bagatta M, De Nicola R, Iori R, Ioannides C (2011) Up-regulation of cytochrome P450 and phase II enzyme systems in rat precision-cut rat lung slices by the intact glucosinolates, glucoraphanin and glucoerucin. Lung Cancer 71:298–305

Sanlier N, Guler SM (2018) The benefits of Brassica vegetables on human health. J Hum Heal Res 1:1–13

Sharma A, Rai PK (2018) Assessment of bioactive compounds in *Brassica juncea* using chromatographic techniques. J Pharmacogn Phytochem 7: 1274–1277

Sharma A, Sharma A, Yadav P, Singh D (2016) Isothiocyanates in Brassica: potential anti-cancer agents. Asian Pacific J Cancer Prev 17:4507–4510

Sharma AK (1979) Powdery mildew diseases of some crucifers from Jammu and Kashmir State. Indian J Mycol Plant Pathol 9:29–32

Sharma TR, Singh BM (1992) Transfer of resistance to *Alternaria brassicae* in *Brassica juncea* through interspecific hybridization among Brassicas. J Genet Breed 46:373–378

Shaw FJF, Ajrekar SL (1915) The genus Rhizoctonia in India. Botanical Series 7: 77. Mem Dep Agric New Delhi, India

Shigesaburo T (1967) A plant exploration in Brassica and allied genera. Tohoku J Agric Res 17:249–277

Singh DP, Sangwan VP, Pannu RK, Chaudhary BD (1996) Comparison of osmotic adjustment in leaves and siliquae of oilseed Brassicae. Indian J Plant Physiol 1:284–285

Singh M, Chauhan JS, Kumar PR (2003) Response of different rapeseed-mustard varieties for growth, yield and yield component under irrigated and rainfed condition. Indian J Plant Physiol 8:53–59

Song KM, Osborn TC, Williams PH (1988) Brassica taxonomy based on nuclear restriction fragment length polymorphisms (RFLPs). Theor Appl Genet 75:784–794

Spect CE, Diederichsen A (2001) Brassica. In: Hanelt P, Buttner R, Mansfeld R (eds) Mansfeld's Encyclopedia of agricultural and horticultural crops (except ornamentals). Springer, Berlin, Heidelberg, pp 1453–1456

Swati, Sehwag S, Das M (2015) A brief overview: Present status on utilization of mustard oil and cake. Indian J Trad Knowl 14:244–250

Trachootham D, Zhang H, Zhang W, Feng L, Du M, Zhou Y, Chen Z, Pelicano H, Plunkett W, Wierda WG (2008) Effective elimination of fludarabine-resistant CLL cells by PEITC through a redox-mediated mechanism. Blood 112:1912–1922

Vaughan J, Hemingway JS, Schofield HJ (1963) Contributions to a study of variation in *Brassica juncea* Coss. & Czern. Bot J Linn Soc 58:435–447

Vaughan JG, Gordon EI (1973) A taxonomic study of *Brassica juncea* using the techniques of electrophoresis, gas-liquid chromatography and serology. Ann Bot 37:167–184

Warwick SI, Francis A, Al-Shehbaz IA (2006) Brassicaceae: species checklist and database on CD-Rom. Plant Syst Evol 259:249–258

Wittstock U, Burow M (2010) Glucosinolate breakdown in Arabidopsis: mechanism, regulation and biological significance. Arab. Book/American Soc Plant Biol 8

Woods DL, Capcara JJ, Downey RK (1991) The potential of mustard (*Brassica juncea* (L.) Coss) as an edible oil crop on the Canadian Prairies. Can J Plant Sci 71:195–198

Xu K, Thornalley PJ (2001) Involvement of glutathione metabolism in the cytotoxicity of the phenethyl isothiocyanate and its cysteine conjugate to human leukaemia cells in vitro. Biochem Pharmacol 61: 165–177

Xu Z, Wang S, Ji H, Zhang Z, Chen J, Tan Y, Wintergerst K, Zheng Y, Sun J, Cai L (2016) Broccoli sprout extract prevents diabetic cardiomyopathy via Nrf2 activation in db/db T2DM mice. Sci Rep 6:1–12

2 Botanical Descriptions of *Brassica juncea*: Taxonomy, Cytology, Cytogenetics and Phylogenetic Relationships

Arun Kumar, V. V. Singh, and Priyamedha

Abstract

The genus *Brassica* is one of the most important plant families within family *Crucriferae* (*Brassicaceae*) which includes economically important cultivated crop species, ornamentals, with many wild and weedy relatives. Cytology and cytogenetics has played significantly a prominent role in terms of elucidating genetic architecture of the crop species. Cytological researches on Brassicas initiated with the confirmation of somatic chromosome numbers in various crop species between 1916 and 1930, followed by genomic analysis of *Brassica* by Morinaga and U (1928–1935) that gave acceleration to another group of researchers to look more passionately into genome homology, phylogenetic relationships and investigations of chromosome evolution in the family *Brassicaceae*. With the development to tissue culture techniques, viz. ovary culture and embryo rescue and protoplast fusion in the 1950s that enabled research workers to make enormous progress in the detailed investigations of genome homologies on the basis of chromosome pairing analyses. Form a cytology and cytogenetic perspective in the period from 1960 to 1980s mostly lead to a concentrated effort undertaken for collection and classification of botanical taxa of the *Brassicaceae*. One of the important contributions was made by Harberd (Bot J Linn Soc 65:1–23, 1972). His studies led to the classification of cytodemes describing homologous genomes. It is now well known that there is a huge amount of genome homology or homoeology throughout the entire *Brassica* coeno-species, as well as from a plant breeding point of view, we there upon have the opportunity to broaden genetic base through introgression of unique genes of economic important traits and/or alleles well beyond the species barrier. A remarkable advance in *Brassica* cytology and cytogenetics during the last two to three decades is the use of molecular markers. In situ hybridization has been successfully used for identification and characterization of individual chromosomes and assimilation of genetic and physical maps. Using another molecular cytogenetic tool like Genomic in situ hybridization (GISH), A, B and C genome chromosomes were differentiated in all of polyploid species. GISHis also used to identify alien/unusual chromosome introgressions in many interspecific/or intergeneric wide hybrids. Linking of cytology, cytogenetic and molecular cytogenetics with other molecular biological techniques; for example, DNA sequencing and protein expression analysis have bridged the gap

A. Kumar · V. V. Singh (✉) · Priyamedha
ICAR-Directorate of Rapeseed-Mustard Research, Sewar, Bharatpur, Rajasthan, India

C. Kole and T. Mohapatra (eds.), *The Brassica juncea Genome*, Compendium of Plant Genomes,
https://doi.org/10.1007/978-3-030-91507-0_2

between chromosomes and linkage groups and resulted in a plenty of new and valuable information in the *Brassicaceae* family.

2.1 Introduction

The *Crucriferae* (*Brassicaceae*) family is a very important plant family which includes a number of economically important cultivated species such as oleiferous (oil yielding), leafy (vegetable yielding), rapiferous (rapeseed yielding), ornamental flowers and forage crop forms. Three monogenomic diploid species, specifically *B. rapa* L. (2*n* = 20, AA genome), *B. nigra* (L.) Koch (2*n* = 16, BB) and *B. oleracea* L. (2*n* = 18, CC) and their allo tetraploid species, viz. *B. napus* L. (2*n* = 38, AACC), *B. juncea* (L.) Czern. & Coss. (2*n* = 36, AABB) and *B. carinata* A. Braun (2*n* = 34, BBCC) are represented by the genus *Brassica* (Kaneko and Bang 2014).

The genesis of cytogenetical research in Brassicas was initiated by determination of somatic chromosome number in *Brassica rapa* (syn. *B. campesteris*) by Takamine (1916). Notwithstanding that Sageret (1826) and Herbert (1847) reported intergeneric hybrids in *Raphanussativus* × *B. oleracea*, *B. napus* × *B. rapa*, respectively, much earlier. Two researchers, Karpechenko (1927) and Manton (1932), also lead the way in cytogenetical investigations; the former synthesized *Raphanobrassica*, and the latter determined the mitotic and meiotic chromosome number for number of species. The pioneer works of Moringa (1928, 1934a) and Nagaharu (1935) investigated the meiotic behaviour of chromosomes in hybridization between different Brassica species and elucidated genetic architecture of the crop species.

A comprehensive range of wild relatives and weed species related to crop Brassicas and possessing extensive genetic diversity and variability occur in the Mediterranean region, particularly in Spain, Morocco and Algeria and have a wide distribution from Mediterranean to north-west of India. Mizushima (1950, 1968), Harbard (1972, 1976) and Harbard and McArthur (1980) extensively investigated this germplasm which included determination of chromosome number, hybridization among different species and studies on the chromosome pairing in interspecific or intergeneric hybrids to interpret homology and determine phylogenetic relationships among them. Takahata and Hinata (1983) and Warwick and Black (1997) extended this study further by adding a few more genera. While early phase of cytogenetic researches was mainly in or around polyploidy breeding, the later attention shifted to expanding of genetic variability, introgressing nuclear genes that conferred desirable agronomic traits or cytoplasmic genes for inducing male sterility and chromosome addition lines to locate genes on specific chromosomes.

In *Brassica* due to meagre size of chromosomes, lack of unique cytological landmarks and not being amenable to pachytene investigations were largely a hindrance to cytogenetical analyses. With the development of tissue culture techniques, viz. ovary culture and embryo rescue and protoplast fusion in the 1950s made excessive growth in the study of genome homologies based on chromosome pairing analyses. Following the initial developments, *Brassica* cytogenetical researches from elucidating complex species relationships in the *Brassica* crop species and wild relatives have achieved amazing improvements. These developments from classical cytogenetics to some recent achievements can be divided in two phases and will be discussed in *Brassica* crop.

2.2 Taxonomy

Brassica oilseed crops belong to the family *Brassicaceae* (Syn. *Crucifarae*). The family presently has about 3709 species and 338 genera (Warwick et al. 2006) (Table 2.1) and is one of the ten most important plant families in terms of economic importance (Rich 1991). And within *Brassicaceae*, the genus *Brassica* is one of the ten core genera. The genera display excessive plant diversity and have a number of wild and weed species related to crop *Brassica* in nature.

Table 2.1 Taxonomic classification of Family *Brassicaceae*

Kingdom	Plantae—Plants
Subkingdom	Tracheobionta—Vascular plants
Superdivision	Spermatophyta—Seed plants
Division	Magnoliophyta—Flowering plants
Class	Magnoliopsida—Dicotyledons
Subclass	Dilleniidae
Order	Capparales
Family	*Brassicaceae*—Mustard family

The *Brassica* crops are extensively used as a source of oil, vegetable, condiments and fodder. Those of particular importance are *Brassica juncea, B. rapa* syn. *B. campestris, B. napus* and *B. carinata* as source of edible oil; *B. oleracea* as cole-crops and *B. nigra* as seed condiments. The genus *Raphanus* is bred for its edible root, *Sinapis* as antecedent of oil and condiments and *Eruca* as a source of green salad, vegetable and non-edible industrial oil.

2.3 Morphological Description

The family consists mostly of herbaceous plants with annual, biennial or perennial lifespans. The leaves are alternate (hardly opposite), at times organized as basal rosettes, in very rarely shrub crucifers of Mediterranean, their leaves most often in terminal rosettes, and sometimes coriaceous and evergreen. Leaves pinnately incised and exstipulate. The flowers of this family is exceptionally uniform and four free saccate sepals and four clawed free petals. They can be dissymmetric or sometimes zygomorphic, with a classic cross-like structure (hence the name '*Cruciferae*'). Flower possess six stamens, four longer (as long as the petals) and are organized in a cross-like, and two stamens are shorter, hence known as Tetradynamous flower. The pistil comprises of two fused carpels with very short style and stigma is bilobed. Superior ovary, unilocular and become bilocular with development of false septum known as replum. The flowers are ebracteate with racemose inflorescences, often apically corymb-like in florescences also encountered. Pollination is facilitated by entomogamy, nectar glands present at the base of the four longer stamens and stored on the sepals (Fig. 2.1). The fruit is a capsule named as siliqua (plural siliquae). It opens by two valves, which are the modified carpels; the seeds are attached to replum and dehisce from below to upward. The siliqua sometimes dissect apart at constrictions present between the segments of the seeds, forming a loment (e.g. *Raphanus*); it may expel the seeds completely (e.g. Cardamine) or can evolve in a sort of samara (e.g. Isatis).

Family *Brassicaceae* is categorized on the basis of the existence of conduplicate cotyledons, (cotyledons folded around the radical lengthwise) and/or two-segmented fruits (siliquae) that contains seed in one or both segments and only simple hairs, if present (Gomez-Campo 1980,

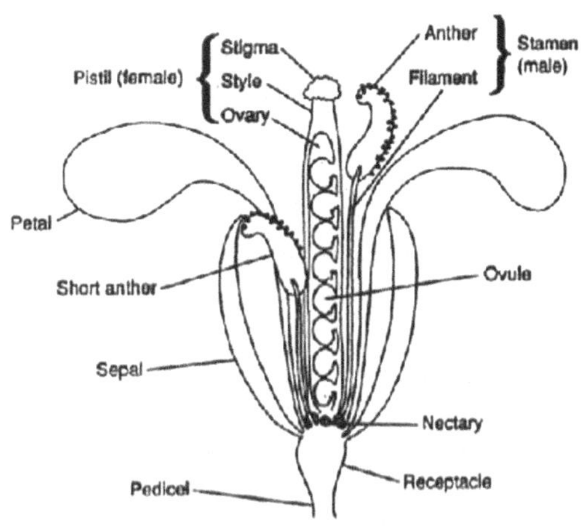

Fig. 2.1 A typical flower of family *Brassicaceae*

1999; Al-Shehbaz 1985). The following three important features separate the family *Brassicaceae* from all other plant families:

1. The stamens in tetradynamous condition (4 + 2), four long and two short stamens in each flower.
2. The flowers have four petals that form a cross.
3. The pod, called siliqua, has false septum known as replum (thin translucent inner membrane), separating two sides of siliqua and to which the seeds are attached.

B. juncea, B. rapa, Brassica napus and *B. carinata* are the most common species of genus *Brassica* cultivated for edible oil production worldwide. The brief description is as under:

Brassica juncea: It is known as mustard. The plants are tall (90–200 cm), erect and much branched. The leaf is dilated from base and stalked, broad and pinnatified. The fruits are known as siliquae and range about 2–6.5 cm in length, increases with short and stout beak. The seed colour is brown or dark brown with rough seed coat.

Brassica rapa: It is known as rapeseed and have many important oilseed varieties of toria, yellow sarson and brown sarson. The plant is shorter than mustard and plant height is between 45 and 150 cm; the stem is largely covered with a wax deposit. Plants can be distinguished from mustard plant by the characters of the leaves. In rape, the leaves are sessile (lower leaf lamina surrounding the stem completely or partially), glabrous and hairy. Fruits are thicker in comparison to *B. juncea* and are laterally compressed, with a beak one-third to half their length. Seeds are yellow or brown with smooth seed coat.

Varying amount of self- and cross-pollination and self- and cross-incompatibility occur in different species and strains. *B. juncea* is self-fertile and largely self-pollinated. Due to insects about 4–14%, cross-pollination may take place. In *B. rapa*, var. yellow sarson and toria form of brown sarson are self-fertile and largely self-pollinated; although 5–12% cross-pollination may occur in yellow sarson due to insect pollination. The *lotni* type of brown sarson and toria are cross-pollinated as a result of self-sterility. It has been found that in toria, the pollen tubes require 24 to 48 h to reach the ovule in self-fertilization in comparison to just 5 h with cross-pollination. This shows that the sterility may be due to self-incompatibility of pollen and stigma.

Brassica napus: *B. napus* is known by the name rape, rapeseed or canola, a mainly cultivated crop of Europe, Canada, China and Australia. In Asian countries, this crop is mostly cultivated as winter rapeseed. This species is divided into two sub-species which includes swedes (*B. napus* ssp. *napobrassica*) and *B. napus* ssp. *napus* which comprises both winter and spring oilseed crops. The seed is dark in colour and no natural yellow-seeded forms are known to occur. Mediterranean region is believed to be place of origin where both the parents *B. campestris* and *B. oleracea* grow together in wild form (Sinskaia 1928).

Brassica carinata: *B. carinata* is commonly known as Abyssinian or Ethiopian mustard. It is comparatively slow growing than other *Brassica* species, and it is mostly cultivated in the Ethiopia and border areas in East Africa. *B. carinata* has several agronomically important traits that are rare in other brassica oilseed crops such as non-dehiscent siliquae and a much more developed and aggressive root system, heat, drought tolerance, tolerance to various biotic and abiotic stresses. East African plateau is regarded as its place of origin.

2.4 *Brassica* Cytology and Cytogenetics

2.4.1 The First Phase (1916–1970)

The dawn in the area of *Brassica* cytogenetic research was the investigations on somatic chromosomes by several workers (Takamine 1916; Manton 1932; Catcheside 1934).

Takamine in 1916 reported the mitotic chromosome number in *Brassica rapa* (syn. *B. campesteris*); Karpechenko (1927) investigated the chromosome number of *B. nigra*, *B. oleracea* and *B. carinata*. During the same period, Karpechenko (1927) synthesized *Raphanobrassica* following hybridization between *Raphanus sativus* ($2n = 18$) × *B. oleracea* ($2n = 18$), and that became the first example of a new species obtained through experimental hybridization. However, Sageret (1826) and Herbert (1847) though reported intergeneric and intergeneric hybrids in *Raphanussativus* × *B. oleracea*, *B. napus* × *B. rapa*, respectively, much earlier. Jaretzky (1932) and Manton (1932) carried out intensive surveys of the family *Brassicaceae* for somatic chromosomes and published number of new chromosome numbers of the family.

Morinaga (1928, 1929a, b, c, 1931, 1933, 1934a, b) began his experiments in 1920s with hybridization between all six crop species in various combinations and investigated chromosome pairing/behaviour of the hybrids. He proposed that *Brassica* crop species group comprises three diploid monogenomic species viz. *B. nigra* (L.) Koch ($n = 8$, B), *B. oleracea* L. ($n = 9$, C) and *B. rapa* L. ($n = 10$, A) and three high chromosome allot etraploid species, viz. *B. napus* L. ($n = 19$, AC), *B. juncea* (L.) Czern. &Coss. ($n = 18$, AB), and *B. carinata* A. Braun ($n = 17$, BC). These species are digenomic and their evolution took place in natural environment via convergent alloploid evolution with any two of the diploid species and also assigned genome symbols to these species. At the same time, Nagaharu U investigated detailed cytogenetic relationships of *Brassica* crop species, which is now commonly referred to as U triangle (U 1935) (Fig. 2.2). He further extended the *Brassica* genome by studying the chromosome pairing/behaviour in interspecific hybrids between *B. carinata* × *B. oleracea*, *B. carinata* × *B. nigra*, *B. napus* × *B. carinata*, *B. juncea* × *B. carinata*, *B. rapa* × *B. oleracea*, *B. napus* × *B. oleracea* and *B. napus* × *B. rapa*. Nagaharu U verified the allopolyploid origin of three high chromosome species viz. *B. carinata*, *B. juncea* and *B. napus* and confirmed its validity by experimentally synthesizing *B. napus* from its parents, i.e. *B. rapa* and *B. oleracea*.

Studies on karyotype investigations on somatic chromosomes of *Brassica* species were initiated in 1934 by Catcheside and later followed by Alam (1936), Richharia (1937) and Sikka (1940). These studies distinguished chromosomes into long, medium, small and very small with median, subterminal and terminal constrictions. Since *Brassica* chromosomes are very small in size, lack of visible cytological landmarks, the studies were mostly confined to frequency of various chromosomes types, number of satellite chromosomes and nucleoli in different genomes (Prakash 2010). The advancement of ovary culture and embryo rescue techniques in the 1950s helped colossal progress in the study of genome homologies based on chromosome pairing analyses. Additionally, technological advances in optical equipment and microscopy brought a great improvement in cytological techniques in general, and based on these techniques, Robbelen (1960) was the first to report the detailed cytological descriptions of *Brassica* somatic chromosome structure. He for the first time, analysed pachytene chromosomes in *B. nigra*, *B. oleracea* and *B. rapa*. The chromosomes revealed differentiation into proximal heterochromatic and distal euchromatic segments. Individual chromosomes within the genomes were identified by the number, size and distribution pattern of the heterochromatic segments near the centromeres (Prakash et al. 2009).

2.4.2 The Second Phase (1970 to Date)

Researches on related wild germplasm of *Brassica* crop

From a cytogenetics as well as cytological approaches, the second phase starts with the researches on wild germplasm. A range of wild species related to crop Brassicas and possessing extensive genetic variability was found in the Mediterranean region—particularly in Spain, Morocco and Algeria and have a broad distribution from Mediterranean to the north-west of

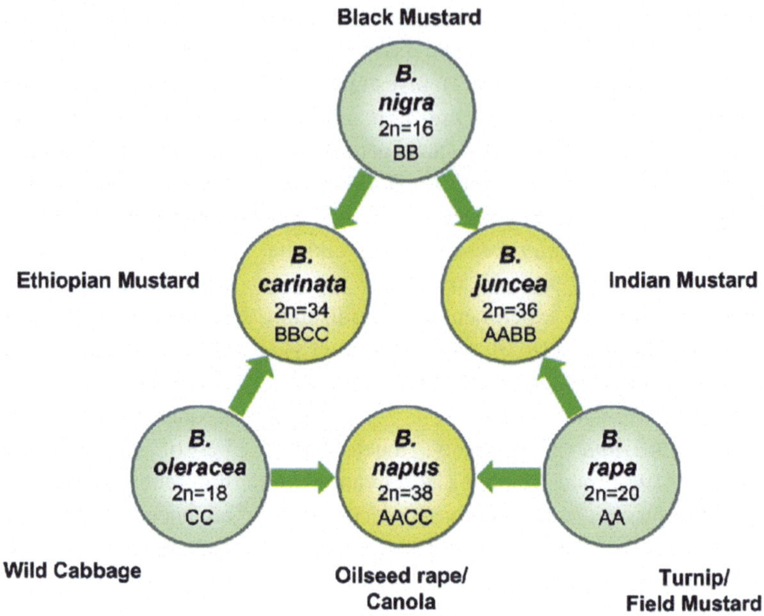

Fig. 2.2 The 'Triangle of U' (Nagaharu 1935) showing the genetic relationships between the important species of genus *Brassica* (*Source* viswiki.com/en/triangle of U)

India. Mizushima (1950), Harberd (1972, 1976) and Harberd and McArthur (1980) extensively investigated this germplasm which included determination of somatic chromosome number, hybridization among different species and studying the chromosome pairing in interspecific or intergeneric hybrids to interpret homology and determine phylogenetic relationships among them. Wild germplasm is a rich pool of many useful agronomic traits of economic importance, such as resistance to pests and insects including abiotic stresses as well as having mitochondrial genome variability which can be useful for inducing cytoplasmic male sterility of variable origin in crop species.

2.4.3 Cytogenetic Structure of *Brassica* Cenospecies

Harberd (1972) was the first to propose the concept based on his extensive cyto-taxonomical investigations on Brassica crop and related wild species. Harberd classified the germplasm which is known as Brassica cenospecies into cytodemes or crossing groups. His study included bivalent pairing between a huge number of species that eventually led to the classification of cytodemes describing homologous genomes. This study was further extended by Takahata and Hinata (1983) and Warwick and Black (1997) who added a few more genera. At present, 63 cytodemes are identified in the wild germplasm and comprise both diploids and tetraploids. All these species are related to crop Brassicas and capable of exchanging genetic material among them (Fig. 2.3). Coeno-species comprise 13 genera, viz. *Brassica, Diplotaxis, Eruca, Erucastrum, Coincya, Enarthrocarpus, Sinapis, Sinapidendron, Trachystoma, Raphanus, Hirschfeldia, Moricandia, Rytidocarpus* and *Pseuderucaria* and closely corresponds to subtribe *Brassicineae* of Schulz (1936), in addition to two genera from subtribe *Raphaninae* and three genera from subtribe *moricandinae.* The minimum chromosome number in coeno-species, $n = 7$, is represented by seven cytodemes. Harberd (1972) was in the opinion that cytodeme with $n = 14$ or higher chromosome numbers should be accredited to polyploidy. According to this opinion, 43 cytodemes are diploids where every chromosome number from $n = 7$ to $n = 13$ is illustrated. However, variations in isozyme numbers of a broad range of taxa in the tribe Brassicacineae suggested that genera with $n = 14$ to 18 may not be necessarily polyploids of $n = 7$ to 13

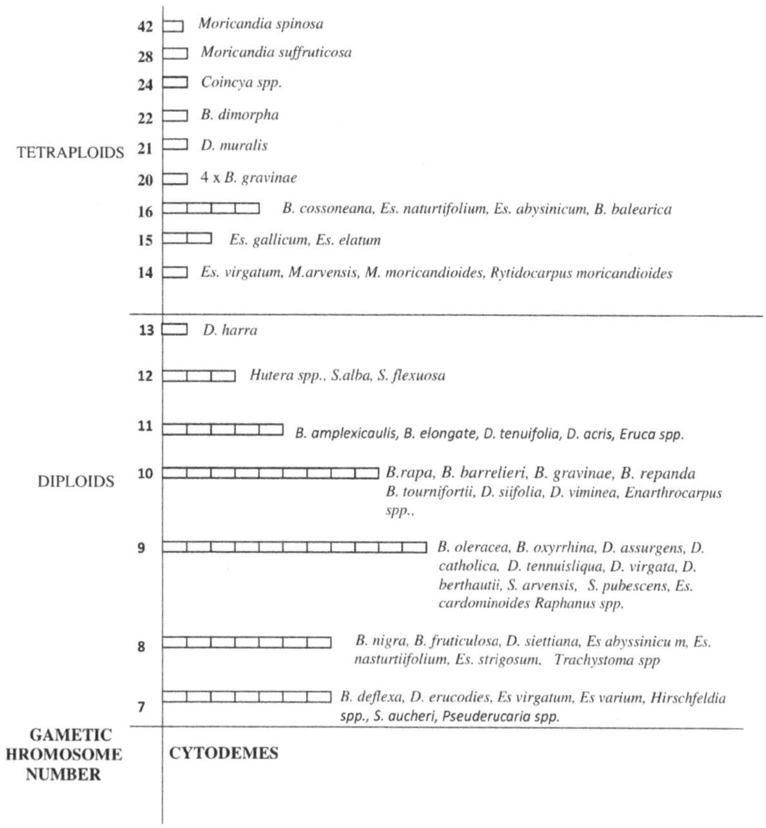

Fig. 2.3 Cytodemes in *Brassica* coeno-species (*B* = *Brassica*; *D* = *Diplotaxis*: *Es*. = *Erucastrum*; *S* = *sinapis*, *M* = *moricandia*) (*Source* Prakash et al. 2009 Plant Breeding Rev 3:21–187)

genomes. About half of the cytodemes have haploid chromosome number $n = 9$ and $n = 10$. Polyploidy also played a role as both auto- and allopolyploids which are represented by 20 cytodemes. This polyploidy level is overtaken by only in some of *Moricandia spinosa* ($2n = 84$, $x = 6$) accessions and *Brassica repanda* ($2n = 160$, $x = 8$) in general. The genus *Moricandia* appears to be polyploid.

2.4.4 *Brassica* Karyotypes

During nineties, karyotype studies were conducted based on somatic chromosome numbers, especially on diploid species by Olin-Fatih and Heneen (1992); Cheng et al. (1995) and Fukui et al. (1998). Meiotic chromosome was investigated in detail by Mackowiak and Heneen (1999) and Koo et al. (2004), and their morpho-types have been prepared using different methods of staining. For characterization of individual chromosomes and construction of karyotypes, mitotic prometaphase and meiotic diakinesis stages of meiosis offer much better possibilities. With the beginning of 1990s, use of fluorescence in situ hybridization (FISH) with ribosomal DNA probes has further helped in generating chromosome markers. Mitotic chromosomes of A and C genomes are phenotypically similar and difficult to distinguish as was established by Olin-Fatih and Heneen (1992). However, now with the use of techniques like FISH and genomic in situ hybridization (GISH), it is possible to identify individual chromosomes of A, B and C genomes and also to match chromosomes with corresponding counterparts in alloploid species with considerable accuracy (Snowdon et al. 2002a Kamisugi et al. 1998; Howell et al. 2002). All these studies are very fruitful in integrating genetic maps, arising from study of molecular markers, with physical maps on the basis of cytomorphic investigations. Many research papers appeared since 1934 characterizing the karyotypes of *Brassica* species. The general observations can be summarized in the way that chromosomes of A genome are morphologically more diverse, B genome chromosomes are generally uniform and are difficult to identify at individual label and C genome chromosomes are poorly differentiated in morphological terms and size, and these chromosomes show variable degree of condensation in hetero- and euchromatin (Olin-Fatih 1994).

2.5 Wide Hybridization and Introgression

Wide or distance hybridization in Brassicas dates back to early nineteenth century when Sageret (1826) first reported hybrid *Raphanus-sativus* × *B. oleracea* and Herbert (1847) reported interspecific hybrid between *B. napus* and *B. rapa*. These researches in the area of cytogenetics by the above researchers encouraged others to shift towards wide hybridization. Though, the initial attempts of hybridizations were for elucidating genomic homoeology. But in later years, focus shifted towards utilization of wide hybridization for creation of genetic variability, introgression of novel/unique genes or alleles for economic agronomic traits or for exploiting cytoplasmic genes for inducing male sterility. Chromosome addition lines have been produced for locating specific genes on chromosomes and for construction of genetic maps (Prakash et al. 2009). From the last 40 years, the development of in vitro techniques such as sequential ovary and embryo culture and protoplast fusion have enabled fruitful growth in the investigations of genome homologies based on chromosome associations analyses and obtaining a large number of sexual and somatic hybrids.

Although wide hybridizations in Brassica nowadays is very common, utilization of species of secondary and tertiary gene pools will be favourable in terms of creating unique genetic variability for distinct biotic and abiotic stresses as well as present changing climatic conditions. Some intensive investigations were made by Harberd and McArthur (1980), using these secondary and tertiary germplasm and reported about 50 wide hybrids in which a majority were intergeneric hybrids. In sexual hybrids, meiotic behaviour of chromosome is highly disorganized, particularly when both parents are diploid due to the inability of a homologous pairing, they remain mostly as univalents. But occasionally, pairing occurs and also forms bivalents in a very low frequency. Bivalents are mostly rod-shaped, monochiasmates and rarely ring-shaped chromosome with multiple chiasmata are observed. Multivalent chromosomes if encountered in diploid hybrids are observed very rarely. However, a variable number of bivalents and frequent multivalents in the form of trivalents and quadrivalents are observed in combination of triploid (tetraploid × diploid) and tetraploid (tetraploid × tetraploid species). A close relationship between mean chromosome number and mean bivalent frequency at three ploidy levels was observed by Harberd and McArthur (1980) (Table 2.2).

In diploid hybrids, high chromosome associations have been reported in several combinations; for example, in *B. tournefortii* × *B. rapa* 1IV + 1III + 4II (Kumar et al. 2015a); *B. rapa* *B. fruticulosa* 1III + 8II (Kumar et al. 2013) in sexual hybrids, *E. cardaminoides* × *B. oleracea* ($2n$ = 18, 1IV + 1III + 1 II, Mohanty 1996), *Enarthrocarpus lyratus* × *B. rapa* ($2n$ = 20, 2III + 4II, Gundimeda et al. 1992), *Diplotaxis erucoides* × *B. nigra* ($2n$ = 15, 6II, Quiros et al. 1988) and *Erucastrum canariense* × *B. oleracea* ($2n$ = 18, 8II, Harberd and McArthur 1980). Higher chromosome associations have been observed in triploid and tetraploid hybrids; few examples include. *Brassica carinata* × *B. tournefortii* (1IV, 2III, Choudhary and Joshi 2012), *B juncea* × *Diplotaxis virgata* (1 IV, 2 III, Inomata 2003), *Diplotaxis viminea* × *B. napus* (2 IV, Mohanty 1996) and *B. napus* × *Hirschfeldia incana* (1 IV, Kerlan et al. 1993). Hybrids between the diploids were absolutely pollen and seed sterile while triploid

Table 2.2 Mean chromosome number and bivalent frequency at three ploidy levels in the tribe *Brassiceae*

Hybrid	Mean chromosome number, 2*n*	Mean bivalent frequency
Diploids	18.7	2.9
Triploids	25.6	6.2
Tetraploids	34.3	10.2

and tetraploid hybrids had a little pollen and seed fertility. The observation of bivalents in various pollen mother cells (PMCs) in the hybrids could be interpreted to be due to archaic homology within the chromosomes of the same genome (autosyndesis). However, observations of higher chromosome associations in terms of tetravalents or trivalents, can be attributed to allosyndesis indicating partial homoeology or segmental allopolyploidy between the two parental genomes (Prakash et al. 2009). Occasional laggards, including late disjunction of bivalents, intermixing of chromatids at one pole, bridge fragment configuration are recorded at anaphase I and II (Kumar et al. 2015a). The possibility of occurrence of such phenomenon results from chiasma formation within a heterozygous inversion. Such chiasmata indicates true homology and pairing between the genomes involved is being reported by Attia and Robbelen (1986).

Polyploidy is recognized as another outstanding feature in the evolution of higher plants and adaptation, and polyploids have often been frequently selected during the evolution of crop plants. From a plant breeding point of reference, induction of polyploidy may generate new genetic recombinations, which provides more variability in breeding material for breeder to utilize in crop improvements.

Induction of autotetraploidy using colchicine and cyto-morphological evaluation and the genetic analysis of polyploids have been a subject of interest among geneticists and breeders for a long time as revealed by various studies (Otto and Whitton 2000; Reiseberg 2001; Ramsey and Schemske 2002; Mable 2003). It is now well established that in the genus *Brassica,* three low-chromosome monogenomic diploids, *B. nigra*, *B. oleracea*, and *B. rapa* (syn. *B. campestris*) and three high-chromosome digenomics *B. carinata*, *B. juncea* and *B. napus* evolved in nature through convergent alloploid evolution between any two of the diploid species. This genomic relationship of the six Brassicas (Nagaharu 1935) has been substantiated by molecular analysis of nuclear and chloroplast DNA and by GISH (Snowdon 2007). As reported by Osborn (2004) that polyploidy plays important role in contributing morphological variation via several mechanisms, it induces the significant variation of dosage-regulated gene expression and affect flowering time variation through the gene *FLC* (*Flowering Locus C*) in *Brassica*. Trigenomic hexaploids through interspecific hybridization between *B. napus* and *B. nigra* by colchicine treatment has been developed by Pradhan et al. (2010) in *Brassica*. These hexaploids showed enhancement in some of the phenotypic characters such as size of flowers, pollen and stomata increased pollen viability and hexaploid chromosome count. These hexaploid genotypes can be used as a genetic stock for crosses with other Brassica hexaploid genotypes from the same or different sources.

Successful autotetraploidy has been generated in *B. fruticulosa* Cyr. sub sp. *fruticulosa* (2*n* = 16 FF) wild relative of Brassica using aqueous colchicine by the cotton-swab method, and their subsequent utilization in interspecific hybridization with *B. juncea* by Kumar et al. (2015b, 2018). The synthesized plants showed unique enhancement in several phenotypical and floral aspects making them more robust. From cytological analysis point of view, studies revealed that quadrivalent frequency was low which ranged between 2.2 and 3.1 and bivalent frequency was high that ranged from 6 to 12 per cell. Univalent frequency ranging from 3.4 to 4.1 was characteristicof the colchicine-induced tetraploids that have been reported by Kumar et al. (2015b). The synthesized autotetraploids plants may have opportunities to be helpful in hybridization programme with *Brassica* and other *Brassica* species for introgression of economic traits, especially mustard aphid in crop improvement in Brassicas.

2.6 Phylogenetic Relationships

Theodosius Dobzhansky (1900–1975), the famous biologist, said that nothing in biology makes sense except in the light of evolution and molecular phylogenetic analyses which lead to exceptional understandings in the relationships of plants and other green plants at various levels that formed basis in a reclassification of the angiosperms, particularly on the analysis of different genes such as*rbc*L, *atp*B of the cpDNA and the 18S rDNA reviewed by Haider (2013). Phylogenetic information has come up with new view point on the evolution of polyploid genomes such as those of Brassicas. U-triangle *Brassica* species serves the best model systems for the study of polyploidy as given by Maluszynska and Hasterok (2005). The botanical relationship between the *Brassica* species is well established as result of several taxonomic investigations carried out in the 1930s by Morinaga (1929a, 1934a, b; 1928) and Nagaharu (1935) as discussed earlier.

Chromosome analysis has further suggested an early evolution from a common progenitor species with a basic chromosome number of $n = 6$, and that diploid *Brassica* species, with $n = 8$, 9 and 10 resulted from secondary balanced polyploidy (Robbelen 1960). Truco et al. (1996), studied the intra-genomic homology of the Brassica genome. They found inter-genomic conserved reasons with extensive recording among the genomes. Eighteen linkages from all three species could be associated on the basis of homologues segments, based on at least three homologues conservation which was also observed for some of the chromosomes of A, B and C genomes. A possible chromosome phylogenetic pathway based on an ancestral genome of at least five and no more than seven chromosomes were drawn from the inter-relationship observed. These duplication and rearrangement have been involved in the formation of Brassica genome from a similar ancestral genome. Lazaro and Aguinagalde (1996) studied the phylogenetic relationship among the wild related species of the *B. oleracea* with the help of random amplified polymorphic DNA (RAPD). These genetic markers, which allowed the different taxa to be distinguished, were used to investigate the phylogenetic and evaluative relationship of the wild *Brassica* ($n = 9$) species. Three clearly separated braches are shown, the western group, the silican group and the Aegean group. Maximum diversity was detected in the Asian group and highest similarity coefficient was shown by the silican group. Results were consistent with previous photochemical analysis. Yang et al. (1998) studied the phylogenetic position of *Raphanus* in relation to *Brassica* species based on 58 rRNA spacer sequence data. Based on restriction fragment length polymorphism (RFLP) analysis, the evolutionary lineages for *Brassica* diploid species have been purposed. These two lineages are the *nigra* and the second one, the *rapa*/*oleracea* lineage. The phylogenetic relationship of *Raphanus* species to this lineage is still not clear since the chloroplast and mitochondrial DNA genomic restriction site variation reflected that *Raphanus* is more close to the *rapa/olaracea* lineage, whereas nuclear RFLP and other lines of evidence reflected that the *Raphanus* is close to *nigra* lineage. Here, evidence is presented of the intergenic spacer of nuclear 58 rRNA to support that *Raphanus* is more related to *nigra* linage than to *rapa*/*olaracea*.

For inferring phylogenetic relationships among *Brassica* species, basic features of the nuclear genome are need to be considered here. For instance, it has been reported by various researchers that the diploid cultivated species of *Brassica* are secondary polyploids (Gómez-Campo and Prakash 1999) as part of their genome is represented in duplicate and/or in triplicate by Quiros et al. (1987). These species are considered that they evolved through changes in chromosomes and genes from a prototype with a basic chromosome number of five reported by Sikka (1940) or six as reported by Prakash and Hinata (1980) though there are no known *Brassica* species in nature with genomes of less than

$n = 7$ chromosomes based on the cytogenetic studies such as karyotypic analysis. On the basis of these studies, a partial genomic homology relationship among the genomes A, B and C has been reported by different researchers based on cytogenetic studies like by Morinaga (1934a, b), Nagaharu (1935), Attia and Robbelen (1986) and Attia et al. (1987), studies on nRFLP by Song et al. (1988) and Hosaka et al. (1990), and RAPD (Quiros et al. 1991).

2.7 Molecular Cytogenetics

A remarkable advance in Brassica cytogenetics is the use of molecular markers. During the last two to three decades, in situ hybridization has successfully been attempted for reliable description and characterization of individual chromosome to construct the karyotypes of U triangle species, to determine the genomic components of alloploid species and assimilation of genetic and physical maps. Using another molecular cytogenetic tool like GISH, A, B and C genome chromosomes were differentiated in alloploid species (Snowdon et al. 2002; Snowdon 2007; Maluszynska and Hasterok 2005). GISH was also used to identify alien chromosome introgressions in many interspecific/or intergeneric wide hybrids. Such as in *B. napus* × *Eruca sativa* (Fahleson et al. 1997); *Sinapis arevensis* (Snowdon et al. 2000) and in *Crambe abyssinica* by Wang et al. in 2004.

Fluorescence in situ hybridization (FISH) was, for the first time, used in Brassica crop to determine the number of rDNA loci in U triangle species by Malusyzynska and Heslop-Harrison in 1993. The utility of this approach in all the diploid and alloploid species have amply been demonstrated. Now, with advancement of multicolour FISH with 45S rDNA and 5S rDNA is being increasingly used in Brassica molecular cytogenetics. Individual chromosomes of all the U triangle species have been characterized in detail, and their karyotypes are now well documented (Snowdon et al. 2002; Maluszynska and Hasterok 2005). Translocations and rearrangements has been explained well with use of in situ hybridization (ISH) technique for forecasting of ancestral karyotypes in *Brassicaceae* using ISH technique with painting of whole chromosomes with bacterial artificial chromosomes (BACs) by Mandakova and Lysak in 2008. Using this technique, Lysak et al. (2007) also provided additional information for a hypothesized whole-genome triplication in the common ancestor that includes species of *Brassica* and *Sinapis* (tribe *Brassiceae*).

Now, BAC libraries are the most common approach used as sources of large-insert DNA clones in Brassica molecular cytogenetics, and BAC clones have efficiently utilized in assigning 9 linkage groups in *B. oleracea* by Howell et al. in 2002. Lot of progress has been made in cytomolecular mapping of *B. rapa* by Brassica Genome project (Yang 2005; Lim et al. 2006) and to analyse the structure of centromeric region (Lim et al. 2007) has been critically reviewed by Prakash et al. (2009). As the genomic studies on Arabidopsis (Crucifer) progresses, which is closely related to Brassicas, may serve the base in for a new insight in Brassica genome and may play critical role in the direction of evolutionary studies of the *Brassicaceae*. Studies on comparative genomics between Arabidopsis and *Brassica* species have elucidated evolutionary processes. Arabidopsis has now become a model plant in the area of experimental biology in particular in the family of *Brassicaceae* (Prakash 2010).

2.8 Conclusion

In Brassica, due to small size of chromosomes, lack of unique cytological land marks and not being amenable to pachytene investigations were a large hindrance to cytogenetical analyses; even then, many eminent researchers have contributed a lot to Brassica cytology and in cytogenetics. Initiation started with determination of somatic chromosome number in early nineteenth century followed by genome analysis by Morinaga and Nagaharu U, leading to unraveling the genetic architecture of this crop. The enhanced use of wild relatives related to *Brassica* germplasm for incorporating new and novel genes has become a common programme in *Brassica* crop improvement. Incorporation of advance tissue culture techniques like protoplast regeneration, fusion generated number of somatic hybrids. Unprecedented developments made in molecular cytogenetics, FISH, GISH and of sequence information facilitated the detailed investigations in meiosis and meiotic behaviour of chromosomes. Molecular cytogenetics aided in physical and genetic maps as well as played an important role in in addressing chromosome rearrangement, duplication and deletion occurring which otherwise are difficult to identify. With progress of next-generation sequencing (NGS) technologies, and Arabidopsisas model plant of Brassica family in the area of experimental biology will further improve crop improvement programmes in near future.

References

Alam Z (1936) Cytological studies of some Indian oleiferous *Cruciferae* III. Ann Bot 50:85–102

Al-Shehbaz IA (1985) The genera of *Brassiceae* (*Cruciferae*; *Brassicaceae*) in the southeastern United States. J Arnold Arbor 66:279–351

Attia T, Robbelen G (1986) Cytogenetic relationship within cultivated *Brassica* analyzed in amphihaploids from the three diploid ancestors. Can J Genet Cytol 28:323–329

AttiaT BC, Robbelen G (1987) Digenomic triploids for an assessment of chromosome relationships in the cultivated diploid *Brassica* species. Genome 29(2):326–330

Catcheside DG (1934) The chromosomal relationships in the swede and turnip groups of *Brassica*. Ann Bot 48:601–633

Cheng BF, Heneen WK, Chen BY (1995) Mitotic karyotypes of *Brassica campestris* and *Brassica alboglabra* and identification of the *B. alboglabra* chromosome in an addition line. Genome 38:313–319

Choudhary BR, Joshi P (2012) Crossability of *Brassica carinata* and *B. tournefortii*, and cytomorphology of their F_1 hybrid. Cytologia 77:453–458

Fahleson J, Lagercrantz U, Mouras A, Glimelius K (1997) Characterization of somatic hybrids between *Brassica napus* and *Eruca sativa* using species-specific repetitive sequences and genomic in situ hybridization. Plant Sci 123:133–142

Fukui K, Nakayama S, Ohmido N, Yoshiaki H, Yamabe M (1998) Quantitative karyotyping of three diploid *Brassica* species by imaging methods and localization of 45S rDNA loci on the identified chromosomes. TheorAppl Genet 96:325–330

Gómez-Campo C, Prakash S (1999) Origin and Domestication. In: Gómez-Campo C (ed) Biology of *Brassica* coenospecies. Elsevier, Oxford and New York

Gómez-Campo C (1980) Morphology and morphotaxonomy of the tribe *Brassiceae*. In: Tsunoda S, Hinata K, Gómez-Campo C (eds) Brassica crops and wild allies: biology and breeding. Japan Scientific Society Press, Tokyo, pp 3–31

Gundimeda HR, Prakash S, Shivanna KR (1992) Intergeneric hybrids between *Enarthrocarpuslyratus*, a wild species and crop brassicas. TheorAppl Genet 83:655–662

Haider N (2013) Recent advances in phylogenetic relationships among the U-triangle six *Brassica* L. species. In: CP Malik, Sanghera GS, Wani SH, Sharma P (eds) Crop improvement: biotechnological, physiological and nanotechnological approaches. Aavishkar Publishers, Jaipur, India, p 134

Harberd DJ (1976) Cytotaxonomic studies of *Brassica* and related genera. In: Vaughan JG, MacLeod AJ, Jones MG (eds) The biology and chemistry of the *Cruciferae*. Academic Press, London, pp 47–68

Harberd DJ, McArthur ED (1980) Meiotic analysis of some species and genus hybrids in the *Brassiceae*. In: Tsunoda S, Hinata K, Gomez-Campo C (eds) Brassica crops and wild allies: biology and breeding. Japan Scientific Society Press, Tokyo, pp 65–87

Harberd DJ (1972) A contribution to cytotaxonomy of *Brassica* (*Cruciferae*) and its allies. Bot J Linn Soc 65:1–23

Herbert W (1847) On hybridization amongst vegetables. J HortSoc 2:81–107

Hosaka K, Kianian SF, Mcgrath JM, Quiros CF (1990) Development and chromosomal localization of genome-specific DNA markers of *Brassica* and the evolution of amphidiploids and n=9 diploid species. Genome 33(1):131–142

Howell BC, Barker GC, Jones GH, Kearsey MJ, King GJ, Kop EP, Ryder CD, Teakle GR, Vicente JG, Armstrong SJ (2002) Integration of the cytogenetic and genetic linkage maps of *Brassica oleracea*. Genetics 161:1225–1234

Inomata N (2003) Production of intergeneric hybrids between *Brassica juncea* and *Diplotaxis virgate* through ovary culture and the cytology and crossability of their progenies. Euphytica 133:57–64

Jaretzky R (1932) BeziehungenzwischenChromosomenzahi und SystematikbeidenCruciferen. JahrbWiss Bot 76:485–527

Kamisugi Y, Nakayama S, O'Neil CM, Mathias RJ, Trick M, Fukui K (1998) Visualization of the *Brassica* self-incompatibility S-locus on identified oilseed rape chromosomes. Plant MolBiol 38:1081–1087

Kaneko Y, Bang SW (2014) Interspecific and intergeneric hybridization and chromosomal engineering of *Brassicaceae* crops. Breed Sci 64(1):14–22

Karpechenko GD (1927) Polyploid hybrids of *Raphanus-sativus* L. *Brassica oleracea* L. Bull Appl Bot Genet Plant Breed 7:305–410

Kerlan MC, Chevre AM, Eber F (1993) Interspecific hybrids between a transgenic rapeseed (*Brassica napus*) and related species; cytogenetical characterization and detection of the transgene. Genome 36:1099–1106

Koo DH, Plaha P, Lim YP, Hur Y, Bang JW (2004) A high resolution karyotype of *Brassica rapa* ssp. *pekinensis* revealed by pachytene analysis and multicolour fluorescence in situ hybridization. TheorAppl Genet 109:1346–1352

Kumar A, Singh BK, Singh VV, Chauhan JS (2013) Cytomorphological and molecular evidences of synthesis of interspecific hybrids between *Brassica rapa* and *B.fruticulosa* through hybridization. Aust J Crop Sci 7:849–854

Kumar A, Singh BK, Meena HS, Singh VV, Singh YP, Singh D (2015a) Cytomorphological and molecular characterization of F1 hybrids between *B. tournefortii* and *B. rapa*. Cytologia 80:317–326

Kumar A, Singh BK, Meena HS, Ram B, Singh VV, Singh D (2015b) Cytogenetical Investigations in colchicine-induced tetraploids of *Brassica fruticulosa* —an important wild relative of cultivated Brassicas. Cytologia 80(2):223–230

Kumar A, Meena HS, Ram B, Priyamedha, Sharma A, Yadav S Singh VV, Rai PK (2018) Some Cytomorphological evidence for synthesis of interspecific hybrids between *Brassica juncea* and autotetraploid *B. fruticulosa*. Cytologia 83(4):421–426

Lazaro A, Aguinagalde I (1996) Phylogenetic relationship between the wild taxa of the *Brassica oleracea* L. group (2n=18) using random amplified polymorphic DNA assay. SciHort 65 (4): 219–227

Lim YP, Plaha P, Choi SR, Uhm T, Hong CP, Bang JW, Hur YK (2006) Toward unraveling the structure of *Brassica rapa* genome. Physiol Plant 126:585–591

Lim KB, Yang TJ, Hwang YJ, Kim JS, Park JY, Kwon SJ, Kim JA, Choi BS, Lim MH, Jin M, Kim HI, Jong H, Bancroft I, Lim YP, Park BS (2007) Characterization of the centromere and pericentromere retrotransposons in *Brassica rapa* and their distribution in related *Brassica* species. Plant J 49:173–183

Lysak MA, Cheung K, Kitschke M, Bures P (2007) Ancestral chromosomal blocks are triplicated in *Brassiceae* species with varying chromosome number and genome size. Plant Physiol 145:402–410

Mable BK (2003) Breaking down taxonomic barriers in polyplody research. Trends Plant Sci 8:582–590

Mackowiak M, Heneen WK (1999) Meiotic karyotypes of the B genomes of *Brassica nigra* and *B. carinata*. Hereditas 130:131–135

Maluszynska J, Hasterok RJ (2005) Identification of individual chromosomes and parental genomes in *Brassica juncea* using GISH and FISH. Cytogenet Genome Res 109:310–314

Maluszynska J, Heslop-Harrison JS (1993) Physical mapping of rDNA loci in *Brassica* species. Genome 36:774–781

Mandakova T, Lysak MA (2008) Chromosomal phylogeny and karyotype evolution in x=7 crucifer species (*Brassicaceae*). Plant Cell 20:2559–2570

Manton I (1932) Introduction to the general cytology of the *Cruciferae*. Ann Bot 46:509–556

Mizushima U (1950) Karyogenetic studies of species and genus hybrids in the tribe *Brassiceae Cruciferae*. Tohoku J Agr Res 1:1–14

Mizushima U, Tsunda S (1968) A plant exploration in *Brassica* and allied genera. Tohoku J Agri Res 17:249–277

Mohanty A (1996) Hybridization between crop Brassicas and some of their wild allies. Ph.D. Thesis, Univ Delhi, India

Morinaga T (1928) Prelimary note on interspecific hybridization in *Brassica*. Proc Imp Acad 4:620–622

Morinaga T (1929a) Interspecific hybridization in Brassica. I. The cytology of F_1 hybrids of *B. napella* and various other species with 10 chromosomes. Cytologia 1:16–27

Morinaga T (1929b) Interspecific hybridization in *Brassica*. III. The cytology of F_1 hybrids of *B. cernua* and *B. napella*. J Dept Agr Kyushu Imp Univ 2:199–206

Morinaga T (1929c) Interspecific hybridization in Brassica. II. The cytology of F_1 hybrids of *B. cernua* and various other species with 10 chromosomes. Jan J Bot 4:277–289

Morinaga T (1931) Interspecific hybridization in *Brassica*. IV. The cytology of F_1 hybrids of *B. carinata* and some other species with 10 chromosomes. Cytologia 3:77–83

Morinaga T (1933) Interspecific hybridization in *Brassica*. V. The cytology of F_1 hybrids of *B.carinata* and *B. alboglabra*. Japan J Bot 6:467–475

Morinaga T (1934) Interspecific hybridization in *Brassica*. VI. The cytology of F_1 hybrids of *B. juncea* and *B. nigra*. Cytologia 6:62–67

Morinaga T (1934) On the chromosome number of *Brassica juncea* and *B. napus*, on the hybrid between the two, and on offspring line of the hybrid. Jpn J Genet 9:161–163

Nagaharu U (1935) Genome analysis in *Brassica* with special reference to the experimental formation of *B. napus* and peculiar mode of fertilization. Jpn J Bot 7:389–452

Olin-Fatih M (1994) A new method for differential staining of *Brassica* metaphase chromosomes, and karyotypes of *B. campestris*, *B. oleracea*, and *B. napus*. Hereditas 120:253–259

Olin-Fatih M, Heneen WK (1992) C-banded karyotypes of Brassica campestris *B. oleracea* and *B. napus*. Genome 35:583–589

Osborn TC (2004) The contribution of polyploidy to variation in *Brassica* species. Physiol Planta 121 (4):531–536

Otto SP, Whitton J (2000) Polyploidy incidence and evolution. Annu Rev Genet 34:297–309

Pradhan A, Julie A, Plummer MN, Nelson WA, Cowling GY (2010) Successful induction of trigenomichexaploid *Brassica* from a triploid hybrid of *B. napus* L. and *B. nigra* (L.) Koch. Euphytica 176:87–98

Prakash S (2010) Brassica cytogenetics—a historical journey and my personal reminiscence. Chin J Oil Crop Sci 32:163–172

Prakash S, Hinata K (1980) Taxonomy, cytogenetics and origin of *Brassica*, a review. Per Bot 55:1–57

Prakash S, Bhat SR, Quiros CF, Kirti PB, Chopra VL (2009) *Brassica* and its close allies: cytogenetics and evolution. Plant Breed Rev 3:21–187

Quiros CF, Ochoa O, Douches DS (1988) Exploring the role of x=7 species in *Brassica* evolution: hybridization with *B. nigra* and *B. oleracea*. J Hered 79:351–358

Quiros CF, Hu J, This P, Chevre AM, Delseny M (1991) Development and chromosomal localization of genome-specific markers by polymerase chain reaction in *Brassica*. Theor Appl Genet 82(5):627–632

Quiros CF, Ochoa O, Kianian SF, Douches D (1987) Analysis of the *Brassica-oleracea*genome by the generation of *B. campestris-oleracea* chromosome addition lines—characterization by isozymes and rDNA genes. Theor Appl Genet 74(6):758–766

Ramsey J, Schemske DW (2002) Neopolyploidy in flowering plants. Annu Rev Ecol Evol Syst 33:589–639

Reiseberg LH (2001) Polyploid evolution: keeping the peace at genomic reunions. Curr Biol 11:R925–R928

Rich TCG (1991) Crucifers of Great Britain and Ireland. Botanical Society of the British Isles, London, p 336

Richharia RH (1937) Cytological investigations of *Raphanussativus*, *Brassica oleracea* and their F_1 hybrids. J Genet 34:45–55

Robbelen G (1960) Beitra ̈gezurAnalyse des *Brassica*-Genomes. Chromosoma 11:205–228

Sageret M (1826) Considerations sur la production des variants et des varie´tie´sen general, et sur celled de la famille de Cucurbitacee´senparticulier. Ann Sci Nat 8:94–314

Schulz OE (1936) *Cruciferae*. In: Engler A, Prantl P (eds) Die natu ̈rlichenPflanzenfamilien. Wilhelm Engelmann, Leipzig, pp 17b: 227–658

Sikka SM (1940) Cytogenetics of *Brassica* hybrids and species. J Genet 40:441–509

Sinskaia EN (1928) The oleiferous plants and root crops of the family *cruciferae*. Bull App/bot Genet Pl Breed 175:3–166

Snowdon RJ (2007) Cytogenetics and genome analysis in *Brassica* crops. Chrom Res 15:85–95

Snowdon RJ, Winter H, Diestal A, Sacristan MD (2000b) Development and characterization of *Brassica napus–Sinapisarvensis* addition lines exhibiting resistance to *Leptosphaeriamaculans*. Theor Appl Genet 101:1008–1014

Snowdon R J, Friedrich T, Friedt W, Kohler W (2002a) Identifying the chromosomes of the A- and C-genome diploid *Brassica* Species *B. rapa* (syn. *campestris*) and *B. oleracea* in their amphidiploid *B. napus*. Theor Appl Genet 104:533–538

Song KM, Osborn, TC, Williams PH (1988) *Brassica* taxonomy based on nuclear restriction fragment length polymorphisms (RFLPs). 1. Genome evolution of diploid and amphidiploid species. Theor Appl Genet 75(5):784–794

Takahata Y, Hinata K (1983) Studies on cytodemes in the subtribe *Brassicineae*. Tohoku J Agr Res 33:111–124

Takamine N (1916) U ̈ber die rubenden und die pra ̈synaptischen Phasen der Reduktionsteilung. Bot Mag Tokyo 30:293–303

Truco MJ, Hu J, Sandowski J, Quiros CF (1996) Inter-and intra-genomic homology of the *Brassica* genomes: implications for their origin and evolution. Theor Appl Genet 93 (8): 1225-1233

Wang YP, Snowdown RJ, Rudloff E, Wehling P, Friedt W, Sonntag K (2004) Cytogenetic characterization and fae1 gene variation in progenies from asymmetric somatic hybrids between *Brassica napus* and *Crambeabyssinica*. Genome 47:724–731

Warwick SI, Black LD (1997) Phylogenetic implications of chloroplast DNArestriction site variation in subtribes *Raphaninae* and *Cakilinae* (*Brassicaceae*). Can J Bot 75:960–973

Warwick SI, Francis A, Al-Shehbaz IA (2006) *Brassicaceae*: species checklist and database on CD-Rom. Plant Syst Evol 259:249–258

Yang YW, Tseng PF, Tai PY, Chang- CJ, Yang YW, Tseng PF, Tai PY, Chang CJ (1998) Phylogenetic position of *Raphanus* in relation to Brassica species based on 5 s r RNA Speacer sequence data. Bot Bull Acad Sin 39(3):153–160

Yang T (2005) The Korea Brassica genome Project: a glimpse of the Brassica genome based on comparative genome analysis with Arabidopsis. In: Plant and Animal Genomes 13th Conf California State Univ California, 6:138–145

Chemical Composition of Oil and Cake of *Brassica juncea*: Implications on Human and Animal Health

3

Sanjula Sharma, Manju Bala, Gurpreet Kaur, Saad Tayyab, and Shevin Rizal Feroz

Abstract

Oilseed Brassicas have become the third important edible oilseed crop next to soybean and palm in the world. The four main Brassica oilseed species, *B. napus, B. rapa, B. juncea* and *B. carinata* are widely cultivated for oil, condiment and vegetable purposes. *B. juncea* (L.) Czern, commonly referred to as Indian mustard is a major oilseed crop in South Asia, while *B. napus* and *B. rapa* are popular in Europe, Canada and Australia. Oil extracted from *B. juncea* using expeller/*kachi ghani* has always been favored as cooking oil in India due to its interesting chemical properties. Its oil is a perfect blend of saturated and unsaturated fatty acids along with bioactive components such as phytosterols and tocopherols. The meal or cake left after oil extraction from mustard seeds is rich in minerals, vitamins and high-quality proteins. Despite its nutritional richness, oil and meal of the traditional *B. juncea* varieties are considered inferior in quality as they contain a very high amount of undesirable long-chain fatty acids, namely erucic acid ($C_{22:1}$) (40–57%) in oil and deleterious glucosinolates (GLSs) (50–120 μ moles/g) in seed meal. Intake of high erucic acid has been associated with cardiac injury in cattle and experimental rodents, whereas the presence of a high concentration of GLSs with goitrogenic effects limits the use of mustard meal to ruminants only. Based on animal trials, a statutory limit of < 2% erucic acid in oil and < 30 μmoles/g of GLSs in a meal has been set by many countries, initially for rapeseed varieties which have been registered under the trade name of 'canola' or '00' quality and is also being implemented for mustard. The quantitative reduction of erucic acid is balanced by the increase in desirable oleic and linoleic acid in the mustard oil. In India, since 90% of the area under rapeseed-mustard is planted with *B. juncea*

S. Sharma (✉) · G. Kaur
Oilseeds Section, Department of Plant Breeding and Genetics, Punjab Agricultural University, Ludhiana, Punjab 141004, India
e-mail: drsanjula@pau.edu

G. Kaur
e-mail: gurpreetkaur@pau.edu

M. Bala
Food Grains and Oilseed Processing Division, ICAR-Central Institute of Post-Harvest Engineering and Technology, Ludhiana, Punjab 141004, India
e-mail: manju.bala1@icar.gov.in

S. Tayyab
Biomolecular Research Group, Biochemistry Programme, Institute of Biological Sciences, Faculty of Science, Universiti Malaya, 50603 Kuala Lumpur, Malaysia

S. R. Feroz
Department of Biological Sciences and Biotechnology, Faculty of Science and Technology, Universiti Kebangsaan Malaysia, 43600 Bangi, Selangor, Malaysia
e-mail: shevin@ukm.edu.my

C. Kole and T. Mohapatra (eds.), *The Brassica juncea Genome*, Compendium of Plant Genomes,
https://doi.org/10.1007/978-3-030-91507-0_3

varieties, improvement of its oil and meal quality is of prime importance and hence, led to the development of canola or '00' quality *B. juncea* varieties which have been fairly successful and can be considered excellent for food and feed purpose.

3.1 Introduction

Nutrition has a long-term impact on health. Nutritive food has always been considered equally significant as medicine to cure any disease or build immunity. Dietary fats, carbohydrates and proteins constitute a macronutrient group that acts as an energy source in the human diet. Intake of dietary fats derived from plant sources (oilseeds and nuts) and animal sources (butter, meat, fish, poultry and egg) should be moderate and lie between 25 and 35% of calories in a balanced diet as it aids in the prevention of heart disease and stroke. Oilseeds are dual purpose crops as their derivatives namely oil and meals, are in great demand globally.

Oilseed Brassicas are an important group of oilseed crops consisting of rapeseed and mustard varieties. Rapeseed (*B. rapa* and *B. napus*) and mustard (*B. juncea*) together constitute the world's third major vegetable oil source after soybean and oil palm (Chand et al. 2021). *Brassica* spp. cultivated in India for edible oils are *B. juncea* (Indian mustard), *B. rapa* (ssp. Toria), *B. napus* (gobhi sarson) and *B. carinata* (karan rai) with variation in their oil content from 31.81–42.13%, 36.5–40.8%, 34.52–45.09% and 25–29%, respectively (Sharafi et al. 2015; Sen et al. 2018; Sra et al. 2019). Rapeseed-mustard is mainly grown in the states of Rajasthan, Haryana, Uttar Pradesh, Madhya Pradesh, West Bengal and Gujarat (India Agristat 2020). At one time in India a significant area was under *B. rapa*, but nowadays, it is dominated by *B. juncea* (Indian mustard) with 90% acreage (Jat et al. 2019). *B. juncea,* which belongs to the Brassicaceae or Cruciferae family, is commonly referred to as Indian mustard, Chinese mustard, or Oriental mustard. Indian mustard seeds are known for their therapeutic properties including antioxidant, anticancer, antimicrobial, antidiabetic, anti-obesity and anti-inflammatory activities (Thirumalai et al. 2011; Engels et al. 2012; Oh et al. 2016; Mazumder et al. 2016; Le et al. 2020).

Oil content in Indian mustard varieties varies from 27.57 to 40.23% (Sra et al. 2019), and thus about 59.77–72.43% of seed weight constitute the meal. The oil obtained from Indian mustard (*B. juncea*) is a popular cooking oil in certain regions of India and its fatty acid profile is very close to the profile recommended by global health agencies. After extraction of oil from mustard seeds, the remaining portion is known as meal or cake as per the residual oil content in it. Both meal and cake are rich in protein (32–48%) (Sadeghi and Bhagya 2009) but differ in their oil content which is 1–4% in the meal (Klein-Hessling 2007) and 10–14% in cakes (Mahoonak and Swamylingappa 2007). In India, at present, the mustard meal is fed to the ruminants. The amino acid composition of the mustard meal proteins has been reported to be comparable with milk protein casein (Sen and Bhattacharyya 2000). Due to the nutritionally rich amino acid composition of its proteins, the mustard meal has also been exploited for the preparation of protein isolates and concentrates for human use (Tan et al. 2011). Besides proteins, the meal is also a rich source of minerals and vitamins.

Despite its nutritional richness, there are concerns about the use of mustard oil and meal for food and feed purposes (Chauhan et al. 2002; Sadeghi and Bhagya 2009). The desirable fatty acid profile of this oil is antagonized by the presence of a high amount of erucic acid ($C_{22:1}$) (40–57%), which is detrimental to mammalian health and lowers its utility as an edible oil but makes it industrially important (Singh et al. 2014). On the other hand, the presence of antinutritional compounds *viz.* glucosinolates (GLSs), phytates, sinapine, tannin and fiber makes mustard meal unfit for consumption and limits its use as a source of protein in food products.

Initially, in the 1970s, with the aim to improve seed oil and meal quality, the Brassica improvement program led to the development of *B. napus* and *B. rapa* varieties with reduced erucic acid ($C_{22:1}$) (<2%) in oil and GLSs (<30 μmoles/g) in seed meal. Such varieties were registered under the trademark 'Canola', also referred to as double zero or '00' by the Canola Council of Canada. Later in 2003, canola type mustard (*B. juncea)* varieties were developed in Canada (Potts et al. 2003). Being major producer of *B. juncea,* India has developed and released many international standard canola quality mustard varieties. Some of them are now available for commercial cultivation. Farmers still prefer growing traditional mustard; hence, framing appropriate policies and market support is needed to convince farmers to adopt the shift from traditional mustard to canola mustard. Efforts by the government agencies and others are underway to promote contract farming for enhancing coverage and production of both single zero or '0' and double zero '00' quality mustard in India.

3.2 Composition and Characteristics of Mustard Oil

Brassica oils are mainly composed of triacylglycerols making up about 98% of the oils. The remaining 2% is non-glyceride fractions constituted by sterols, sterol esters, tocopherols, pigments (chlorophylls and carotenoids) and waxes (Booth 2004). *B. juncea* is one of India's primary sources of edible oil with tremendous variability in oil content ranging from 20.4 to 49.30% (Table 3.1), which denotes its potential to be exploited in breeding programs for developing new cultivars with raised oil content.

3.2.1 Fatty Acids

The nutritional and industrial value of Brassica oil, like other vegetable oils, is determined by its fatty acid profile, made up of distinct carbon chain length and number of double bonds. Literature analysis showed that the average amount of saturated fatty acid (SFA), namely palmitic acid ($C_{16:0}$) and stearic acid ($C_{18:0}$) together constituted 1.8–10.1% of total fatty acids, monounsaturated fatty acid (MUFA) like oleic acid ($C_{18:1}$) varied from 7.30–66.70%, and polyunsaturated fatty acid (PUFA) such as linoleic acid (ω6; $C_{18:2}$) and linolenic acid (ω3; $C_{18:3}$) ranged from 11.54 to 32.20% and 5.84–23.92%, respectively. Very long-chain fatty acids (VLCFAs) like eicosenoic acid ($C_{20:1}$) was found in the range of 0.73–14.5%, whereas based on erucic acid ($C_{22:1}$) content, *B. juncea* accessions have been categorized into low erucic acid (LEA) accessions with <2% erucic acid (0.0–2.3%) and high erucic acid (HEA) accessions with >2% of erucic acid (8.50–55.5%) (Table 3.1). As a dietary component, heart-friendly oil should be low in SFA (<10%), and erucic acid (<2%), high in MUFA (oleic acid), and possess an ideal PUFA ω6:ω3 ratio (5–10:1) (Kale 2007; WHO 2018).

Among all the available oils, mustard oil is a perfect blend of SFA and unsaturated fatty acid (UFA), which helps in balancing cholesterol levels. Rapeseed-mustard oil possesses a low level of SFA (<10%) in comparison to other edible oils such as olive (15.03–20.48%), soybean (13.5–18.3%), sesame (14.00–36.43%), rice bran (15.40–24.77%), palm (12.30–49.45%), groundnut (10.7–19.2%) and sunflower oil (8.50–12.36%) (Orsavova et al. 2015; Awogbemi et al. 2019). A high level of SFA in edible oils is nutritionally undesirable because it increases the concentration of harmful low-density lipoproteins (LDL). Various research groups demonstrated the effectiveness of MUFA-rich vegetable oils in reducing the harmful LDL cholesterol level without any effect on beneficial high-density lipoprotein (HDL) cholesterol in the blood (Mensink and Katan 1989; Rakow and Raney 2003). In addition to this, high oleic acid is more resistant than PUFA toward oxidation at ambient storage and high temperatures (Frankel 2012). Traditional mustard germplasm contain oleic acid in the range of 7.30–20.39%; alternatively, canola quality mustard possesses 43.71–66.70% (Table 3.1). Mustard oil is not permitted for use as an edible oil in many developed countries around the world due to the presence of

Table 3.1 Range of oil content and fatty acid composition in seeds of *B. juncea* cultivars

B. juncea cultivars		Fatty acid profile (%)						References
		SFA	MUFA	PUFA		VLCFA		
	Oil (%)	Palmitic $C_{16:0}$ + Stearic $C_{18:0}$	Oleic $C_{18:1}$	Linoleic $C_{18:2}$	Linolenic $C_{18:3}$	Eicosenoic $C_{20:1}$	Erucic $C_{22:1}$	
LEA	ND	7.50	47.20	32.20	8.40	1.80	0.60	Ostrikov et al. (2020)
	ND	6.05	43.71	32.09	11.19	2.53	2.30	Simakova et al. (2019)
	36.90–39.00	ND					0.00	Rout et al. (2018)
	ND	5.70	62.20–66.70	13.80–15.80	11.80–14.10	1.20–1.50	0.20–0.30	Wijesundera et al. (2008)
HEA	40.70–49.30	ND					20.50–55.50	Rout et al. (2018)
	ND	5.73	10.16	15.58	11.70	5.48	51.18	Dorni et al. (2018)
	31.81–42.13	4.14	15.95	19.37	20.29	0.73	39.55	Sen et al. (2018)
	40.95–41.98	4.00	14.72–18.69	14.38–15.11	5.84–7.14	ND	50.60–51.35	Sharif et al. (2017)
	ND	5.50	17.00	13.00	9.00	14.50	8.50	Khansili and Rattu (2017)
	36.70	1.80	15.70	17.60	19.40	3.20	42.30	Lee et al. (2015)
	20.4–31.8	4.80	12.46–16.69	17.46–22.39	18.35–22.80	ND	23.75–38.19	Sharafi et al. (2015)
	40.30	3.82	13.8	16.38	13.21	4.80	47.96	Singh et al. (2014)
	38.45	4.38	18.32	23.57	23.92	ND	29.81	Al-Jasass and Al-Jasser (2012)
	36.32	5.10	20.24	21.36	11.56	12.10	23.90	Abul-Fadl et al. (2011)
	39.76	4.17	20.39	18.63	14.76	ND	41.89	Chhokar et al. (2008)
	36.90	10.10	13.94	11.54	10.59	7.05	35.99	Kanrar et al. (2006)
	31.20	3.90	7.30	12.80	10.60	10.20	40.40	Lionneton et al. (2004)

ND Not determined; *LEA* Low in erucic acid; *HEA* High in erucic acid; *SFA* Saturated fatty acid; *MUFA* Monounsaturated fatty acid; *PUFA* Polyunsaturated fatty acid; *VLCFA* Very long-chain fatty acid

high level of erucic acid (40–57%), which is considered detrimental to health because of its slow tendency to oxidize resulting in its accumulation in arterial lining leading to myocardial lipidosis in laboratory rats (Charlton et al. 1975). Internationally, oils with <2% erucic acid are recommended for consumption purpose. Mustard varieties with oils low in erucic acid (<2%) as canola oil or '0' have been developed namely, Pusa Karishma, Pusa Mustard 21, Pusa Mustard 22, RLC1, Pusa Mustard 24, RLC2, Pusa Mustard 29, Pusa Mustard 30, Pusa Mustard 32, RLC3, PDZ1, RCH1 (Table 3.2). The first source identified for LEA content in *B. juncea* was 'Zero Erucic Mustard' (ZEM) (Kirk and Oram 1978), which was extensively used in Brassica breeding programs for the development of LEA cultivars.

The two PUFAs in Brassica oil, linoleic (ω6) and linolenic acid (ω3), are also known as essential fatty acids (EFAs), as they cannot be synthesized in mammals and are obtained from the diet. EFAs are involved in synthesis of prostaglandins in the human body that regulate the cardiovascular and reproductive systems and play a role in immune function (Horrobin 1986; Singh et al. 1997; Rastogi et al. 2004; Endo and Arita 2016). *Brassica juncea* oil contains a considerable amount of EFAs, including linoleic acid ($C_{18:2}$) (13.80–32.20% in LEA accessions; 11.54–23.57% in HEA accessions) and linolenic acid ($C_{18:3}$) (8.40–14.10% in LEA and 9.00–23.92% in HEA accessions) with a nearly ideal balanced ratio of two PUFAs, 1.1–3.8 in LEA accessions and 0.9–2.1 in HEA (Table 3.1). Although linolenic acid is another EFA, its presence in the oil makes it highly susceptible to oxidation leading to oil rancidity and off-flavor (Sharafi et al. 2015). Striking the right balance between the two EFAs *viz.,* linoleic (ω6) and linolenic acid (ω3) is considered nutritionally important. It is recommended to be in the ratio of 5–10:1 or lower to prevent heart disease, cancer and autoimmune diseases (ICMR 2010). In consideration of the above reports, reduction in erucic ($C_{22:1}$) and linolenic ($C_{18:3}$) acids and increase in oleic ($C_{18:1}$) and linoleic ($C_{18:2}$) acids has become an essential objective in Brassica improvement programs. The reduction in the erucic acid content increases the oleic acid content and imparts a better ω6:ω3 ratio in seed oil (Jagannath et al. 2011). The canola type mustard oil being free of erucic acid has a high level of MUFA (61% oleic acid) and an intermediate level of PUFA (32%) with a nearly favorable balanced ratio of ω6/ω3 (Kaur et al. 2019). However, the high erucic acid level in *B. juncea* and *B. rapa*, has shown potential biofuel production and industrial applications such as softener in textiles, as well as in the manufacturing of surfactants, plasticizers, surface coatings and pharmaceuticals (Hossain et al. 2019). Erucic acid has a low flash point, good combustion and lubricating qualities, all of which makes it a valuable component for biodiesel (Zanetti et al. 2009).

3.2.2 Phytosterols

Recently, improvement in the content and composition of some minor non-glyceride fractions such as carotenoids, phytosterols and tocopherols gained importance due to their health benefits. Pieces of evidence support that phytosterols play an important role as an anti-inflammatory (Yuan et al. 2019) and anti-carcinogenic agent (Ramprasath and Awad 2015) and have therapeutic effects on dementia (Shuang et al. 2016). It is well-documented that plant sterols increase cholesterol-lowering efficacy. An intake of phytosterols at an effective dose of 2 g/day leads to significant reductions (8–15%) in harmful LDL cholesterol (Cabral and Klein 2017). Sterols are high in a range of vegetable oils such as rice bran oil (1891.82 mg/100 g), corn oil (990.94 mg/100 g), rapeseed oil (893.84 mg/100 g), sesame oil (637.60 mg/100 g) and sunflower oil (253.25 mg/100 g) (Yang et al. 2019). Members of Brassicaceae are rich in phytosterols which comprise sitosterol followed by campesterol, brassicasterol and avenasterol (Teh and Möllers 2016). Brassicasterol is a typical phytosterol found only in this family. Limited studies are available on free individual sterol content in *B. juncea* seeds (Table 3.3). A negative association was detected between the content of erucic acid and phytosterols in a doubled haploid population of winter oilseed rape recommending for the conversion of high erucic acid lines to zero erucic acid resulting in an increase in phytosterol content (Amar et al. 2008). Aside from the domestic demand in India,

Table 3.2 Oil content, fatty acid composition and glucosinolate content in single- and double low varieties of mustard in India

B. juncea variety		Oil (%)	Fatty acid profile (%)						GLS (μmoles/g meal)	Year of release
			SFA	MUFA	PUFA		VLCFA			
			Palmitic $C_{16:0}$ + Stearic $C_{18:0}$	Oleic $C_{18:1}$	Linoleic $C_{18:2}$	Linolenic $C_{18:3}$	Eicosenoic $C_{20:1}$	Erucic $C_{22:1}$		
Low in erucic acid '0'	Pusa Karishma	38.00	4.30	42.30	32.60	18.50	0.05	0.85	55.40	2004
	Pusa Mustard 21	35.60	4.60	42.10	38.20	13.30	1.20	0.70	70.50	2007
	Pusa Mustard 22	36.00	6.20	42.60	35.30	13.90	0.98	1.10	62.80	2007
	RLC 1	37.80	6.80	42.80	28.40	18.30	0.50	1.50	95.00	2007
	Pusa Mustard 24	35.60	8.40	39.80	36.30	13.40	3.70	0.11	57.20	2008
	RLC 2	37.60	4.30	35.70	35.80	22.70	0.77	0.77	58.90	2012
	Pusa Mustard 29	37.20	6.20	41.00	34.30	15.90	2.20	1.00	78.20	2013
	Pusa Mustard 30	37.70	5.80	40.50	30.50	18.90	2.60	1.30	81.80	2013
	Pusa Mustard 32	38.00	5.50	34.70	36.70	20.60	4.90	1.30	54.80	2020
Canola '00'	RLC 3	41.50	7.60	42.30	35.30	11.50	1.00	0.50	15.00	2015
	PDZ 1	40.00	7.40	50.30	32.80	7.80	0.50	0.60	25.00	2017
	RCH1	39.40	6.80	41.00	32.00	16.70	0.20	1.50	25.00	2019

SFA Saturated fatty acid; *MUFA* Monounsaturated fatty acid; *PUFA* Polyunsaturated fatty acid; *VLCFA* Very long-chain fatty acid; *GLS* Glucosinolate

Table 3.3 Phytosterol composition (mg/100 g) in *B. juncea* seed oil

	Brassicasterol	Campesterol	Stigmasterol	β-Sitosterol	γ-Sitosterol	Δ^5-Avenasterol	Δ^7-Avenasterol	Clerosterol	Stigmasten-3,5- diene	Cycloartenol	References
B. juncea	ND	ND	30.7	182.4	181.2	ND	ND	ND	72.1	ND	Sharma et al. (2018)
	134.0	359.0	ND	599.0	ND	20.0	13.0	6.0	ND	ND	Jham et al. (2009)
	135.0	353.0	12.0	451.0	ND	31.0	ND	ND	ND	18.0	Mortuza (2006)
	192.0	236.0	ND	572.0	ND	ND	ND	ND	ND	ND	Appelqvist et al. (1981)

ND Not detected

Table 3.4 Tocopherol composition (μg/g) in the seed oil of *B. juncea* varieties/cultivars

	α-Tocopherol	β-Tocopherol	γ-Tocopherol	δ-Tocopherol	Total	References
B. juncea	89.56	ND	290.34	ND	–	Shrestha and Meulenaer (2014)
	76.13	ND	179.77	3.89	260.44	Li et al. (2012)
	75.51	ND	318.30	16.01	409.82	Vaidya and Choe (2011)
	200	1	585	14	800.00	Jham et al. (2009)
	76–335	4–31	163–579	ND	256–928	Richards et al. (2007)
	14.96–19.05	Traces	45.79–66.51	2.43–3.83	67.92–87.71	Mortuza, (2006)

ND Not detected

there is enormous scope for the export of bioactive compounds in the Asia Pacific and China for the manufacture of steroids and pharmaceutical products. Globally, the phytosterols market exceeded USD 780 million in 2020 and is expected to grow at over 9% Compound Annual Growth Rate (CAGR) between 2021 and 2027. Mo et al. (2013) detected high levels of brassicasterol (48.8 mg/100 g) in canola oil compared to other edible oils. Piironen et al. (2000) also reported brassicasterol in the range of 55 to 73 mg/100 g. Therefore, identification of phytosterols rich *B. juncea* lines could give an added value to the crop.

3.2.3 Tocopherols

Another main component of the non-glyceride fractions in vegetable oils is tocopherol (Vitamin E). Tocopherols are gaining a lot of attention these days because of their potential antioxidant properties and important role in avoiding aging-related diseases (Conti et al. 2016), cardiovascular disease (Mathur et al. 2015) and Alzheimer's disease (Gugliandolo et al. 2017). Although the concentration of tocopherols in oil is relatively low (700–1200 ppm), they still act as antioxidants and significantly contribute to the oxidative stability of oil (Ratnayake and Daun 2004; Gunstone 2011). The tocopherol content accounted for 800 ppm in mustard oil which was higher than that of palm (642 ppm), sunflower (546 ppm) and safflower (413 ppm) oils (Jham et al. 2009). Tocopherols consist of α-, β-, γ- and δ-Tocopherols, and their structural characteristics determine their metabolic fate and biological activities. The α-Tocopherol is biologically the most active form of vitamin E, while γ-Tocopherol is essential for oxidative stability of vegetable oils particularly at higher temperatures (Seppanen et al. 2010). The major tocopherols in *B. juncea* varieties/cultivars are presented in Table 3.4. Brassica oils have low α-Tocopherol but a high concentration of γ-Tocopherols. The relative concentration of α- and γ-Tocopherols is dependent on the γ-Tocopherol methyltransferase (γ-TMT), which catalyses a methylation step from γ-Tocopherol to α-Tocopherol (Bergmuller et al. 2003). Over expression of γ-TMT gene in *B. juncea* plants resulted in about six fold increase in α-Tocopherol content over the control plants (Yusuf and Sarin 2007). A significant positive correlation was noticed between α-tocopherol and the sum of oleic and linoleic acid, on the other hand, hardly any association existed between γ-Tocopherol and fatty acids, indicating another possible function of α-Tocopherol beyond its antioxidant property (Li et al. 2012). A significant varietal effect on tocopherol content in rapeseed genotypes has been reported (Wang et al. 2012; Fritsche et al. 2012), which allows for further investigation into the breeding of tocopherol enhanced cultivars.

3.3 Composition and Characteristics of Mustard Cake/Meal

The seed coat contributes a significant proportion of the total seed volume. The hulls represent about 19% of whole seed in mustard and other *Brassica* species (Cilly et al. 1978). Mustard meal left after oil extraction comprises hulls and embryos, with the hull contributing to about 30% of the meal weight. The hull is composed of cellulose, hemicelluloses, pentosans and lignin. When the hull is compared with the embryo fraction, it contains more fiber and low protein content. Besides fiber and proteins, the hull of the rapeseed is reported to have ash, phenolic compounds and GLSs content (Naczk et al. 1998). Total carbohydrates, crude protein, oil, crude fiber and ash content of 23.9%, 32.3%, 13.2%, 14.0% and 5.4% for mustard cake (Mahoonak and Swamylingappa 2007) and 40.8%, 39.9%, 2.4%, 6.8% and 6.9% for a mustard meal (Daghiar and Mian 1976) have been reported (Table 3.5). The composition of oilseed meal is affected by the variety and seed quality, oil extraction method, storage and environmental conditions. Mustard seed meal/cake is rich in protein and is used as a feed to ruminants. However, the protein and fiber digestibility is very low in hulls, and therefore mustard meal cannot be utilized for feeding the non-ruminants.

3.3.1 Carbohydrates

Total carbohydrate content constitutes about 50% of the mustard meal composition, including fiber. The crude fiber in general consists of cellulose, hemicellulose, pentosans and lignin of the cell walls. Most of the fiber in *Brassica* species is present in hulls, while a tiny amount is present in the embryo part. Sadeghi et al. (2006) reported a crude fiber content of 10% in *B. juncea* meal. However, the crude fiber content in the range of 5.12–12.06% and 6.3–18.9% has been observed by Bala and Singh (2013) and Chauhan and Kumar (2011), respectively, for Indian genotypes of *B. juncea*. Simbaya et al. (1995) reported that meal from brown seeded *B. juncea* contained more dietary fiber (35.1%) on a dry weight basis as compared to the meal from yellow seeded *B. juncea* (27.8%). Therefore, the mustard meal obtained from the dark- colored seed coat has been associated with the high fiber and tannin content. In contrast, meal from yellow seeded varieties shows reduced fiber content. In the case of mustard seed meal, free sugars content is more while pectin, soluble polysaccharides, and cellulose content are less compared to rapeseed meal (Sindhu Kanya and Kantharaj Urs 1983; Siddique and Woods 1977), as shown in Table 3.6. Sucrose content of 1.68%, has been reported for Indian varieties. Free sugars present in the mustard meal have been reported to be stachyose, raffinose, melibiose, galactose, glucose and fructose, having a concentration of 2.10%, 1.20%, 0.68%, 1.80%, 0.96% and 2.10%, respectively (Sindhu Kanya and Kantharaj Urs 1983). However, sucrose content of 7.0 and 8.3% on a dry weight basis was reported for the meals obtained from brown and yellow seeded mustard, respectively, while oligosaccharide content of 2% was reported in both meal types (Simbaya et al. 1995).

High amounts of fiber may negatively influence protein digestibility (Bell 1993) and bioavailability of minerals such as manganese and zinc (Kies and Umoren 1989). The use of mustard meal is limited in the diets of non-ruminant animals because of the presence of a high amount of fiber (Slominski et al. 1994). Therefore, dehulling is considered one option to remove the fiber content (Naczk and Shahidi 1990; Matthaus 1998; Mohapatra et al. 2004). Dehulling reduced the fiber content up to 3–6% in the mustard meal (Sadeghi et al. 2006; Cilly et al. 1978; Mahoonak and Swamylingappa 2007). Treatment of meal using enzymes also lower the fiber content as Musigwa et al. (2021) observed the inclusion of multi-carbohydrase enzyme in the broiler diets reduced the effect of fiber. Although reduction of fiber content could be achieved by removing the hull before oil extraction, it is not feasible technology to be adopted by the industry due to the losses of oil during the dehulling process (Khajali and Slominski 2012).

Table 3.5 Composition (%) of mustard cake/meal

B. juncea	Moisture	Crude protein	Oil	Dietary fiber	Ash	Crude fiber	Carbohydrates	References
Whole brown mustard seed flour	4.98 ± 0.42	32.48 ± 0.74	36.32 ± 0.27	6.34 ± 0.13	3.88 ± 0.36	NE	16.49 ± 0.77	Abul-Fadl et al. (2011)
Mustard meal (Dehulled defatted)	8.00 ± 0.3	48.00 ± 0.5	1.00 ± 0.2	NE	4.20 ± 0.2	3.00 ± 0.3	35.80 ± 0.5	Sadeghi et al. (2006)
Mustard cake	11.20 ± 0.2	32.30 ± 0.8	13.20 ± 0.3	NE	5.40 ± 0.1	14.00 ± 0.5	23.90	Mahoonak and Swamylingappa (2007)
Mustard meal (Low GLS)	NE	45.00–47.2	0.40–0.60	25.55–27.90	NE	NE	NE	Newkirk et al. (1997)
Mustard meal	10.00	39.90	2.40	NE	6.90	6.80	40.80	Daghiar and Mian (1976)

NE Not estimated

Table 3.6 Carbohydrate components (%) of mustard seed meal

Component	Mustard seed meal	*B. campestris*
Cellulose	2.80	7.00
Polysaccharides	2.20	4.50
Pectic substances	6.50	14.50
Sucrose	1.68	2.26–7.49
Free sugars	12.6	5–10
Composition of sugars		
Stachyose	2.10	1.52–2.43
Raffinose	1.20	0.31–0.34
Melibiose	0.68	–
Galactose	1.80	–
Glucose	0.96	0.28–0.40
Fructose	2.10	0.15–0.51

Source Cited from Nagaraj (2009), Sindhu Kanya and Kantharaj Urs (1983), Siddiqui and Woods (1977)

3.3.2 Proteins

The dry matter analysis of mustard meal showed high crude protein content of 32–48%, indicating their suitability as protein supplements (Ramachandran et al. 2007). Mustard cakes are generally used for feeding ruminants (cattle and buffaloes), and with the development of canola varieties of mustard, it can also find feed applications to non-ruminants. As the mustard meal is rich in protein, its use for food supplementation has been tried through the preparation of isolates and concentrates by various workers using different extraction techniques (Sadeghi et al. 2006; Das et al. 2009). Newkirk et al. (1997) reported that compared to canola type *B. napus* meal and canola type *B. juncea* meal possesses more protein and less fiber and showed ether extract content similar to *B. napus*. Removal of hull increased the protein concentration from 39 to 46% in the defatted mustard meal and other allied *Brassica* species (Cilly et al. 1978). Dehulling and defatting of mustard seed increased the protein content from 22 g/100 g to 48 g/100 g (Mahoonak and Swamylingappa 2007). The amino acid profile of mustard cake presents it as a rich source of glutamic acid, arginine, leucine and aspartic acid (Table 3.7). Mustard proteins like rapeseed proteins have a good balance of essential amino acids possess higher levels of methionine and cystine (the sulphur amino acids) than soybean protein. The proteins have the highest amount of glutamic acid (20.67 g/100 g). Protein digestibility of about 92% and 79% mustard protein have been detected in five weeks and 20-month-old rats, respectively (Gilani and Sepehr 2003). The mustard protein isolate showed better digestibility of proteins (92.4%) than a mustard meal (80.6%). Both mustard meal and protein isolate showed similar amino acid profiles with differences in concentration of some amino acids.

Mustard seed storage proteins exist in two classes *viz.* globulins (legumin-type; 11S or 12S or cruciferin) and albumins (napin-type; 2S or napins) (Wanasundara 2011). The composition of storage proteins in terms of the type of protein and polypeptides is genetically controlled, but their proportions may vary due to environmental effects (Norton 1989). Mustard proteins consist of two fractions–high molecular weight (12S) and low molecular weight, which constitutes about 25% and 70%, respectively (Rao et al. 1978). Venkatesh and Rao (1988) isolated and characterized low molecular weight protein (22.9 kDa). Aluko et al. (2005) reported that the molecular masses of the protein bands of the *B juncea* meal proteins showed bands in the molecular weight ranging

Table 3.7 Amino acid composition of the mustard meal (mg/100 g protein)

Amino acid	Dehulled defatted mustard meal	Mustard seed flour	Mustard meal
Alanine	4.34 ± 0.07	4.54	–
Arginine	9.82 ± 0.05	7.76	2.80
Aspartic acid	7.13 ± 0.10	8.69	–
Cysteine as cysteic acid	2.40 ± 0.02	–	–
Glutamic acid	20.67 ± 0.20	20.15	–
Glycine	5.20 ± 0.07	5.04	–
Histidine	2.79 ± 0.02	2.61	1.10
Isoleucine	3.71 ± 0.05	4.54	1.70
Leucine	7.59 ± 0.10	8.00	2.90
Lysine	5.30 ± 0.20	5.54	1.90
Methionine	2.52 ± 0.01	–	-
Phenylalanine	4.57 ± 0.09	–	1.60
Proline	5.62 ± 0.04	4.19	–
Serine	4.45 ± 0.05	4.94	–
Threonine	4.38 ± 0.06	4.04	1.60
Tryptophan	1.87 ± 0.07	–	0.60
Tyrosine	2.37 ± 0.11	–	–
Valine	5.27 ± 0.07	5.66	2.10
Methionine + Cysteine	–	5.89	1.90
Phenylalanine + Tyrosine	–	8.41	–
References	Sadgehi et al. (2006)	Abul-Fadl et al. (2011)	Chowdhury et al. (2012)

from 12 to 80 kDa. Similar results have been shown by Rao and Rao (1981), who found proteins in the range of 11–70 kDa. The *B. juncea* proteins are of low molecular weight, recognized as proteins with low antigenicity (Drew 2004) and are considered better as compared to other plant-derived proteins.

3.3.3 Minerals

In wild mustard meal, calcium and phosphorus content of 0.78% and 1.06%, while sodium, potassium, magnesium, iron, manganese, zinc content of 185, 212, 257, 28, 9.2 and 10.2 mg/kg DM has been reported (Daghir and Mian 1976). The mineral content of mustard meal with regard to calcium and phosphorus is comparable with rapeseed meal, for which 0.64% and 1.03% (at 8.5% moisture basis) values have been reported (Bell and Keith 1991), whereas these values are higher than in soybean meal. Copper, manganese and zinc content of 7.0, 35.3 and 55.3 mg/kg dry matter (DM), respectively has been observed in Indian mustard cake (Jankowski et al. 2014).The group further reported that the fertigation of sulphur did not affect the nutrient composition of mustard cake.

3.3.4 Glucosinolates

Glucosinolates (β-thioglucoside-*N*-hydroxysulfates) (GLSs), are secondary metabolites found in almost all plants of the family Brassicaceae and help in protecting plants from pathogen attack (Sharma and Gupta 2020). These compounds are present in considerable amounts in seed meal of *B. juncea* (45–130.1 μmol g^{-1}) (Akhatar et al. 2020). Due to their adverse effects on the thyroid, they fall under the category of antinutrients (Walker and Booth 2001). On the other hand, they are also known for their various health-promoting activities (Latté et al. 2011). After the discovery of sinigrin in mustard (*B. juncea*), over 200 GLSs are known in Brassica crops. Among these, glucoiberin, epiprogoitrin, sinigrin, progoitrin, glucoraphanin, gluconapin, glucobrassicanapin, glucoerucin, glucobrassicin, etc. are the most prevalent (Wittstock and Halkier 2002). The mustard meal contains higher levels and different kinds of GLSs compared with rapeseed meal. Usually, intact GLSs are non-toxic; however, when the seed is crushed in the presence of moisture, myrosinase hydrolyses the GLSs to release glucose and sulphate and various toxic compounds such as isothiocyanates, thiocyanates and nitriles. The enzyme myrosinase, also called thioglucosidase (EC 3.2.3.1), exists in a separate compartment within the mustard seed. GLSs can be aliphatic, aromatic and indole type, based on their precursor amino acids (Clarke 2010; Bala et al. 2015). However, aliphatic GLSs are the prominent ones in *B. juncea* (Yadav and Rana 2018). Newkirk et al (1997) reported that *B. juncea* canola meal had relatively higher amounts of aliphatic GLSs (24.2 mmol/g), 3-butenyl glucosinolates (21.2 mmol/g) and lesser concentration of indole GLSs (3.9 mmol/g) than meals from *B. napus* and *B. rapa*. Different compounds, namely, progoitrin (0.3 μmol g^{-1}), glucoraphanin (0.1 μmol g^{-1}), sinigrin (104.3 μmol g^{-1}), glucoalyssin (0.2 μmol g^{-1}), gluconapin (0.8 μmol g^{-1}), 4-hydroxyglucobrassicin (5.1 μmol g^{-1}), glucoerucin (0.2 μmol g^{-1}), glucobrassicin (0.4 μmol g^{-1}), 4-methoxyglucobrassicin (0.5 μmol g^{-1}), gluconasturtin (0.7 μmol g^{-1}), neoglucobrassicin (0.2 μmol g^{-1}) have been detected with total GLSs content of 112.77 μmol g^{-1} in mustard meal (Park et al. 2019). In the mustard seed meal, 2-propenyl glucosinolate (sinigrin) has been reported as the single largest compound which upon hydrolysis yields allylisothiocyanate (AITC), responsible for imparting pungency in mustard oil and meal. Oxazolidine-2-thione, derived from progoitrin, represents a vital member of harmful compounds (Rosa et al. 1997). Accumulation of this compound in the oilseeds is responsible for the harmful effect, *viz.* goiter and other antinutritional effects on animals. Inhibition of iodine uptake and iodine binding to thyroglobulin by oxazolidine-2-thione and thiocyanate anions may result in goiter (hypertrophy of the thyroid gland) (Burel et al. 2001; De Groef et al. 2006). Lowered thyroid hormones (T3 and T4) indicating iodine deficiency have been reported in growing calves fed with Brassica meal containing diets showing a positive correlation between plasma T3 levels and calves growth rates. This deficiency was significantly overcome with copper and iodine-supplemented Brassica meal diet (Tripathi et al. 2001). Bell (1993) reported that aliphatic GLSs are primarily goitrogenic while indole GLSs have a lesser concern. Some of the adverse effects of GLSs degradation products on the livestock can be described as behavioral changes (Rodriguez et al. 1997), interference with thyroid metabolism leading to goiter (Spiegel et al. 1993), gastrointestinal mucosa irritation and liver damage (Martland et al. 1984), growth and fertility impairment (Schone et al. 1997; Ahlin et al. 1994). Other animals, such as pigs, rabbits and fishes, are also susceptible to GLSs poisoning (Tripathi and Mishra 2017). Impairment in sexual maturity, conception, reduced live piglet birth rate, hypothyroidism, inadequate feed intake and retarded growth are seen in pigs (Lee et al. 2020). Growth retardation and the increased mortality rate have been noticed in rabbits fed with a high GLSs-rich diet (Tripathi et al. 2003). GLSs levels higher than 10 mmol/kg in the diet had significantly reduced the growth rate and daily weight gain in broiler chicks (Mawson et al. 1994). The enhanced time interval between one calving to the subsequent calving and poor productive performance are some of the features noticed in cattle upon GLSs intake (Katamoto et al. 2001). In animal diets, when the mustard meal is used as a major dietary component the concentration of

GLSs and their degradation products being very high may result in negative effects of these compounds. Further, the varieties of mustard meal containing less than 30 μmol g^{-1} of GLSs have been designated as a canola meal, a trade name given by the Canola Council of Canada. The differences in the levels of GLSs in canola varieties of rapeseed and mustard exist, and values are more for *B. juncea* than *B. napus* or *B. rapa*, although these are within the defined limit (Bell 1993). It has been reported that low GLSs cultivars possess lower levels of aliphatic GLSs, and therefore, the meal has been widely accepted for use in animal feeds. Although various negative effects of GLSs and their degradation products have been reported, mustard meal-seed derived products (GLS and their degradation products) when used in small concentrations have shown health-promoting and chemopreventive effects. Nowadays, many studies (Kwon et al. 2020; Mazumder et al. 2016; Augustine and Bisht 2016) have supported their positive effects, as discussed in Sect. 3.4.

3.3.5 Phenolic Compounds

Phenolic compounds are ubiquitously distributed secondary metabolites found in most plants. Phenolic acids, flavonoids, anthocyanins, stilbenes, lignans, lignins, tannins etc., are the different classes of these compounds (Lattanzio et al. 2006). Compared to other oilseed meals, rapeseed-mustard contains ten to thirty folds higher content of phenolic compounds (Kozlowska et al. 1990). These are mainly found in the embryo, while the hull contains only minor amounts. Phenolic compounds may contribute to the dark color, bitter taste and astringency of mustard seed and meals, but they have been known to possess antioxidant activity (Khattab et al. 2010; Bala et al. 2011). Mayengbam et al. (2014) studied the scavenging activity of the various phenolic extracts from whole seed mustard, the cotyledon and hulls extract and found that scavenging activity of mustard cotyledon was more than whole seed, and least activity was observed in mustard hulls. In mustard meal, the total phenolic acid content of 1668.5 mg/100 g has been reported, in which free, esterified and insoluble bound forms are represented as 108.1, 1538, and 22.4 mg/100 g, respectively (Shahidi and Naczk 1992). Esterified phenolic acids comprise about 90% of the total phenolic compounds in mustard meal; the remaining exist in free and insoluble bound form. The total phenolic acid content of 1711 to 2100 mg gallic acid equivalent/100 g has been reported for a meal of Indian mustard (*B. juncea*) varieties (Table 3.8). Among the phenolic compounds, sinapine, sinapoyl glucose and free sinapic acid have been reported as major compounds in mustard meal (Thiyam et al. 2006). Engels et al. (2012) has reported sinapine, kaempferol-sinapoyl-trihexoside, sinapic acid, sinapoyl-hexoside, disinapoyl-dihexoside, disinapoyl-hexoside, trisinapoyl-dihexoside, sinapoyl conjugate(s) and sinapic acid in mustard meal extracts. Among all compounds, sinapine was found in abundance, while free sinapic acid was in traces. Sinapoyl glucose (glucopyranosyl sinapate) has been reported as the most active antioxidative component (Wanasundara et al. 1994). Sinapic acid (3,5-dimethoxy-4-hydroxycinnamic acid) is

Table 3.8 Antinutrients of mustard cake/meal

Component	Mustard meal	References
Phenolic compounds (%)	1.7–2.1	Sadeghi et al. (2006), Bala et al. (2011)
Phytic acid (%)	1.8–5.6	Sadeghi et al. (2006), Sharma et al. (2019)
Tannins (%)	1.0–1.4	Cilly et al. (1978)
Fiber (%)	5.12–18.9	Chauhan and Kumar (2011), Bala and Singh (2013)
Glucosinolates (μmole/g seed meal)	50–120	Chauhan et al. (2007)

considered as the precursor of sinapine and other esters (Chalas et al. 2001). The sinapic acid content of 960–1503 mg sinapic acid equivalents per 100 g has been reported for rapeseed canola varieties, and similar quantities have been observed for the mustard meal by Engels et al. (2012). Sinapic acid concentration may vary depending on the time of preparation of extract. Extracts prepared with hot water showed higher amounts of sinapine and only a small quantity of sinapic acid (Dubie et al. 2013). However, keeping the extract in water at 20 °C for seven days resulted in a decrease in sinapine and increased the sinapic acid content. Chadni et al. (2021) reported that the extraction yield for sinapic acid (13.22 μmol/g of DM) and sinapine (15.73 μmol/g DM) was maximum at pH 12 and pH 2, respectively. The dietary importance of the mustard meal is affected by their presence as these compounds and/or their oxidized products decrease the bioavailability of proteins and essential amino acids. Sinapine has a bitter flavor, shows adverse effects on feed intake with diets containing mustard meal, and found that body weight gain, feed intake and feed efficiency decreased with the increasing levels of mustard meal (Tangtaweewipat et al. 2004). The removal/reduction of sinapine content may improve the flavor, taste and food characteristics of mustard seeds and meal. Different groups have suggested ways to remove or reduce sinapine (Wojciechowski et al. 1994), but no method is economically feasible (Wang et al. 1998).

3.3.6 Tannins

Tannins are polyphenolic compounds, localized in the endosperm. Tannin content was found to be 1.0–1.4% in raya (*B. juncea*), toria (*B. campestris* var. toria), yellow and brown sarson (*B. campestris* var. sarson), 1.74% in taramira (Cilly et al. 1978) and 1.5% in rapeseed meal (Bell 1995). The tannin content in *B. juncea* seeds and meal is significantly lower than those of canola (*B. napus*) seed, cake and meals. Press cake and meal of canola varieties had condensed tannin contents of 1.12–1.32 g/100 g and 0.59–1.23 g/100 g, respectively, while for mustard seed meal, it was 0.46 g/100 g and 0.79 g/100 g, respectively (Khattab et al. 2010). Canola and rapeseed hulls have been reported to contain up to 6% tannins. Tannins possess antioxidant properties (Amarowicz 2007) as crude tannins extracted from hull fraction of canola showed significantly ($P \leq 0.025$) more antioxidant activity than those from rapeseed (Amarowicz et al. 2000). However, tannins may also show antinutritional properties. It has been reported that when tannins rich fraction of rapeseed was fed to laying hen, egg tainting was observed, which could be due to inhibition of trimethylamine oxidase enzyme activity which converts trimethylamine to a water-soluble, odorless oxide (Shahidi and Naczk 1992).

3.3.7 Phytates

Phytic acid or phytate, named myo-inositol-1, 2, 3, 4, 5, 6-hexakis-dihydrogen phosphate, exists as assorted salts of iron, calcium, zinc etc. (Yiu et al. 1982). It constitutes up to 10% of the dry weight of Brassica oilseeds (Yiu et al. 1983) and accounts for about 36–70% of the phosphorus content in Brassica meals (Broz and Ward 2007). Phytate accumulation is mainly confined to within the embryo throughout seed development and maturation (Dang et al. 2013). Moreover, it is found inside protein bodies of mustard seed in the form of crystalline globoids of 0.5–2.8 μm in size. Phytate levels of 2.0–4.0% for the whole mustard seed, 2.0–5.6% for the defatted mustard meal, and <1.0% to 9.8% for the protein isolates prepared using different methods have been reported (Sadeghi et al. 2006; Sadeghi and Bhagya 2009). The phytic acid content of mustard seed (2.99 g/100 g) and meal (2.45 g/100 g) was found to be similar to canola seeds and the meal (Khattab et al. 2010). Phytates bind with metal ions such as iron, calcium, magnesium and zinc due to their strong chelating properties (Kies et al. 2006). They are known to lower protein and starch digestibility, and reduce the bioavailability of the minerals and amino acids (Noureddini and Dang 2008) and thus act as an antinutritional substance in the diets of monogastric animals and humans. However, with the help of phytase activity of the ruminal microflora, ruminants can utilize phytate phosphorous (P), and that is why in India, the mustard meal is given to ruminants only. Monogastric animals cannot utilize dietary phytic acid because of the absence of the enzyme phytase. Effective reduction of phytic acid can be made using enzymatic and non-enzymatic treatments (Greiner and Konietzny 2006). Phytase can hydrolyse phytate to inositol and inorganic phosphorus. There are reports showing enhanced digestibility of phytate-associated phosphorus with the addition of phytase to the diets of monogastric animals (Pontoppidan et al. 2007) and improved growth performance. The development of mustard varieties with low levels of phytic acid can be an alternative approach to lower phytate levels and improve the quality of rapeseed-mustard meal (Eifler et al. 2021).

3.4 Pharmacological Significance of *B. juncea*

3.4.1 Antioxidant Activity

In a recent investigation, Kwon et al. (2020) compared the antioxidant activities of two different *B. juncea* cultivars (Dolsan and Jeongseon). The Jeongseon cultivar was found to possess superior antioxidant potential based on 2,2-diphenyl-1-picrylhydrazyl (DPPH) radical scavenging activity and ferric reducing antioxidant power (FRAP). This property was attributed to its higher total phenol and flavonoid content. A similar observation was reported by Oh et al. (2016), who found that the *B. juncea* seeds possessed more potent antioxidants than the leaves, and there was a direct correlation between antioxidant capacity and the levels of phenol and flavonoid. Nevertheless, it should be noted that polyphenolic phytochemicals alone do not account for all of the antioxidant properties of *B. juncea,* as it also contains high concentrations of other compounds with known antioxidant properties such as vitamin C, vitamin A and GLSs (Park et al. 2017).

3.4.2 Anticancer Activity

One of the well-studied metabolites of *B. juncea* showing anticancer activity is the GLS sinigrin, the precursor of AITC (Mazumder et al. 2016). In a rat model of carcinogen-induced hepatotoxicity, treatment with sinigrin inhibited liver tumor cells proliferation by inducing apoptosis (Jie et al. 2014). The effects of sinigrin were also tested in vitro on HL60 human promyelocytic leukemia cells, where it exhibited potent tumor inhibitory activity with a half-maximal inhibitory concentration (IC_{50}) value of 2.71 μM (Lozano-Baena et al. 2015). The antiproliferative activity of *B. juncea* against different cancer cell types is summarized in Table 3.9. Many studies have also focused on the therapeutic properties of AITC.

Using University of Michigan-Urothelial Carcinoma-3 (UM-UC-3) normal human bladder carcinoma cells, Bhattacharya et al. (2010) demonstrated that AITC strongly inhibited cancer growth (IC_{50} value = 2.7 μM) without affecting normal bladder epithelial cells. A study by Tripathi et al. (2015) involving a human non-small cell lung carcinoma cell line (H1299) provided insight into the anticancer mechanisms of AITC. It was shown that AITC mediated its effects by inducing replication stress in cancer cells through the generation of fork-stalling DNA lesions. *Brassica juncea* also contains significant amounts of another GLS glucobrassicin, and its degradation product, indole-3-carbinol (I3C), which is well-documented for its anticancer activities (Augustine and Bisht 2016). Its mechanism of action revealed that I3C stimulated the expression of tumor suppressive micro RNA in Michigan Cancer Foundation-7 (MCF7) human breast cancer cells (Hargraves et al. 2016). Apart from breast cancer cells, the anticancer activity of I3C has also been demonstrated in various cell lines of colon cancer (Megna et al. 2016) and hepatocellular carcinoma (Wang et al. 2015).

Table 3.9 Antiproliferative activity of *B. juncea* against human cancer cell lines

Cancer cell type	Efficacy (μg/mL)	Extracting solvent	References
A549, human lung carcinoma	54.35[a]	Ethyl acetate	Bassan et al. (2018)
	80.16[a]	Dichloromethane	Bassan et al. (2018)
	45% at 60[b]	Methanol/Ethyl acetate	Li et al. (2019)
HCT116, human colorectal carcinoma	61.50[a]	Ethyl acetate	Bassan et al. (2018)
	78.86[a]	Dichloromethane	Bassan et al. (2018)
	81.1[a]	Ethanol	Tian and Deng (2020)
HeLa, human cervix adenocarcinoma	67.25[a]	Ethyl acetate	Bassan et al. (2018)
	68.56[a]	Dichloromethane	Bassan et al. (2018)
MCF7, human breast adenocarcinoma	32.93[a]	Ethyl acetate	Bassan et al. (2018)
	43.10[a]	Dichloromethane	Bassan et al. (2018)
MDA-MB-231, human breast adenocarcinoma	37.16[a]	Ethyl acetate	Bassan et al. (2018)
	51.14[a]	Dichloromethane	Bassan et al. (2018)
PC-3, human prostate adenocarcinoma	54.73[a]	Ethyl acetate	Bassan et al. (2018)
	65.23[a]	Dichloromethane	Bassan et al. (2018)

[a] IC_{50} value
[b] Percentage of cell growth inhibition

3.4.3 Antimicrobial Activity

Brassica juncea seed extracts have been shown to exhibit selective antibacterial activity against several food borne pathogens such as *Staphylococcus aureus*, *Listeria monocytogenes*, *Salmonella enterica, Bacillus subtilis*, *Escherichia coli* and *Pseudomonas fluorescens* (Engels et al. 2012). The antagonistic effect toward these bacteria has been partly attributed to the presence of sinapic acid, one of the most abundant phenolic compounds in *B. juncea* seeds (Nicácio et al. 2021). Surprisingly, sinapic acid did not negatively affect the growth and activity of the beneficial *Lactobacillus plantarum* (Engels et al. 2012). The antimicrobial property of sinapic acid is also evident from its ability to inhibit mycotoxin production by toxigenic strains of fungi (Kulik et al. 2017). These compounds also demonstrated a synergistic effect when combined with conventional antibiotics, thereby improving their efficacy (Dias et al. 2012). It is thought that sinigrin and AITC are part of the plant's defense system that helps protect them against parasitic threats (Melrose 2019). As the dominant isothiocyanate of *B. juncea*, the inhibitory effects of AITC against bacterial and fungal pathogens have been widely reported and are summarized in Table 3.10. Remarkably, the antibacterial potency of AITC has been described as being similar to that of vancomycin to treat antibiotic-resistant bacteria (Melrose, 2019). Experimental evidence suggests that AITC produces its antimicrobial effects by disrupting cell membrane integrity and inhibiting the activity of the enzymes involved in DNA biosynthesis and energy metabolism (Romeo et al. 2018). As such, AITC is valuable to the food industry, where it has found application as components of antimicrobial films to prevent food spoilage caused by bacteria and fungi contamination (Gao et al. 2017; Maruthupandy and Seo 2019). The antifungal properties of *B. juncea* have also been attributed to proteins and peptides. Napin, a 15 kDa protein isolated from *B. juncea* seeds, was shown to repress the mycelial growth of *Aspergillus* and *Fusarium* fungi. Besides, napin exhibited inhibitory activity against several clinically significant bacterial species (Munir et al. 2019). Another antifungal protein found in *B. juncea* seeds is juncin (18.9 kDa), which is effective against *Helmintho sporiummaydis*,

Table 3.10 Microbial species sensitive to growth inhibitory effect of AITC

Organism type	Species	References
Gram-positive bacteria	*Bacillus cereus*	Clemente et al. (2016)
	Listeria monocytogenes	Blažević et al. (2019), Liu and Yang (2010), Olaimat and Holley (2013)
	Staphylococcus aureus	Blažević et al. (2019), Clemente et al. (2016), Dias et al. (2014), Liu and Yang (2010)
Gram-negative bacteria	*Acinetobacter baumannii*	Blažević et al. (2019)
	Campylobacter jejuni	Dufour et al. (2012)
	Escherichia coli	Blažević et al. (2019), Clemente et al. (2016), Liu and Yang (2010), Luciano and Holley (2009)
	Pseudomonas aeruginosa	Blažević et al. (2019), Kaiser et al. (2017)
	Salmonella enterica	Clemente et al. (2016), Liu and Yang (2010)
	Salmonella typhimurium	Blažević et al. (2019)
	Vibrio parahaemolyticus	Liu and Yang (2010)
Fungi	*Aspergillus niger*	Suhr and Nielsen (2003)
	Aspergillus flavus	Clemente et al. (2019), Otoni et al. (2014)
	Aspergillus parasiticus	Lopes et al. (2018), Quiles et al. (2015)
	Botryotinia fuckeliana	Clemente et al. (2019)
	Candida albicans	Blažević et al. (2019)
	Endomyces fibuliger	Suhr and Nielsen (2003)
	Eurotiumrepens	Suhr and Nielsen (2003)
	Penicillium corylophilum	Suhr and Nielsen (2003)
	Penicillium roqueforti	Clemente et al. (2019), Suhr and Nielsen (2003)

Fusarium oxysporum and Mycosphaerella arachidicola (Ye and Ng 2009).

3.4.4 Anti-obesity/Antidiabetic Activities

The role of natural products, including those present in *B. juncea*, in managing abnormal metabolic conditions such as diabetes and obesity, has been widely acknowledged. A recent study demonstrated that both ethanol extract of *B. juncea* and purified sinigrin reduced lipid accumulation in a cell line derived from mouse 3T3 cells (3T3-L1) adipocytes (Kwon et al. 2021). Given the high phenolic content of *B. juncea,* it is plausible that these effects may also be modulated by flavonoids, whose anti-obesity and antidiabetic properties, as well as their mechanisms of action, have been expounded in detail (Hossain et al. 2016). It has been reported that administration of seed extract of *B. juncea* to streptozotocin-induced diabetic rats (Thirumalai et al. 2011) led to a dose-dependent increase in the rat's serum insulin level.

3.4.5 Anti-inflammatory Activity

The anti-inflammatory potential of *B. juncea* has been proven in vivo by Xian et al. (2018), who studied the therapeutic effects of *B. juncea* seed extract in mouse models of acute and chronic inflammation induced by tetradecanoylphorbol-acetate, arachidonic acid and croton oil. The treated mice exhibited noticeably reduced signs of inflammation in the affected organ due to decreased expressions of tumor necrosis factor-alpha (TNF-α), interleukin-6 (IL-6) and interleukin-1 beta (IL-1β) compared to control mice. This effect was expected since *B. juncea* contains a wide variety of phenolic compounds long valued as natural alternatives in treating inflammation (Ambriz-Pérez et al. 2016). Indeed, the anti-inflammatory effects of sinapic acid present in high concentrations in *B. juncea* have been demonstrated using both in vitro and in vivo models (Yun et al. 2008). In RAW 264.7 cells, sinapic acid dose-dependently blocked nuclear factor kappa-light chain enhancer of activated B cells (NF-kB) activation, thus suppressing the production of cyclooxygenase-2 (COX-2), TNF-α and IL-1β. This is believed to be responsible for its effect in inhibiting serotonin and carrageenan-induced paw edema in mice and rats, and its effect was comparable to that of indomethacin and ibuprofen. A recent study by Le et al. (2020) provided further evidence on the relationship between the anti-inflammatory activity of *B. juncea* and its phenolic profile. Fermentation of *B. juncea* with *Lactobacillus plantarum* reportedly enhanced its anti-inflammatory effects in RAW 264.7 cells, attributed to the dephosphorylation of the mitogen-activated protein kinase (MAPK) signaling pathway. The fermented *B. juncea* extract's improved activity over the raw extract was credited to increase in its total phenolic and total flavonoid content upon fermentation. Among the many phenolic compounds identified, caffeic acid, rutin and ferulic acid showed the most significant increase in concentration.

3.5 Potential Benefits and Threats to Human and Animal Health

Human and animal health has become an essential subject in the modern world. Traditional medicinal uses of *B. juncea* seeds and oils have been known for a long time. Several reviews showing phytochemistry, therapeutic potential and adverse effects of *B. juncea* are available in the literature (Cartea and Velasco 2008; Kumar et al. 2011; Rahman et al. 2018; Prieto et al. 2019; Tian and Deng 2020). Figure 3.1 shows some of these activities of *B. juncea* extracts to treat pathological conditions.

GLSs, present in *B. juncea* have attracted the attention of scientists in evaluating their impact on human and animal health. The possible degradation products of GLSs upon myrosinase (β-thioglucosidase) action are isothiocyanates, oxazolidine-2-thiones, thiocyanates, nitriles and epithionitriles (Blažević et al. 2020). Isothiocyanates constitute the major group of hydrolytic products of GLSs, which are beneficial to human health (Vig et al. 2009). The isothiocyanate derivatives present in *B. juncea* essential oil comprises of AITC (73.4%), 1-butene-4-

Fig. 3.1 Therapeutic and biological activities of *Brassica juncea* extracts

isothiocyanate (19.1%), 2-phenylethyl-isothiocyanate (1.2%) and 3-methylthiopropyl-isothiocyanate (0.7%) (Singh et al. 2017). They possess novel therapeutic properties such as biocidal, chemopreventive, anti-inflammatory and anti-mutagenic effects (Wu et al. 2017; Rampal et al. 2017; Grundemann and Huber 2018), antioxidant activity, inhibitory action against mitosis and stimulating apoptosis in human tumor cells, bactericidal, fungicidal and nematicidal activities (Sonderby et al. 2010; Meyer et al. 2011; Barba et al. 2016). 3-butenyl isothiocyanates, present in the seed extract of *B. juncea* L. displayed strong cytotoxic activities against human cancer cell lines (Arora et al. 2016).

3.6 Concluding Remarks and Future Prospects

Brassica oilseeds are among the successful crops whose seed quality has been significantly improved. *B. juncea* also known as Indian mustard is widely used for its oil and meal quality. In terms of nutritional properties, the fatty acid profile of this oil is a perfect blend of different fatty acids along with bioactive compounds such as tocopherols and phytosterols. Also, the meal obtained after extraction of oil is a rich source of proteins, minerals and vitamins. However, the benefits from this crop are not realized to the fullest due to the presence of nutritionally undesired components *viz.* erucic acid in oil and glucosinolates in the meal. To overcome these shortfalls, breeding strategies have been applied to reduce these two compounds resulting in the successful development of highly productive *B. juncea* varieties with oil and meal quality closer to that of the internationally accepted canola (*B. napus*), which is considered healthy for food and feed purpose. The development of canola quality mustard varieties in India has also helped in achieving 40–50% of oleic acid and 30–36% linoleic acid, which are considered desirable for good cooking oil. From future perspective, possibilities exist for identifying Indian mustard genotypes rich in bioactive compounds. It is envisaged that the export of these bioactive compounds may contribute to the national GDP. Research is needed to understand the unknown mechanism of action through which the GLS sinigrin exerts its therapeutic effects. Efforts are also required for the effective dissemination of the canola quality mustard. Further strengthening the linkage among farmer-industry-institution could increase the coverage area under canola quality mustard, thus ensuring the availability of high-quality mustard oil for the Indian population.

References

Abul-Fadl MM, El-Badry N, Ammar MS (2011) Nutritional and chemical evaluation for two different varieties of mustard seeds. World Appl Sci J 15 (9):1225–1233

Ahlin KA, Emanuelson M, Wiktorson H (1994) Rapeseed products from double-low cultivars as feed for dairy cows: effects of long-term feeding on thyroid function, fertility and animal health. Acta Veter Scand 35 (1):37–53

Akhatar J, Singh MP, Sharma A, Kaur H, Kaur N, Sharma S, Bharti B, Sardana VK, Banga S (2020) Association mapping of seed quality traits under varying conditions of nitrogen application in *Brassica juncea* L. Czern. & Coss. Front Genet 11:744

Al-Jasass FM, Al-Jasser MS (2012) Chemical composition and fatty acid content of some spices and herbs under Saudi Arabia conditions. Sci World J, 859892

Aluko RE, McIntosh T, Katepa MF (2005) Comparative study of the polypeptide profiles and functional properties of *Sinapis alba* and *Brassica juncea* seed meals and protein concentrates. J Sci Food Agri 85 (11):1931–1937

Amar S, Becker HC, Möllers C (2008) Genetic variation and genotype× environment interactions of phytosterol content in three doubled haploid populations of winter rapeseed. Crop Sci 48(3):1000–1006

Amarowicz R (2007) Tannins: the new natural antioxidants. Eur J Lipid Sci Technol 109:549–551

Amarowicz R, Naczk M, Shahidi F (2000) Antioxidant activity of crude tannins of canola and rapeseed hulls. J Amer Oil Chem Soc 77(9):957–961

Ambriz-Perez DL, Leyva-Lopez N, Gutierrez-Grijalva EP, Heredia JB (2016) Phenolic compounds: natural alternative in inflammation treatment. A review. Cogent Food Agri 2:1131412

Appelqvist LAD, Kornfeld AK, Wennerholm JE (1981) Sterols and steryl esters in some *Brassica* and *Sinapis* seeds. Phytochemistry 20(2):207–210

Arora R, Kumar R, Mahajan J, Vig AP, Singh B, Singh B, Arora S (2016) 3-Butenyl isothiocyanate: a hydrolytic product of glucosinolate as a potential cytotoxic agent against human cancer cell lines. J Food Sci Technol 53:3437–3445

Augustine R, Bisht N (2016) Biofortification of oilseed *Brassica juncea* with the anti-cancer compound glucoraphanin by suppressing *GSL-ALK* gene family. Sci Rep 5:18005

Awogbemi O, Onuh EI, Inambao FL (2019) Comparative study of properties and fatty acid composition of some neat vegetable oils and waste cooking oils. Int J Low Carbon Technol 14(3):417–425

Bala M, Kumar S, Singh L (2011) Antioxidant potential of rapeseed–mustard seed meal extracts. Indian J Agric Biochem 24(1):55–59

Bala M, Singh M (2013) Non-destructive estimation of total phenol and crude fiber content in intact seeds of rapeseed–mustard using FTNIR. Ind Crops Prod 42:357–362

Bala M, Tushir S, Tyagi SK, Gupta RK (2015) Antinutrients in oilseed brassica: uses and potential applications. Anim Nutr Technol 15(2):295–310

Barba FA, Nikmaram N, Roohinejad S, Khelfa A, Zhu Z, Koubaa M (2016) Bioavailability of glucosinolates and their breakdown products: impact of processing. Front Nutr 3:1–12

Bassan P, Bhushan S, Kaur T, Arora R, Arora S, Vig AP (2018) Extraction, profiling and bioactivity analysis of volatile glucosinolates present in oil extract of *Brassica juncea* var. Raya. Physiol Mol Biol Plants 24:399–409

Bell JM (1993) Factors affecting the nutritional value of canola meal: a review. Can J Anim Sci 73(4):689–697

Bell JM (1995) Meal and by-product utilization in animal nutrition. In: Kimber DS, McGregor DI (eds) Brassica oilseeds: production and utilization. Cab International, Wallingford, Oxford, UK, pp 301–337

Bell JM, Keith MO (1991) A survey of variation in the chemical composition of commercial canola meal produced in Western Canadian crushing plants. Can J Anim Sci 71(2):469–480

Bergmuller E, Porfirova S, Dormann P (2003) Characterization of an Arabidopsis mutant deficient in γ-tocopherol methyltransferase. Plant Mol Biol 52 (6):1181–1190

Bhattacharya A, Tang L, Li Y, Geng F, Paonessa JD, Chen SC, Zhang Y (2010) Inhibition of bladder cancer development by allyl isothiocyanate. Carcinog 31:281–286

Blažević I, Dulovic A, Maravic A, CikesCulic V, Montaut S, Rollin P (2019) Antimicrobial and cytotoxic activities of *Lepidium latifolium* L. Hydrodistillate, extract and its major sulfur volatile allyl isothiocyanate. Chem Biodivers 16:1800661

Blažević I, Montaut S, Burcul F, Olsen CE, Burow M, Rollin P, Agerbirk N (2020) Glucosinolate structural diversity, identification, chemical synthesis and metabolism in plants. Phytochemistry 169:112100

Booth EJ (2004) Extraction and refining. In: Gunstone E (ed) Rapeseed and canola oil: production, processing, properties and uses. Blackwell, Oxford, UK, pp 17–36

Broz J, Ward NE (2007) The role of vitamins and feed enzymes in combating metabolic challenges and disorders. J Appl Poult Res 16(1):150–159

Burel C, Boujard T, Kaushik SJ, Boeuf G, Mol KA, Geyten SV, Darras VM, Kuhn ER, Predet-Balade B, Querat B, Quinsac A, Krouti M, Ribaillier D (2001) Effects of rapeseed meal-glucosinolates on thyroid metabolism and feed utilization in rainbow trout. Gen Comp Endocrinol 124:343–358

Cabral CE, Klein MRST (2017) Phytosterols in the treatment of hypercholesterolemia and prevention of cardiovascular diseases. Arq Bras Cardiol 109(5):475–482

Cartea ME, Velasco P (2008) Glucosinolates in *Brassica* foods: bioavailability in food and significance for human health. Phytochem Rev 7:213–229

Chadni M, Flourat AL, Reungoat V, Mouterde LM, Allais F, Ioannou I (2021) Selective extraction of sinapic acid derivatives from mustard seed meal by acting on pH: toward a high antioxidant activity rich extract. Molecules 26(1):212

Chalas J, Claise C, Edeas M, Messaoudi C, Vergnes L, Abella A, Lindenbaum A (2001) Effect of ethyl esterification of phenolic acids on low density lipoprotein oxidation. Biomed Pharmacother 55:54–60

Chand S, Patidar OP, Chaudhary R, Saroj R, Chandra K, Meena VK, Limbalkar OM, Patel MK, Pardeshi PP, Vasisth P (2021) Rapeseed-mustard breeding in India: Scenario, achievements and research needs. In: Brassica breeding and biotechnology. IntechOpen

Charlton KM, Corner AH, Davey K, Kramer JK, Mahadevan S, Sauer FD (1975) Cardiac lesions in rats fed rapeseed oils. Can J Comp Med 39(3):261

Chauhan JS, Kumar S (2011) Assessment of oil and seed meal quality parameters of rapeseed–mustard group of crops. Indian J Agric Sci 81:140–144

Chauhan JS, Singh M, Bhadauria VPS, Kumar A (2007) Genetic analysis of glucosinolate content in Indian mustard (*Brassica juncea L.*). In: Proceedings of the 12th international rapeseed congress. Science Press USA Inc., Wuhan, China, pp 167–169

Chauhan JS, Tyagi MK, Kumar PR, Tyagi P, Singh M, Kumar S (2002) Breeding for oil and seed meal quality in rapeseed-mustard in India—a review. Agric Rev 23(2):71–92

Chhokar V, Beniwal V, Kumar A, Rana JS (2008) Lipid content and fatty acid composition of mustard (*Brassica juncea* L.) during seed development. Asian J Exp Chem 3(1–2):6–9

Chowdhury MK, Tacon AG, Bureau DP (2012) Digestibility of amino acids in Indian mustard protein concentrate and Indian mustard meal compared to that of a soy protein concentrate in rainbow trout and Atlantic salmon. Aquaculture 356:128–134

Cilly VK, Lodhi GN, Ichhponani JS (1978) Nutritive value of mustard cake derived from seeds of different

genetic make up for growing chicks. J Agric Sci 91 (3):543–550
Clarke DB (2010) Glucosinolates, structures and analysis in food. Anal Methods 2:310–325
Clemente I, Aznar M, Nerín C (2019) Synergistic properties of mustard and cinnamon essential oils for the inactivation of foodborne moulds*in vitro* and on Spanish bread. Int J Food Microbiol 298:44–50
Clemente I, Aznar M, Silva F, Nerín C (2016) Antimicrobial properties and mode of action of mustard and cinnamon essential oils and their combination against foodborne bacteria. Innov Food Sci Emerg Technol 36:26–33
Conti V, Izzo V, Corbi G, Russomanno G, Manzo V, De Lise F, Di Donato A, Filippelli A (2016) Antioxidant supplementation in the treatment of aging-associated diseases. Front Pharmacol 7:24
Daghir NJ, Mian NA (1976) Mustard seed meal as a protein source for chickens. Poult Sci J 55(5):1699–1703
Dang J, Yan W, Bock C, Nokhrina K, Keller W, Georges F (2013) Perturbing the metabolic dynamics of myo-inositol in developing *Brassica napus* seeds through *in vivo* methylation impacts its utilization as phytate precursor and affects downstream metabolic pathways. BMC Plant Biol 13:84
Das R, Bhattacherjee C, Ghosh S (2009) Preparation of mustard (*Brassica juncea* L.) Protein isolate and recovery of phenolic compounds by ultrafiltration. Ind Eng Chem Res 48:4939–4947
De Groef B, Decallonne BR, Geyten SV, Darras VM, Bouillon R (2006) Perchlorate versus other environmental sodium/iodide symporter inhibitors: potential thyroid-related health effects. Eur J Endocrinol 155:17–25
Dias C, Aires A, Bennett RN, Rosa EA, Saavedra MJ (2012) First study on antimicriobial activity and synergy between isothiocyanates and antibiotics against selected Gram-negative and Gram-positive pathogenic bacteria from clinical and animal source. Med Chem 8:474–480
Dias C, Aires A, Saavedra MJ (2014) Antimicrobial activity of isothiocyanates from cruciferous plants against methicillin-resistant *Staphylococcus aureus* (MRSA). Int J Mol Sci 15:19552–19561
Dorni C, Sharma P, Saikia G, Longvah T (2018) Fatty acid profile of edible oils and fats consumed in India. Food Chem 238:9–15
Drew M (2004) Canola protein concentrate as a feed ingredient for salmonid fish. In: Cruz Suárez LE, Ricque M, Nieto López MG, Villarreal D, Scholz UY, González M (eds) Advances en Nutritión Acuícola VII, Memorias del VII Simposium Internacional de Nutrición Acuícola. Hermosillo, Sonora, Mexico
Dubie J, Stancik A, Morra M, Nindo C (2013) Antioxidant extraction from mustard (*Brassica juncea*) seed meal using high-intensity ultrasound. J Food Sci 78:542–548
Dufour V, Alazzam B, Ermel G, Thepaut M, Rossero A, Tresse O, Baysse C (2012) Antimicrobial activities of isothiocyanates against *Campylobacter jejuni* isolates. Front Cell Infect Microbiol 2:53
Frankel EN (2012) Lipid oxidation, 2nd edn. Elsevier, Oily Press
Eifler J, Wick JE, Steingrobe B, Mollers C (2021) Genetic variation of seed phosphorus concentration in winter oilseed rape and development of a NIRS calibration. Euphytica 217:53
Endo J, Arita M (2016) Cardioprotective mechanism of omega-3 polyunsaturated fatty acids. J Cardiol 67 (1):22–27
Engels C, Schieber A, Ganzle MG (2012) Sinapic acid derivatives in defatted oriental mustard (*Brassica juncea* L.) seed meal extracts using UHPLC-DAD-ESI-MSN and identification of compounds with antibacterial activity. Eur Food Res Technol 234:535–542
Fritsche S, Xingxing W, Li J, Benjamin S, Friedrich J, Kopisch-Obuch JE, Gunhild L, Dreyer F, Friedt W, Meng J, Jung C (2012) A candidate gene-based association study of tocopherol content and composition in rapeseed (*Brassica napus*). Front Plant Sci 3:129
Gao H, Fang X, Chen H, Qin Y, Xu F, Jin TZ (2017) Physiochemical properties and food application of antimicrobial PLA film. Food Control 73:1522–1531
Gilani GS, Sepehr E (2003) Protein digestibility and quality in products containing antinutritional factors are adversely affected by old age in rats. J Nutr 133 (1):220–225
Greiner R, Konietzny U (2006) Phytase for food application. Food Technol Biotechnol 44(2):125–140
Grundemann C, Huber R (2018) Chemoprevention with isothiocyanates—from bench to bedside. Cancer Lett 414:26–33
Gugliandolo A, Bramanti P, Mazzon E (2017) Role of vitamin E in the treatment of Alzheimer's disease: evidence from animal models. Int J Mol Sci 18 (12):2504
Gunstone F (2011) Vegetable oils in food technology: composition, properties and uses. Wiley.
Hargraves KG, He L, Firestone GL (2016) Phytochemical regulation of the tumor suppressive micro RNA, miR-34a, by p53-dependent and independent responses in human breast cancer cells. Mol Carcinog 55:486–498
Horrobin DF (1986) Essential fatty acid and prostaglandin metabolism in Sjögren's syndrome, systemic sclerosis and rheumatoid arthritis. Scand J Rheumatol Suppl 61:242–245
Hossain MK, Dayem AA, Han J, Yin Y, Kim K, Saha SK, Yang GM, Choi HY, Cho SG (2016) Molecular mechanisms of the anti-obesity and anti-diabetic properties of flavonoids. Int J Mol Sci 17:569
Hossain Z, Johnson EN, Wang L, Blackshaw RE, Gan Y (2019) Comparative analysis of oil and protein content and seed yield of five Brassicaceae oilseeds on the Canadian prairie. Ind Crop Prod 136:77–86
ICMR (2010) Nutrient requirements and recommended dietary allowances for Indians: a report of the expert group of the Indian Medical Research. ICMR, New Delhi, India, pp 93–106

India Agristat (2020) http://www.indiaagristat.com/table/agriculture-data/2/total-oilseeds/19582/1067842/data.aspx. Dated 24 July 2021

Jagannath A, Sodhi YS, Gupta V, Mukhopadhyay A, Arumugam N, Singh I, Pental D (2011) Eliminating expression of erucic acid-encoding loci allows the identification of "hidden" QTL contributing to oil quality fractions and oil content in *Brassica juncea* (Indian mustard). Theor Appl Genet 122(6):1091–1103

Jankowski KJ, Budzyński WS, Klasa A (2014) Concentrations of copper, zinc and manganese in the roots, straw and oil cake of white mustard (*Sinapis alba* L.) and Indian mustard (*Brassica juncea* (L.) Czern. et Coss.) depending on sulphur fertilization. Plant Soil Environ 60(8):364–371

Jat RS, Singh VV, Sharma P, Rai PK (2019) Oilseed brassica in India: demand, supply, policy perspective and future potential. OCL 26:8

Jham GN, Moser BR, Shah SN, Holser RA, Dhingra OD, Vaughn SF, Berhow MA, Winkler-Moser JK, Isbell TA, Holloway RK, Walter EL, Natalino R, Anderson JC, Stelly DM (2009) Wild Brazilian mustard (*Brassica juncea* L.) seed oil methyl esters as biodiesel fuel. J Amer Oil Chem Soc 86:917–926

Jie M, Cheung WM, Yu V, Zhou Y, Tong PH, Ho JW (2014) Anti-proliferative activities of sinigrin on carcinogen-induced hepatotoxicity in rats. PLoS One 9(10):e110145

Kaiser SJ, Mutters NT, Blessing B, Günther F (2017) Natural isothiocyanates express antimicrobial activity against developing and mature biofilms of *Pseudomonas aeruginosa*. Fitoterapia 119:57–63

Kale V (2007) In souvenir National seminar on changing vegetable oil scenario: issues and challenges before India. Indian Society of Oilseeds Research, Directorate of oilseeds Research, Hyderabad, India, pp 81–86

Kanrar S, Venkateswari J, Dureja P, Kirti PB, Chopra VL (2006) Modification of erucic acid content in Indian mustard (*Brassica juncea*) by upregulation and downregulation of the *Brassica juncea* Fatty Acid Elongation1 (Bjfae1) gene. Plant Cell Rep 25:148–155

Katamoto H, Nishiguchi S, Harada K, Ueyama I, Fujita T, Watanabe O (2001) Suspected oriental mustard (*Brassica juncea*) intoxication in cattle. Vet Rec 149 (7):215–216

Kaur R, Sharma AK, Rani R, Mawlong I, Rai PK (2019) Medicinal qualities of mustard oil and its role in human health against chronic diseases: a review. Asian J Dairy Food Res 38(2):98–104

Khajali F, Slominski BA (2012) Factors that affect the nutritive value of canola meal for poultry. Poult Sci J 91(10):2564–2575

Khansili N, Rattu G (2017) A comparative study of hidden characteristics of canola and mustard oil. Int J Chem Stud 5(3):632–635

Khattab R, Eskin M, Aliani M, Thiyam U (2010) Determination of sinapicacid derivatives in canola extracts using high-performance liquid chromatography. J Amer Oil Chem Soc 87:147–155

Kies AK, Kemme PA, Sebek LBJ, Van Diepen JTM, Jongbloed AW (2006) Effect of graded doses and a high dose of microbial phytase on the digestibility of various minerals in weaner pigs. J Anim Sci 84 (5):1169–1175

Kies C, Umoren J (1989) Inhibitors of copper bioutilization: fiber, lead, phytate and tannins. Adv Exp Med Biol 258:81–93

Kirk JTO, Oram RN (1978) Mustards as possible oil and protein crops for Australia [Indian mustard, *Brassica juncea*, white mustard, *Sinapsis alba*]. J Austral Inst Agric Sci 44:143–156

Klein-Hessling H (2007) Value canola meal before using it. World Poult 23:24–28

Kozlowska H, Naczk M, Shahidi F, Zadernowski R (1990) Phenolic acids and tannins in rapeseed and canola. Canola and Rapeseed. Springer, Boston, MA, pp 193–210

Kulik T, Stuper-Szablewska K, Bilska K, Buśko M, Ostrowska-Kołodziejczak A, Załuski D, Perkowski J (2017) Sinapic acid affects phenolic and trichothecene profiles of *F. culmorum* and *F. graminearum sensustricto*. Toxins 9:264

Kumar V, Thakur AK, Barothia ND, Chatterjee SS (2011) Therapeutic potentials of *Brassica juncea*: an overview. TANG Int J Genuine Trad Med 1(1):2

Kwon HY, Choi SI, Park HI, Choi SH, Sim WS, Yeo JH, Cho JH, Lee OH (2020) Comparative analysis of the nutritional components and antioxidant activities of different *Brassica juncea* cultivars. Foods 9:840

Kwon HY, Choi SI, Han X, Men X, Jang GW, Choi YE, Lee OH (2021) Antiobesity effect of *Brassica juncea* cultivated in Jeongseon with optimized sinigrin content using 3T3-L1 adipocytes. J Food Biochem 45:13650

Lattanzio V, Lattanzio VM, Cardinali A (2006) Role of phenolics in the resistance mechanisms of plants against fungal pathogens and insects. Phytochemistry 661(2):23–67

Latté KP, Appel KE, Lampen A (2011) Health benefits and possible risks of broccoli—an overview. Food Chem Toxicol 49:3287–3309

Le B, Anh PTN, Yang SH (2020) Enhancement of the anti-inflammatory effect of mustard kimchi on RAW 264.7 macrophages by the *Lactobacillus plantarum* fermentation-mediated generation of phenolic compound derivatives. Foods 9:181

Lee JW, Kim IH, Woyengo TA (2020) Toxicity of canola-derived glucosinolate degradation products in pigs—a review. Animals 10:2337

Lee ST, Radu S, Ariffin A, Ghazali HM (2015) Physico-chemical characterization of oils extracted from noni, spinach, lady's finger, bitter gourd and mustard seeds, and copra. Int J Food Prop 18:2508–2527

Li Y, Hussain N, Zhang L, Chen X, Ali E, Jiang L (2012) Correlations between tocopherol and fatty acid components in germplasm collections of Brassica oilseeds. J Agric Food Chem 61(1):34–40

Lionneton E, Aubert G, Ochatt S, Merah O (2004) Genetic analysis of agronomic and quality traits in mustard (*Brassica juncea*). Theor Appl Genet 109 (4):792–799

Liu TT, Yang TS (2010) Stability and antimicrobial activity of allyl isothiocyanate during long-term storage in an oil-in-water emulsion. J Food Sci 75:445–451

Li W, Yang H, Li C, Tan S, Gao X, Yao M, Zheng Q (2019) Chemical composition, antioxidant activity andantitumor activity of tumorous stem mustard leaf and stem extracts. CyTA – J Food 17:272–279

Lopes LF, Bordin K, de Lara GH, Saladino F, Quiles JM, Meca G, Luciano FD (2018) Fumigation of Brazil nuts with allyl isothiocyanate to inhibit the growth of *Aspergillus parasiticus* and aflatoxin production. J Sci Food Agric 98:792–798

Lozano-Baena MD, Tasset I, Obregón-Cano S, de Haro-Bailon A, Muñoz-Serrano A, Alonso-Moraga Á (2015) Antigenotoxicity and tumor growing inhibition by leafy *Brassica carinata* and sinigrin. Molecules 20:15748–15765

Luciano FB, Holley RA (2009) Enzymatic inhibition by allyl isothiocyanate and factors affecting its antimicrobial action against *Escherichia coli* O157: H7. Int J Food Microbiol 131:240–245

Mahoonak AS, Swamylingappa B (2007) A new method for preparation of non-toxic, functional protein hydrolysate from commercial mustard cake. Qual Nutr Process Trade 142

Martland MF, Butler EJ, Fenwick GR (1984) Rapeseed induced liver haemorrhage reticulolysis and biochemical changes in laying hens: The effects of feeding high and low glucosinolate meals. Vet Sci Res 36:298–309

Maruthupandy M, Seo J (2019) Allyl isothiocyanate encapsulated halloysite covered with polyacrylate as a potential antibacterial agent against food spoilage bacteria. Mater Sci Eng 105:110016

Mathur P, Ding Z, Saldeen T, Mehta JL (2015) Tocopherols in the prevention and treatment of atherosclerosis and related cardiovascular disease. Clin Cardiol 38:570–576

Matthaus B (1998) Effect of dehulling on the composition of antinutritive compounds in various cultivars of rapeseed. Fett/lipid 100:295–301

Mawson R, Heaney RK, Zdunczyk Z, Kozlowska H (1994) Rapeseed meal-glucosinolates and their antinutritional effects. Part 3 Animal growth and performance. Nahrung 38:167–177

Mayengbam S, Aachary A, Thiyam-Holländer U (2014) Endogenous phenolics in hulls and cotyledons of mustard and canola: a comparative study on its sinapates and antioxidant capacity. Antioxidants 3:544–558

Mazumder A, Dwivedi A, du Plessis J (2016) Sinigrin and its therapeutic benefits. Molecules 21:416

Megna BW, Carney PR, Nukaya M, Geiger P, Kennedy GD (2016) Indole-3-carbinol induces tumor cell death: function follows form. J Surg Res 204:47–54

Melrose J (2019) The glucosinolates: a sulphur glucoside family of mustard anti-tumour and antimicrobial phytochemicals of potential therapeutic application. Biomedicines 7:62

Mensink R, Katan M (1989) Dietary fat guidelines for men and women. The Lancet 333:790–791

Meyer SLF, Zasada IA, Orisajo SB, Morra MJ (2011) Mustard seed meal mixtures: management of *Meloidogyne incognita* on pepper and potential phytotoxicity. J Nematol 43:7–15

Mo S, Dong L, Hurst WJ, van Breemen RB (2013) Quantitative analysis of phytosterols in edible oils using APCI liquid chromatography–tandem mass. Lipids 48(9):949–956

Mohapatra M, Narain M, Sarkar BC (2004) Quality characteristics of dehulled rapeseed. J Food Sci Technol 41:194–195

Mortuza MG (2006) Tocopherol and sterol content of some rapeseed/mustard cultivars developed in Bangladesh. Pak J Biol Sci 9:1812–1816

Munir A, Iqbal S, Khaliq B, Saeed Q, Hussain S, Shah KH, Ahmad F, Mehmood S, Ali Z, Munawar A, Saeed MQ, Mehmood S, Asharf MY, Akrem A (2019) In silico studies and functional characterization of a napin protein from seeds of *Brassica juncea*. Int J Agric Biol 22:1655–1662

Musigwa S, Cozannet P, Morgan N, Swick RA, Wu SB (2021) Multi-carbohydrase effects on energy utilization depend on soluble non-starch polysaccharides-to-total non-starch polysaccharides in broiler diets. Poult Sci 100:788–796

Naczk M, Amarowicz R, Sullivan A, Shahidi F (1998) Current research developments on polyphenolics of rapeseed/canola: a review. Food Chem 62:489–502

Naczk M, Shahidi F (1990) Carbohydrates of canola and rapeseed. Canola and Rapeseed. Springer, Boston, MA, pp 211–220

Nagaraj G (2009) Oilseeds: properties, processing, products and procedures. New India Publishing

Newkirk RW, Classen HL, Tyler RT (1997) Nutritional evaluation of low glucosinolate mustard meals (*Brassica juncea*) in broiler diets. Poult Sci 76(9):1272–1277

Nicácio AE, Rodrigues CA, Visentainer JV, Maldaner L (2021) Evaluation of the quencher method for the determination of phenolic compounds in yellow (*Brassica alba*), brown (*Brassica juncea*), and black (*Brassica nigra*) mustard seeds. Food Chem 340:128162

Norton G (1989) Nature and biosynthesis of storage proteins. In: Robbelen G, Downey RK, Ashri A (eds) Oil crops of the world. McGraw-Hill, New York, USA, pp 105–191

Noureddini H, Dang J (2008) Degradation of phytase in Distillers' grains and corn gluten feed by *Aspergillus niger* phytase. Appl Biochem Biotechnol 159:11–23

Oh S, Kim K, Choi M (2016) Antioxidant activity of different parts of Dolsan leaf mustard. Food Sci Biotechnol 25:1463–1467

Olaimat AN, Holley RA (2013) Effects of changes in pH and temperature on the inhibition of *Salmonella* and

Listeria monocytogenes by allyl isothiocyanate. Food Control 34:414–419

Orsavova J, Misurcova L, Ambrozova JV, Vicha R, Mlcek J (2015) Fatty acids composition of vegetable oils and its contribution to dietary energy intake and dependence of cardiovascular mortality on dietary intake of fatty acids. Int J Mol Sci 16:12871–12890

Ostrikov AN, Kleymenova NL, Bolgova IN, Kopylov MV (2020) Gas chromatographic analysis of the fatty acid composition of mustard oil obtained by cold pressing (method). Emir J Food Agri 32(5):391–396

Otoni CG, Soares NDFF, da Silva WA, Medeiros EAA, Baffa Junior JC (2014) Use of allyl isothiocyanate-containing sachets to reduce *Aspergillus flavus* sporulation in peanuts. Package Technol Sci 27:549–558

Park SY, Jang HL, Lee JH, Choi Y, Kim H, Hwang J, Seo D, Kim S, Nam JS (2017) Changes in the phenolic compounds and antioxidant activities of mustard leaf (*Brassica juncea*) kimchi extracts during different fermentation periods. J Food Sci Biotechnol 26:105–112

Park W, Lee YH, Kim KS, Cha YL, Moon YH, Song YS, Kwon DE, Lee JE (2019) The optimal mixing ratio of *Brassica napus* and *Brassica juncea* meal improve nematode *Meloidogyne hapla* effects. Plant Signal Behav 14:1678369

Piironen V, Lindsay DG, Miettinen TA, Toivo J, Lampi AM (2000) Plant sterols: biosynthesis, biological function and their importance to human nutrition. J Sci Food Agri 80:939–966

Pontoppidan K, Pettersson D, Sandberg AS (2007) *Peniophoralycii* phytase is stable and degrades phytate and solubilises minerals in vitro during simulation of gastrointestinal digestion in the pig. J Sci Food Agri 87:2700–2708

Potts DA, Rakow GW, Males DR, Woods DL (2003) The development of canola-quality *Brassica juncea*. Can J Plant Sci 83:117–118

Prieto MA, Lopez CJ, Simal-Gandara J (2019) Glucosinolates: molecular structure, breakdown, genetic, bioavailability, properties and healthy and adverse effects. Adv Food Nutr Res 90:305–350

Quiles JM, Manyes L, Luciano F, Manes J, Meca G (2015) Influence of the antimicrobial compound allyl isothiocyanate against the *Aspergillus parasiticus* growth and its aflatoxins production in pizza crust. Food Chem Toxicol 83:222–228

Rahman M, Khatun A, Liu L, Barkla BJ (2018) Brassicaceae mustards: traditional and agronomic uses in Australia and New Zealand. Molecules 23:231

Rakow G, Raney JP (2003) Present status and future perspectives of breeding for seed quality in Brassica oilseed crops. In: Proceedings of the 11th international rape seed congress, Copenhagen, Denmark, pp 181–185

Ramachandran S, Singh SK, Larroche C, Soccol CR, Pandey A (2007) Oil cakes and their biotechnological applications—a review. Bioresour Technol 98 (10):2000–2009

Rampal G, Thind TS, Arora R, Vig AP, Arora S (2017) Synergistic anti-mutagenic effect of isothiocyanates against varied mutagens. Food Chem Toxicol 109:879–887

Ramprasath VR, Awad AB (2015) Role of phytosterols in cancer prevention and treatment. J AOAC Int 98 (3):735–738

Rao AG, Rao MN (1981) Comparative study of the high molecular weight protein fraction of mustard (*B. juncea*) and rapeseed (*B. campestris*). Int J Peptide Protein Res 18(2):154–161

Rao AG, Urs MK, Rao MN (1978) Studies on the proteins of mustard seed (*B. juncea*). Can Inst Food Sci Technol J 11(3):155–161

Rastogi T, Reddy KS, Vaz M, Spiegelman D, Prabhakaran D, Willett WC, Ascherio A (2004) Diet and risk of ischemic heart disease in India. Amer J Clin Nutr 79(4):582–592

Ratnayake WMN, Daun JK (2004) Chemical composition of canola and rapeseed oils. In: Gunstone FD (ed) Rapeseed and Canola Oil. Blackwell Publishing Limited, Oxford, UK, pp 37–73

Richards A, Wijesundera C, Salisbury P (2007) Genotype and growing environment effects on the tocopherols and fatty acids of *Brassica napus* and *B. juncea*. J Amer Oil Chem Soc 85(2):159–168

Rodriguez RA, Maciel MG, de Ochoteco MB (1997) Blindness in Holando Argentine calves due to ingestion of turnip (*Brassica campestris*). Vet Argent 14:601–605

Romeo L, Iori R, Rollin P, Bramanti P, Mazzon E (2018) Isothiocyanates: an overview of their antimicrobial activity against human infections. Molecules 23:624

Rosa EAS, Heaney RK, Fenwick GR, Portas CAM (1997) Glucosinolates in crop plants. Hort Rev 19:199–215

Rout K, Yadav BG, Yadava SK, Mukhopadhyay A, Gupta V, Pental D, Pradhan AK (2018) QTL landscape for oil content in *Brassica juncea*: analysis in multiple bi-parental populations in high and "0" erucic background. Front Plant Sci 9:1448

Sadeghi MA, Bhagya S (2009) Effect of recovery method on different property of mustard protein. World J Dairy Food Sci 4(2):100–106

Sadeghi MA, Rao AA, Bhagya S (2006) Evaluation of mustard (*Brassica juncea*) protein isolate prepared by steam injection heating for reduction of antinutritional factors. LWT-Food Sci Technol 39(8):911–917

Schone F, Rudolph B, Kirchheim U, Knapp G (1997) Counteracting the negative effects of rapeseed and rapeseed press cake in pig diets. British J Nutr 78 (6):947–962

Sen M, Bhattacharyya DK (2000) Nutritional effects of a mustard seed protein on detoxified with aqueous isopropanol in young rats. J Lipid Sci Technol 102:727–733

Sen R, Sharma S, Kaur G, Banga SS (2018) Near-infrared reflectance spectroscopy calibrations for assessment of oil, phenols, glucosinolates and fatty acid content in the intact seeds of oilseed Brassica species. J Sci Food Agric 98(11):4050–4057

Seppanen CM, Song Q, Csallany AS (2010) The antioxidant functions of tocopherol and tocotrienol homologues in oils, fats, and food systems. J Amer Oil Chemist Soc 87(5):469–481

Shahidi F, Naczk M (1992) An overview of the phenolics of canola and rapeseed: chemical, sensory and nutritional significance. J Amer Oil Chem Soc 69(9):917–924

Sharafi Y, Majidi MM, Goli SAH, Rashidi F (2015) Oil content and fatty acids composition in Brassica species. Int J Food Prop 18(10):2145–2154

Sharif RH, Paul RK, Bhattacharjya DK, Ahmed KU (2017) Physicochemical characters of oilseeds from selected mustard genotypes. J Bang Agril Univ 15 (1):27–40

Sharma A, Aacharya M, Punetha H, Sharma S, Kumari N, Rai PK (2019) Biochemical characterization and correlations in *Brassica juncea* genotypes. Int J Curr Microbiol Appl Sci 8(1):2408–2417

Sharma S, Gupta N (2020) Defense signaling in plants against micro-creatures: do or die. Indian Phytopathol 73:605–613

Sharma A, Rai PK, Prasad S (2018) GC–MS detection and determination of major volatile compounds in *Brassica juncea* L. leaves and seeds. Microchem J 138:488–493

Shrestha K, De Meulenaer B (2014) Effect of seed roasting on canolol, tocopherol, and phospholipid contents, maillard type reactions, and oxidative stability of mustard and rapeseed oils. J Agric Food Chem 62(24):5412–5419

Shuang R, Rui X, Wenfang L (2016) Phytosterols and dementia. Plant Foods Human Nutr 71(4):347–354

Siddiqui IR, Wood PJ (1977) Carbohydrates of rapeseed: a review. J Sci Food Agri 28(6):530–538

Simakova I, Volf E, Strizhevskaya V, Popova O, Kozyreva V, Karagulova E (2019) Blends of unrefined vegetable oils for functional nutrition. Agron Res 17 (4):1761–1768

Simbaya J, Slominski BA, Rakow G, Campbell LD, Downey RK, Bell JM (1995) Quality characteristics of yellow-seeded Brassica seed meals: protein, carbohydrate, and dietary fiber components. J Agric Food Chem 43(8):2062–2066

Sindhu Kanya TC, Kantharaj Urs M (1983) Carbohydrate composition of mustard (*Brassica juncea*) seed meal. J Food Sci Technol 20(3):124–126

Singh RB, Niaz MA, Sharma JP, Kumar R, Rastogi V, Moshiri M (1997) Randomized, double-blind, placebo-controlled trial of fish oil and mustard oil in patients with suspected acute myocardial infarction: the Indian experiment of infarct survival. Cardiovasc Drugs Ther 11(3):485–491

Singh BK, Bala M, Rai PK (2014) Fatty acid composition and seed meal characteristics of Brassica and allied genera. Natl Acad Sci Lett 37(3):219–226

Singh S, Das SS, SinghG PM, Schuff S, Catalan CAN (2017) Comparison of chemical composition, antioxidant and antimicrobial potentials of essential oils and oleoresins obtained from seeds of *Brassica juncea* and *Sinapis alba*. MOJ Food Process and Technol 4 (4):113–120

Slominski BA, Campbell LD, Guenter W (1994) Carbohydrates and dietary fibre components of yellow and brown seeded canola. J Agric Food Chem 42:704–707

Sonderby IE, Geu-Flores F, Halkier BA (2010) Biosynthesis of glucosinolates—gene discovery and beyond. Trends Plant Sci 15(5):283–290

Spiegel C, Bestetti G, Rossi G, Blum JW (1993) Feeding of rapeseed presscake meal to pigs: effects on thyroid morphology and function and on thyroid hormone blood levels, on liver and on growth performance. J Vet Med 40:45–57

Sra SK, Sharma M, Kaur G, Sharma S, Akhatar J, Sharma A, Banga SS (2019) Evolutionary aspects of direct or indirect selection for seed size and seed metabolites in *Brassica juncea* and diploid progenitor species. Mol Biol Rep 46:1227–1238

Suhr KI, Nielsen PV (2003) Antifungal activity of essential oils evaluated by two different application techniques against rye bread spoilage fungi. J Appl Microbiol 94:665–674

Tan SH, Mailer RJ, Blanchard CL, Agboola SO (2011) Extraction and residual antinutritional components in protein fractions of *Sinapis alba* and *Brassica napus* oil-free meals. In: 17th Australian Research Assembly on Brassicas, pp 107–114

Tangtaweewipat S, Cheva-Isarakul B, Sangsrijun P (2004) The use of mustard meal as a protein source in broiler diets. Songklanakarin J Sci Technol 26 (1):23–30

Teh L, Möllers C (2016) Genetic variation and inheritance of phytosterol and oil content in a doubled haploid population derived from the winter oilseed rape Sansibar XOase cross. Theor Appl Genet 129:181–199

Thirumalai T, Therasa SV, Elumalai EK, David E (2011) Hypoglycemic effect of *Brassica juncea* (seeds) on streptozotocin induced diabetic male albino rat. Asian Pac J Trop Biomed 1:323–325

Thiyam U, Stöckmann H, ZumFelde T, Cwarz K (2006) Antioxidative effect of the main sinapic acid derivatives from rapeseed and mustard oil by-products. Eur J Lipid Sci Tech 108:239–248

Tian Y, Deng F (2020) Phytochemistry and biological activity of mustard (*Brassica juncea*): a review. CyTA —J Food 18:704–718

Tripathi K, Hussein UK, Anupalli R, Barnett R, Bachaboina L, Scalici J, Rocconi RP, Owen LB, Piazza GA, Palle K (2015) Allyl isothiocyanate induces replication-associated DNA damage response in NSCLC cells and sensitizes to ionizing radiation. Oncotarget 6:5237–5252

Tripathi MK, Agrawal IS, Sharma SD, Mishra DP (2001) Effect of untreated, HCl treated or copper and iodine supplemented high glucosinolate mustard (*Brassica juncea*) meal on nutrient utilization, liver enzymes, thyroid hormones and growth of calves. Anim Feed Sci Technol 92:73–85

Tripathi MK, Mishra AS (2017) Prospects and problems of dietary glucosinolates in animal feeding. Adv Diary Res 5:180

Tripathi MK, Mishra AS, Misra AK, Prasad R, Mondal D, Jakhmola RC (2003) Effect of graded levels of high glucosinolate mustard (*Brassica juncea*) meal inclusion on nutrient utilization, growth performance, organ weight and carcass composition of growing rabbits. World Rabbit Sci 11(4):211–226

Vaidya B, Choe E (2011) Stability of tocopherols and lutein in oil extracted from roasted or unroasted mustard seeds during the oil oxidation in the dark. Food Sci Biotechnol 20:193–199

Venkatesh A, Rao AA (1988) Isolation and characterization of low molecular weight protein from mustard (*Brassica juncea*). J Agric Food Chem 36:1150–1155

Vig AP, Rampal G, Thind TS, Arora S (2009) Bio-protective effects of glucosinolates—a review. LWT-Food Sci Technol 42(10):1561–1572

Walker KC, Booth EJ (2001) Agricultural aspects of rape and other *Brassica* products. Eur J Lipid Sci Technol 103:441–446

Wanasundara JP (2011) Proteins of Brassicaceae oilseeds and their potential as a plant protein source. Crit Rev Food Sci Nutr 51(7):635–677

Wanasundara U, Amarowicz R, Shahidi F (1994) Isolation and identification of an antioxidative component in canola meal. J Agric Food Chem 42(6):1285–1290

Wang AS, Hu P, Hollister EB, Rothlisberger KL, Somenahally A, Provin TL, Hons FM, Gentry TJ (2012) Impact of Indian mustard (*Brassica juncea*) and flax (*Linumus itatissimum*) seed meal applications on soil carbon, nitrogen, and microbial dynamics. Appl Environ Soil Sci, 1–14

Wang SX, Oomah BD, Ian McGregor D, Downey RK (1998) Genetic and seasonal variation in the sinapine content of seed from Brassica and Sinapis species. Can J Plant Sci 78:395–400

Wang X, He H, Lu Y, Ren W, Teng K, Chiang CL, Yang Z, Yu B, Hsu S, Jacob ST, Ghoshal K (2015) Indole-3-carbinol inhibits tumorigenicity of hepatocellular carcinoma cells via suppression of microRNA-21 and upregulation of phosphatase and tensin homolog. Biochim Biophys Acta (BBA)—Mol Cell Res 1853:244–253

WHO (2018) Draft guidelines on saturated fatty acid and trans fatty acids. Intake for adults and children, pp 1–103

Wijesundera C, Ceccato C, Fagan P, Shen Z (2008) Seed roasting improves the oxidative stability of canola (*B. napus*) and mustard (*B. juncea*) seed oils. Eur J Lipid Sci Technol 110(4):360–367

Wittstock U, Halkier BA (2002) Glucosinolate research in the *Arabidopsis era*. Trends Plant Sci 7:263–270

Wojciechowski A, Kott L, Beversdorf W (1994) Content of sinapine in haploid embryos of rapeseed (*Brassica napus*) from microspore in in vitro culture. Rośliny Oleiste—Oilseed Crop 15:105–110

Wu X, Sun J, Haytowitz DB, Harnly JM, Chen P, Pehrsson PR (2017) Challenges of developing a valid dietary glucosinolate database. J Food Compos Anal 64:78–84

Xian YF, Hu Z, Ip SP, Chen JN, Su ZR, Lai XP, Lin ZX (2018) Comparison of the anti-inflammatory effects of *Sinapis alba* and *Brassica juncea* in mouse models of inflammation. Phytomedicine 50:196–204

Yadav M, Rana JS (2018) Quantitative analysis of Sinigrin in *Brassica juncea*. J Pharmacogn Phytochem 7:948–954

Yang R, Xue L, Zhang L, Wang X, Qi X, Jiang J, Li P (2019) Phytosterol contents of edible oils and their contributions to estimated phytosterol intake in the Chinese diet. Foods 8:334

Ye X, Ng TB (2009) Isolation and characterization of juncin, an antifungal protein from seeds of Japanese Takana (*Brassica juncea* Var. integrifolia). J Agric Food Chem 57(10):4366–4371

Yiu SH, Altosaar I, Fulcher RG (1983) The effects of commercial processing on the structure and microchemical organization of rapeseed. Food Struct 2(2):7

Yiu SH, Poon H, Fulcher RG, Altosar I (1982) Microscopic structure and chemistry of rapeseed and its products. Food Struct 1:135–143

Yuan L, Zhang F, Shen M, Jia S, Xie J (2019) Phytosterols suppress phagocytosis and inhibit inflammatory mediators via ERK pathway on LPS-triggered inflammatory responses in RAW264. 7 macrophages and the correlation with their structure. Foods 8 (11):582

Yun KJ, Koh DJ, Kim SH, Park SJ, Ryu JH, Kim DG, Lee JY, Lee KT (2008) Anti-inflammatory effects of sinapic acid through the suppression of inducible nitric oxide synthase, cyclooxygenase-2, and proinflammatory cytokines expressions via nuclear factor-κB inactivation. J Agric Food Chem 56:10265–10272

Yusuf MA, Sarin NB (2007) Antioxidant value addition in human diets: genetic transformation of *Brassica juncea* with γ-TMT gene for increased α-tocopherol content. Transgen Res 16(1):109–113

Zanetti F, Vamerali T, Mosca G (2009) Yield and oil variability in modern varieties of high-erucic winter oilseed rape (*Brassica napus L.* var. oleifera) and Ethiopian mustard (*Brassica carinata A. Braun*) under reduced agricultural inputs. Ind Crops Prod 30 (2):265–270

4 Genepools of Brassica

Hariom Kumar Sharma, V. V. Singh, Arun Kumar, H. S. Meena, Pankaj Sharma, and P. K. Rai

Abstract

The Brassicaceae is a plant family of great economic significance since human civilization as plants of this family were used as vegetables, condiments, and edible oils. The Brassica genetic resources can be grouped into primary, secondary, and tertiary genepool based on extent of crossability between them. The primary genepool including landraces, obsolete cultivars, modern cultivars, and advanced breeding lines and genetic stocks of corresponding species of the U-triangle, which means each species has its own primary gene pool. Secondary genepool includes all species that can be crossed with the crop and are coenospecies. The species and genera of different cytodemes of Brassica coenospecies constitutes tertiary gene pool. Wild Brassica and its related genera are packed with novel genes of agronomic importance, biotic and abiotic stress tolerant, oil and seed quality and male sterility. So, their proper characterization for agro-morphological, physiological, and biochemical traits is very important to understand their genetic value and for planning of long-term breeding programs for Brassica crops. These species are source for disease resistance including white rust (*Brassica maurorum* and *Eruca versicaria* ssp. *sativa*), Alternaria blight (*Brassica fruticulose, Trachystom aballii),* blackleg/phoma disease (*Sinapis arvensi, Sinapis alba*, *Thlaspi arvense, B. tournefortii*), powdery mildew (*B. oleracea*), clubroot disease (*B. rapa, B. oleracea* and *B. napus).* Some are source of C3–C4 intermediate photosynthetic system (*Moricandia, Diplotaxis* species), high erucic acid (*Crambe abyssinica)* and cytoplasmic male sterility (*Sinapis incana,* and *Diplotaxis siifolia*). The gene transfer between *Brassica* and allied genera can be done with some limitations despite difference in chromosome numbers through conventional and biotechnological approaches.

4.1 Introduction

Brassicaceae (Cruciferae) is a plant family of great economic significance since human civilization as plants of this family were used as vegetables, condiments, and edible oils. It contains over 3,700 species in 338 genera, having significant genetic diversity (Demeke et al. 1992; Warwick et al. 2009; Hayward 2012). The taxonomic studies of Brassica were conducted by many researchers since 1700 and still it is continued to add new genera and species in the family (Linnaeus 1753; De Candolle 1821;

H. K. Sharma · V. V. Singh · A. Kumar · H. S. Meena · P. Sharma (✉) · P. K. Rai
ICAR-Directorate of Rapeseed-Mustard Research, Bharatpur 321303, India

C. Kole and T. Mohapatra (eds.), *The Brassica juncea Genome*, Compendium of Plant Genomes,
https://doi.org/10.1007/978-3-030-91507-0_4

Hooker 1862; Baillon 1871; Prantl 1891; Schulz 1919, 1936; Beilstein et al. 2006; Branca and Cartea 2011). Crop Brassicas including six *Brassica* species are distributed around the world and used for various purposes. *Brassica carinata* (cultivated in Ethiopia and North-east Africa), *B. napus* (cultivated in Europe and North America), and *B. juncea* (mainly cultivated in South- and South-East Asia) are amphidiploids originated through natural hybridization between the diploid species *B. rapa, B. nigra,* and *B. oleracea* (UN 1935; Branca and Cartea 2011). The genus *Brassica* includes morphologically diverse species with variable uses (Dixon 2007; Warwick et al. 2009). *B. carinata, B. juncea, B. napus,* and the diploid *B. rapa* are naturally oliferous species. These species have good combination of saturated and unsaturated fatty acids along with secondary metabolites, glucosinolates in moderate to high amount in seed oil and erucic acid and protein content in the seed meal in moderate to high proportions (Downey and Röbbelen 1989). *B. nigra* is mainly used as condiments and *B. oleracea* is cultivated for vegetable purpose due to its enormous morphological diversity. Different *Brassica* oilseeds species predominate in different regions of the world. *B. juncea* and *B. rapa* mainly distributed predominantly in the warm sub-tropical regions, while distribution of *B. napus* and *B. rapa, B. carinata* is limited in cooler temperate regions in Ethiopia and northeast Africa, whereas *B. nigra* is grown in Europe and Asia (Hammer et al. 2013). Other genera in Brassicaceae are valued for condiments (*Sinapis, Eruca*), vegetable (*Eurca, Raphanus, Diplotaxis*), industrial uses (*Eruca, Crambe, Sinapis, Lepidium, Camelina, Thlaspi*), and food/fodder (*Orychophragmu, Eruca*) (Warwick et al. 2009).

The Brassica genetic resources can be grouped into primary, secondary, and tertiary genepool on the basis of extent of gene exchange possible between different species and genera. Each crop *Brassica* species represents the primary genepool. Further several hybridization and cytological experiments have been carried out to investigate and assign other brassica species to primary, secondary, and tertiary genepool for their effective utilization in Brassica improvement program. Many cytogenetic investigations, genomic and molecular marker-based studies helped to identify the basic genomes of Brassica crops and possible evolution of Brassica and allied genera (Prakash and Hinata 1980; Hosaka et al. 1990; Slocum et al. 1990; Song et al. 1990; Branca and Cartea 2011). The basic chromosome number of some of the genera in tribe Brassiceae is presented in Table 4.1. One can refer Warwick et al. (2009) for detailed Karyotype study of these species.

4.2 Taxonomy

4.2.1 Family Brassicaceae

The Brassicaceae, is of tremendous significance in terms of economic and scientific importance. It includes 3709 species and 338 genera (Warwick et al. 2006). The increasing importance of *Arabidopsis* and *Brassica* species as model organisms in the plant sciences has greatly advanced research into the systematics, taxonomy, evolution, and development of the entire family, including the cultivated taxa and their wild relatives (Koch and Al-Shehbaz 2009).

4.2.2 Tribe Brassiceae

This tribe Brassiceae consists of nearly 46 poorly defined genera and about 230 species of economically important plants in the family (e.g., species of *Brassica, Eruca, Raphanus, Sinapis*). The tribe has been studied thoroughly for molecular, taxonomic, and evolutionary research (Tsunoda et al. 1980; Gómez-Campo 1999; Warwick and Sauder 2005; Koch and Al-Shehbaz 2009; Kagale et al. 2014; Xue et al. 2020.). In family Brassicaceae, the tribe Brassiceae is most diverse in terms of morphotypes and economic values (Gómez-Campo 1999). The crop Brassicas have tremendous diversity and variations in end use as these are sources of oil, vegetables, condiments, fodder, and industrial applications. Among these crop Brassicas,

Table 4.1 Common name and chromosome number of some of species of the Brassica genepool

Scientific name	Common name/description	*n*
Brassica rapa	Bird rape	10 (A)
subsp. *campestris* (L.) A.R. Clapham	Summer turnip rape, wild turnip rape	10 (A)
subsp. *oleifera* (DC.) Metzg	Winter turnip rape	10 (A)
var. *brown sarson*	Brown sarson	10 (A)
var. *toria*	Toria	10 (A)
var. *yellow sarson*	Yellow sarson	10 (A)
subsp. *japonica*	Pot herb mustard	10 (A)
subsp. *chinensis* (L.) Hanelt	Pak-choi or bok choy, Chinese mustard, Chinese broccoli	10 (A)
subsp. *pekinensis* (Lour.) Hanelt	Petsai, Chinese cabbage	10 (A)
subsp. *rapifera*	Turnip	10 (A)
subsp. *nipposinica* (L.H. Bailey) Hanelt Curled mustard	Curled mustard	10 (A)
subsp. *parachinensis*	–	10 (A)
Brassica nigra (L.) W.D.J. Koch.	Black mustard, Banarasi rai	8 (B)
Brassica oleracea L.	–	9 (C)
var. *viridis* L.	Kale, collard	9 (C)
var. *botrytis* L.	Cauliflower	9 (C)
var. *capitata* L.	Cabbage	9 (C)
var. *gongylodes* L.	Kohlrabi	9 (C)
var. *gemmifera* (DC.) Zenker	Brussels sprouts	9 (C)
var. *italic* Plenck	Broccoli	9 (C)
var. *sabauda*	Savoy cabbage	9 (C)
subsp. *Alboglabra* L.H. Bailey	Chinese kale, kailan	9 (C)
Brassica juncea (L.) Czern.	Indian mustard, brown mustard	18 (AB)
B. juncea Coss.	Oilseed type of India and China	18 (AB)
var. *napiformis* Baile	Enlarged turnip like root, China	18 (AB)
var. *foliosa* Bailey	*Hsueh li hung*, glabrous leaves, China	18 (AB)
var. *japonica* Bailey	*Naganszkaai*, glabrous leaves, China	18 (AB)
var. *crispifolia* Bailey	*Azanina*, salads and as ornamentals, USA	18 (AB)
var. *integrifolia* (Stokes) Kitam	Entire succulent leaves, China, and Asia	18 (AB)
var. *rugosa* (Roxb) Kitam	Extremely succulent with large leaves	18 (AB)
var. *bulbifolia* Mas	*Ta hsintsai*, Succulent stems and elongated internodes	18 (AB)
Brassica napus L.	Rapeseed, Canola	19 (AC)
subsp. *oleifera* (Delile) Sinskaya	Summer oilseed rape, Canola	19 (AC)
B. napus f. biennis (Schübl. &G. Martens) Thel.	Winter oilseed rape, Winter canola	19 (AC)
var. *pabularia* (DC.) Rchb.	Rape-kale	19 (AC)
var. *napobrassica* (L.) Rchb.	Rutabaga, swede	19 (AC)

(continued)

Table 4.1 (continued)

Scientific name	Common name/description	*n*
Brassica carinata A. Braun	Abyssinian mustard, Ethiopian mustard,	17 (BC)
B. tournefortii	African mustard, Sahara mustard	10 (T)
Capsella bursa - pastoris	Shepherds purse	8, 16
Diplotaxis erucoides	White Wall Rocket	7
Eruca sativa Mill	Garden rocket, Rocket salad	11 (E)
Erucastrum gallicum	Dog mustard, Hairy rocket	15
Orychophragmus violaceus	Chinese violet cress	12
Raphanus raphanistrum L.	Wild radish, Jointed charlock	9 (R)
Raphanus sativus L.	Radish	9 (R)
Sinapis alba	White mustard	12 (Sal)
Hirschfeldia incana	Hoary mustard	7 (H)

Source Yarnell (1956), Dixon (2007), Prakash et al. (2009), OCED (2016), Warwick (2009)

Brassica napus, *B. rapa,* and *B. juncea* are main source of edible canola oil, and *B. oleracea* most commonly used as vegetable known as cole crops while *B. nigra* is consumed in the form condiment. The other genera are used for edible roots (*Raphanus*), condiments (*Sinapis*), industrial oil (*Crambe*), and green salads (*Eruca* and *Diplotaxis*). Many other species (*Sinapis arvensis, Raphanus raphanistrum, B. rapa*) have become naturalized weeds around the world. Hence, these species and other allied genera of tribe Brassiceae, can be good donors for many stresses faced by Brassica crops.

The geographical distribution range of tribe is primarily in the Mediterranean region mainly Algeria, Morocco, and Spain, where 40 genera are either endemic or exhibit maximum diversity, adjacent southwestern Asia including India and Pakistan, and South Africa, except four species of *Cakile* which are native to North America (Hedge 1976; Gómez-Campo 1999; Koch and Al-Shehbaz 2009; Warwick et al. 2009). The tribe Brassiceae currently contains 242 species in 48 genera, 20 of which are monotypic (Table 4.2). Chloroplast-based studies found that most of the genera in the tribe evolved through polyphyletic origin following *rapa* and *nigra* lineage while some genera (e.g., *Cakile, Vella, Crambe*) are monophyletic (Warwick and Black 1997b; Francisco-Ortega et al. 2002; Gómez-Campo 1999; Warwick and Hall 2009). These genera exhibit tremendous fruit diversity, hence it is used as main characters in their delimitation.

Tribal members are morphologically characterized by having conduplicate cotyledons (i.e., the cotyledons longitudinally folded around the radicle in the seed), and/or transversely segmented fruits that have seeds or rudimentary ovules in both segments (heteroarthrocarpic) (Appel 1999) and, if present, only simple trichomes or hairs (Gómez-Campo 1980, 1999; Al-Shehbaz 1985). The first two features are unknown elsewhere in the family. The few exceptions to this character combination are the genera *Ammosperma* and *Pseuderucaria*, neither of which has the conduplicate cotyledons or the segmented fruits (Warwick and Hall 2009).

On basis of morphological characters, Schulz (1919, 1923, 1936) also recognized seven subtribes: *Brassicinae, Cakilinae, Moricandiinae, Raphaninae, Savignyinae, Vellinae,* and *Zillinae*. The fruit characters are the most reliably used structures for the proper identification of genera and species (Hedge 1976; Al-Shehbaz 1985). The prime characteristics of Brassicinae are elongated (siliquose) dehiscent fruits, presence of median nectaries, and usually seeded beaks. The subtribe *Brassicinae* includes ten core genera namely, *Brassica, Coincya, Diplotaxis, Eruca, Erucastrum, Raphanus, Sinapidendron, Sinapis,*

Table 4.2 Genera of the tribe Brassiceae (no. species in brackets) (*Source* Warwick et al. 2000)

Genera	Genera	Genera	Genera
Ammosperma (2)	*Diplotaxis* (32)	*Henophyton* (2)	*Raffenaldia* (2)
Brassica (39)	*Douepea* (2)	*Kremeriella* (1)	*Raphanus* (3)
Cakile (6)	*Enarthrocarpus* (5)	*Moricandia* (8)	*Rapistrum* (2)
Carrichtera (1)	*Eremophyton* (1)	*Morisia* (1)	*Rytidocarpus* (1)
Ceratocnemum (1)	*Eruca* (4)	*Muricaria* (1)	*Savignya* (1)
Chalcanthus (1)	*Erucaria* (10)	*Orychophragmus* (2)	*Schouwia* (1)
Coincya (6)	*Erucastrum* (25)	*Otocarpus* (1)	*Sinapidendron* (4)
Conringia (6)	*Fezia* (1)	*Physorhynchus* (2)	*Sinapis* (5)
Cordylocarpus (1)	*Foleyola* (1)	*Pseuderucaria* (2)	*Succowia* (1)
Crambe (34)	*Fortuynia* (2)	*Pseudofortuynia* (1)	*Trachystoma* (3)
Crambella (1)	*Guiraoa* (1)	*Psychine* (1)	*Vella* (7)
Didesmus (2)	*Hemicrambe* (3)	*Quezeliantha* (1)	*Zilla* (2)

Hirschfeldia, and *Trachystoma* (Gomez-Campo 1999; Warwick et al. 2009). A complete list of genera, species, and subspecies in the tribe Brassiceae, is provided by Warwick et al. (2000). Morphological, cytological, hybridization, isozyme, and molecular studies in tribe Brassiceae helped in understanding species relationships within the tribe and also identified new relationships between genera and species (Warwick et al. 2009).

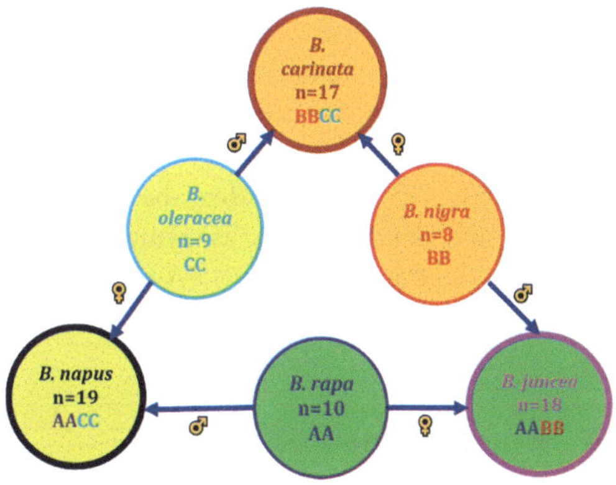

Fig. 4.1 U's triangle of Crop Brassica Species showing their genetic relationships [*Source* Morinaga (1934), UN (1935), Purty et al. (2008)]

4.2.3 Phylogenetics of Tribe Brassiceae

Because of its economic and scientific significance, extensive molecular-based phylogenetic studies have been conducted on members of the tribe. Earlier studies focused on *Brassica* crops and relatives. Relationships between the three diploid *Brassica* crop species [*B. nigra* ($n = 8$, BB), *B. rapa* (syn. *campestris*) ($n = 10$, AA), and *B. oleracea* ($n = 9$, CC)] and related amphidiploid species [*B. napus* ($n = 19$, AACC), *B. carinata* ($n = 17$, BBCC), and *B. juncea* ($n = 18$, AABB)] were first proposed by Morinaga in 1934 and UN in 1935 (Fig. 4.1).

It was proposed that *B. juncea*, *B. napus,* and *B. carinata* were natural amphidiploid hybrids derived from hybridization of the diploid species *B. nigra*, *B. oleracea*, and *B. rapa, B. napus* has been derived from multiple hybridization events hence considered as polyphyletic origin, with *B. oleracea* as one of several maternal ancestors (Allender and King 2010; Chalhoub et al. 2014).

Many phylogenetic studies have been conducted to understand the evolution of Brassica and allied genera and it was concluded that they might have evolved from a common ancestor ($n = 6$) with increase in chromosome numbers which is evident from the partial homology of A, B, and C-genomes (Prakash and Hinata 1980; Song et al. 1990). The U triangle of origin of crop Brassica was further confirmed at molecular level by use of restriction site data from the chloroplast DNA (cpDNA) by Palmer et al. (1983) and Erickson et al. (1983) which was further supported by using nuclear RFLP markers (Song et al. 1988). Later hypothesis of monophyletic origin of genus *Brassica* was disagreed by Yanagino et al. (1987) and Song et al. (1990) by study of cpDNA and nuclear RFLP of *Brassica* species of tribe.

Recent morphological, hybridization, and molecular data sets have provided support for alternative subtribal and generic circumscriptions. Various molecular studies have been done for phylogenetic analysis in tribe Brassiceae. Molecular study includes restriction site analysis of cpDNA (Warwick and Black 1993, 1994, 1997a; Warwick et al. 1992), cpDNA and mitochondrial DNA (mtDNA) RFLP data (Pradhan et al. 1992), ITS (internal transcribed spacers) ITS-1 and ITS-2 of nuclear DNA, and the 5.8 rRNA gene sequence-based phylogenetic studies (Crespo et al. 2000; Francisco-Ortega et al. 1999, 2002), S-locus related gene *SLR1* (Inaba and Nishio 2002), *trn*L-*trn*F based phylogeny (Lysak et al. 2005), ITS and ITS/*trn*L-based study (Warwick and Sauder 2005), isozyme duplication studies for *Pgm*-2 and *Tpi*-1 (Anderson and Warwick 1999).

4.3 Brassica Coenospecies and Cytodemes

Brassica coenospecies are the part of tribe Brassiceae whose members are closely related to crop *Brassica* species and are potentially capable of inter-crossing with them. Concept of coenospecies was proposed by Harberd (1972) based on cyto-taxonomical investigations on Brassica crop and allied wild genera and species. According to him, nine genera (*Brassica, Coincya, Diplotaxis, Eruca, Erucastrum, Hirschfeldia, Sinapis, Sinapidendron,* and *Trachystoma*) from subtribe *Brassicinae* and two genera (*Enarthrocarpus* and *Raphanus*) from subtribe *Raphaninae* constitute Brassica coenospecies (Prakash et al. 2009). Later on, Harberd (1972) classified different coenospecies of Brassiceae into 38 cytodemes or crossing groups (Table 4.3). A cytodeme is a group consisting of any number of species or genera that have the same chromosome number, and crosses between them always yield fertile hybrids. Furthermore, Harberd (1976), Takahata and Hinata (1983), Warwick and Black (1991), and Warwick and Black (1997a) extended the number of cytodemes upto 63 including 14 genera (Prakash et al. 1999, 2009). The extent of hybridization between cytodemes is limited which can be enhanced by application of some of the biotechnological techniques. Hybridization success rate can also be manipulated changing the direction of the cross to break unilateral incompatibility (Katche et al. 2019).

Based on the cpDNA-based phylogenetic analysis and phenetic clustering, genera and species of *Brassica coenospecies* divided in to two evolutionary lineages viz., *nigra* and *rapa/oleracea* (Warwick and Black 1991; Pradhan et al. 1992; Prakash et al. 2009) (Table 4.4). Recently six or seven lineages are currently recognized in the tribe Brassiceae, some of which are consistent with traditional subtribal delimitations (Warwick and Hall 2009). Maximum 16 genera followed *Nigra* lineage which is also known as *Sinapis* lineage. While, 10 genera followed *Rapa/Oleracea* lineage also referred as the *Brassica* lineage (Warwick and Hall 2009).

4.4 Brassica Genepool

The Russian geneticist Aleksandr Sergeevich Serebrovskii introduced the term *genofond* ("gene fund" in English) in 1926 and called it to the complete set of different genes found within a group of organisms. The *genofond,* term was later named as "genepool," into English, which represents a reservoir of genetic diversity that assist organisms to cope up to a changing environment and that can be tapped by scientists for plant breeding and crop improvement. Two pioneers of the crop diversity conservation movement, Jack Harlan and Jan de Wet, divided the genepool into three categories (primary, secondary, and tertiary) (Harlan and de Wet 1971). In their system, crop wild relative (CWR) species were classified into groups based on cross compatibility with their cultivated species. In this system, wild relatives can be in the crop's

Table 4.3 Cytodemes in Brassica coenospecies

Chromosome no. (n)	Coenospecies
7	*Brassica deflexa, Diplotaxis erucoides, Erucastrum virgatum, E. varium, Sinapis aucheri, Hirschfeldia incana, Pseuderucaria* spp.
8	*B. nigra, B. fruiticulosa* (+*maurorum* + *spinescens*), *Diplotaxis siettiana, Erucastrum abyssinicum, E. nasturtiifolium* (+*leucanthum*), *E. strigosum, Trachystoma* spp.
9	*B. oleracea* and 8 wild Mediterranean allied species, *B. oxyrrhina, Diplotaxis assurgens, D. catholica, D. tenuisiliqua, D. virgata, D. berthautii, Erucastrum cardaminoides*, (+ *canariense* + *ifniense*), *Raphanus* all species and subspecies, *Sinapis arvensis* (+ *allioni*), *S. pubescens*
10	*B. tournefortii, B. barrelieri, B. gravinae, B. repanda* (+ *desnottesii*), *B. rapa* (+ many cultivated subspecies), *Diplotaxis siifolia, D. viminea, Enarthrocarpus* spp., *Sinapidendron* spp.
11	*B. souliei, B. elongate, D. acris, Diplotaxis* (+ *pitardiana*); *Eruca* spp.
12	*Coincya* spp. (syn. *Hutera* and *Rhynchosinapis*), *Sinapis alba, S. flexuosa*
13	*Diplotaxis harra*, (+ several subsps.)
14	*Erucastrum virgatum* (subsp. *pseudosinapis*)[a], *Moricandia arvensis*[a], *M. moricandioides*[a], *Rytidocarpus moricandioides*[a]
15	*Erucastrum gallicum*[b], *E. elatum*[b]
16	*B. cossoniana* (4x)[a], North African subspecies, *B. balearica*[b], *Erucastrum nasturtiifolium*[a], *E. abyssinicum*[a]
17	*Brassica carinata*[b]
18	*Brassica juncea*[b]
19	*Brassica napus*[b]
20	*Brassica gravinae*[a]
21	*Diplotaxis muralis*[b]
22	*Brassica dimorpha*[a]
24	*Coincya* spp. (4x)[a]
28	*Moricandia suffruticosa*[a]
42	*Moricandia spinosa*[a]
80?	*Brassica repanda*[a]

Note [a] Autopolyploid cytodemes. [b] Allopolyploid cytodemes (*Source* From Prakash et al. 1999)

primary, secondary, or tertiary genepools (https://www.cwrdiversity.org/about/what-is-a-genepool/). Genepool concept was further elaborated by Spillane and Gepts (2001). Genepool of Brassica can also be classified into three categories on the basis of classification of Harlan and de Wet (1971) as follows:

4.4.1 Primary Genepool (GP-1)

GP-1 refers to group of germplasm in which hybridization leads to successful sexual recombination. It includes true biological species. Crossing between members of this genepool, is easy; with fertile hybrids along with good chromosome pairing; gene segregation is approximately normal therefore gene transfer is generally easy. The biological species includes all the wild and weedy form of the species which evolved spontaneously in nature and cultivated forms as well (Harlan and de Wet 1971). Hence, GP1 includes landraces, cultivars in current use, obsolete cultivars, and advanced breeding lines. This category also includes the conspecific wild progenitor (Dobzhansky 1937; Mayr 1963).

Table 4.4 Genera and species of Brassica coenospecies in *nigra* and *rapa/oleracea* lineage

Nigra lineage	*Rapa/Oleracea* lineage
Group I *Brassica nigra* (8), *B. fruticulosa* (8) *Sinapis arvensis* (9), *S. alba* (12) *Diplotaxis ibicensis* (8), *D. siettiana* (8), *Erucastrum littoreum* (16), *Trachystoma balii* (8) **Group II** *B. tournefortii* (10), *B. procumbens* (9), *D. brachycarpa* (9), *E. varium* (7), *E. virgatum* (7), *Hirschfeldia incana* (7), *S. pubescens* (9) **Group III** *E. canariense* (9), *E. brevirostre* (9) *D. assurgens* (9), *D. siifolia* (10), *D. berthautii* (10), *D. virgata* (9), *D. catholica* (9), *Sinapidendron* spp. (10) **Group IV** *Coincya* spp. (12)	**Group I** *B. rapa* (10), *B. oleracea* (9), *B. deflexa* (7), *B. barrelieri* (10), *B. oxyrrhina* (9), *D. cossoneana* (9), *D. erucoides* (7), *E. abyssinicum* (16), *E. strigosum* (8), *E. nasturtifolium* (8), *S. aucheri* (7), *Enarthrocarpus lyratus* (10), *Raphanus* spp. (9) **Group II** *D. harra* (13), *Eruca* spp. (11), *D. tenuifolia* (11), *Rytidocarpu smoricandiodes* (14) **Group III** *Moricandia arvensis* (14); *M. moricandiodes* (14) *M. suffruticosa* (28) **Group IV** *B. gravinae* (10), *B. repanda* (10) *D. viminea* (10) **Group V** *B. elongata* (11)

Note Values in parenthesis indicate haploid chromosome no. of the species (*Source* Warwick and Black 1991, 1997a; Prakash et al. 2009)

It also includes cytodeme or crossing group which have same chromosome numbers and readily cross with each other (Harberd 1972; Hammer et al. 2013). However, cytodeme of different *Brassica* species presented in Table 4.3 forms the primary genepool of respective *Brassica* species. Each cultivated *Brassica* species have their own primary genepool which includes different subspecies and morphological varieties.

The genome-A is carried by *B. rapa* (Syn. *B. campestris*) which has many variants like Chinese cabbage, sarson turnip, turnip greens, turnip, and turnip rape crops. On the basis of morphology these are assigned the leafy, rapifera, and oleifera types. The Chinese cabbage is used as salad in Asia; sarson turnip is a minor crop in Europe and in New Zealand; turnip greens and turnip tops have culinary uses in Portugal and north Spain and turnip rape is widely distributed and cultivated for oil production in the North America as oilseed crop (McNaughton 1995; Padilla et al. 2005; Branca and Cartea 2011).

The genome-B is possessed by *B. nigra* which is also known as black mustard. It is used as condiment since Middle Ages. Naturally it is distributed as weed in Europe. *B. oleracea*, which contains genome-C is the most diversified *Brassica* species which evolved into many several botanical varieties and related crops by domestication processes. These include var. *acephala* (kale), var. *botrytis* (cauliflower), var. *capitata* (cabbage), var. *gemmifera* (Brussel's sprout), var. *gongylodes* (kohlrabi), var. *italica* (broccoli), and var. *sabauda* (Savoy cabbage) (Linnaeus 1753; Lamarck 1784; De Candolle 1821; Branca and Cartea 2011).

4.4.2 Secondary Genepool (GP-2)

GP-2 refers to germplasm group in which crossing with the cultivated species are difficult. However, it can be achieved using conventional plant breeding methods due to some biological barriers. Hybrids may be weak or partially sterile, chromosomes may pair poorly and there may be differences in ploidy levels. This includes nonconspecific wild relatives of the cultivated species, as well as other cultivated species belonging to the same genus. This group includes all biological species that can be crossed with the crop species and will form an experimentally defined coenospecies (Harlan and de Wet 1971; Spillane and Gepts 2001). This is corresponding to the coenospecies or syngameon, and is very useful and readily defined entity where adequate investigation of interspecific hybrid production between the crop and related species has been carried out (Smartt 1970). The term Brassica coenospecies (more or less second genepool) includes all species that can share their genes with the important cultivated Brassica species (Prakash et al. 1999; Hammer et al. 2013). This genepool can be utilized, with some efforts of the plant breeder or geneticist. Once a fertile hybrid is obtained the recombination potential is high in these hybrids because of genomic affinity.

4.4.3 Tertiary Genepool (GP-3)

Crossing of GP1 to members of this group tend to produce anomalous, lethal or completely sterile hybrids. Gene transfer is extremely difficult or can be possible only by advanced breeding or biotechnological techniques. Special techniques like, embryo rescue, cell culture and tissue culture, protoplast fusion, chromosome engineering, chromosome doubling, bridging crosses, genetic engineering would be required to produce the viable fertile hybrid plant to transfer genes for trait of interest from this genepool to GP-1. This category includes all the distant wild relatives including different genera of the cultivated from in which gene transfer through sexual recombination is very difficult. Tertiary genepool (GP3) can be more or less equated with the comparium, i.e., the grouping of all these coenospecies between which hybridization is possible. The enlarged tertiary genepool of Brassica includes the related genera *Coincya, Diplotaxis, Eruca, Erucastrum, Hirschfeldia, Sinapis, Sinapidendron, Trachystoma, Enarthrocarpus, Raphanus, Moricandia, Pseuderucaria,* and *Rytidocarpus* (Hu et al. 2009; Hammer et al. 2013). Introgressions from some of these genera have been reported through somatic hybridization for transfer of different traits of interest (Chatterjee et al. 1988; Kirti et al. 1992; Kirti et al. 1995; Muller et al. 2001; Hu et al. 2009; Hammer et al. 2013; Kumari et al. 2018; Kumari et al. 2020a, b).Please check the clarity of the sentence "This category includes all the distant wild relatives including different genera of the cultivated from which gene transfer through sexual recombination very difficult."Corrected

Branca (2008) divided genetic resources of Brassica crops in to three genepools based on their crossability. The primary genepool includes landraces, obsolete cultivars, modern cultivars, and advanced breeding lines and genetic stocks of corresponding species of the U-triangle, which means each species has its own primary gene pool. The secondary genepool is represented by the other species of the U-triangle, since their relationship has been confirmed by various phylogenetic and cytogenetic studies. Finally, the tertiary genepool includes *Brassica coenospecies* including species and genera grouped into different cytodemes based on their crossability and cytological studies. Genera in cytodems include *Brassica, Coincya, Diplotaxis, Eruca, Erucastrum, Hirschfeldia, Sinapis, Sinapidendron, Enarthrocarpus,* and *Raphanus* (Cartea et al. 2011). A list of selected members of the primary, secondary, and tertiary gene pools is presented in Table 4.5.

4.5 Brassica Genepool as Gene Reservoir

Wild and related genera of Brassica species are reservoirs of novel genes which possess some traits of interest including gene for agronomic importance, biotic and abiotic stress tolerant, oil and seed quality and male sterility. However, an understanding of the genetic potential of wild relatives is important for planning of long-term breeding programs for these crops (Warwick et al. 2009). The potential some of these wild relatives of Brassica is presented in Table 4.6. An exhaustive list of these traits was summarized by Warwick et al. (2009). These species are source for disease resistance including white rust (*Brassica maurorum* and *Eruca versicaria* ssp. *sativa*) to Alternaria blight (*Brassica fruticulose, Trachystom aballii*), blackleg/Phoma disease (*Sinapis arvensi, Sinapis alba, Thlaspi arvense, B. tournefortii*), powdery mildew (*B. oleracea*), clubroot disease (*B. rapa,*

Table 4.5 Selected members of the primary, secondary, and tertiary genepools of Brassica crops

Genepool	Genepool species
Primary genepool (GP1)	*Brassica carinata A. Braun* (Ethiopian mustard), *Brassica juncea* (L.) *Czern* (Indian mustard, brown mustard) *Brassica napus* ssp. *napus* (oilseed rape, fodder rape) *B. napus* ssp. *napobrassica* (swede) *B. napus* ssp. *napus* var. *pabularia* (leaf rape, kale)
Secondary genepool (GP2)	*Brassica nigra* *Brassica oleracea* (includes crop varieties), *B. alboglabra, B. bourgeaui, B. cretica, B. hilarionis, B. incana, B. insularis, B. macrocarpa, B. montana, B. rupestris, B. villosa)* *Brassica rapa* (includes wild and cultivated varieties)
Tertiary genepool (GP3)	*Brassica fruticulosa* *Brassica gravinae* *Brassica maurorum* *Brassica oxyrrhina* *Brassica repanda* (include *B. desnottesii, B. nudicaulis, B. saxatilis*) *Brassica souliei* (*syn. B. amplexicaulis*) *Brassica tournefortii* *Coincya* spp. (include all species in the genus) *Crambe abyssinica* *Diplotaxis acris* *Diplotaxis assurgens* *Diplotaxis berthautii* *Diplotaxis catholica* *Diplotaxis cossoniana* *Diplotaxis erucoides* *Diplotaxis harra* (include *D. crassifolia, D. gracilis, D. hirtum, D. lagascana*) *Diplotaxis muralis* *Diplotaxis siettiana* (include *D. ibicensis*) *Diplotaxis siifolia* *Diplotaxis tenuifolia* *Diplotaxis tenuisiliqua* *Diplotaxis viminea* *Diplotaxis virgata* *Enarthrocarpus* ssp. (includes *E. lyratus, E. pterocarpus, E. strangulatus*) *Eruca* spp. (includes *E. vesicaria, E. sativa, E. pinnatifida)* *Erucastrum abyssinicum* *Erucastrum canariense* (includes *E. cardaminoides*) *Erucastrum elatum* *Erucastrum gallicum* *Erucastrum nasturtiifolium* (includes *E. leucanthum*) *Erucastrum strigosum* *Erucastrum varium* *Erucastrum virgatum* *Onchophragmus violaceus* *Raphanus* ssp. (includes *R. raphanistrum, R. sativus, R. caudatus, R. maritimus, R. landra*) *Sinapidendron* spp. (include *S. angustifolium, S. frutescens, S. rupestre*) *Sinapis alba* (includes *S. dissecta*) *Sinapis arvensis* (include *S. allioni, S. turgida*) *Sinapis aucheri* (syn. *Raphanus aucheri*) *Sinapis flexuosa* *Sinapis pubescens* (include *S. aristidis, S. boivinii, S. indurata*) *Trachystoma labasii* *Trachystoma* spp. (include *T. aphanoneurum, T. ballii*)

Note Secondary and tertiary genepool species listed together are members of the same cytodeme, i.e., they share a single diploid chromosome number and are generally fully interfertile (*Source* Snowdon et al. 2007)

Table 4.6 List of Brassica genepool species as a source of biotic and abiotic stress tolerance (*Source* Warwick et al. 2009)

Trait	Species
Abiotic stress	
Salt tolerance	*Crambe maritime, Eruca vesicaria* subsp. *sativa, Lesquerella fendleri, Lobularia maritime, Raphanus raphanistrum* ssp. *Thellungiella salsuginea,*
Cold tolerance	*Barbarea vulgaris, Descurainia sophia, Thellungiella salsuginea, Thlaspi arvense*
Drought tolerance	*B. carinata, B. tournefortii, Diplotaxis acris, D. harra, Enarthrocarpus strangulates, Eruca vesicaria, Erucaria boveana,* subsp. *sativa, Moricandia* spp.
Disease resistance	
White rust—*Albugo candida*	*B. carinata, B. nigra, B. maurorum, Eruca vesicaria* subsp. *sativa, Raphanus sativus*
Black leaf spot/ Alternaria spp.—*Alternaria brassicae, A. brassicicola, A. raphani*	*Alliaria petiolata, Brassica nigra, B. elongate, B. fruticulosa, B. maurorum, B. nigra, B. souliei, Camelina sativa, Capsella bursa-pastoris, D. catholica, D. erucoides, D. tenuifolia, Eruca vesicaria* subsp. *sativa, Hemicrambe fruticulosa, Neslia paniculata, Raphanus sativus, Sinapis alba*
Blackleg—*Leptosphaeria maculans* [=*Phoma lingam*]	*Arabidopsis thaliana, B. carinata, B. nigra, B. elongata, B. fruticulosa, B. insularis, B. atlantica, Camelina sativa, Diplotaxis muralis, D. tenuifolia, Eruca vesicaria, Hirschfeldia incana, Raphanus raphanistrum, Raphanus sativus, Sinapis alba, S. avensis, Sisymbrium loeselii, Thlaspi arvense*
Downy mildew (*Peronospora parasitica)*	*Eruca vesicaria*
Clubroot (*Plasmodiophora brassicae)*	*Arabidopsis thaliana, Armoracia rusticana, Capsella bursa-pastoris, Raphanus* spp.
Sclerotinia stem rot (*Sclerotinia sclerotiorum)*	*Capsella bursa-pastoris, Eruca vesicaria* subsp. sativa, *Erucastrum gallicum*
Black rot (*Xanthomonas campestris)*	*Alliaria petiolata, Barbarea vulgaris, B. nigra, Erysimum hieraciifolium, Matthiola incana, B. carinata, B. nigra*
Insect resistance	
Flea beetles (*Phyllotreta cruciferae* and *P. striolata)*	*Arabidopsis thaliana, B. incana, B. villosa, Camelina sativa, Capsella bursa-pastoris, Crambe abyssinica, Crambe hispanica, Sinapis alba, Thlaspi arvense*
Diamond-back moth (*Plutella xylostella*)	*Barbarea vulgaris, Crambe abyssinica, Raphanus raphanistrum*
Cabbage aphid (*Brevicoryne brassicae)*	*Brassica fruticulosa, B. spinescens, B. cretica, B. incana, B. macrocarpa, B. villosa, Eruca vesicaria* subsp. *sativa, Sinapis alba*
Mustard aphid (*Lipaphis erysimi*)	*B. carinata, B. nigra, Eruca vesicaria* subsp. *sativa*
Cabbage white fly (*Aleyrodesproletella)*	*B. cretica, B. fruticulosa, B. incana, B. insularis, B. spinosa, B. villosa*
Mustard sawfly (*Athalia proxima*)	*Camelina sativa*

B. oleracea and *B. napus).* Some species are resistant/tolerant to pod shattering and heavy metals (*B. juncea).* Some are source of C3–C4 intermediate photosynthetic system (*Moricandia, Diplotaxis* species), high erucic acid (*Crambe abyssinica),* and cytoplasmic male sterility (*Sinapis incana, Diplotaxis siifolia*) (Katche et al. 2019). The gene transfer between *Brassica* and allied genera is possible with some limitations despite differences in chromosome numbers (Hammer et al. 2013). On the basis of intercrossing and cytological studies genome interrelationships between *Brassica* and allied genera were proposed (Fig. 4.2) (Mizushima 1980; Quezada-Martinez et al. 2021). This can act as guide for wide hybridization in Brassica coenospecies.

The utilization of tertiary genepool species is difficult because of less recombination potential and higher genetic load. These hurdles can be overcome by consistent pre-breeding. The recombination frequency can be increased by generating more backcross progeny. Introgression from secondary and tertiary genepool is more usually associated with genetic load (load of undesirable alleles) caused by undesirable linkages which make selections difficult during segregating generations, which in term increases the number of breeding cycles (Khadi et al. 2003). Therefore, pre-breeding or germplasm enhancement is very important activity for exploitation of untapped potential of wild relatives of Brassica. With scientific advancement, genetic engineering tools and marker-assisted selection (MAS) proved to be a boon for distant hybridization. However, there is a need for identification of markers for different traits. Hence MAS provides new dimension for utilization of Brassica germplasm. An exhaustive information on interspecific and intergeneric hybridization among Brassica crops and related species, was presented by Warwick et al. (2000, 2009).

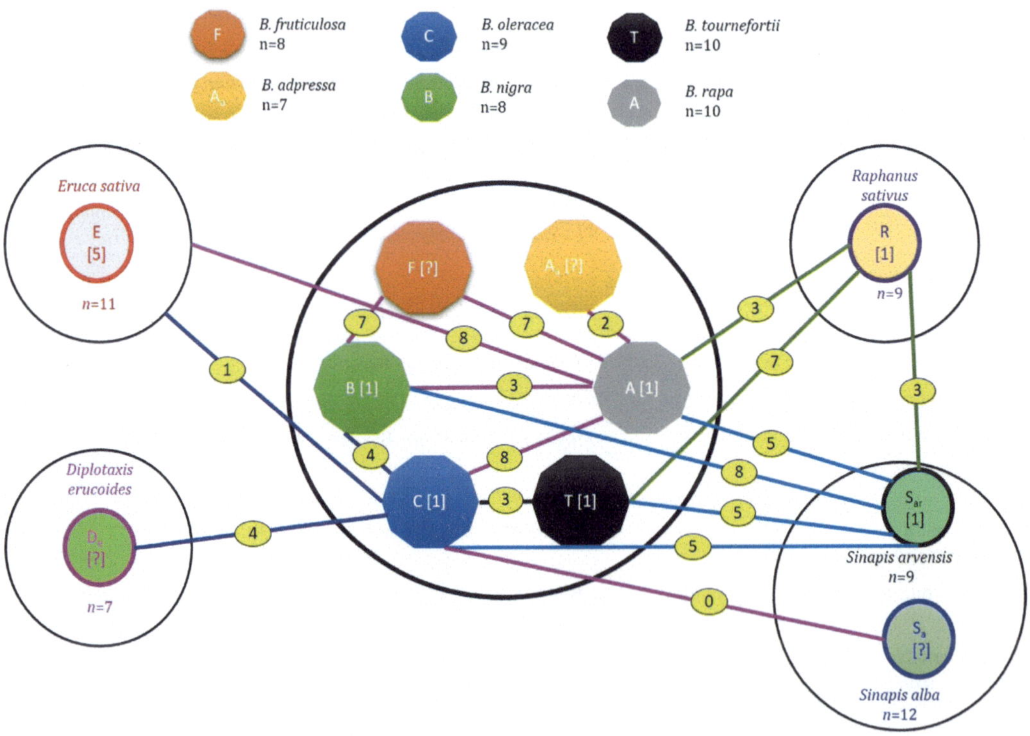

Fig. 4.2 Genome inter-relationships in *Brassica* and allied genera. Autosyndetic bivalents observed in haploids are denoted by numbers in brackets whereas numbers on lines represent the maximum number of bivalents observed in hybrids between the two species (*Source* Mizushima 1980; Quezada-Martinez et al. 2021)

4.6 Conservation of Brassica Genepool

To protect valuable genepool of Brassicaceae, the collection and conservation programs started worldwide. During the 1970s, wild germplasm of Brassica was extensively collected and conserved (Gomez-Campo et al. 2006, 2007). The Universidad Polite´cnica of Madrid (UPM), Spain maintains the one of the largest collections of wild *Brassica* species and allied genera. The Plant Germplasm Bank-UPM holds 600 crucifer accessions of rare and endangered species widespread in the western Mediterranean area (http://www.etsia.upm.es/ANTIGUA/DEPARTAMENTOS/biologia/documentos/GC-2000-Int.htm). Information on standard techniques and protocols for long-term effective preservation and germplasm regeneration were provided by Gladis and Hammer (1990, 2003). Presently Brassica germplasm are conserved across the globe in seed genebanks of many countries (China, India, United Kingdom, United States, Netherland, Spain, and Germany), and are constantly enriched by further collections through plant exploration from different geographical regions with more diversity based on identified gaps (Punjabi et al. 2019). These countries hold around 60% of total worldwide collections of crucifers (Singh and Sharma 2007). Large national collections, in USA, China, and India, are generally maintained as a network of active regional collections and a set of it is conserved in situ or ex situ in gene banks (Knee et al. 2011). Germplasm from these genebanks are exchanged across the globe for research purpose after following proper guidelines of germplasm exchange and signing of Material Transfer Agreement (MTA) between indenter and seed supplier organizations.

References

Allender CJ, King GJ (2010) Origins of the amphidiploid species *Brassica napus* L. investigated bychloroplast and nuclear molecular markers. BMC Plant Biol 10:54

Al-Shehbaz IA (1985) The genera of Brassiceae (Cruciferae; *Brassicaceae*) in the south eastern United States. J Arnold Arbor Harv Univ 66:279–351

Anderson JK, Warwick SI (1999) Chromosome number evolution in the tribe Brassiceae (*Brassicaceae*): evidence from isozyme number. Plant Syst Evol 215:255–285

Appel O (1999) The so-called 'beak', a character in the systematics of *Brassicaceae*? Bot Jahr Syst Pflanzengesch Pflanzengeogr 121:85–98

Baillon HE (1871) Cruciferes. Historie Des Plantes (paris) 3(188–195):248

Beilstein MA, Al-Shehbaz IA, Kellog EA (2006) Brassicaceae phylogeny and trichome evolution. Amer J Bot 93:607–619

Branca F, Cartea E (2011) Brassica. In: Kole C (ed) Wild crop relatives: genomic and breeding resources, oilseeds. Springer, Berlin, Heidelberg, pp 17–36

Branca F (2008) Cauliflower and broccoli. In: Prohens J, Nuez F (eds) Vegetables I: Asteraceae, Brassicaceae, Chenopodicaceae, and Cucurbitaceae. Handbook of crop breeding. Springer, Berlin, pp 151–186

Cartea ME, Lema M, Francisco M, Velasco P (2011) Basic information on vegetable Brassica crops. In: Sadowski J, Kole C (eds) Genetics, genomics and breeding of vegetable brassicas. Science Publishers, Enfield, pp 1–33

Chalhoub B, Denoeud F, Liu S, Parkin IAP, Tang H, Wang X et al (2014) Early allopolyploid evolution in the post-Neolithic *Brassica napus* oilseed genome. Science 345:950–953

Chatterjee G, Sikdar SR, Das S, Sen SK (1988) Intergeneric somatic hybrid production through protoplast fusion between *Brassica juncea* and *Diplotaxis muralis*. Theor Appl Genet 76:915–922

Crespo MB, Lledo MD, Fay MF, Chase MW (2000) Subtribe Vellinae (Brassiceae, *Brassicaceae*): a combined analysis of ITS nrDNA sequences and morphological data. Ann Bot (london) 86:53–62

De Candolle AP (1821) Cruciferae. Syst Nat 2:139–700

Demeke T, Adams RP, Chibbar R (1992) Potential taxonomic use of random amplified polymorphic DNA (RAPD)—a case study in *Brassica*. Theor Appl Genet 84:990–994

Dixon GR (2007) Vegetable Brassicas and related Crucifers. CAB International, Willingford, UK, p 416

Dobzhansky T (1937) Genetics and the origin of species. Colombia University Press, New York

Downey RK, Röbbelen G (1989) *Brassica* species. In: Röbbelen G, Downey RK, Ashri A (eds) Oil crops of the world. McGraw-Hill, New York, pp 339–362

Erickson LR, Straus NA, Beversdorf WD (1983) Restriction patterns reveal origins of chloroplast genomes in *Brassica* amphidiploids. Theor Appl Genet 65:201–206

Francisco-Ortega J, Fuertes-Aguilar J, Kim SC, Santos-Guerra A, Crawford DJ, Jansen RK (2002) Phylogeny of the Macaronesian endemic *Crambe* section *Dendrocrambe* (*Brassicaceae*) based on internal transcribed spacer sequences of nuclear ribosomal DNA. Amer J Bot 89:1984–1990

Franciso-Ortega J, Fuertes-Aguilar J, Gómez-Campo C, Santos-Guerra A, Jansen RK (1999) Internal

transcribed spacer sequence phylogeny of *Crambe* L. (*Brassicaceae*): molecular data revealed two old world disjunctions. Mol Phylogenet Evol 11:361–380
Gladis T, Hammer K (2003) Die *Brassica-oleracea*-Gruppe. Schriften des Vereins zur Erhaltung der Nutzpflanzenvielfalt 1, 79 pp. + Appendix. VEN, Lennestadt
Gladis T, Hammer K (1990) Die Gaterslebener *Brassica* Kollektion – eine Einführung. Kulturpflanze 38:121–156
Gómez-Campo C (1980) Morphology and morphotaxonomy of the tribe Brassiceae. In: Tsunoda S, Hinata K, Gómez-Campo C (eds) Brassica crops and wild allies. Japan Science Societies Press, Tokyo, pp 3–31
Gómez-Campo C (1999) Taxonomy. In: Gómez-Campo C (ed) The biology of *Brassica coenospecies*. Elsevier Science B.V, Amsterdam, pp 3–32
Gomez-Campo C (1999) Biology of Brassica coenospecies. Elsevier Science, Amsterdam, p 489
Gomez-Campo C, Aguinagalde I, Ceresuela J, Lazaro A, Martinez-Laborde J (2006) Erosion of genetic resources within seed genebanks: the role of seed containers. Seed Sci Res 16:291–294
Gomez-Campo C, Aguinagalde I, Arus P, Jimenez-Aguilar C, Lazaro A, Martın-Clemente JP, Parra-Quijano M, Sanchez Yelamo MD, Simonetti E, Torres E, Torcal L, Tortosa ME (2007) Geographical distribution and conservation status of *Brassica montana* in NE Spain. Cruciferae Newsl 27:32–34
Hammer K, Gladis T, Laghetti G, Pignone D (2013) The wild and the grown-remarks on *Brassica*. Intl J AgriSci 3(6):453–480
Harberd DJ (1972) A contribution to cytotaxonomy of Brassica (Cruciferae) and its allies. Bot J Linn Soc 65:1–23
Harberd DJ (1976) Cytotaxonomic studies of Brassica and related genera. In: Vaughan JG, MacLeod AJ, Jones MG (eds) The biology and chemistry of the cruciferae. Academic Press, London, pp 47–68
Harlan JR, De Wet JMJ (1971) Toward a rational classification of cultivated plants. Taxon 20:509–517
Hayward A (2012) Introduction-Oilseed Brassicas. In: Edwards E, Batley J, Parkin I, Kole C (eds). Genetics, genomics and breeding of oilseed brassicas. CRC Press, Taylor & Francis Group. ISBN-13: 978-1-4398-8335-8
Hedge IC (1976) A systematic and geographical survey of the old world cruciferae. In: Vaughn JG, MacLeod AJ, Jones BMG (eds) The biology and chemistry of the cruciferae. Academic Press, London, pp 1–45
Hooker JD (1862). In: Bentham G, Hooker JD (eds) Genera Plantarum, vol 1. Lovell Reed, London, pp 57–102
Hosaka K, Kianian SF, McGrath JM, Quiros CF (1990) Development and chromosomal localization of genome specific DNA markers of Brassica and evolution of amphidiploids and n=9 diploid species. Genome 33:131–142
Hu Q, Li Y, Mei D (2009) Introgression of genes from wild crucifers. In: Gupta SK (ed) Biology and breeding of Crucifers. CRC Press, Boca Raton, FL, pp 261–283
Inaba R, Nishio T (2002) Phylogenetic analysis of Brassiceae based on the nucleotide sequences of the S-locus related gene, *SLR1*. Theor Appl Genet 105:1159–1165
Kagale S, Robinson SJ, Nixon J, Xiao R, Huebert T, Condie J, Kessler D, Clarke WE, Edger PP, Links MG, Sharpe AG, Parkin IAP (2014) Polyploid evolution of the brassicaceae during the cenozoic era. Plant Cell 26:2777–2791
Katche E, Quezada-Martinez D, Katche EI, Vasquez-Teuber P, Mason AS (2019) Interspecific hybridization for Brassica crop improvement. Crop Breed Genet Genom 2019,1:e190007. https://doi.org/10.20900/cbgg20190007
Khadi BM, Kulkarni VN, Ansingkar AS, Singh VV, Gotmore V (2003) Gene pool concept for diploid cultivated cotton. In: World cotton research conference-3, Cape Town, South Africa, pp 214–225
Kirti PB, Narasimhulu SB, Prakash S, Chopra VL (1992) Somatic hybridization between *Brassica juncea* and *Moricandia arvensis* by protoplast fusion. Plant Cell Rep 11(5–6):318–321
Kirti PB, Mohapatra T, Khanna H, Prakash S, Chopra VL (1995) *Diplotaxis catholica* + *Brassica juncea* somatic hybrids: molecular and cytogenetic characterization. Plant Cell Rep 14:593–597
Knee EM, Rivero L, Crist D, Grotewold E, Scholl R (2011) Germplasm and molecular resources. In: Genetics and genomics of the Brassicaceae. Springer, New York, NY, pp 437–467
Koch MA, Al-Shehbaz IA (2009) Molecular systematics and evolution. In: Gupta S (ed) Biology and breeding of Crucifers. CRC Press, Boca raton, FL, pp 1–18
Kumari P, Bisht DS, Bhat SR (2018) Stable, fertile somatic hybrids between *Sinapis alba* and *Brassica juncea* show resistance to *Alternaria brassicae* and heat stress. Plant Cell Tiss Organ Cult 133:77–86. https://doi.org/10.1007/s11240-017-1362-9
Kumari P, Singh KP, Bisht D, Kumar S (2020a) Somatic hybrids of *Sinapis alba* + *Brassica juncea*: study of backcross progenies for morphological variations, chromosome constitution and reaction to *Alternaria brassicae*. Euphytica 216:93. https://doi.org/10.1007/s10681-020-02629-3(0123456789(),-volV)(01234567
Kumari P, Singh KP, Kumar S, Yadava DK (2020b) Development of a yellow-seeded stable allohexaploid Brassica through inter-generic somatic hybridization with a high degree of fertility and resistance to *Sclerotinia sclerotiorum*. Front Plant Sci 11:575591. https://doi.org/10.3389/fpls.2020.575591
Lamarck JBA (1784) "Chou". Encyclopédie Methodique Botanique. I. Paris, France
Linnaeus C (1753) Species plantarum II. Stockholm, Sweden, 561 p
Lysak ML, Koch MA, Pecinka A, Schubert I (2005) Chromosome triplication found across the tribe Brassiceae. Genome Res 15:516−525

Mayr E (1963) Animal species and evolution. Harvard University Press, Cambridge, Massachusetts

McNaughton IH (1995) Turnip and relatives. *Brassica napus* (Cruciferae). In: Smartt J, Simmonds NW (eds) Evolution of crop plants, Chap 17. Longman, London, pp 62–68

Mizushima U (1980) Genome analysis in *Brassica* and allied genera. In: Tsunoda S, Hinata K, Gomez-Campo C (eds) Brassica crops and wild allies: biology and breeding. Japan Scientific Societies Press, Tokyo, pp 89–106

Morinaga T (1934) Interspecific hybridization in *Brassica*. VI. The cytology of F1 hybrids of *B. juncea* and *B. nigra*. Cytologia 6:62–67

Müller J, Sonntag K, Rudloff E (2001) Somatic hybridisation between *Brassica* spp. and *Raphanus sativus*. Acta Hort 560:219–220. https://doi.org/10.17660/ActaHortic.2001.560.41

OECD (2016) Brassica crops (*Brassica* species). In: Safety assessment of transgenic organisms in the environment, Volume 5: OECD Consensus Documents. OECD Publishing, Paris

Padilla G, Cartea ME, Rodríguez VM, Ordás A (2005) Genetic diversity in a germplasm collection of *Brassica rapa* subsp. rapa L. from northwestern Spain. Euphytica 145:171–180

Palmer JD, Shields CR, Cohen DB, Orton TJ (1983) Chloroplast DNA evolution and the origin of amphidiploid *Brassica* species. Theor Appl Genet 65:181–189

Panjabi P, Yadava SK, Kumar N, Bangkim R, Ramchiary N (2019) Breeding *Brassica juncea* and *B.* rapa for sustainable oilseed production in the changing climate: progress and prospects. In: Kole C (ed) Genomic designing of climate-smart oilseed crops, oilseeds. Springer Nature, Cham, Switzerland

Pradhan AK, Prakash S, Mukhopadhyay A, Pental D (1992) Phylogeny of Brassica and allied genera based on variation in chloroplast and mitochondrial DNA patterns: molecular and taxonomical classifications are incongruous. Theor Appl Genet 85:331–340

Prakash S, Hinata K (1980) Taxonomy, cytogenetics and origin of crop brassicas—a review. Oper Bot 55:1–57

Prakash S, Takahata Y, Kirti PB, Chopra VL (1999) Cytogenetics. In: Gomez-Campo C (ed) Biology of *Brassica coenospecies*. Elsevier Science, Amsterdam, pp 59–106

Prakash S, Bhat SR, Quiros CF, Kirti PB, Chopra VL (2009) Brassica and its close allies: cytogenetics and evolution. Plant Breeding Reviews 31:21–167

Prantl K (1891) Cruciferae. In: Engler A, Prantl K (eds) Die Naturlichen Pflanzenfamilien. Wilhelm Englmann, Leipzig, pp 145–208

Purty RS, Kumar G, Singla-Pareek SL, Pareek A (2008) Towards salinity tolerance in *Brassica*: an overview. Physiol Mol Biol Plants 14:39–49

Quezada-Martinez D, Nyarko CPA, Sarah V, Schiessl SV, Mason AS (2021) Using wild relatives and related species to build climate resilience in *Brassica* crops. Theor Appl Genet. https://doi.org/10.1007/s00122-021-03793-3

Schulz OE (1919) Cruciferae-Brassicaceae. Part 1. Subtribes I. Brassicinae and II. Raphaninae. In: Engler A (ed) Das Pflanzenreich, vol 68–70. Wilhelm Engelmann, Leipzig, Germany, pp 1–290

Schulz OE (1936) Cruciferae. In: Engler A, Harms H (eds) Die naturlichen Pflanzenfamilien. Wilhelm Engelmann, Leipzig, Germany, pp 227–658

Schulz OE (1923) Cruciferae–Brassiceae. Part II. Subtribes Cakilinae, Zillinae, Vellinae, Savignyinae, and Moricandiinae. In: Engler A (ed) Pflanzenr. 82–85 (IV. 105). Verlag von Wilhelm Engelmann, Leipzig, pp 1–100

Singh R, Sharma SK (2007) Evaluation, maintenance, and conservation of germplasm. In: Gupat SK, Delseney M, Kader JC (eds) Advances in botanical research, incorporating advances in plant pathology, rapeseed breeding, vol 35. Academic Press, Elsevier, CA, USA, pp 466–481

Slocum MK, Figdore SS, Kennard WC, Suzuki JY, Osborn TC (1990) Linkage arrangement of restriction fragment length polymorphism loci in *Brassica oleracea*. Theor Appl Genet 80:57–64

Smartt J (1970) Evolving gene pools in crop plants. Euphytica 30(1981):415–418

Snowdon R, Lühs W, Friedt W (2007) *Brassica* Oilseeds. In: Singh RJ (eds) Genetic resources, chromosome engineering, and crop improvement, vol 4. CRC Press, Taylor & Francis Group, Broken Sound Parkway NW, Suite 300, Boca Raton, pp 195–230

Song KM, Osborne TC, Williams PH (1988) *Brassica* taxonomy based on nuclear restriction fragment length polymorphisms (RFLPs). 1. Genome evolution of diploid and amphidiploid species. Theor Appl Genet 75:784–794

Song KM, Osborn TC, Williams PH (1990) Brassica taxonomy based on nuclear restriction fragment length polymorphism (RFLPs). 3. Genome relationship in Brassica and related genera and the origin of *B. oleracea* and *Brassica rapa* (syn. *campestris*). Theor Appl Genet 79:497–506

Spillane C, Gepts P (2001) Evolutionary and genetic perspectives on the dynamics of crop genepools. In: Cooper HD, Spillane C, Hodgkin T (eds) Broadening the genetic base of crop production. CABI Publishing in association with Food and Agriculture Organization of the United Nations and International Plant Genetic Resources Institute, pp 25–70

Takahata Y, Hinata K (1983) Studies on cytodemes in the subtribe Brassicineae. Tohoku J Agric Res 33:111–124

Tsunoda S, Hinata K, Gómez-Campo C (1980) Brassica crops and wild allies. Japan Scientific Societies Press, Tokyo, pp 1–354

UN (1935) Genome analysis in Brassica with special reference to the experimental formation of *B. napus* and peculiar mode of fertilization. Jpn J Bot 7:389–452

Warwick SI, Black LD (1991) Molecular systematics of Brassica and allied genera (subtribe Brassicinae, Brassiceae)-chloroplast genome and cytodeme congruence. Theor Appl Genet 82:81–92

Warwick SI, Black LD (1993) Molecular relationships in subtribe Brassicinae (Cruciferae, tribe Brassiceae). Can J Bot 71:906–918

Warwick SI, Black LD (1994) Evaluation of the subtribes Moricandiinae, Savignyinae, Vellinae and Zillinae (*Brassicaceae*, tribe Brassiceae) using chloroplast DNA restriction site variation. Can J Bot 72:1692–1701

Warwick SI, Black LD (1997a) Phylogenetic implications of chloroplast DNA restriction site variation in subtribes Raphaninae and Cakilinae (*Brassicaceae*, tribe Brassiceae). Can J Bot 75:960–973

Warwick SI, Black LD (1997b) Molecular phylogenies from theory to application in *Brassica* and allies (tribe Brassiceae, *Brassicaceae*). Opera Bot 132:159–168

Warwick SI, Sauder C (2005) Phylogeny of tribe Brassiceae (*Brassicaceae*) based on chloroplast restriction site polymorphisms and nuclear ribosomal internal transcribed spacer and chloroplast *trn*L intron sequences. Can J Bot 83:467–483

Warwick SI, Black LD, Aguinagalde I (1992) Molecular systematics of *Brassica* and allied genera (subtribe Brassicinae, Brassiceae) chloroplast DNA variation in the genus *Diplotaxis*. Theor Appl Genet 83:839–850

Warwick SI, Francis A, Al-Shehbaz IA (2006) Brassicaceae: species checklist and database on CD-Rom. Pl Syst Evol 259:249–258

Warwick SI, Hall JC (2009) Phylogeny of Brassica and wild relatives. In: Gupta SK (ed) Biology and breeding of crucifers. CRC Press Taylor & Francis Group, 6000 Broken Sound Parkway NW, Suite 300, Boca Raton, FL, pp 18–36

Warwick SI, Francis A, La Flèche J (2000) Guide to the wild germplasm of Brassica and allied crops (tribe Brassiceae, Brassicaceae) [online], 2nd edn. Agriculture and Agri-Food Canada Research Branch Publication, ECORC, Ottawa, Ontario, Canada. Contribution No. 991475. Available from http://www.brassica.info/information.htm. Accessed April 2008

Warwick SI, Francis A, Gugel RK (2009) Guide to wild germplasm of *Brassica* and allied crops (tribe Brassiceae, Brassicaceae), 3rd edn. Available from: http://www.brassica.info/info/publications/guidewild/Guide_ed.3_Introd_16July2009.pdf. Accessed 2019 July 18

Xue JY, Wang Y, Chen M, Dong S, Shao Z-Q, Liu Y (2020) Maternal inheritance of U's triangle and evolutionary process of brassica mitochondrial genomes. Front Plant Sci 11:805. https://doi.org/10.3389/fpls.2020.00805

Yanagino T, Takahata Y, Hinata K (1987) Chloroplast DNA variations among diploid species in *Brassica* and allied genera. Jpn J Genet 62:119–125

Yarnell SH (1956) Cytogenetics of the vegetable crops: II. Crucifers. Bot Rev 22(2):81–166

Genetic Diversity Characterization and Population Structure in *Brassica juncea*

5

D. K. Yadava, Rashmi Yadav, Harinder Vishwakarma, Yashpal, Sangita Yadav, Navinder Saini, and Sujata Vasudev

Abstract

Brassica is a significant group of oilseed crops with an extensive evolutionary history. Various markers have played a vital role in understanding the origin and evolution of Brassica species. The molecular markers are better than the morphological traits in discriminating *B. juncea* genotypes. Modern scientific advances are helping in the development of high-quality Brassica genomes. Genomics and diversity study will lead to Brassica improvement; a multi-disciplinary plant breeding approach will ensure the development of resilient Brassica varieties. An attempt has been made in this chapter to give a comprehensive account of the diversity characterization and population structure studies using the latest techniques in *Brassica juncea*. Efforts have been made to highlight the variability available in *Brassica* spp. to be used in Brassica improvement programs.

D. K. Yadava (✉)
ADG (Seeds), Indian Council of Agricultural Research, Krishi Bhawan, New Delhi 110012, India

R. Yadav · H. Vishwakarma
Division of Germplasm Evaluation, ICAR-National Bureau of Plant Genetic Resources, Pusa Campus, New Delhi 110012, India

S. Yadav
Division of Seed Science and Technology, Indian Agricultural Research Institue, New Delhi 110012, India

Yashpal · N. Saini · S. Vasudev
Division of Genetics, Indian Agricultural Research Institue, New Delhi 110012, India

5.1 Introduction

Breeding disease-resistant, high-yielding vegetable, and oilseed Brassica crops having superior seed, and oil quality is the most imperative objective for the breeders. Assessment of population structure, genetic diversity, and phylogenetic relationships of *Brassica* spp. is indispensable for the sustainable breeding programs for the development of new breeding lines. Population structure is defined by the organization of genetic variation. Combined effects of evolutionary processes like recombination, mutation, genetic drift, demographic history, and natural selection design the population structure. By simple definition, genetic diversity is the occurrence of a difference in a selected allele among a population. The phylogenetic relationship reflects the timeline, wherein the species shared common ancestry. It is of utmost importance to understand this genetic phenomenon for the optimum utilization of the crop germplasm. Available genetic diversity is the foundation and is vital in plant improvement. Genetic diversity assessment would be of great value to know variability in cultivars, segregating population,

C. Kole and T. Mohapatra (eds.), *The Brassica juncea Genome*, Compendium of Plant Genomes,
https://doi.org/10.1007/978-3-030-91507-0_5

and introgression of the desirable gene across genus/species (Smith 1984; Cox et al. 1986; Barrett and Kidwell 1998; Thompson and Nelson 1998). Conservation of the diverse gene pools and wild genotypes is a critical step in preserving the variability for future breeding efforts. Elite varieties, landraces, and wild species are key resources of useful variation.

Brassica of the Brassicaceae (Cruciferae) family is a major source of edible oil and vegetables (Table 5.1). The antiquity of crops belonging to the genus *Brassica* is manifested from references in ancient literature of the Indian, Chinese, Greek, and Roman civilizations (Prakash and Hinata 1980; Gómez-Campo and Prakash 1999a). Several studies of this family are available. For example, studies conducted by Tournefort (1700), Linnaeus (1753) are of great practical importance. Similarly, studies by De Candolle (1821), Baillon (1871), and Prantl (1891) have laid the foundation. The most comprehensive study on family Brassicaceae was attempted by Schulz (1919, 1936), a German school teacher (Hedge 1976; Prakash and Hinata 1980; Gómez-Campo 1999b). Schulz identified 19 tribes, of which Brassicaceae is one of them, which was further divided into 7–9 subtribes (Gómez-Campo 1980, 1999b). Genus *Brassica* is the most important group of family Brassicaceae. A majority of the species related to *Brassica* is wild and weedy.

Six cultivated species in genus Brassica consist of, *B. carinata* ($2n = 34$, BBCC), *B. juncea* ($2n = 36$, AABB), and *B. napus* ($2n = 38$, AACC), the tetraploid species, while *B. campestris/rapa* ($2n = 20$, AA), *B. nigra* ($2n = 16$, BB), and *B. oleracea* ($2n = 18$, CC) are diploid species. The tetraploids have naturally evolved by hybridization between the constituent diploid species (Prakash et al. 2009a, b; Yadava et al. 2012; Kaur et al. 2014). Chromosomal rearrangements like translocations, inversions, deletions, duplications, and transposon activation that occurred during interspecific hybridization, between different subgenomes of these polyploid *Brassica* species, resulted in genome reshuffling (Zou et al 2010, 2011). For the creation of novel genotypes, the genetic diversity of important oilseed crops has been used (Chen et al. 2011). Understanding the genetics behind variation among three sets (AA, BB, CC) of *Brassica* subgenomes would provide deep insights for Brassica breeding programs utilizing favorable allele introgression from allied species (Zou et al. 2016).

Brassica juncea L. Czern. and Coss., known as Oriental, Brown, or Indian mustard, is first described in Sanskrit texts during 3000 BC (Hemingway 1995). *B. juncea* spread to Europe as a medicinal crop in the middle ages. In the present day, *B. juncea* is used worldwide as an oilseed, a condiment, and a vegetable (Edwards et al. 2007). In India, *B. juncea* is the principal winter oilseed crop, covering above 90% of the area under rapeseed–mustard crops (Yadava et al. 2012). Indian mustard is recognized for cultivation under drought and hot conditions (Chen et al. 2013). However, its genetic base, restricted by two major constraints of polyploidy and domestication, is gradually manipulated by vigorous plant breeding activities (Chauhan et al. 2011; Banuelos et al. 2013). Breeders may improve the productivity of this important oilseed crop, by combining the knowledge of Brassica genetics and plant breeding (Gupta et al. 2015). Superior alleles for disease resistance, higher yield potential, and other favorable traits are being explored in the primary gene pools and the related species for transfer into the superior agronomic base in cultivated germplasm with the prime aim of imparting stability, wider adaptability, stress tolerance, improved oil quality and seed traits, and increased productivity to meet the existing and emerging nutritional and industrial needs (Banga and Banga 2016).

5.2 Genetic Resources in Brassicas

Brassicas are cultivated worldwide, and it is therefore certain that rapid environmental changes occurring globally would affect the crop yield (Francisco et al. 2017). Increasing temperature due to raised CO_2 is resulting in a drier atmosphere and a high rate of evapotranspiration (Ficklin and Novick 2017). Environment resilient

Table 5.1 Taxonomic components of *Brassica* and related genera and their usage

Botanical name	Common name	Usage
B. nigra	Black mustard	Condiment (seed)
B. oleracea		
var. *acephala*	Kale	Vegetable, fodder (leaves)
var. *capitata*	Cabbage	Vegetable (head)
var. *sabauda*	Savoy cabbage	Vegetable (terminal buds)
var. *gemmifera*	Brussels sprouts	Vegetable (head)
var. *gongylodes*	Kohlrabi	Vegetable, fodder (stem)
var. *botrytis*	Cauliflower	Vegetable (inflorescence)
var. *italica*	Broccoli	Vegetable (inflorescence)
var. *fruticosa*	Branching bush kale	Fodder (leaves)
var. *alboglabra*	Chinese kale	Vegetable (stem, leaves)
B. rapa		
spp. *oleifera*	Turnip rape	Oilseed
var. brown sarson	Brown sarson	Oilseed
var. yellow sarson	Yellow sarson	Oilseed
var. toria	Toria	Oilseed
ssp. *rapifera*	Turnip fodder	Vegetable (root)
ssp. *chinensis*	Bok choi	Vegetable (leaves)
ssp. *pekinensis*	Chinese cabbage	Vegetable, fodder (head)
ssp. *nipposinica*	–	Vegetable (leaves)
ssp. *parachinensis*	–	Vegetable (leaves)
B. carinata	Ethiopian mustard	Vegetable, oilseed
B. juncea	Mustard	Oilseed, vegetable
B. napus		
spp. *oleifera*	Rapeseed	Oilseed
spp. *rapifera*	Rutabaga	Swede fodder
Eruca sativa	Rocket, taramira	Vegetable, nonedible oilseed
Raphanus sativus	Radish	Vegetable, fodder
Sinapis alba	White mustard	Oilseed

Brassica varieties are perhaps the best strategy to ensure sustainability. The future crop should have resilience against both abiotic stresses (heat, drought, and salinity), as well as biotic stresses. Bebber et al. (2013) used a combination of observation data and mathematical equations to hypothesize that pathogens and pests affecting global crops are moving @ of 2.7 km/year toward poles due to rising global temperature. Pathogens such as *Pseudocercosporella capsellae* (causative agent of white leaf spot), *Albugo candida* (white rust), *Alternaria brassicae* (Alternaria blight), *Plasmodiophora brassicae* (club root), *Leptosphaeria maculans* (black leg or stem canker), and *Sclerotinia sclerotiorum* (causative agent of Sclerotinia stem rot) of Brassica oilseed and vegetable crops, adversely affect crop development and yield (Murray and Brennan 2012).

Evaluation of germplasm needs collaboration between multiple disciplines, i.e., biochemistry, breeding, pathology, entomology, physiology, etc. Globally, approximately, 39,532 accessions of different *Brassica* spp. and its relatives have been conserved which include 3127 in USDA gene bank (https://www.ars-grin.gov/), 11,119 in Indian National Gene Bank, ICAR-National Bureau of Plant Genetic Resources (http://www.nbpgr.ernet.in/Research_Projects/Base_Collection_in_NGB.aspx), and 1387 in Chinese gene bank (https://www.genesys-pgr.org/a/v2QAYWY43R9).

The characterization of the Brassica germplasm conserved at ICAR-NBPGR, National Gene Bank is carried out for genetic diversity status based on agronomic traits. Potential distinct genotypes are also identified by estimation of the magnitude of genetic variability existing for different characters (Table 5.2).

Evaluation of rapeseed–mustard genotypic performance under different environments and measurement of stability parameters can be used to fulfill effective selection for the improvement of yield. Based on the morphological and biochemical phenotyping, a good number of donors have been identified for various quality traits in India (Table 5.3).

5.3 Abiotic Stress Tolerant Germplasm Identified in India

Systematic efforts to identify and utilize the trait-specific germplasm for various abiotic stresses have not been made in oilseed crops which is evident from the fluctuations in area, production, and productivity of these crops over years. During past decade, efforts have been initiated to identify trait-specific germplasm in mustard which has been summarized as under (Yadava et al. 2019).

Salt tolerant lines/varieties: CS-52, CS-54, CS-234-2-2, CS-56, RH-8814 (IC-401570), CS-58, CS-1100-1-2-3-5-1, CS-1500-1-2-2-2-1, BPR-540-6.

Drought tolerant lines/varieties: RH-406, RH-781, RH-819, RB-50, RH-725, RVM-2, Pant Rai-20, RGN-298, Aravali, Geeta, Shivani, DRM-541-44.

Heat tolerance lines/varieties: Pusa Vijay (NPJ-93), Pusa Mustard 25 (NPJ-112), Pusa Mustard 27 (EJ-17), Pusa Mustard 28 (NPJ-124), BPR-540, BPR-541-4, BPR-543-2, BPR-549-9, Pant

Table 5.2 Range of some of the important agro-morphological traits in important Brassica germplasm accessions and wild relatives

Characters	Crop species								
	Brassica juncea	*B. rapa* var. *yellow sarson*	*B. rapa* var. *toria*	*B. napus*	*B. chinensis*	*B. nigra*	*Crambe* spp.	*Eruca sativa*	*Lepidium sativum*
Plant height (cm)	69–241.8	51.4–185.2	35.4–167.3	96.26–220.1	79.2–264.9	72.67–257.6	59–83.30	43–132	60.2–259.8
Main shoot length (cm)	23.8–112.4	22.8–93.6	16.6–75	23.4–92.0	23.4–79	13.30–71.35	30.6–43.3	14.9–77.6	8.5–65.8
Siliques on main shoot	10.0–86.8	6.3–78.3	13.3 -79.6	12.67–96.67	10.00–67.8	8.00–74.00	29.8–70	9–43.4	4.8–62.6
Siliqua length (cm)	2.4–6.5	2.4–7.9	2.4–7.7	2.50–7.60	2.46–5.5	0.50–5.30	0.39–1.34	0.75–3.8	0.20–4.30
Seed per siliqua	4.3–25.8	8.6–53.5	10.8–25.6	8.02–33.0	6.20–21.7	3.00–19.70	7.10–14.2	5.6–23.8	2–5
Seed weight/plant	1.22–205.2	1.8–112	0.84–124.2	0–129	0.4–37.6	1.4–129.52	2–17	3.6–40	5.12–40
Days to maturity	112–132	110–171	61–130	112–193	122–18	111–190	133–168	124–160	123–152.5
Oil content (%)	24.1–46.11	35.1–47.0	35.7–45.6	27.85–43.13	29.34–45.22	20.34–39.39	33.83–35.8	28.24–38.39	23.68–28.26

Table 5.3 Different donors for quality traits in *Brassica juncea* identified in India

Quality trait	Donor
Low glucosinolate (<30 μmol/g defatted seed meal)	NUDH-YJ-1, NUDH-YJ-2
	HNS-99(0E)3, NUDB-9
Low erucic acid (<2%)	LES-17-1, Pusa Karishma (LES-39)[a], LES-1-27 (Pusa Mustard 21)[a], LET-17 (Pusa Mustard 22)[a], LET-18 (Pusa Mustard 24)[a], Pusa Mustard 29, Pusa Mustard 30, YSRL-9-18-23, TERI (0E) M21 (IC-296684)[b], ELM-079[a], ELM-123, PRQ-2005-1 (IC-546946)[b], Pusa Mustard 32[a]
Low erucic acid (<2%) and low glucosinolate (<30 μmol/g defatted seed meal)	BJ-1058, NUDH-YJ-5 (IC-296507)[b], Heera (IC-296501)[b], Pusa Double Zero Mustard 31[a], RLC 3[a], Pusa Double Zero Mustard 33[a]
High oleic and linoleic acid	PM 30[a], TERI GZ-05 (IC-404233)[b]
High oil content (~45%)	NDYR-8 (IC-296689)[b], NDYR-10 (IC-296689)[b]

[a]Released varieties
[b]Registered germplasm

Rai-18, RH-406, RGN-229, RGN-236, RGN-298.

Frost tolerant varieties: RH-819, RGN-48.
In addition, 35 genetic stocks of different *B. juncea* spps. with specific traits for different biotic and abiotic stresses and quality have been registered with the ICAR-NBPGR for use by the Brassica breeding programs in the country (Table 5.4).

5.4 Molecular Techniques for Studying Genetic Diversity

Major steps are taken to conserve genetic resources by establishing gene banks, which are now gaining attention the world over. But for efficient conservation practices, understanding of population structure and genetic diversity polymorphism is important. Studies on genetic diversity and population structure are important. Many methods are available for diversity identification in germplasm and breeding lines. These methods are highly reliant on morphological data, pedigree data, biochemical data, and DNA marker-based data (Mohammadi and Prasanna 2003). Various markers based on morphological, biochemical, and cytological data were successfully used (Raney et al. 1995; Rimmer et al. 1995). Different techniques can be used to detect polymorphisms at the DNA level including non-PCR-based approaches, e.g., restriction fragment length polymorphism (RFLP), the first technology developed for polymorphism detection at the sequence level. Specific probe/enzyme combinations result in highly reproducible patterns for an individual. However, mutation can cause variation in the restriction patterns between individuals can arise when mutations in the DNA sequence result in changed restriction sites. RFLP analysis was used earlier in the development of genetic maps and has been effectively used in the assessment of genetic diversity (Deu et al. 1994). Thermal cycler and polymerase chain reaction led to the development of a number of techniques for the screening of genetic diversity. The techniques are based on the use of a single random primer, in a PCR reaction on genomic DNA and result in the amplification of several distinct DNA fragments. A number of closely related techniques based on this principle were developed simultaneously and are collectively referred to as multiple arbitrary amplicon profiling (MAAP). Randomly amplified polymorphic DNA (RAPD) analysis became the most extensively used technique, where the primers are usually 10 or 20-mers. The amplification products are separated on agarose gels in the presence of ethidium bromide and visualized under ultraviolet light (Williams et al. 1990). Arbitrary primed PCR (AP-PCR) (Welsh and McClelland 1991) and DNA amplification

Table 5.4 Trait-specific genetic stocks/donors of *Brassica juncea* registered with ICAR-NBPGR, New Delhi, India

Trait	Accession
Restorer gene for cytoplasmic male sterile (CMS) trachy with recombinant mitochondrial genome	IC296682, IC296683
Cytolplasmic male steile (CMS) lines	IC296701, IC296702, IC296703, IC296704, IC296705, IC296706, IC296707
Yellow seeded with high oil content	IC296689, IC296690, IC296829, IC443623
Early maturity (117 d), zero erucic acid, and yellow seed color	IC296684
Low glucosinolate content ($\sim$17 µmol/g seed), oil with low erucic acid (0.12%)	IC296501, IC296507
Resistant to salinity	IC401570
Tolerant to blight (*Alternaria brassicae* and *A. brassicola*)	IC546946
High water use efficiency, thermo-toleranace at the terminal stage, salinity tolerance at the juvenile stage	IC0583386
Tolerance to salinity at 10 ds/m at seedling stage	IC0584669
Only for thermo-tolerance at juvenile stage	IC0589778
Low glucosinolate content in seed, high oil content (45%), resistance to white rust, yellow seed coat color	IC0595268
Salinity tolerance at juvenile stage, thermo-tolerance at juvenile stage	IC0593927, IC595525
Salinity tolerant up to Eciw 12 dS/m. High level of alkalinity tolerant up to pH 9.4. High seed yield (20.9 q/ha) and oil content (39%) under salt stress condition	IC0511389
Drought tolerance (high water use efficiency under rainfed conditions)	IC0598624
Moricandia system-based CMS line. Resistant to white rust disease	IC0622804
High tolerance to salinity (ECe 12 dS/m) and alkalinity (pH 9.4)	IC0624502
White rust resistant	IC0598622, IC0598623
Tetralocular silique	IC296741
Bold seed (1000 seed weight:6.10 g)	IC573439
White petal color	IC0593926
Early maturing	IC296827, IC296828

fingerprinting (DAF) (Caetano-Annollés et al. 1991a, b) are two other techniques that branched out of PCR but differ from RAPDs in the stringency conditions, primer length, and detection of the fragments. In all cases, polymorphisms are detected as the presence/absence of bands and result from sequence differences in one or both of the primer binding sites. The opposite approach to arbitrary primed PCR is to design primers to target specific regions of the genome. The targeted amplified product can be compared on a gel to the corresponding product from another individual, but the resolution achievable will only detect differences in fragment size resulting from any base-pair changes. In order to resolve all the possible sequence differences, it is necessary to sequence the entire fragment, either manually or using an automated DNA sequencer. Once this has been done, sequences from different individuals can be aligned, any differences detected, and the data entered and analyzed in statistical packages.

High-throughput sequencing has helped to create dense genetic linkage maps for diverse agricultural crops (Bus et al. 2012; Delourme et al. 2013). In addition to the complete resolution, these high-density maps help identify molecular markers linked with trait(s) of interest. Genome-wide association study (GWAS) facilitates the discovery of chromosomal rearrangements as a consequence of genome evolution. GWAS also helps in the identification of novel alleles for genetic advance within and among crop species (Parkin 2011; Sharma et al. 2014). A number of genetic linkage maps have been generated, largely focusing on commercially important Brassica species.

Recently, genotyping-by-sequencing (GBS), genetic diversity panels are being utilized for genome-wide association studies. A better potent approach for studying plant species is associative transcriptomics (AT), because it utilizes transcribed sequences. RNA-sequencing is used to measure the transcript abundance, and to identify gene sequence variants (including hemi-single nucleotide polymorphisms (SNPs), which are unique to polyploid species) (Harper et al. 2012).

5.4.1 Analysis of Population Structure

Population constitution can be predicted by various algorithms, e.g., Bayesian Markov chain Monte Carlo model (MCMC). STRUCTURE software is used to calculate the number of sub-population present within a population (Pritchard et al. 2000). Principal component analysis (PCA) is used for the clustering of the groups in a population. Single-nucleotide polymorphism (SNPs) is used to discover transition or transversion mutations taking place during natural selection. The amount of information obtained from SNP marker system may be highly informative (PIC > 0.05), moderate informative (PIC 0.25–0.5), and less informative (PIC value < 0.25). PIC stands for polymorphism information content which indicates the genetic properties of SNPs present in a population. The various studies on population structure and genetic relationship in different *Brassica* spps. are presented in the next section.

5.4.1.1 Advances in Genetic Diversity and Relationship in *Brassica juncea*

Identification of the genetic bases for trait variation, beneficial alleles, and molecular markers for the selection of elite cultivars is a key aim in pre-breeding. The use of RAPD is of great success in many crop species for knowing genetic diversity relationships (DosSantos et al. 1994; Thormann et al. 1994; Shengwu et al. 2003; Sabina et al. 2006). Additional techniques branching out from PCR such as amplified fragment length polymorphism (AFLP) (Qi et al. 2008), inter-retrotransposon amplified polymorphism (IRAP) (Novakova et al. 2009), simple sequence repeat (SSR) (Verma et al. 2021), and inter-simple sequence simple repeat (ISSR) (Faisal et al. 2021) are being extensively used to study genetic variability among diverse *B. juncea* (Karp 1997). SSR markers-based approach used to study the effect of stem rot disease by 69 strains of *Sclerotinia sclerotiorum* on *B. juncea* revealed that all the isolates were genetically diverse (Sharma et al. 2018). Studies based on isozymes, morphological traits, the protein present in seed, and DNA markers are well known (Paterson et al. 1991; Shengwu et al. 2003). A study of forty-two genotypes of *Brassica* species for genetic diversity using RAPD, ISSR, and 5′-anchored simple sequence repeat (ASSR) markers revealed a high degree of interspecific polymorphism (Kalita et al. 2007). DNA profiles could not identify all the *B. juncea* genotypes independently however, collective DNA profile consisting of 125 markers could differentiate individual Brassica genotypes. Absence of intra-varietal heterogeneity for a set of five *B. juncea* varieties suggested utility of ASSR profiles in varietal identification and differentiation.

A study was carried out for evaluating genomic sequence-tagged microsatellite site (STMS) markers developed earlier in *Brassica napus*, *B. oleracea*, *B. rapa*, and *B. nigra* for their use in *B. juncea* and *B. carinata* (Yadava et al. 2009). 100 STMS markers used revealed high cross-transferability in *Brassica* species including *B. carinata* (91.6%) and *B. juncea* (87.5%) suggesting the prospect of utilizing these markers for genome analysis where no such markers are available. Interspecific polymorphism was much higher than the intraspecific polymorphism. The polymorphic markers are observed to play significant role in establishing the distant relationships among six different species of *Brassica* and two closely related genera.

Brassica juncea genotypes including varieties/pure lines were evaluated for genetic diversity from different agro-climatic zones of India and few exotic genotypes (Australia, Poland and China). SSR markers specific to the A and B genomes were used along with the phenotypic data on 12 yield and yield contributing traits. Based on principal component analysis (PCoA) it was found that the grouping of genotypes based on SSR marker data was more realistic than that based on the phenotypic data. Hence, SSR markers are superior to quantitative traits in discriminating *B. juncea* genotypes (Vinu et al. 2013).

SSR markers were used for genetic diversity evaluation in a germplasm collection of 59 accessions leafy mustard (*B. juncea* var. *rugosa*). 482 alleles were identified by 55 SSRs and the number of alleles varied from 1 to 8 with an average of 3.11 alleles per marker. 78.70% (122) SSRs resulted in polymorphic amplicons. PIC value ranged from 0.32 to 0.77 with a mean of 0.44 per SSR locus. Dendrogram analysis grouped all the 59 accessions into two major clusters based on both agro-morphological traits and SSR markers. However, three subpopulations/ subgroups were predicted by population STRUCTURE analysis (Sharma et al. 2020).

Resynthesized Brassica genomes, by artificially hybridizing diploid donors have poor breeding value, as a result of linkage drag associated with the parental genotypes. Gupta et al. (2015) resynthesed Brassica genomes through allotetraploid hybridization. A and B genomes in *B. napus* and *B. carinata*, respectively, were combined to resynthesize *B. juncea*. The progeny of allotetraploid hybridisation was analyzed based on the DNA polymorphism generated using 108 locus-specific SSR primers. These SSR markers helped to identify different pools of allelic diversity. The progenies with determinate plant growth habits aligned closely with the B genome of *B. carinata*. The indeterminate group showed a greater genetic affinity with the *B. juncea*.

Therefore, inherent breeding strategies combined with genetic diversity would allow a more accurate prediction of heterosis. Besides, generating new variabilities of higher economic importance, the method of developing amphiploidy is another great option for establishing heritable variation across *Brassica* species (especially digenomes) (Gupta et al. 2015).

Identifying molecular markers in linkage disequilibrium with trait-controlling loci is now a well-known tool (Atwell et al. 2010; Cockram et al. 2010; Tian et al. 2011; Zhao et al. 2011). However, these markers are getting limited in use now due to time and cost restrictions. Innovative and superior techniques are being used in this day and age. A genome-wide association study (GWAS) is defined as any study of genetic variation across the entire genome that is designed to identify genetic associations with observable traits such as yield, height, and disease. Genetic variants across the genomes of many individuals tested by GWAS to identify genotype–phenotype associations. GWAS seek to identify the single nucleotide polymorphisms (SNPs, pronounced "snips") that are common to the genome, and to determine how these polymorphisms are distributed across different populations. Among molecular markers, SNPs are the most robust markers and are ubiquitously used. SNPs have high heritability, biallelic nature, and stable distribution. Genotyping-by-sequencing (GBS), which is to identify genetic variants and quickly genotype samples, reduces genome complexity by using restriction enzymes to divide the genome into fragments whose ends are sequenced on short-read sequencing platforms. GWAS and GBS are among the latest trends for studying genetic diversity and population structure.

The genetic architecture of oil, protein, and glucosinolates was evaluated by conducting a genome-wide association study (GWAS), using 92 diverse genotypes for two levels of nitrogen (N) application. Genotyping-by-sequencing

(GBS) identified 66,835 loci, spread over 18 chromosomes revealing high genetic diversity, and phenotypic variations, and 16, 23, and 27 loci found associated with oil, protein, and glucosinolates, respectively. Functional candidate genes orthologs (21no.) related to the biosynthesis of oil, protein, and glucosinolates were also identified (Akhatar et al. 2020).

Agronomic traits are evaluated by numerous quantitative trait loci (QTLs) that can have pleiotropic effects. A multiparent advanced generation intercross (MAGIC) population of *B. juncea* was derived from eight *B. juncea* lines. The diversity in agronomic and quality traits was used for the assessment of multiple QTLs for complex agronomic traits in *B. juncea*. Twenty-two QTLs for nine seed-related traits were identified including one each for oil content, seed number per silique, and thousand-seed weight (Zhao et al. 2021).

Timely flowering and maturity are important for agronomic adaptation and productivity of Indian mustard (*B. juncea*). These traits involve complex genetic and environmental interactions. A population of 92 diverse genotypes of mustard was used to analyze under deficient (N75), normal (N100), or excess (N125) nitrogen application. It was observed that the lower nitrogen induced early flowering and maturity in most genotypes, whereas high nitrogen delayed both. GBS identified 406,888 SNPs. 282 marker-trait associations were identified.

Annotation of genomic region (50 kb within the peak SNPs) led to the prediction of 30 potential genes involved in light signal perception, floral meristem identity, flowering regulation, circadian rhythm, GA_3 pathways, plant growth, and development. More than one copy of each gene, i.e., *AP1, AGL24, FRI, FVE, GNC,* and *GID1A* related to above-mentioned processes were present. *FLC* and *CO* were predicted on chromosomes A02 and B08, respectively (Akhatar et al. 2021).

Genome-wide identification of receptor-like kinases *(RLK)* and receptor-like proteins *(RLP)* genes was done by using the RGAugury pipeline (Yang et al. 2021). Though several candidate genes were identified in the *B. juncea* genome, however, only nine *RLKs* and two *RLPs* were observed in *B. juncea*.

Phenotypic variations for pod length, number of seeds per pod, and pod shattering were assessed in large collections of *B. juncea* and *B. rapa* and *B. nigra* germplasm. Pod shatter resistance was defined as the energy required for rupturing a mature dry pod. Rupture energy (RE) was found to be in the range of 3.3–11.0 mJ in *B. juncea*. Genome = wide association studies with select sets of *B. juncea* and *B. rapa* lines predict the role of FRUITFULL, MANNASE7, and NAC secondary wall thickening promoting factor (NST2) in the genetic regulation of shatter resistance in *B. juncea* (Kaur et al. 2020). Similarly, 300 *BjuNAC* genes were identified in *B. juncea* var. tumida, of which 278 were mapped to specific chromosomes (Jiang et al. 2021).

Genotyping-by-sequencing approach (DArT-seq) was conducted to understand subgenome differentiation in *B. juncea* in comparison to other A genome-carrying Brassica species (*B. rapa* and *B. napus*). A dense genetic linkage map of *B. juncea* was constructed using an F_2 population derived from Sichuan Yellow/purple mustard and population. The map integrated 3329 DArT-seq markers on 18 linkage groups and covered 1579 cM with an average density of two markers per cM. Comparative analyses among the three A subgenomes revealed 30 potential inversion events across large segments and 20 potential translocation events. Population genetics of Brassica species carrying three A-genome suggest that *B. juncea* is distinct in terms of genomic diversity, and/or has originated from a different progenitor of *B. napus* (Zou et al. 2016).

5.5 Conclusion

Large number of accessions of *Brassica* spps. with wider genetic variability for different traits of economic importance are available globally. The need of the day is to have well-characterized trait-specific germplasm with genomic information which can be used in precise breeding programs with ensured introgression of the gene(s)/

QTL of interest. Efforts in this direction have been made, but more emphasis has been given into *B. napus* and *B. rapa*. The spp. like *B. juncea* and other oilferous *Brassica* spps have not been fully characterized and genomic resources associated with these species are scanty. Likewise, though the population structure information in different *Brassica* spp. is available, but further efforts are required for generating information for use in crop improvement programs of various *Brassica* spps.

Author Contribution DY conceptualized and contributed to reviewing the manuscript. RY and HV contributed to experimental material and preparation of the manuscript. YP, SY, and NS coordinated and revised the manuscript. SV revised the original draft manuscript. All authors read the manuscript and agreed with its content.

References

Akhatar J, Goyal A, Kaur N, Atri C, Mittal M et al (2021) Genomewide association analyses to understand genetic basis of flowering and plant height under three levels of nitrogen application in *Brassica juncea* (L.) Czern & Coss. Sci Rep 11:4278. https://doi.org/10.1038/s41598-021-83689

Akhatar J, Singh MP, Sharma A, Kaur H, Kaur N, Sharma S, Bharti B, Sardana VK, Banga SS (2020) Association mapping of seed quality traits under varying conditions of nitrogen application in *Brassica juncea* L. Czern & Coss. Front Genet 11:744. https://doi.org/10.3389/fgene.2020.00744

Atwell S, Huang YS, Vilhjalmsson BJ, Willems G, Horton M et al (2010) Genome-wide association study of 107 phenotypes in *Arabidopsis thaliana* inbred lines. Nature 465:627–631

Baillon HE (1871) Historie des plantes—cruciferes. Paris 3(188):195–248

Banga SS, Banga S (2016) Genetic diversity and germplasm patterns in *Brassica juncea*. In: Rajpal VR, Rama Rao S, Raina SN (eds) Gene pool diversity and crop improvement. Springer, Berlin, pp 163–186. https://doi.org/10.1007/978-3-319-27096-8_5

Banuelos GS, Dhillon KS, Banga SS (2013) Oilseed brassicas. In: Singh BP (ed) Biofuel crops: production, physiology and genetics. CABI, Wallingford, pp 339–368

Barrett BA, Kidwell KK (1998) AFLP-Based genetic diversity assessment among wheat cultivars from the pacific Northwest. Crop Sci 38(5):1261–1271. https://doi.org/10.2135/cropsci1998.0011183X003800050025x

Bebber DP, Ramotowski MAT, Gurr SJ (2013) Crop pests and pathogens move polewards in a warming world. Nat Clim Change 3:985–988. https://doi.org/10.1038/nclimate1990

Bus A, Hecht J, Huettel B, Reinhardt R, Stich B (2012) High-throughput polymorphism detection and genotyping in *Brassica napus* using nextgeneration RAD sequencing. BMC Genomics 13:281

Caetano-Anollés G, Bassam BJ, Gresshoff PM (1991a) DNA amplification fingerprinting using very short arbitrary oligonucleotide primers. Biotechnology 9:553–557

Caetano-Anollés G, Bassam BJ, Gresshoff PM (1991b) DNA amplification fingerprinting: a strategy for genome analysis. Plant Mol Biol Rep 9:294–307

Chauhan JS, Singh KH, Singh VV, Kumar S (2011) Hundred years of rapeseed-mustard breeding in India: accomplishments and future strategies. Indian J Agric Sci 81:1093–1109

Chen S, Nelson MN, Chèvre A-M, Jenczewski E, Li Z, Mason AS et al (2011) Trigenomic bridges for Brassica improvement. Crit Rev Plant Sci 30 (6):524–547

Chen S, Wan Z, Nelson MN, Chauhan JS, Redden R et al (2013) Evidence from genome-wide simple sequence repeat markers for a polyphyletic origin and secondary centers of genetic diversity of *Brassica juncea* in China and India. J Hered 104:416–427

Cockram J, White J, Zuluaga DL, Smith D, Comadran J et al (2010) Genome-wide association mapping to candidate polymorphism resolution in the unsequenced barley genome. Proc Natl Acad Sci USA 107:21611–21616

Cox MC, Qualset CO, Rains DW (1986) Genetic variation for nitrogen assimilation and translocation in Wheat. III. Nitrogen translocation in relation to grain yield and protein. Crop Sci 26(4):737–740. https://doi.org/10.2135/cropsci1986.0011183X002600040022x

De Candolle AP (1821) Cruciferae. Syst Nat 2:139–700

de Tournefort JP (1700) Institutiones rei herbariae editio altera, vol 1, pp 219–227

Delourme R, Falentin C, Fomeju BF, Boillot M, Lassalle G et al (2013) Highdensity SNP-based genetic map development and linkage disequilibrium assessment in *Brassica napus* L. BMC Genomics 14:120

Deu M, Gonzalez de Leon D, Glaszmann JC, Degremont I, Chanereau J, Lanaud C, Hamon P (1994) RFLP diversity in cultivated sorghum in relation to racial differentiation. Theor Appl Genet 88:838–844

DosSantos JB, Nienhuis J, Skroch P, Tivang J, Slocum M (1994) Comparison of RAPD and RFLP genetic markers in determining genetic similarity among *Brassica oleracea* L., genotypes. Theor Appl Genet 87:909–915

Edwards D, Salisbury PA, Burton WA, Hopkins CJ, Batley J (2007) In: Kole OC (ed) Indian Mustard in genome mapping and molecular breeding in plants. Springer, Berlin, pp 179–210

Faisal M, Abdel-Salam EM, Alatar AA, Qahtan AA (2021) Induction of somatic embryogenesis in *Brassica juncea* L. and analysis of regenerants using ISSR-

PCR and flow cytometer. Saudi J Biol Sci 28 (1):1147–1153. https://doi.org/10.1016/j.sjbs.2020.11.050
Ficklin DL, Novick KA (2017) Historic and projected changes in vapor pressure deficit suggest a continental-scale drying of the United States atmosphere. J Geophys Res Atmos 122:2061–2079. https://doi.org/10.1002/2016jd025855
Francisco M, Tortosa M, Martínez-Ballesta M, Velasco P, García-Viguera C et al (2017) Nutritional and phytochemical value of Brassica crops from the agri-food perspective. Ann Appl Biol 170:273–285. https://doi.org/10.1111/aab.12318
Gómez-Campo C (1980) Morphology and morphotaxonomy of the tribe Brassiceae. In: Tsunoda S, Hinata K, Gómez-Campo C (eds) Brassica crops and wild allies: biology and breeding. Japan Scientific Society Press, Tokyo, pp 3–31
Gómez-Campo C, Prakash S (1999a) Origin and domestication. In: Gómez-Campo C (ed) Develoments in plant genetics and breeding. Elsevier, Amsterdam, pp 33–58
Gómez-Campo C (1999b) Seedless and seeded beaks in the tribe Brassiceae (Cruciferae). Cruciferae Newsl 21:11–12
Gupta M, Gupta S, Kumar H, Kumar N Banga SS (2015) Population structure and breeding value of a new type of *Brassica juncea* created by combining A and B genomes from related allotetraploids. Theor Appl Genet 128: 221–234
Harper AL, Trick M, Higgins J, Fraser F, Clissold L, Wells R, Hattori C, Werner P, Bancroft I (2012) Associative transcriptomics of traits in the polyploid crop species *Brassica napus*. Nat Biotechnol 30:798–802
Hedge IC (1976) A systematic and geographical survey of the Old World Cruciferae. In: Vaughan JG, MacLeod AJ, Jones MG (eds) The biology and chemistry of the cruciferae. Academic Press, London, pp 1–45
Hemingway JS (1995) The mustard species: condiment and food ingredient use and potential as oilseed crops. In: Kimber D, McGregor DI (eds) *Brassica* oilseeds—production and utilization. CAB International, Oxon, pp 373–383
Jiang L, Sun Q, Wang Y, Chang P, Kong H, Luo C, He X (2021) Genome-wide identification and characterization of NAC genes in *Brassica juncea* var. tumida. Peer J 9:e11212. https://doi.org/10.7717/peerj.11212
Kalita MC, Mohapatra T, Dhandapani A, Yadava DK, Srinivasan K, Mukherjee AK, Sharma RP (2007) Comparative evaluation of RAPD, ISSR and Anchored-SSR markers in the assessment of genetic diversity and fingerprinting of oilseed *Brassica* genotypes. J Plant Biochem Biotechnol 16(1):41–48
Karp A (1997) Reproducibility testing of RAPD, AFLP and SSR markers in plants by a network of European laboratories. Mol Breed 3:381–390
Kaur P, Banga S, Kumar N, Gupta S, Akhtar J, Banga S (2014) Polyphyletic origin of *Brassica juncea* with *B. rapa* and *B. nigra* (Brassicaceae) participating as cytoplasm donor parents in independent hybridization events. Am J Bot 101(7):1157–1166. https://doi.org/10.3732/ajb.1400232
Kaur J, Akhatar J, Goyal A, Kaur N, Kaur S, Mittal M, Kumar N, Sharma H, Banga S, Banga SS (2020) Genome wide association mapping and candidate gene analysis for pod shatter resistance in *Brassica juncea* and its progenitor species. Mol Biol Rep 47(4):2963–2974. https://doi.org/10.1007/s11033-020-05384-9. PMID: 32219770
Linnaeus C (1753) Species plantarum, vol II. Stockholm, p 561
Mohammadi SA, Prasanna BM (2003) Analysis of genetic diversity in crop plants—salient statistical tools and considerations. Crop Sci 43(4):1235–1248. https://doi.org/10.2135/cropsci2003.1235
Murray GM, Brennan JP (2012) The current and potential costs from diseases of oilseed crops in Australia. Grains research and development corporation. Available online at: https://grdc.com.au/__data/assets/pdf_file/0021/82641/grdcreportdiseasecostoilseedspdf.pdf.pdf
Novakova A, Simackova K, Barta J, Curn V (2009) Variety identification by molecular markers based on retrotransposon analyses. Czech J Genet Plant Breed 45:1–10
Parkin I (2011) Chasing ghosts: comparative mapping in the Brassicaceae. In: Bancroft I, Schmidt R (ed) Genetics and genomics of the Brassicaceae. Springer, New York, pp 153–170
Paterson AH, Tanksley SD, Sorrells ME (1991) DNA markers in plant improvement. Adv Agron 46:39–90
Prakash S, Hinata K (1980) Taxonomy, cytogenetics and origin of crop Brassicas, a review. Opera Bot 55:1–57
Prakash S, Bhat SR, Quiros CF, Kirti PB, Chopra VL (2009a) *Brassica* and its close allies: cytogenetics and evolution. Plant Breed Rev 31:21–187
Prakash S, Bhat SR, Fu T (2009b) Wild germplasm and male sterility. In: Gupta SK (ed) Biology and breeding of crucifers. CRC Press, Taylor & Francis, USA, pp 113–127
Prantl K (1891) Cruciferae. In: Engler A, Prantl K (ed) Die natürlichen pflanzenfamilien. Wilhelm Englmann, Leipzig, pp 145–208
Pritchard JK, Stephens M, Donnelly P (2000) Inference of population structure using multilocus genotype data. Genetics 155:945–959
Qi XH, Yang JH, Zhang MF (2008) AFLP-based genetic diversity assessment among Chinese vegetable mustards (*Brassica juncea* (L.) Czern.). Genet Resour Crop Evol 55:705–711. https://doi.org/10.1007/s10722-007-9278-6
Raney JP, Rakow G, Olson T (1995) Development of zero erucic, low linolenic *Brassicajuncea* utilizing interspecific crossing. In: Proceedings of the 9th international rapeseed congress. Cambridge, UK, pp 413–415
Rimmer SRR, Scarth PBE, McVetty D, Gugel WR, Swartz GS (1995) Breeding for resistance to blackleg

(Stem canker) in Western Canada. Blackleg Newsl 5:11–15

Sabina Y, Shahidul I, Nasiruddin K, Samsul A (2006) Molecular characterization of potato germplasm by random amplified plolymorphic DNA markers. Biotechnology 5:27–31

Schulz OE (1919) Cruciferae-brassiceae. Part I: brassicinae and raphaninae. In: Engler A (ed) Das pflanzenreich. Wilhelm Engelmann, Leipzig, pp 68–70

Schulz OE (1936) Cruciferae. In: Engler A, Prantl P (eds) Die natürlichen pflanzenfamilien. Wilhelm Engelmann, Leipzig, pp 227–658

Sharma A, Li X, Lim YP (2014) Comparative genomics of Brassicaceae crops. Breed Sci 64(1):3–13

Sharma P, Samkumar A, Rao M, Singh VV, Prasad L, Mishra DC, Bhattacharya R, Gupta NC (2018) Genetic diversity studies based on morphological variability, pathogenicity and molecular phylogeny of the *Sclerotinia sclerotiorum* population from Indian mustard (*Brassica juncea*). Front Microbiol 9:1169. https://doi.org/10.3389/fmicb.2018.01169

Sharma D, Nanjundan J, Singh L, Singh SP, Parmar N, Sujith Kumar MS, Singh KH, Mishra AK, Singh R, Verma KS, Thakur AK (2020) Genetic diversity in leafy mustard (*Brassica juncea* var. *rugosa*) as revealed by agro-morphological traits and SSR markers. Physiol Mol Biol Plants 26(10):2005–2018. https://doi.org/10.1007/s12298-020-00883-2

Shengwu H, Ovesná J, Kučera L, Kučera V Vyvadilová M (2003) Evaluation of genetic diversity of *Brassica napus* germplasm from China and Europe assessed by RAPD markers. Plant Soil Environ 49(3):106–113

Smith JSC (1984) Genetic variability within U.S. hybrid maize: multivariate analysis of isozyme data. Crop Sci 24(6):1041–1046. https://doi.org/10.2135/cropsci1984.0011183X002400060009x

Thompson JA, Nelson RL (1998) Core set of primers to evaluate genetic diversity in soybean. Crop Sci 38:1356–1362

Thormann CE, Ferreira ME, Carmago LEA, Tivang JG, Osborn TC (1994) Comparison of RFLP and RAPD markers to estimate genetic relationships within and among cruciferous species. Theor Appl Genet 88:973–980

Tian F, Bradbury PJ, Brown PJ, Hung H, Sun Q et al (2011) Genome-wide association study of leaf architecture in the maize nested association mapping population. Nat Genet 43:159–162

Verma K, Tiwari S, Tripathi N (2021) Analysis of genetic diversity among *Brassica juncea* genotypes using morpho-physiological and SSR markers. Int J Curr Microbiol Appl Sci 10(1). 10.1108-1117.https://doi.org/10.20546/ijcmas.2021.1001.134

Vinu V, Singh N, Vasudev S, Yadava DK, Kumar S, Naresh S, Bhat SR, Prabhu KV (2013) Assessment of genetic diversity in *Brassica juncea* (Brassicaceae) genotypes using phenotypic differences and SSR markers. Rev Biol Trop 61(4):1919–1934. PMID: 24432543

Welsh J, McClelland M (1991) Genomic fingerprinting using arbitrary primed PCR and a matrix of pairwise combinations of primers. Nucleic Acids Res 19:5275–5279

Williams JGK, Kubelik AR, Livak KJ, Rafalski JA, Tingey SV (1990) DNA polymorphisms amplified by arbitrary primers are useful as genetic markers. Nucleic Acids Res 18:6531–6535

Yadava DK, ParidaSK DVK, Varshney A, Ghazi IA, Sujata V, Mohapatra T (2009) Cross-transferability and polymorphic potential of genomic STMS markers of *Brassica* species. J Plant Biochem Biotechnol 18 (1):29–36

Yadava DK, Vasudev S, Naveen S, Mohapatra T, Prabhu KV (2012) Breeding major oil crops: present status and future research needs. Gupta SK (ed) Technological innovations in major world oil crops, pp 17–51

Yadava DK, Choudhury PR, Yadav R (2019) Genetic resources of oilseed crops adaptable to climate change: issues and way forward. In: Proceedings and recommendations of the satellite symposium on 'Dryland agrobiodiversity for adaptation to climate change. Rajasthan, India

Yang H, Bayer PE, Tirnaz S, Edwards D, Batley J (2021) Genome-wide identification and evolution of receptor-like kinases (RLKs) and receptor like proteins (RLPs) in *Brassica juncea*. Biology 10:17

Zhao H, Yan W, Kunjiang Yu, Wang T, Khattak AN, Tian E (2021) QTL identification for nine seed-related traits in *Brassica juncea* using a multiparent advanced generation intercross (MAGIC) population. Czech J Genet Plant Breed 57(1)9–18. https://doi.org/10.17221/73/2020-CJGPB

Zhao K, Tung CW, Eizenga GC, Wright MH, Ali ML et al (2011) Genome-wide association mapping reveals a rich genetic architecture of complex traits in *Oryza sativa*. Nat Commun 2:467

Zou J, Zhu J, Huang S, Tian E, Xiao Y, Fu D et al (2010) Broadening the avenue of intersubgenomic heterosis in oilseed Brassica. Theor Appl Genet 120(2):283

Zou J, Fu D, Gong H, Qian W, Xia W, Pires JC et al (2011) De novo genetic variation associated with retrotransposon activation, genomic rearrangements and trait variation in a recombinant inbred line population of *Brassica napus* derived from interspecific hybridization with *Brassica rapa*. Plant J 68 (2):212–224

Zou J, Hu D, Peifa L, Raman H, Liu Z et al (2016) Colinearity and divergence of the A subgenome of *Brassica juncea* compared with other Brassica species carrying different A subgenomes. BMC Genomics 17:18 ff. https://doi.org/10.1186/s12864-015-2343-1ff. ffhal-02634780

Classical Genetics and Traditional Breeding in *Brassica juncea*

6

Gurpreet Kaur, V. V. Singh, K. H. Singh, Priyamedha, Indu Rialch, Mehak Gupta, and S. S. Banga

Abstract

Brassica juncea, known as the Indian mustard, was initially confined to Indian sub-continent. It is becoming a crop choice for drier ecologies in Australia, Canada and northern United States. The increase in seed yield, oil content, resistance to biotic and abiotic stress and canola quality are the major breeding objectives. Both conventional and induced variations have been used to meet stated objectives. However, response to directed selection for productivity-related traits has been low due to a narrow genetic base of germplasm used for varietal development. This is inspite of the availability of a large number of germplasm accessions in many gene banks. Genetic diversity analysis has divided *B. juncea* germplasm into east European and Indian gene pools. Efforts have also been made to broaden the base of genetic diversity through the resynthesis of *B. juncea* and alien introgressions. Hybrids based on cytoplasmic male sterility have been developed, but there is a need to further enhance the level of yield heterosis in these hybrids. Integration of biotechnological and genomics tools like doubled haploid, soma clonal variation, molecular markers, genetic engineering and gene-editing is expected to accelerate the pace of varietal development and sustainability of productivity by incorporating genes for disease and insect resistance in agronomically superior cultivars.

6.1 Introduction

The crop Brassicas belongs to the family Brassicaceae earlier known as Cruciferae. It includes over 300 genera and 3,709 species (Warwick et al. 2006). This family also includes six crop Brassica species, which are cultivated throughout the world as oilseed, vegetables, condiments, and fodder crops (Dixon 2007). Of these, *Brassica rapa* (AA genome, n = 10), *Brassica nigra* (BB, n = 8) and *Brassica oleracea* (CC, n = 9) are diploids, whilst *Brassica juncea* (AABB, n = 18), *Brassica napus* (AACC, n = 19), and *Brassica carinata* (BBCC, n = 17) are allotetraploids. *B. juncea* and *B. napus* are major

G. Kaur (✉) · I. Rialch · M. Gupta · S. S. Banga
Department of Plant Breeding and Genetics, Punjab Agricultural University, Ludhiana 141004, India
e-mail: gurpreetkaur@pau.edu

I. Rialch
e-mail: indudehal@pau.edu

M. Gupta
e-mail: mehakgupta@pau.edu

S. S. Banga
e-mail: nppbg@pau.edu

V. V. Singh · K. H. Singh · Priyamedha
ICAR-Directorate of Rapeseed-Mustard Research, Bharapur, Rajasthan 321303, India
e-mail: vijay.singh4@icar.gov.in

C. Kole and T. Mohapatra (eds.), *The Brassica juncea Genome*, Compendium of Plant Genomes,
https://doi.org/10.1007/978-3-030-91507-0_6

edible and industrial oilseed crops and both winter and spring forms are cultivated in different regions of the world. Winter type of *B. napus* predominates in Europe, China, and eastern United States, whilst spring forms are grown in Canada and Australia. In China, winter type of *B. juncea* is grown but it is now being replaced by high yielding *B. napus*. Spring types of *B. juncea* are mainly confined to the Indian sub-continent. More recently, it has also become an option for drier ecologies of Australia, Canada and in northern United States. Globally, oilseed Brassicas ranks third after soybean and palm. During 2019–20, the area, production and yield of rapeseed and mustard in the world were 36.11 million hectares (m ha), 71.81 million tonnes and 1988 kg/ha, respectively (USDA 2021). India ranks fourth in the edible oil sector after the US, China, and Brazil and accounts for 19.8% and 9.8% of the total acreage and production, respectively (USDA 2021).

Brassica juncea known as Indian mustard in the trade is the principal winter oilseed crop of India. It accounted for about 75–80% of the 6.85 m ha under rapeseed-mustard crops during the 2019–20 crop season. Average productivity for rapeseed-mustard crop in India is 1331 kg/ha (SOPA 2021), which is lower than the international productivity levels recorded in *B. napus*. Crop breeding programme is now focussed on broadening the genetic base of this crop by accumulating superior alleles for productivity, diseases and insect-pest resistance from primary and secondary gene pools. It will not only help increase productivity but also impart stability and wider adaptability to the crop. This chapter describes the genetic diversity, attempts at broadening the genetic base by exploiting related species, trait variation and genetics, conventional and heterosis breeding to develop varieties and hybrids.

6.2 Geographic and Genetic Origins

Brassica juncea (AABB) is an ancient polyploid originated from independent multiple hybridizations between diploid *B. rapa* (AA) and *B. nigra* (BB) followed by polyploidy (Prakash and Hinata 1980; Kaur et al. 2014). Multiple hybridizations may have occurred in the sympatric distribution area of middle East and its adjoining area as well as in eastern Europe and China (Burkill 1930; Olsson 1960; Vaughan et al. 1963; Axelsson et al. 2000; Spect and Diederichsen 2001; Prakash et al. 2009). Based on diversity, Vavilov (1949) proposed Afghanistan and neighbouring territories as the primary origin centre, with Asian Minor, central or western China, and eastern India as secondary origin centres. Separate origins have been hypothesized for the Chinese foliage and Indian oleiferous forms of *B. juncea* (Vaughan et al. 1963; Vaughan 1977). Duhoon and Koppar (1998) have opined an Indian origin for this crop. Indian and oriental mustard were reported to be close to progenitors *B. rapa* and *B. nigra,* respectively (Vaughan et al. 1963; Vaughan 1977; Prakash et al. 2009). Taxonomic, morphological and molecular investigations have confirmed multiple independent hybridization events with cytoplasm from *B. rapa* (Uchimiya and Wildman 1978; Banga et al. 1983; Erickson et al. 1983; Palmer et al. 1983; Warwick and Black 1991; Pradhan et al. 1992). Sequential variation of internal transcribed spacer regions of ribosomal DNA has allowed the grouping of vegetable *B. juncea* of Chinese origin along with its parental species into two clades. One clade contains vegetables forms with homology with *B. nigra*, whereas as other clade contained only oilseed forms of *B. juncea*. These clearly contradicted the theory of uni-directional hybridization regarding origin of *B. juncea* (Qi et al. 2007). Polyphyletic origin of *B. juncea* with cytoplasms from both diploid species, *B. rapa* and *B. nigra* have also been confirmed by SSR markers specific to cytoplasmic organelle. Two major plasmotypes, with one group represented *B. nigra* lineage, whereas and other group included species/genotypes belonging to *B. rapa*/*B. oleracea* lineage were proposed (Kaur et al. 2014). Most landraces of east European origin shared the cytoplasmic homology with *B. nigra* indicated that *B. nigra* may have contributed cytoplasm of east European *B. juncea.* Chen et al. (2013) also reported the two groups based on SSR markers, with group I mainly comprised east European types with few genotypes from India, whereas genotypes of Indian, Australian and Chinese origin clustered in other groups.

6.3 Genetic Diversity

Studies on the genetic diversity have been conducted at the cytological, molecular, biochemical and morphological levels in *B. juncea*. Extensive literature available on genetic diversity of Brassica species is based on morphological traits but most of these studies lacked a global diverse collection and were conducted on small and indigenous collections (Shalini et al. 2000; Ghosh and Gulati 2002; Goswami and Behl 2006; Singh and Srivastava, 2006; Kumar et al. 2007; Lodhi et al. 2013; Shekhawat et al. 2014; Singh et al. 2018). Diversity analysis in a germplasm set comprising resynthesized *B. juncea* (Prakash 1973) and genotypes from India and east European origins indicated that the maximum divergence existed between Indian and east European genotypes (Pradhan et al. 1993). East European genotypes possessed variation for yellow seed colour, resistance to white rust, seed quality traits, oil content along with morphological traits like branching and pod number (Pradhan et al. 2003). Traits like early flowering, dwarf height, long siliquae, bold seed and resistance to pod shatter can be exploited from Indian germplasm pool (Ramchiary et al. 2007b). Singh et al. (2014) reported significant variability for all agronomic traits. However, leaf and phenological traits contributed maximum towards variability. A broad range of dispersion for maturity duration may be attributed to the adoption of Indian mustard under diverse environmental conditions prevailing in distinct ecogeographies, broadly representing sub-temperate, sub-tropical and semi-arid climatic conditions. Bulk of the crop is cultivated under sub-tropical and semi-arid regions of India. Varieties bred at centres located in these regions tended to be grouped together representing medium plant height, maturity and broad leaf characteristics. Genotypes of tropical conditions possess short plant stature as well as maturity. Traits like oil content show limited variation and necessitates utilization of diverse sources for expanding the range of variation. Similarly, diversity in maturity duration was also observed, whilst assessing the 60 germplasm lines representing 27 Indian, 8 Chinese and 25 Australian genotypes using morpho-physiological and quality characters (Meena et al. 2014a). Many Chinese accessions were late in maturity and failed to set seed under Indian conditions. On the basis of quality characteristics, Indian genotypes in general possessed higher erucic acid (31.5–52.5%) and low oleic acid (6.0–18.1%) as compared Australian genotypes with low erucic acid (<2%) and high oleic acid (29.7–58.4%). Similarly, glucosinolates (GLSs) content in Indian germplasm was higher (75–136.6 μ moles/g defatted seed meal) than the germplasm from Australia (up to 30 μ moles/g defatted seed meal). Phylogenetic analysis with molecular markers further substantiated the presence of the two gene pools (Jain et al. 1994; Srivastava et al. 2001; Burton et al. 2004). Polymorphism generated by genome specific SSR markers also revealed a broad genetic diversity, with oilseeds forms being grouped into two clusters. Central and western Indian genotypes clustered in group I along with the germplasm from eastern China, whereas group II comprised genotypes from Europe, Australia, northern and eastern India along with genotypes from central and western China (Chen et al. 2013). In another study, population structure analysis based on molecular markers grouped a diverse germplasm set of 122 geographically diverse accessions into two clusters (Kaur et al. 2014). A considerable number of Indian, Chinese, and Australian genotypes formed a group distinct from some genotypes Indian and East Europe origin. Few studies included both morphological as well as molecular datasets on geographically diverse accessions. Majority divided the *B. juncea* germplasm into two major clusters. 59 accessions of leafy mustard (*B. juncea* var. rugosa) could be divided into two groups on the basis of genetic distance generated with the help of morphological and SSR markers, whereas population structure analysis revealed three sub populations (Sharma et al. 2020). Many studies have been attempted to find correlation between genetic and geographic diversities (Gupta et al. 1991) as well as genetic diversity and heterosis (Gupta et al. 1991; Krishna and Ghose 1992). Sodhi et al. (2006) found maximum heterosis between genotypes belonging to Indian and East European pools. In some cases, intermediate diversity was found to be important for heterosis (Ghosh and Gulati 2002), however, lack of association between genetic distance and heterosis was also reported (Thakur and Zarger 1989; Ali et al.1995; Acharya and Swain 2003). Diverse genotypes were also observed from the same geographical region (Verma et al. 2000).

6.4 Broadening the Genetic Base

Continual exploration and effective utilization of novel genetic variation by crop breeders is imperative to meet the immediate and future challenges of food and nutritional security. Genetic manipulation of conventional gene pool by introducing variability from progenitor species following resynthesis approach or introgressing variation from the relative species are the alternate options when traits of interest do not exist in a crop species (Prakash et al. 2009; Meyer et al. 2012; Zhan et al. 2017; Gupta and Banga 2020).

6.4.1 Resynthesized *B. juncea*

Considerable efforts have been made in the past to reconstruct Brassica allopolyploids by combining their constituent genomes from related species (Prakash et al. 2009; Pires and Gaeta 2011). Resynthesized *B. juncea* was first generated with an aim to experimentally verify its origin from interspecific crossing between then putative genome donor species namely *B. rapa* and *B. nigra* (Nagaharu 1935). Subsequently, the prime objective shifted to meet the necessity of augmenting the available genetic variation or reintroduce some specific traits from diploid progenitor species. *B. juncea* allotetraploids were repeatedly resynthesized through reciprocal hand pollinations between the selected genetic variants of the diploid progenitors (Frandsen 1943; Ramanujam and Srinivasachar 1943; Olsson 1960; Prakash 1973; Song et al. 1993, 1995; Axelsson et al. 2000; Srivastava et al. 2001, 2004). However, resynthesized genotypes using *B. rapa* as female parent were easier to develop (Song et al. 1993). Most resynthesized genotypes displayed meiotic instability, reduced pollen viability and seed set in early generations, which considerably improved upon selection in advanced generations (Olsson 1960). Hassan and Rahman (2018) developed clubroot resistant *B. juncea* by resynthesizing it using a resistant genotype of *B. rapa* in crossing with two susceptible lines of *B. nigra.* Concurrently, the purpose was also to elicit novel genetic and phenotypic variability ensuing from intergenomic recombination between the chromosomes of constituent parental species and other polyploidy associated consequences (Song et al. 1995). Undoubtedly, resynthesized genotypes were recognized as good diversity conduits as revealed by allozyme and molecular marker studies to genetically enrich crop Brassica (Song et al. 1993; Srivastava et al. 2004). Bansal et al. (2012) reported high heterosis for seed yield in crosses between resynthesized lines and crop cultivars of *B. juncea*. But, despite enormous efforts, resynthesized genotypes could not be established for direct as release as crop cultivars or parents of mustard hybrids. Apparent reasons for low breeding value of resynthesized genotypes could be the associated genetic or phenotypic instabilities and linkage drag (Bansal et al. 2009; Banga and Kaur 2009). More innovative sort of genetic resources named derived *B. juncea* were developed by combining A genome of *B. napus* with B genome of *B. carinata* (Gupta et al. 2015). Authors confirmed the superior agronomic performance and excellent heterosis inducing potential of the developed germplasm. They reasoned that two of the three diploid progenitor species (i.e. *B. nigra* and *B. oleracea*) employed in alloploids resynthesis were never domesticated or bred as oilseed crops. Whereas, three Brassica allopolyploids have been specially evolved or bred as oilseed crops in different ecological habitats. Additionally, the approach tended to generate a type of genetic variation distinct from what was available in primary gene pool or in resynthesized germplasm utilizing diploid progenitors. A new class of *B. juncea* showing determinacy was also identified from the produced variation. Gene for determinacy (*Sdt1*-an orthologue of Arabidopsis terminal flower 1 (*TFL1*) was mapped and cloned by the same research group (Kaur and Banga 2015). This gene was also transferred to various diverse male sterile and restorer genotypes of *B. juncea* for development of productive determinate hybrids. Currently, some of the determinate mustard hybrids are under evaluation phase.

6.4.2 Alien Introgressions

Partial introgression of alien genome components from related wild species is another approach to augment genetic diversity of primary gene pool for desired traits (Kaneko and Bang 2014; Li and Wang 2017). Brassica coenospecies (an assemblage of species from 14 genera-*Diplotaxis, Brassica, Eruca, Erucastrum, Hirschfeldia, Coincya, Sinapis, Sinapidendron, Trachystoma, Rytidocarpus, Moricandia, Pseuderucaria, Raphanus, Enarthocarpus*) along with other species of the Brassicaceae family constitute a goldmine of genetic diversity and trait novelty that can be explored to enrich the germplasm base of cultivated Brassicas (Gomez-Campo 1999a, b; Warwick et al. 2000; Prakash et al. 2009). These are extremely valuable genetic resources for diverse nuclear and cytoplasmic encoded traits like resistance to various insect-pests, diseases, salinity, water and heat stresses (Prakash et al. 2009). Stable and diverse alloplasmic CMS lines could be synthesized by substituting the cytoplasm of crop species with the wild species for hybrid seed production in rapeseed-mustard crops (Banga et al. 2015). There are two important prerequisites that must be met for the effective exploitation of wild relatives for crop upgradation—(1) synthesis of interspecific hybrids after overcoming pre- and post-fertilization barriers and (2) transfer of alien chromatin from wild to crop genome via homoeologous recombination or chance chromosome breakages and fusions (Gupta and Banga 2020). One can initially attempt some of the simple known approaches such as the including a broader sample of germplasm variation to use as parents of the wide hybrids, reciprocal crossing, bud and stump pollination, use of mentor pollen and application of various growth regulators to improve hybrid seed set (Batra et al. 1990; Li et al. 1995; Ahuja et al. 2003; Tonosaki et al. 2013). If post fertilization obstacles persist, then tissue culture interventions such as embryo rescue and somatic hybridization can be employed to overcome species barriers (Harberd and McArthur 1980; Zenkteler 1990; Garg et al. 2007; Kaneko et al. 2009; Mohanty et al. 2009). In Brassicas, a broad range of interspecific, intergeneric and even intertribal hybrids have been obtained using somatic fusion techniques (Lelivelt and Krens 1992; Lelivelt et al. 1993; Narasimhulu et al. 1994; Skarzhinskaya et al.1996; Hu et al. 2002). Achieving integration of alien genome components from wild relatives into stable karyotypes of cultivated Brassica species is the next major constraint. Many times, only monosomic or disomic addition lines with desired traits were achieved after repeated backcrossing, no introgression lines could be recovered. Very few cases are found in literature where introgressions for desired traits have been demonstrated in the recipient genome background of crop species. For example, euploid *B. juncea* carrying genomic fragments from two wild species *B. fruticulosa* and *E. cardaminoides* has been established showing varying levels of resistance to Sclerotinia stem rot (Rana et al. 2017, 2019; Atri et al. 2019). Introgression genotypes of *B. juncea* with alien segments from *B. fruticulosa* showed resistance to aphid infestation (Kumar et al. 2011; Atri et al. 2012). It may also be possible to manipulate recently identified pairing control locus (*BnaPh1*) in Brassica as is done for *Ph1* locus in wheat to promote homoeologous recombination between alien and crop chromosomes (Higgins et al. 2021). Severe linkage drag (negative effects associated with desirable targeted genes) is the next problem that prevents practical utilization of introgressed germplasm for crop improvement. Linkage drag is primarily caused by physical linkages of unwanted chromatin with the introgressed chromosome fragments bearing genes controlling the trait of interest. It is difficult to get rid of this type of linkage drag if recombination is suppressed between crop and wild genomes in the introgression regions. However, mutagenesis can be deployed to overcome linkage drag. Of the several established *B. juncea–B. fruticulosa* introgression lines, three ILs were highly resistant to aphid infestation. However, these exhibited poor agronomic potential, possibly due to uncontrolled introductions of unwanted genes along with desirable aphid resistance genes. Cytogenetic and in silico studies also revealed large segment substitutions in these introgression lines. These alien fragment substitutions could be physically disrupted by irradiating seeds of introgression lines with a heavy dose of gamma irradiation (200kR) (Agrawal et al. 2021). Another potential approach that can be followed to obviate the linkage drag is the marker assisted introgression of targeted fragments from the wild species. It will be advantageous to genetically map and clone the targeted genes in the wild species itself. Prior knowledge on the genetic basis of desirable trait to be introgressed may increase the chances of recovering the introgression lines with least possible linkage drag.

6.5 Breeding Objectives

Major objectives of mustard improvement programme are:

6.5.1 High Seed and Oil Yield

High seed and oil yield are the key breeding objectives, which have been sought to be achieved by broad basing the hybridization programme, resynthesis of *B. juncea* and development of CMS based hybrids.

6.5.2 Early Maturity

Early maturity is an important objective for multiple cropping sequences. Early maturing toria, yellow sarson, brown sarson and mustard are required for north eastern and eastern parts of the country depending upon agro-ecology and crop sequence.

6.5.3 Oil and Meal Quality

Quality of oil is defined by fatty acid composition, whereas quality of seed meal is determined by higher quality and quantity of seed protein with minimum amount of anti-nutritional components. Mustard oil is characterized by high amount of erucic acid (50%) in its oil. Consumption of higher amount of mustard oil having high erucic acid (>2%) is arguably associated with myocardial infractions, lipidosis, etc. (Ackman et al. 1977). Therefore, breeding for low erucic acid conforming to international norms is one of the major objectives. The presence of high GLSs content in seed meal is growth inhibiting factor in animals and poultry when used as feed (Tripathi and Mishra 2007). As per international accepted quality norms, it is desirable to reduce erucic acid content below 2% and glucosinolate content in seed meal < 30 micro moles/g defatted seed meal.

6.5.4 Resistance to Biotic and Abiotic Stresses

Current mustard cultivars are susceptible to many biotic (white rust, stem rot, *Alternaria* blight, powdery and downy mildew) and abiotic (drought, heat, salinity) stresses. *Alternaria* blight is one of the most important diseases of this crop. The yield losses due to *Alternaria* infections varied between 11 and 100% depending upon stage of crop, environmental conditions and management strategies (Kolte 1985; Tewari and Conn 1993; Verma and Saharan 1994; Seidle et al. 1995; Meah et al.

2002; Prasad et al. 2003; Mondal et al. 2007). Shrivelling of siliquae and seeds, reduction of seed size; oil content are some of the major manifestations of *Alternaria* infections. No source of resistance to this disease is available in cultivated germplasm. White rust, caused by *Albugo candida,* is prevalent in many parts of the world, including India. Many Albugo pathotypes have been reported but situation regarding the domineering races is still unclear (Dev et al. 2020). Host resistance is available in Brassica germplasm for white rust. Stem rot is a serious threat, particularly in mono-cropping areas. A wide host range of a pathogen complicates breeding for resistant cultivars. Mustard aphid (*Lipaphis erysimi*) is a major pest and has been reported to cause 10–90% yield losses depending upon phenological stage of infestation and environmental conditions (Rohilla et al. 2004; Rana 2005; Ahuja et al. 2009; Kular and Kumar 2011; Deka et al. 2017; Kumar 2017).

Drought, heat, frost and salinity are important abiotic stresses, which affect rapeseed-mustard production. Rain fed area suffers from water stress at one or more phenological stages depending upon moisture availability in the root zone. Yield reductions due to water stress may range from 17 to 94% in this crop (Chauhan et al. 2007b). High temperature stresses at terminal and crop establishment stages also cause significant yield losses. Salinity reduces crop yields of oilseed Brassica through reduced seedling emergence and yield losses due to impaired physiological or metabolic processes.

6.6 Germplasm Variation and Trait Genetics

6.6.1 Yield and Its Components

Yield is a complex trait with many contributing traits. These include: number of siliqua bearing branches, length of siliqua, seeds per siliqua, seed size, etc. These contribute directly or indirectly towards overall yield. Hence, these traits can be used as indirect measures of selection for yield (Smith 1936; Hazel 1943). Productivity can be improved by substituting bottle-neck loci with more efficient alleles (Banga and Banga 2009). Genetics of yield contributing traits are summarized in Table 6.1.

6.6.2 Biotic Stresses

6.6.2.1 White Rust

Genotypic (host) variability exists for white rust resistance. Many donor sources, BIOYSR, NRCDR 515, have been registered for white rust resistance (Chauhan et al. 2011b). Resistance to white rust has also been observed in wild Brassicaceae species including: *Brassica fruticulosa, Crambe abyssinica* and *Thlaspi arvense. Brassica tournefortii* and some species of genus *Diplotaxis* and *Sinapis* are moderately resistant (Saharan et al. 1988; Dang et al. 2000). Monogenic dominant resistance was reported in *B. nigra*, *B. rapa*, *B. juncea* and *B. carinata* (Singh et al. 2021). It was also possible to map locus AcB1-A4.1 to a linkage group A4, and locus AcB1-A5.1 to linkage group A5 in *B. juncea*. In both instances, closely linked flanking markers were identified based on synteny between *Arabidopsis* and *B. juncea* (Panjabi-Massand et al. 2010).

6.6.2.2 Stem Rot

Stem rot (*S. sclerotiorum*) is emerging as a devastating disease causing severe losses in *B. juncea*. No variability for stem rot has been reported in the cultivated germplasm of Brassica. The maximum level of field tolerance in rapeseed was reported in Chinese *B. napus* cultivars Zhongyou 821 and Zhongshuang No. 9 (Li et al. 1999; Wang et al. 2004). Additionally, partial resistance has been identified in some of the genotypes, ZY004 and 06-6-3792 of *B. napus* (Zhao et al. 2006; Gyawali et al. 2016; Wu et al. 2016) and *B. juncea* from China, Australia (Li et al. 2009) and India (Singh et al. 2008). Navabi et al. (2010a, b) reported a usable level of resistance in *B. nigra* and *B. carinata* lines. Resistance to Sclerotinia in oilseed Brassicas has been reported to be governed by a complex interplay of multiple minor genes (Disi et al.

Table 6.1 Inheritance of economically important traits of *B. juncea*

Traits		Inheritance	References
Yield and its attributed traits	Plant height Number of primary branches Number of secondary branches Length of main shoot Number of seeds per siliqua Length of siliqua 1000-seed weight	Additive	Yadav et al. (1974), Asthana and Pandey (1977), Kumar and Singh (1987), Chauhan (1987), Yadav et al. (1992), Malkhandale (1993), Pradhan et al. (1993), Singh et al. (1996b), Kumar et al. (1998), Zhang et al. (2010)
	Days of 50% flowering Number of seeds per siliqua Seed yield per plant Oil content	Non-additive	Labana et al. (1975), Labana et al. (1987), Yadav et al. (1992), Pradhan et al. (1993), Kumar et al. (1998)
Biotic stresses	White rust	Monogenic dominant gene	Mukherjee et al. (2001), Varshney et al. (2004)
		Monogenic di-allelic (AcB1-A4.1 and AcB1-A5.1)	Panjabi-Massand et al. (2010)
	Alternaria blight	Quantitative (*PR1, PR2, PR3, NPR1, PDF1.2* genes for resistance)	Nayanakantha et al. (2016)
	Stem rot	Multiple loci	Rana et al. (2017, 2019), Atri et al. (2019)
Abiotic stresses	Drought and high temperature	Non additive gene action	Cheema and Sadaqat (2004)
	Salinity	Additive and non-additive gene action	Thakral and Prakash (1998)
Quality traits	Oil content	Polygenic additive	Cheung et al. (1998), Mahmood et al. (2006), Yadava et al. (2012)
	Oil quality: Erucic acid	Two additive genes	Kirk and Hurlstone (1983)
		Digenic and additive (*FAE1* genes)	Gupta et al. (2004)
	Seed meal quality: Glucosinolate content	Two to eight genes (Polygenic additive)	Love et al. (1990), Stringam and Thiagarajah (1995), Sodhi et al. (2002), Chauhan et al. (2007a), Ramchiary et al. (2007a)

2014). Moderate resistance was identified in *B. rupestris*, *B. incana*, *B. insularis* and *B. villosa* (Mei et al. 2011, 2013). Additionally, a promising high level of resistance to Sclerotinia has been documented in the related wild crucifer species *B. fruticulosa*, *B. oxyrrhina*, *B. parachinensis*, *B. tournefortii* (Uloth et al. 2013), *E. cardaminoides*, *E. abyssinicum, D. tenuisiliqua* (Garg 2010), *Capsella bursa-pastoris* (Chen et al. 2007). *B. fruticulosa* and *E. cardaminoides* have been utilized for transfer of genes in cultivated background of *B. juncea.*

Two sets of introgression lines, *B. juncea–B. fruticulosa and B. juncea–E. cardaminoides* have been developed, which revealed significant variation for stem rot. GWAS on both set of introgressions studies led to identification of candidate genes. Annotation of candidate genes revealed the role of anti-fungal proteins, metabolites, hypersensitive reaction and signal transduction pathways, TIR-NBS-LRR class, Chitinase, Malectin/receptor-like protein kinase, defensin-like (DEFL), desulfoglucosinolate sulfotransferase protein and lipoxygenase (Rana et al. 2017, 2019; Atri et al. 2019).

6.6.2.3 *Alternaria* Blight

High level of resistance against A. *brassicicola* and *A. brassicae* have not been identified amongst the cultivated Brassica species. However, related species such as *B. alba*, *B. carinata*, *B. desnottessi*, *B. elongata*, *B. fruticulosa*, *B. maurorum* and *B. spinescens* have been found resistant against *A. brassicae* (Brun et al. 1987a, b; Tewari and Conn 1993; Hansen and Earle 1997; Chrungu et al. 1999; Sharma et al. 2002). Some of the wilds of the *Brassicaceae* family viz. *Alliaria petiolata*, *Barbarea vulgaris*, *Capsella bursa-pastoris*, *Coincya* spp., *D. catholica*, *D. berthautii*, *D. creacea*, *D. erucoides*, *D. tenuifolia*, *E. gallicum*, *E. vesicaria* subsp. *sativa*, *Hemicrambe fruticulosa*, *H. matronalis*, *N. paniculata*, *R. sativus* and *S. arvensis* (Conn and Tewari 1986; Conn et al. 1988; Tewari 1991; Tewari and Conn 1993; Sharma et al. 2002; Warwick 2011) have shown resistance to *Alternaria.*

6.6.2.4 Aphid

Inter-varietal hybridization, induced mutagenesis, or autotetraploidy failed to provide significant resistance against aphid. Kumar et al. 2011 identified *B. fruticulosa* as resistant to *L. erysimi* after screening of diverse array of wild crucifers. *B. fruticulosa* have been used to develop *B. juncea–B. fruticulosa* introgression lines (IL) carrying genes for aphid resistance (Chandra et al. 2004; Kumar et al. 2011). These ILs demonstrated resistance against mustard aphid (Atri et al. 2012). Of these, 3 ILs have been selected with high resistance to *L. erysimi.* Molecular-cytogenetic analysis has demonstrated large translocation from *B. fruticulosa* to terminal region of chromosomes A05, B02, B03 and B04 (Agrawal et al. 2021). These ILs revealed antibiosis and antixenosis as a mechanism of resistance against *L. erysimi* under artificial infestation conditions (Palial et al. 2021). Response to resistance has also been related with synthesis of high level of phenols and low amount of GLSs (Palial et al. 2018).

6.6.3 Abiotic Stresses

6.6.3.1 Drought

Production losses may range from 17 to 94% due to drought (Mohammad et al. 1990; Chauhan et al. 2007b). It affects photosynthesis, dry matter accumulation, yield components, oil and protein content (Larcher 2003; Ohashi et al. 2006; Sinaki et al. 2007; Nasri et al. 2008; Ashraf and Harris 2013). Several physiological parameters such as transpiration cooling, epicuticular wax on the leaves, osmotic adjustment, thicker leaves, canopy temperature and drought susceptibility index (Chaudhary et al.1989; Kumar 1990; Singh et al. 1996a; Kumar and Singh 1998; Singh et al. 2003; Chauhan et al. 2007b; Sharma et al. 2011) were reported to be associated with drought tolerance. Genotypic differences for stomatal conductance, osmotic adjustment and oxalic acid exist in Brassicas (Ashraf and Mehmood 1990; Kumar and Elston 1992; Kumar and Singh 1998). In India, several varieties, including Aravali, PBR 97, RL91-27, Pusa Bahar, Pusa Bold, RH 781, RH 819, RGN 48, RB 50, Shivani, Vaibhav and PBR 378, have been released for cultivation under rainfed conditions.

6.6.3.2 Temperature

Heat, especially terminal heat stress, is the second most important stress for mustard. The crop plants entail different heat stresses during growing phase and the extent of exposure depends upon the varying climatic conditions. Due to changed climatic conditions, the weather extremes are increasing than before during crop growing season and increased temperature in the near future is expected to be adverse for winter crops like *B. juncea* (Qian et al. 2018). Two genetic stocks, BPR 541-4 and BPR 543-2, have been registered with the National Bureau of Plant Genetic Resources (NBPGR) for high-thermo-tolerance at terminal and juvenile stages, respectively (Chauhan et al. 2011b). Several varieties of Indian mustard, viz. Kranti, Pusa Agrani, Urvashi, NRCDR 02, Pusa Mustard 25 and Pusa Mustard 27, showing good thermo-tolerance during crop establishment stage have been commercialized.

6.6.3.3 Salinity

Saline soils and low-quality brackish irrigation water are additional issues for rapeseed-mustard growers. *B. juncea* and other allotetraploid Brassicas are more salt tolerant than their diploid progenitors (Nazir et al. 2001; Kumar et al. 2009). Photosynthetic rate, antioxidant enzyme activity, shoot Na + level, relative cell membrane permeability appeared to be associated with salt tolerance (Khayat et al. 2010; Munir et al. 2013). *B. juncea* germplasm have been screened to identify salt tolerant varieties (Siddiqui et al. 2009; Hayat et al. 2011). A strong correlation was observed between the physiological responses of the different species under salt stress with salt overly sensitive (SOS) pathway-related genes (Kumar et al. 2009). Of these, the *SOS2* gene was found to be upregulated when Brassica plants were exposed to salt stress. *B. juncea* was the only exception wherein high levels of *SOS2* expression were found even under normal conditions (Kumar et al. 2009). RH 8814 and BPR 541-4 are amongst genetic stocks registered for salinity-tolerance. Several salt tolerant varieties CS 52, CS 54, CS 56, CS 58, CS 60 and Narendra Rai 1 have been recommended for cultivation. Other salt tolerant varieties include: Varuna, TH 68, RH 30, Pusa Bold, Kranti, CS-4, CS-15, Pant Rai 2030, PR 1002, RH 7818, DIRA 337, BM-1, LL-84, P-15 and KS-51 (Ashraf and McNeilly 2004).

6.6.4 Oil Content

Improving the oil content and quality of *B. juncea* has always been a priority. The oil content of modern cultivars ranges from 37 to 42%. The switch to yellow-seeded varieties may result in an additional increase. Yellow-seeded varieties of *B. rapa, B. juncea,* and *B. carinata* are available. The genetics of oil content is complicated. In addition to maternal and environmental effects, it is influenced by a large number of genes with additive effects (Banga and Banga 2009; Weselake et al. 2009; Wang et al. 2010). Manipulation of photosynthetic activity in the silique is one of the genetic alterations used to boost oil content and fatty acid synthesis pathway (Hua et al. 2011). Apetalous and erectophile genotypes are thought to attract more photosynthetically active radiation, and so these show higher oil content than the conventional varieties (Fray et al. 1996).

6.6.5 Oil and Meal Quality

Modification in content of erucic acid in Indian mustard began with the discovery of mutants, ZEM I and ZEM II in *B. juncea* (Kirk and Oram 1981). Erucic acid in mustard was later found to be controlled by the genotype of embryo and showed digenic inheritance with additive effects (Gupta et al. 2004; Bhat et al. 2002). Attempts to reduce levels of meal GLSs content in mustard seeds got impetus from the development of low GLSs line, BJ 1058 in Canada. It was derived from interspecific cross of high GLSs mustard genotype with Low GLSs *B. rapa*. Aliphatic GLSs in the east European lines was controlled by two loci, whilst contribution of six to eight loci for GLSs content have been reported for Indian types. Maternal influence on the heredity of GLSs complicates breeding for low GLSs using the backcross method.

6.7 Breeding Methods

Indian mustard is a largely self-pollinated crop, though cross pollination up to 18 per cent has been reported (Labana and Banga 1984; Bhajan et al. 1991; Abraham 1994). Various selection methods including pure line selection, mass selection have been utilized. In pure line selection, progeny test is conducted to judge the value of selected plant, so a new variety is expected to be the progeny of a single, self-fertilized homozygous plant. In contrast to pure line selection, progeny is not tested in mass selection and phenotypically superior plants are harvested and bulked. Mass selection with some modifications is also followed in mustard.

6.7.1 Pedigree Method

The most common method for cultivar development in Indian mustard is pedigree selection. This involves hybridization between two or more parents and selection of superior plants in F_2 generation on the basis of plant phenotypes. Selected plants are advanced through plant to progeny rows in subsequent generations (F_3–F_5). Best progenies are bulked in F_6–F_7 and evaluated in multilocation replicated yield trials. It is possible to initiate replicated progeny testing in early generations due to high seed multiplication of this crop.

Different sorts of pedigree procedures aided yield improvements in conventional Brassica species (Downey and Rimmer 1993). Selection in the segregating progenies from bi-parental, multiple crosses or bulk-pedigree methods are commonly followed. Visual selection in F_2 on a single-plant basis for seed yield and oil content has proven to be ineffective, largely due to large $G \times E$ interaction rather than a preponderance of dominance genetic variance (Brown 1995). Delaying the first backcross to F_2 generation is frequently advantageous since a recombination cycle affords the opportunity to break certain unwanted linkages (Robbelen and Nitsch 1975). Delaying the selection until F_3 or later generations or through manipulation of simple characters or combination of characters was also found more effective. However, the environmental influence on yield components (Thurling and Depittayanan 1992) modulates the response to selection.

6.7.2 Recurrent Selection

It is rarely followed in mustard due to the inability to ensure random mating. However, it is possible to carry out recurrent selection programmes aided by genetic male sterility. The bulked seed (selected genotypes + GMS line) is grown in isolation for random mating. Seed is harvested only from selected male sterile plants. This cycle is repeated 2–3 times to get improved population with increased frequency of desirable alleles. In the end, male sterile gene is bred out of the population. The improved population can be used as directed as a cultivar or as a source for deriving new inbred lines.

6.7.3 Backcross Breeding

The backcross method is used for transferring specific gene(s) into an agronomically superior variety that is deficient in one or a few characteristics by repeated backcrossing to the recurrent parent of a hybrid accompanied by selection for specific character(s) being transferred from the donor parent. In Brassica, backcross method is used for transfer of disease resistance, transfer of low erucic and glucosinolate content and incorporation of male sterility and fertility restorer genes.

6.7.4 Mutation Breeding

Spontaneous and induced mutations cause changes at chromosomal level and nucleotide level. Physical mutagens consisting of ionizing radiation cause changes at chromosomal level including deletion, breakage, etc., whereas chemical mutagens produce changes at nucleotide level. Mutations have been successfully utilized for improvement of various qualitative and quantitative traits in Brassicas (Robbelen 1990; Bhatia et al. 1999; Choudhery and Jambulkar 2016). Remarkable success was achieved in tailoring fatty acids composition in the oil. Notable successes include high C18:1 (Auld et al.1992; Rucker and Robbelen 1997) and low C18:3 (Robbelen and Rakow 1970; Rakow 1973; Robbelen and Nitsch 1975) mutants in *B. napus*. LEA (Low erucic acid) mutants were produced due to loss of functional activity of enzyme fatty acid elongase (FAE 1). The enzyme loss its functional activity due to replacement of serine by phenylalanine aminoacid at position 282 and caused LEA. Similarly, single mutation for high C18:1 has been reported in *B. rapa* (Tanhuanpaa et al. 1998), *B. napus* (Hu et al. 2006) and *B. juncea* (Ripley et al. 2014). These caused premature termination of polypepitide and led to high C18:1. Dwarf mutants were induced as a result of mutations in the biosynthetic pathway of gibberellin (GA). GA caused degradation of DELLA protein and promotes the growth. Dwarfing mutant, *Brrga1-d*, has single NT substitution in the DELLA protein and as a result GA could not interact with DELLA protein and hence led to formation of dwarf mutants (Muangprom and Osborn 2004). Various modified traits through the use of both physical as well as chemical mutagens are summarized in Table 6.2. Over 31 mutant varieties have been released over the world. India contributed eight, Bangladesh, China and Sweden each have contributed five, whilst Canada, Japan, the former

Table 6.2 Trait modification through use of mutagens

Trait	Mutagens	Mutants	References
Morphological traits			
Chlorophyll	EMS	11 mutants (albina, viridis, xantha, maculata, tigrina, albo-xantha, xantho-alba, variegated, xantho-viridis, virido-xantha, virido-alba)	Gupta et al. (2012)
Height	Y rays	Dwarf and yellow seeded	Choudhary and Jambhulkar (2016)
Flower	EMS, ENU, EtBr	Male sterile	Bhat et al. (2001)
	EMS, Y rays	Male sterile	Chauhan and Singh (1998)
Siliqua	X rays	Appressed pod mutants	Rai (1958) Nayar and George (1969)
	Y rays	Bunching and appressed pod mutants	Singh and Sareen (2004)
	EMS,ENU, EtBr	Tri and tetralocular siliqua, non shattering	Bhat et al. (2001)
Seed colour	S radio isotope	Yellow colour	Nayar (1968)
	P radio isotope	Yellow colour	Nayar (1968)
Root traits	Y rays	Taller root	Basak and Prasad (2004)
Productivity related traits			
Seed yield	Y rays and EMS	Early, high dwarf and high grain yield	Khatri et al. (2005)
		Yield and its components traits	Mahla et al. (1990, 1991)
Oil content	Y rays EMS	Mutants with 3% more oil content Mutants with 1–5% more oil content	Khatri et al. (2005)
Quality traits			
Oleic acid	EMS	High-oleic acid mutant (70%)	Ripley et al. (2014)

Soviet Union and Pakistan each have contributed two varieties (Choudhary and Jambulkar 2016).

6.7.5 Somaclonal Variation

A somaclone of Varuna namely Bio-902 (Pusa Jai Kisan) of Indian mustard has been released as a variety (Katiyar 1997).

6.8 History and Achievements of Varietal Development in India

Improvement work of rapeseed-mustard crops was first initiated at Pusa (Bihar) during last century. These researches primarily involved collection and purification of the land races. Pioneering work at Layallpur (now Faisalabad in Pakistan) later led to the development of Indian mustard variety RL 18 in 1937 through pure line selection. Several high-yielding varieties of Indian mustard, such as Laha 101, Varuna, Durgamani and Patan Mustard, were subsequently produced through selection and released for commercial cultivation between 1947 and 1967. Eighties heralded a very successful era of intervarietal hybridization and led to the release of 112 varieties till 2021. During this period, mutation breeding has also been exploited and 13 varieties were commercialized. 77 novel genetic stocks of rapeseed-mustard for plant architectural traits (dwarf, long main shoot, tetralocular siliquae), quality characteristics (LEA, low GLSs, High OA, Low linolenic acid), disease resistance (white rust), abiotic stress (WUE, temperature, salinity), seed characteristics (yellow, bold seed) CMS and restorer have been registered with National Bureau of Plant Genetic Resources (NBPGR 2022).

6.9 Heterosis Breeding in Indian Mustard

Exploitation of heterosis is one of the most useful contributions of genetics for the improvement of global agriculture. Heterosis or hybrid vigour was first discovered in eighteen century when self and cross fertilization studies were carried out in vegetable species (Darwin 1876). It is the deviation of F_1 hybrids from mid parental value of a given trait (Hayes et al. 1955). If F_1 hybrid expresses superior performance relative to the better parent then it is called heterobeltiosis. Heterosis relative to the standard check variety is described as standard or commercial heterosis. Hybrids offer an opportunity for mobilizing greater genetic variation and heterotic response. Commercial viability of the hybrids depends upon three factors: yield advantage of hybrids against commercial checks (standard heterosis), efficient pollination control mechanism and economics of hybrid seed production. Indian mustard is well suited for hybrid development as floral structure permits out crossing up to 18% along with demonstrated yield advantages. In *B. juncea* heteosis from 13 to 100% have been reported (Yadava et al 1974; Banga and Labana 1984a; Dhillon et al. 1990; Kumar et al. 1990; Thakur and Bhateria 1993; Baishakh et al. 1994; Rai 1995; Verma et al. 1998). Heterosis was generally higher in hybrids involving geographically diverse parents as compared to closely related genotypes (Pradhan et al. 1993). However, the highest heterosis usually happens till an optimum level of divergence between parents. Genetic instability occurs when the parents are more diverse, and are reproductively isolated due to variation in days to flowering. Genetic diversity between parents and SCA (specific combining ability) determines the probability of superior hybrid performance in *B. juncea* (Gupta et al. 1991). Heterobeltiosis for seed yield and its attributing traits has been observed for the parental combinations showing high SCA (Vaghela et al. 2011; Yadava et al. 2012; Meena et al. 2014b; Adhikari et al. 2017; Rai et al. 2017; Singh et al. 2020). Association between the genetic diversity and combining ability of the parents has been observed for various traits including oil and seed yield as well as its components (Thakur and Zarger 1989; Gupta et al. 1991; Monalisa et al. 2005). These results indicate that low/average GCA with high SCA along with medium genetic diversity between the parents provide high heterosis in hybrids.

6.9.1 Pollination Control Mechanisms

Genetic male sterility, cytoplasmic-genetic male sterility systems, chemical hybridizing agents and genetically engineered male sterility system have been reported in *B. juncea.*

6.9.1.1 Genetic Male Sterility (GMS)

Several examples of spontaneously occurring systems of genetic male sterility have been reported in mustard (Badwal and Labana 1983; Banga and Labana 1985). Male sterility, in all these cases, was inherited in monogenic recessive fashion. None of these could be used in hybrid breeding due to the difficulties in maintenance of male sterility.

6.9.1.2 Cytoplasmic Male Sterility (CMS)-Fertility Restorer Systems

CMS is the result of cytoplasmic-nuclear gene interaction. This generally occurs when CMS-inducing mitochondrial genome interacts with a nuclear genome lacking *Rf* genes. Such combination may occur in nature or by artificial hybridizations including both sexual as well as somatic hybridizations. The natural occurring mutants arise due to intragenic rearrangement of mt DNA that create new open reading frames (ORFs) (Dufay et al. 2007). Identification of candidate gene is easier or possible where CMS is produced by recent activity as compared to somatic hybrids. In somatic hybrids, genes can be predicted through differential segregation analysis (Kim et al. 2016; Makarenko et al. 2019). In contrast to normal alloplasmic genes, where difference between normal and cytoplasmic DNA (Arimura et al. 2018; Kim et al.2019; Luo et al. 2013) could predict genes for CMS. In general, all cytoplasmic sterilizing genes shares some common features such as these arise as a consequence of recombination, co transcribed along with existing *mt* gene and translated product has transmembrane domain (Kim et al. 2019; Tang et al. 2017).

F_1 hybrid seed production necessitates complete restoration of male fertility to ensure self-seed set. Fertility restorer (*Rf*) genes for sterilizing cytoplasm resulting from intraspecific variation are generally found within the primary gene pool of the species. Whereas fertility restorer genes are mostly required to be introgressed from cytoplasmic donor species for CMS of alloplasm origin. For example. *Rf* genes from for CMS systems *ogu*, *trachy*, *mori, can*, *lyr* and *tour* were introgressed from cytoplasmic donor species (Heyn 1976; Kirti et al. 1997, 1998; Prakash et al. 2001; Banga et al. 2003; Janeja et al. 2003). Introgression of *Rf* from cytoplasm donor species is generally associated large alien segment substitutions. For example, introgression of *Rf* gene from radish substituted a major portion of C09 chromosome. This introgression also harboured some negative associated traits as well (Heyn 1976; Delourme and Eber 1992; Delourme et al. 1998; Hu et al. 2008; Tanaka et al. 2012). *Rf* gene was then transferred to *B. juncea* and it also possessed large segment as of unimproved *B. napus* (Primard-Brisset et al. 2005; Tian et al. 2014). Gudi et al. (2020) also reported transfer as well as of physical mapping of *Rf* gene in *B. juncea*. Introgression (108 kb) was located to the distal end of chromosome A09. In silico map of A09 indicated many homeologous exchanges between A and C chromosomes as well. KASPar markers for marker assisted transfer for *Rf* was also reported. *Rf* gene for *ogu* CMS showed linkage drag with genes synthesis for GLSs in *B. napus* (Delourme et al. 1995; Primard-Brisset et al. 2005). Fertility restorers are mostly inherited as sporophytic traits, but gametophytic control of fertility restoration was also reported from CMS systems based on *M. arvensis*, *D. erucoides* and *D. berthautii* cytoplasms (Kaur et al. 2004; Bhat et al. 2005). *Rf* gene for *ogura* CMS are well characterized at molecular level. The *Rf* gene caused modification in the processing of RNA ato restore fertility (L'Homme and Brown 1993; Singh and Brown 1993). Interestingly, *Rf* gene of *mori* CMS could restore the fertility of *cath, eru, ber* CMS system. (Bhat et al. 2005, 2006, 2008), as all these CMS systems possess common gene *orf108* upstream of *atpA*. These gene co transcribed and led to formation of bicystronic transcript. Presence of restorer gene caused modification in the processing of transcript and Bi-cistronic *orf108-atpA* transcript cleaved just downstream of the start codon of *orf108*, led to the production of only single transcript of *atpA* (Ashutosh et al. 2008; Kumar et al. 2012) and restored fertility.

CMS-Rf systems could be broadly classified into three types based on its origin:

Natural Variations as Source of CMS

Naturally occurring male sterile plants were observed in a Japanese radish variety (Ogura 1968). Interaction of recessive nuclear gene *ms ms* with cytoplasm caused male sterility. This system was not utilized in exploiting heterosis in radish due to lack of fertility restorer genes. The male sterilizing cytoplasm was later transferred to European radish, where restorer genes were available (Bannerot et al. 1977; Bonnet 1977). This male sterilizing cytoplasm was also transferred to *B. napus* (Bannerot et al. 1977) as well as in *B. juncea* (Labana and Banga 1989). Primary CMS lines based on *ogura* system could not be used in hybrid seed production due to leaf chlorosis, low nectar production and the absence of fertility restoring genes in crop brassica species. Chlorosis and low nectar production were subsequently rectified through somatic hybridization by substituting radish chloroplasts with those from *B. napus* (Jarl and Bornman 1988; Menczel et al. 1987; Pelletier et al. 1983) or *B. juncea* (Kirti et al. 1995a). Sterilizing genes for *ogu* CMS from radish are well characterized. Sterility is caused due to gene *orf 224* which co transcribed with *atp6*. In the presence of *Rf* gene, processing of RNA is affected and caused restoration of fertility. (L'Homme and Brown 1993; Singh and Brown 1993). Cell fusions between alloplasmic CMS of ogura and euplasmic *B. napus* resulted in recombination of mt DNA. Detailed genome analysis of recombined mt DNA identified the *orf 138* up stream of *orf 8/atp 8* as a cause of male sterility. Both the genes are co transcribed. A deletion of 39 bases and nucleotide variants (up to 9) has been reported to cause sterility action (Yamagishi and Terachi 2001). Ogura cytoplasm could not cause male sterility in *A. thaliana* (Duroc et al. 2006). Polima (*pol*) CMS also resulted from spontaneous mitochondrial mutation in *B. napus* (Fu 1981; Liu et al. 1987). The sterility was caused by interaction of sterilizing cytoplasm with one or two pairs of recessive nuclear genes (Liu et al. 1987). The use of this system as male sterility appeared to breakdown at higher temperatures. Other examples of spontaneously occurring CMS include *nap* CMS (L'Homme et al. 1997), 681A (Liu et al. 2005) in *B. napus* and *hau* (Wan et al. 2008) in *B. juncea*. Male sterility in *hau* CMS occurred due to gene *orf 288* and co transcribed with *atp 6* (Jing et al. 2012). Male sterility in *B. juncea* CMS system was found due to new gene *Orf 220*. It shared homology with *orf 222* and *orf 224* (Yang et al. 2005). CMS system 126-1 arose spontaneously from DH population of resynthesized *B. napus* line ISN 706. This CMS was designated as '126-1'. This was transferred to *B. juncea*. The male sterility could be maintained only by few specific genotypes of mustard, whereas most mustard germplasm lines acted as restorer of fertility (Sodhi et al. 2006). Uniqueness of 126-1 has been demonstrated using mt DNA using southern hybridization.

CMS Systems Based on Alien Cytoplasm Substitution

(i) **Allopmasmic CMS line development through sexual hybridization**

Male sterility was reported from cytoplasmic substitutions following interspecific or intergeneric hybridizations. Interspecific hybridization between *B. rapa* and *B. juncea* led to the male sterility. This CMS was subsequently transferred to tuber mustard (*B. juncea* var. *tumida*) (Zhang et al. 2003), leaf mustard (*B. juncea* var. *multiceps*) (Yang et al. 2005), and stem mustard (*B. juncea*) (Yang et al. 2009). Substitution of cytoplasm from eight wild species: *B. tournefortii, B. oxyrrhina D. berthautii, D. catholica, D. erucoides, D. siifolia, E. lyratus, E.*

canariense have been reported to cause cytoplasmic male serilty in *B. juncea* (Prakash and Chopra 1990; Rao et al. 1994; Landgren et al. 1996; Malik et al. 1999; Banga et al. 2003; Pathania et al. 2003; Prakash et al. 2001; Kumar et al. 2012).

***Tournefortii system* (*tour*)** Male sterile plants were initially identified in *B. juncea* by Rawat and Anand (1979). Comparison of cp DNA profiles between *B. juncea* and allied Brassica species indicated that *B. tournefortii* was the most probable cause of male sterility and source of the cytoplasm (Pradhan et al. 1991). Sterility inducing cytoplasm was subsequently transferred to *B. napus* (Banga et al. 1994) and *B. carinata* (Anand 1987) Fertility restoration system is stable in *B. napus,* whilst in *B. juncea* is influenced by the genotype as well as photoperiod. On set of long days caused reversion of fertility restoration.

***Oxyrrhina* system (*oxy*)**: Synthetic alloploid developed between from *B. oxyrrhina* and *B. campestris* was repeated backcrossed with *B. juncea* to produce *oxyrrhina* CMS system in Indian mustard (Prakash and Chopra 1990).

***Siifolia* system (*siif*)**: Substitution of *B. juncea* nucleus in cytoplasm of *D. siifolia* cytoplasm led to male sterility (Rao et al. 1994). Male sterility was stable over the range of environmental conditions.

***Moricandia arvensis* system (*mori*)**: The alloplasmic CMS was developed through cytoplasmic substitution of *B. juncea* cytoplasm with *Moricandia arvensis*. The male sterility is characterized by reduced anther and cholortic plants. Chlorosis was subsequently rectified through somatic hybridization (Prakash et al. 1995). Genes for fertility restoration were introgressed in *B.juncea* by generating the monosomic addition line of *M. arvensis* in *B. juncea* CMS (Prakash et al. 1998).

***Enarthrocarpus lyratus system* (*lyr*)**: The backcross substitution of *B. rapa* genome in cytoplasm of *E. lyratus* resulted in male sterile genotype. The CMS plant had petaloid anthers. Euplasmic *B. juncea* plants acted as maintainers of male sterility. Fertility restorer genes could be introgressed from *E. lyratus* (Banga et al. 2003).

***Canariense system* (*can*)**: Transfer of *Erucastruum canariense* cytoplasm in the nuclear background of *B. juncea* resulted in the male sterility. Male sterility was stable over the range of environmental conditions (Prakash et al. 2001). The *Rf* genes could be introgressed from cytoplasm donor species. Restoration is complete and acted in monogenic dominant manner.

Erucoides system (eru): Transfer of *B. juncea* nucleus in cytoplasmic background of *D. erucoides* led to development of a stable male sterility system male sterility (Bhat et al. 2006). Fertility restorer of *M. arvensis* system could restore the fertility of this system as well. Restoration was found to be monogenic and gemetophytic.

***Berthautii* system (*ber*)**: This CMS system was developed through backcross substitution of *B. juncea* nucleus into the cytoplasm of *D. berthautii* (Bhat et al. 2006). Fertility restorer of *M. arvensis* system could restore the fertility of this system as well. Restoration was monogenic and gemetophytic (Bhat et al. 2008).

Catholica system (*cath*): Transfer of *B. juncea* nucleus in the cytoplasmic background of *D. catholica* produced this male sterility. *Rf* gene were later introgressed by hybridizing CMS line with fertile plants derived in the progeny of somatic hybrid of *B. juncea* and *D. cathoilca*.

(ii) **Allopmasmic CMS line development through somatic hybridization**

Wild species which were not amenable to sexual hybridization were exploited through somatic hybridization to produce male sterile lines. Somatic hybrids have been produced with *Sinapis alba*, *D. catholica*, *M. arvensis* and *Trachystoma ballii* (Prakash et al. 1995; Kirti et al. 1993, 1995a, b, c).

Sinapis alba system: CMS plants of *B. juncea* were developed following backcrossing of somatic hybrid (*S. alba* + *B. juncea*; $2n = 60$; S^{al}

S^{al} AABB) with *B. juncea*. The CMS plants were normal green and bore flowers have good nectaries. Molecular analysis indicated recombined mitochondrial genome, whereas cp DNA was derived only from *S. alba* (Prakash et al.1995).

***Diplotaxis catholica* (*cath*)**: This system was derived from somatic hybrid *D. catholica* + *B. juncea* ($2n = 54$; D^C D^C AABB). Flowers had reduced anthers with excellent nectaries. Somatic hybrids had both recombined mitochondrial and chloroplast genome (Prakash et al. 1995). Mitochondrial gene, *cox I-2* was suggested to cause male sterility in this CMS system (Pathania et al. 2007).

***Trachystoma ballii system* (*trachy*)**: The CMS plants were selected from regenerated plants from somatic hybrid of *T. ballii* + *B. juncea* ($2n = 522$; TTAABB). Male sterile plants had petaloid anthers and recombined mt DNA as well as cp DNA (Kirti et al. 1995c). Fertility restorer genes have been introgressed from *T. ballii.*

Penalties Associated with CMS System

CMS lines based on alien cytoplasm substitutions are known produce hybrids with productivity lower than the hand bred hybrids based on same parental combination. This yield penalty may result from adverse effect of alien cytoplasm has been reported in *B. juncea* (Kaur et al. 2004; Katiyar et al. 2007; Chamola et al. 2013). In some instances, negative effect of cytoplasm could be reversed by the restorer gene like in *T. ballii,* whereas, in *D. catholica*, only partial improvement could be achieved.

Six alloplasms (*tournefortii, siifolia, ogura, moricandia, oxyrrhina* and *trachystoma*) in different genetic backgrounds along with their maintainer lines were assessed for their floral and agronomic traits. All CMS except *trachy* were at par with their maintainers for height, harvest index, yield, oil and protein content (Singh and Srivastava 2006).

6.9.1.3 Chemical Hybridizing Agents

Chemicals that can induce sterility or check pollen spread can be an alternative to genetic/ cytoplasmic systems of sterility. For Indian mustard, limited studies have conducted to assess the role of male gametocides. Ethrel has been reported to cause male sterility in Indian mustard. However, commercial use of Ethrel was not recommended due to shorter period of gametocidal effect and reduced female fertility (Banga and Labana 1983b, b). Other gametocidal agents, synthetic detergent (Surf excel, Nirma), gibberellic acid, benzotriazole, sodium arsenate, azetidine 3-corboxylate, maleic hydrazide, pendimethalin and arsenic trioxide were also evaluated in mustard with limited male sterilizing effects (Chauhan et al. 2007c, 2010, 2011a; Singh and Chauhan 2001, 2003, 2004; Chauhan and Singh 2002; Lavania and Chauhan 2006). Along with male sterility, CHA also caused reduction in plant height, number of branches, seed set, seed size and yield/plant (Singh and Chauhan 2001, 2004; Chauhan and Singh 2002). Most CHAs interfered with the development of tapetum, anther and microspore (Chauhan and Kinoshita 1982).

6.9.1.4 Genetic Engineered Male Sterility

Genetic engineering approaches has also been used to develop male sterility/restorer system in *B. juncea.* The barnase-barstar system is most prominent genetically engineered male sterility used for hybrid seed production. First male sterile plants were developed in rapeseed by a transgenic approach through transforming it with barnase gene from bacteria (*Bacillus amyloliquefaciense*) using a tapetum-specific promoter (TA29) from tobacco (Mariani et al. 1990). Likewise, genetically engineered restorer plants were developed by transferring the barstar gene in TA29-barnase plants (Mariani et al. 1992). Both barnase and barstar genes expressed in F_1 offspring's when male sterile and restorer transgenic plants are hybridized. These genes were expressed in anther of hybrid plants due to the use of tapetum-specific promoter. A dominant herbicide resistant gene called bar (from *Streptomycin hygroscopicus*), conferring resistance to the herbicide phosphinothricin/glufosinate (Basta), was also used in conjunction with genetically engineered male sterility. Only male sterile segregants could survive herbicide spray at seedling stage owing to the bar gene expression. This method allowed the development of completely male sterile progenies in hybrid seed production blocks. TA29-barnase male sterility system was also developed developed in *B. juncea*. For this TA 29 construct was modified by the insertion of a spacer DNA fragment in the middle of the barnase gene and the promoter of the bar gene (driven by CaMV 35S promoter; Jagannath et al. 2001). This fragment acted as a functional insulator for tissue-specific expression of barnase gene and for agronomically superior male sterile lines. Fertility restoring genotype was also developed using the barstar gene and TA29 promoter sequence (Jagannath et al. 2002). Both the transgenics (male sterile and restorer) could be used for hybrid seed production in *B. juncea*. Another barnase/barstar transgenic system involved using ALS^{dm} as selection marker. This gene conferred resistance to Pursuit herbicide (Ray et al. 2007). Pursuit herbicide is less expensive compared to Basta. Barnase/barstar system provides an excellent approach for hybrid seed production in Indian mustard.

6.9.1.5 Commercial Hybrids in Mustard

Systematic and coordinated efforts for developing hybrids resulted in the release of many CMS based hybrids. These include NRCHB 506 (*mori* cytoplasm), DMH 1(based on CMS 126-1), Coral 432/PAC 432, PHR 126 and RCH1 (all based on *ogu* cytoplasm). DMH-11 is first transgenic hybrid of India was developed through barnase-barstar system by Delhi University. Unfortunately, it was not commercialized due to biosafety concerns from environmental activists.

References

Abraham V (1994) Rate of out crossing in Indian mustard (*Brassica juncea* L.). Eucarpia Cruciferae News Lett 16:69–70

Acharya NN, Swain D (2003) Hybrid performance in relation to genetic divergence in Indian mustard [*Brassica juncea* (L.) Czern. & Coss.]. Crop Res 25:312–315

Ackman RG, Eaton CA, Sipos JC, Loew FW, Hancock D (1977) Comparison of fatty acid from high levels of erucic acid of RSO and partially hydrogenated fish oil in non-human primate species in a short termexploratory study. Nutr Diet 25:170–185

Adhikari S, Pathak S, Joshi D, Pant U, Singh AK, Bhajan R (2017) Heterosis analysis for seed yield and other component traits in Indian mustard [*Brassica juncea* (L.) Czern. & Coss.]. Intl J Curr Microbiol App Sci 6(10):1157–1162

Agrawal N, Gupta M, Atri C, Akhatar J, Kumar S, Heslop-Harrison JS, Banga SS (2021) Anchoring alien chromosome segment substitutions bearing gene(s) for resistance to mustard aphid in *Brassica juncea–B. fruticulosa* introgression lines and their possible disruption through gamma irradiation. Theor Appl Genet. https://doi.org/10.1007/s00122-021-03886-z

Ahuja I, Bhaskar PB, Banga SK, Banga SS, Rakow G (2003) Synthesis and cytogenetic characterization of intergeneric hybrids of *Diplotaxis siifolia* with *Brassica rapa* and *B. juncea*. Plant Breed 122:447–449

Ahuja I, Rohloff J, Bones AM (2009) Defence mechanisms of Brassicaceae: implications for plant insect interactions and potential for integrated pest management—a review. Agron Sustain Dev 30(2):311–348. https://doi.org/10.1051/agro/2009025

Ali M, Copeland LO, Elias SG, Kelly JD (1995) Relationship between genetic distance and heterosis for yield and morphological traits in winter canola (*Brassica napus* L.). Theor Appl Genet 91:118–121

Anand IJ (1987) Breeding hybrids in rapeseed and mustard. In: Proceedings of international rapeseed conference, Poland, pp 79–85

Arimura S, Yanase S, Tsutsumi N, Koizuka N (2018) The mitochondrial genome of an asymmetrically cell-fused rapeseed, *Brassica napus*, containing a radish-derived cytoplasmic male sterility-associated gene. Genes Genet Syst 93(4):143–148

Ashraf M, Harris PJC (2013) Photosynthesis under stressful environments: an overview. Photosynthetica 51:163–190. https://doi.org/10.1007/s11099-013-0021-6

Ashraf M, McNeilly T (2004) Salinity tolerance in Brassica oilseeds. Crit Rev Plant Sci 23(2):157–174

Ashraf M, Mehmood S (1990) Response of four Brassica species to drought stress. Environ Exp Bot 30:93–100

Ashutosh KP, Sharma PC, Prakash S, Bhat SR (2008) A novel *orf108* co-transcribed with *atpA* gene is associated with cytoplasm male sterility in *Brassica juncea* carrying *Moricandia arvensis* cytoplasm. Plant Cell Physiol 49:284–289

Asthana AN, Pandey UK (1977) Combining ability and rank correlations in a diallel cross of Indian mustard *B. juncea*. Exp Agri 13:71–79

Atri C, Kumar B, Kumar H, Kumar S, Sharma S, Banga SS (2012) Development and characterization of *Brassica juncea–fruticulosa* introgression lines exhibiting resistance to mustard aphid (*Lipaphis erysimi* Kalt). BMC Genet 13:104. https://doi.org/10.1186/1471-2156-13-104

Atri C, Akhatar J, Gupta M, Gupta N, Goyal A, Rana K, Kaur R, Mittal M, Sharma A, Singh MP, Sandhu PS, Barbetti MJ, Banga SS (2019) Molecular and genetic analysis of defensive responses of *Brassica juncea–B. fruticulosa* introgression lines to *Sclerotinia* infection. Sci Rep 9:1–12

Auld DL, Heikkinen MK, Erickson DA, Sernyk JL, Romero JE (1992) Rapeseed mutants with reduced level of polyunsaturated fatty acids and increased levels of oleic acid. Crop Sci 32:657–662

Axelsson T, Bowman CM, Sharpe AG, Lyndiate DJ, Lagercrantz U (2000) Amphidiploid *Brassica juncea* contains conserved progenitor genomes. Genome 43:679–788

Badwal SS, Labana KS (1983) Male sterility in Indian mustard. In: International conference genetics, New Delhi, p 514

Baishakh BS, Panda S, Nayak BK (1994) Heterosis for yield in Indian mustard. Environ Ecol 12(3):572–574

Banga SS, Labana KS (1983) Production of F_1 hybrids using ethrel-induced male sterility in Indian mustard [*B. juncea* (L.) Coss.]. J Agric Sci Camb 101:453–455

Banga SS, Labana KS (1984a) Heterosis in Indian mustard (*Brassica juncea* (L.) Coss.). Z Pflanzenzucht 92:61–70

Banga SS, Labana KS (1984b) Ethrel-induced male sterility in Indian mustard [*B. juncea* (L.) Coss.]. Z Pflanzenzucht 92:232–233

Banga SS, Labana KS (1985) Male sterility in Indian mustard IV. Genetics of MS 4. Can J Genet Cytol 27:487–490

Banga SS, Deol JS, Banga SK (2003) Alloplasmic male-sterile *Brassica juncea* with *Enarthrocarpus lyratus* cytoplasm and the introgression of gene(s) for fertility restoration from cytoplasm donor species. Theor Appl Genet 106:1390–1395

Banga SK, Kumar P, Bhajan R, Singh D, Banga SS (2015) Genetics and breeding. In: Kumar A, Banga SS, Meena PD, Kumar PR (eds) *Brassica* oilseeds: breeding and management. CABI, UK, pp 11–41

Banga SS, Banga SK (2009) Crop improvement strategies in rapeseed-mustard. In: Hegde DM (ed) Vegetable oils scenario: approaches to meet the growing demands. ISOR, Hyderabad, pp 13–35

Banga SS, Kaur N (2009) An alternate procedure for resynthesis of *Brassica juncea.* In : Proceedings of 16th Australian Res Assembly Brassicas, Ballarat Victoria, pp 1–4

Banga SS, Labana KS (1983a) Male sterility in Indian mustard II Genetics and Cytology of MS-1. In: Proceedings of 6th international rapeseed conference, Paris, pp 349–353

Banga SS, Banga SK, Labana KS (1983) Nucleo-cytoplasmic interaction in *Brassica.* In: Proceedings of the 6th international rapeseed congress, Paris, 17–19, pp 602–606

Banga SS, Banga SK, Sandha GS (1994) Hybrids in oilseed rape based on *tour* cytoplasm. In: Verma MM, Virk DS, Chahal GS (eds) Proceedings of symposium on heterosis breeding in crop plants-theory and application. Ludhaiana, pp 123–129

Bansal P, Kaur P, Banga SK, Banga SS (2009) Augmenting genetic diversity in *Brassica juncea* through its resynthesis using purposely selected diploid progenitors. Int J Plant Breed 3:41–45

Bansal P, Banga S, Banga SS (2012) Heterosis as investigated in terms of polyploidy and genetic diversity using designed *Brassica juncea* amphiploid and its progenitor diploid species. PLoS ONE 7: e29607. https://doi.org/10.1371/journal.pone.0029607

Bannerot HL, Boulidard L, Chupeau Y (1977) Unexpected difficulties met with the radish cytoplasm in Brassica oleracea. Eucarpia Crucifers Newsl 2:16

Basak S, Prasad C (2004) Screening for root mutant in turnip (*Brassica rapa L.*). Cruciferae Newslett 25:9–10

Batra V, Prakash S, Shivanna KR (1990) Intergeneric hybridization between *Diplotaxis siifolia*, a wild species and crop brassicas. Theor Appl Genet 80:537–541

Bhajan R, Chauhan YS, Kumar K (1991) Natural cross pollination in Indian mustard. Cruciferae Newslett 14:24–25

Bhat SR, Haque A, Chopra VL (2001) Induction of mutations for cytoplasmic male sterility and some rare phenotypes in Indian mustard (*Brassica juncea L.*). Indian J Genet 61:335–340

Bhat MA, Gupta ML, Banga SK, Raheja RK, Banga SS (2002) Erucic acid heredity in *Brassica juncea* (L.) Czern. & Coss.—some additional information. Plant Breed 121:456–458

Bhat SR, Vijayan P, Ashutosh K, Dwivedi K, Prakash S (2006) *Diplotaxis erucoides* induced cytoplasmic male sterility in *Brassica juncea* is rescued by the *Moricandia arvensis* restorer genetic and molecular analysis. Plant Breed 125:150–155

Bhat SR, Kumar P, Prakash S (2008) An improved cytoplasmic male sterile (*Diplotaxis berthautii*) *Brassica juncea*—identification of restorer and molecular characterization. Euphytica 159:145–152

Bhat SR, Prakash S, Kirti PB, Dinesh kumar V, Chopra VL (2005) A unique introgression from *Moricandia arvensis* confers male fertility upon two different cytoplasmic male-sterile lines of *Brassica juncea.* Plant Breed 124:117–120

Bhatia CR, Nichterlein K, Maluszynski M (1999) Oilseed cultivars development from induced mutations and mutation altering fatty acid composition. Mut Breed 11:1–36

Bonnet A (1977) Breeding in France of a radish hybrid obtained by use of cytoplasmic male sterility. Eucarpia Cruciferae Newslett 2:5–6

Brown J (1995) Early generation selection in spring canola (*Brassica napus* L.). In: Proceedings of 9th international rapeseed congress, Cambridge, UK, pp 699–701

Brun H, Plessis J, Renard M (1987a) Resistance of some crucifers to *Alternaria brassicae* (Berk.) Sacc. In: Proceedings of 7th international rapeseed congress, Poznan, pp 1222–1227

Brun H, Renard M, Plessis J (1987b) Resistance of some crucifers to *Alternaria brassicae* (Berk.) Sacc. In: Proceedings of 7th international rapeseed congress, Poznan, pp 247

Burkill IH (1930) The Chinese mustards in the Malaya Peninsula. Gardners Bull 5:99–117

Burton W, Ripley V, Potts D, Salisbury P (2004) Assessment of genetic diversity in selected breeding lines and cultivars of canola quality *Brassica juncea* and their implications for canola breeding. Euphytica 136:181–192

Chamola R, Balyan HS, Bhat SR (2013) Effect of alien cytoplasm and fertility restorer genes on agronomic and physiological traits of *Brassica juncea* (L) Czern. Plant Breed 132:681–687

Chandra A, Gupta ML, Banga SS, Banga SK (2004) Production of an interspecific hybrid between *Brassica fruticulosa* and *B. rapa*. Plant Breed 123:497–498

Chaudhary BD, Singh P, Singh P, Kumar P (1989) Inheritance studies of plant water relations in *Brassica juncea*. Biol Planta 31:202–206

Chauhan YS (1987) Genetic analysis for yield and its components in Indian mustard. Indian J Agric Res 21:77–82

Chauhan SVS, Kinoshita T (1982) Chemically induced male sterility in angiosperm. Seiken Ziho 30:54–75

Chauhan SVS, Singh NK (1998) Gamma ray and EMS induced male sterility mutants in *Brassica juncea*. J Cytol Genet 33:29–34

Chauhan SVS, Singh V (2002) Detergent-induced male sterility and bud pollination in *Brassica juncea (L)*. Curr Sci 82(8):918–920

Chauhan JS, Singh M, Bhadauria VPS, Kumar A, Meena ML (2007a) Genetic analysis of glucosinolate content in Indian mustard (*Brassica juncea* L.). Indian J Genet 67:411–413

Chauhan JS, Tyagi MK, Kumar A, Nashaat NI, Singh M, Singh NB, Jakhar ML, Welham SJ (2007b) Drought effects on seed yield and its components in Indian mustard (*Brassica juncea* L.). Plant Breed 126:399–402

Chauhan SVS, Agnihotri DK, Singh V (2007c) Some new male gametocides—a review. Theor Appl Genet, 210–221

Chauhan SVS, Gupta HK, Singh V (2011a) Efficacy of a novel ester azetidine 3-corboxylate as chemical hybridizing agents for *Brassica juncea* (L.) Czern. & Coss. J Indian Bot Soc 90:139–141

Chauhan JS, Singh KH, Singh VV, Kumar S (2011b) Hundred years of rapeseed-mustard breeding in India: accomplishments and future strategies. Indian J Agric Sci 81(12):1093–1109

Chauhan SVS, Gupta HK, Singh V (2010) Sodium arsenate A metabolic inhibitor used as a chemical hybridizing agent for *Brassica juncea* (L.) Czern & Coss. Int J Plant Reprod Biol 2(1):93–98

Cheema KL, Sadaqat HA (2004) Potential and genetic basis of drought tolerance in canola (*Brassica napus*): I. Generation mean analysis for some phenological and yield components. Int J Agric Biol 6:74–81. http://www.fspublishers.org/ijab/index_ijab.jsp

Chen HF, Wang H, Li ZY (2007) Production and genetic analysis of partial hybrids in intertribal crosses between Brassica species (*B. rapa, B. napus*) and *Capsella bursa-pastoris*. Plant Cell Res 26:1791–1800

Chen S, Wan Z, Nelson MN, Chauhan J, Redden R, Burton WA, Lin P, Salisbury PA, Fu T, Cowling WA (2013) Evidence from genome-wide simple sequence repeat markers for a polyphyletic origin and secondary centers of genetic diversity of *Brassica juncea* in China and India. J Hered 104:416–427

Cheung WY, Landry BS, Raney P, Rakow GFW (1998) Molecular mapping of seed quality traits in *Brassica juncea* L. Czern. and Coss. Acta Hort 459:139–147

Choudhary V, Jambhulkar SJ (2016) Genetic improvement of rapeseed mustard through induced mutations. In: Gupta SK (ed) Breeding oilseed crops for sustainable production. Academic Press, San Diego, pp 370–390

Chrungu B, Verma N, Mohanty A, Pradhan A, Shivanna KR (1999) Production and characterization of interspecific hybrids between *Brassica maurorum* and crop brassicas. Theor Appl Genet 98:608–613

Conn KL, Tewari JP (1986) Hyper sensitive reaction induced by *Alternaria brassicae* in *Eruca sativa*, an oil-yielding crucifer. Can J Plant Pathol 8:348

Conn KL, Tewari JP, Dahiya JS (1988) Resistance to *Alternaria brassicae* and phytoalexin-elicitation in rapeseed and other crucifers. Plant Sci 55:21–25

Dang JK, Sangwan MS, Mehta N, Kaushik CD (2000) Multiple disease resistance against four fungal foliar diseases of rapeseed-mustard. Indian Phytopathol 53 (4):455–458

Darwin CR (1876) The effects of cross and self fertilization in the vegetable kingdom. John Murray, London

Deka AC, Goswami NK, Sharma I (2017) Biocontrol prospects of entomopathogenic fungi for management of mustard aphid (*Lipaphis erysimi* Kalt.) on rapeseed-mustard. Adv Appl Sci Res 8(4):21–29

Delourme R, Eber F (1992) Linkage between an isozyme marker and a restorer gene in radish cytoplasmic male sterility of rapeseed (*Brassica napus L*). Theor Appl Genet 85:222–228

Delourme R, Foisset N, Horvais R, Barret P, Champagne G, Cheung WY, Renard M (1998) Characterisation of the radish introgression carrying the *Rfo* restorer gene for the Ogu-INRA cytoplasmic male sterility in rapeseed (*Brassica napus* L.). Theor Appl Genet 97:129–134

Delourme R, Eber F, Renard M (1995) Breeding double low restorer lines in radish cytoplasmic male sterility of rapeseed (*Brassica napus* L). Cambridge, UK, pp 6–8

Dev D, Tewari AK, Upadhyay P, Daniel GR (2020) Identification and nomenclature of *Albugo candida* pathotypes of Indian origin causing white rust disease of rapeseed-mustard. Eur J Plant Pathol 158:987–1004

Dhillon SS, Labana KS, Banga SK (1990) Studies on heterosis and combining ability in Indian mustard (*Brassica juncea* [L.] Coss.). J Res PAU 27:1–8

Disi J, Mei O, Wei D, Ding Y, Qian W (2014) Inheritance of leaf and stem resistance to *Sclerotinia sclerotiorum* in a cross between *Brassica incana* and *Brassica oleracea* var. alboglabra. J Agric Sci 152:146–152

Dixon GR (2007) Vegetable Brassicas and related crucifers. CABI, Wallingford, UK; Cambridge, Massachusetts, USA

Downey RK, Rimmer SR (1993) Agronomic improvement in oilseed Brassicas. Adv Agron 50:1–65

Dufay M, Touzet P, Maurice S, Cuguen J (2007) Modelling the maintenance of male-fertile cytoplasm in a gynodioecious population. Heredity 99(3):349–356

Duhoon SS, Koppar MN (1998) Distribution, collection and conservation of biodiversity in cruciferous oilseeds in India. Genet Resour Crop Evol 45:317–323

Duroc Y, Gaillard C, Hiard S, Tinchant C, Berthom R, Pelletier G, Budar F (2006) Nuclear expression of a cytoplasmic male sterility gene modifies mitochondrial morphology in yeast and plant cells. Plant Sci 170:755–767

Erickson LR, Straus NA, Beversdorf WD (1983) Restriction patterns reveal origins of chloroplast genomes in *Brassica amphiploids*. Theor Appl Genet 65:201–206

Frandsen KJ (1943) The experimental formation of *Brassica juncea* Czern & Coss. Dansk Bot Ark 11:1–17

Fray J, Evans EJ, Lydiate DJ, Arthur AE (1996) Physiological assessment of apetalous flowers and erectophile pods in oilseed rape (*Brassica napus*). J Agric Sci 127:193–200

Fu TD (1981) Production and research on rapeseed in the peoples of Republic in China. Cruciferae Newsl 6:6–7

Garg H (2010) High level of resistance to *Sclerotinia sclerotiorum* in introgression lines derived from hybridization between wild crucifers and the crop Brassica species *B. napus* and *B. juncea*. Field Crop Res 117:51–58

Garg H, Banga S, Bansal P, Atri C, Banga SS (2007) Hybridizing *Brassica rapa* with wild crucifers *Diplotaxis erucoides* and *Brassica maurorum*. Euphytica 156:417–424

Ghosh SK, Gulati SC (2002) Parental diversity to realize maximum hetrosis in Indian mustard (*Brassica juncea* (L.) Czern & Coss). Indian J Genet 62:25–28

Gomez-Campo C (1999a) Biology of *Brassica* coenospecies. Elsevier Science, Amsterdam, p 486

Gomez-Campo C (1999b) Seedless and seeded beaks in the tribe *Brassiceae*. Cruciferae Newslett 21:11–13

Goswami PK, Behl RK (2006) Genetic divergence in Indian mustard. Ann Agric Res 27:187–190

Gudi S, Atri C, Goyal A, Kaur N, Akhtar J, Mittal M, Kaur P, Kaur G, Banga SS (2020) Physical mapping of introgressed chromosome fragment carrying the fertility restoring (Rfo) gene for *Ogura* CMS in *Brassica juncea* L. Czern & Coss. Theor Appl Genet 133:2949–2959

Gupta VP, Sekhon MS, Satija DR (1991) Studies on genetic diversity, heterosis and combining ability in Indian mustard. Indian J Genet 51:448–453

Gupta V, Mukhopadhyay A, Arumugam N, Sodhi YS, Pental D, Pradhan AK (2004) Molecular tagging of erucic acid trait in oilseed mustard (*Brassica juncea*) by QTL mapping and SNPs in *FAE 1* gene. Theor Appl Genet 108:743–749

Gupta S, Tyagi DVS, Rawat RS (2012) Induced chlorophyll mutations and mutagenic effectiveness in Indian mustard (*Brassica juncea* L. Czern & Coss). Progress Agr 12:128–131

Gupta M, Gupta S, Kumar H, Kumar N, Banga SS (2015) Population structure and breeding value of a new type of *Brassica juncea* created by combining A and B genomes from related allotetraploids. Theor Appl Genet 128(2):221–234

Gupta M, Banga SS (2020) Exploiting alien genetic variation for germplasm enhancement in *Brassica* oilseeds. In: Kang MS (ed) Quantitative genetics, genomics and plant breeding, CABI, UK, pp 338–384

Gyawali S, Harrington M, Durkin J, Horner K, Parkin IAP, Hegedus DD (2016) Microsatellite markers used for genome-wide association mapping of partial resistance to *Sclerotinia sclerotiorum* in a world collection of *Brassica napus*. Mol Breed 36:72

Hansen LH, Earle ED (1997) Somatic hybrids between *Brassica oleracea* (L.) and *Sinapis alba* (L.) with resistance to *Alternaria brassicae* (Berk.) Sacc. Theor Appl Genet 94:1078–1085

Harberd DJ, McArthur ED (1980) Meiotic analysis of some species and genus hybrids in the Brassiceae. In: Tsunoda S, Hinata K, Gomez-Campo C (eds) Brassica crops and wild allies. Japan Scientific Societies Press, Tokyo, pp 65–87

Hasan MJ, Rahman H (2018) Resynthesis of *Brassica juncea* for resistance to *Plasmodiophora brassicae* pathotype 3. Breed Sci 68:385–391

Hayat S, Mir BA, Wani AS, Hasan SA, Irfan M, Ahmad A (2011) Screening of salt tolerant genotypes of *Brassica juncea* based on photosynthetic attributes. J Plant Interact 6:53–60

Hayes HK, Immer IR, Smith OC (1955) Methods of plant breeding. McGraw Hill, pp 52–65

Hazel LN (1943) The genetic basis for constructing selection indexes. Genetics 28:476–490

Heyn FW (1976) Transfer of restorer genes from *Raphanus* to cytoplasmic male sterile *Brassica napus*. Cruciferae Newslett 1:15–16

Higgins EE, Howell EC, Armstrong SJ, Parkin IA (2021) A major quantitative trait locus on chromosome A9, *BnaPh1*, controls homoeologous recombination in *Brassica napus*. New Phytol 229:3281–3293

Hu Q, Andersen S, Dixelius C, Hansen L (2002) Production of fertile intergeneric somatic hybrids between *Brassica napus* and *Sinapis arvensis* for the enrichment of the rapeseed gene pool. Plant Cell Rep 21:147–152

Hu X, Sullivan-Gilbert M, Gupta M, Thompson SA (2006) Mapping of loci controlling oleic and linolenic acid contents and development of *fad2* and *fad3* allele-specific marker in canola. Theor Appl Genet 113:497–501

Hu X, Sullivan-Gilbert M, Kubik T, Danielson J, Hnatiuk N, Marchione W, Greene T, Thompson S (2008) Mapping of the *Ogura* fertility restorer gene *Rfo* and development of *Rfo* allele-specifc markers in canola (*Brassica napus L*). Mol Breed 22:663–674

Hua W, Li RJ, Zhan GM, Liu J, Li J, Wang XF, Liu GH, Wang HZ (2011) Maternal control of seed oil content in *Brassica napus*: the role of silique wall photosynthesis. Plant J. https://doi.org/10.1111/j.1365-313X.2011.04802.x

Jagannath A, Bandyopadhyay P, Arumugam N, Gupta V, Burma PK, Pental D (2001) The use of a spacer DNA fragment insulates the tissue-specific expression of a cytotoxic gene (*barnase*) and allows high frequency generation of transgenic male sterile lines in *Brassica juncea* L. Mol Breed 8:11–23

Jagannath A, Arumugam N, Gupta V, Pradhan AK, Burma PK, Pental D (2002) Development of transgenic barstar lines and identification of a male sterile (*barnase*)/restorer (*barstar*) combination for heterosis breeding in Indian oilseed mustard (*Brassica juncea*). Curr Sci 82:46–52

Jain A, Bhatia S, Banga SS, Prakash S, Lakshmikumaran M (1994) Potential use of random amplified polymorphic DNA (RAPD) technique to study the genetic diversity in Indian mustard (*Brassica juncea*) and its relationship to heterosis. Theor Appl Genet 88:116–122

Janeja HS, Banga SS, Lakshmikumaran M (2003) Identification of AFLP markers linked to fertility restorer genes for tournefortii cytoplasmic male-sterility system in *Brassica napus*. Theor Appl Genet 107:148–154

Jarl CI, Bornman CH (1988) Correction of chlorophyll-defective, male sterile winter oilseed rape (*Brassica napus*) through organelle exchange: phenotypic evaluation of progeny. Hereditas 108:97–102

Jing B, Heng S, Tong D, Wan Z, Fu T et al (2012) A male sterility-associated cytotoxic protein Orf288 in *Brassica juncea* causes aborted pollen development. J Exp Bot 63:1285–1295

Kaneko Y, Bang SW (2014) Interspecific and intergeneric hybridization and chromosomal engineering of Brassicaceae crops. Breeding Sci 64:14–22

Kaneko Y, Bang SW, Matsuzawa Y (2009) Distant Hybridization. In: Gupta SK (ed) Biology and breeding of crucifers. Taylor & Francis Group, New York, pp 207–247

Katiyar RK (1997) Pusa Jai Kisan: a new high-yielding mustard variety developed through tissue culture. Indian Farming 45:6–7

Katiyar RK, Chamola R, Yadav M (2007) Effect of alien cytoplasm on expression of heterosis and productivity of Indiana mustard (*Brassica juncea*) hybrids. Indian J Agric Sci 77(3):158–161

Kaur G, Banga SK, Gogna K, Joshi S, Banga SS (2004) *Moricandia arvensis* cytoplasm based system of cytoplasmic male sterility in *Brassica juncea*: reappraisal of fertility restoration and agronomic potential. Euphytica 138:271–276

Kaur H, Banga SS (2015) Genetic divergence in novel determinate variants of resynthesized Indian mustard [*Brassica juncea* (L.) Czern. & Coss.]. Indian J Genet 75(2):260–263

Kaur P, Banga S, Kumar N, Gupta S, Akhatar J, Banga SS (2014) Polyphylectic origin of *Brassica juncea* with *B. rapa* and *B. nigra* (Brassicaceae) participating as cytoplasm donor parents in independent hybridization events. Amer J Bot 101(7):1157–1166

Khatri A, Khan IA, Siddiqui MA, Raza S, Nizamani GS (2005) Evaluation of high yielding mutants of *Brassica juncea* Cv S-9 developed through gamma rays and EMS. Pak J Bot 37:279–284

Khayat PN, Jamaati-e-Somarin S, Zabihi-e-Mahmoodabad R, Yari A, Khayatnezhad M, Gholamin R (2010) Screening of salt tolerance canola cultivars (*Brassica napus* L.). World Appl Sci J 10:817–820

Kim B, Kim K, Yang TJ, Kim S (2019) Identification of a gene responsible for cytoplasmic male-sterility in onions (*Allium cepa L*) using comparative analysis of mitochondrial genome sequences of two recently diverged cytoplasms. Theor Appl Gen 132(2):313–322

Kim B, Kim K, Yang TJ, Kim S (2016) Completion of the mitochondrial genome sequence of onion (*Allium cepa* L.) containing the CMS-S male-sterile cytoplasm and identification of an independent event of the ccmF N gene split. Curr Gen 62(4):873–885

Kirk JTO, Hurlstone CG (1983) Variation and inheritance of erucic acid content in *Brassica juncea*. J Plant Breed 90:331–338

Kirk JTO, Oram RN (1981) Isolation of erucic acid free lines of *Brassica juncea*: Indian mustard now a potential oilseed crop in Australia. J Aust Inst Agric Sci 47:51–52

Kirti PB, Narasimhulu SB, Mohapatra T, Prakash S, Chopra VL (1993) Correction of chlorophyll deficiency in alloplasmic male sterile *Brassica juncea* through recombination between chloroplast genomes. Genet Res 62(1):11–14

Kirti PB, Mohapatra T, Prakash S, Chopra VL (1995) Development of a stable cytoplasmic male sterile line of *Brassica juncea* from somatic hybrid *Trachystoma ballii* + *Brassica juncea*. Plant Breed 114:434–438

Kirti PB, Mohapatra T, Khanna H, Prakash S, Chopra VL (1995) *Diplotaxis catholica* + *Brassica juncea* somatic hybrids: molecular and cytogeneric characterization. Plant Cell Rep 14:593–597

Kirti PB, Banga SS, Prakash S, Chopra VL (1995a) Transfer of *Ogu* cytoplasmic male sterility to *Brassica juncea* and improvement of male sterile through somatic cell fusion. Theor Appl Genet 91:517–521

Kirti PB, Baldev A, Gaikwad K, Bhat SR, Dineshkumar V, Prakash S, Chopra VL (1997) Introgression of a gene restoring fertility to CMS (*Trachystoma*)

Brassica juncea and the genetics of restoration. Plant Breed 116:1179–1182

Kirti PB, Prakash S, Gaikwad K, Dinesh Kumar V, Bhat SR, Chopra VL (1998) Chloroplast substitution overcomes leaf chlorosis in a Moricandiaarvensis-based cytoplasmic male sterile *B. juncea*. Theor Appl Genet 97:1179–1182

Kolte SJ (1985) Diseases of annual edible oilseed crops, vol 2. CRC Press, Boca Raton, FL, p 135

Krishna P, Ghose SK (1992) Heterosis in relation to genetic divergence in rapessed and mustard. J Oilseeds Res 9:169–174

Kular JS, Kumar S (2011) Quantification of avoidable yield losses in oilseed *Brassica* caused by insect pests. J Plant Prot Res 51:38–43

Kumar S (2017) Assessment of avoidable yield losses in crop brassicas by insect-pests. J Entomol Zool Stud 5:1814–1818

Kumar A, Elston J (1992) Genotypic differences in leaf water relations between *Brassica juncea* and *B. napus*. Ann Bot 70:3–9. https://doi.org/10.1093/oxfordjournals.aob.a088436

Kumar P, Singh MM (1987) Inheritance of seed weight in Indian mustard. J Oilseeds Res 4:297–299

Kumar A, Singh DP (1998) Use of physiological indices as a screening technique for drought tolerance in oil seeds *Brassica* species. Ann Bot 81:413–420

Kumar PR, Arora RK, Singh NP, Yadav RC, Kumar P (1990) A study of heterosis in Indian mustard. Acta Agron Hung 39(1–2):137–143

Kumar M, Singh KP, Singh D (2007) Genetic divergence analysis for seed yield, its parameters and oil content in Indian mustard [*Brassica juncea* (L.) Czern & Coss]. Res Crops 8:183–185

Kumar G, Purty RS, Sharma MP, Singla-Pareek SL, Pareek A (2009) Physiological responses among *Brassica* species under salinity stress show strong correlation with transcript abundance for SOS pathway-related genes. J Plant Physiol 166:507–520

Kumar S, Atri C, Sangha MK, Banga SS (2011) Screening of wild crucifers for resistance to mustard aphid, *Lipaphis erysimi* (Kaltenbach) and attempt at introgression of resistance gene(s) from *Brassica fruticulosa* to *Brassica juncea*. Euphytica 179:461–470

Kumar P, Vasupalli N, Srinivasan R, Bhat SR (2012) An evolutionarily conserved mitochondrial *orf108* is associated with cytoplasmic male sterility in different alloplasmic lines of *Brassica juncea* and induces male sterility in transgenic *Arabidopsis thaliana*. J Exp Bot 63:2921–2932

Kumar P, Thakral NK, Yadav TP, Lekhraj, Chandra N (1998) Inheritance of yield attributing traits In Indian mustard (*Brassica juncea* L.) Cruciferae Newslt 20:73–74

Kumar C (1990) Drought management: research on rapeseed and mustard. In: Proceedings of Indo-Swedish symposium. The Swedish University of Agricultural Sciences, Uppsala, Sweden, 4–6 Sept 1989, pp 44–48

L'Homme Y, Brown GG (1993) Organizational differences between cytoplasmic male sterile and male fertile Brassica mitochondrial genomes are confined to a single transposed locus. Nucleic Acids Res 21 (8):1903–1909

L'HommeY SRJ, LiXQ HA, Brown GG (1997) Brassica nap cytoplasmic male sterility is associated with expression of a mtDNA region containing a chimeric gene similar to the pol CMS-associated *orf224* gene. Curr Genet 31:325–335

Labana KS, Banga SS (1984) Floral biology in Indian mustard (*Brassica juncea* L.). Genetica Agraria 38:131–138

Labana KS, Banga SK (1989) Transfer of Ogura cytoplasmic male sterility of *Brassica napus* into genetic background of *Brassica juncea*. Crop Improv 16:82–83

Labana KS, Badwal SS, Chaurasia BD (1975) Heterosis and combining ability analysis in (*Brassica juncea* (L.) Czern & Coss.). Crop Improv 2:46–51

Labana KS, Chaurasia BD, Sangha AS (1987) Inheritance of oil content in Indian mustard. Indian J Agric Sci 57:539–543

Landgren M, Zetterstrand M, Sundberg E, Glimelius K (1996) Alloplasmic male-sterile Brassica lines containing *B. tournefortii* mitochondria express an ORF 3′ of the *atp6* gene and a 32 kDa protein. Plant Mol Biol 32:879–890

Larcher W (2003) Physiological plant ecology: ecophysiology and stress physiology of functional groups. Springer Science & Business Media

Lelivelt CLC, Krens FA (1992) Transfer of resistance to the beet cyst nematode (*Heterodera schachtii* Schm.) into the *Brassica napus* L. gene pool through intergeneric somatic hybridization with *Raphanus sativus* L. Theor Appl Genet 83:887–894

Lelivelt CLC, Leunissen EHM, Frederiks HJ, Helsper JPFG, Krens FA (1993) Transfer of resistance to the beet cyst nematode (*Heterodera schachtii* Schm.) from *Sinapis alba* L. (white mustard) to the *Brassica napus* L. gene pool by means of sexual and somatic hybridization. Theor Appl Genet 85:688–696

Levania M, Chauhan SVS (2006) Efficacy of three chemical hybridizing agents in *Brassica juncea* var. Pusa Bold. J Cyto Genet 7:141–148

Li ZY, Wang YP (2017) Cytogenetics and germplasm enrichment in *Brassica* allopolyploids in China. J Integr Agric 16:2698–2708

Li Z, Liu HL, Luo P (1995) Production and cytogenetics of intergeneric hybrids between *Brassica napus* and *Orychophragmus violaceus*. Theor Appl Genet 91:131–136

Li CX, Liu SY, Sivasithamparam K, Barbetti MJ (2009) New sources of resistance to Sclerotinia stem rot caused by *Sclerotinia sclerotiorumin* Chinese and Australian *Brassica napus* and *B. juncea* germplasm screened under Western Australian conditions. Australasian Plant Path 38:149–152

Li Y, Chen J, Bennett R, Kiddle G, Wallsgrove R (1999) Breeding, inheritance, and biochemical studies on

Brassica napus cv. Zhongyou 821: tolerance to *Sclerotinia sclerotiorum* (stem rot). In: Proceedings of the 10th international rapeseed congress, Chap, Canberra, ACT Vol 2, p 230
Liu Z, Guan C, Zhao F, Chen S (2005) Inheritance and mapping of a restorer gene for the rapeseed cytoplasmic male sterile line 681A. Plant Breed 124:5–8
Liu H, Fu T, Yang X (1987) Discovery and studies on Polima CMS line. In: Proceedings of 7th international rapeseed congress, Pozman, Poland, pp 69–78
Lodhi B, Thakral NK, Singh D, Avtar R, Bahadur R (2013) Genetic diversity analysis in Indian mustard (*Brassica juncea*). J Oilseed Brassica 4:57–60
Love HR, Rakow G, Raney JP, Downey RK (1990) Genetic control of 2-propenyl and 3-butenyl glucosinolate synthesis in mustard. Can J Plant Sci 70:425–429
Luo D, Xu H, Liu Z, Guo J, Li H et al (2013) A detrimental mitochondrial-nuclear interaction causes cytoplasmic male sterility in rice. Nature Genet 45:573
Mahla SVS, Mor BR, Yadava JS (1990) Effect of mutagen on yield and its component characters in mustard. Haryana Agric Univ J Res 20:259–264
Mahla SVS, Mor BR, Yadava JS (1991) Mutagen induced polygenic variability in some mustard (*Brassica juncea L*) varieties and their hybrids. J Oilseed Res 8:173–177
Mahmood T, Rahman MH, Stringam GR, Yeh F, Good AG (2006) Identification of quantitative trait loci (QTL) for oil and protein contents and their relationships with other seed quality traits in *Brassica juncea*. Theor Appl Genet 113:1211–1220
Makarenko MS, Usatov AV, Tatarinova TV, Azarin KV, Logacheva MD, Gavrilova VA, Horn R (2019) Characterization of the mitochondrial genome of the MAX1 type of cytoplasmic male-sterile sunflower. BMC Plant Biol 19(1):51
Malik M, Vyas P, Rangaswamy NS, Shivanna KR (1999) Development of two new cytoplasmic male sterile lines of *Brassica juncea* through wide hybridization. Plant Breed 118:75–78
Malkhandale JD (1993) Diallel analysis for some metric traits in Indian mustard. Ann Plant Physiol 7:247–250
Mariani C, De Beuckeleer M, Truettner J, Leemans J, Goldberg RB (1990) Induction of male sterility in plants by a chimaeric ribonuclease gene. Nature 347:737–741
Mariani C, Gossele V, De Beuckeleer M, De Block M, Goldberg RB, De Greef W, Leemans J (1992) A chimaeric ribonuclease-inhibitor gene restores fertility to male sterile plants. Nature 357:384–387
Meah MB, Hau B, Siddiqua MK (2002) Relationships between disease parameters of Alternaria-blight (*Alternaria brassicae*) and yield of mustard. J Plant Dis Protec 109:243–251
Meena H, Ram B, Kumar A, Brijendra SP, Meena V, Singh SD (2014) Heterobeltiosis and standard heterosis for seed yield and important traits in *Brassica juncea*. J Oilseed Brassica 5(2):134–140
Meena CP, Chauhan JS, Singh M, Meena ML, Singh KH (2014) Evaluation of Indigenous and Exotic Germplasm of Indian mustard [*Brassica juncea* (L.) Czern & Coss] for morpho-physiological and quality characters. Indian J Plant Genet Resour 27:118–124
Mei J, Qian L, Disi JO, Yang X, Li Q, Li J, Frauen M, Cai D, Qian W (2011) Identification of resistant sources against *Sclerotinia sclerotiorum* in Brassica species with emphasis on *B. oleracea*. Euphytica 177:393–399
Mei J, Ding Y, Lu K, Wei D, Liu Y, Disi JO, Li J, Liu L, Liu S, McKay J, Qian W (2013) Identification of genomic regions involved in resistance against *Sclerotinia sclerotiorum* from wild *Brassica oleracea*. Theor Appl Genet 126:549–556
Menczel L, Morgan A, Brown S, Maliga P (1987) Fusion mediated combination of Ogura-type cytoplasmic male sterility with *Brassica napus* plastids using X-irradiated CMS protoplasts. Plant Cell Rep 6:98–101
Meyer RS, DuVal AE, Jensen HR (2012) Patterns and processes in crop domestication: an historical review and quantitative analysis of 203 global food crops. New Phytol 196:29–48
Mohammad A, Olsson AI, Sevensk H (1990) Drought management. In: Olsson I, Kumar PR (eds) Research on rapeseed and mustard. In: Proceedings of the Indo-Swedish symposium, Swedish University of Agricultural Sciences, Uppsala, Sweden, 4–6 Sept 1989, pp 49–67
Mohanty A, Chrungu B, Verma N, Shivanna KR (2009) Broadening the genetic base of crop brassicas by production of new intergeneric hybrid. Czech J Genet Plant Breed 45:117–122
Monalisa P, Singh NB, Singh NG, Laishram JM (2005) Genetic divergence and combining ability in relation to heterosis in Indian mustard [*Brassica juncea* (L.) Czern and Coss] for seed yield, its attributes and oil yield. Indian J Genet 65:302–304
Mondal KK, Bhattacharya R, Koundal K, Chatterjee S (2007) Transgenic Indian mustard (*Brassica juncea*) expressing tomato glucanase leads to arrested growth of *Alternaria brassicae*. Plant Cell Rep 26:247–252
Muangprom A, Osborn TC (2004) Characterisation of dwarf gene in *Brassica rapa*, including the identification of a candidate gene. Theor Appl Genet 108:1378–1384
Mukherjee AK, Mohapatra T, Varshney A, Sharma R, Sharma RP (2001) Molecular mapping of a locus controlling resistance to *Albugo candida* in Indian mustard. Plant Breed 120:483–497
Munir S, Hussain E, Bhatti KH, Nawaz K, Hussain K, Rashid R, Hussain I (2013) Assessment of inter-cultivar variations for salinity tolerance in winter radish (*Raphanus sativus L*) using photosynthetic attributes as effective selection criteria. World Appl Sci J 21:384–388
Nagaharu U (1935) Genomic analysis in Brassica with special reference to the experimental formation of *B. napus* and peculiar mode of fertilization. Jpn J Bot 7:389–452

Narasimhulu SB, Kirti PB, Bhatt SR, Prakash S, Chopra VL (1994) Intergeneric protoplast fusion between *Brassica carinata* and *Camelina sativa*. Plant Cell Rep 13:657–660

Nasri M, Khalatbari M, Zahedi H, Paknejad F, Moghadam HRT (2008) Evaluation of micro and macro elements in drought stress condition in cultivars of rapeseed (*Brassica napus* L.). Am J Agric Biol Sci 3(3):579–583

Navabi ZK, Strelkov SE, Good AG, Thiagarajah MR, Rahman MH (2010) Brassica B-genome resistance to stem rot (*Sclerotinia sclerotiorum*) in a doubled haploid population of *Brassica napus* × *Brassica carinata*. Can J Plant Pathol 32:237–246

Navabi ZK, Parkin IA, Pires JC, Xiong Z, Thiagarajah MR, Good AG, Rahman MH (2010a) Introgression of B-genome chromosomes in a doubled haploid population of *Brassica napus* x *B. carinata*. Genome 53:619–629

Nayanakantha NMC, Rawat S, Ali S, Grover A (2016) Differential expression of defense-related genes in *Sinapis alba* and *Brassica juncea* upon the infection of *Alternaria brassicae*. Trop Agric Res 27:123–136. https://doi.org/10.4038/tar.v27i2.8161

Nayar GG, George KP (1969) X-ray induced early flowering, appressed pod mutant in *Brassica juncea* Coss. In: Proceedings of the symposium on radiation and radiomimetic substances in mutation breeding, Sept 26–29, Bombay, India, pp 409–413

Nayar GG (1968) Seed colour mutation in *Brassica juncea* induced by radioactive phosphorus. P Sci Cult 34:421–422

Nazir N, Ashraf M, Rasul E (2001) Genomic relationships in oilseeds Brassicas with respect to salt tolerance-photosynthetic capacity and ion relations. Pak J Bot 33:483–501

NBPGR (2022) Germpalsm registration information system. National bureau of plant genetic resources. http://www.nbpgr.ernet.in:8080/registration/InventoryofGermplasm.aspx

Ogura H (1968) Studies on the new male sterility in Japanes radish with special reference to the utilization of this sterility towards the practical raising of hybrid seed. Mem Fac Agric Kagoshima Uni 6:40–75

Ohashi Y, Nakayama N, Saneoka H, Fujita K (2006) Effects of drought stress on photosynthetic gas exchange, chlorophyll fluorescence and stem diameter of soybean plants. Biol Plant 50(1):138–141

Olsson G (1960) Species crosses within the genus Brassica I. Artificial *Brassica juncea* Coss. Hereditas 4:171–222

Palial S, Kumar S, Sharma S (2018) Biochemical changes in *Brassica juncea*-fruticulosa introgression lines after *Lipaphis erysimi* (Kaltenbach) infestation. Phytoparasitica 46:499–509

Palial S, Kumar S, Atri C, Sharma S, Banga SS (2021) Antixenosis and antibiosis mechanisms of resistance to turnip aphid, *Lipaphis erysimi* (Kaltenbach) in *Brassica juncea-fruticulosa* introgression lines. J Pest Sci. https://doi.org/10.1007/s10340-021-01418-8

Palmer JD, Shields CR, Cohen DB, Orton TJ (1983) Chloroplast DNA evolution and the origin of amphiploid Brassica species. Theor Appl Genet 65:181–189

Panjabi-Massand P, Yadava SK, Sharma P, Kaur A, Kumar A, Arumugam N, Sodhi YS, Mukhopadhyay A, Gupta V, Pradhan AK, Pental D (2010) Molecular mapping reveals two independent loci conferring resistance to *Albugo*. Theor Appl Genet 121:137–145

Pathania A, Bhat SR, Dinesh Kumar V, Asutosh PS, Chopra VL (2003) Cytoplasmic male sterility in alloplasmic *Brassica juncea* carrying *Diplotaxis catholica* cytoplasm: molecular characterization and genetics of fertility restoration. Theor Appl Genet 107:455–461

Pathania A, Kumar R, Dinesh Kumar V, Asutosh, Dwivedi KK, Kirti PB, Prakash S, Chopra VL, Bhat SR (2007) A duplication of coxI gene is associated with CMS (*Diplotaxis catholica*) *Brassica juncea* derived from somatic hybridization with *Diplotaxis catholica*. J Genet 86:93–101

Pelletier G, Primard C, Vedel F, Chétrit P, Rémy R, Rousselle P, Renard M (1983) Intergeneric cytoplasmic hybridization in Cruciferae by protoplast fusion. Genet Mol Biol 191:244–250

Pires JC, Gaeta RT (2011) Structural and functional evolution of resynthesized polyploids. In: Schmidt R, Bancroft I (eds) Genetics and genomics of the Brassicaceae. Springer, New York, pp 195–214

Pradhan AK, Mukhopadhyay A, Pental D (1991) Identification of the putative cytoplasmic donor of a CMS system in *Brassica juncea*. Plant Breed 106:204–208

Pradhan AK, Prakash S, Mukhopadhyay A, Pental D (1992) Phylogeny of *Brassica* and allied genera based on variation in chloroplast and mitochondrial DNA patterns: molecular and taxonomic classifications are incongruous. Theor Appl Genet 85:331–340

Pradhan AK, Sodhi YS, Mukhopadhayay A, Pental D (1993) Hetrerosis breeding in Indian mustard [*Brassica juncea* (L.) Czern. & Coss.] Analysis of component characters contributing to heterosis for yield. Euphytica 69:219–229

Pradhan AK, Gupta V, Mukhopadhyay A, Sodhi YS, Arumugam N, Pental D (2003) A high-density linkage map in *Brassica juncea* using AFLP and RFLP markers. Theor Appl Genet 106:607–614

Prakash S (1973) Artificial synthesis of *Brassica juncea* Coss. Genetica 44:249–263

Prakash S, Hinata K (1980) Taxonomy, cytogenetics and origin of crop Brassicas: a review. Opera Bot 55:1–57

Prakash S, Kirti PB, Bhat SR, Gaikwad K, Kumar VD, Chopra VL (1998) A *Moricandia arvensis* based cytoplasmic male sterility and fertility restoration system in *Brassica juncea*. Theor Appl Genet 97:488–492

Prakash S, Ahuja I, Upreti HC, Kumar DV, Bhat SR, Kirti PB, Chopra VL (2001) Expression of male sterility in alloplasmic *Brassica juncea* with *Erucastrum canariense* cytoplasm and development of a fertility restoration system. Plant Breed 120:479–482

Prakash S, Bhat SR, Quiros CF, Kirti PB, Chopra VL (2009) Brassica and its close allies: cytogenetics and evolution. Plant Breed Rev 31:21–187

Prakash S, Chopra VL (1990) Male sterility caused by cytoplasm of *Brassica oxyrrhina* in *B. campestris* and *B. juncea*. Theor Appl Genet 79(2):285–287

Prakash S, Kirti PB, Chopra VL (1995) Cytoplasmic male sterility systems other than *ogu* and *polima*-current status. In: Proceedings of 9th international rapeseed congress, Cambridge 1:44–48

Prasad R, Saxena D, Chandra S (2003) Yield losses by Alternaria blight in promising genotypes of Indian mustard. Indian Phytopathol 56:205–206

Primard-Brisset C, Poupard JP, Horvais R, Eber F, Pelletier G, Renard M, Delourme R (2005) A new recombined double low restorer line for the Ogu-INRA CMS in rapeseed (*Brassica napus L*). Theor Appl Genet 111:736–746

Qi XH, Zhang MF, Yang JH (2007) Molecular phylogeny of Chinese vegetable mustard (*Brassica juncea*) based on the internal transcribed spacers (ITS) of nuclear ribosomal DNA. Genet Resour Crop Evol 54:1709–1716

Qian B, Jing Q, Bélanger G, Shang J, Huffman T, Liu J, Hoogenboom G (2018) Simulated canola yield responses to climate change and adaptation in Canada. Agron J 110:133–146

Rai UK (1958) X-ray induced appressed pod mutant in *Brassica juncea*. Sci Cult 24:46–47

Rai B (1995) Development of hybrid cultivars in oilseed Brassica: status and strategies. J Oilseed Res 12 (2):239–244

Rai SK, Sandhu R, Jat L, Kumar A, Kiran U, Mukhtar S, Kohli S, Rai GK (2017) Study of heterosis and combining ability for yield and its component traits in *Brassica juncea* L. Intl J Curr Microbiol App Sci 6 (12):2570–2579

Rakow G (1973) Selection of linoleic and linolenic fatty acid in rapeseed with mutation breeding [In German]. J Plant Breed 69:62–68

Ramanujam S, Srinivasachar D (1943) Cytogenetical investigations in Brassica and the artificial synthesis of *Brassica juncea*. Indian J Genet 3:73–88

Ramchiary N, Bisht NC, Gupta V, Mukhopadhyay A, Arumugam N, Sodhi YS, Pental D, Pradhan AK (2007a) QTL analysis reveals context-dependent loci for seed glucosinolate trait in the oilseed *Brassica juncea*: importance of recurrent selection backcross (RSB) scheme for the identification of 'true' QTL. Theor Appl Genet 116:77–85

Ramchiary N, Padmaja KL, Sharma S, Gupta V, Sodhi YS, Mukhopadhyay A, Arumugam N, Pental D, Pradhan AK (2007b) Mapping of yield influencing QTL in Brassica juncea: implications for breeding of a major oilseed crop of dryland areas. Theor Appl Genet 115:807–817

Rana J (2005) Performance of *Lipaphis erysimi* (Homoptera: Aphididae) on different Brassica species in a tropical environment. J Pest Sci 78:155–160

Rana K, Atri C, Gupta M, Akhatar J, Sandhu PS, Kumar N, Jaswal R, Barbetti MJ, Banga SS (2017) Mapping resistance responses to *Sclerotinia* infestation in introgression lines of *Brassica juncea* carrying genomic segments from wild Brassicaceae *B. fruticulosa*. Sci Rep 7:1–12

Rana K, Atri C, Akhatar J, Kaur R, Goyal A, Singh MP, Kumar N, Sharma A, Sandhu PS, Kaur G, Barbetti MJ, Banga SS (2019) Detection of first marker trait associations for resistance against *Sclerotinia sclerotiorum* in *Brassica juncea–Erucastrum cardaminoides* introgression lines. Front Plant Sci 10:1015

Rao GU, Batra VS, Prakash S, Shivanna KR (1994) Development of a new cytoplasmic male sterility system in *Brassica juncea* through wide hybridization. Plant Breed 112:171–174

Rawat DS, Anand IJ (1979) Male sterility in Indian mustard. Indian J Genet 39:412–415

Ray K, Bisht N, Pental D, Burma P (2007) Development of barnase/barstartransgenics for hybrid seed production in Indian oilseed mustard (*Brassica juncea* L. Czern. & Coss.) using a mutant acetolactate synthase gene conferring resistance to imidazolinone-based herbicide 'Pursuit'. Curr Sci 93(10):1390–1396

Ripley VL, Thompson SA, Ehlert ZC (2014) Omega-9 quality *Brassica juncea*. US Patent 8,637,740, 28 Jan 2014

Robbelen G (1990) Mutation breeding for quality improvement a case study for oilseed crops. Mutation Breeding 6:1–43

Robbelen G, Nitsch A (1975) Genetical and physiological investigations on mutants for polyeonic fatty acid in rapeseed *Brassica napus* L.I. Selection and description of new mutants. Z Pflanzuchtg 75:93–105

Robbelen G, Rakow G (1970) Selection for fatty acids in rapeseed. In: Proceedings of the international conference on rapeseed. Sainte-Adele, Canada, pp 476–490

Rohilla HR, Bhatnagar P, Yadav PR (2004) Chemical control of mustard aphid with newer and conventional insecticides. Indian J Entomol 66:30–32

Rucker B, Robbelen G (1997) Mutants of *Brassica napus* with altered seed lipid fatty acid composition. In: Proceedings of the 12th international symposium. plant lipids. Kluwer Academic Publisher Dordrecht, The Netherlands, pp 316–318

Saharan GS, Kaushik CD, Kaushik JC (1988) Sources of resistance and epidemiology of white rust of mustard. Indian Phytopathol 41:96–99

Seidle E, Rude S, Petrie A (1995) Final report: the effect of Alternaria black spot of canola on seed quality and seed yield and studies on disease control. Agric Dev Fund Canada, Saskatoon, p 41

Shalini S, Sheriff RA, Kulkarni RS, Venkantarmana P (2000) Correlation and path analysis of Indian mustard germplasm. Res Crops India 1:226–229

Sharma G, Kumar VD, Haque A, Bhat SR, Prakash S, Chopra VA (2002) Brassica coenospecies: a rich reservoir for genetic resistance to leaf spot caused by *Alternaria brassicae*. Euphytica 125:411–417

Sharma P, Sardana V, Banga S, Salisbury PA, Banga SS (2011) Morpho-physiological responses of oilseed rape (*Brassica napus* L.) genotypes to drought stress. In: Proceedings of 13th international rapeseed congress, 5–9 June 2011, Prague, Czech Republic, pp 238–239

Sharma D, Nanjundan J, Singh L, Singh SP, Parmar N, Sujith Kumar MS, Singh KH, Mishra AK, Singh R, Verma KS, Thakur AK (2020) Genetic diversity in leafy mustard (*Brassica juncea* var. *rugosa*) as revealed by agro-morphological traits and SSR markers. Physiol Mol Biol Plants 26(10):2005–2018

Shekhawat N, Jadeja GC, Singh J, Ramesh (2014) Genetic diversity analysis in relation to seed yield and its component traits in Indian mustard (*Brassica juncea* Czern. & Coss.). Bioscan 9:713–717

Siddiqui MH, Mohammad F, Khan MN (2009) Morphological and physio-biochemical characterization of *Brassica juncea* (L.) Czern. & Coss. genotypes under salt stress. J Plant Interact 4:67–80

Sinaki JM, Heravan EM, Ra AHS, Noormohammadi G, Zarei G (2007) The effects of water deficit during growth stages of canola (*Brassica napus* L.). Am Eurasian J Agric Environ Sci 4:417–422

Singh M, Brown GG (1993) Characterization of expression of a mitochondrial gene region associated with the Brassica "Polima" CMS: developmental influences. Curr Genet 24:316–322

Singh V, Chauhan SVS (2001) Benotriazole-A new chemical hybridizing agents for *Brassica juncea* (L.). J Cyt Genet 2:81–83

Singh V, Chauhan SVS (2003) Bud pollination and hybrid seed production in detergent induced male sterile plants of *Brassica juncea* (L.). Plant Breed 122 (5):421–425

Singh V, Chauhan SVS (2004) Evaluation of three chemical hybridizing agents in *Brassica juncea* (L.) Czern. & Coss. Brassica 6(3 & 4):71–73

Singh KP, Sareen PK (2004) Induced podding mutants of Indian mustard (*Brassica juncea* L. Czern. & Coss.). Cruciferae Newsl 25:17

Singh KH, Srivastava KK (2006) Characterization of different cytoplasmic male sterility systems in Indian mustard (*Brassica juncea* L. Czern. & Coss.). Plant Breed 125(1):72–76

Singh DP, Sangwan VP, Pannu RK, Chaudhary BD (1996a) Comparison of osmotic adjustment in leaves and siliquae of oilseed Brassicae. Indian J Plant Physiol 1:284–285

Singh JN, Yadav MC, Sheik IA (1996b) Genetical studies for yield and oil content in *Brassica juncea* (L.) Czern. & Coss. Indian J Genet 56:299–304

Singh M, Chauhan JS, Kumar PR (2003) Response of different rapeseed-mustard varieties for growth, yield and yield component under irrigated and rainfed condition. Indian J Plant Physiol 8:53–59

Singh R, Singh D, Sivasithamparam SK, Yadav N, Salisbury P (2008) Management of Sclerotinia rot of oilseed Brassicas: a focus on India. Brassica 10:1–27

Singh KH, Shakya R, Mahawar RR (2014) Genetic diversity and patterns of variation among Indian mustard [*Brassica juncea* (L.) Czern. & Coss.] varieties. SABRAO J Breed Genet 46:329–339

Singh KH, Shakya R, Nanjundan R, Thakur AK, Singh K, Singh KK (2018) Genetic diversity assessment and characterization of Indian mustard (Brassicaceae) varieties using agro-morphological traits. Indian J Plant Genet Resour 36(1):44–50

Singh RK, Singh SK, Singh A (2020) Heterosis and combining ability analysis in Indian mustard (*Brassica juncea*). Int J All Res Educ Sci Methods 8(12):1618–1623

Singh KP, Kumari P, Rai PK (2021) Current status of the disease-resistant gene(s)/QTLs, and strategies for improvement in *Brassica juncea*. Front Plant Sci 12:295

Skarzhinskaya M, Landgren M, Glimelius K (1996) Production of intertribal somatic hybrids between *Brassica napus* L. and *Lesquerella fendleri* (Gray) Wats. Theor Appl Genet 93:1242–1250

Smith FH (1936) A discriminate function for plant selection. Ann Eugen 7:240–250. https://doi.org/10.1111/j.1469-1809.1936.tb02143.x

Sodhi YS, Mukhopadhyay A, Arumugam N, Verma JK, Gupta V, Pental D, Pradhan AK (2002) Genetic analysis of total glucosinolate in crosses involving a high glucosinolate Indian variety and a low glucosinolate line of *Brassica juncea*. Plant Breed 121:508–511

Sodhi YS, Chauhan A, Verma JK, Arumugam N, Mukhopadhyay A, Gupta V, Pental D, Pradhan AK (2006) A new cytoplasmic male sterility system for hybrid seed production in Indian oilseed mustard (*Brassica juncea*). Theor Appl Genet 114:93–99

Song KM, Tang K, Osborn TC (1993) Development of synthetic *Brassica* amphidiploids by reciprocal hybidization and comparision to natural amphidiploids. Theor Appl Genet 86:811–821

Song KM, Lu P, Tang KL, Osborn TC (1995) Rapid genome change in synthetic polyploids of *Brassica* and its implications for polyploid evolution. Proc Natl Acad Sci USA 92:7719–7723

SOPA (2021) India oilseeds-Area, production and productivity. The Soybean Processors Association of India. http://www.sopa.org/india-oilseeds-area-production-and-productivity

Spect CE, Diederichsen A (2001) Brassica. In: Hanelt P (ed) Mansfeld's encyclopedia of agricultural and horticultural crops. Springer, Berlin, pp 1453–1456

Srivastava A, Gupta V, Pental D, Pradhan AK (2001) AFLP-based genetic diversity assessment amongst agronomically important natural and some newly synthesized lines of *Brassica juncea*. Theor Appl Genet 102:193–199

Srivastava A, Mukhopadhyay A, Arumugam M, Gupta V, Verma JK, Pental D, Pradhan AK (2004) Resynthesis of *Brassica juncea* through interspecific crosses between *B. rapa* and *B. nigra*. Plant Breed 123:204–206

Stringam GR, Thiagarajah MR (1995) Inheritance of alkenyl glucosinolates in traditional and microspore-derived doubled haploid populations of *Brassica juncea* (L.) Czern & Coss. In: Proceedings of 9th international rapeseed congress, 4–7 July 1995. Cambridge, UK, GCIRC, pp 804–806

Tanaka Y, Tsuda M, Yasumoto K, Yamagishi H, Terachi T (2012) A complete mitochondrial genome sequence of Ogura-type malesterile cytoplasm and its comparative analysis with that of normal cytoplasm in radish (*Raphanus sativus* L). BMC Genom 13:352

Tang H, Zheng X, Li C, Xie X, Chen Y et al (2017) Multi-step formation, evolution, and functionalization of new cytoplasmic male sterility genes in the plant mitochondrial genomes. Cell Res 27(1):130–146

Tanhuanpaa P, Vikki J, Vihinen M (1998) Mapping and cloning of *FAD2* gene to develop allele-specific PCR for oleic acid in spring turnip rape (*Brassica rapa* ssp. oleifera). Mol Breed 4:543–550

Tewari JP (1991) Structural and biochemical bases of the black spot disease of crucifers. Adv Struct Biol 1:325–349

Tewari JP, Conn KL (1993) Reaction of some wild crucifers to *Alternaria brassicae*. Bull OILB SROP 16:53–58

Thakral NK, Prakash K (1998) Combining ability for vigour characters under normal and saline environments in Indian mustard [*Brassica juncea* (L.) Czern. & Coss.]. J Oilseed Res 15:234–237

Thakur HL, Bhateria S (1993) Heterosis and inbreeding depression in Indian mustard. Indian J Genet 53 (1):60–65

Thakur HL, Zarger MA (1989) Heterosis in relation to genetic diversity and specific combining ability in Indian mustard. Indian J Genet 51:448–453

Thurling N, Depittayanan V (1992) EMS Induction of early flowering mutants in spring rape (*Brassica napus*). Plant Breed 108:177–184

Tian E, Roslinsky V, Cheng B (2014) Molecular marker-assisted breeding for improved Ogura CMS restorer line (Rfo) and mapping of the restorer gene (*Rfo*) in *Brassica juncea*. Mol Breed 34:1361–1371

Tonosaki K, Michiba K, Bang SW, Kitashiba H, Kaneko Y, Nishio T (2013) Genetic analysis of hybrid seed formation ability of *Brassica rapa* in intergeneric crossings with *Raphanus sativus*. Theor Appl Genet 126:837–846

Tripathi MK, Mishra AS (2007) Glucosinolates in animal nutrition: a review. Anim Feed Sci Technol 132:1–27

Uchimiya H, Wildman SG (1978) Evolution of fraction 1 protein in relation to origin of amphidiploid Brassica species and other members of the Cruciferae. J Hered 69:299–303

Uloth M, You M, Finnegan P, Banga S, Banga S, Sandhu P, Yi H, Salisbury P, Barbetti M (2013) New sources of resistance to Sclerotinia sclerotiorum for crucifer crops. Field Crops Res 154:40–52. https://doi.org/10.1016/j.fcr.2013.07.013

USDA (2021) Oilseeds: world markets and trade. United States Department of Agriculture. https://www.fas.usda.gov/data/oilseeds-world-markets-and-trade

Vaghela PO, Thakkar DA, Bhadauria HS, Sutariya DA, Parmar SK, Prajapati DV (2011) Heterosis and combining ability for yield and its component traits in Indian mustard [*Brassica juncea* (L.)]. J Oilseed Brassica 2:39–43

Varshney A, Mohapatra T, Sharma RP (2004) Development and validation of CAPS and AFLP markers for white rust resistance gene in *Brassica juncea*. Theor Appl Genet 109:153–159. https://doi.org/10.1007/s00122-004-1607-y

Vaughan G (1977) A multidisciplinary study of the taxonomy and origin of Brassica crops. Bioscience 27:35–40

Vaughan JG, Hemingway JS, Schofield HJ (1963) Contributions to a study of variation in *Brassica juncea* Coss. & Czern. J Linn Soc Bot 58:435–447

Vavilov NI (1949) The origin, variation, immunity and breeding of cultivated plants. Chron Bot 13:1–364

Verma OP, Singh HP, Singh PV (1998) Heterosis and inbreeding depression in Indian mustard. Cruciferae Newsl 20:75–76

Verma OP, Kushwaha GD, Singh HP (2000) Heterosis in relation to genetic diversity in Indian mustard. Cruciferae Newsl 22:93–94

Verma PR, Saharan GS (1994) Monograph on Alternaria diseases of crucifers. Saskatoon Research Centre, Technical Bulletin 1994–6E, Agriculture and Agri-Food Canada, Saskatoon, SK, Canada, p 162

Wan Z, Jing B, Tu J, Ma C, Shen J, Yi B, Wen J, Huang T, Wnag X, Fu T (2008) Genetic characterization of a new cytoplasmic male sterility system (hau) in *Brassica juncea* and its transfer to *B. napus*. Theor Appl Genet 116:355–362

Wang HZ, Liu GH, Zheng YB, Wang XF, Yang Q (2004) Breeding of the *Brassica napus* cultivar Zhongshuang 9 with high-resistance to *Sclerotinia sclerotiorum* and dynamics of its important defense enzyme activity. Sci Agri Sin 37:23–28

Wang XF, Liu GH, Yang Q, Hua W, Liu J, Wang HZ (2010) Genetic analysis on oil content in rapeseed (*Brassica napus* L.). Euphytica 173:17–24

Warwick SI, Black LD (1991) Molecular systematics of Brassica and allied genera (Subtribe Brassicinae Brassicae) chloroplast genome and cytodeme congruence. Theor Appl Genet 82:81–92

Warwick SI, Beckie HJ, Thomas AG, McDonald T (2000) The biology of Canadian weeds. 8. *Sinapis arvensis*. L. (updated). Can J Plant Sci 80:939–961

Warwick SI, Gugel RK, McDonald T, Falk KC (2006) Genetic variation of Ethiopian mustard (*Brassica carinata* A. Braun) germplasm in western Canada. Genet Resour Crop Evol 53:297–312

Warwick SI (2011) Brassicaceae in agriculture. In: Schmidt R, Bancroft I (eds) Genetics and genomics of the Brassicaceae. Plant genetics and genomics: crops and models. Springer, New York, pp 33–65

Weselake RJ, Taylor DC, Rahman MH, Shah S, Laroche A, McVetty PBE, Harwood JL (2009) Increasing the flow of carbon into seed oil. Biotechnol Adv 27:866–878

Wu J, Zhao Q, Liu S, Shahid M, Lan L (2016) Genome-wide association study identifies new loci for resistance to Sclerotinia stem rot in *Brassica napus*. Front Plant Sci 7:1418

Yadav OP, Yadava TP, Kumar P (1992) Combining ability studies for seed yield, its component characters and oil content in Indian mustard (*Brassica juncea* (L.) Czem. & Coss.). Oilseed Res 9:14–20

Yadava TP, Singh H, Gupta VP, Rana RK (1974) Heterosis and combining ability in raya for yield and its components. Indian J Genet 34:684–686

Yadava DK, Singh N, Vasudev S, Singh S, Giri SC, Dwivedi VK, Prabhu KV (2012) Combining ability and heterobeltiosis for yield and yield contributing traits in Indian mustard (*Brassica juncea*). Indian J Agri Sci 82(7):563–567

Yamagish H, Terachi T (2001) Intra- and inter-specific variations in the mitochondrial gene *orf 138* of Ogura-type male sterile cytoplasm from *Raphanus sativus* and *Raphanus raphanistrum*. Theor Appl Genet 103:725–732

Yang JH, Zhang MF, Yu JQ, Zhang S, Wang T, Chen ZJ (2005) Identification of alloplasmic cytoplasmic male-sterile line of leaf mustard synthesized by intra-specific hybridization. Plant Sci J 168:865–871

Yang JH, Huai Y, Zhang MF (2009) Mitochondrial *atpA* gene is altered in a new *orf220*-type cytoplasmic male-sterile line of stem mustard (*Brassica juncea*). Mol Biol Rep 36:273–280

Zenkteler M (1990) *In vitro* fertilization of ovules of some species of Brassicaceae. Plant Breed 105:221–228

Zhan ZX, Nwafor CC, Hou ZK, Gong JF, Zhu B, Jiang YF, Zhou YM, Wu JS, Piao Z, Tong Y, Liu C, Zhang C (2017) Cytological and morphological analysis of hybrids between *Brassicoraphanus*, and *Brassica napus* for introgression of clubroot resistant trait into *Brassica napus* L. PLoS One 12:e0177470

Zhang MF, Chen LP, Wang BL, Yang JH, Chen ZJ, Yutaka H (2003) Characterization of *atpA* and *orf220* genes distinctively present in a cytoplasmic male-sterile line of tuber mustard. J Hortic Sci Biotechnol 78:837–841

Zhang L, Yang G, Liu P, Hong D, Li S, He Q (2010) Genetic and correlation analysis of silique-traits in *Brassica napus* L. by quantitative trait locus mapping. Theor Appl Genet 122(1):21–31

Zhao J, Becker H, Zhang D, Zhang Y, Ecke W (2006) Conditional QTL mapping of oil content in rapeseed with respect to protein content and traits related to plant development and grain yield. Theor Appl Genet 133:33–38

7 Heteroploidy in *Brassica juncea*: Basics and Applications

Mehak Gupta and Surinder S. Banga

Abstract

Heteroploidy is a condition in which cells, tissues and whole organisms exhibit a chromosome number that constitutes a deviation from the species constant. And, individuals with aberrant chromosome number are termed as heteroploids. Changes in chromosome numbers may result from polyploidy and aneuploidy or dysploidy. These arise due to spontaneous or induced errors of cell division and also through hybridizations across species domains. Aberrant meiosis may lead to the formation of aneuploids with change in one or a few numbers of chromosomes relative to euploid form of the species (e.g., trisomics and monosomics). Upward or downward shifts in ploidy along with the gains and losses of single chromosomes can have strong evolutionary consequences, if these are stabilized and selected for by natural selection. Reticulate evolution through hybridization, introgression and polyploidization has been the most potent force of evolution in angiosperms as it facilitated the establishment of new euploid species by bringing together highly differentiated gene pools. The emergence of a new species adds to the genetic and landscape diversity through shifts in allele frequencies and niche specialization. Heteroploids have been extensively synthesized for their use in genetic, evolutionary and plant breeding investigations. Aneuploid stocks such as monosomics and trisomics have allowed mechanistic processes underlying homoeologous chromosome pairing, pairing regulation and structural chromosomal rearrangements. These were also used as potent tools for assigning genes to the chromosomes in the pre-genomic era. Chromosome addition lines and double ditelosomic stocks have been used in the development and refinement of a draft assembly of highly complex and redundant wheat genome by allowing flow-sorted isolation, sequencing and de novo assembly of each individual chromosome or chromosome arms. A wide variety of heteroploids (including haploids, aneuploids, autopolyploids and novel polyploids) have been developed in Brassicas for genetic studies and development of synthetic species. Haploids are most commonly produced through anther or microspore culture for use in instant production of pure lines and cutting short of breeding cycles in various crop *Brassica* species. Doubled-haploid populations are considered ideal for genetic mapping as well.

M. Gupta · S. S. Banga (✉)
Department of Plant Breeding and Genetics,
Punjab Agricultural University,
Ludhiana, Punjab, India
e-mail: nppbg@pau.edu

M. Gupta
e-mail: mehakgupta@pau.edu

C. Kole and T. Mohapatra (eds.), *The Brassica juncea Genome*, Compendium of Plant Genomes,
https://doi.org/10.1007/978-3-030-91507-0_7

7.1 Introduction

The chromosome number is a species constant, and it is a key genomic characteristic of an organism or lineage in sexually reproducing species. There are two chromosomally distinct phases in all eukaryotes. The first one is the gametic phase (n), where gametes are developed through the process of reductional division or meiosis in the germ cells, which decreases the chromosome number from diploid to a haploid state. The second somatic phase ($2n$) arises from equational division or mitosis in the fertilized zygotic cells after the fusion of male and female gametes. The alternation of these two phases ensures a chromosome number continuum from one generation to another. Fertilization of male and female gametes (n) restores the chromosome number in the zygote ($2n$) vital for the continuation of a species. Meiosis encompasses DNA replication followed by two rounds of cell division to produce four genetically dissimilar daughter cells, each carrying only one representative copy of homologous chromosomes (Crismani et al. 2013). Whereas, during mitosis all replicated chromosomes of a species partition equally to produce two daughter cells genetically identical to each other and with their parent cell. These exhibit the same number of chromosomes, as in the original cell (Walczak et al. 2010). The development of an organism from a zygote to functional adult is ensued through mitotic cell divisions. Two copies of every chromosome known as homologous chromosomes are present in all somatic cells. The homologous chromosomes are morphologically similar to the same gene content, but the allelic composition can vary. Each step of mitosis and meiosis is controlled by an intricate network of genes to check mis-division of chromosomes or development of chromosomally imbalanced cells (Koduru and Rao 1981; Pagliarini 2000; Hamant et al. 2006; Musacchio and Salmon 2007; Mercier et al. 2015; Lambing and Heckmann 2018). However, anomalous meiosis does occur, albeit at a low frequency. It can occur spontaneously or is induced through environmental stresses or experimental manipulations. This results in cells with deviant chromosome numbers of a species. The condition is termed as heteroploidy, and the cells, tissues and whole organisms exhibiting aberrant chromosome numbers are known as heteroploids. These are of two types: (1) aneuploids that display the change in one or a few numbers of chromosomes relative to euploid form of the species (e.g., trisomies and monosomies) and (2) euploids that show an alteration in one or more sets of genomes (such as haploids, triploids and hexaploids).

Brassicaceae is a large angiosperm family comprising 338 genera and 3709 species (Warwick et al. 2006; Wang et al. 2011). *Brassica* is the most important genus in this family carrying multiple species which are used for vegetables, oils, condiments, spices and fodder. Of major economic value, there are three diploid species {*Brassica rapa* (AA, $2n = 20$); *B. nigra* (BB, $2n = 16$) and *B. oleracea*, (CC, $2n = 18$)} and three derived allotetraploid species {*B. juncea* (AABB, $2n = 36$); *B. carinata* (BBCC, $2n = 34$) and *B. napus* (AACC, $2n = 38$)}. Allotetraploid species arose by pairwise natural hybridization followed by chromosome doubling among three diploid species. The cytogenetic and evolutionary relationships among the six main domesticated species of *Brassica* are traditionally represented as U's triangle (Nagaharu 1935). These three allotetraploid species have also been resynthesized repeatedly by Brassica researchers via hand pollination and somatic hybridization between preferred genotypes of known progenitor species (Prakash et al. 2009). Apart from these, a wide variety of heteroploids (including haploids, aneuploids and novel polyploids) have been developed in Brassicas for various purposes. In this chapter, we have discussed methods and utility and several other important aspects of Brassica heteroploids available so far. Heteroploidy is a wide topic, and hence, its narrative could not be limited to *B. juncea* alone, so we have included the literature from related *Brassica* species. In some cases, references from other crops are also included to convey the message.

7.2 Aneuploidy

Aneuploidy refers to the condition in which the chromosome number or chromosome constitution of an individual organism deviates from its wild type or a diploid state by only one or a few numbers of whole chromosomes. The organisms depicting such chromosome changes are known as aneuploids that can have a lesser, greater or the same number of chromosomes as that of the parental euploid. Aneuploids can be of intraspecific or interspecific origin depending upon species specificity of missing, added or substituted chromosomes. These can be from their own gametic set or from different species.

7.2.1 Intraspecific Aneuploidy

The intraspecific aneuploid conditions, i.e., $2n - 1$ and $2n - 2$ are called monosomic and nullisomic, respectively, as only one or both copies of some specific chromosome are missing from the regular somatic chromosome complement. Similarly, the other two common types of intraspecific aneuploids are $2n + 1$ and $2n + 2$, called trisomic and tetrasomic, respectively, in which extra copies (either one or two of a specific chromosome), are present in addition to normal two homologous copies of all chromosomes of a chromosome set. If the organism exhibits added or deleted copies from two pairs of homologous chromosomes, then accordingly, it will be called double tri-/tetrasomic or double mono-/nullisomic. Nullisomic-tetrasomics are another special kind of intraspecific aneuploids in which chromosome number remains the same as exist in regular euploid organism of the species, but one chromosome will be in tetrasomic state, and some other chromosome of same somatic complement will be in nullisomic condition. Different pathways exist for the production of intraspecific aneuploids.

7.2.1.1 Asynapsis, Desynapsis and Non-disjunction

Perturbations during synapsis and non-disjunction of chromosomes are mainly responsible for generating intraspecific aneuploids. Aberrations during synapsis lead to the failure of homologous chromosomes to pair (asynapsis) or separate prematurely (desynapsis). Both lead to univalent formation at metaphase-I which may either be lost or be randomly distributed to daughter cells, producing chromosomally unbalanced or aneuploid gametes. Disjunction is the separation of homologous chromosomes or sister chromatids and distribution to opposite poles during karyokinesis. Deviation from this cellular process leads to meiotic non-disjunction, where the chromosomes fail to disjoin properly at the first or second division. This may lead to unilateral transfer of two chromosomes or chromatids at one pole and none to the other, leading to the production of duplication ($n + 1$) and deficiency ($n - 1$) gametes, in a scenario where a single pair of chromosome or chromatid fails to disjoin. Fertilization of the $n - 1$ megaspore with n bearing pollen will lead to a monosomic ($2n - 1$) zygote. Similarly, the fusion of $n + 1$ and n gametes will yield a trisomic individual with $2n + 1$ somatic chromosome complement. The selfing of monosomic and trisomic plants may lead to the generation of nullisomic and tetrasomic plants, respectively, for the corresponding chromosomes. These anomalies (asynapsis, desynapsis and non-disjunction) spontaneously occur at low frequencies under the influence of environmental factors such as temperature and humidity (Ramanna and Hermsen 1979; Koduru and Rao 1981; Singh 2002; Rao and Kumar 2003; Singhal and Kumar 2010; Kumar and Singhal 2011; Singhal et al. 2012). Such anomalies can also be induced by the application of chemical (ethyl methane sulfonate) and physical mutagens (X-rays) (Carter 1978; Singh and Ikehashi 1981; Martin et al. 1997; Tepperberg et al. 1999; Higgins et al. 2004; Naseem and Kumar 2013). Meiotic gene mutations that interfere with normal synaptonemal complex formation, homologous recombination and segregation events may also trigger asynapsis, desynapsis or non-disjunction (Armstrong et al. 2002; Higgins et al. 2004; Nonomura et al. 2004; Pradillo et al. 2007; Jiang et al 2009). These cellular abnormalities have been observed in newly evolved auto- or allopolyploids (Mestiri et al. 2010; Chester et al. 2012; Zhang et al. 2013). Autopolyploids contain more than two copies of homologous chromosomes that produce trivalents (in triploids) or quadrivalents (in tetraploids) or various other chromosome conformations. These complex pairing associations often fail to segregate or disjoin unevenly to yield aneuploid gametes. Shen et al. (2006) could establish a complete series of primary trisomics (from Tri-1 to Tri-10) of *B. rapa* through microspore culture in tetraploid Chinese cabbage. A complete set of primary trisomic lines of *B. campestris* syn. *B. rapa* (Hill et al. 1986; Tang et al. 1988) and *B. oleracea* var. *capitata* (Zhang et al. 2006) was also isolated from the progenies developed through hybridization between autotriploid and diploid individuals. These were categorized on the basis of chromosome size, centromere position (arm ratio) and presence or absence of satellite regions. Aneuploid plants also occurred in the selfed progenies of newly developed or young allopolyploids (Yang 1964; Tokumasu 1976; Choudhary et al. 2000). Homoeologous pairing during early generations in the induced polyploids, precipitates non-paired, multi-paired chromosomes and aberrant chromosome segregation. This type of pairing leads to the development of abnormal sporad and aneuploid progeny (Leflon et al. 2010; Szadkowski et al. 2010; Cui et al. 2012). Wen et al. (2010) observed four nullisomics ($2n = 36$), two nullihaploids ($2n = 18$) and one tetrasomic haploid ($2n = 20$) in a microspore-derived (MD) progeny from a resynthesized *B. napus*. These aneuploids may have resulted from non-disjunction of one bivalent or the distribution of two univalents to the same pole during microsporogenesis. Molecular analysis indicated the loss of one A-genome chromosome from one nullihaploid. The remaining five nullisomic/nullihaploid seemed to have lost the different C-genome chromosome(s). Selfing and crossing of haploid with diploids can also be a good source of aneuploids as chromosomes tend to go to one pole (one or two chromosomes are left behind and not included in daughter cells) or separate randomly to two poles during meiosis producing unbalanced spores.

7.2.1.2 Interspecific Hybridization

Partial to complete loss of the genome of one donor species may occur in wide hybrids developed following crossing of Brassica crop species with wild crucifers, and aneuploids of both intraspecific and interspecific origin are common in hybrid progeny. The preferential loss of some C-genome chromosomes from the somatic chromosome complement of *B. napus* has been repeatedly observed in its intergeneric crosses with *Capsella bursa-pastoris*, *Lesquerella fendleri* and *Isatis indigotica* (Chen et al. 2007; Du et al. 2008; Tu et al. 2010). Hua and Li (2006) isolated one *B. napus* nullisomic ($2n = 36$) of unknown chromosome specificity among mixoploid hybrids obtained by crossing *B. napus* with *Orychophragmus violaceus*. Later, the lost chromosome was determined as C2 by RNA sequencing of the nullisomic individual (Zhu et al. 2015). Similarly, some other Brassica aneuploids of uncertain chromosome identity have been recognized in mixoploids of Brassica diploids and allotetraploids with *O. violaceus* (Li et al. 1995, 1998; Li and Heneen 1999; Li and Ge 2007; Liu and Li 2007; Xu and Li 2007). There is also a report for the identification of nullitetrasomics lines of *B. napus* with 22 A-genome and 16 C-genome chromosomes and aneuploids ($2n = 37$ or 39) with variable A- or C-genome chromosomes with a C-genome-specific BAC (BoB014O06) fluorescent in situ hybridization (FISH) probe in an advanced generation of an intergeneric hybridization between *B. rapa* and *O. violaceus* with accidental pollination by *B. napus* (Xu et al. 2019). Four *B. campestris* (*B. rapa*) trisomic plants were isolated among the progeny of one selfed *B. campestris*-alboglabra monosomic addition line (Chen et al. 1992). Further analysis revealed the trivalent frequencies of 33.3% and 20% and transmission rates of 30.6% and 32.4% of the extra chromosome through the ovule in trisomics 2891 and 2881, respectively (Cheng et al. in 1993). Fan and Tai (1985) isolated two types of monosomics of *B. napus,* one (mono-1) from a hybrid between *Diplotaxis muralis* and *Brassica napus* and second (mono-2) from a hybrid between Polima cytoplasmic male sterile

(CMS) *B. napus* and *B. juncea* both backcrossed to *B. napus*. Disomic to monosomic plants occurred in 1:1 ratio instead of theoretically expected 1:2:1 ratio of disomic: monosomic: nullisomic plants in the progenies of two monosomics (Chang et al. 1987). This deviant ratio was more due to greater functionality of n gametes than $n - 1$ gametes. A nullisomic low glucosinolate *B. juncea* breeding line missing one pair of B-genome chromosomes was also identified from a BC_1F_3 plant of an interspecific cross between high glucosinolate Indian *B. juncea* line 60,143 and *B. rapa* canola strain CZY (Cheng et al. 2001).

7.2.1.3 Uses of Intraspecific Aneuploids

Monosomics and nullisomics are usually lethal in diploids, but they are better tolerated in allopolyploid species as homoeologous chromosomes of closely related species present in the allopolyploids can functionally compensate for a missing homolog pair (Sears 1954; Siegel and Amon 2012). Intraspecific aneuploids have been used prominently in wheat (*Triticum aestivum* L.), where a diverse range of whole chromosome aneuploids such as monosomics, nullisomics and compensating nullitetrasomics and ditelosomics have been established (Sears 1954; Gupta et al. 2005). Allosyndetic pairing suppressor locus (*Ph1*) of bread wheat was detected, following the development of aneuhaploids and nullisomics, deficient for Chromosome 5B. These lines showed meiotic pairing that was different from euhaploids and euploid chromosome complement of the species (Riley and Chapman 1958; Riley et al. 1959). Nullitetrasomics cytogenetic stocks were also used to explore homoeologous relationships among wheat chromosomes (Sears 1952; Brewer et al. 1969) and gene dosage balance (Zhang et al. 2019b). The results revealed that two additional copies of a chromosome can genetically compensate for the absence of a homoeologous chromosome pair. The availability of complete series of double ditelosomic stocks of the hexaploid wheat cultivar Chinese Spring facilitated the development of a draft assembly of highly complex and redundant wheat genome by allowing flow-sorted isolation, sequencing and de novo assembly of each individual chromosome arm (Mayer et al. 2014). However, only trisomics were useful in diploids (Khush et al. 1984; Blakeslee et al. 1922). These cytogenetic stocks have been used for assigning a gene to a particular chromosome or arm, to test the independence of linkage groups and to associate linkage groups with individual chromosomes in right orientation in rice (Singh et al. 1996; Dong et al. 2001), soybean (Gardner et al. 2001; Zhou et al. 2003; Zou et al. 2006), foxtail millet (Wang et al. 2002, 2007; Gao et al. 2003). Zhang et al. (2009) located a recessive gene controlling the $2n$ gamete formation to Chromosome 4 (chromosomes sorted using morphometric features of chromosomes) of Chinese cabbage by exploiting full series of primary trisomics in crossing with diploid Chinese cabbage genotype Bp058 which could produce $2n$ male gametes. Transcriptome analysis of aneuploids has provided novel insights into gene expression disturbances caused by loss or gain of individual chromosomes. Aneuploidy disturbed the gene dosage balance system in the chromosome complement of a species, especially for the genes located on missing and additional chromosomes. This made the aneuploids appear phenotypically distinct, mostly with reduced fertility and impaired fitness, from regular diploids (Henry et al. 2010; Siegel and Amon 2012). More obvious phenotypes are identified in monosomic than in trisomic individuals (Siegel and Amon 2012). Surprisingly, aneuploids displayed rather severe alterations in gene expression compared to the whole genome changes in polyploids (Birchler and Newton 1981; Guo et al. 1996). The phenotypic deviations in aneuploids are known to result from changed gene expression levels caused by global imbalance of the entire regulatory system precipitated by altered copy numbers (cis-effects) of missing chromosomes and unaltered disomic chromosomes (*trans*-effects) (Birchler and Veitia 2007; Huettel et al. 2008; Zhang et al. 2017; Rey et al. 2018; Zhu et al. 2018). Zhu et al. (2015) detected more differentially expressed genes (DEGs) in monosomic than nullisomic for C2 chromosome in *B. napus* ($2n = 38$, AACC). Nullisomic plant was earlier in maturity and showed a greater reduction in plant height as compared to the euploid donor. The trans-acting effects of missing one or two copies of C2 chromosome were more pronounced than cis-effects for majority of involved genes. Overall, higher gene expression was detected in the aneuploids though lower gene expression occurred on homoeologous A2 chromosome.

7.2.2 Interspecific Aneuploids Including Alien Addition and Substitution Lines

7.2.2.1 Monosomic and Disomic Alien Addition Lines

The addition of one copy of an alien chromosome from a related species to the normal somatic chromosome complement of a cultivated species is termed as monosomic alien addition line. If both homologous copies of an alien chromosome are added, then recipient is termed as disomic alien addition line. These lines can be produced by consecutive backcrossing of interspecific hybrids or synthetic amphiploids to species in the background of which alien chromosome additions are required. They act as intermediate or bridging materials in a route of gene introgression from one species to another during alien introgression program (Fu et al. 2012). These also serve as powerful genetic stocks to examine the genome organization of a species by partitioning it into individual chromosomes units. These lines have been utilized in numerous aspects of plant genome research such as alien gene localization to specific individual donor chromosome(s) (Chen et al. 1992; Geleta et al. 2012), to examine the heterologous gene expression, analysis of homoeologous relationships between alien chromosome(s) and corresponding orthologous recipient genome and inducing crop-wild genome recombination (Pertuze et al. 2003; Cifuentes and Benavente 2009; Molnar and Molnar 2010). Alien addition lines are also useful for microdissection and microcloning for chromosome-specific library construction (Kynast et al. 2004; Bento et al. 2010; Shim et al. 2010). A wide variety of monosomic alien addition lines (MAALs) have been available in many crop species, i.e., wheat (Kong et al. 2008), oat (Kynast et al. 2001), rice (Multani et al. 2003), potato (Ali et al. 2001), cucumber (Chen et al. 2004), tobacco (Chen et al. 2002), sugarbeet (Gao et al. 2001) and Brassica crop species (Gupta and Banga 2020). An effective set of alien addition lines requires all chromosomes of the donor genome is added individually in the genetic background of recipient genome and characterized based on chromosome landmarks of each alien chromosome or appropriate set of FISH probes or molecular markers specific for each pair of donor chromosomes. There are only a few reports for such attempts in Brassicas as it has not been possible to identify individual Brassica chromosomes until recently (Xu et al. 2016; Agrawal et al. 2020). Early studies aimed at the development of chromosome addition lines were directed at dissecting crop Brassica genomes. McGarth and Quiros (1990) analyzed the chromosome complement of *B. oleracea* by adding seven of the nine C-genome chromosomes in the background of *B. rapa* (syn. *B. campestris*). These were characterized by morphology, pollen and seed fertility and transmission frequency of chromosome-specific markers. Transmission frequency of chromosome additions was influenced by the genotype of the recurrent parent. This et al. (1990) used hyperploid derivatives of *Diplotaxis erucoides* × *B. nigra* hybrid combinations to develop seven out of the eight possible monosomic addition lines for *B. nigra* (genome B). No phenotypic changes were obvious, but plants carrying extra chromosomes were slow growing. *B. nigra* chromosomes were recognized by size, heterochromatin distribution and restriction fragment length polymorphism (RFLP) markers. Chevre et al. (1991) could develop *B. napus*-*B. nigra* disomic addition lines and characterized six synteny groups, representing six of the eight *B. nigra* chromosomes with the help of isozyme, fatty acid and RFLP markers. This study was followed by the successful addition of four *B. alboglabra* chromosomes in the background of *B. rapa* (Chen et al. 1997).

The added chromosomes of *B. alboglabra* chromosomes were distinguished by random amplified polymorphic DNA (RAPD) marker analysis or morphological landmarks of mitotic chromosomes. Backcrossing *B. rapa* to *B. juncea* facilitated isolation of plants carrying addition of three A-genome Chromosomes (6, 7 and 8) in the background of *B. nigra* (Kapoor et al. 2011). These chromosome addition stocks were characterized by chromosome counts, pollen and seed fertility and chromosome-specific simple sequence repeat (SSR) markers. Varied transmission of A-genome chromosomes was indicated. A-genome Chromosomes 3 and 4 were the first to get eliminated followed by Chromosome 10. It is possible that Chromosomes 6, 7 and 8 had higher transmission frequency, and these were better tolerated by the *B. nigra* genome. Srinivasan et al. (1998) developed monosomic chromosome addition lines of *Brassica oxyrrhina* in the background of *B. campestris* (syn. *B. rapa*) carrying *B. oxyrrhina* cytoplasm. These were characterized through cytology and molecular (RAPD) marker analysis, and seven synteny groups were identified. Meiotic studies revealed the homeology of four *B. oxyrrhina* chromosomes (Synteny groups 1, 3, 5 and 6) with those of *B. rapa*. *B. napus-Crambe abyssinica* monosomic addition lines ($2n = 39$, AACC with added chromosome from *C. abyssinica*) were isolated from the F_2 progeny of the asymmetric somatic hybrid between two species (Wang et al. 2006). The added chromosome from *C. abyssinca* was clearly distinguished by genomic in situ hybridization (GISH). Zhu et al. (2016) extracted the ancestral type of *B. rapa* (Restituted *B. rapa* Oro; RBR Oro) from the allotetraploid *B. napus* donor "Oro" by inducing the preferential loss of the C-genome chromosomes in intertribal crosses of *B. napus* with *Isatis indigotica* ($2n = 14$) (Tu et al. 2010). Subsequently, a complete series of monosomic alien addition lines with each of the nine sub-genome chromosomes of *B. oleracea* was obtained in the background of extracted *B. rapa* genome by backcrossing restituted *B. rapa* RBR with *B. napus* "Oro" (Zhu et al. 2016). Huo et al. (2019) exploited this alien addition set to affiliate 54.68% (44.11 Mb) of the 80.67 Mb of non-anchored scaffolds belonging to the C-subgenome to the corresponding chromosomes of *B. napus*, using multiple comparisons of expression values of genes located on unanchored scaffolds in the MAALs and RBR Oro, thus improving the genome sequence assembly of *B. napus*.

7.2.2.2 Alien Substitution Lines

Replacement of one or both copies of a homologous pair of chromosomes of cultivated species with the corresponding homoeologous chromosomes of wild species allows the synthesis of monosomic and disomic alien chromosome substitution lines, respectively. These substitution lines are very specific as alien chromosome must genetically compensate for the loss of substituted host chromosome. Such genetic stocks are useful for establishing homoeologous relationships between chromosomes of crop and wild genomes. Systematic syntheses of alien substitution lines require hybridization between a known chromosome addition line as pollinator with monosomic aneuploid set of the crop species. The produced F_1s with the same chromosome number as of euploid crop species can either be selfed or be backcrossed to same addition line to yield monosomic or disomic alien substitution line. Unfortunately, it has not been possible to purposefully develop chromosome substitution lines in the absence of fully characterized aneuploid stocks. However, random chromosome substitutions can occur in crossing between two allotetraploids sharing one genome. Banga (1988) identified chromosome substitutions for at least three A-genome chromosomes, which were replaced by C-genome homeologs. These genetic stocks were identified in F_7 generation of an interspecific cross between *B. juncea* and *B. napus*. These were meiotically stable and were characterized by certain specific traits like bolting habit, leaf aberrations and variation for erucic acid content. The same group could develop and recognize three whole chromosome (C1, C3 and C8) substitutions and 13 major C-genome segmental substitutions in *B. juncea* genotypes developed by hybridizing *B. napus* with *B. carinata* (Gupta et al. 2015, 2016b). C-genome chromosome substitutions were confirmed through genotyping with Brassica Illumina 60 K Infinium SNP array. Subsequent molecular assays with B-genome-specific primers suggested that C-genome chromosome (s) were likely to have replaced B-genome chromosome(s). C1 was the most common substituting chromosome, while substituted B chromosome(s) appeared random (Gupta et al. 2016b). Zhu et al. (2018) also reported a monosomic substitution line of restituted *B. rapa* (RBR Oro from *B. napus* "Oro") in the progenies of monosomic alien addition line C1 (of RBR) after pollination with RBR (Zhu et al. 2018). FISH analysis with C-genome-specific probe and polymerase chain reaction (PCR) amplification with C1 chromosome-specific gene markers confirmed the presence of C1 chromosome, but its monosomic substitution with A1 chromosome was revealed through RNASeq analysis. Substitution line showed low gene expression levels along Chromosome A1, thus confirming the absence of one copy of A1 chromosome. Both cis-effects DEGs (along altered chromosome) and trans-effects DEGs (across disomic chromosomes) were prevalent in addition and substitution lines induced by addition and substitution of one copy of Chromosome C1. Although aneuploids of both intra-and inter-specific origin have been obtained many times in Brassica, these could not be used to understand genome structure, homoeologous relationships, linkage group assignment and gene localization in crop Brassica as seen in other crop species. This has been primarily due to the small size and lack of karyological features in Brassica chromosomes, which hindered their reliable and consistent classification. However, recently developed oligo fluorescent in situ hybridization (FISH) techniques can facilitate development and characterization of various aneuploid stocks, especially alien addition and substitution lines at cytogenetic level for their applications in pre-breeding research programs (Agrawal et al. 2020). Such stocks are important to validate genome assemblies developed purely on the basis of in silico analysis.

7.3 Whole Ploidy Manipulations

7.3.1 Haploids

A sporophyte with gametic chromosome number is known as a haploid. It is a generalized term and applied unequivocally to haploids extracted from both diploid and polyploid species. However, the haploids obtained from diploid and polyploid species are also termed as monoploids/monohaploids and polyhaploids, as they carry only one or multiple sets of chromosomes. Polyhaploids can be further categorized into autopolyhaploids and allopolyhaploids/amphihaploids depending upon the origin of polyploid species from which they are derived. Haploids with an exact gametic chromosome complement or one representative (copy) of all chromosomes of a euploid can be called as euhaploid. In contrast, aneuhaploid carries either extra doses of or show deficiency for one or more chromosomes from the haploid chromosome set. An aneuhaploid with chromosome composition $2n = n + 1$, where an additional chromosome is added from its own gametic set, is known as disomic haploid due to disomic condition for one and monosomic state for all other chromosomes. On the other hand, an aneuhaploid {chromosome constitution $2n = n + 1^*$; (1^* represents an alien chromosome)} having an additional chromosome from a different species is termed an addition haploid. Similarly, an aneuhaploid missing for one chromosome of the gametic complement is known as nullisomic haploid ($2n = x - 1$). An aneuhaploid may also have one or more alien chromosomes substituted for one or more chromosomes of its own gametic set, leading to chromosome composition ($2n = x - 1 + 1^*$ or $2n = x - 1 - 1 + 1^* + 1^*$) and the chromosome number equivalent to euhaploid plants. These types of aneuhaploids are denoted as substitution haploids. Sometimes, the term doubled haploid is used as synonym for dihaploid; in the true sense, both are completely different from each other. Dihaploids are autopolyhaploids of autotetraploid species {such as haploids obtained from

Solanum tuberosum L. ($2n = 4x = 48$)} and do not show complete homozygosity like doubled haploids.

7.3.1.1 Synthesis

Haploidization is a powerful tool for use in the study of evolution, cytogenetics, genetics and genomics and practical plant breeding (Dunwell 2010; Murovec and Bohanec 2011). Spontaneous occurrence of haploids is rare (Bhojwani and Razdan 1996). Blakeslee et al. (1922) was the first to identify two haploid individuals of *Datura stramonium* ($2n = 24$) with twelve chromosomes in their somatic cells from the anomalous plants produced following cold treatment. Later, a haploid plant of *Triticum compactum* ($2n = 42$) was also obtained in Washington Agricultural Research station, Pullman in 1925 (Gaines and Aase 1926). Extremely low frequency of naturally occurring haploids prompted scientists to use induced parthenogenesis (mainly *via* anther culture), pollen irradiation, temperature shocks, sparse pollination, seed selection with twin embryos, alien cytoplasm substitutions, distant hybridizations, haploid inducer stocks and gene engineering (e.g., *CENH3*—centromere-specific histone 3) to increase the frequency of haploids in different crops (Kasha and Maluszynski 2003; Touraev et al. 2009; Britt and Kuppu 2016; Wang et al. 2019). Distant crossing and anther/microspore culture are now used in practical plant breeding. Haploid induction results from distant hybridization in which the chromosomes of one parental species are preferentially eliminated from the developing hybrid embryos (Ishii et al. 2016). Such uniparental chromosome elimination was first observed in crosses between *Nicotiana tobacum* and *Nicotiana sylvestris* (Clausen and Mann 1924). This was followed by the high-frequency induction of barley haploids in the cross between *Hordeum vulgare* and its wild relative *H. bulbosum* (Kasha and Kao 1970). This method is now popular as "The *Hordeum bulbosum* method" (Devaux 2003). Sanei et al. (2011) later demonstrated that loss of *CENH3*, a centromere-specific histone variant, from the *H. bulbosum* chromosomes preceded the elimination of *bulbosum* chromosomes from the hybrid embryos. Haploids of wheat and oats are now commonly produced using chromosome elimination strategy by pollinating wheat and oats with maize. Some other grass relatives pearl millet, sorghum, barley and teosinte were also used (Laurie and Bennett 1986, 1988; Laurie 1989; Ohkawa et al. 1992; Riera-Lizarazu and Mujeeb-Kazi 1993; Niroula and Bimb 2009; Marcinska et al. 2013; Polgári et al. 2014). Few instances of partial to complete genome exclusion of one parental species from the hybrid embryos were also reported in the family Brassicaceae. Tu et al. (2009) identified a completely sterile *B. rapa* haploid plant in the progenies of *B. rapa* × *I. indigotica* ($2n = 14$) combination. In another study, hybrids obtained from two combinations of interspecific hybridization involving *B. rapa* and *B. napus* as female with *C. bursa-pastoris* ($2n = 4x = 32$) produced mixoploids. These included plants with haploid chromosome number of *B. napus* (i.e., $2n = 19$) and disomic chromosome constitution of *B. rapa* ($2n = 20$) and *B. napus* ($2n = 38$) along with variable number of chromosomes, some with whole chromosome additions from *C. bursa-pastoris* were identified (Chen et al. 2007). Partial to complete genome elimination was also observed in crosses involving *O. violaceus* ($2n = 24$) with all cultivated Brassica diploid and allotetraploid species (Li and Ge 2007). Fu et al. (2018) attained a remarkably high frequency (40–99%) of maternal doubled haploids of *B. napus* by using synthetic Brassica allo-octaploid (AAAACCCC, $2n = 8x = 76$) as the pollen donor. The method appears very advantageous as simultaneous duplication of Brassica chromosomes also occurred along with genome elimination of synthetic species. It can also be attempted in other *Brassica* species to induce concurrent haploidy and doubled haploidy. In vivo haploid induction by inducer lines Stock 6 and more recently its derivatives are also widely used for inbred line development in maize (Coe 1959; Zhang et al. 2008; Trentin et al. 2020). Studies now suggest that chromosome fragmentation in sperm cells and selective elimination of uniparental chromosomes after fertilization are likely causes of haploid induction in maize (Zhao et al. 2013; Li et al. 2017). More recent studies have linked the haploid induction in maize (*Zea mays*) to a frame-shift mutation in *MATRILINEAL* (*MTL*), a pollen-specific phospholipase. It was also shown that novel edits in *MTL* of the inducer line led to a 6.7% haploid induction rate (Gilles et al. 2017; Kelliher et al. 2017; Liu et al. 2017). Mutations in *MTL* gene were also able to induce haploids in wheat and rice (Yao et al. 2018; Liu et al. 2020). Availability of transformation and gene editing technologies are now facilitating the development of more efficient haploid inducer stocks. *CENH3* has been variously used for this purpose. It is a centromere-specific histone 3 variant which epigenetically specifies the centromere by recruiting constitutive centromere proteins that next recruit overlying kinetochore proteins (Palmer et al. 1987; Zhong et al. 2002; Musacchio and Desai 2017; Kixmoelle et al. 2020). It possesses an *N*-terminal tail domain and a more conserved C-terminal histone fold domain (HFD). Ravi and Chan (2010) developed modified *CENH3* (GFP-tailswap) in *Arabidopsis thaliana*. The tail domain of *CENH3* was replaced with the histone 3.3 tail domain and fused with green fluorescent protein (GFP). HFD of *CENH3* gene was retained unmodified. Crossing of these genetically transformed plants with wild types led to the elimination of chromosomes of transformed plants in the hybrids. This led to the induction haploids of wild type at high frequencies (Ravi and Chan 2010). In subsequent studies, replacement of the endogeneous *CENH3* of *Arabidopsis thaliana* with heterologous *CENH3* from *B. rapa* and *Lepidium oleraceum* also induced haploids following crossing with wild-type plants (Maheshwari et al. 2015; Britt and Kuppu 2016). Recent investigations have suggested that even point mutations in the *CENH3* α-N-helix or centromere targeting domain (*CATD*) can also produce haploids (Karimi-Ashtiyani et al. 2015; Kuppu et al. 2015, 2020). Alterations to *CENH3* gene show impaired or reduced *CENH3* loading during meiosis due to reduction in effective centromere size or weakening of the centromere function in gametophytes of haploid inducer lines (Wang and Dawe 2018). When these lines are crossed to wild-type plants, the progeny shows centromere size disparity or imbalance, that leads to targeted destruction of the smaller or weaker centromeres by natural clearing mechanisms of the cell, thus leaving behind the chromosomes of only wild type. This concept of "centromere size model" may also explain chromosome elimination in many, if not all, interspecific combinations. Chromosomes of small-centromere species (sorghum, maize, pearl millet, adlay millet and perennial ryegrass) are preferentially eliminated from the hybrid embryo while retaining chromosomes of oat, barley and wheat carrying large centromeres in wide crosses. This may be due to failure of small centromeres to expand to the necessary size because of insufficient protein loading in a larger genome environment during the early stages of embryogenesis (Wang et al. 2014). Recently, Wang et al. (2021) have now simplified the method of haploid induction by demonstrating use of heterozygous haploid inducer line for a *CENH3* null mutation. Lines with *CENH3* in heterozygous condition are vigorous and can be multiplied easily. Normal seed production is a major issue in the homozygous *CENH3* mutant. Diluted effect of *CENH3* in the heterozygous plant in the *CENH3* genotype of the gametophyte during the postmeiotic cell divisions that precede gamete formation is expected to trigger genome elimination from the F_1 progeny upon crossing with wild type. This simple in vivo seed-based system can also be applied in other crops lines for haploid production of both maternal and paternal origin as *CENH3* is ubiquitous in all eukaryotes. The technique has the potential to replace the highly genotype dependent and cost-demanding tissue culture mediated in vitro techniques of haploid induction in the future.

In vitro culture of immature male or female gametophyte under controlled conditions remains the method of choice to induce haploids (Dunwell 2010) in spite of the forementioned innovations. Anther and microspore culture techniques are now exploited in more than 250 different plant species including trees, vegetables and cereals for androgenic haploid production

(Maluszynski et al. 2003; Germana 2006; Wedzony et al. 2009; Hazarika et al. 2013). Microspore culture has been found more efficient and reliable inducer of embryogenesis and embryo formation (Siebel and Pauls 1989). Gametophyte culture techniques owe their origin to the induction of embryo-like haploids from anther culture of *Datura innoxia* (Guha and Maheshwari 1964). It has been also possible to induce haploids by culturing young ovule and gynoecium in onion and sugarbeet (Murovec and Bohanec 2011). Brassica crops are very amenable to in vitro manipulations and respond well to microspore embryogenesis (Lionneton et al. 2001). In genus *Brassica*, first microspore culture mediated haploid induction was demonstrated in *B. napus* (Lichter 1982). Since then, haploids have been produced in other *Brassica* species using isolated microspores (Babbar et al. 2004). During microspore culture, they essentially involve their reprogramming microspores to initiate sporophytic development under controlled in vitro conditions. Microspore cell undergoes microspore embryogenesis to develop into an embryo-like structure that on "germination" results in a haploid plantlet or microspore can directly give rise to haploid plant through organogenesis (Aionesei et al. 2005). The frequency of embryogenesis following microspore culture is influenced by genotype and growth conditions of donor plants, development stage of microspores, culture environment, type and pH of culture and regeneration media, hormones in the media and stress factors (Dwivedi et al. 2015). Genotypic differences for responsive to embryogenesis are now very well documented (Cloutier et al. 1995; Chuong et al. 1988; Ajisaka et al. 1999; Ferrie et al. 1999; Chanana et al. 2005; Shumilina et al. 2020). Detection of genes and the development of allelic markers associated with microspore embryogenesis are important for germplasm screening and transfer of relevant genes to poor responders. Valdes et al. (2018) have identified 13 quantitative trait Loci (QTLs) and 19 possible candidate genes located on A01, A02, A05, A10, C04 and C06 chromosomes for association with microsporogenesis in *B. napus*. Many of the identified candidate genes had roles in the induction, development and maintenance of apical and subapical meristems. Selection of appropriate bud size for microspore embryogenesis is also important to get the best response from the plant individual, which again varies species-wise and with respect to genotype (Gil-Humanes and Barro 2009; Gu et al. 2014). Microspores undergoing late uninucleate to early binucleate stages of Brassica are considered the best to change their fate from gametogenesis to embryogenesis. Alia et al. (2008) suggested that buds of size 3.0–3.1 mm give better response for microspore embryogenesis in *B. juncea*. Agarwal et al. (2006) reported that an antiauxinic (p-chlorophenoxyisobutyricacid (PCIB) has a promotional effect on microspore embryogenesis by converting several multicellular grains or preglobular embryos into fully developed embryos when culture is supplemented with 20 μM of PCIB after 1 day of culture initiation at 35 °C. A significant increase in yield of microspore embryos has been observed upon pretreatment of flower buds at 10 °C and addition of activated charcoal in the culture medium at medium pH levels in the turnip genotypes studied (Shumilina et al. 2020).

7.3.1.2 Applications

Haploids are usually weak and sterile as they possess only one partner of homologous pairs of chromosomes that segregate unevenly to yield non-viable aneuploid gametes. However, these have diverse applications following the production of doubled haploids (DH) either spontaneously (by meiotic non-reduction) or with the use of mitotic inhibitors like colchicine, oryzalin, amiprophosmethyl, trifluralin and pronamide. The resulting homozygous fertile lines have proved advantageous in plant breeding and basic research. Haploids have been used for establishing intergenomic and intragenomic chromosome homologies, which are difficult to be detected in the diploids due to preferential pairing between homologous chromosomes. Occurrence of bivalents in haploids of Brassica diploids (*B. rapa, B. nigra* and *B. oleracea*) established these as secondary balanced polyploids (Prakash 1973a; Armstrong and Keller

1982). Allosyndetic pairing between chromosomes of A and C-genomes of Brassica was also demonstrated in polyhaploids of allopolyploid *Brassica* species, establishing greater synteny between A- and C-genome compared to the B-genome (Attia et al. 1987; Yang et al. 1994). Pairing regulator (*PrBn*) for homoeologous pairing control in *B. napus* was also detected during a study of F_1 haploid population developed between high- and low-pairing genotypes of *B. napus* (Jenczewski et al. 2003). *PrBn* was later mapped to Chromosome C9 of *B. napus* (Liu et al. 2006). Higgins et al. (2021) have now identified a more reliable QTL (*BnaPh1*) on *Bna*A9 and two minor QTLs. This QTL explained a substantially large chunk of variation (32–58%) for homoeologous pairing in a doubled-haploid population derived from crossing between natural and synthetic *B. napus*. The immediate transformation of heterozygous allele combinations into homozygous state in only one generation is the most important use of doubled-haploid technology in plant breeding. Furthermore, fertile doubled-haploid lines can be produced in species where strong self-incompatibility barriers are predominant. Haploid technology has significantly reduced the timeframe to fix the genetic variation from three to four years required in the traditional method of recurrent self-pollination to only one year, when using off-season nurseries (Prigge and Melchinger 2012). Haploid-derived homozygous lines can be directly released as cultivars in self-pollinated crops or used as inbreds in hybrid cultivar development in several Brassica crops (Ferrie and Möllers 2011; Shmykova et al. 2016; Watts et al. 2020). Haploids have also found extensive usage in quantitative genetic studies. Doubled-haploid-derived populations are used extensively for constructing genetic linkage maps, QTL analysis and marker-assisted breeding using small sized populations (Rahman and de Jimenez 2016). This is due to the high probability of getting the desirable genotype, i.e., 1/2n in haploid method (as only two types of genotypes will occur for a pair of alleles, A and a, with the frequency of 1/2 AA and 1/2 aa) in contrast to 1/4n by conventional method (here, three forms of genotypes will appear with the frequency of 1/4 AA, 1/2 Aa, 1/4 aa) when *n*-number of loci are segregating. However, linkage maps constructed from DH populations may be less densely saturated or a QTL may be mapped to a large genomic region than localized in recombinant inbred lines (RILs). This is due to lesser opportunities of recombination (just only one) to shuffle parental linkages in DH populations in comparison to many cycles of recombination in RILs produced by successive self-pollination. Complete homozygosity of doubled-haploid lines facilitates a more precise phenotyping across multiple years and locations and better correlation of genotype to phenotype. So, QTL effects are better estimated in DH populations during marker trait association studies (Yan et al. 2017). A dense genetic linkage map of allohexaploid Brassica has been prepared from SNPs derived from restriction-site associated DNA (RAD) sequencing of 146 F_1 doubled-haploid progenies and two synthesized allohexaploid parents (Yang et al. 2018). DH mapping populations from microspore culture of F_1s between the genotypes representing Indian and east European gene pool and between Chinese vegetable and Indian genotypes (Tumida × Varuna) have been developed and used in *B. juncea* for mapping of various important traits like seed weight (Dhaka et al. 2017), oil content (Rout et al. 2018), high pod density and other yield contributing traits (Ramchiary et al. 2007; Yadava et al. 2012), oil and seed meal quality (Gupta et al. 2004; Rout et al. 2015) and resistance to white rust (Arora et al. 2019; Bhayana et al. 2019). Relatively, high genetic variance and increased trait heritability due to 100% homozygosity facilitate enhanced response to selection in doubled-haploid germplasm. The integration of molecular markers with doubled-haploid technology further helps in quick fixation of multiple genes in any desired background (Chaikam et al. 2019). Doubled-haploid genotypes are also ideal candidates for accomplishing the whole genome sequencing of a species. For instance, sequencing for *B. rapa* (Chiifu-401–42) (Wang et al. 2011), *B. nigra* (YZ12151, Sangam) (Yang et al. 2016; Paritosh et al. 2021) and *B.*

napus (Darmor-bzh) (Chalhoub et al. 2014) were carried out with doubled-haploid lines. Identification of mutants following chemical or physical mutagenesis is very efficient in haploids as both diploid and recessive mutations can be immediately identified due to the hemizygous state of all loci and their immediate fixation in the homozygous state after chromosome doubling. Isolated microspores are preferred as starting material for mutagenic treatment as being single cells, these will not lead to chimera's production which is otherwise a major issue when seeds are used as base materials for mutagenesis. Varied saturated fatty acid proportions were observed in doubled-haploid lines of *B. napus* and *B. juncea*, developed following microspore mutagenesis (Ferrie et al. 2008). Microspores or microspore-derived embryos are useful materials for plant genetic transformation and developing homozygous transgenics in a short time span. Homozygous transgenic *B. napus* were produced expressing the *uidA* gene encoding β-glucuronidase (GUS) and the *bar* gene as a marker of resistance to phosphinotricin by *Agrobacterium tumefaciens-mediated* transformation and chromosome doubling of microspore-derived embryos (Cegielska-Taras et al. 2008). Doubled-haploid *B. napus* with stably inherited transgene [a barley cDNA for the thaumatin-like protein Hv-TLP8 driven by the cauliflower mosaic virus promoter (CaMV 35S)] exhibiting enhanced resistance to clubroot pathogen (*Plasmodiophora brassicae*) was also developed *via* agrobacterium-mediated transformation (Reiss et al. 2009).

7.3.2 Polyploidy

It is defined as the presence of more than two sets of chromosomes in a nucleus. Polyploidy is recognized as an invaluable force driving plant evolution and speciation (Madlung 2013; Zhang et al. 2019a; Nieto Feliner et al. 2020). Although polyploidy is ubiquitously distributed in the plant kingdom, it has been considered more important for evolution and diversification of higher plants such as angiosperms and ferns than lower plants such as liverworts, hornworts and gymnosperms (Husband et al. 2013; Schneider et al. 2017). Two types of polyploids are known on the basis of the origin of progenitor species. These are autopolyploids and allopolyploids (Kihara and Ono in 1926). Autopolyploids arose from intraspecific genome duplication of structurally similar or identical genomes, whereas allopolyploids owe their origin to interspecific hybridization between two or more species, followed by chromosome doubling of non-homologous genomes. Stebbins (1947) later added a new class of polyploids; "segmental" allopolyploids to define those polyploids that carry the majority of chromosomal segments in common in its component genomes and show varying numbers of multivalents during meiosis (Stebbins 1947). They are considered intermediates between true autopolyploids and allopolyploids. The current concept of segmental allopolyploidy includes both transitional autopolyploids and allopolyploids. These depict a blend of auto- and allopolyploid segments after fixation of duplication-deletion events precipitated by homoeologous exchanges during allopolyploidization (Sun et al. 2017; Bertioli et al. 2019; Mason and Wendel 2020). Allopolyploids were previously perceived to be more important than autopolyploids in terms of contribution to species diversification and niche exploitation due to the benefits of fixed heterosis and intergenomic interactions with allopolyploidy (Hegarty and Hiscock 2008; Bansal et al. 2012). However, recent data have supported the importance of autopolyploids in the natural populations and their significance in shaping available biodiversity (Parisod et al. 2010a; Barker et al. 2016). Both autopolyploids (e.g., potato, watermelon, strawberries, alfalfa) and allopolyploids (such as wheat, soybean, rapeseed and mustard, cotton, tobacco) are key contributors for human nutrition (Leitch and Bennett 1997; Chen 2007; Dubcovsky and Dvorak 2007; Pignatta et al. 2010; Chalhoub et al. 2014; Van de Peer et al. 2017). Plant species like *Arabidopsis* (Arabidopsis Genome Initiative 2000), maize (Gaut and Doebley 1997) and rice (Zhang et al. 2005), previously assumed to be strict diploid species, are now considered as "paleopolyploids," i.e., species that have experienced ancient polyploidy events and later rediploidized through extensive genome reshuffling and gene fractionation to become functional diploids (Nieto Feliner et al. 2020). Deep genome sequencing and analysis technologies are continuously evidencing multiple (both ancient and recent) episodes of polyploidization in almost all investigated plant species (Cheng et al. 2018; Ren et al. 2018; Zhang et al. 2019a). A model plant species *Arabidopsis thaliana*, placed in the Brassicaceae plant family, has experienced many incidents of genome wide duplication events across its evolutionary timeframe (Jiao et al. 2011). *Brassica* species also underwent the same polyploidy history as experienced by *A. thaliana*, but with an extra lineage-specific whole genome triplication event (i.e., paleohexaploid event) after divergence from *Arabidopsis thaliana* about 15–20 million years ago (Lysak et al. 2005; Couvreur et al. 2010; Wang et al. 2011; Cheng et al. 2013; Yang et al. 2016). The paleohexaploid event has been interpreted to be very crucial for promoting intra- and inter-species diversity in three elementary diploid species of *Brassica* through extensive genome restructuring, karyotype changes and biased gene fractionation in three triplicated sub-genomes (Wang et al. 2011). The three diploid species of *Brassica* hybridized multiple times in all possible diploid pairs to produce three cultivated allopolyploids (*B. juncea*, AABB, $2n = 36$; *B. carinata*, BBCC, $2n = 34$; and *B. napus*, AACC, $2n = 38$) (Gomez-Campo and Prakash 1999, Flannery et al. 2006). Apparently, the present day Brassica allopolyploids have undergone both ancient and recent events of genome duplication. These allopolyploids occupy a greater range of ecological habitats than their diploid progenitors.

7.3.2.1 Synthetic Polyploids

Polyploidy has proven to be a highly dynamic and continuous evolutionary activity that cyclically provided novel evolutionary substrates for species diversification, phenotypic novelty and adaptation to wide ecological range (Hohmann et al. 2015; Tank et al. 2015; Soltis and Soltis 2016; Mandakova et al. 2017; Van de Peer et al. 2017; Jiao 2018). Advantages associated with natural polyploidy fascinated the Brassica breeders to artificially synthesize polyploids for exploiting this phenomenon for trait diversification. Plant geneticists used synthetic polyploidy to model the genetic and genomic events responsible for genome stabilization, establishment and phenotypic innovation in neopolyploids. All natural allotetraploids (*B. juncea*, *B. napus* and *B. carinata*) have been repeatedly resynthesized by sexual and somatic hybridizations of selected genotypes of progenitor species (Prakash 1973b; Sarla and Raut 1988; Jourdan and Salazar 1993; Srivastava et al. 2004; Song et al. 1995; Axelsson et al. 2000; Schranz and Osborn 2000, 2004; Pires et al. 2004; Albertin et al. 2006; Lukens et al. 2006; Gaeta et al. 2007; Hasan and Rahman 2018) to directly produce productive forms or donors for desired traits. They are also recognized as excellent genetic models to reveal the immediate genetic and genomic consequences of de novo and recurrent polyploidization due to exact known parentage, which is not possible in existing polyploids (Samans et al. 2018). Allotetraploids of *B. juncea* and *B. napus* were also resynthesized by hybridizing non-parental but related allotetraploids (Gupta et al. 2015; Chatterjee et al. 2016). Reports abound on synthesis of novel polyploids in Brassicaceae family including both autopolyploid and allopolyploid species. *Raphanobrassica* developed through intergeneric hybridization between *Raphanus sativus* and *B. oleracea* was the first synthetic allopolyploid generated in the family (Karpechenko 1928). Many more allotetraploids were also synthesized by hybridizing cultivated crop Brassica diploids and wild Brassicaceae species. These do not exist in nature and were synthesized only to establish genome relatedness or use as bridging species to transfer cytoplasmic or nuclear encoded traits to crop *Brassica* species (Gupta and Banga 2020). There is also lot of interest in the development of new synthetic Brassica allohexaploids (AABBCC; $2n = 54$). These were synthesized to combine the useful traits of three genomes of *Brassica* and expand the gene pool of Brassica (reviewed in Gaebelein and Mason 2018). It is expected that allohexaploids will outcompete the diploids and allotetraploid species of *Brassica* in seed yield and oil quality and will show greater adaptation to wide geographical range including increased resistance to pests and diseases. Three potential approaches have been attempted to produce allohexaploids by using species from the U's triangle, i.e., by crossing allotetraploids and with diploid species followed by chromosome doubling (Pradhan et al. 2010; Tian et al. 2010; Gupta et al. 2016a; Zhou et al. 2016; Gaebelein et al. 2019a; Mwathi et al. 2020), two step hybridization between three elementary diploid species (Cui et al. 2012) and hybridization among three allotetraploid species (Mason et al. 2011, 2012; Mwathi et al. 2017). Most of these allohexaploids are unstable, and loss of chromosomes was observed during various generations of selfing following their initial synthesis. Allohexaploids developed through cyclic hybridization between three allotetraploids could not attain a chromosome number and complement expected from a trigenomic Brassica allohexaploid (Mwathi et al. 2017). Till date, crossing *B. carinata* with *B. rapa* followed by chromosome doubling is the exploited cross-combination to produce allohexaploids in crop Brassica (Iwasa 1964; Takeda and Takahata 1996; Meng et al. 1998; Rahman 2001; Jiang et al. 2007; Tian et al. 2010). Only the allohexaploid synthesized through hybridization between *B. carinata* and a specific genotype of *B. rapa* (Gupta et al. 2016a) has proven to be stable over several generations of selfing. Attempts are underway to identify the pairing regulating gene(s) involved in this combination. Some other novel allohexaploid *Brassica* species possessing third genome from wild species (*Sinapsis alba*, *B. fruticulosa* and *B. maurorum*) have also been generated by somatic hybridization or reciprocal crossing between Brassica allotetraploids and wild species (Chen et al. 2012; Yao et al. 2012; Kumari et al. 2018, 2020). They can serve as potential bridging species to transfer traits of interest from these wild aliens into cultivated Brassicas.

7.3.2.2 Diploidization of Neo-Polyploids

A neo-polyploid (whether autopolyploid or allopolyploid) must maintain or attain meiotic stability and niche specialization for its emergence as a new species (Comai 2005; Baduel et al. 2018; Pele et al. 2018). In neo-autopolyploids, complex pairing or multivalents are formed due to the presence of more than two similar copies of each chromosome in a single nucleus (Lloyd and Bomblies 2016). The complex pairing associations often irregularly disjoin and produce unbalanced spores. Low pollen viability and severe meiotic abnormalities were common in synthetic autooctaploids of *B. rapa, B. nigra*, *B. oleracea* (Liu 2014) and *Arabidopsis thaliana* (Wang et al. 2010). However, such complex associations are uncommon in natural or established autopolyploids. Only bivalents are formed in autotetraploid *A. arenosa*. Multivalents are not formed as there is an evolutionary reduction in the number of crossovers that can be formed. Only one crossover per homolog is formed in the natural accessions of the species (Carvalho et al. 2010; Cifuentes et al. 2010; Pecinka et al. 2011; Yant et al. 2013; Lloyd and Bomblies 2016; Spoelhof et al. 2017). However, in nascent allopolyploids, the presence of two or more chromosome sets with varying degree of ancestral similarity may confound the ability of chromosomes to recognize correct partners to synapse and recombine with. This trigger illegitimate exchanges due to multiple chromosome configurations at metaphase-I. Result is the production of aneuploid gametes and reduced fertility (Ramsey and Schemske 2002). Level of illegitimate or homoeologous recombination in an allopolyploid is governed by two factors: first physical divergence between homoeologous chromosomes and second genetic control of pairing. Resynthesized *B. napus* is meiotically unstable and shows much higher frequency of homoeologous recombination through pairing between A- and C-genome chromosomes (Gaeta et al. 2007; Szadkowski et al. 2011; Xiong et al. 2011), which share large regions of synteny. Xiong et al. (2021) have provided strong evidence for high frequency of homoeologous chromosome pairing as tetravalent, exchanges, homoeologous chromosome replacement and unresolved pairing of 45S rDNA containing chromosomes. These caused a greater number of chromosome breakages and non-homologous centromere associations in resynthesized *B. napus* compared to the natural *B. napus*. These were inferred by phenotyping meiosis-I through repetitive DNA and BAC-FISH probes. They also confirmed that these meiotic errors are mainly exhibited by homoeologues that share synteny along entire chromosome length. In contrast, synthetic *B. juncea* (Axelsson et al. 2000; Srivastava et al. 2004) and *B. carinata* (Sarla and Raut 1988; Jourdan and Salazar 1993) display comparatively a much stable meiotic behavior due to poor sequence similarity of B-genome to A- and C-genomes (Li et al. 2017). Similarly, an intergeneric allotetraploid *Brassicoraphanus* (AARR; $2n = 38$) developed from a cross between *B. rapa* and *Raphanus sativus* (RR; $2n = 18$) showed stable meiosis and high fertility, probably due to little or no homoeologous pairing in the absence of large syntenic regions between the two participating genomes (Park et al. 2020). However, largely unstructured sub-genomes and a few homoeologous exchange events have been recognized at genome microscale in established *B. napus* cultivars during its comparative genome analysis with diploid progenitor species (Chalhoub et al. 2014). This supported the idea that genetic variants for high meiotic stability or low homoeologous exchanges must exist in the true progenitor genotypes of present *B. napus* cultivars. Otherwise, two sub-genomes of *B. napus* should be structurally diverged enough to be recognized distinct from actual involved genomes through interchanges between homoeologous chromosomes during cytological diploidization (Soltis and Soltis 2000; Wendel 2000). Meiotic investigations in synthetic allohexaploids of Brassica created from different cross-combinations *B. carinata* × *B. rapa* (Tian et al. 2010; Gupta et al. 2016a), *B. juncea* × *B. oleracea* (Zhou et al. 2016; Mwathi et al. 2020), *B. napus* × *B. nigra* (Gaebelein et al. 2019a) and (*B. napus* × *B. carinata*) × *B. juncea* (Gaebelein et al. 2019b; Mwathi et al. 2017) showed that meiotic stability is the net outcome of interactions between A-, B- and C-sub-genomes, in addition to few undocumented genetic loci. For instance, out of many allohexaploids synthesized from various genotypes of *B. carinata and B. rapa* parental species, only those hexaploids that involved a specific genotype of *B. rapa* remained meiotically stable and produced mostly euploid progenies in selfing generations (Gupta et al. 2016a). A major genetic locus called *PrBn* was identified in allohaploids of *B. napus* controlling homoeologous recombination (Jenczewski et al. 2003). This locus was later mapped to Chromosome *Bna*C9 (Liu et al. 2006). A role for six minor effect QTLs on Chromosomes *Bna*A1, *Bna*C1, *Bna*C3 and *Bna*C6 was also indicated. However, diploid parental lines used in forementioned study produced regular bivalents (Jenczewski et al. 2003). Knockout of one copy of meiosis gene *MSH4* prevents A–C pairing in Brassica allohaploids, although this gene was considered an unlikely candidate for the stabilization of *B. napus* due to its presence in two functional copies in natural *B. napus* (Gonzalo et al. 2019). A major QTL (*BnaPh1*) has been recently recognized on Chromosome A9 (Higgins et al. 2021). It explained 32–58% of the observed variation for homoeologous recombination. Role of two minor QTLs was also indicated.

7.3.2.3 Genome Evolution of a Neo-Polyploid

Polyploidy sets in motion a myriad of genetic and epigenetic alterations depend upon the genome background of involved species. Such alterations are critical for reconciling two or more genomes brought together in a single nucleus (McClintock 1984). These changes have been investigated more widely in allopolyploids than autopolyploids. Homoeologous exchanges, chromosomal rearrangements, deletions and duplications are all responsible for chromosome instability and reduced fertility in neo-allopolyploids. However, at the other end of the horizon, all these contribute to genetic and phenotypic diversities over the time as all of these are subject to the natural selection (Gaeta and Pires 2010; Lashermes et al. 2014; Rousseau-Gueutin et al. 2017; Stein et al. 2017; Hurgobin et al. 2018; Samans et al. 2018). Mispairing between ancestrally related but differentiated chromosomes during meiosis results in homoeologous reciprocal translocations (HRTs) and homoeologous non-reciprocal translocations (HNRTs) (Gaeta and Pires 2010). While HRTs are balanced, HNRTs are unbalanced. Reciprocal translocations involve swapping the pieces of DNA between two sub-genomes. In contrast, HNRT results from the loss of one homoeologous segment and concomitant duplication of the other segment. Homoeologous exchanges are a major cause of gene copy number variation in *Brassica napus* varieties (Hurgobin et al. 2018). Large structural alterations such as aneuploidy, inter- and intragenomic rearrangements, deletions, duplications, a wide variety of genetic and epigenetic events like specific DNA sequence elimination (Ozkan and Feldman 2009; Khasdan et al. 2010), rDNA loci changes, (Kovarik et al. 2008), gene conversion (Salmon et al. 2010), activation of transposons and induced insertional mutagenesis (Parisod et al. 2010b; Vicient and Casacuberta 2017) cause reprogramming of transcriptomes, proteomes and metabolomes in the neo-polyploids (Albertin et al. 2006; Feldman and Levy 2012; Wendel et al. 2016; Hurgobin et al. 2018). All these events contribute to the diploid like meiotic behavior in newly formed allopolyploids. Meiotic analysis of 32 lines of resynthesized *Brassica napus* involving different genotypes of *B. rapa* and *B. oleracea* showed marked changes in number and positioning of rDNA loci in early generations (S0–S3) from parental rDNA loci patterns. Loss of 35S rDNA loci specifically occurred when *B. rapa* was the maternal parent (Sosnowska et al. 2020). The genome size of allopolyploid is usually reduced from the predicted total of the genomes of parental species during the process of polyploidization (Boyko et al. 1984, 1988; Ma and Gustafson 2005; Paterson et al. 2012). In Brassicas, a decline of nearly 1.072% DNA was observed in resynthesized Brassica genotypes derived from the interspecific crosses following five cycles of polyploidization (Gupta et al. 2014). Genome downsizing during polyploidization occurs mainly due to the loss of repetitive DNA sequences and not the gene content per se (Buggs et al. 2012). Sudden doubling of all genome content generates multiple copies of the genes paralogous to each other. Polyploid individuals may take advantage of this beneficial situation by adopting different gene fractionation pathways (Feldman and Levy 2012). Relaxed selection pressure on duplicated genes due to their functional redundancy may allow one copy either to mutate (either genetically or epigenetically) to become nonfunctional (pseudogenization), or to evolve a new function (neofunctionalization). Nonfunctional gene copy may be lost subsequently if it does not provide any selective advantage to polyploid organisms. Allopolyploid may also retain both copies of the genes without adopting new function by partitioning the ancestral functional role of gene among duplicate copies (sub-functionalization) (Grover et al. 2012). Sub-genome dominance is very common among allopolyploids. It refers to gene fractionation bias and expression-level dominance between homoeologous/paralogous genes from different sub-genomes of an allopolyploid (Zhang et al. 2019a). This phenomenon is widely observed and discussed in *Arabidopsis*, *Gossypium*, *Zea* and *Brassica* species (Chang et al. 2010; Yoo et al. 2013; Cheng et al. 2016). It is speculated that dominant sub-genome undergoes less gene fractionation during polyploidization and, therefore, exhibits higher gene density. Dominant sub-genome also shows fewer numbers and density of TEs and ensures minimal changes in their gene complement caused by epigenetic alterations. Expression level of the genes in the dominant sub-genome is also higher as compared to their paralogs from the recessive sub-genome(s) (Bottani et al. 2018). Sub-genome dominance is the outcome of varied sub-genome stability and nucleolar dominance of the interacting genomes. A hierarchy of sub-genome stability, i.e., B > A > C is known among the sub-genomes of *Brassica* species (Li et al. 2020). Intergenomic conflict may also arise in newly sexually formed polyploids due to altered cytonuclear interactions as they carry maternal and paternal nuclear chromosomes from both progenitor species, but inherit cytoplasm only from the female parents. It is hypothesized that the maternal nuclear genes are preferentially retained over the paternally inherited copies to avoid cytoplasm-nuclear conflict (Sharbrough et al. 2017; Gong et al. 2012). Other biases may be also be important for the fractionation of multicopy genes to different fates. For example, genes acting independently or less connected with others, i.e., involved in the functional categories of DNA-repair, recombination, enzyme activity, kinase activity, transport, tRNA ligation and defense are preferentially retained as single copies (Blanc and Wolfe 2004; Samans et al. 2017). In contrast, genes functioning in complexes or interacting with other genes associated with the basic cellular machinery, nucleotide sugar metabolism and regulatory functions are preferentially preserved in multiple copies (Sankoff et al. 2010; Pont et al. 2011; Freeling et al. 2015). Increased allele dosage and fixed heterosis (due to fixation of alternate alleles in paralogs) help to increase phenotypic variance of traits and the emergence of novel phenotypes in neo-allopolyploids. Epigenetic changes such as DNA methylation, histone modifications and RNA interference are also prevalent in neo-polyploids (Zhang et al. 2015; Ding and Chen 2018). Approximately 1.7% of the fragments underwent DNA methylation changes upon genome doubling in autotetraploid *B. rapa* with respect to diploid donor (Xu et al. 2017). Xu et al. (2009) reported a higher frequency of alterations, i.e., 4.09% in gene expression and 6.84% in DNA methylation among resynthesized *B. napus* compared to its diploid progenitors. C-genome-specific gene expression and methylation changes were significantly higher than those of A-genome-specific alterations. The resynthesized *B. juncea* revealed a higher internal methylation (40.4%) level than the natural *B. juncea* (17.1%), but external methylation was almost the same (Gupta et al. 2019). Genetic and epigenetic alterations in neo-polyploids cause extensive repatterning of duplicate gene expression which may lead to transgressive upregulation/downregulation, unequal parental contributions and gene silencing (Yoo et al. 2014). Transgressive upregulation or downregulation occurs when cumulative expression level of duplicate gene pairs is statistically higher or lower in polyploids than that sum total of individual gene expression in the two diploid progenitors. Li et al. (2020) noted the effects of natural hybridization and polyploidization on gene expression dynamics in natural *B. napus* relative to diploid progenitors. They used RNA-Seq data from stems, leaves, flowers and siliques and found 48, 29.7 and 22.3% homoeologous gene pairs exhibited additive expression, expression-level dominance and transgressive expression. Non-stochastic and non-additive expression were also observed at protein level in polyploids with expression-level dominance veering toward a particular sub-genome (Shen et al. 2014).

7.3.2.4 Phenotypic Novelty in Polyploids

Polyploidy increases the range of trait values represented by parents and may also drive novel characters absent in progenitor species (Leitch et al. 2008). Polyploid plants generally possess enlarged cells and may have increased organ size and vigorous plant growth habit compared to their diploid counterparts (Ramsey and Ramsey 2014). However, these effects may be genotype or tissue specific and may vary for the level of ploidy (Abel and Becker 2007; Doyle and Coate 2019; Rashid et al. 2021). At many instances, autopolyploids of Brassica diploids and tetraploids are reported to have equal or reduced vegetative biomass relative to their donors (Olsson 1960a; Olsson 1960b; Abel and Becker 2007; Yin et al. 2020). This may be attributed to inbreeding depression owing to increased homozygosity in autopolyploids. However, Rashid et al. (2021) have reported different effects of autopolyploidization in *B. rapa* and *B. oleracea*. *B. rapa* displayed enlarged vegetative and reproductive organs, whereas *B. oleracea* did not show increased organ size except for seed size. Novel functions were also identified in some synthetic autopolyploids. Increased pollen apertures were also noted in synthetic autoallooctaploids of *B. napus* (Yin et al. 2020). These were helpful in inducing maternal doubled haploids of *B. napus* when used as pollinator with allotetraploid *B. napus* (Fu et al. 2018). Allopolyploids also showed vigorous plant growth, probably due to fixed heterosis obtained through intergenomic hybridization (Quijada et al. 2006; Udall et al. 2006; Bansal et al. 2012). Derived *B. juncea* and *B. napus* populations extracted from single genotypes of non-progenitor species showed substantial variation for morphological and yield related traits and high heterosis with natural *B. juncea* and *B. napus* genotypes (Gupta et al. 2015; Chatterjee et al. 2016). A novel phenotype, i.e., determinate growth habit also appeared in a subset of the population of derived *B. juncea* (Gupta et al. 2015; Kaur and Banga 2015). HNRTs arising due to homoeologous exchanges were associated with QTLs for yield (Osborn et al. 2003), disease resistance (Zhao et al. 2006), flowering time variation (Pires et al. 2004; Stein et al. 2017) and glucosinolates metabolism (Hurgobin et al. 2018) in *B. napus*.

References

Abel S, Becker HC (2007) The effect of autopolyploidy on biomass production in homozygous lines of *Brassica rapa* and *Brassica oleracea*. Plant Breed 126:642–643

Agarwal PK, Agarwal P, Custers JB, Liu CM, Bhojwani SS (2006) PCIB an antiauxin enhances microspore embryogenesis in microspore culture of *Brassica juncea*. Plant Cell Tiss Org Cult 86:201–210

Agrawal N, Gupta M, Banga SS, Heslop-Harrison JP (2020) Identification of chromosomes and chromosome rearrangements in crop Brassicas and *Raphanus sativus*: a cytogenetic toolkit using synthesized massive oligonucleotide libraries. Front Plant Sci 11:598039

Aionesei T, Touraev A, Heberle-Bors E (2005) Pathways to microspore embryogenesis. In: Palmer CE, Keller WA, Kasha KJ (eds) Haploids in crop improvement II. Springer, Berlin, pp 11–34

Ajisaka H, Kuginuki Y, Shiratori M, Ishiguro K, Enomoto S, Hirai M (1999) Mapping of loci affecting the culturing efficiency of microspore culture of *Brassica rapa* L. syn. *campestris* L. using DNA polymorphism. Breed Sci 49:187–192

Albertin W, Balliau T, Brabant P, Chevre AM, Eber F, Malosse C, Thiellement H (2006) Numerous and rapid nonstochastic modifications of gene products in newly synthesized *Brassica napus* allotetraploids. Genetics 173:1101–1113

Ali SNH, Ramanna MS, Jacobsen E, Visser RGF (2001) Establishment of a complete series of a monosomic tomato chromosome addition lines in the cultivated potato using RFLP and GISH analyses. Theor Appl Genet 103:687–695

Ali MM, Mian MAK, Custers JBM, Khuram MMH (2008) Microspore culture and the performance of microspore derived doubled haploid in *Brassica juncea* (L.). Bang J Agric Res 33:571–578

Armstrong KC, Keller WA (1982) Chromosome pairing in haploids of *Brassica oleracea*. Can J Genet Cytol 24:735–739

Armstrong SJ, Caryl AP, Jones GH, Franklin FCH (2002) *Asy 1*, a protein required for meiotic chromosome synapsis, localizes to axis-associated chromatin in *Arabidopsis* and *Brassica*. J Cell Sci 115:3644–3655

Arora H, Padmaja KL, Paritosh K, Mukhi N, Tewari AK, Mukhopadhyay A, Gupta V, Pradhan AK, Pental D (2019) *BjuWRR1*, a CC–NB–LRR gene identified in *Brassica juncea*, confers resistance to white rust caused by *Albugo candida*. Theor Appl Genet 132:2223–2236

Attia T, Busso C, Robbelen G (1987) Digenomic triploid for an assessment of chromosome relationships in the cultivated diploid *Brassica* species. Genome 29:326–330

Axelsson T, Bowman CM, Sharpe AG, Lydiate DJ, Lagercrantz U (2000) Amphidiploids *Brassica juncea* contains conserved progenitor genomes. Genome 43:679–688

Babbar SB, Agarwal PK, Sahay S, Bhojwani SS (2004) Isolated microspore culture of *Brassica*: an experimental tool for developmental studies and crop improvement. Indian J Plant Biotechnol 3:185–202

Baduel P, Bray S, Vallejo-Marin M, Kola F, Yant L (2018) The "polyploid hop": shifting challenges and opportunities over the evolutionary lifespan of genome duplications. Front Ecol Evol 6:117

Banga SS (1988) C-genome chromosome substitution lines in *Brassica juncea*. Genetica 77:81–86

Bansal P, Banga SK, Banga SS (2012) Heterosis as investigated in terms of polyploidy and genetic diversity using designed *Brassica juncea* amphiploid and its progenitor diploid species. PLoS ONE 7(2): e29607. https://doi.org/10.1371/journal.pone.0029607

Barker MS, Arrigo N, Baniaga AE, Li Z, Levin DA (2016) On the relative abundance of autopolyploids and allopolyploids. New Phytol 2:391–398

Bento M, Gustafson P, Viegas W, Silva M (2010) Genome merger: from sequence rearrangements in triticale to their elimination in wheat-rye addition lines. Theor Appl Genet 121:489–497

Bertioli DJ, Jenkins J, Clevenger J, Dudchenko O, Gao D, Seijo G et al (2019) The genome sequence of segmental allotetraploid peanut *Arachis hypogaea*. Nat Genet 51:877–884

Bhayana L, Paritosh K, Arora H, Yadava SK, Singh P, Nandan D, Mukhopadhyay A, Gupta V, Pradhan AK, Pental D (2019) A mapped locus on LG A6 of *Brassica juncea* line Tumida conferring resistance to white rust contains a CNL type R gene. Front Plant Sci 10:1690

Bhojwani SS, Razdan MK (1996) Plant tissue culture: theory and practice. Elsevier, Amsterdam, pp 177–201

Birchler JA, Newton KJ (1981) Modulation of protein levels in chromosomal dosage series of maize: the biochemical basis of aneuploid syndromes. Genetics 99:247–266

Birchler JA, Veitia RA (2007) The gene balance hypothesis: from classical genetics to modern genomics. Plant Cell 19:395–402

Blakeslee AF, Belling J, Farnham ME, Bergner AD (1922) A haploid mutant in the Jimson weed, *Datura stramonium*. Science 55:646–647

Blanc G, Wolfe KH (2004) Functional divergence of duplicated genes formed by polyploidy during *Arabidopsis* evolution. Plant Cell 16:1679–1691

Bottani S, Zabet NR, Wendel JF, Veitia RA (2018) Gene expression dominance in allopolyploids: hypotheses and models. Trends Plant Sci 23:393–402

Boyko EV, Badaev NS, Maximov NG, Zelenin AV (1984) Does DNA content change in the course of *Triticale* breeding? Cereal Res Commun 12:99–100

Boyko EV, Badaev NS, Maximov NG, Zelenin AV (1988) Regularities of genome formation and organization in cereals. I. DNA quantitative changes in the process of allopolyploidization. Genetika 24:89–97

Brewer GJ, Sing CF, Sears ER (1969) Studies of isozyme patterns in nullisomic-tetrasomic combinations of hexaploid wheat. Proc Natl Acad Sci USA 64 (4):1224–1229

Britt AB, Kuppu S (2016) Cenh3: an emerging player in haploid induction technology. Front Plant Sci 7:357

Buggs RJA, Renny-Byfield S, Chester M, Jordon-Thaden IE, Viccini LF, Chamala S, Leitch AR, Schnable PS, Barbazuk WB, Soltis PS, Soltis DE (2012) Next-generation sequencing and genome evolution in allopolyploids. Amer J Bot 99:372–382

Carter CR (1978) The cytology of *Brachycome* 8. The inheritance, frequency and distribution of B chromosomes in *B. dichromosomatica* ($n = 2$), formerly included in *B. lineariloba*. Chromosoma 67:109–121

Carvalho A, Delgado M, Barao A, Frescatada M, Ribeiro E, Pikaard CS, Viegas W, Neves N (2010) Chromosome and DNA methylation dynamics during meiosis in the autotetraploid *Arabidopsis arenosa*. Sex Plant Reprod 23:29–37

Cegielska-Taras T, Pniewski T, Szala L (2008) Transformation of microspore–derived embryos of winter oilseed rape (*Brassica napus* L.) by using *Agrobacterium tumefaciens*. J Appl Genet 49:343–347

Chaikam V, Molenaar W, Melchinger AE, Boddupalli PM (2019) Doubled haploid technology for line development in maize: technical advances and prospects. Theor Appl Genet 132:3227–3243

Chalhoub B, Denoeud F, Liu S, Parkin IA, Tang H et al (2014) Early allopolyploid evolution in the post-Neolithic *Brassica napus* oilseed genome. Science 345:950–953

Chanana NP, Dhawan V, Bhojwani SS (2005) Morphogenesis in isolated microspore cultures of *Brassica juncea*. Plant Cell Tiss Org Cult 83:169–177

Chang R, Tai W, Fan Z (1987) Cytogenetic studies of aneuploids in rapeseed. II. Morphology, fertility, and cytology of self-pollinated progenies of monosomic plants of *Brassica napus* L. Genome 29:174–179

Chang PL, Dilkes BP, McMahon M, Comai L, Nuzhdin SV (2010) Homoeolog-specific retention and use in allotetraploid *Arabidopsis suecica* depends on parent of origin and network partners. Genome Biol 11:R125

Chatterjee D, Banga SK, Gupta M, Bharti S, Salisbury PA, Banga SS (2016) Resynthesis of *Brassica napus* through hybridization between *B. juncea* and *B. carinata*. Theor Appl Genet 129:977–990

Chen ZJ (2007) Genetic and epigenetic mechanisms for gene expression and phenotypic variation in plant polyploids. Annu Rev Plant Biol 58:377–406

Chen BY, Simonsen V, Lannrr-Herrera C, Heneen WK (1992) A *Brassica campestris-alboglabra* addition line and its use for gene mapping, intergenomic gene transfer and generation of trisomics. Theor Appl Genet 84:592–599

Chen B, Cheng B, Jorgensen R, Heneen WK (1997) Production and cytogenetics of *Brassica campestris-alboglabra* chromosome addition lines. Theor Appl Genet 94:633–640

Chen CC, Chen SK, Liu MC, Kao YY (2002) Mapping of DNA markers to arms and sub-arm regions of *Nicotiana sylvestris* chromosomes using aberrant alien addition lines. Theor Appl Genet 105:8–15

Chen JF, Luo XD, Qian CT, Jahn MM, Staub JE, Zhuang FY, Lou QF, Ren G (2004) Cucumis monosomic alien addition lines: morphological, cytological, and genotypic analyses. Theor Appl Genet 108:1343–1348

Chen HF, Wang H, Li ZY (2007) Production and genetic analysis of partial hybrids in intertribal crosses between *Brassica* species (*B. rapa*, *B. napus*) and *Capsella bursa-pastoris*. Plant Cell Rep 26:1791–1800

Chen JP, Ge XH, Yao XC, Li ZY (2012) Genome affinity and meiotic behaviour in trigenomic hybrids and their doubled allohexaploids between three cultivated *Brassica* allotetraploids and *Brassica fruticulosa*. Genome 55:164–171

Cheng BF, Olin-Fatih M, Heneen WK (1993) Cytogenetic studies on *Brassica campestris* primary trisomics. Hereditas 119:143–148

Cheng BF, Seguin-Swartz G, Somers DJ, Rakow G (2001) Low glucosinolate *Brassica juncea* breeding line revealed to be nullisomic. Genome 44:738–741

Cheng F, Mandakova T, Wu J, Xie Q, Lysak MA, Wang X (2013) Deciphering the diploid ancestral genome of the mesohexaploid *Brassica rapa*. Plant Cell 25:1541–1554

Cheng F, Sun C, Wu J, Schnable J, Woodhouse WR, Liang J, Cai C, Freeling M, Wang X (2016) Epigenetic regulation of subgenome dominance following whole genome triplication in *Brassica rapa*. New Phytol 211:288–299

Cheng F, Wu J, Cai X, Liang J, Freeling M, Wang X (2018) Gene retention, fractionation and subgenome differences in polyploid plants. Nat Plants 4:258–268

Chester M, Gallagher JP, Symonds VV, Cruz da Silva AV, Mavrodiev EV, Leitch AR, Soltis PS, Douglas ES (2012) Extensive chromosomal variation in a recently formed natural allopolyploid species, *Tragopogon miscellus* (Asteraceae). Proc Natl Acad Sci USA 109:1176–1181

Chevre AM, This P, Eber F, Deschamps M, Renard M, Delseny M, Quiros CF (1991) Characterization of disomic addition lines *Brassica napus-Brassica nigra* by isozyme, fatty acid, and RFLP markers. Theor Appl Genet 81:43–49

Choudhary BR, Joshi P, Singh K (2000) Synthesis, morphology and cytogenetics of *Raphanofortii* (TTRR, $2n = 38$): a new amphidiploid of hybrid *Brassica tournefortii* (TT, $2n = 20$) 9 *Raphanus caudatus* (RR, $2n = 18$). Theor Appl Genet 101:990–999

Chuong PV, Deslauriers C, Kott LS, Beversdorf WD (1988) Effects of donor genotype and bud sampling on microspore culture of *Brassica napus*. Can J Bot 6:1653–1657

Cifuentes M, Benavente E (2009) Wheat-alien metaphase I pairing of individual wheat genomes and D genome chromosomes in interspecific hybrids between *Triticum aestivum* L. and *Aegilops geniculata* Roth. Theor Appl Genet 119:805–813

Cifuentes M, Eber F, Lucas MO, Lode M, Chevre AM, Jenczewski E (2010) Repeated polyploidy drove different levels of crossover suppression between homoeologous chromosomes in *Brassica napus* allohaploids. Plant Cell 22:2265–2276

Clausen RE, Mann MC (1924) Inheritance in *Nicotiana tabacum*: V. The occurrence of haploid plants in interspecific progenies. Proc Natl Acad Sci USA 10:121–124

Cloutier S, Cappadocia M, Landry BS (1995) Study of microspore-culture responsiveness in oilseed rape (*Brassica napus* L.) by comparative mapping of a F_2 population and two microspore–derived populations. Theor Appl Genet 91:841–847

Coe EH (1959) A line of maize with high haploid frequency. Amer Nat 93:381–382. https://doi.org/10.1086/282098

Comai L (2005) The advantages and disadvantages of being polyploid. Nat Rev Genet 6:836–846

Couvreur TLP, Franzke A, Al-Shehbaz IA, Bakker FT, Koch MA, Mummenhoff K (2010) Molecular phylogenetics, temporal diversification, and principles of evolution in the mustard family (Brassicaceae). Mol Biol Evol 27:55–71

Crismani W, Portemer V, Froger N, Chelysheva L, Horlow C, Vrielynck N, Mercier R (2013) MCM8 is required for a pathway of meiotic double-strand break repair independent of DMC1 in *Arabidopsis thaliana*. PLoS Genet 9:e1003165

Cui C, Ge X, Gautam M, Kang L, Li Z (2012) Cytoplasmic and genomic effects on meiotic pairing in *Brassica* hybrids and allotetraploids from pair crosses of three cultivated diploids. Genetics 191:725–738

Devaux P (2003) The *Hordeum bulbosum* (L.) method. In: Maluszynski M, Kasha K, Forster BP, Szarejko I (eds) Doubled haploid production in crop plants: a manual. Springer, Netherlands, pp 15–19

Dhaka N, Rout K, Yadava SK, Sodhi YS, Gupta V, Pental D, Pradhan AK (2017) Genetic dissection of seed weight by QTL analysis and detection of allelic variation in Indian and east European gene pool lines of *Brassica juncea*. Theor Appl Genet 130:293–307

Ding M, Chen ZJ (2018) Epigenetic perspectives on the evolution and domestication of polyploid plant and crops. Curr Opin Plant Biol 42:37–48

Dong YJ, Tsuzuki E, Terao H (2001) Trisomic genetic analysis of aroma in three Japanese native rice varieties (*Oryza sativa* L.). Euphytica 117:191–196

Doyle JJ, Coate JE (2019) Polyploidy, the nucleotype, and novelty: the impact of genome doubling on the biology of the cell. Int J Plant Sci 180:1–52

Du XZ, Ge XH, Zhao ZG, Li ZY (2008) Chromosome elimination and fragment introgression and recombination producing intertribal partial hybrids from *Brassica napus* × *Lesquerella fendleri* crosses. Plant Cell Rep 27:261–271

Dubcovsky J, Dvorak J (2007) Genome plasticity a key factor in the success of polyploid wheat under domestication. Science 316:1862–1866

Dunwell JM (2010) Haploid in flowering plants: origin and exploitation. Plant Biotechnol J 8:377–424

Dwivedi SL, Britt AB, Tripathi L, Sharma S, Upadhyaya HD, Ortiz R (2015) Haploids: constraints and opportunities in plant breeding. Biotechnol Adv 33 (6):812–829

Fan Z, Tai W (1985) A cytogenetic study of monosomics in *Brassica napus* L. Can J Genet Cytol 27:683–688

Feldman M, Levy AA (2012) Genome evolution due to allopolyploidization in wheat. Genetics 192(3):763–774

Ferrie AMR, Möllers C (2011) Haploids and doubled haploids in *Brassica* spp. for genetic and genomic research. Plant Cell Tiss Org Cult 104:375–386

Ferrie AMR, Taylor DC, Mackenzie SL, Keller WA (1999) Microspore embryogenesis of high sn-2 erucic acid *Brassica oleracea* germplasm. Plant Cell Tiss Org Cult 57:79–84

Ferrie AMR, Taylor DC, MacKenzie SL, Rakow G, Raney JP, Keller WA (2008) Microspore mutagenesis of *Brassica* species for fatty acid modifications: a preliminary evaluation. Plant Breed 127:501–506

Flannery ML, Mitchell FJG, Coyne S, Kavanagh TA, Burke JI, Salamin N, Dowding P, Hodkinson TR (2006) Plastid genome characterization in *Brassica* and Brassicaceae using a new set of nine SSR. Theor Appl Genet 113:1221–1231

Freeling M, Scanlon MJ, Fowler JE (2015) Fractionation and sub-functionalization following genome duplications: mechanisms that drive gene content and their consequences. Curr Opin Genet Dev 35:110–118

Fu S, Lv Z, Qi B, Guo X, Li J, Liu B, Han F (2012) Molecular Cytogenetic characterization of wheat-*Thinopyrum elongatum* addition, substitution and translocation lines with a novel source of resistance to wheat *Fusarium* head blight. J Genet Genom 39:103–110

Fu S, Yin L, Xu M, Li Y, Wang M, Yang J, Tingdong F, Wang J, Shen J, Ali A, Zou Q, Yi B, Wen J, Tao L, Kang Z, Tang R (2018) Maternal doubled haploid production in interploidy hybridization between *Brassica napus* and *Brassica* allooctaploids. Planta 247:113–125

Gaebelein R, Mason AS (2018) Allohexaploids in the genus *Brassica*. Crit Rev Plant Sci 37(5):422–437

Gaebelein R, Alnajar D, Koopmann B, Mason AS (2019a) Hybrids between *Brassica napus* and *B. nigra* show frequent pairing between the B and A/C genomes and resistance to blackleg. Chrom Res 27 (3):221–236

Gaebelein R, Schiessl SV, Samans B, Batley J, Mason AS (2019b) Inherited allelic variants and novel karyotype changes influence fertility and genome stability in *Brassica* allohexaploids. New Phytol 223(2):965–978

Gaeta RT, Pires CJ (2010) Homoeologous recombination in allopolyploids: the polyploid ratchet. New Phytol 186:18–28

Gaeta RT, Pires JC, Iniguez-Luy F, Leon E, Osborn TC (2007) Genomic changes in resynthesized *Brassica napus* and their effect on gene expression and phenotype. Plant Cell 19:3403–3417

Gaines EF, Aase HC (1926) A haploid wheat plant. Amer J Bot 13(6):373–385

Gao D, Guo D, Jung C (2001) Monosomic addition lines of *Beta corolliflora* Zoss in sugar beet: cytological and molecular-marker analysis. Theor Appl Genet 103:240–247

Gao ZH, Wang RQ, Mao LP (2003) Chromosome location of dwarf gene in foxtail millet. Acta Agron Sin 29:152–154

Gardner ME, Hymowitz T, Xu SJ, Hartman GL (2001) Physical map location of the Rps1-k allele in soybean. Crop Sci 41:1435–1438

Gaut BS, Doebley JF (1997) DNA sequence evidence for the segmental allotetraploid origin of maize. Proc Natl Acad Sci USA 94(13):6809–6814

Geleta M, Heneen WK, Stoute AI, Muttucumaru N, Scott RJ, King GJ, Kurup S, Bryngelsson T (2012) Assigning *Brassica* microsatellite markers to the nine C-genome chromosomes using *Brassica rapa* var. *trilocularis–B. oleracea* var. *alboglabra* monosomic alien addition lines. Theor Appl Genet 125:455–466

Germana MA (2006) Doubled haploid production in fruit crops. Plant Cell Tiss Org Cult 86:131–146

Gil-Humanes J, Barro F (2009) Production of doubled haploids in *Brassica*. In: Touraev A, Forster BP, Jain SM (eds) Advances in haploid production in higher pants. Springer, Dordrecht, pp 65–73

Gilles LM, Khaled A, Laffaire JB, Chaignon S, Gendrot G, Laplaige J, Berges H, Beydon G, Bayle V, Barret P, Comadran J, Martinant JP, Rogowsky PM, Widiez T (2017) Loss of pollen-specific phospholipase NOT LIKE DAD triggers gynogenesis in maize. EMBO J 36(6):707–717

Gomez-Campo C, Prakash S (1999) The phylogeny of *Brassica* and allied genera. Domestication of cultivated Brassicas and allies. In: Gomez-Campo C (ed) Biology of *Brassica* coenospecies. Elsevier, Amsterdam, pp 33–58

Gong L, Salmon A, Yoo MJ, Grupp KK, Wang Z, Paterson AH, Wendel JF (2012) The cytonuclear dimension of allopolyploid evolution: an example from cotton using RUBISCO. Mol Biol Evol 29:3023–3036

Gonzalo A, Lucas MO, Charpentier C, Sandmann G, Lloyd A, Jenczewski E (2019) Reducing *MSH4* copy number prevents meiotic crossovers between non-homologous chromosomes in *Brassica napus*. Nat Commun 10:2354. https://doi.org/10.1038/s41467-019-10010-9

Grover CE, Gallagher JP, Szadkowski EP, Yoo MJ, Flagel LE, Wendel JF (2012) Homoeolog expression bias and expression level dominance in allopolyploids. New Phytol 196:966–971

Gu H, Sheng X, Zhao Z, Yu H, Wang J (2014) Initiation and development of microspore embryogenesis and plant regeneration of *Brassica nigra*. In Vitro Cell Dev Biol Plant 50:534–540

Guha S, Maheshwari SC (1964) In vitro production of embryos from anthers of *Datura*. Nature 204:497

Guo M, Davis D, Birchler JA (1996) Dosage effects on gene expression in a maize ploidy series. Genetics 142:1349–1355

Gupta M, Banga SS (2020) Exploiting alien genetic variation for germplasm enhancement in *Brassica* oilseeds. In: Kang MS (ed) Quantitative genetics, genomics and plant breeding. CABI, UK, pp 338–384

Gupta V, Mukhopadhyay A, Arumugam N, Sodhi YS, Pental D, Pradhan AK (2004) Molecular tagging of erucic acid trait in oilseed mustard (*Brassica juncea*) by QTL mapping and single nucleotide polymorphisms in *FAE1* gene. Theor Appl Genet 108:743–749

Gupta P, Kulwal P, Rustgi S (2005) Wheat cytogenetics in the genomics era and its relevance to breeding. Cytogenet Genome Res 109:315–327

Gupta M, Atri C, Banga SS (2014) Cytogenetic stability and genome size variations in newly developed derived *Brassica juncea* allopolyploid lines. J Oilseeds Brassica 5(2):118–127

Gupta M, Gupta S, Kumar H, Kumar N, Banga SS (2015) Population structure and breeding value of a new type of *Brassica juncea* created by combining A and B genomes from related allotetraploids. Theor Appl Genet 128(2):221–234

Gupta M, Atri C, Agarwal N, Banga SS (2016a) Development and molecular-genetic characterization of a stable *Brassica* allohexaploid. Theor Appl Genet 129:2085–2100

Gupta M, Mason AS, Batley J, Bharti S, Banga S, Banga SS (2016b) Molecular-cytogenetic characterization of C-genome chromosome substitution lines in *Brassica juncea* (L.) Czern and Coss. Theor Appl Genet 129:1153–1166

Gupta S, Sharma N, Akhatar J, Atri C, Kaur J, Kaur G, Banga SS (2019) Analysis of epigenetic landscape in a recombinant inbred line population developed by hybridizing natural and resynthesized *Brassica juncea* (L.) with stable C genome introgressions. Euphytica 215:174

Hamant O, Ma H, Cande WZ (2006) Genetics of meiotic prophase I in plants. Annu Rev Plant Biol 57:267–302

Hasan MJ, Rahman H (2018) Resynthesis of *Brassica juncea* for resistance to *Plasmodiophora brassicae* pathotype 3. Breed Sci 68(3):385–391. https://doi.org/10.1270/jsbbs.18010

Hazarika RR, Mishra KV, Chaturvedi R (2013) In vitro haploid production—a fast and reliable approach for crop improvement. In: Tuteja N, Gill SS (eds) Crop improvement under adverse conditions. Springer, New York, pp 171–212

Hegarty MJ, Hiscock SJ (2008) Genomic clues to the evolutionary success of polyploid plants. Curr Biol 18 (10):435–444

Henry IM, Dilkes BP, Miller ES, Burkart-Waco D, Comai L (2010) Phenotypic consequences of aneuploidy in *Arabidopsis thaliana*. Genetics 186:1231–1245

Higgins JD, Armstrong SJ, Franklin FCH, Jones GH (2004) The Arabidopsis MutS homolog *AtMSH4* functions at an early step in recombination: evidence for two classes of recombination in *Arabidopsis*. Genes Dev 18:2557–2570

Higgins EE, Howell EC, Armstrong SJ, Parkin IA (2021) A major quantitative trait locus on chromosome A9, *BnaPh1*, controls homoeologous recombination in *Brassica napus*. New Phytol 229:3281–3293

Hill CB, Tang K, Williamsp H (1986) Development of primary trisomics of rapid-cycling *Brassica campestris*. Crucferae Newsl 11:30

Hohmann N, Eva MW, Lysak MA, Koch MA (2015) A time-calibrated road map of Brassicaceae species radiation and evolutionary history. Plant Cell 27:2770–2784

Hua Y, Li Z (2006) Genomic *in situ* hybridization analysis of *Brassica napus* × *Orychophragmus violaceus* hybrids and production of *B. napus* aneuploids. Plant Breed 125:144–149

Huettel B, Kreil DP, Matzke M, Matzke AJ (2008) Effects of aneuploidy on genome structure, expression, and interphase organization in *Arabidopsis thaliana*. PLoS Genet 4:e1000226

Huo DA, Zhu B, Tian GF, Du XY, Guo J, Cai MX (2019) Assignment of unanchored scaffolds in genome of *Brassica napus* by RNA-seq analysis in a complete set of *Brassica rapa-Brassica oleracea* monosomic addition lines. J Integr Agri 18:1541–1546

Hurgobin B, Golicz AA, Bayer PE, Chan CKK, Tirnaz S, Dolatabadian A, Schiessl SV, Samans B, Montenegro JD, Parkin IAP, Pires JC, Chalhoub B, King GJ, Snowdon R, Batley J, Edwards D (2018) Homoeologous exchange is a major cause of gene presence/absence variation in the amphidiploid *Brassica napus*. Plant Biotechnol J 16:1265–1274

Husband BC, Baldwin SJ, Suda J (2013) The incidence of polyploidy in natural plant populations: major patterns and evolutionary processes. In: Leitch IJ, Greilhuber J, Doležel J, Wendel JF (eds) Plant genome diversity 2: physical structure, behaviour and evolution of plant genomes. Springer, Wien, pp 255–276

Ishii T, Karimi-Ashtiyani R, Houben A (2016) Haploidization *via* chromosome elimination: means and mechanisms. Annu Rev Plant Biol 67:421–438

Iwasa S (1964) Cytogenetic studies on the artificially raised trigenomic hexaploid hybrid forms in the genus *Brassica*. J Fac Agric Kyushu Univ 13:309–352

Jenczewski E, Eber F, Grimaud A, Huet S, Lucas MO, Monod H, Chevre AM (2003) *PrBn*, a major gene controlling homoeologous pairing in oilseed rape (*Brassica napus*) haploids. Genetics 164:645–653

Jiang Y, Tian E, Li R, Chen L, Meng J (2007) Genetic diversity of *Brassica carinata* with emphasis on the interspecific crossability with *B. rapa*. Plant Breed 126:487–491

Jiang H, Wang F-F, Wu Y-T, Zhou X, Huang X-Y, Zhu J, Gao J-F, Dong R-B, Cao K-M, Yang Z-N (2009) Multipolar Spindle 1 (*MPS1*), a novel coiled-coil protein of *Arabidopsis thaliana*, is required for meiotic spindle organization. Plant J 59:1001–1010

Jiao Y (2018) Double the genome, double the fun: Genome duplications in angiosperms. Mol Plant 11 (3):357–358

Jiao Y, Wickett NJ, Ayyampalayam S, Chanderbali AS, Landherr L, Ralph PE, Tomsho LP, Hu Y, Liang H, Soltis PS, Soltis DE, Clifton SW, Schlarbaum SE, Schuster SC, Ma H, Leebens-Mack J, Depamphilis CW (2011) Ancestral polyploidy in seed plants and angiosperms. Nature 473:97–100

Jourdan PS, Salazar E (1993) *Brassica carinata* resynthesized by protoplast fusion. Theor Appl Genet 86:567–572

Kapoor R, Kaur G, Banga SK, Banga SS (2011) Generation of *Brassica rapa*—*B. nigra* chromosome addition stocks: cytology and microsatellite (SSR) based characterization. New Biotechnol 28:407–417

Karimi-Ashtiyani R, Ishii T, Niessen M, Stein N, Heckmann S, Gurushidze M, Banaei-Moghaddam AM, Fuchs J, Schubert V, Koch K, Weiss O, Demidov D, Schmidt K, Kumlehn J, Houben A (2015) Point mutation impairs centromeric CENH3 loading and induces haploid plants. Proc Natl Acad Sci USA 112:11211–11216

Karpechenko GD (1928) Polyploid hybrids of *Raphanus sativus* L. × *Brassica oleracea* L. Z Ind Abst Vererbgsl 48:1–85

Kasha KJ, Kao KN (1970) High frequency haploid production in barley (*Hordeum vulgare* L). Nature 225:874

Kasha KJ, Maluszynski M (2003) Production of doubled haploids in crop plants an introduction. In: Maluszynski M, Kasha KJ, Forster BP, Szarejko I (eds) Doubled haploid production in crop plants: a manual. Springer, Netherlands, pp 1–4

Kaur H, Banga SS (2015) Genetic divergence in novel determinate variants of resynthesized Indian mustard (*Brassica juncea* (L.) Czern & Coss). Indian J Genet 75(2):260–263

Kelliher T, Starr D, Richbourg L, Chintamanani S, Delzer B, Nuccio ML, Green J, Chen Z, McCuiston J, Wang W, Liebler T, Bullock P, Martin B (2017) MATRILINEAL, a sperm–specific phospholipase, triggers maize haploid induction. Nature 542:105–109

Khasdan V, Yaakov B, Kraitshtein Z, Kashkush K (2010) Developmental timing of DNA elimination following allopolyploidization in wheat. Genetics 185:387–390

Khush CS, Singh RJ, Sur SC, Librojo AL (1984) Primary trisomics of rice: origin, morphology, cytology and use in linkage mapping. Genetics 107:141–163

Kihara H, Ono T (1926) Chromosomenzahlen und systematische gruppierung der Rumex arten. Zeitschr Zellf Mikrosk Anat 4:475–481

Kixmoeller K, Allu PK, Black BE (2020) The centromere comes into focus: from CENP-A nucleosomes to kinetochore connections with the spindle. Open Biol 10:200051

Koduru PRK, Rao KM (1981) Cytogenetics of synaptic mutants in higher plants. Theor Appl Genet 59:197–214

Kong F, Wang H, Cao A, Qin B, Ji J, Wang S, Wang X (2008) Characterization of *T. aestivum-H. californicum* chromosome addition lines DA2H and MA5H. J Genet Genom 35:673–678

Kovarik A, Dadejova M, Lim YK, Chase MW, Clarkson JJ, Knapp S, Leitch AR (2008) Evolution of rDNA in *Nicotiana* allopolyploids: a potential link between rDNA homogenization and epigenetics. Ann Bot 101(6):815–823

Kumar P, Singhal VK (2011) Chromosome number, male meiosis and pollen fertility in selected angiosperms of the cold deserts of Lahaul Spiti and adjoining areas (Himachal Pradesh, India). Plant Syst Evol 297:271–297

Kumari P, Bisht DS, Bhat SR (2018) Stable, fertile somatic hybrids between *Sinapis alba* and *Brassica juncea* show resistance to *Alternaria brassicae* and heat stress. Plant Cell Tiss Org Cult 133:77–86

Kumari P, Singh KP, Bisht D, Kumar S (2020) Somatic hybrids of *Sinapis alba* + *Brassica juncea*: study of backcross progenies for morphological variations, chromosome constitution and reaction to *Alternaria brassicae*. Euphytica 216:93. https://doi.org/10.1007/s10681-020-02629-3

Kuppu S, Tan EH, Nguyen H, Rodgers A, Comai L, Chan SW, Britt AB (2015) Point mutations in centromeric histone induce post–zygotic incompatibility and uniparental inheritance. PLoS Genet 11:e1005494

Kuppu S, Ron M, Marimuthu MPA, Li G, Huddleson A, Siddeek MH, Terry J, Buchner R, Shabek N, Comai L, Britt AB (2020) A variety of changes, including CRISPR/Cas9 mediated deletions, in CENH3 lead to haploid induction on outcrossing. Plant Biotechnol J 18:2068–2080

Kynast RG, Riera-Lizarazu O, Vales MI, Okagaki RJ, Maquieira S, Chen G, Ananiev EV, Odland WE, Russell CD, Stec AO, Livingston SM, Zaia HA, Rines HW, Phillips RL (2001) A complete set of maize individual chromosome additions to the oat genome. Plant Physiol 125:1216–1227

Kynast RG, Okagaki RJ, Galatowitsch MW, Granath SR, Jacobs MS, Stec AO, Rines HW, Phillips RL (2004) Dissecting the maize genome by using chromosome addition and radiation hybrid lines. Proc Natl Acad Sci USA 101:9921–9926

Lambing C, Heckmann S (2018) Tackling plant meiosis: from model research to crop improvement. Front Plant Sci 9:829. https://doi.org/10.3389/fpls.2018.00829

Lashermes P, Combes MC, Hueber Y, Severac D, Dereeper A (2014) Genome rearrangements derived from homoeologous recombination following allopolyploidy speciation in coffee. Plant J 78:674–685

Laurie DA (1989) The frequency of fertilization in wheat × pearl millet crosses. Genome 32:1063–1067

Laurie DA, Bennett MD (1986) Wheat × maize hybridization. Can J Genet Cytol 28:313–316

Laurie DA, Bennett MD (1988) The production of haploid wheat plants from wheat × maize crosses. Theor Appl Genet 76:393–397

Leflon M, Grandont L, Eber F, Huteau V, Coriton O, Chelysheva L, Jenczewski E, Chevre AM (2010) Crossovers get a boost in *Brassica* allotriploid and allotetraploid hybrids. Plant Cell 22:2253–2264

Leitch IJ, Bennett MD (1997) Polyploidy in angiosperms. Trends Plant Sci 2:470–476

Leitch IJ, Hanson L, Lim KY, Kovarik A, Chase MW, Clarkson JJ, Leitch AR (2008) The ups and downs of genome size evolution in polyploid species of *Nicotiana* (Solanaceae). Ann Bot 101(6):805–814

Li ZY, Ge XH (2007) Unique chromosome behavior and genetic control in *Brassica* × *Orychophragmus* wide hybrids: a review. Plant Cell Rep 26:701–710

Li Z, Heneen WK (1999) Production and cytogenetics of intergeneric hybrids between the three cultivated *Brassica* diploids and *Orychophragmus violaceus*. Theor Appl Genet 99:694–704

Li Z, Liu H, Luo P (1995) Production and cytogenetics of intergeneric hybrids between *Brassica napus* and *Orychophragmus violaceus*. Theor Appl Genet 91:131–136

Li Z, Wu JG, Liu Y, Liu HL, Heneen WK (1998) Production and cytogenetics of intergeneric hybrids *Brassica juncea* × *Orychophragmus violaceus* and *B. carinata* × *O. violaceus*. Theor Appl Genet 96:251–265

Li X, Meng D, Chen S, Luo H, Zhang Q, Jin W, Yan J (2017) Single nucleus sequencing reveals spermatid chromosome fragmentation as a possible cause of maize haploid induction. Nat Commun 8:991

Li M, Wang R, Wu X, Wang J (2020) Homoeolog expression bias and expression level dominance (ELD) in four tissues of natural allotetraploid *Brassica napus*. BMC Genomics 21:330

Lichter R (1982) Induction of haploid plants from isolated pollen of *Brassica napus*. Z Pflanzenphysiol 105:427–434

Lionneton E, Beuret W, Delaitre C, Ochatt S, Rancillac M (2001) Improved microspore culture and doubled haploid plant regeneration in the brown condiment mustard (*Brassica juncea*). Plant Cell Rep 20:126–130

Liu F (2014) Creation of autopolyploidy germplasm of *Brassica rapa*, *Brassica oleracea* and *Brassica nigra*. Hebei Agricultural University, Hebei, China

Liu M, Li ZY (2007) Genome doubling and chromosome elimination with fragment recombination leading to the formation of *Brassica rapa*-type plants with genomic alterations in crosses with *Orychophragmus violaceus*. Genome 50:985–993

Liu Z, Adamczyk K, Manzanares-Dauleux M, Eber F, Lucas MO, Delourme R, Chevre AM, Jenczewski E (2006) Mapping *PrBn* and other quantitative trait loci responsible for the control of homoeologous chromosome pairing in oilseed rape (*Brassica napus* L.) haploids. Genetics 174:1583–1596

Liu C, Li X, Meng D, Zhong Y, Chen C, Dong X, Xu X, Chen B, Li W, Li L, Tian X, Zhao H, Song W, Luo H, Zhang Q, Lai J, Jin W, Yan J, Chen S (2017) A 4-bp insertion at ZmPLA1 encoding a putative phospholipase A generates haploid induction in maize. Mol Plant 10(3):520–522

Liu H, Wang K, Jia Z, Gong Q, Lin Z, Du L, Pei X, Ye X (2020) Efficient induction of haploid plants in wheat by editing of *TaMTL* using an optimized Agrobacterium-mediated CRISPR system. J Exp Bot 71:1337–1349

Lloyd A, Bomblies K (2016) Meiosis in autopolyploid and allopolyploid *Arabidopsis*. Curr Opin Plant Biol 30:116–122

Lukens LN, Pires JC, Leon E, Vogelzang R, Oslach L, Osborn TC (2006) Patterns of sequence loss and cytosine methylation within a population of newly resynthesized *Brassica napus* allopolyploids. Plant Physiol 140:336–348

Lysak MA, Koch MA, Pecinka A, Schubert I (2005) Chromosome triplication found across the tribe Brassiceae. Genome Res 15(4):516–525

Ma XF, Gustafson JP (2005) Genome evolution of allopolyploids: a process of cytological and genetic diploidization. Cytogenet Genome Res 109:236–249

Madlung A (2013) Polyploidy and its effect on evolutionary success: old questions revisited with new tools. Heredity 110:99–104

Maheshwari S, Tan EH, West A, Franklin FC, Comai L, Chan SW (2015) Naturally occurring diferences in CENH3 affect chromosome segregation in zygotic mitosis of hybrids. PLoS Genet 11:e1004970

Maluszynski M, Kasha KJ, Forster BP, Szarejko I (2003) Doubled haploid production in crop plants: a manual. Springer, Netherlands, p 428

Mandakova T, Li Z, Barker MS, Lysak MA (2017) Diverse genome organization following 13 independent mesopolyploid events in Brassicaceae contrasts with convergent patterns of gene retention. Plant J 91:3–21

Marcinska I, Nowakowska A, Skrzypek E, Czyczyło-Mysza I (2013) Production of double haploids in oat (*Avena sativa* L.) by pollination with maize (*Zea mays* L.). Cent Eur J Biol 8:306–313

Martin R, Ernst S, Rademaker A, Barclay L, Ko E, Summers N (1997) Chromosomal abnormalities in sperm from testicular cancer patients before and after chemotherapy. Hum Genet 99:214–218

Mason AS, Wendel JF (2020) Homoeologous exchanges, segmental allopolyploidy, and polyploid genome evolution. Front Genet 11:1014

Mason AS, Nelson MN, Yan G, Cowling WA (2011) Production of viable male unreduced gametes in *Brassica* interspecific hybrids is genotype specific and stimulated by cold temperatures. BMC Plant Biol 11:103

Mason AS, Yan G, Cowling WA, Nelson MN (2012) A new method for producing allohexaploid *Brassica* through unreduced gametes. Euphytica 186:277–287

Mayer KF, Rogers J, Dolezel J, Pozniak C, Eversole K, Feuillet C et al (2014) A chromosome-based draft sequence of the hexaploid bread wheat (*Triticum aestivum*) genome. Science 345:1251788. https://doi.org/10.1126/science.1251788

McClintock B (1984) The significance of responses of the genome to challenge. Science 226:792–801

McGrath JM, Quiros CF (1990) Generation of alien chromosome addition lines from synthetic *Brassica napus*: morphology, cytology, fertility, and chromosome transmission. Genome 33:374–383

Meng J, Shi S, Gan L, Li Z, Qu X (1998) The production of yellow-seeded *Brassica napus* (AACC) through crossing interspecific hybrids of *B. campestris* (AA) and *B. carinata* (BBCC) with *B. napus*. Euphytica 103:329–333

Mercier R, Mezard C, Jenczewski E, Macaisne N, Grelon M (2015) The molecular biology of meiosis in plants. Annu Rev Plant Biol 66:297–327

Mestiri I, Chague V, Tanguy AM, Huneau C, Huteau V, Belcram H, Coriton O, Chalhoub B, Jahier J (2010) Newly synthesized wheat allohexaploids display progenitor-dependent meiotic stability and aneuploidy but structural genomic additivity. New Phytol 186:86–101

Molnar I, Molnar LM (2010) GISH reveals different levels of meiotic pairing with wheat for individual *Aegilops biuncialis* chromosomes. Bilo Plant 54:259–264

Multani DS, Khush GS, delos Reyes BG, Brar DS (2003) Alien genes introgression and development of monosomic alien addition linesfrom *Oryza latifolia* Desv. to rice *Oryza sativa* L. Theor Appl Genet 107:395–405

Murovec J, Bohanec B (2011) Haploids and doubled haploids in plant breeding. In: Abdurakhmonov IY (eds) Plant breeding. InTech, Rijeka, Croatia, pp 87–106

Musacchio A, Desai A (2017) A molecular view of kinetochore assemby and function. Biology 6:5

Musacchio A, Salmon ED (2007) The spindle-assembly checkpoint in space and time. Nat Rev Mol Cell Biol 8 (5):379–393

Mwathi MW, Gupta M, Atri C, Banga SS, Batley J, Mason AS (2017) Segregation for fertility and meiotic stability in novel *Brassica* allohexaploids. Theor Appl Genet 130:767–776

Mwathi MW, Gupta M, Quezada D, Pradhan A, Batley J, Mason AS (2020) Fertile allohexaploid *Brassica* hybrids obtained from crosses between *B. oleracea* and *B. juncea* via ovule rescue and colchicine treatment of cuttings. Plant Cell Tiss Org Cult 140:301–313

Nagaharu U (1935) Genome analysis in *Brassica* with special reference to the experimental formation of B. napus and peculiar mode of fertilization. Jpn J Bot 7:389–452

Naseem S, Kumar G (2013) Induced desynaptic variation in poppy (*Papaver somniferum* L.). Crop Breed Appl Biotechnol 13:363–366

Nieto Feliner G, Casacuberta J, Wendel JF (2020) Genomics of evolutionary novelty in hybrids and polyploids. Front Genet 11:792

Niroula RK, Bimb HP (2009) Overview of wheat × maize system of crosses for dihaploid induction in wheat. World Appl Sci J 7(8):1037–1045

Nonomura K, Nakano M, Fukuda T, Eiguchi M, Miyao A, Hirochika H, Kurata N (2004) The novel gene homologous pairing aberration in rice meiosis I of rice encodes a putative coiled-coil protein required for homologous chromosome pairing in meiosis. Plant Cell 16:1008–1020

Ohkawa Y, Suenaga K, Ogawa T (1992) Production of haploid wheat plants through pollination of sorghum pollen. Jpn J Breed 42(4):891–894

Olsson G (1960a) Species crosses within the genus *Brassica*. I. Artificial *Brassica juncea* coss. Hereditas 46:171–223

Olsson G (1960b) Species crosses within the genus *Brassica*. II. Artificial *Brassica napus* L. Hereditas 46:351–386

Osborn TC, Butruille DV, Sharpe AG, Pickering KJ, Parkin IAP, Parker JS, Lydiate DJ (2003) Detection and effects of a homoeologous reciprocal transposition in *Brassica napus*. Genetics 165:1569–1577

Ozkan H, Feldman M (2009) Rapid cytological diploidization in newly formed allopolyploids of the wheat (*Aegilops-Triticum*) group. Genome 52:926–934

Pagliarini MS (2000) Meiotic behavior of economically important plant species: the relationship between fertility and male sterility. Genet Mol Biol 23:997–1002

Palmer DK, O'Day K, Wener MH, Andrews BS, Margolis RL (1987) A 17–kD centromere protein (CENP–A) copurifies with nucleosome core particles and with histones. J Cell Biol 104(4):805–815

Parisod C, Alix K, Just J, Petit M, Sarilar V, Mhiri C, Ainouche M, Chalhoub B, Grandbastien MA (2010a) Impact of transposable elements on the organization and function of allopolyploid genomes. New Phytol 186:37–45

Parisod C, Holderegger R, Brochmann C (2010b) Evolutionary consequences of autopolyploidy. New Phytol 186:5–17

Paritosh K, Yadava SK, Singh P, Bhayana L, Mukhopadhyay A, Gupta V, Bisht NC, Zhang J, Kudrna DA, Copetti D, Wing RA, Bhaskar V, Lachagari R, Pradhan AK, Pental D (2021) A chromosome–scale assembly of allotetraploid *Brassica juncea* (AABB) elucidates comparative architecture of the A and B genomes. Plant Biotechnol J 19:602–614

Park HR, Park JE, Kim JH, Shin H, Yu SH, Son S et al (2020) Meiotic chromosome stability and suppression of crossover between non-homologous chromosomes in *xBrassicoraphanus*, an intergeneric allotetraploid derived from a cross between *Brassica rapa* and *Raphanus sativus*. Front Plant Sci 11:851

Paterson AH, Wendel JF, Gundlach H, Guo H, Jenkins J, Jin D, Llewellyn D et al (2012) Repeated polyploidization of *Gossypium* genomes and the evolution of spinnable cotton fibres. Nature 492:423–427

Pecinka A, Fang W, Rehmsmeier M, Levy AA, Scheid OM (2011) Polyploidization increases meiotic recombination frequency in *Arabidopsis*. BMC Biol 9:24. https://doi.org/10.1186/1741-7007-9-24

Pele A, Rousseau-Gueutin M, Chevre AM (2018) Speciation success of polyploid plants closely relates to the regulation of meiotic recombination. Front Plant Sci 9:907

Pertuze RA, Ji Y, Chetelat RT (2003) Transmission and recombination of homoeologous *Solanum sitiens* chromosomes in tomato. Theor Appl Genet 107:1391–1401

Pignatta D, Dilkes BP, Yoo SY, Henry IM, Madlung A, Doerge RW et al (2010) Differential sensitivity of the *Arabidopsis thaliana* transcriptome and enhancers to the effects of genome doubling. New Phytol 186:194–206

Pires JC, Zhao J, Schranz ME, Leon EJ, Quijada PA, Lukens LN, Osborn TC (2004) Flowering time divergence and genomic rearrangements in resynthesized *Brassica* polyploids (Brassicaceae). Biol J Lin Soc 82:675–688

Polgári D, Cseh A, Szakacs E, Jager K, Molnar-Lang M, Sagi L (2014) High frequency generation and characterization of intergeneric hybrids and haploids from new wheat-barley crosses. Plant Cell Rep 33:1323–1331

Pont C, Murat F, Confolent C, Balzergue S, Salse J (2011) RNA-seq in grain unveils fate of neo- and paleopolyploidization events in bread wheat (*Triticum aestivum* L.). Genome Biol 12:119

Pradhan A, Plummer JA, Nelson MN, Cowling WA, Yan G (2010) Trigenomic hybrids from interspecific crosses between *Brassica napus* and *B. nigra*. Crop Pasture Sci 61:464–474

Pradillo M, Lopez E, Romero C, Sanchez-Moran E, Cunado N, Santos JL (2007) An analysis of univalent segregation in meiotic mutants of *Arabidopsis thaliana*: a possible role for synaptonemal complex. Genetics 175:505–511

Prakash S (1973a) Haploidy in *Brassica nigra* Koch. Euphytica 22:613–614

Prakash S (1973b) Non-homologous meiotic pairing in the A and B genomes of *Brassica*: its breeding significance in the production of variable amphidiploids. Genet Res Camb 21:133–137

Prakash S, Bhat SR, Quiros CF, Kirti PB, Chopra VL (2009) *Brassica* and its close allies: cytogenetics and evolution. Plant Breed Rev 31:21–187

Prigge V, Melchinger AE (2012) Production of haploids and doubled haploids in maize. In: Loyola-Vargas VM, Ochoa-Alejo N (eds) Plant cell culture protocols. Humana Press, Totowa, NJ, pp 161–172

Quijada PA, Udall JA, Lambert B, Osborn TC (2006) Quantitative trait analysis of seed yield and other complex traits in hybrid spring rapeseed (*Brassica napus* L.): 1. Identification of genomic regions from winter germplasm. Theor Appl Genet 113:549–561

Rahman MH (2001) Production of yellow-seeded *Brassica napus* through interspecific crosses. Plant Breed 120:463–472

Rahman M, de Jimenez MM (2016) Behind the scenes of microspore-based double haploid development in *Brassica napus*: a review. J Plant Sci Mol Breed 5:1

Ramanna MS, Hermsen JGT (1979) Unique meiotic behavior in F_1 plants from a cross between a non-tuberous and a tuberous *Solanum* species in section Petota. Euphytica 28:9–15

Ramchiary N, Padmaja KL, Sharma S, Gupta V, Sodhi YS, Mukhopadhyay A, Arumugam N, Pental D, Pradhan AK (2007) Mapping of yield influencing QTL in *Brassica juncea*: implications for breeding of a major oilseed crop of dryland areas. Theor Appl Genet 115:807–817

Ramsey J, Ramsey TS (2014) Ecological studies of polyploidy in the 100 years following its discovery. Phil Trans Roy Soc B 369:20130352. https://doi.org/10.1098/rstb.2013.0352

Ramsey J, Schemske DW (2002) Neopolyploidy in flowering plants. Annu Rev Ecol Syst 33:589–639

Rao SR, Kumar A (2003) Cytological investigations in a synaptic variant of *Anogeissus sericea* var. *sericea* Brandis (Combretaceae), an important hardwood tree of Rajasthan. Bot J Linn Soc 142:103–109

Rashid H, Bowei C, Yang Z, Qingyong Y, Xianhong G, Zaiyun L (2021) Asymmetrical effects of autopolyploidization on organ size and gene expression in *Brassica rapa* and *B. oleracea*. Sci Hort 282:109991

Ravi M, Chan SW (2010) Haploid plants produced by centromere-mediated genome elimination. Nature 464:615–618

Reiss E, Schubert J, Scholze P, Kramer R, Sonntag K (2009) The barley thaumatin-like protein Hv-TLP8 enhances resistance of oilseed rape plants to *Plasmodiophora brassicae*. Plant Breed 128:210–212

Ren R, Wang H, Guo C, Zhang N, Zeng L, Chen Y, Ma H, Qi J (2018) Widespread whole genome duplications contribute to genome complexity and species diversity in angiosperms. Mol Plant 11:414–428

Rey E, Abrouk M, Keeble-Gagnere G, Karafiatova M, Vrana J, Balzergue S, Taconnat LS, Brunaud V, Magniette MM, Endo TR, Bartos J (2018) Transcriptome reprogramming due to the introduction of a barley telosome into bread wheat affects more barley genes than wheat. Plant Biotechnol J 16:1767–1777

Riera-Lizarazu O, Mujeeb-Kazi A (1993) Polyhaploid production in the Triticeae: wheat × Tripsacum crosses. Crop Sci 33(5):973–976

Riley R, Chapman V (1958) Genetic control of the cytologically diploid behaviour of hexaploid wheat. Nature 182:713–715

Riley R, Chapman V, Kimber G (1959) Genetic control of chromosome pairing in intergeneric hybrids with wheat. Nature 183:1244–1246

Rousseau-Gueutin M, Morice J, Coriton O, Huteau V, Trotoux G, Negre S et al (2017) The impact of open pollination on the structural evolutionary dynamics, meiotic behavior and fertility of resynthesized allotetraploid *Brassica napus* L. G3-Genes Genom Genet 7:705–717

Rout K, Sharma M, Gupta V, Mukhopadhyay A, Sodhi YS, Pental D, Pradhan AK (2015) Deciphering allelic variations for seed glucosinolate traits in oilseed mustard (*Brassica juncea*) using two bi-parental mapping populations. Theor Appl Genet 128:657–666

Rout K, Yadav BG, Yadava SK, Mukhopadhyay A, Gupta V, Pental D, Pradhan AK (2018) QTL landscape for oil content in *Brassica juncea*: analysis in multiple bi-parental populations in high and "0" erucic background. Front Plant Sci 9:1448

Salmon A, Flagel L, Ying B, Udall JA, Wendel JF (2010) Homoeologous non-reciprocal recombination in polyploid cotton. New Phytol 186:123–134

Samans B, Chalhoub B, Snowdon RJ (2017) Surviving a genome collision: genomic signatures of allopolyploidization in the recent crop species. Plant Genome 10. https://doi.org/10.3835/plantgenome2017.02.0013

Samans B, Snowdon R, Mason AS (2018) Homoeologous exchanges and gene losses generate diversity and differentiate the *B. napus* genome from that of its ancestors. In: Liu S, Snowdon R, Chalhoub B (eds) The *Brassica napus* genome. Springer, Cham, pp 131–148

Sanei M, Pickering R, Kumke K, Nasuda S, Houben A (2011) Loss of centromeric histone H3 (CENH3) from centromeres precedes uniparental chromosome

elimination in interspecific barley hybrids. Proc Natl Acad Sci USA 108:E498-505
Sankoff D, Zheng C, Zhu Q (2010) The collapse of gene complement following whole genome duplication. BMC Genomics 11:313. https://doi.org/10.1186/1471-2164-11-313
Sarla N, Raut RN (1988) Synthesis of *B. carinata* from *B. nigra* × *B. oleracea* hybrids obtained by ovary culture. Theor Appl Genet 76:846–849
Schneider H, Liu HM, Chang YF, Ohlsen D, Perrie LR, Shepherd L et al (2017) Neo-and paleopolyploidy contribute to the species diversity of *Asplenium*-the most species-rich genus of ferns. J Syst Evol 55:353–364
Schranz ME, Osborn TC (2000) Novel flowering time variation in the resynthesized polyploid *Brassica napus*. J Hered 91:242–246
Schranz ME, Osborn TC (2004) De novo variation in life-history traits and responses to growth conditions of resynthesized polyploid *Brassica napus* (Brassicaceae). Amer J Bot 91:174–183
Sears ER (1952) Homoeologous chromosomes in *Triticum aestivum*. Genetics 37:624
Sears ER (1954) The aneuploids of common wheat. Agric Exp State Res Bull 572:1–58
Sharbrough J, Conover JL, Tate JA, Wendel JF, Sloan DB (2017) Cytonuclear responses to genome doubling. Amer J Bot 104:1277–1280
Shen SX, Hou XL, Zhang CH (2006) A study on obtaining primary trisomics by the isolated microspore culture of autotetraploid Chinese cabbage. Acta Hort Sin 33:1209–1214
Shen D, Suhrkamp I, Wang Y, Liu S, Menkhaus J, Verreet JA, Fan L, Cai D (2014) Identification and characterization of micro RNAs in oilseed rape (*Brassica napus*) responsive to infection with the pathogenic fungus *Verticillium longisporum* using *Brassica* AA (*Brassica rapa*) and CC (*Brassica oleracea*) as reference genomes. New Phytol 204:577–594
Shim J, Panaud O, Vitte C, Mendioro MS, Brar DS (2010) RDA derived *Oryza minuta*-specific clones to probe genomic conservation across *Oryza* and introgression into rice (*O. sativa* L.). Euphytica 176:269–279
Shmykova NA, Shumilina DV, Suprunova TP (2016) Doubled haploid production in *Brassica* L. species. Russ J Genet Appl Res 6:68–77
Shumilina D, Kornyukhin D, Domblides E, Soldatenko A, Artemyeva A (2020) Effects of genotype and culture conditions on microspore embryogenesis and plant regeneration in *Brassica rapa* ssp. *rapa* L. Plants 9 (2):278
Siegel JJ, Amon A (2012) New insights into the troubles of aneuploidy. Annu Rev Cell Dev Biol 28:189–214
Siebel J, Pauls KP (1989) A comparison of anther and microspore culture as a breeding tool in *Brassica napus*. Theor Appl Genet 78:473–479
Singh RJ (2002) Plant cytogenetics. CRC Press, London, p 398
Singh RJ, Ikehashi H (1981) Monogenic male sterility in rice: induction, identification and inheritance. Crop Sci 21:286–289
Singh K, Multani D, Khush GS (1996) Secondary trisomics and telotrisomics of rice: origin, characterization, and use in determining the orientation of chromosome map. Genetics 143:517–529
Singhal VK, Kumar P (2010) Variable sized pollen grains due to impaired male meiosis in the cold desert plants of Northwest Himalayas (India). In: Kaiser BJ (ed) Pollen: structure, types and effects. Nova Science Publishers Inc., New York, pp 101–126
Singhal VK, Kaur M, Himshikha KP, Gupta RC (2012) High pollen sterility and 2n pollen grains in an asynaptic 4× cytotype ($2n = 48$) of *solanum nigrum* L. Cytologia 77(3):333–342
Soltis PS, Soltis DE (2000) The role of genetic and genomic attributes in the success of polyploids. Proc Natl Acad Sci 97:7051–7057
Soltis PS, Soltis DE (2016) Ancient WGD events as drivers of key innovations in angiosperms. Curr Opin Plant Biol 30:159–165
Song KM, Lu P, Tang KL, Osborn TC (1995) Rapid genome change in synthetic polyploids of *Brassica* and its implications for polyploid evolution. Proc Natl Acad Sci USA 92:7719–7723
Sosnowska K, Majka M, Majka J, Bocianowski J, Kasprowicz M, Książczyk T, Szała L, Cegielska-Taras T (2020) Chromosome instabilities in resynthesized *Brassica napus* revealed by FISH. J Appl Genet 61:323–335
Spoelhof J, Soltis P, Soltis D (2017) Pure polyploidy: Closing the gaps in autopolyploid research: pure polyploidy. J Syst Evol 55:340–352
Srinivasan K, Malathi V, Kirti P, Prakash S, Chopra VL (1998) Generation and characterisation of monosomic chromosome addition lines of *Brassica campestris*—*B. oxyrrhina*. Theor Appl Genet 97:976–981
Srivastava PS, Narula A, Srivastava S (2004) Plant biotechnology and molecular markers. Springer, Dordrecht, p 400
Stebbins GL (1947) Types of polyploids: their classification and significance. Adv Genet 1:403–429
Stein A, Coriton O, Rousseau-Gueutin M, Samans B, Schiessl SV, Obermeier C, Parkin IAP, Chevre AM, Snowdon RJ (2017) Mapping of homoeologous chromosome exchanges influencing quantitative trait variation in *Brassica napus*. Plant Biotechnol J 15:1478–1489
Sun Y, Wu Y, Yang C, Sun S, Lin X, Liu L (2017) Segmental allotetraploidy generates extensive homoeologous expression rewiring and phenotypic diversity at the population level in rice. Mol Ecol 26:5451–5466
Szadkowski E, Eber F, Huteau V, Lode M, Huneau C, Belcram H, Coriton O, Manzanares-Dauleux MJ, Delourme R, King GJ, Chalhoub B, Jenczewski E, Chevre AM (2010) The first meiosis of resynthesized *Brassica napus*, a genome blender. New Phytol 186:102–112

Szadkowski E, Eber F, Huteau V, Lode M, Coriton O, Jenczewski E, Chebvre AM (2011) Polyploid formation pathways have an impact on genetic rearrangements in resynthesized *Brassica napus*. New Phytol 191:884–894

Takeda T, Takahata Y (1996) Production of alloplasmic Chinese cabbage using synthesized trigenomic hexaploid (AABBCC) in *Brassica*. Acta Hort 407:151–154

Tang K, Hill CB, Williamsp H (1988) Development of trisomics of rapid-cycling *Brassica rapa*. Eucarpia Cruciferae Newsl 13:60–61

Tank DC, Eastman JM, Pennell MW, Soltis PS, Soltis DE, Hinchliff CE, Brown JW, Sessa EB, Harmon LJ (2015) Nested radiations and the pulse of angiosperm diversification: increased diversification rates often follow whole genome duplications. New Phytol 207(2):454–467

Tepperberg JH, Moses MJ, Nath J (1999) Colchicine effects on meiosis in the male mouse II. Inhibition of synapsis induction of nondisjunction. Mutat Res 429:93–105

This P, Ochoa O, Quiros CF (1990) Dissection of the *Brassica nigra* genome by monosomic addition lines. Plant Breed 105:211–220

Tian E, Jiang Y, Chen L, Zou J, Liu F, Meng J (2010) Synthesis of a *Brassica* trigenomic allohexaploid (*B. carinata_B. rapa*) de novo and its stability in subsequent generations. Theor Appl Genet 121:1431–1440

Tokumasu S (1976) The increase of seed fertility of *Brassicoraphamus* through cytological irregularity. Euphytica 25:463–470

Touraev A, Forster BP, Jain SM (2009) Advances in haploid production in higher plants. Springer, Netherlands, p 348

Trentin HU, Frei UK, Lubberstedt T (2020) Breeding maize maternal haploid inducers. Plants 9:614

Tu YQ, Sun J, Ge X, Li ZY (2009) Chromosome elimination, addition and introgression in intertribal partial hybrids between *Brassica rapa* and *Isatis indigotica*. Ann Bot 103:1039–1048

Tu YQ, Sun J, Ge XH, Li ZY (2010) Production and genetic analysis of partial hybrids from intertribal sexual crosses between *Brassica napus* and *Isatis indigotica* and progenies. Genome 53:146–156

Udall JA, Quijada PA, Lambert B, Osborn TC (2006) Quantitative trait analysis of seed yield and other complex traits in hybrid spring rapeseed (*Brassica napus* L.): 2. Identification of alleles from unadapted germplasm. Theor Appl Genet 113:597–609

Valdes A, Clemens R, Mollers C (2018) Mapping of quantitative trait loci for microspore embryogenesis-related traits in the oilseed rape doubled haploid population DH4069 × express 617. Mol Breed 38:65

Van de Peer Y, Mizrachi E, Marchal K (2017) The evolutionary significance of polyploidy. Nat Rev Genet 18:411–424

Vicient CM, Casacuberta JM (2017) Impact of transposable elements on polyploid plant genomes. Ann Bot 120(2):195–207

Walczak CE, Cai S, Khodjakov A (2010) Mechanisms of chromosome behaviour during mitosis. Nat Rev Mol Cell Biol 11(2):91–102

Wang N, Dawe RK (2018) Centromere size and its relationship to haploid formation in plants. Mol Plant 11:398–406

Wang RQ, Gao JH, Mao LP, Du RH (2002) Chromosome location of the male-sterility and yellow seedling gene in line 1066A of foxtail millet. Acta Bot Sin 44:1209–1212

Wang Y, Sonntag K, Rudloff E, Wehling P, Snowdon RJ (2006) GISH analysis of disomic *Brassica napus-Crambe abyssinica* chromosome addition lines produced by microspore culture from monosomic addition lines. Plant Cell Rep 25:35–40

Wang RQ, Gao ZH, Guan ZB, Mao LP (2007) Chromosome location and linkage analysis of a few agronomically important traits in foxtail millet. Acta Agron Sin 33:9–14

Wang Y, Jha AK, Chen R, Doonan JH, Yang M (2010) Polyploidy-associated genomic instability in *Arabidopsis thaliana*. Genesis 48:254–263

Wang X, Wang H, Wang J, Sun R, Wu J, Liu S, Bai Y, Mun JH, Bancroft I, Cheng F, Huang S, Li X, Hua W, Wang J, Wang X, Freeling M, Pires JC, Paterson AH, Chalhoub B, Wang B et al (2011) *Brassica rapa* genome sequencing project consortium. The genome of the mesopolyploid crop species *Brassica rapa*. Nat Genet 43(10):1035–1039

Wang K, Wu YF, Zhang WL, Dawe RK, Jiang JM (2014) Maize centromeres expand and adopt a uniform size in the genetic background of oat. Genome Res 24:107–116

Wang S, Jin W, Wang K (2019) Centromere histone H3- and phospholipase-mediated haploid induction in plants. Plant Methods 15:1–10

Wang N, Gent JI, Dawe RK (2021) Haploid induction by a maize cenh3 null mutant. Sci Adv 7:eabe2299. https://doi.org/10.1126/sciadv.abe2299

Warwick SI, Francis A, Al-Shehbaz IA (2006) Brassicaceae: species checklist and database on CD-Rom. Plant Syst Evol 259:249–258

Watts A, Sankaranarayanan S, Raipuria RK, Watts A (2020) Production and application of doubled haploid in *Brassica* improvement. In: Wani SH, Thakur AK, Khan YJ (eds) *Brassica* improvement. Springer, Cham, pp 67–84

Wedzony M, Forster BP, Zur I, Golemiec E, Szechynska-Hebda M, Dubas E, Gotebiowska G, Wedzony M (2009) Progress in doubled haploid technology in higher plants. In: Touraev A, Forster BP, Jain SM (eds) Advances in haploid production in higher plants. Springer, Dordrecht, pp 1–33

Wen J, Zeng XH, Pu YY, Qi LP, Li ZY, Tu JX, Ma CZ, Shen JX, Fu TD (2010) Meiotic nondisjunction in resynthesized *Brassica napus* and generation of aneuploids through microspore culture and their characterization. Euphytica 173:99–111

Wendel JF (2000) Genome evolution in polyploids. Plant Mol Biol 42:225–249

Wendel JF, Jackson SA, Meyers BC, Wing RA (2016) Evolution of plant genome architecture. Genome Biol 17:37. https://doi.org/10.1186/s13059-016-0908-1

Xiong Z, Gaeta RT, Pires JC (2011) Homoeologous shuffling and chromosome compensation maintain genome balance in resynthesized allopolyploid *Brassica napus*. Proc Natl Acad Sci USA 108:7908–7913

Xiong Z, Gaeta RT, Edger PP, Cao Y, Zhao K, Zhang S, Pires JC (2021) Chromosome inheritance and meiotic stability in allopolyploid *Brassica napus*. G3-Genes Genom Genet 11(2):jkaa011. https://doi.org/10.1093/g3journal/jkaa011

Xu CY, Li ZY (2007) Origin of new *Brassica* types from single intergeneric hybrid between *B. rapa* and *Orychophragmus violaceus* by rapid chromosome evolution and introgression. J Genet 86:249–257

Xu Y, Zhong L, Wu X, Fang X, Wang J (2009) Rapid alterations of gene expression and cytosine methylation in newly synthesized *Brassica napus* allopolyploids. Planta 229:471–483

Xu Z, Xie B, Wu T, Xin X, Man L, Tan G, Xiong Z (2016) Karyotyping and identifying all of the chromosomes of allopolyploid *Brassica juncea* using multicolor FISH. Crop J 4:266–274

Xu Y, Zhang W, Chen G, Wang J (2017) DNA methylation alteration is a major consequence of genome doubling in autotetraploid *Brassica rapa*. Arch Biol Sci 69:689–697

Xu C, Huang Q, Ge X, Li Z (2019) Phenotypic, cytogenetic, and molecular marker analysis of *Brassica napus* introgressants derived from an intergeneric hybridization with *Orychophragmus*. PLoS ONE 14 (1):e0210518

Yadava SK, Arumugam N, Mukhopadhyay A, Sodhi YS, Gupta V, Pental D, Pradhan AK (2012) QTL mapping of yield–associated traits in *Brassica juncea*: meta-analysis and epistatic interactions using two different crosses between east European and Indian gene pool lines. Theor Appl Genet 125:1553–1564

Yan G, Liu H, Wang H, Lu Z, Wang Y, Mullan D, Hamblin J, Liu C (2017) Accelerated generation of selfed pure line plants for gene identification and crop breeding. Front Plant Sci 8:1786

Yang SJ (1964) Numerical chromosome instability in *Nicotiana* hybrids I. Interplant variation among offspring of amphiploids. Genetics 50:745–756

Yang GS, Robbelen G, Fu TD (1994) Effects of B genome on chromosome pairing among the 3 homologous genomes A, B and C in *Brassica*. Huazhong Agric Univ 13:111–117

Yang J, Liu D, Wang X, Ji C, Cheng F, Liu B, Hu Z, Chen S, Pental D, Ju Y, Yao P, Li X, Xie K, Zhang J, Wang J, Liu F, Ma W, Shopan J, Zheng H, Mackenzie SA, Zhang M (2016) The genome sequence of allopolyploid *Brassica juncea* and analysis of differential homoeolog gene expression influencing selection. Nat Genet 48:1225–1232

Yang S, Chen S, Zhang K, Li L, Yin Y, Gill RA, Yan G, Meng J, Cowling WA, Zhou W (2018) A high-density genetic map of an allohexaploid *Brassica* doubled haploid population reveals quantitative trait loci for pollen viability and fertility. Front Plant Sci 9:1161

Yant L, Hollister JD, Wright KM, Arnold BJ, Higgins JD, Franklin FCH, Bomblies K (2013) Meiotic adaptation to genome duplication in *Arabidopsis arenosa*. Curr Biol 23:2151–2156

Yao XC, Ge XH, Li ZY (2012) Different fertility and meiotic regularity in allohexaploids derived from trigenomic hybrids between three cultivated *Brassica* allotetraploids and *B. maurorum*. Plant Cell Rep 31:781–788

Yao L, Zhang Y, Liu C, Liu Y, Wang Y, Liang D, Liu J, Sahoo G, Kelliher T (2018) OsMATL mutation induces haploid seed formation in indica rice. Nat Plants 4:530–533

Yin L, Zhu Z, Luo X, Huang L, Li Y, Mason AS, Yang J, Ge X, Long Y, Wang J, Zou Q, Tao L, Kang Z, Tang R, Wang M, Fu S (2020) Genome wide duplication of allotetraploid *Brassica napus* produces novel characteristics and extensive ploidy variation in self-pollinated progeny. G3-Genes Genom Genet 10:3687–3699

Yoo MJ, Szadkowski E, Wendel JF (2013) Homoeolog expression bias and expression level dominance in allopolyploid cotton. Heredity 110:171–180

Yoo MJ, Liu X, Pires JC, Soltis PS, Soltis DE (2014) Nonadditive gene expression in polyploids. Annu Rev Genet 48:485–517

Zhang Y, Xu GH, Guo XY, Fan LJ (2005) Two ancient rounds of polyploidy in rice genome. J Zhejiang Univ Sci B 6:87–90

Zhang CH, Zhu HY, Li XF, Shen SX, Chen XP, Hong MAN, Xuan SX (2006) Obtaining and cytological identification of a set of primary trisomics in cabbage. Agric Sci China 5:655–660

Zhang Z, Qiu F, Liu Y, Ma K, Li Z, Xu S (2008) Chromosome elimination and in vivo haploid production induced by Stock 6-derived inducer line in maize (*Zea mays* L.). Plant Cell Rep 27:1851–1860

Zhang CH, Li XF, Shen SX, Yuan H, Xuan SX (2009) Determination of n+1 gamete transmission rate of trisomics and location of gene controlling 2n gamete formation in Chinese cabbage (*Brassica rapa*). J Integr Plant Biol 51(1):29–34

Zhang H, Bian Y, Gou X, Zhu B, Xu C, Qi B, Li N, Rustgi S, Zhou H, Han F, Jiang J, Wettstein D, Liu B (2013) Persistent whole-chromosome aneuploidy is generally associated with nascent allohexaploid wheat. Proc Natl Acad Sci USA 110:3447–3452

Zhang J, Liu Y, Xia EH, Yao QY, Liu XD, Gao LZ (2015) Autotetraploid rice methylome analysis reveals methylation variation of transposable elements and their effects on gene expression. Proc Natl Acad Sci USA 112(50):E7022–E7029. https://doi.org/10.1073/pnas.1515170112

Zhang A, Li N, Gong L, Gou X, Wang B, Deng X, Li C, Dong Q, Zhang H, Liu B (2017) Global analysis of gene expression in response to whole-chromosome aneuploidy in hexaploid wheat. Plant Physiol 175:828–847

Zhang K, Wang X, Cheng F (2019a) Plant polyploidy: origin, evolution, and its influence on crop domestication. Hortic Plant J 5(6):231–239

Zhang R, Geng S, Qin Z, Tang Z, Liu C, Liu D, Song G, Li Y, Zhang S, Li W, Gao J, Han X, Li G (2019b) The genome-wide transcriptional consequences of the nullisomic-tetrasomic stocks for homoeologous group 7 in bread wheat. BMC Genomics 20:29. https://doi.org/10.1186/s12864-018-5421-3

Zhao J, Udall JA, Quijada PA, Grau CR, Meng J, Osborn TC (2006) Quantitative trait loci for resistance to *Sclerotinia sclerotiorum* and its association with a homoeologous non-reciprocal transposition in *Brassica napus* L. Theor Appl Genet 112:509–516

Zhao X, Xu X, Xie H, Chen S, Jin W (2013) Fertilization and uniparental chromosome elimination during crosses with maize haploid inducers. Plant Physiol 163:721–731

Zhong CX, Marshall JB, Topp C, Mroczek R, Kato A, Nagaki K, Birchler JA, Jiang J, Dawe RK (2002) Centromeric retroelements and satellites interact with maize kinetochore protein CENH3. Plant Cell 14:2825–2836

Zhou JJ, Sing RJ, Lee SJ (2003) Assignment of molecular linkage groups to soybean chromosomes by primary trisomics. Theor Appl Genet 107:745–750

Zhou J, Chen T, Cheng C, Xianhong G, Zaiyun L (2016) Distinct subgenome stabilities in synthesized *Brassica* allohexaploids. Theor Appl Genet 129:1257–1271

Zhu B, Shao Y, Pan Q, Ge X, Li Z (2015) Genome-wide gene expression perturbation induced by loss of C2 chromosome in allotetraploid *Brassica napus* L. Front Plant Sci 6:763

Zhu B, Tu Y, Zeng P, Ge X, Li Z (2016) Extraction of the constituent subgenomes of the natural allopolyploid rapeseed (*Brassica napus* L.). Genetics 204(3):1015–1027

Zhu B, Xiang Y, Zeng P, Cai B, Huang X, Ge X, Weng Q, Li Z (2018) Genome-wide gene expression disturbance by single A1/C1 chromosome substitution in *Brassica rapa* restituted from natural *B. napus*. Front Plant Sci 9:377. https://doi.org/10.3389/fpls.2018.00377

Zou JJ, Singh RJ, Lee J, Xu SS, Hymowitz T (2006) SSR markers exhibit trisomic segregation distortion in soybean [*Glycine max* (L.) Merr.]. Crop Sci 46:1456–1461

8 Introgression Breeding in *Brassica juncea*

Helen Behn and Annaliese S. Mason

Abstract

Introgression breeding refers to the transfer of one or more genomic segments from one genome to another, usually between species, for crop improvement. *Brassica juncea* is a major vegetable and oilseed crop with a large number of close and distant relatives. Many of these relatives contain useful agronomic traits and can be hybridized with varying degrees of difficulty with *B. juncea*. Previously, introgression breeding in *B. juncea* ($2n$ = AABB) has been successfully carried out to introduce novel genetic diversity and heterosis from progenitor diploid species *B. rapa* ($2n$ = AA) and *B. nigra* ($2n$ = BB) as well as closely related allotetraploids *B. carinata* ($2n$ = BBCC) and *B. napus* ($2n$ = AACC). Additional traits of agronomic importance have been introgressed from wider relatives such as *Brassica fruticulosa, Erucastrum cardaminoides* and *Sinapis arvensis* which do not share a genome in common with *B. juncea*, to produce *B. juncea* lines with resistance to mustard aphid, Sclerotinia disease and auxin herbicide. Several hybrid breeding systems (cytoplasmic male sterility plus fertility restorer lines) have also been developed via hybridization with species such as *Moricandia arvensis*, *Trachystoma ballii* and various *Diplotaxis* species. Introgression breeding can be challenging, but shows excellent potential for the introduction of highly valuable qualitative traits such as biotic stress resistances from wild relatives into *B. juncea*. In future, increasing accessibility and availability of resources such as genome sequences of wild relatives, allowing genotyping of early generation hybrids and development of genetic markers linked to specific traits, is expected to facilitate introgression breeding for crop improvement of *B. juncea*.

8.1 Introduction

Brassica juncea is a temperate crop species, grown worldwide but particularly across Asia as a root, stem and leafy vegetable, as an oilseed crop, and as a condiment (mustard). Common names for this species include "brown mustard," "leaf mustard," "Indian mustard" and "rapeseed-mustard"; "root mustard" is also less commonly grown as a vegetable-type. A member of the *Brassica* genus, which also includes rapeseed (*B. napus*), cabbage (*B. oleracea*) and turnip (*B. rapa*), among other crop types, *B. juncea* is

H. Behn · A. S. Mason (✉)
Plant Breeding Department, University of Bonn, Katzenburgweg 5, 53115 Bonn, Germany
e-mail: annaliese.mason@uni-bonn.de

H. Behn
e-mail: hbehn@uni-bonn.de

C. Kole and T. Mohapatra (eds.), *The Brassica juncea Genome*, Compendium of Plant Genomes,
https://doi.org/10.1007/978-3-030-91507-0_8

related to many other species of agricultural importance.

Brassica juncea is an allotetraploid species formed from hybridization between progenitor species *B. rapa* (2*n* = AA = 20 chromosomes) and *B. nigra* (2*n* = BB = 16 chromosomes). The hybridization event between these two parent species most likely occurred about 10,000 years ago, making this a relatively young species (Qiao et al. 2020). Although reports exist of multiple cytoplasm types in *B. juncea* (Kaur et al. 2014b), it has been proposed that a single hybridization event gave rise to this species (Qiao et al. 2020), followed by diversification and differentiation into different germplasm pools through selective breeding for specific morphotypes and geographic isolation. As a result of these processes two well-defined genepools of *B. juncea* have developed—Indian and east European (Chen et al. 2011). Hybrids between the two genepools are heterotic for yield (Paritosh et al. 2014). However, despite the relatively high genetic and morphological diversity present in *B. juncea*, this species as a recent allotetraploid is thought to contain only a subset of the genetic variation and diversity present in parental species *B. nigra* and *B. rapa*.

The *Brassica* crop species in general provide an excellent example of the targeted use of interspecific hybridization for crop improvement (Katche et al. 2019). This is particularly true of the *Brassica* "Triangle of U" species, where *B. rapa*, *B. nigra* and *B. oleracea* (which are diploids with the A, B and C genomes, respectively) were found to be the extant progenitors of the allotetraploid species *B. juncea* (A and B genomes), *B. napus* (A and C genomes) and *B. carinata* (B and C genomes). This relationship provides an excellent means of transferring traits from one species to the other via various methods (Chen et al. 2011; Mason and Chèvre 2017; Katche et al. 2019).

Possibly the most commonly used method for improvement of *B. juncea* via introgression breeding (or just to increase the genetic and phenotypic variation available for further selection) is production of resynthesized *B. juncea* via various methods. The extant progenitor species *B. rapa* and *B. nigra* can be readily cross-pollinated by hand to produce interspecific hybrids (Fitzjohn et al. 2007), following which application of chromosome-doubling agents such as colchicine can be used to produce allotetraploids (AABB) from the allohaploid (AB) hybrids. Resynthesis can also be carried out via hybridization between *B. napus* and *B. carinata* to produce CCAB hybrids, followed by chromosome doubling to CCCCAABB karyotypes which lose the C genomes in successive generations of self-pollination to produce AABB types (Banga and Kaur 2009).

Hybridization between *B. juncea* and the allotetraploid species *B. napus* and *B. carinata* (with which it shares a subgenome in common) is also relatively straightforward: in the first generation AABC or BBAC hybrids are produced, which are then usually backcrossed to *B. juncea* coupled with selection to produce introgressions with the trait of interest. For desirable genetic variants located in another species but in effectively the same genome, this close relationship potentially enables even quantitative traits to be readily transferred via homologous recombination (Mason and Chèvre 2017). Transfer of many alleles simultaneously for the improvement of quantitative traits in *B. juncea* may also be possible with the closely related C genome, since the A and C genomes readily pair and exchange segments in many interspecific hybrid types (Mason et al. 2010, 2014; Gupta et al. 2016).

For more distantly-related species, more traditional introgression breeding approaches need to be undertaken: following the production of interspecific hybrids (possibly requiring biotechnological interventions like embryo rescue), recombinant chromosomes between the *B. juncea* genomes and the wild relative genomes need to be recovered, requiring homoeologous exchange in the interspecific hybrid followed by backcrossing to eliminate the additional undesirable chromosome segments. This latter method is of course better for qualitative, strongly heritable traits, and can be facilitated by methods such as marker-assisted selection (for review see Quezada-Martinez et al. 2021).

In general, there exists excellent potential in *B. juncea* for introgression breeding approaches, and this method has previously been highly successful in obtaining introgression lines carrying desirable traits from a range of related and not-so-related species. In this review, we will introduce the various methods for producing introgression lines, with reference to previous studies with respect to the results of interspecific hybridization attempts, various cross combinations, and results in terms of chromosome inheritance and subsequent agricultural outcomes.

8.2 Introgressions from Progenitor Species *B. rapa* and *B. nigra*

8.2.1 Methods and Approaches

The introgression of genomic regions from progenitor species *B. rapa* ($2n = 2x =$ AA) and *B. nigra* ($2n = 2x =$ BB) can be carried out via either direct hybridization or via resynthesis. In the direct hybridization approach, *B. juncea* is hybridized with either *B. rapa* or *B. nigra*. Both of these cross combinations are usually successful in producing hybrid progeny without the need for biotechnological intervention, but in both cases *B. juncea* has been found to generally be a better maternal parent than either of the diploid species, with more chance of success in producing hybrid seeds if *B. juncea* is the maternal parent in this cross (Fitzjohn et al. 2007). The expected genomic constitution of the hybrids is AAB from the cross *B. juncea* × *B. rapa*, and BBA from the cross *B. juncea* × *B. nigra*. However, due to the frequent occurrence of unreduced gametes in crosses between species and ploidy levels in the genus *Brassica* (Mason and Pires 2015), it may also be possible to directly recover AABB types (or AABBB, or AAAB) from these crosses (U N 1935; Heyn 1977; Mason et al. 2011). Resulting AAB and BBA hybrids from these direct crossing events can in any case then be backcrossed to the parent *B. juncea* in order to start the process of selection and restoration of euploid (AABB) chromosome complements, as required for establishing new *B. juncea* cultivars.

In the resynthesis approach, *B. rapa* and *B. nigra* are hybridized to produce AB hybrids, which are then chromosome-doubled to produce synthetic AABB types. The cross combination *B. rapa* × *B. nigra* is not generally as successful as that of the *B. juncea* × *B. rapa* or *B. juncea* *B. nigra* crosses, but many studies have nonetheless reported hybrid seed production. In this case, *B. rapa* is by far the preferred maternal parent in this cross combination, as *B. nigra* tends to more readily abort interspecific hybrid seeds. Following production of interspecific hybrids (which are usually AB, although AAB and BBA have also been observed), a chromosome-doubling agent such as colchicine is used to produce synthetic allotetraploid lines (AABB). Although direct hybridization of the AB hybrid to *B. juncea* is theoretically possible, allohaploid hybrids are generally highly sterile (Ramsey and Schemske 1998), and hence recovering seeds from these lines may be challenging. Despite the increased difficulty in generating interspecific hybrids via resynthesis (Fitzjohn et al. 2007), this approach may be preferred because it removes the need to select for euploidy ($2n =$ AABB $= 36$ chromosomes) in resulting progeny from the *B. rapa* × *B. nigra* hybrids, and produces homozygous lines which can be directly crossed to *B. juncea* for further selection or for hybrid breeding approaches (Bansal et al. 2012).

8.2.2 Novel Genetic Diversity and Heterosis

Hybridization of *B. juncea* with *B. rapa* or *B. nigra* via either approach has frequently been carried out with the goal of broadening the genetic basis of the *B. juncea* crop germplasm for further crop improvement. Additionally, some studies have investigated the heterotic potential of various resynthesized lines in hybrid breeding, as the genetic distance between "natural" and "resynthesized" *B. juncea* is much greater than that which would be expected between any two *B. juncea* lines within the cultivated germplasm pool. Liu et al. (2021) screened the genetic diversity among 83 *B. nigra*, 16 *B. juncea*, and other *Brassica* accessions using 42 simple sequence repeat (SSR) markers (Liu et al. 2021). It was found that *B. nigra* had rich genetic diversity and could be classified into four subgroups with varying degrees of genetic relationship to *B. juncea*. This study revealed a great potential of *B. nigra* in widening genetic diversity of *B. juncea*, particularly using *B. nigra* from divergent germplasm groups (Liu et al. 2021). Bansal et al. (2009) aimed to broaden the genetic diversity of *B. juncea* by hybridization of genetically diverse germplasm of the progenitor species *B. rapa* and *B. nigra* followed by colchicine-induced chromosome doubling. Resulting synthetic lines were subsequently assayed for genetic diversity using microsatellite markers, and resynthesis was found to generate significant variability (Bansal et al. 2009). In a novel resynthesis approach incorporating the A genome from *B. rapa* and the B genome from *B. carinata* ($2n$ = BBCC), Wei et al. (2016) created an allohexaploid (AABBCC) and crossed it with *B. juncea* to generate pentaploid hybrids (AABBC) (Wei et al. 2016). Subsequent self-pollination led to complete loss of the C genome and a new-type *B. juncea* which was found to be genetically stable at the F_6 generation. Hybrids between newly synthesized and traditional *B. juncea* showed considerable potential for heterosis in terms of increased seed yield (Wei et al. 2016).

8.3 Introgressions from the Related A Subgenome in *B. napus* and the Related B Subgenome in *B. carinata*

8.3.1 Methods and Approaches

The evolutionary relationships between the allotetraploid species *B. juncea* (AABB), *B. napus* (AACC) and *B. carinata* (BBCC) facilitate both hybridization and genome introgression between these three species. The subgenomes in each species are homologous, not identical; both different interspecific hybridization events between progenitor diploid accessions (e.g., between *B. rapa* and *B. nigra* compared to between *B. rapa* and *B. oleracea*), a complex history of domestication and selection and possibly even homoeologous exchanges between the two subgenomes in each allopolyploid have led to differentiation and potentially useful genetic and phenotypic variation present in each of these species (Chalhoub et al. 2014; Yang et al. 2016; Song et al. 2021). Introgression into *B. juncea* from the shared A genome in *B. napus* or the B genome in *B. carinata* can readily be achieved by homologous recombination, and with a lower frequency from the C genome via homeologous recombination (usually with the A genome) (Mason and Chèvre 2017).

Hybridization between *B. juncea* and *B. napus* and *B. juncea* and *B. carinata* can be readily carried out via hand-pollination (Mason et al. 2011). Both crossing directions are possible; some slight preference for *B. juncea* as a maternal parent seems to be present in terms of success rate (Fitzjohn et al. 2007), but as with all crosses this is also strongly genotype-dependent. Crosses between *B. juncea* and *B. napus* usually result in the production of AABC hybrids, while crosses between *B. juncea* and *B. carinata* usually result in the production of BBAC hybrids. However, other genome compositions have also been observed, including aneuploidy (missing chromosomes from a *B. carinata* parent gamete) and AAABCC compositions from the combination of an unreduced gamete from *B. napus* with a reduced gamete from *B. juncea* (Mason et al. 2011). AABC and BBAC hybrid types can be readily backcrossed to either parent to restore *B. juncea* types, with fertility generally improving with subsequent backcross generations (Frello et al. 1995; Schelfhout et al. 2006). Improvement in fertility is putatively coupled to successive loss of undesired C genome chromosomes.

8.3.2 Novel Genetic Diversity and Heterosis

As with *B. rapa* and *B. nigra*, hybridizations with *B. napus* and *B. carinata* have been frequently carried out in order to introduce novel genetic diversity into *B. juncea*, and to explore the potential for heterosis and hybrid breeding using this material. Schelfhout et al. (2008) made reciprocal crosses between *Brassica napus* cv. Mystic (canola) and *B. juncea* JN29 (near canola quality), and self-pollinated and backcrossed the F_1 hybrids in all possible combinations to the parent plants (Schelfhout et al. 2008). $F_{2:3}$ and $BC_1S_{0:1}$ progeny showed transgressive segregation for agronomic and quality traits in two contrasting field environments, and several *B. juncea*-type progeny had improved seed chemical composition (lower total seed glucosinolates and higher % oleic acid) than the *B. juncea* parent. They concluded that self-pollination of interspecific hybrids between canola-quality *B. napus* and *B. juncea* has the potential to greatly enhance genetic diversity in canola-quality progeny of both species, without the loss of donor alleles that normally occurs with repeated backcrossing.

A novel resynthesis method to produce *B. juncea* via the cross between *B. napus* and *B. carinata* was pioneered by Prof. Surinder Banga at Punjab Agricultural University. This method involves cross-pollination between *B. napus* and *B. carinata* to produce CCAB hybrids, followed by colchicine treatment to induce chromosome doubling to produce CCCCAABB types. Interestingly, self-pollination of these lines then results in complete loss of the C genome chromosomes via tetravalent elimination, thus producing AABB-type plants which represent a novel resynthesis of *B. juncea* (Banga and Kaur 2009). Gupta et al. (2014) analyzed 62 progeny resulting from this approach for cytogenetic stability and genome size changes in the S_5 and S_6 generations, and confirmed the viability of this resynthesis concept for *B. juncea* improvement (Gupta et al. 2014). These new resynthesized hybrids showed excellent potential for hybrid breeding in terms of heterosis after hybridization with natural *B. juncea* (Gupta et al. 2015). Some progeny even showed novel, desirable agronomic traits such as determinate inflorescences (Kaur et al. 2014a). Further molecular cytogenetic investigation also revealed that some of these lines contained C genome chromosome substitutions (whole C genome chromosomes which had replaced homoeologous *B. juncea* genome chromosomes), as well as C-genome introgressions, with potential value for trait introgression and increased heterosis in crosses with natural *B. juncea* (Gupta et al. 2016). Interestingly, the resynthesized *B. juncea* lines were found to have a higher internal methylation level (40.4%) than the natural *B. juncea* cultivars (17.1%) (Gupta et al. 2019), possibly suggesting gene methylation as an additional source of variation which may affect trait phenotypes.

8.3.3 Cytoplasmic Male Sterility and Restorer Lines for Hybrid Breeding Systems

Hybridization between *B. napus* and *B. juncea* has also been done to transfer established traits or introgressions from other species. In particular, "Ogura" cytoplasmic male sterility (CMS) is a system that has been developed in *B. napus* and then transferred to *B. juncea*. This CMS trait was first identified in *Raphanus sativus* (radish). Male-sterile *B. napus* lines carrying the radish cytoplasm and subsequently a fertility restorer line were produced by targeted hybridization breeding and introgression of a radish locus into chromosome C9 of *B. napus* (Pellan-Delourme and Renard 1988; Delourme et al. 1998). Due to the amount of work required to build this complete hybrid breeding system, which is still a highly efficient and widely used hybrid seed production system today, direct transfer of the ogu radish cytoplasm into *B. juncea* from *B. napus* via somatic fusion has been attempted (Kirti et al. 1995). An alternative approach was the transfer of the fertility restorer (*Rfo*) gene in *B. napus* to *B. juncea* via hybridization, and subsequent improvement of this system (Gudi et al. 2020). Tsuda et al. (2014) also explored the possibility of indirectly introgressing traits from *B. napus* to related Brassicaceae species through *B. juncea* as well as direct introgression from *Brassica napus* to *B. juncea* (Tsuda et al. 2014).

Similarly, Arumugam et al. (1996) synthesized hexaploid hybrids (AABBCC) by somatic cell hybridization between hygromycin-resistant *B. juncea* (AABB) carrying "tour" cytoplasm and phosphinotricin-resistant, normal *B. oleracea* (CC). This bridging material was backcrossed to all of the diploid and allopolyploid *Brassica* species in order to transfer "tour" CMS derived from *Brassica tournefortii* (Arumugam et al. 1996).

8.3.4 Transgene Escape

Due to regulatory issues around transgenic crops, the possibility of unintended introgression from transgenic lines to crops has also been studied in the context of hybridization between crop rapeseed and wild *B. juncea*. Tang et al. (2018) studied transgene introgression from transgenic rapeseed (*B. napus*) to different varieties of *B. juncea* (Tang et al. 2018), and Frello et al. (1995) studied the inheritance of rapeseed (*Brassica napus*) specific random amplified polymorphic DNA (RAPD) markers and a transgene in the cross *B. juncea* × (*B. juncea* × *B. napus*) (Frello et al. 1995). Guan et al. (2020) evaluated the introgression of transgenic *B. napus* into wild *B. juncea* and found a loss of C-genome-specific markers during transgene introgression (Guan et al. 2020).

8.4 Introgressions from Outside the A, B and C Genome Species

8.4.1 Methods and Approaches

Hybridization between *B. juncea* and species outside the "Triangle of U" has also frequently been carried out in order to introgress useful traits for *B. juncea* crop improvement. In this case, following production of interspecific hybrids, repeated backcrossing to *B. juncea* is usually carried out in order to eliminate all chromosomes from the wild species genome with the exception of the trait locus of interest. This trait locus will ideally be introgressed into the host *B. juncea* genome such that only a small chromosomal segment is present on a *B. juncea* chromosome, to avoid linkage drag (negative agronomic effects contributed by adjacent genetic loci on the same segment resulting from the wild species). This approach can be challenging due to the low frequencies of crossovers usually observed in the interspecific hybrids, but has been used to great effect in both production of CMS hybrid breeding systems and introgression of resistances to mustard aphid and Sclerotinia disease. Mithila and Hall (2013) also introgressed auxinic herbicide (dicamba) resistance into *B. juncea* by crossing with a resistant *Sinapis arvensis* (old name *B. kaber*; wild mustard) genotype and rescuing the embryo rescue followed by backcrosses to *B. juncea* (Mithila and Hall 2013).

8.4.2 Cytoplasmic Male Sterility Systems for Hybrid Breeding

CMS systems in *Brassica juncea* have been developed via wide hybridization with a variety of species to result in either male sterility or the development of a complete hybrid breeding system (CMS lines plus fertility restorer lines). Rao et al. (1994) developed a CMS system in *Brassica juncea* through wide hybridization with *Diplotaxis siifolia* (Rao et al. 1994), while Banga et al. (2003) transferred CMS to *Brassica juncea* by introducing its nucleus into the cytoplasm of *Enarthrocarpus lyratus* ($2n = 20$; EE). Banga et al. (2003) and Prakash et al. (2001) examined the expression of male sterility in alloplasmic *Brassica juncea* with *Erucastrum canariense* cytoplasm and the development of a fertility restoration system (Prakash et al. 2001; Banga et al. 2003). Somatic fusion to produce hybrids with *Trachystoma ballii* followed by repeated backcrossing to *B. juncea* was first carried out to produce CMS lines (Kirti et al. 1995), and subsequently to introgress a fertility restorer gene (Kirti et al. 1997). Pathania et al. (2003) studied CMS and fertility restoration in alloplasmic *Brassica juncea* carrying *Diplotaxis catholica* cytoplasm (Pathania et al. 2003). *Moricandia arvensis* has frequently been investigated and utilized: Prakash et al. (1998) studied a *Moricandia arvensis*-based CMS and fertility restoration system in *Brassica juncea* (Prakash et al. 1998), while Kirti et al. (1998) examined effects of chloroplast substitution on leaf chlorosis in a *Moricandia arvensis*-based cytoplasmic male sterile *B. juncea* (Kirti et al. 1998). Kaur et al. (2004) investigated fertility restoration in *B. juncea* CMS lines based on the cytoplasm from *Moricandia arvensis* (Kaur et al. 2004), and Ashutosh et al. (2007) developed a CMS line of *Brassica juncea* through somatic hybridization with *Moricandia arvensis* and introgressed the fertility restorer gene into *B. juncea*. The authors identified AFLP markers linked to the male fertility restorer gene and converted them into SCAR markers (Ashutosh et al. 2007). Bhat et al. (2005) studied restoration of male fertility to two CMS *B. juncea* lines carrying either *M. arvensis* or *Diplotaxis catholica* cytoplasm by crossing them with a *B. juncea* line carrying an introgression from *Moricandia arvensis* (Bhat et al. 2005). Restoration was found to be monogenic. Bhat et al. (2006) developed an alloplasmic CMS line of *Brassica juncea* by repeated backcrossing of the sexual hybrid (*Diplotaxis erucoides* × *Brassica rapa*) with *B. juncea* cv. "Pusa Bold" and studied fertility rescue by the *Moricandia arvensis* restorer (Bhat et al. 2006). In each of the *mori, erucoides* and *catholica* CMS systems altered expression of *atpα* at the floral stage was associated with CMS and a common restorer was able to restore male fertility, suggesting *atpα* may be involved in causing CMS in these lines (Bhat et al. 2006). Bhat et al. (2008) reported the development of an improved CMS system of *B. juncea* carrying cytoplasm of the wild species *Diplotaxis berthautii* and restoration of fertility using *Moricandia arvensis* and *D. catholica*-based alloplasmic CMS systems of *B. juncea* (Bhat et al. 2008).

8.4.3 Resistance to Mustard Aphid

Mustard aphid (*Lipaphis erysimi*) is a major pest in *Brassica* oilseeds, but limited resistance exists within the crop germplasm against this pest (Bhatia et al. 2011). Kumar et al. (2011) screened a number of wild crucifers for resistance traits and found *Brassica fruticulosa* and *Brassica montana* to be the most promising (Kumar et al. 2011). Atri et al. (2012) synthesized an artificial amphiploid from *B. fruticulosa* × *B. rapa* var. brown sarson for use as a bridge to transfer resistance traits (Atri et al. 2012). Introgression lines generated by self-pollination showed euploid chromosome numbers (2n = 36), normal meiosis and high pollen grain fertility. *B. juncea-fruticulosa* introgression lines may prove to be a very powerful breeding tool for aphid resistance related quantitative trait locus (QTL)/gene discovery and fine mapping of the desired genes/QTLs to facilitate marker-assisted transfer of identified gene(s) for mustard aphid resistance in the background of commercial mustard genotypes (Atri et al. 2012). *Brassica juncea* lines carrying genomic introgressions from *B. fruticulosa* have subsequently investigated for mustard aphid resistance and found to be resistant (Kumar et al. 2011). Palial et al. (2018) also identified changes in biochemical constituents after aphid infestation in three *Brassica juncea fruticulosa* introgression lines (already reported resistant to *Lipaphis erysimi*) in terms along with *B. fruticulosa* (resistant parent) in comparison with *B. juncea* var PBR −210 (susceptible parent) and *B. rapa* ecotype brown sarson BSH-1 (susceptible check) (Palial et al. 2018).

8.4.4 Resistance to *Sclerotinia sclerotiorum*

Sclerotinia wilt or stem stripe is a disease with serious impact on Brassica crop production. Garg et al. (2010) introgressed resistance to *Sclerotinia sclerotiorum* into *B. napus* and *B. juncea* by hybridization with three wild crucifer species: *Erucastrum cardaminoides*, *Diplotaxis tenuisiliqua* and *E. abyssinicum* (Garg et al. 2010). Introgression of segments of the genome from *E. cardaminoides* into cultivated lines of *B. juncea* was achieved by hybridizing the wild relative with *B. rapa* or *B. nigra* followed by chromosome doubling to give a synthetic amphidiploid. This amphidiploid was then backcrossed to *B. juncea*. Introgression lines showed much higher levels ($P < 0.001$) of resistance compared with the *B. juncea* germplasm. Rana et al. (2019) developed a set of 96 *Brassica juncea–Erucastrum cardaminoides* introgression lines (ILs): marker-trait associations and genomic regions associated with Sclerotinia stem rot (*Sclerotinia sclerotiorum*) resistance from wild Brassicaceae species *E. cardaminoides* were established (Rana et al. 2019). Rana et al. (2017) mapped resistance responses to Sclerotinia infection in introgression lines of *Brassica juncea* carrying genomic segments from wild Brassicaceae *B. fruticulosa* (Rana et al. 2017), and associated high levels of resistance against *S. sclerotiorum* with particular segments. Atri et al. (2019) also analyzed defensive responses of *Brassica juncea—B. fruticulosa* introgression lines to stem rot caused by *Sclerotinia sclerotiorum* (Atri et al. 2019): over 40% of introgression lines showed higher levels of resistance.

8.5 Facilitating Introgression Breeding

Introgression breeding in *B. juncea* has successfully led to the production of introgression lines with increased genetic diversity as well as useful agronomic traits, as outlined above in previous sections. However, introgression breeding can be challenging: successful production of interspecific hybrids can be difficult with more distantly-related species, and recovering small chromosome introgression regions carrying the agronomic trait of interest in the crop genome background usually requires substantial breeding efforts to obtain and select for recombinant chromosomes, which may result from very rare crossover events in the hybrids (Quezada-Martinez et al. 2021). In the case of *B. juncea*, the availability of several related species which share the A or B genome or subgenome in common greatly facilitates this process: in hybrids which share a genome (e.g., AAB, AABC genome structure), homologous chromosome pairing can be used to transfer genetic loci directly, and may hence allow for large-scale introduction of novel genetic variation or potentially even QTLs (reviewed by Mason and Chèvre 2017).

Within the Brassiceae tribe, many species are known to successfully hybridize with *B. juncea* (reviewed by Fitzjohn et al. 2007). Successful interspecific hybridization is usually correlated to some extent with phylogenetic divergence between species: however, although the Brassiceae is thought to be a monophyletic clade, genera within the Brassiceae are highly polyphyletic (Al-Shehbaz et al. 2006). Based on ITS sequence comparisons (https://brassibase.cos.uni-heidelberg.de/), "A genome species" *Brassica juncea*, *B. napus* and *B. rapa* form one subgroup which is very closely related to the "C genome species" cytodeme of *B. oleracea*, *B. incana*, *B. rupestris*, *B. insularis*, *B. montana*, *B. macrocarpa*, *B. villosa* and *B. cretica*, while the B genome closely groups with *Sinapis arvensis*, *Trachystoma labasii* and *Guiraoa arvensis* (Koch et al. 2012, 2018; Kiefer et al. 2014), making these species potentially a good choice for introgressions. Another 30–60 species fall into a "relatively close" phylogenetic position to the A and B genomes, including species in the large polyphyletic genera *Erucastrum* and *Diplotaxis*, as well as other *Brassica* species such as *B. fruticulosa* (Koch et al. 2012, 2018; Kiefer et al. 2014). Although phylogenetic proximity may not always be directly related to hybridization or introgression success, prior knowledge of these relationships may be helpful in selecting the best target species for introgression attempts, if multiple options are available.

In recent years, due to the increasing accessibility and reduced costs of whole genome sequencing and other genomics approaches, huge numbers of complex crop and wild relative genomes are being sequenced and assembled, providing a robust platform for genomics approaches to breeding (Bevan et al. 2017). This is of major benefit for introgression breeding: if genome sequence data or even transcriptomics data is available for the wild relative of interest, then this allows molecular markers to be rapidly developed. This can be done via comparison of the genomic regions of interest between *B. juncea* and the wild relative to find variants of interest, or via comparison of previously identified markers in *B. juncea* for potential cross-genome use. Marker-assisted tracking of genomic introgressions can be hugely beneficial for early selection of plants which are carrying the desired genomic segment. As well, markers can be used to identify whether whole addition chromosomes or recombinant chromosomes are present: large numbers of putative introgression lines can be screened with a marker at the top and bottom of the wild relative chromosome containing the target locus of interest, and individuals where only one of the two marker loci is present identified for further investigation. Genomic or transcriptomic information can also allow the detailed elucidation of the genomic introgression fragments in terms of size and location (He et al. 2017), and may even allow identification of possible candidate genes if the genomic introgression is small enough.

8.6 Conclusions

Introgression breeding for crop improvement in *Brassica juncea* has been successfully carried out to introduce novel genetic diversity from *B. rapa*, *B. nigra*, *B. napus* and *B. carinata*, and with wider relatives to develop hybrid breeding systems (cytoplasmic male sterile and fertility restorer lines) and to introgress specific traits of agronomic importance (Fig. 8.1). These include resistances to major diseases and pests afflicting *B. juncea* crops such as mustard aphid infestation

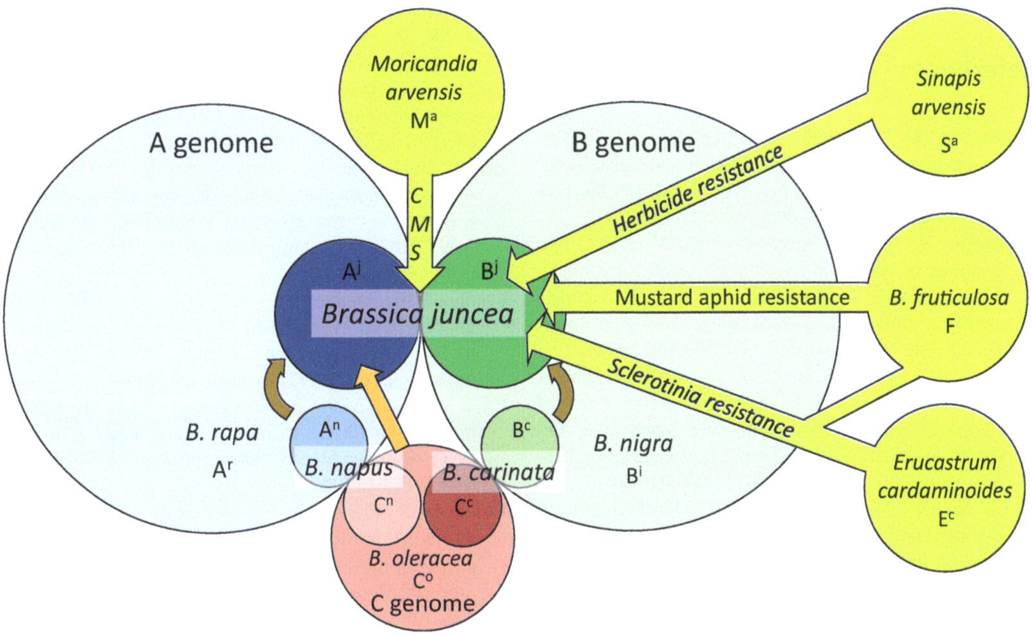

Fig. 8.1 Example introgressions into *Brassica juncea* from related species. *B. juncea* is an allotetraploid with 2n = AABB; introgression can hence readily be achieved between *B. juncea* and relative species which share subgenomes in common (*B. rapa*, *B. nigra*, *B. napus* and *B. carinata*; A genome = blue, B genome = green; brown arrows), as well as between the closely related A and C genomes (C genome = red, orange arrow indicates possible introgression of C-genome segments into the A genome; introgression from the C to the B is possible but much more difficult). Wider introgressions (yellow) have successfully transferred CMS from *Moricandia arvensis*, herbicide resistance from *Sinapis arvensis*, mustard aphid resistance from *B. fruticulosa*, and *Sclerotinia* resistance from *B. fruticulosa* and *Erucastrum cardaminoides*, as some examples

and Sclerotinia disease, as well as introgression of herbicide resistance. Despite these impressive results, introgression breeding has a great deal of as-yet-unexplored potential. Both the ready introgressions between subgenomes permitted by the genomic similarity between *B. juncea* and the other "Triangle of U" species as well as the extremely wide range of wild relatives for which hybridization approaches are possible suggest numerous possibilities for improvement. The ongoing development of additional genetic and genomic resources (genome references for the wild species, markers and germplasm collections) as well as new biotechnological approaches (hopefully including targeted transfer of identified genes directly between species using transformation, pending regulatory approval) can only lead to immense benefits for *B. juncea*, an important and diverse vegetable, oilseed and condiment crop around the world.

References

Al-Shehbaz IA, Beilstein MA, Kellogg EA (2006) Systematics and phylogeny of the Brassicaceae (Cruciferae): an overview. Plant Syst Evol 259:89–120. https://doi.org/10.1007/s00606-006-0415-z

Arumugam N, Mukhopadhyay A, Gupta V et al (1996) Synthesis of hexaploid (AABBCC) somatic hybrids: a bridging material for transfer of "tour" cytoplasmic male sterility to different *Brassica* species. Theor Appl Genet 92:762–768. https://doi.org/10.1007/BF00226099

Ashutosh SPC, Prakash S, Bhat SR (2007) Identification of AFLP markers linked to the male fertility restorer gene of CMS (*Moricandia arvensis*) *Brassica juncea* and conversion to SCAR marker. Theor Appl Genet 114:385–392. https://doi.org/10.1007/s00122-006-0444-6

Atri C, Kumar B, Kumar H et al (2012) Development and characterization of *Brassica juncea*—*fruticulosa* introgression lines exhibiting resistance to mustard aphid (*Lipaphis erysimi* Kalt). BMC Genet 13

Atri C, Akhatar J, Gupta M et al (2019) Molecular and genetic analysis of defensive responses of *Brassica juncea*—*B. fruticulosa* introgression lines to *Sclerotinia* infection. Sci Rep 9:1–12. https://doi.org/10.1038/s41598-019-53444-3

Banga SS, Kaur N (2009) An alternate procedure for resynthesis of *Brassica juncea*. In: 16th Australian research assembly on Brassicas. Ballarat Victoria, pp 1–4

Banga SS, Deol JS, Banga SK (2003) Alloplasmic male-sterile *Brassica juncea* with *Enarthrocarpus lyratus* cytoplasm and the introgression of gene(s) for fertility restoration from cytoplasm donor species. Theor Appl Genet 106:1390–1395. https://doi.org/10.1007/s00122-002-1180-1

Bansal P, Kaur P, Banga SK, Banga SS (2009) Augmenting genetic diversity in *Brassica juncea* through its resynthesis using purposely selected diploid progenitors. Int J Plant Breed, 41–45

Bansal P, Banga S, Banga SS (2012) Heterosis as investigated in terms of polyploidy and genetic diversity using designed *Brassica juncea* amphiploid and its progenitor diploid species. PLoS ONE 7. https://doi.org/10.1371/journal.pone.0029607

Bevan MW, Uauy C, Wulff BBH et al (2017) Genomic innovation for crop improvement. Nature 543:346–354. https://doi.org/10.1038/nature22011

Bhat SR, Prakash S, Kirti PB et al (2005) A unique introgression from *Moricandia arvensis* confers male fertility upon two different cytoplasmic male-sterile lines of *Brassica juncea*. Plant Breed 124:117–120. https://doi.org/10.1111/j.1439-0523.2005.01054.x

Bhat SR, Vijayan P, Ashutosh et al (2006) *Diplotaxis erucoides*-induced cytoplasmic male sterility in *Brassica juncea* is rescued by the *Moricandia arvensis* restorer: genetic and molecular analyses. Plant Breed 125:150–155. https://doi.org/10.1111/j.1439-0523.2006.01184.x

Bhat SR, Kumar P, Prakash S (2008) An improved cytoplasmic male sterile (*Diplotaxis berthautii*) *Brassica juncea*: identification of restorer and molecular characterization. Euphytica 159:145–152. https://doi.org/10.1007/s10681-007-9467-6

Bhatia V, Uniyal PL, Bhattacharya R (2011) Aphid resistance in *Brassica* crops: Challenges, biotechnological progress and emerging possibilities. Biotechnol Adv 29:879–888. https://doi.org/10.1016/j.biotechadv.2011.07.005

Chalhoub B, Denoeud F, Liu S et al (2014) Early allopolyploid evolution in the post-neolithic *Brassica napus* oilseed genome. Science 345(80-):950–953. https://doi.org/10.1126/science.1253435

Chen S, Nelson MN, Chèver AM et al (2011) Trigenomic bridges for *Brassica* improvement. CRC Crit Rev Plant Sci 30:524–547. https://doi.org/10.1080/07352689.2011.615700

Delourme R, Horvais R, Barret P et al (1998) Mapping the radish introgression carrying the *rfo* restorer gene used for the Ogu-inra cytoplasmic male sterility in rapeseed (*Brassica napus* L.). Acta Hortic 459:135–138. https://doi.org/10.17660/ActaHortic.1998.459.14

Fitzjohn RG, Armstrong TT, Newstrom-Lloyd LE et al (2007) Hybridisation within *Brassica* and allied genera: evaluation of potential for transgene escape. Euphytica 158:209–230. https://doi.org/10.1007/s10681-007-9444-0

Frello S, Hansen KR, Jensen J, Jørgensen RB (1995) Inheritance of rapeseed (*Brassica napus*)-specific RAPD markers and a transgene in the cross *B. juncea* × (*B. juncea* × *B. napus*). Theor Appl Genet 91:236–241. https://doi.org/10.1007/BF00220883

Garg H, Atri C, Sandhu PS et al (2010) High level of resistance to *Sclerotinia sclerotiorum* in introgression lines derived from hybridization between wild crucifers and the crop *Brassica* species *B. napus* and *B. juncea*. F Crop Res 117:51–58. https://doi.org/10.1016/j.fcr.2010.01.013

Guan ZJ, Shen BC, Wei W et al (2020) Loss of C-genome-specific markers during transgene introgression from *Brassica napus* to wild *Brassica juncea*. Nord J Bot 38:1–5. https://doi.org/10.1111/njb.02794

Gudi S, Atri C, Goyal A et al (2020) Physical mapping of introgressed chromosome fragment carrying the fertility restoring (*Rfo*) gene for Ogura CMS in *Brassica juncea* L. Czern Coss Theor Appl Genet 133:2949–2959. https://doi.org/10.1007/s00122-020-03648-3

Gupta M, Atri C, Banga S (2014) Cytogenetic stability and genome size variations in newly developed derived *Brassica juncea* allopolyploid lines. J Oilseed Brassica 5:118–127

Gupta M, Gupta S, Kumar H et al (2015) Population structure and breeding value of a new type of *Brassica juncea* created by combining A and B genomes from related allotetraploids. Theor Appl Genet 128:221–234. https://doi.org/10.1007/s00122-014-2423-7

Gupta M, Mason AS, Batley J et al (2016) Molecular-cytogenetic characterization of C-genome chromosome substitution lines in *Brassica juncea* (L.) Czern and Coss. Theor Appl Genet 129:1153–1166. https://doi.org/10.1007/s00122-016-2692-4

Gupta S, Sharma N, Akhatar J et al (2019) Analysis of epigenetic landscape in a recombinant inbred line population developed by hybridizing natural and re-synthesized *Brassica juncea* (L.) with stable C-genome introgressions. Euphytica 215:1–11. https://doi.org/10.1007/s10681-019-2498-y

He Z, Wang L, Harper AL et al (2017) Extensive homoeologous genome exchanges in allopolyploid crops revealed by mRNAseq-based visualization. Plant Biotechnol J 15:594–604. https://doi.org/10.1111/pbi.12657

Heyn FW (1977) Analysis of unreduced gametes in the Brassiceae by crosses between species and ploidy levels. Z Pflanzenzuechtung 78:13–30

Katche E, Quezada-Martinez D, Katche EI (2019) Interspecific hybridization for *Brassica* crop improvement. Crop Breed Genet Genom 1–32. https://doi.org/10.20900/cbgg20190007

Kaur G, Banga SK, Gogna KPS et al (2004) *Moricandia arvensis* cytoplasm based system of cytoplasmic male sterility in *Brassica juncea*: reappraisal of fertility restoration and agronomic potential. Euphytica 138:271–276. https://doi.org/10.1023/B:EUPH.0000047098.33909.b7

Kaur H, Gupta S, Kumar N et al (2014a) Progression of molecular and phenotypic diversification in resynthesized *Brassica juncea* (L) gene pool with determinate inflorescence. Euphytica 199:325–338. https://doi.org/10.1007/s10681-014-1133-1

Kaur P, Banga S, Kumar N et al (2014b) Polyphyletic origin of *Brassica juncea* with *B. rapa* and *B. nigra* (Brassicaceae) participating as cytoplasm donor parents in independent hybridization events. Am J Bot 101:1157–1166. https://doi.org/10.3732/ajb.1400232

Kiefer M, Schmickl R, German DA et al (2014) BrassiBase: introduction to a novel knowledge database on brassicaceae evolution. Plant Cell Physiol 55:1–9. https://doi.org/10.1093/pcp/pct158

Kirti P, Banga S, Prakash S, Chopra V (1995) Transfer of Ogu cytoplasmic male sterility to *Brassica juncea* and improvement of the male sterile line through somatic cell fusion. Theor Appl Genet 91:517–521

Kirti PB, Baldev A, Gaikwad K et al (1997) Introgression of a gene restoring fertility to CMS (*Trachystoma*) *Brassica juncea* and the genetics of restoration. Plant Breed 116:259–262. https://doi.org/10.1111/j.1439-0523.1997.tb00992.x

Kirti PB, Prakash S, Gaikwad K et al (1998) Chloroplast substitution overcomes leaf chlorosis in a *Moricandia arvensis*-based cytoplasmic male sterile *Brassica juncea*. Theor Appl Genet 97:1179–1182. https://doi.org/10.1007/s001220051007

Koch MA, Kiefer M, German DA et al (2012) BrassiBase: Tools and biological resources to study characters and traits in the Brassicaceae-version 1.1. Taxon 61:1001–1009. https://doi.org/10.1002/tax.615007

Koch MA, German DA, Kiefer M, Franzke A (2018) Database taxonomics as key to modern plant biology. Trends Plant Sci 23:4–6. https://doi.org/10.1016/j.tplants.2017.10.005

Kumar S, Atri C, Sangha MK, Banga SS (2011) Screening of wild crucifers for resistance to mustard aphid, *Lipaphis erysimi* (Kaltenbach) and attempt at introgression of resistance gene(s) from *Brassica fruticulosa* to *Brassica juncea*. Euphytica 179:461–470. https://doi.org/10.1007/s10681-011-0351-z

Liu J, Rana K, McKay J, et al (2021) Investigating genetic relationship of *Brassica juncea* with *B. nigra* via virtual allopolyploidy and hexaploidy strategy. Mol Breed 41. https://doi.org/10.1007/s11032-020-01197-7

Mason AS, Chèvre AM (2017) Optimization of recombination in interspecific hybrids to introduce new genetic diversity into oilseed rape (*Brassica napus* L.). Polyploidy Hybrid Crop Improv 431–444. https://doi.org/10.1201/9781315369259

Mason AS, Pires JC (2015) Unreduced gametes: meiotic mishap or evolutionary mechanism? Trends Genet 31:5–10. https://doi.org/10.1016/j.tig.2014.09.011

Mason AS, Huteau V, Eber F et al (2010) Genome structure affects the rate of autosyndesis and allosyndesis in AABC, BBAC and CCAB *Brassica* interspecific hybrids. Chromosom Res 18:655–666. https://doi.org/10.1007/s10577-010-9140-0

Mason AS, Nelson MN, Yan G, Cowling WA (2011) Production of viable male unreduced gametes in *Brassica* interspecific hybrids is genotype specific and stimulated by cold temperatures. BMC Plant Biol 11. https://doi.org/10.1186/1471-2229-11-103

Mason AS, Batley J, Bayer PE et al (2014) High-resolution molecular karyotyping uncovers pairing between ancestrally related *Brassica* chromosomes. New Phytol 202:964–974. https://doi.org/10.1111/nph.12706

Mithila J, Hall JC (2013) Transfer of auxinic herbicide resistance from *Brassica kaber* to *Brassica juncea* and *Brassica rapa* through embryo rescue. Vitr Cell Dev Biol Plant 49:461–467. https://doi.org/10.1007/s11627-013-9515-y

Palial S, Kumar S, Sharma S (2018) Biochemical changes in the *Brassica juncea-fruticulosa* introgression lines after *Lipaphis erysimi* (Kaltenbach) infestation. Phytoparasitica 46:499–509. https://doi.org/10.1007/s12600-018-0686-2

Paritosh K, Gupta V, Yadava SK et al (2014) RNA-seq based SNPs for mapping in *Brassica juncea* (AABB): Synteny analysis between the two constituent genomes A (from *B. rapa*) and B (from *B. nigra*) shows highly divergent gene block arrangement and unique block fragmentation patterns. BMC Genomics 15:1–14. https://doi.org/10.1186/1471-2164-15-396

Pathania A, Bhat SR, Dinesh Kumar V et al (2003) Cytoplasmic male sterility in alloplasmic *Brassica juncea* carrying Diplotaxis catholica cytoplasm: molecular characterization and genetics of fertility restoration. Theor Appl Genet 107:455–461. https://doi.org/10.1007/s00122-003-1266-4

Pellan-Delourme R, Renard M (1988) Cytoplasmic male sterility in rapeseed (*Brassica napus* L.): female fertility of restored rapeseed with "Ogura" and cybrids cytoplasms. Genome 30:234–238

Prakash S, Kirti PB, Bhat SR et al (1998) A *Moricandia arvensis*—based cytoplasmic male sterility and fertility restoration system in *Brassica juncea*. Theor Appl Genet 97:488–492. https://doi.org/10.1007/s001220050921

Prakash S, Ahuja I, Upreti HC et al (2001) Expression of male sterility in alloplasmic *Brassica juncea* with *Erucastrum canariense* cytoplasm and the development of a fertility restoration system. Plant Breed 120: 479–482. https://doi.org/10.1046/j.1439-0523.2001.00627_x

Qiao J, Zhang X, Chen B et al (2020) Comparison of the cytoplastic genomes by resequencing: insights into the genetic diversity and the phylogeny of the agriculturally important genus *Brassica*. BMC Genomics 21:1–15. https://doi.org/10.1186/s12864-020-06889-0

Quezada-Martinez D, Addo Nyarko CP, Schiessl SV, Mason AS (2021) Using wild relatives and related species to build climate resilience in *Brassica* crops. Theor Appl Genet. https://doi.org/10.1007/s00122-021-03793-3

Ramsey J, Schemske DW (1998) Pathways, mechanisms, and rates of polyploid formation in flowering plants. Annu Rev Ecol Syst 29:467–501. https://doi.org/10.1146/annurev.ecolsys.29.1.467

Rana K, Atri C, Gupta M et al (2017) Mapping resistance responses to *Sclerotinia* infestation in introgression lines of *Brassica juncea* carrying genomic segments from wild Brassicaceae *B. fruticulosa*. Sci Rep 7:1–12. https://doi.org/10.1038/s41598-017-05992-9

Rana K, Atri C, Akhatar J et al (2019) Detection of first marker trait associations for resistance against *Sclerotinia sclerotiorum* in *Brassica juncea–Erucastrum cardaminoides* introgression lines. Front Plant Sci 10. https://doi.org/10.3389/fpls.2019.01015

Rao GU, Batra-Sarup V, Prakash S, Shivanna KR (1994) Development of a new cytoplasmic male-sterility system in *Brassica juncea* through wide hybridization. Plant Breed 112:171–174. https://doi.org/10.1111/j.1439-0523.1994.tb00666.x

Schelfhout CJ, Snowdon R, Cowling WA, Wroth JM (2006) Tracing B-genome chromatin in *Brassica napus* × *B. juncea* interspecific progeny. Genome 49:1490–1497. https://doi.org/10.1139/G06-103

Schelfhout CJ, Wroth JM, Yan G, Cowling WA (2008) Enhancement of genetic diversity in canola-quality *Brassica napus* and *B. juncea* by interspecific hybridisation. Aust J Agric Res 59:918–925. https://doi.org/10.1071/AR07425

Song X, Wei Y, Xiao D et al (2021) *Brassica carinata* genome characterization clarifies U's triangle model of evolution and polyploidy in *Brassica*. Plant Physiol 186:388–406. https://doi.org/10.1093/plphys/kiab048

Tang T, Chen GM, Bu CP et al (2018) Transgene introgression from *Brassica napus* to different varieties of *Brassica juncea*. Plant Breed 137:171–180. https://doi.org/10.1111/pbr.12567

Tsuda M, Ohsawa R, Tabei Y (2014) Possibilities of direct introgression from *Brassica napus* to *B. juncea* and indirect introgression from *B. napus* to related Brassicaceae through *B. juncea*. Breed Sci 64:74–82. https://doi.org/10.1270/jsbbs.64.74

U N (1935) Genome analysis in *Brassica* with special reference to the experimental formation of *B. napus* and peculiar mode of fertilization. Jpn J Bot 7:389–452

Wei Z, Wang M, Chang S et al (2016) Introgressing subgenome components from *Brassica rapa* and *B. carinata* to *B. juncea* for broadening its genetic base and exploring intersubgenomic heterosis. Front Plant Sci 7. https://doi.org/10.3389/fpls.2016.01677

Yang J, Liu D, Wang X et al (2016) The genome sequence of allopolyploid *Brassica juncea* and analysis of differential homoeolog gene expression influencing selection. Nat Genet 48:1225–1232. https://doi.org/10.1038/ng.3657

In Vitro Culture for Micropropagation, Somatic Embryogenesis, Somatic Mutation, and Somatic Hybridization in *Brassica juncea*

9

Simone Treccarichi, Giulio Flavio Rizzo, and Ferdinando Branca

Abstract

Development of *Brassica juncea* in vitro regeneration protocol can be a useful tool for breeders to select genotypes having interesting traits which can be related to the oil production which is the main goal for an oilseed crop. Several techniques have been mentioned for their possible application in breeding programs, starting from the explant and finishing with regenerated plant acclimatation. Explant regeneration process needs the optimization of protocol for the medium composition, its environmental conditions and plant selection. Regeneration of explant is carried out via organogenesis through the excised organs in vitro regeneration or via somatic embryogenesis (SE) which consist in the production of somatic embryos following the typical stages of the embryo's development (from "globular" to "heart" stage). Organogenesis is also studied to improve the efficiency of *Rhizobium radiobacter* gene mediated transformation protocol for the infected explant regeneration. Breeding programs are also focused on the production of homozygous pure lines through androgenesis generating doubled haploid (DH) lines, homozygous at each locus. In contrast, it is possible to enhance the heterozygosity and genetic variability through the somaclonal variation of mutant hybrids which have different traits compared to the parental plant. Protoplast fusion is a technique which allows the production of hybrid and cybrid combinations of species that are sexually incompatible facilitating the intergeneric and the intraspecific hybridization without genetic transformation. In this chapter these techniques have been evaluated and described comparing several methods and results.

9.1 Introduction

During the last decades Brassica plant tissue regeneration has been improved and optimized by organogenesis and somatic embryogenesis utilizing various explants, as such as roots, hypocotyl, cotyledons, leaf petiole, shoots, and inflorescence portions (Cardoza and Stewart 2004). The setting up of in vitro plant regeneration methodologies for *Brassica* species is influenced by several factors such as culture medium composition, temperature, light and photoperiod, organs utilized for explants and

S. Treccarichi · G. F. Rizzo · F. Branca (✉)
Department of Agriculture, Food and Environment (Di3A), University of Catania, 95131 Catania, Italy
e-mail: fbranca@unict.it

S. Treccarichi
e-mail: simone.treccarichi@phd.unict.it

G. F. Rizzo
e-mail: giulio.rizzo@phd.unict.it

C. Kole and T. Mohapatra (eds.), *The Brassica juncea Genome*, Compendium of Plant Genomes,
https://doi.org/10.1007/978-3-030-91507-0_9

genotype which shows different propagation reactivity. Establishment of efficient in vitro regeneration protocols for *Brassica* species is a crucial starting point for improving their economical value also in order to optimize new biotechnological protocols for setting desirable traits of the plant (Ravanfar et al. 2017).

In vitro culture allows selection in shorter time than in conventional breeding, increasing plants homozygous for economic locus by establishing diploaploids or by somaclonal mutants of the donor plant. Plant variability can be enhanced through the somaclonal variation and the epigenetic genetic traits of interest can be detected performing molecular biology techniques, such as the marker assisted selection (MAS). Tissue culture exploits the cellular totipotency which is the ability of a single cell to be able to differentiate into embryonic or extraembryonic cell by the stimulation of environmental conditions and/or hormones.

Conventional breeding is a long and costly process which requires space and work to analyze phenotypes showing interesting economic traits; biotechnologies can reduce time to select improved breeding lines in agronomic traits.

Since some decades Brassica breeding activities have been improved to enhance androgenesis protocols by anther or pollen culture, in order to obtain pure lines (homozygous at each locus) for hybrids' F1 production.

In organic breeding it is not possible to use cytoplasmic male sterility (CMS) lines such as the Ogura-type (Ogura 1968), which was individuated in a rapeseed cultivar and is widely utilized for *Brassica oleracea* F_1 hybrids production such as for broccoli, cauliflower and cabbage (Branca 2008).

CMS system in Brassica has many limitations including the difficult availability of restorer lines (R), vulnerability to pest and diseases, difficulty in maintenance of each line, non-synchronization of flowering among the *Brassica* species, unstable sterility over environments and incomplete fertility restoration (Sinha et al. 2016; Singh et al. 2017). The diploid *Brassica* species are *B. rapa*, *B. nigra*, and *B. oleracea* and they represent the U triangle (Nagaru 1935) model with A, B, and C genomes, respectively. The above cited species show self-incompatibility (SI) while *B. napus* (AACC, $n = 19$), *B. juncea* (AABB, $n = 18$), and *B. carinata* (BBCC, $n = 17$), exhibit self-compatibility (Kitashiba and Nasrallah 2014). SI system is carried out at the surface of stigma epidermal cells through the failure of incompatible pollen grains to germinate producing pollen tubes (Kitashiba and Nasrallah 2014).

In the present chapter the main methods of in vitro tissue cultures and their alternatives have been described focusing on their possible application in breeding programs to enhance commercial, nutraceutical, and technological traits of *B. juncea* oilseed crop.

9.2 Organogenesis

In vitro plant tissue culture can stimulate organogenesis in differentiating cells by the application of exogenous phytohormones and the cells explant competence which influence tissue formation (Sugiyama 1999). In vitro organogenesis processes determines complex cell morphological modification which induces *de novo* tissues production (Thorpe and Biondi 1981).

George and Rao (1980) cultivated hypocotyl segments and cotyledons, excised from 8-day-old seedlings of *B. juncea* var. Rai-5, on Murashige and Skoog's basal medium (Murashige and Skoog 1962) supplemented with 1 mg/L naphthalene acetic acid (NAA), 1 mg/L benzyladenine (BA) and containing 2% sucrose; shoots were produced on the cotyledons while the hypocotyls gave only roots. *B. juncea* var. Rai-5 was subject of other trials in which in vitro yellow seeded variants were selected and analyzed, finding 2% more of oil production than yellow seeds (George and Rao 1983).

In *B. juncea* cv RLK-81–1 Sharma et al. (1990) studied shoot and root regeneration from excised 5 days old seedling cotyledons and they observed the maximum adventitious shoot buds differentiation by Murashige and Skoog medium (MS) containing 5 μM 6-Benzyladenine (6-BA).

Organogenesis developed only in the petiole's cut end in contact with the medium while in the leaf surface there did not occur any organogenesis. (Sharma et al. 1990) also studied the effect of sucrose and glucose on shoot buds differentiation and they observed the best results by using glucose 3% (about 95% of shoots regenerated) and the worst one by using sucrose 1% (about 5% of shoots regenerated quickly turned brown without much further growth). Organogenesis can occur starting from callus arised from hypocotyl and cotyledon explants cultured on MS medium; callus formation carried out mostly in cotyledon explant than to hypocotyle; callus treatment with indole acetic acid (IAA) (0.2 mg^{-1}) and kinetinen (2.0 mg^{-1}) showed significant differences in the callus induction, fresh and dry weight of callus (Ratan et al. 2001).

Bano et al. (2010) evaluated the in vitro regeneration response of cotyledons, hypocotyls, and roots explants of three *B. juncea* genotypes (UCD-635, RL-18 and NIFA-RAYE) in MS medium supplemented with auxins and cytokinins. The hormones concentration (BAP /NAA 0.5 mg L^{-1} and Kin 2.0 mg L^{-1}/IAA 0.2 mg L^{-1} and Kin 3.0 mg L^{-1}/IAA 0.5 mg L^{-1}) and their application played an important role for the differentiation of the callus cells, induced by cotyledons explants, producing shoots efficiently of the UCD-635 and RL-18 genotypes.

Paladi et al. (2017) studied the effect of silver nitrate ($AgNO_3$) which modulates the organogenesis in *B. juncea* inducing high frequency of shoots and several changes in antioxidant defense; transcript profile of hormone synthesis was analyzed signaling the activity of SOD, CAT and EIN2 genes involved in the oxidative stress response and in the ethylene synthesis (Paladi et al. 2017).

Kashyap et al. (2019) studied the effect of the formulated organic medium composed by vermicompost (30%) extracts integrated with 20 mg/L coelomic fluid in combination with 1 mg/L BA and 0.1 mg/L NAA in comparison to standard MS medium; the organic medium showed higher cotyledons' survival rate than MS medium, 98% and 63% respectively.

Cytokinins plays an important role in in vitro hypocotyl callus regeneration; callus explant in MS medium supplemented with 98.9 μmol/L,2,4-dichlorophenoxyacetic acid (2,4-D) and 4.5×10^{-3} μmol/L kinetin, induced the upregulation of transcriptomic factors involved in cytokinins pathways (ARR genes) and in others plant hormone signaling pathways such as auxins, ethilens, and brassinosteroids (Lu et al. 2020).

Brassica juncea shoot regeneration was studied by *Rhizobium radiobacter* gene transformation which needs the tumor tissue stem's explant regeneration; *R. radiobacter* gene mediated transformation requires an efficient tissue regeneration protocol and several authors studied the effect of the concentration and combination of different hormones in MS medium for shoot regeneration from callus (Mathews et al. 1985, 1990; Guo et al. 2005). Cotyledons provided high shoot regeneration frequency than leaves, 67.9% and 52.4% respectively, using 4.54 μM of cytokinin-like herbicide thidiazuron (TDZ) combined with 5.37 μM NAA in the medium; shoots induction and differentiation in all cytokinins used was enhanced in combination of auxin (Guo et al. 2005) and further studies have been conducted to enhance the regeneration from cotyledons preparing explants for *R. radiobacter* genetic transformation (Bhuiyan et al. 2011).

Pental et al. (1993) studied the effect of genotype on shoots regeneration from hypocotyl explants comparing 12 *B. juncea* cultivars grown in different regions of the world (India, Australia and Commonwealth of Independent States—CIS) used for the *R. radiobacter*. mediated genetic transformation; the Indian cultivar RLM 198 showed higher in vitro performances than the other cultivars in MS medium supplemented with BAP 1 mg/1 and 2,4-D 0.05 mg/L stained with 20 μM of $AgNO_3$. The $AgNO_3$ effect was also studied by analyzing in vitro shoot regeneration from several kinds of explants including cotyledons, hypocotyl, petiole obtained from three different *B. juncea* genotypes (I39/1, AB79/1, J99). From each explant the transverse thin cell layers (tTCLs) (400–500 μm thin) were isolated; tTCLs of cotyledons gave the best results from the formed callus regeneration, followed by petiole and hypocotyl (Aoun et al. 2008).

High regeneration efficiency is the main focus for *R. radiobacter* mediated transformation; *B. juncea* (Aari canola) showed about 70% efficiency from cotyledon explant (Fig. 9.1) (Farooq et al. 2019). *R. radiobacter* mediated transformation requires the optimization of regeneration protocol to produce several shoots from induced calli (the best result was 87% callus frequency on MS supplemented with 4 μM BA, after 4 weeks) to insert interesting traits into *B. juncea* genome. Transformation occurred in *R. radiobacter* binary vector plasmid having *Euonymusalatus* diacylglycerol acetyltransferase (*Ea*DAcT) involved in the fatty acids' modulation and *bar* gene, carrier of herbicide resistance to select transgenic recombinant mutants (Naeem et al. 2020).

Organogenesis performing protocol can also provide a valid tool to screen and investigate biotic stress resistance such as white rust disease, caused by *Albugo candida*; *B. juncea* resistant mutant genotypes (C66 and C69) were compared with the susceptible *B. ca juncea* var. Varuna using various hormonal compositions (Chatterjee et al. 2021).

9.3 Somatic Embryogenesis

Somatic embryogenesis (SE) is a process derived from non-zygotic cells without vascular connection with the original tissue which evolves a bipolar structure resembling a zygotic embryo through a series of typical stages (Fig. 9.2) (von Arnold et al. 2002). SE can occur directly from the removed cells or indirectly through the callus formation (Williams and Maheswaran 1986). Microspores, ovules, zygotic and somatic embryos, and seedlings represent the most used target involved in the SE process. In *Brassica* genus several researches have been conducted aiming to reduce time-consuming and costly manipulation of the explants cultivated through the organogenesis process (Lichter 1989). SE stages can be categorized to five steps which are:

1. Initiation of embryogenic cultures by culturing the starting explant on medium supplemented with plant grown regulators (PGRs).
2. Proliferation of embryogenic cultures on medium supplemented with PGRs.
3. Development of somatic embryos in medium without PGRs to inhibit proliferation and stimulate somatic embryo formation and early development.

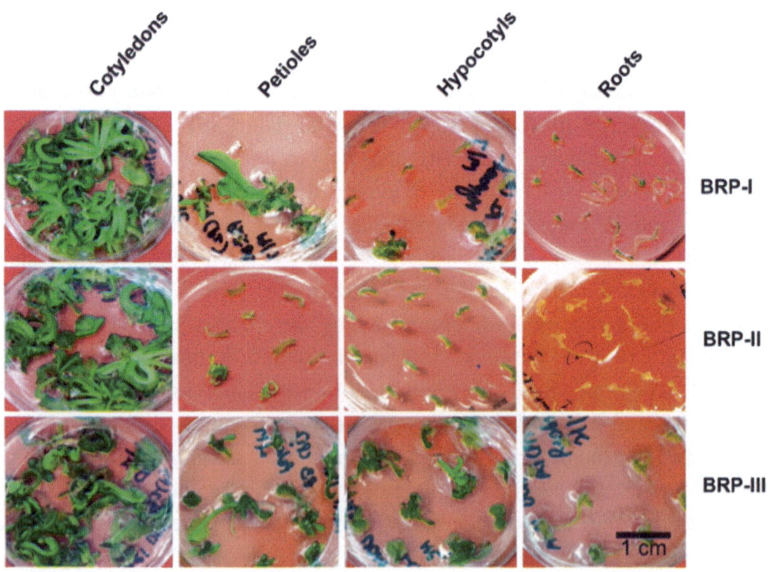

Fig. 9.1 Regeneration of *B. juncea* different from different explants (Farooq et al. 2019)

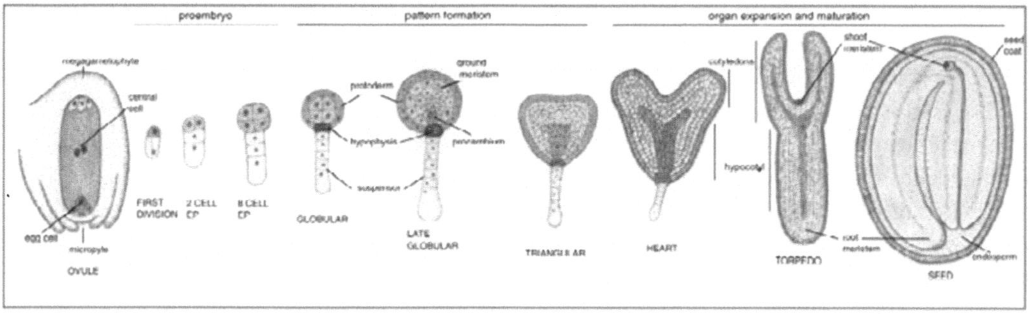

Fig. 9.2 Schematic overview of angiosperm somatic embryo development. Picture was taken from von Arnold et al. (2002)

4. Maturation of somatic embryos by culturing on medium supplemented with ABA and/or reduced osmotic potential.
5. Development of plants on medium without plant growth regulator (PGRs) (Högberg et al. 2001).

As reported in literature, the genotype is the most important factor for embryogenic frequency of most *Brassica* species. During the last 30 years several authors reported studies to evaluate the regeneration response of the excised organs in *Brassica* species and their response to differentiate into embryos starting from excised somatic cells.

Hypocotyl explants of *B. juncea* var. RL.M-198 were used to analyze a simple regeneration protocol based on MS medium supplemented with 2% sucrose, 0.25 mg L^{-1} 2.4-D, 0.5 mg L^{-1} NAA and BAP; this method provides 31% embryos production and each embryo developed about 25 shoots (Kirti and Chopra 1989). In vitro culture of *B. juncea* could provide individuals, after in vitro selection, show tolerant traits to prevent abiotic stress such as sodium chloride level starting from hypocotyl explants (Kirti et al. 1991a). Calli produced from hypocotyl explants differentiated into somatic embryos and then shoots on the same medium of Kirti and Chopra (1989).

Auxins content can significantly induce developmental patterns interfering with the embryo production in its evolution from globular to transition to "heart" embryo stage (Hadfi et al. 1998).

In *B. juncea* cultivar Pusa Jai Kisan embryogenic calli were induced from hypocotyl segments and cotyledons of in vitro germinated seedlings; the highest embryogenic calli frequency was 98% in cotyledons on MS medium supplemented with 2 mg/L 2,4-D. Somatic embryo germination achieved best results by adding 2.6 mg/L ABA to the embryos development medium. Embryo development stages were checked using transmission electron microscopy (TEM) and scanning electron microscopy (SEM) which allowed histological studies concerning the embryos' bipolar structure produced (Akmal et al. 2011).

Faisal et al. (2021) developed and standardized a protocol for direct somatic embryogenesis and plant regeneration of aseptic seedlings derived from unreaped green seeds of *B. juncea* after 7, 14, 21, or 28 days from pollination; immature seeds were disinfected and placed into Nitsch culture medium fortified with sucrose (3% w/v) followed by fructose and maltose. Protocol application depended on the seeds' age and medium composition was an important factor which influenced the number of embryos. The highest number of embryos from immature seeds was collected after 21 days from pollination observing 12.7 embryos per seedling while after 7 days from pollination 4.1 embryos per seedling. Intern simple sequence repeat (ISSR) PCR analysis was performed to check genetic diversity of regenerated plantlets and the ploidy level and flow cytometry; the comparison between these two methods showed no substantial differences between in vitro somatic embryo-derived plants of *B. juncea* and control plants grown to ex vitro conditions (Faisal et al. 2021).

SE was also checked starting from callus and from cell suspension cultures; the most performing callus induction was obtained with MS medium fortified with 3.0 mg L^{-1} 2,4-D. Further regenerations were obtained supplementing 0.5 mg L^{-1} BAP in combination with 0.5–1.0 mg L^{-1} 2,4-D (Shyam et al. 2021b).

9.4 Anther/Microspore Culture and Doubled Haploids

During the last year breeders have needed to obtain homozygous individuals for each locus providing pure lines which can be crossed and used for breeding programs. In vitro anther cultures were firstly successfully performed in *Daturainnoxia* for which haploid embryos were produced (Guha and Maheshwari 1964). Haploid plant production occurs occasionally in natural conditions (Chase 1963) and this process was discovered in *Zeamays* (one haploid plant each 80,000 plants). Narayanaswamy and George (1982) confirmed the possibility to grow sporophytic cells under in vitro artificial stimulation. Haploid technique through microspore embryogenesis can play a significant role to reach homozygosity reducing time (Keller et al. 1987). Pollen structure is characterized by dimorphism that can produce sporophytic and/or gametophytic organs and its type and stage of development are the key elements for DH individuals production. DH methods and technologies could provide a useful tool in breeding self-incompatible, outcrossing lines, which are common in *B. oleracea* complex species (n = 9). Various techniques for microspore/anther culture were developed in *Brassica* species (Palmer et al. 1996).

Malik et al.(2001) used *B. juncea* cv Pusa Bold as a parent in a series of crosses among *Diplotaxis*spp. genotypes and *B. campestris* to produce cytoplasmic male sterile (CMS) lines and to examine the androgenic plantlets produced from microspore cultures; late season donor plantlets (sown during the flowering month) yielded about 8% higher androgenic anthers than the usual season donor plants. Microspores activation, after anther collection and their incubation at 35 °C in darkness conditions for 5 days, occurred in 4–16 days in the responsive genotypes and in 4–28% anthers. Cell division level was equal in about 80% of the activated microspores using KA-77 medium which induced androgenesis only in *B. juncea.* The addition of glutathione and 30 μM $AgNO_3$ in the medium enhanced the percentage of androgenic anthers.

Brassica juncea cultivar Rajat had a significant response in androgenesis producing around 3500 embryos/100 buds. The microspore embryos germinated on B_5 medium with 2% sucrose after 20 days at 5–10 °C; androgenic plantlets had about 52% survival and 74% fertility (Chanana et al. 2005).

Androgenesis was usually successfully induced by high concentration of colchicine (>10 mg/L) for 24 h (Agarwal et al. 2006). In *B. juncea* var. PR-45 higher temperature complemented with an antiauxin 20 μM PCIB application (p-chlorophenoxyisobutyric acid) improved microspore embryogenesis markedly in comparison to the colchicine treatment showing 586 embryos than 119 embryos in the control from 3 mL of microspore suspension dispensed into 60 × 15 mm petri dishes (Agarwal et al. 2006).

Culture medium composition markedly influences the embryos production; the addition of activated charcoal in NLN medium supplemented with 13% (w/v) sucrose and 10 μM silver nitrate induced high microspore embryogenesis level, ranging from 100 to 405 embryos per Petri dish corresponding to 2700–10,935 embryos per 100 buds; embryos were incubated at 4 °C for 10 days and plant microspores produced (mostly haploids) were treated with colchicine having 70% survival and 75% chromosome doubling frequencies (Prem et al. 2008).

The *B. juncea* doubled haploid (DH) plants in comparison to their original parents can achieve improvement in agronomic traits such as plant height and seeds yield and also in the oilseed content and erucic acid concentration (Ali et al. 2009).

9.5 Somatic Protoplast Fusion

Protoplast fusion is a technique which allows the setting up of hybrid and cybrid combinations of species that are sexually incompatible facilitating the gene transfer in sexually incompatible species, to another without genetic transformation. This technology has allowed not only intrageneric hybridizations, but the production of intergeneric hybrids and cybrids (Cardoza and Stewart 2004). Protoplast cell fusion process is also used to integrate cytoplasmic traits by modifying organelle combination, such as chloroplasts and mitochondria and their organellar DNA (Pelletier 1986).

In *Brassica* crops protoplast fusion occurred successfully in *B. rapa* and *B. oleracea* for the production of somatic hybrids to transfer biotic stress resistance genes involved in the bacterial soft rot disease caused by *Erwiniacarotovora* subsp. *carotovora*. Fusion of protoplasts was confirmed by selecting recombinant hybrids through OPA-01 primers (Ren et al. 2000).

Interspecific hybridization could be achieved through mesophyll protoplast fusion using polyethylene glycol to produce interspecific hybrids between *B. juncea* cv. RLM-198 and *B. spinescens*. Hypocotyl derived protoplasts were fused and fusion products were microscopically identified. Putative somatic hybrids were of intermediate phenotype of the parents (Kirti et al. 1991b).

Intergeneric somatic hybridization has been conducted by protoplast fusion originating "*Brassicamoricanda*" hybrids from *B. oleracea* var. *capitata* ($2n$ = 18, CC) (red cabbage) and *Moringacampestris* ($2n$ = 28, MM) as parental, donor line; protoplasts were isolated from *B. oleracea* var. *capitata* hypocotyl calli and from *Moricandia arvensis* cells suspension. (Toriyama et al. 1987). Intergeneric somatic hybridization has been performed to produce hybrids between *B. juncea* ($2n$ = 36, AABB) cv. RLM-198 and *M. arvensis* obtaining shoots with 5% fusion frequency; somatic hybrid pollen was sterile and the *siliqua* were produced by backcrossing with *B. juncea* (Kirti et al. 1992). *Diplotaxiscatholica* ($2n$ = 18) and *B. juncea* somatic hybrids were produced and confirmed by molecular characterization by specific rDNA nuclear wheat probe 18–25 s to detect *Hind* III fragment which is present in both parental species (Kirti et al. 1995).

Intergeneric somatic hybrids which show fertility can be important generic resources concerning biotic or abiotic stress resistance as reported by Kumari et al. (2018) which produced *B. juncea* × *Sinapisalba* hybrid, confirmed by cytological analysis and using 10 simple sequence repeat (SSR) markers to check the hybridization of the regenerated plants analyzing the parents' polymorphic amplicons. Somatic hybrids were screened and selected for their *Alternariabrassicae* resistance and for heat stress tolerance. Further studies were conducted to characterize *B. juncea* and *Sinapisalba* somatic hybrid screening *Sclerotinia sclerotiorum* disease resistance and large number of backcross populations (BC) were produced and BC1 progeny were morphologically very similar (Kumari et al. 2020).

9.6 Somaclonal Variation

Somaclonal variation is an unpredictable in natural process which induces variation in regenerated plants and it can include heritable (genetic) and non-heritable (epigenetic) traits which can be useful compared to the physical and chemical mutagens applied for crops improvement (Jain et al. 1998). In vitro induced somaclonal variation shows cytological abnormalities, phenotypic mutation and changes in genetic regulation through the activation, silencing or mutation in genes involved in qualitative and quantitative traits; epigenetic variation could occur in tissue culture process and quiescent transposable elements and retrotransposons activation indicates this process. Heterochromatin modification is the basis of the phenotypic variation through the modulation of gene function (Kaeppleret al. 2000). Sexual hybridization through protoplast fusion has been tested to bypass genetic barriers among Brassicaceae species related to their fertility, such as chromosome pairing and new nuclear-cytoplasmic combinations, obtaining somatic hybrids and cybrids (Glimelius 1999). Tissue culture exploits the efficiency of mutagenic treatment to induce somaclonal variations; In vitro culture can speed up the breeding program by generating variability, followed by selection (after preliminary screening) and propagation of the desired genotypes (Lestari 2000).

Several of these variations are transient, physiological, and developmental changes while others are a result of epigenetic changes which can induce specific mutations. The variations recovered from somatic tissue cultures, referred to as somaclonal variation which can be exploited in crop improvement (Bhojwani and Dantu 2013).

Somaclonal variation was used in Brassica genotypes identifying male sterile somaclones in *B. juncea*, *B. napus*, and *B. carinata*; in vitro selection was adopted to select mutants having high salt tolerance through the culture in MS medium supplemented with 0.5% sodium and potassium chloride salts for the callus embryogenesis and 1% salts for the stem segments of regenerated plants (Chopra et al. 1989).

Somaclonal variation was observed in *B. juncea* cultivar Prakash in the phenotyping of multiple shoots formation in cotyledonary callus. The main phenotypical variation detected included the number of branches, plants height, siliqua number, 1000 seeds weight and oil content (Jain et al. 1989).

Brassica juncea cultivars RLM 198, RLM 514, RLM 619, and RL 1359 were evaluated for their somaclonal variation and RLM 198 was the best performer showing about 80% average in the cotyledons regeneration process; heritable somaclonal variations were screened for their agronomic traits such as 1000 seeds weight, in vitro ed in vivo germination percentage and speed and seedlings height (Bhatia et al. 2001).

Abiotic stresses tolerance is an important perspective for breeding programs related to the environmental control and phytoremediation, selecting somaclonal mutants having important traits useful for reducing heavy metals' accumulation in the soil; *B. juncea* mutants having enhanced Pb and Cd accumulation were selected and regenerated from metal callus tolerant cells developed in MS medium supplemented with 10–200 μM Cd or Pb (Nehnevajova et al. 2007).

Molecular markers such as random amplified polymorphic DNA (RAPD) and ISSR can be used to detect genetic variation in *B. junceain* in vitro regenerated plants; no difference between banding patterns has been reported comparing regenerated with parental lines (Verma et al. 2016).

Erucic acid content is an important aspect to take into consideration for technological properties related to the oilseed production. Somaclonal variation can be a source of additional variability in the biochemical pathways related to the fatty acids content and accumulation in *B. juncea* PM30 and CS54 genotypes (Shyam et al. 2021a). Putative somaclones of CS54 genotype had 5.48 and 5.52% erucic acid in R0 and R1 regenerants compared to the highest value of the mother plant (41.36%); PM30 regenerant showed a complete absence of erucic acid (Shyam et al. 2021a).

References

Agarwal PK, Agarwal P, Custers JBM Liu C, Bhojwani (2006) PCIB an antiauxin enhances microspore embryogenesis in microspore culture of *Brassica juncea*. Plant Cell Tiss Org Cult 86: 201–210.https://doi.org/10.1007/s11240-006-9108-0

Akmal M, Nafis T, Mirza K, Alam P, Mohammad A, Mujib A, Abdin M (2011) High frequency somatic embryogenesis in mustard crop (*Brassica juncea* L. Cv. Pusa Jai kisan): microscopic and histological analyses. Aust J Crop Sci 5

Ali M, Mian MK, Custers J, Khurram M (2009) Microspore culture and the performance of microspore derived doubled haploid in *Brassica juncea* (L.). Bang J Agric Res 33(4): 571–578. https://doi.org/10.3329/bjar.v33i4.2290

Aoun M, Charles G, Hourmant A (2008) Micropropagation of three genotypes of Indian mustard [{*Brassica juncea*} (L.) Czern.] using seedling-derived transverse thin cell layer (tTCL) explants. https://arxiv.org/abs/0807.1061

Bano R, Khan MH, Khan RS, Rashid H, Munir I, Swati ZA (1990) Callogenesis and organogenesis in three genotypes of *Brassica juncea* affected by explant source. Pak J Bot 42(6): 3925-3932, 2010

Bhojwani SS, Dantu PK (2013) Somaclonal variation. In: Bhojwani SS, Dantu PK (eds) Plant tissue culture: an introductory text. Springer, India. https://doi.org/10.1007/978-81-322-1026-9_12

Branca F (2008) Cauliflower and broccoli. In: Prohens J, Nuez F (eds) Vegetables, vol I. Springer, New York, pp 147–182

Bhatia D, Singh S, Cheema G (2001) Evaluation of plant regeneration and somaclonal variation in *Brassica juncea* L. (Coss and Czern). Cruciferae Newsl 23:23–24

Bhuiyan MSU, Min SR, Jeong WJ, Sultana S, Choi KS, Lim YP, Song WY, Lee Y, Liu JR (2011) An improved method for *Agrobacterium*-mediated genetic transformation from cotyledon explants of *Brassica juncea*. Plant Biotechnol 28:17–23

Cardoza V, Stewart CN (2004) Brassica biotechnology: progress in cellular and molecular biology. In Vitro Cell Dev Biol Plant 40:542–551. https://doi.org/10.1079/IVP2004568

Chanana NP, Dhawan V, Bhojwani SS (2005) Morphogenesis in isolated microspore cultures of *Brassica juncea*. Plant Cell Tiss Org Cult 83:169–177 (2005). https://doi.org/10.1007/s11240-005-4855-x

Chase SS (1963) Androgenesis—its use for transfer of maize cytoplasm. J Hered 54:152–158

Chatterjee A, Nirwan S, Sharma P, Agnihotri A, Shrivastava N (2021) Investigating the in vitro regeneration vigour of *Brassica juncea* var. varuna and its white rust tolerant mutant genotypes: conjugating the conventional and biotechnological approaches. Plant Cell Biotechnol Mol Biol 22(9–10):122–135

Chopra VL, Narashimulu SB, Kirti PB, Prakash S, Anuradha G (1989) Studies on somaclonal variation in *Brassica* sp. and its relevance to improvement to stress tolerance and yield. In: Mujeeb-Kazi, Sitch LA (eds) Review of advances in plant biotechnology 1985–1988, CIMMYT, Mexico and IRRI, The Philippines, pp 220–238

George L, Rao PS (1980) In vitro regeneration of mustard plants (*Brassica juncea* var. RAI-5) on cotyledon explants from non-irradiated, irradiated and mutagen-treated seed. Ann Bot 46(1):107–112. https://doi.org/10.1093/oxfordjournals.aob.a085886

George L, Rao PS (1983) Yellow-seeded variants in in vitro regenerants of mustard (*Brassica juncea* coss var. Rai-5). Plant Sci Lett 30(3): 327–330. ISSN 0304-4211. https://doi.org/10.1016/0304-4211(83)90171-2

Glimelius K (1999) Somatic hybridization. In: Gómez-Campo C (ed) Biology of Brassica coenospecies. Elsevier Science BV, Amsterdam, pp 107–148

Guha S, Maheshwari SC (1964) In vitro production of embryos from anthers of *Datura*. Nature 204:497

Guo DP, Zhu ZJ, Hu XX et al (2005) Effect of cytokinins on shoot regeneration from cotyledon and leaf segment of stem mustard (*Brassica juncea* var. *tsatsai*). Plant Cell Tiss Org Cult 83:123–127. https://doi.org/10.1007/s11240-005-3799-5

Faisal M, Abdel-Salam EM, Alatar AA, Qahtan AA (2021) Induction of somatic embryogenesis in *Brassica juncea* L. and analysis of regenerants using ISSR-PCR and flow cytometer. Saudi J Biol Sci 28 (1):1147–1153. https://doi.org/10.1016/j.sjbs.2020.11.050. Epub 2020 Nov 21. PMID: 33424410; PMCID: PMC7783785

Farooq N, Nawaz MA, Mukhtar Z, Ali I, Hundleby P, Ahmad N (2019) Investigating the in vitro regeneration potential of commercial cultivars of Brassica. Plants (basel) 8(12):558. https://doi.org/10.3390/plants8120558.PMID:31795525;PMCID: PMC6963692

Hadfi K, Speth V, Neuhaus G (1998) Auxin-induced developmental patterns in *Brassica juncea* embryos. Development 125(5):879–887. https://doi.org/10.1242/dev.125.5.879

Högberg K-A, Bozhkov PV, Grönroos R, von Arnold S (2001) Critical factors affecting ex vitro performance of somatic embryo plants of *Piceaabies*. Scand J Res 16:295–304

Jain RK, Sharma DR, Chowdhury JB (1989) High frequency regeneration and heritable somaclonal variation in *Brassica juncea*. Euphytica 40:75–81. https://doi.org/10.1007/BF00023300

Jain SM, Brar DS, Ahloowalia BS (eds) (1998) Somaclonal variation and induced mutations in crop improvement. Kluwer Academic Publishers, UK

Kaeppler SM, Kaeppler HF, Rhee Y (2000) Epigenetic aspects of somaclonal variation in plants. In: Matzke MA, Matzke AJM (eds) Plant gene silencing. Springer, Dordrecht. https://doi.org/10.1007/978-94-011-4183-3_4

Kashyap S, Tharannum S, Taarini R (2019) Influence of formulated organic plant tissue culture medium in the shoot regeneration study of *Brassica juncea* (L.)—Indian mustard. J Plant Biotechnol 46(2):114–118. https://doi.org/10.5010/jpb.2019.46.2.114

Keller WA, Arnison PG, Cardy BJ (1987) Haploids from gametophytic cells—recent developments and future prospects. In: Green CE, Somers DA, Hackett WP, Biesboer DD (eds) Plant tissue and cell culture. Proceedings of the 6th international congress of plant tissue and cell culture. Alan R Liss, New York, pp 223–241

Kirti PB, Chopra VL (1989) A simple method of inducing somatic embryogenesis in *Brassica juncea* (L.) Czern & Coss. Plant Breed 102:73–78. https://doi.org/10.1111/j.1439-0523.1989.tb00317.x

Kirti PB, Hadi S, Kumar PA et al (1991a) Production of sodium-chloride-tolerant *Brassica juncea* plants by in vitro selection at the somatic embryo level. Theor Appl Genet 83:233–237. https://doi.org/10.1007/BF00226256

Kirti PB, Prakash S, Chopra VL (1991b) Interspecific hybridization between *Brassica juncea* and *B. spinescens* through protoplast fusion. Plant Cell Rep 9:639–642

Kirti PB, Narasimhulu SB, Prakash S et al (1992) Somatic hybridization between *Brassica juncea* and *Moricandia arvensis* by protoplast fusion. Plant Cell Rep 11: 318–321 (1992). https://doi.org/10.1007/BF00235090

Kirti PB, Mohapatra T, Khanna H, Prakash S, Chopra VL (1995) *Diplotaxis catholica* + *Brassica juncea* somatic hybrids: molecular and cytogenetic characterization. Plant Cell Rep 14:593–597

Kitashiba H, Nasrallah JB (2014) Self-incompatibility in Brassicaceae crops: lessons for interspecific incompatibility. Breed Sci 64(1):23–37. https://doi.org/10.1270/jsbbs.64.23.PMID:24987288;PMCID: PMC4031107

Kumari P, Bisht DS, Bhat SR (2018) Stable, fertile somatic hybrids between Sinapis alba and *Brassica-juncea* show resistance to *Alternaria brassicae* and heat stress. Plant Cell Tiss Org Cult 133:77–86 (2018). https://doi.org/10.1007/s11240-017-1362-9

Kumari P, Singh KP, Bisht D et al (2020) Somatic hybrids of *Sinapis alba* + *Brassica juncea*: study of backcross progenies for morphological variations, chromosome constitution and reaction to *Alternaria brassicae*. Euphytica 216:93. https://doi.org/10.1007/s10681-020-02629-3

Lestari EG (2016) Combination of somaclonal variation and mutagenesis for crop improvement. J AgroBiogen 8(1):38–44

Lichter R (1989) Efficient yield of embryoids by culture of isolated microspores of different Brassicaceae species. Plant Breed 103:119–123

Lu H, Xu P, Hu K et al (2020) (2020) Transcriptome profiling reveals cytokinin promoted callus regeneration in *Brassica juncea*. Plant Cell Tiss Org Cult 141:191–206. https://doi.org/10.1007/s11240-020-01779-5

Malik MR, Rangaswamy N, Shivanna K (2001) Induction of microspore embryos in a CMS line of *Brassica juncea* and formation of the androgenic plantlets. Euphytica 120:195–203. https://doi.org/10.1023/A:1017502113066

Mathews VH, Bhatia CR, Mitra R, Krishna TG, Rao PS (1985) Regeneration of shoots from *Brassica juncea* (Linn.) Czern and Coss cells transformed by *Agrobacterium tumefaciens* and expression of nopaline dehydrogenase genes. Plant Sci 39:49–54

Mathews VH, Bharathan N, Litz RZ, Rao PS, Bhatia CR (1990) Transgenic plants of mustard *Brassica juncea* L. czern and coss. Plant Sci 72:245–252

Murashige T, Skoog F (1962) A revised medium for rapid growth and bioassay with tabacco tissue culture. Physiol Plant 15:473–497

Naeem I, Munir I, Durrett TP et al (2020) Feasible regeneration and *Agrobacterium*-mediated transformation of *Brassica juncea* with Euonymus alatus diacylglycerol acetyltransferase (EaDAcT) gene. Saudi J Biol Sci 27(5):1324–1332. ISSN 1319-562X. https://doi.org/10.1016/j.sjbs.2019.12.036

Nagaharu U (1935) Genome Analysis in Brassica with Special Reference to the Experimental Formation of B. Napus and Peculiar Mode of Fertilization. Jpn J Bot 7:389–452

Narayanaswamy S, George L (1982) Anther culture. In: Johri BM (ed) Experimental embryology of vascular plants. Springer, New York, pp 79–103

Nehnevajova E, Herzig R, Erismann KH et al (2007) In vitro breeding of *Brassica juncea* L. to enhance metal accumulation and extraction properties. Plant Cell Rep 26:429–437. https://doi.org/10.1007/s00299-006-0264-9

Ogura H (1968) Studies on the new male sterility in Japanese radish with special reference to the utilization of this sterility towards the practical raising of hybrid seeds. Mem Fac Agric Kagoshima Univ 6:39–78

Paladi RK, Rai AN, Penna S (2017) Silver nitrate modulates organogenesis in *Brassica juncea* (L.) through differential antioxidant defense and hormonal gene expression. Sci Hort 226: 261–267. ISSN 0304-4238. https://doi.org/10.1016/j.scienta.2017.08.038

Palmer CE, Keller WA, Arnison PG (1996) Experimental haploidy in *Brassica* species. In: Jain SM, Sopory SK, Veilleux RE (eds) In vitro haploid production in higher plants, vol 2. Kluwer Academic Publishers, Dordrecht, pp 143–172

Pelletier G (1986) Plant organelle genetics through somatic hybridization. Oxford Surv Plant Mol Cell Biol 3:97–121

Pental D, Pradhan AK, Sodhi YS et al (1993) Variation amongst *Brassica juncea* cultivars for regeneration from hypocotyl explants and optimization of conditions for *Agrobacterium*-mediated genetic transformation. Plant Cell Rep 12:462–467. https://doi.org/10.1007/BF00234713

Prem D, Gupta K, Sarkar G et al (2008) (2008) Activated charcoal induced high frequency microspore embryogenesis and efficient doubled haploid production in *Brassica juncea*. Plant Cell Tiss Org Cult 93:269–282. https://doi.org/10.1007/s11240-008-9373-1

Ratan S, Ranwah BR, Singh RP (2001) Tissue culture studies in Indian mustard (*Brassica juncea* L. Czern & Coss). Indian J Agric Res 35:259–262

Ravanfar SA, Orbovic V, Moradpour M, Aziz MA, Karan R, Wallace S, Parajulis S (2017) Improvement of tissue culture, genetic transformation, and applications of biotechnology to Brassica. Biotechnol Genet Eng Rev 33(1):1–25

Ren JP, Dickson MH, Earle ED (2000) Improved resistance to bacterial soft rot by protoplast fusion between *Brassica rapa* and *B. oleracea*. Theor Appl Genet 100:810–819

Sharma KK, Bhojwani SS, Thorpe TA (1990) Factors affecting high frequency differentiation of shoots and roots from cotyledon explants of *Brassica juncea* (L.) Czern. Plant Sci 66:247–253

Sugiyama M (1999) Organogenesis in vitro. Curr Opin Plant Biol 2:61–64

Shyam C, Tripathi MK, Tiwari S, Tripathi N, Solanki RS, Sapre S, Ahuja A, Tiwari S (2021a) In vitro production of somaclones with decreased Erucic acid content in Indian mustard [*Brassica juncea* (Linn.) Czern & Coss]. Plants (Basel) 10(7):1297. https://doi.org/10.3390/plants10071297. PMID: 34202297; PMCID: PMC8309134

Shyam C, Tiwari S, Ahuja A, Tripathi N, Gupta N, Gupta V (2021b) In vitro regeneration from callus and cell suspension cultures in Indian mustard [*Brassica juncea* (Linn.) Czern & Coss]. Intl J Agric Technol 17:1095–1112

Singh RP, Agarwal DK, Sripathy KV, Prasad SR, Kumar SPJ (2017) Varietal and seed replacement in the era of climate change. ICAR-Indian Institute of Seed Science, p 42

Sinha AK, Agarwal DK, Kumar SPJ, Tiwari TN, Chaturvedi AK (2016) Novel technique for precluding hybrid necrosis in bread wheat. Intl J Trop Agri 34 (3):761–765

Thorpe TA, Biondi S (1981) Regulation of plant organogenesis. Advances in cell culture, vol 1. Elsevier, pp 213–239. ISSN 0275-6358, ISBN 9780120079018, https://doi.org/10.1016/B978-0-12-007901-8.50013-X

Toriyama K, Hinata K, Kameya T (1987) Production of somatic hybrid plants, "Brassicomoricandia" through protoplast fusion between *Moricandia arvensis* and *Brassica oleracea*. Plant Sci 48:123–128

Verma V, Thakur AK, Singh BK, Chauhan DK, Singh KH, Chauhan JS (2016) Assessment of genetic fidelity in in vitro regenerated plants of *Brassica juncea* (L.) Czern & Coss. using RAPD and ISSR markers. Indian J Biotechnol 15(1):120–123

von Arnold S, Sabala I, Bozhkov P, Dyachok J, Filonova L (2002) Developmental pathways of somatic embryogenesis. Plant Cell Tiss Org Cult 69:233–249. https://doi.org/10.1023/A:1015673200621

Williams EG, Maheswaran G (1986) Somatic embryogenesis: factors influencing coordinated behavior of cells as an embryogenic group. Ann Bot 57:443–446

Genetic Modification of *Brassica juncea*: Current Scenario and Future Prospects

10

Anjana Rustagi, Ravi Rajwanshi, Deepak Kumar, Neelam Prabha Negi, Paomipem Phazang, Mohd Aslam Yusuf, Arun Vincent Kisku, Nishakant Pandey, Shashi Shekhar, Prerna Chaudhary, Subhash Chandra Prasad, Suchandra Deb Roy, Mukesh Saxena, and Neera Bhalla Sarin

Abstract

In recent years, the development of transgenic *Brassica juncea* with enhanced tolerance to biotic and abiotic stress conditions has been reported by several research groups worldwide. Herein, we present an exhaustive compilation of the studies published on this subject, with a special focus on long noncoding RNAs and development of regulatory genes of *B. juncea* as potential candidates for alleviation of stress. The transgenic approaches for abiotic stress mitigation and disease resistance have also been discussed. The potential of genetically modified *B. juncea* plants as edible vaccines has also been touched upon.

All the authors contributed equally to this chapter.

A. Rustagi
Department of Botany, Gargi College, New Delhi, India

R. Rajwanshi
Discipline of Life Sciences, School of Sciences, Indira Gandhi National Open University, New Delhi, India

D. Kumar
Department of Botany, Institute of Science, Banaras Hindu University, Varanasi, India

N. Prabha Negi
University Institute of Biotechnology, Chandigarh University, Mohali, India

P. Phazang · M. A. Yusuf · A. V. Kisku · N. Pandey · S. Shekhar · P. Chaudhary · S. C. Prasad · S. D. Roy · M. Saxena · N. B. Sarin (✉)
School of Life Sciences, Jawaharlal Nehru University, New Delhi, India

10.1 Introduction

Brassica juncea, Indian mustard, is an economically and industrially important oilseed crop. In India, it is grown predominantly in the states of Gujarat, Haryana, Madhya Pradesh, Rajasthan, Uttar Pradesh, as well as in the South Indian states of Andhra Pradesh, Karnataka, and Tamil Nadu and contributes 28.6% to the total oilseed production (Shekhawat et al. 2012). Primarily cultivated for oil, *B. juncea* has been put to myriad applications in the food and chemical industry and is also used as a biofertilizer. Mustard oilcake is a very nutritious feed for livestock animals. India is the largest exporter of mustard seed meal globally (Thakur et al. 2020). The vegetative parts (leaves) of *B. juncea* plants are consumed as a nutritional delicacy "saag" in North India. Mustard oil is rich in antioxidants (Tian and Deng 2020). It has very good combustion and lubrication properties and is,

C. Kole and T. Mohapatra (eds.), *The Brassica juncea Genome*, Compendium of Plant Genomes,
https://doi.org/10.1007/978-3-030-91507-0_10

therefore, a suitable material for biodiesel production (Fadhil et al. 2020).

The production of mustard crop is severely affected by biotic (leaf blight, white rust, stem rot, powdery mildew, downy mildew, black rot, aphids) and abiotic (drought, salinity, temperature extremes) stress conditions (Grover and Pental 2003; Dutta et al. 2005; Ullah et al. 2012). To meet the ever increasing demand of edible oil, scientists are engaged in developing ways to produce low-input, high-yielding, and better-performing varieties that can withstand environmental fluctuations and expand production across marginal agro-climatic conditions. The main strategies currently adopted for improvement of *Brassica* species are molecular breeding and genetic transformation. Although, more than hundred cultivars of Indian mustard have been developed using the conventional plant breeding program (Chauhan et al. 2011), this approach has fallen short of bringing any tangible breakthrough in certain areas. Conventional breeding has limitations of being time and labor intensive and is often hindered by interspecific barriers with potential of introducing undesirable genetic recombination events and subsequent yield drag. Moreover, there are limited sources of resistance against aphids, Alternaria blight, downy mildew, powdery mildew, stem rot, and *Orobanche* in the *Brassica* germplasm. Most of the pests and diseases are currently managed using insecticides and fungicides, which are expensive, environmentally unsustainable, and undesirable for the consumers. These approaches are also insufficient in combating the new emerging pathotypes. Genetic engineering techniques have been employed for the development of transgenic *B. juncea* germplasm with agronomically desirable traits. Advances in genetic engineering have helped in genetic amendments for integration of value-added traits, surmounting the interspecies, and reproductive barriers.

Molecular responses of plants to environmental cues, both biotic and abiotic, and bearing of these cues on both vegetative and reproductive fitness have been extensively studied. A number of genes participating in such molecular signaling events have been identified, isolated, and modified, and their regulation and function have been investigated. These genes can be easily introduced into a defined recipient for obtaining genetically modified plants with desired traits. Transgenic approaches are valuable for better and sustainable production of *B. juncea*. An efficient and facile in vitro regeneration system are a prerequisite for genetic transformation of plants. Transgenic plants of *Brassica* spp. expressing genes of interest have been generated using different transgenesis methods, for instance via *Agrobacterium tumefaciens*-mediated transformation (Naeem et al. 2020), biolistics (Chen and Beverdorf 1992), and protoplast electroporation (Bergman and Glimelius 1993). The *Agrobacterium*-mediated genetic transformation using hypocotyl explants is the most widely used protocol, which is preferred for its simplicity and cost effectiveness (Barfield and Pua 1991; Pental et al. 1993; Dutta et al. 2005; Das et al. 2006; Yusuf and Sarin 2007; Roy et al. 2009; Shen et al. 2013). This method has been used to transfer many agronomically relevant genes to *B. juncea* for phytoremediation of heavy metals (Zhu et al. 1999), herbicide resistance (Mehra et al. 2000), salt tolerance (Zhang et al. 2001), hybrid seed production (Jagannath et al. 2002), oil quality improvement (Hong et al. 2002; Das et al. 2006), and aphid resistance (Kanrar et al. 2002a, b; Dutta et al. 2005).

In the present chapter, we have elaborated upon various tools and techniques for genetic transformation of *B. juncea* for biotic and abiotic stress tolerance. The application and prospects of the newly developed clustered regularly interspaced short palindromic repeats and CRISPR-associated protein 9 (CRISPR/Cas9)-mediated genome editing tool for crop improvement have also been included. Plants also counter biotic and abiotic stresses by altering the expression of specific transporters or through modulation of epistatic interaction of interlinked genes in different pathways (Yu et al. 2012); we have described developmentally regulated genes of *B. juncea* and their plausible roles in genetic modification. The involvement of long noncoding ribonucleic acids (lncRNAs) in response to different stresses, such as environmental (salinity, drought, high light intensity, temperature extremes), nutritional (nitrogen/phosphate deficiency), hypoxia, and pathogens, has also been elucidated. The potential of transgenic mustard for production of edible vaccines and biopharmaceuticals is also discussed in this chapter.

10.2 Tools and Methods for Genetic Modification in *B. juncea*

Various methods are used to modify the genome of *B. juncea* plants; these include *Agrobacterium*-mediated gene transfer method, floral dip, floral spray, and particle gun bombardment, which are discussed below. The merits and demerits of different methods are depicted in Table 10.1.

10.2.1 *Agrobacterium*-Mediated Gene Transformation

The *Agrobacterium*-mediated gene transfer is the most common method used for genetic transformation of *Brassica* spp. This method has been widely used by scientists to generate stable transgenics (Pental et al. 1993; Rajwanshi and Sarin 2013; Das et al. 2020). Although different tissues, such as cotyledonary leaves, hypocotyls, and pollens, have been used as explants, the hypocotyls are the most suitable explants for tissue culture owing to their amenability to regeneration. Young hypocotyls excised from 3–4-day-old seedlings exhibit optimal regeneration potential (Gasic and Korban 2006). The addition of silver nitrate to the selection medium is imperative to achieve high transformation efficiency.

10.2.2 Floral Dip Method

The floral dip method of transformation is very simple to perform. Floral buds are briefly dipped in a solution of *Agrobacterium*. These buds subsequently mature and set seeds. The transgenic plants are later selected from among the progeny seedlings. Because in vitro regeneration of plant tissue is not required, somaclonal variations can be avoided and the procedure can be performed easily with high success rate. The

Table 10.1 Merits and demerits of different methods used to genetically modify *Brassica juncea*

S. No.	Methods	Merits	Demerits	References
1	*Agrobacterium*-mediated genetic transformation of explants	Easy, minimal equipment and facility required, and less expensive	Chimera formation and somaclonal variation can be performed only in invitro cultures	Eapen (2011), Yusuf and Sarin (2007), Roy et al. (2009)
2	Floral dip method with *Agrobacterium*	Easy to perform; requires less time, overcomes somaclonal variation	Successful only in *B. juncea*	Chhikara et al. (2012) and Wang et al. (2008)
3	Floral spray method with *Agrobacterium*	Easier to perform, tissue culture not required, one successful report in *B. juncea*	Requires mechanical maneuvering of the unopened buds	Aminedi et al. (2019)
4	Particle gun bombardment	Genotype and cultivar independent transformation	Expensive, requires high technical know-how, somaclonal variation, chimera formation, complex DNA, post-transcriptional gene silencing in subsequent generations, multiple copy transgene integration and recombination	Benedikt et al. (1996)

exogenously applied *Agrobacterium* colonizes the deep-seated immature ovaries, wherein cells of female gametophyte are transformed (Clough and Bent 1998). The floral dip transformation method has been harnessed for the improvement of *B. juncea* for phytoremediation (Zhu et al. 1999), herbicide tolerance (Mehra et al. 2000), salt tolerance (Zhang et al. 2001), hybrid seed production (Jagannath et al. 2002), improvement of oil quality (Das et al. 2006; Hong et al. 2002), and aphid resistance (Kanrar et al. 2002a, b; Dutta et al. 2005).

10.2.3 Floral Spray Method

High efficiency (10–30%) of *B. juncea* transformation has been achieved by spraying of *Agrobacterium* cells ($OD_{600} = 1.0$) on floral buds (Aminedi et al. 2019). This method, being simple and highly reproducible, might prove beneficial for functional genomics and transgenic development programs in *B. juncea* and has recently regained priority since the availability of the reference genome sequence of this crop.

10.2.4 Particle Bombardment Method

The particle bombardment device, also known as the gene gun, enables the penetration of the plant cell wall, thereby facilitating the transfer of genetic material containing any gene of interest into the cells. Tungsten or gold microparticles are coated with plasmid DNA and then accelerated under helium pressure and shot into plant tissue placed on a petri plate, under aseptic conditions (Gan 1989). As the microparticles enter the cells, the transgenes are released from the particle surface, allowing their incorporation into the chromosomal DNA of the cells.

This method has been used to transfer genes into single cells of excised globular-stage zygotic *B. juncea* embryos. Bombardment of embryos under optimal conditions does not affect the normal in vitro embryogenesis. Benedikt et al. (1996) were able to express the firefly luciferase gene (*Luc*) in bombarded embryos without affecting the viability of embryos.

10.3 Regulatory Genes for Growth and Development in *B. juncea* and Their Significance in Genetic Manipulation

Plant development is mediated by cellular signaling and communication integrating both environmental and nutritional cues. Molecular events regulating perception and downstream signaling of these cues are regulated at the transcriptional and post-transcriptional levels. (Byrne et al. 2003; Stahl and Simon 2009). Here, we highlight some of the regulatory genes that have been identified through molecular cloning and classical genetics and are found to play important roles in growth and development of *B. juncea.*

10.3.1 *MYB28* (*BjMYB28*) Gene

Glucosinolates are secondary metabolites that are rich in nitrogen and sulfur. These secondary metabolites are specific to Capparales (*Brassica* spp. and *A. thaliana*). In recent years, these have been classified as important because of their potential in improving human health and agriculture (Fahey et al. 2001; Wittstock and Halkier 2002). Some of these have been found to have antipest and antipathogen properties (Halkier and Gershenzon 2006), whereas others are anticarcinogenic. However, there are others that show antinutritional and goitrogenic properties in seed meal (Fahey et al. 1997; Jugeet al. 2007; Cartea and Velasco 2008; Traka and Mithen 2009; Augustine et al. 2013).

The biosynthesis of glucosinolates occurs in three important phases: (1) recruitment of precursor amino acids and elongation of side chain, (2) formation of the core glucosinolate structure, and (3) side group modification (Halkier and Gershenzon 2006). The generation of a variety of glucosinolate compounds is achieved by side chain elongation of the R-group and side group modification (Clarke 2010). Augustine et al.

(2013) identified four *MYB28* homologs in *B. juncea*. They overexpressed the four *BjMYB28* genes in two different genetic backgrounds of the phylogenetically close model plant, *A. thaliana*. They found that the genes encode functional *BjMYB28*, which positively regulates *MAM1* and *MAM3* and *AtSt5b* and *AtST5c*, which are involved in chain elongation and formation of the core structure, respectively. These findings indicated their role in the biosynthesis of aliphatic (Met-derived) glucosinolates. This identification and characterization of *BjMYB28* as well as understanding of their molecular mechanisms have provided a great boost in transgenic approaches. *Brassica juncea* contains relatively high levels of glucosinolates making them antinutritional and less palatable. Hence, developing low-seed glucosinolate (<30 μmol/g DW) lines (Potts et al. 1999) has been a major objective of research. The RNA interference (RNAi) technology has enabled the development of low-seed glucosinolates lines of *B. juncea* (Augustine et al. 2013). The transgene-based RNAi suppression strategy used against *BjMYB28* led to a significant (11.26 mol/g DW) suppression of aliphatic glucosinolates without compromising on the aromatic glucosinolate pool.

10.3.2 *CO, FT, SOC1*, and *LFY* Genes in *B. juncea*

Flowering is generally regulated by an intricate gene network that is receptive to endogenous and exogenous signals (Andres and Coupland 2012; Cho et al. 2017; van Dijk and Molenaar 2017). The reduction of flowering time in *B. juncea* is an important and desired trait because it allows multiple cropping in a growing season or cropping in the regions with a prohibitively short growing season. Previous studies have indicated *CO*, *FT*, *FLC*, *SOC1*, and *LFY* as key regulators of flowering in plants, and *CO* has been identified as the core gene. They belong to the B-box zinc finger family (*BBX*) and are regulated mainly by circadian clock and light signals. The *CO* initiates flowering through its upstream regulators, *GIGAN-TEA* (*GI*), *FLAVIN-BINDING*, *KELCH REPEAT*, *F-BOX 1* (*FKF1*), *CYCLING DOF FACTOR1* (*CDF1*) and downstream genes, *FLOWER-ING LOCUS T* (*FT*) and *TWIN SISTER OF FT* (*TSF*) (Samach et al. 2000; Fornara et al. 2010; Valverde 2011; Mayee and Singh 2016; Tyagi et al. 2018). When floral development is initiated, the downstream target *FT*, owing to its mobile nature, moves to SAM systemically and binds to the *bZIP* transcription factor, *FD*. This, in turn, activates the *MADS* box transcription factors, *SUPPRESSOR OF OVEREXPRESSION OF CONSTANS 1* (*SOC1*) and *APETALA1* (*AP1*), and accelerates floral transition and patterning (Xu et al. 2012; Wagner 2016; Lee et al. 2004; Liu et al. 2007, 2008). External factors like day length and application of gibberellin significantly upregulate the transcription of *SOC1* (Kim and Sung 2014; Samach et al. 2000; Nakamura et al. 2005; Melzer et al. 2008). Thus, *CO-FT-SOC1* is widely accepted as a major network in photoperiodically regulated flowering in plants.

Genome-wide analyzes of *B. juncea* led to the identification of similar flowering regulator genes. A total of 179 genes were identified of which 84 were mapped on the *B. juncea* A subgenome and 79 on the B subgenome. Expression profiling conducted under long day and vernalization conditions revealed that *BjCO*, *BjFT*, and *BjSOC1* have slightly enhanced expression during the early vegetative stage and peaked in the leaf before flowering (Zhaoet al. 2019). Overall, these findings shed light on the regulation of flowering in *B. juncea*. This new evidence on the *CO-FT-SOC1* network in *B. juncea* helped in manipulating the flowering network through genetic approaches. In a study by Tyagi et al. (2018), ectopic expression of *B. juncea*-specific *SOC1* variants resulted in the formation of multiple vegetative rosettes in *A. thaliana*. Taking cue from these findings, they further performed introgression of *B. juncea SOC1*-specific miRNA, which resulted in floral transition and lateral branching, corroborating the role of *SOC1* in flowering. Furthermore, some meristem identity genes including *LFY* (*LEAFY*), the first gene to be expressed prior to flower development and its mutated form, have the most drastic effects on the floral transition (Schultz and Haughn 1991). After turning on the meristem identity switch, the *LFY* gene has an additional role as a transcription factor involved in the activation of floral homeotic genes that specifically identify the organs in the flower (Huala and Sussex 1992; Weigel et al. 1992).The expression of *LFY* was reported to increase with the age of the plant till it reached a threshold level, which is required for transition from the vegetative to the reproductive phase. Overexpression of *AtLEAFY* in *B. juncea* "Varuna" was shown to accelerate the flowering time by approximately a week earlier compared with that in the wild type *B. juncea* (Roy et al. 2009). *LFY* was isolated and cloned from *B. juncea* and its temporal and spatial expression in different cultivars with differences in flowering time were studied with the aim of overexpressing homologs of *BjLFY* in *B. juncea* (Roy et al. 2011).

10.3.3 Sucrose Synthase (*BjSUS*) Genes in *B. juncea*

Sucrose synthase (SuS) is a glycosyl transferase enzyme that plays a key role in sugar metabolism, primarily in sink tissues (Stein and Granot 2019). Genome-wide analysis led to the identification of 14 *SUS* genes in *B. juncea* (Koramutla et al. 2019). All 14 SUS proteins have the sucrose and glycosyltransferase domains, characteristic of SUS in plants (Huber et al. 1996; Hardin and Huber 2004; Li et al. 2015; Wang et al. 2015). The SUS genes in general are characterized into three subgroups, *SUSI*, *SUSII*, and *SUSIII*. Phylogenetic analysis of *BjSUS* genes showed that they belonged to each of the subgroups, thereby supporting the hypothesis that *SUS* genes are only present in higher plants (Hirose et al. 2008).

Ectopic expression of SUS genes in many crop plants showed that they control important agronomic traits. Transgenic expression of potato *SUS* led to enhanced leaf growth and fiber elongation in cotton; in maize, the transgenic plants showed high levels of starch and ADP-glucose in mature seeds. Loss of function of *SUS* genes in several plant species resulted in altered phenotype. For example, suppression of sucrose synthase reduced fiber and seed development in cotton (Ruan et al. 2008), wood density in *Populus* (Gerber et al. 2014), and callose deposition in the sieve elements of *Arabidopsis* (Paul Barratt et al. 2009). In *B. juncea, BjSUS* showed differential expression patterns across a wide range of tissues. The stems, buds, and young pods showed higher levels of *BjSUS1* and *BjSUS2* genes compared to the young or mature leaves. Thus, the differential expression of the *BjSUS* genes across tissues at different time pointed toward spatial and temporal regulation of these genes. Furthermore, their expression was also found to be activated by external application of MeJ, SA, and ET and mechanical wounding in a time-dependent manner. These findings support the centrality of sucrose in energy metabolism, physiological processes, and stress responses. Thus, identifying the *SUS* genes in *B. juncea* and understanding their expression patterns will enable researchers in exploiting the gene functions for developing *B. juncea* crops with desirable agronomical traits.

10.3.4 Auxin Response Factor (*ARF*) Genes in *B. juncea*

A total of 65 *ARF* genes were identified in *B. juncea* "Tumida." These genes are randomly represented in the A and B subgenomes. Furthermore, gene and protein structure analyzes showed that these were similar to those of *A. thaliana*. The *BjARF* genes showed Auxinresp domain at the N-terminal and the protein sequences showed the presence of 9 motifs conserved in all the 65 *BjARFs* identified. Promoter *cis*-element analysis further showed the involvement of *BjARFs* in the expression of most of the plant hormones (Li et al. 2020). Moreover, differential expression patterns of *BjARF* observed during various stages of growth highlighted their role in many physiological activities. Auxin signaling has been known to be pivotal in the growth and development of plants. This important hormone regulates majority of the biological processes (Li et al. 2006; Roosjen et al. 2018). Thus, identification of ARF genes in *B. juncea* will help in detailed dissection of the gene and protein structures subsequently leading to the understanding of their roles as signal molecules.

10.3.5 TRANSPARENT TESTA 1 (*TT1*) Gene in *B. juncea*

Yan et al. (2010) isolated and cloned the full length *BjTT1* gene. They showed that *BjTT1* contains two exons and one intron and encodes a 300 amino acid-long peptide having a conserved WIP domain and a Zn finger. Its expression was detected in the seed coat of *B. juncea* but no significant difference in expression was reported in the yellow-seeded and black-seeded *B. juncea.* In *A. thaliana*, *AtTT1* is reported to be involved in flavonoid synthesis (von Wettstein et al. 1977; Lepiniec et al. 2006). Given the high homology between *BjTT1* and *AtTT1*, the findings also suggest a similar role in *B. juncea*, which needs to be investigated. Keeping in mind the general role of *WIP* domain Zn finger proteins in eukaryotes, this insight into the gene structure and function of *BjTT1* will help in better understanding of the flavonoid biosynthesis in *B. juncea.*

10.4 Genome Editing for Stress Tolerance

Targeted genome editing technologies, especially the CRISPR/Cas9 system provides accurate and efficient ways to produce crops with high yield under biotic and abiotic stress (Abdelrahman et al. 2018). The CRISPR/Cas9-based nucleotide editing using the cytidine deaminase-mediated base editor (the CBE system) efficiently introduced C to T conversion in rapeseed. In *B. napus*, the CBE-based gene modification at position P197 in acetolactate synthasegene (*BnALS1*) conferred tribenuron-methyl resistance whereas P197S substitution in the same gene generated a herbicide-resistant mutant (Wu et al. 2020). A herbicide-resistant mutant of rapeseed has also been developed by Cheng et al. (2021) using A3A-PBE base-editing system by targeting the *ALS* gene. The CRISPR/Cas9 system was also used to study the molecular mechanism of low boron adaptation in *B. napus* by Feng et al. (2020). Knock-out of transcription factor BnaA9. WRKY47 resulted in increased sensitivity under low boron conditions compared with wild type plants. Jiang et al. (2020) carried out gene silencing using the genome editing system to study the role of *ORPHAN GENE* (OG) in the primary metabolism of soluble sugars in *B. rapa*. The two transgenic lines (BrOG1A and BrOG1B) developed using CRISPR/Cas9-mediated mutation resulted in a decrease in fructose, glucose, and total soluble sugar content due to the upregulation of *BrSUSS*, *BrSUS1b*, and *BrSUS3*. In *Brassica*, *SUS* and *OGs* are known to play roles in the primary metabolism of soluble sugars. Although there is limited information on the functional significance of genes in response to abiotic stress, the role of *OGs* in increased tolerance to osmotic stress and soil drought has been studied in *Vigna unguiculata* under specific environmental conditions (Li et al. 2019).

Resistance to pod shattering in brassicas is important for attaining high yield by preventing the shattering of pods under extreme climatic conditions. Zaman et al. (2019) demonstrated multiplex genome editing using CRISPR/Cas9-based knockout of JAGGED (JAG) gene responsible for regulating fruit dehiscence. Of the five homologs of *JAG* mutants, *BnJAG.A08-NUB* mutant line in rapeseed showed significant changes in pod dehiscence zone. The results showed two fold higher resistance to pod shattering in JAG mutant relative to that in wild type plants. The CRISPR/Cas9 system assisted molecular breeding for developing abiotic stress tolerance has recently been studied in greater depth in *B. napus* (Lohani et al. 2020; Razzaq et al. 2021). With the CRISPR/Cas9 approach, a novel susceptibility factor, CRT1a has been identified in *B. napus* infected with *Verticillium longisporum*. The genome-edited plants appeared to develop normally, because only one active

CRT1a locus was targeted, whereas other CRT1a, CRT1b, and CRT3 loci were not affected (Probsting et al. 2020). Since, *B. juncea* is amphidiploid, there may be two or more than two copies of the same gene in its genome. Hence, the use of genome editing technology in the development of stress-tolerant *B. juncea* still remains a challenge and needs to be explored in future research.

10.5 Abiotic Stress Tolerance in *Brassica* Spp. Using Transgenic Approach

Several research groups have been engaged in engineering plants for multiple abiotic stress tolerance, which is highly challenging owing to the genetically multifaceted mechanism of stress response in plants. Besides the conventional approach of plant breeding to enhance abiotic stress tolerance in plants, genetic engineering for improving stress tolerance in different plant species offers an additional approach (Gill et al. 2012).

10.5.1 Factors Affecting Genetic Transformation in *B. juncea*

Effective regeneration of plants entirely depends on the plant genotype used because the in vitro regeneration technique varies with genotypes (Thakur et al. 2020). Gerszberg et al. (2015) compared cotyledons and hypocotyls, from eight cultivars of *B. oleracea* var. capitata for their in vitro regeneration potential. The study revealed the hypocotyl explants of all cultivars to be the best suited for regeneration. Farooq et al. (2019) also evaluated the regeneration potential of four commercial cultivars of *Brassica*, among which *B. juncea* cv. Aari canola exhibited a higher regeneration frequency compared with *B. napus* cv. Westar. They also examined the regeneration efficiency from cotyledons, hypocotyls, roots, and petioles in the four cultivars and found it to be the highest for petiole and cotyledonary explants. The age of the explants has also been shown to determine the success of the genetic transformation in several studies. Bhuiyan et al. (2011a, b) showed better organogenesis from explants taken from 4 to 6-day-old seedlings grown under in vitro conditions. Several workers have recommended 48 h preculture of explants prior to the cocultivation with *Agrobacterium* for higher transformation frequency (Bhuiyan et al. 2011a, b; Ahmed et al. 2017). The concentration of *Agrobacterium* suspension and infection period has also been described as major determinants of the transformation frequency. Infection of *B. juncea* explants with *Agrobacterium* at a density equivalent to 0.01–0.1, absorbance unit (at 600 nm) for a maximum of 30 min has been established as optimal for transformation (Singh et al. 2010). Use of the Positech system is also recommended by several research groups to enhance the transformation efficiency and capture the low positive selection marker gene expressing lines. Sonntag et al. (2004) obtained a transformation frequency of 7.9% using a much higher concentration of mannose (4.5 gL^{-1}) in combination with sucrose (10 gL^{-1}) for the selection of transgenic *B. napus*. Min et al. (2007) reported a transformation frequency of 1.4–3% for Chinese cabbage using 7 gL^{-1} mannose and 2% sucrose. Rajwanshi and Sarin (2013) obtained a transformation efficiency of 4% using a lower concentration of mannose (0.09 gL^{-1}) in combination with glucose (5.0 gL^{-1}) for the selection of putative *B. juncea* transgenics.

10.5.2 Genetic Transformation of Brassica for Abiotic Stress Tolerance

The selection of a suitable promoter is necessary to regulate the expression of a transgene in transgenic plants for enhanced abiotic stress tolerance without any yield penalty. Different studies showed salinity, drought, and abscisic acid (ABA) treatment for inducing the promoter activity, indicating its contribution to stress signaling (Lohani et al. 2020). It is preferable that the stress tolerance associated genes are overexpressed only under stress conditions. Therefore, using stress-inducible promoters for regulating the expression of candidate genes is preferable for developing stress-tolerant plants. Vacuolar processing enzymes (VPEs) are vacuole localized cysteine proteinases and are structurally and evolutionarily linked to human caspases. The *A. thaliana* genome has four VPE genes, namely *α-VPE*, *β-VPE*, *γ-VPE*, and *δ-VPE*. We found that overexpression of *AtγVPE* in *B. juncea* resulted in improved tolerance to abiotic and biotic stress (unpublished data from author's lab). The *in-silico* analysis of *AtγVPE* promoter showed the presence of numerous *cis*-element and TFs binding motifs that are involved in phytohormone response and signaling, regulation of developmental processes, and stress response (Prasad et al. 2018). The promoter: GUS analysis revealed that *AtγVPE* promoter is an inducible promoter. Kaur et al. (2015) isolated a salinity-associated 713 bp *BjSOS2* promoter from *B. juncea* to study the regulation of *BjSOS2* under different abiotic stress conditions and examined the efficacy of the promoter in genetic engineering. Roy et al. (2008) showed that the stress-inducible responsive to desiccation (*rd29A*) promoter regulates gene expression in transgenic plants. Several studies have revealed that salinity, drought, and high temperature stress induce the *rd29A* and *rd29B* genes (Yamaguchi and Shinozaki 1994). Rajwanshi et al. (2016) overexpressed the glyoxalase Igene (*gly I*) in *B. juncea* plants under the regulation of a constitutive (CaMV 35S) as well as a stress-inducible (*rd29A*) promoter and compared the performance of transgenic *B. juncea* under different abiotic stresses. Regulation of *gly I* by constitutive CaMV 35S promoter showed the transgene to be constitutively expressed under stress and non-stress conditions. The study revealed enhanced salinity, drought, and heavy metal stress tolerance of the transgenic plants compared with the untransformed controls. The transgenic plants with *gly I* driven by *rd29A* promoter (*rd29A:gly I*) exhibited improved performance under 200 mM NaCl, 200 mM mannitol, and 5 mM $ZnCl_2$ stress compared with those in which the gene was under *CaMV 35S* promoter (*CaMV 35S:gly I*). Under non-stressed condition, transgenic lines of stress-tolerant *B. juncea* plants having the constitutive *CaMV 35S* promoter showed a higher yield penalty than the other lines expressing the *gly I* gene under the regulation of the stress-inducible *rd29A* promoter.

Identification and characterization of novel miRNAs and genes encoding the same are also important for developing transgenic *Brassica* with enhanced abiotic stress tolerance (Rajwanshi et al. 2014; Rustagi et al. 2020).

A number of studies have been published on the transfer of genes in different *Brassica* species for increasing their abiotic stress tolerance (Table 10.2). Prasad et al. (2000a, b) revealed that transgenic *B. juncea* expressing the choline oxidase gene (*codA*) isolated from *Arthrobacter globiformis* showed up to 200 mM salt tolerance with improved seed germination following exposure to salinity stress. Park et al. (2003) showed that *B. campestris* overexpressing either the trehalose-6-phosphate synthase gene (*otsA*) from *Escherichia coli* or the late embryogenesis abundant (*LEA*) protein gene from *Capsicum annuum* (*CaLEA*) exhibited enhanced salinity, drought, and high temperature tolerance. Rajagopal et al. (2007) showed the potential of an isoform of the vacuolar Na^+/H^+ antiporter isolated from pearl millet (*Pennisetum glaucum*) in conferring salinity tolerance up to 300 mM NaCl in *B. juncea*. Wang et al. (2010) showed that the *Agcod A*gene, when transferred to *B. campestris* conferred enhanced tolerance to extreme temperatures (45 and 1 °C) and up to 300 mM salinity in *B. campestris* with increased betaine accumulation and sustainable photosynthetic rate. Saxena et al. (2011) showed that overexpression of the *Oryza sativa* glyoxalase II gene (*glyII*) in *B. juncea* conferred enhanced salinity tolerance with a higher germination percentage under 200 mM NaCl stress in the transformed plants. Transgenic *Brassica* overexpressing *P. glaucum NHX1* showed tolerance to extreme salinity stress (up to 300 mM NaCl) with enhanced Na^+ accumulation in leaves compared with that in seeds. Ying et al. (2014) showed that the overexpression of genes coding for the NAC

Table 10.2 Genes conferring abiotic stress tolerance in *Brassica* transgenics

Gene name	Source of transgene	Transgenicplant developed	Type of abiotic stress tolerance conferred	References
Thioredoxin-h2 (*AtTrx-h2*)	*A. thaliana*	*B. napus*	Salt (50 and 100 mM NaCl)	Ji et al. (2020)
Expansin-like B1 (*BrEXLB1*)	*B. rapa*	*B. rapa*	Salt (250 mM NaCl), oxidative stress (10 mM H_2O_2), osmotic stress (250 mM D-mannitol), drought (4% PEG-6000)	Muthusamy et al. (2020)
Alternative oxidase (*BnaAOX1b*)	*B. napus*	*B. napus*	Salt (100, 150, and 200 mM NaCl), osmotic stress (300 mM mannitol), ABA treatment (10 μM ABA)	Yang et al. (2019)
Dihydroflavonol-4-reductase (*AtDFR*)	*A. thaliana*	*B. napus*	Salt (200 and 300 mM NaCl	Kim et al. (2017)
SIP1 (6b INTERACTING PROTEIN1) (*BnSIP1-1*)	*B. napus*	*B. napus*	Osmotic pressure (400- and 600-mM mannitol), salt (150 and 200 mM NaCl), ABA treatment (50 and 100 μM ABA)	Luo et al. (2017)
Glyoxalase I (*BjglyI*)	*B. juncea*	*B. juncea*	Salinity (200 mM NaCl), drought (200 mM mannitol), heavy metal (5 mM $ZnCl_2$)	Rajwanshi et al. (2007, 2016)
Late embryogenesis abundant protein gene (*AtLEA4-1*)	*A. thaliana*	*B. juncea*	Drought (20% and 40% PEG-6000, salinity (200 and 400 mM NaCl)	Saha et al. (2016)
ENHANCED DROUGHT TOLERANCE1/HOMEODOMAIN GLABROUS11 (*AtEDT1/HDG11*)	*A. thaliana*	*B. oleracea*	Salt (250 mM NaCl), osmotic stress (25% PEG-6000)	Zhu et al. (2016)
Brass in osteroid biosynthetic gene (*AtDWF4*)	*A. thaliana*	*B. napus*	Drought (withholding water), heat (45 °C)	Sahni et al. (2016)
Aminolevulinate synthase encoding gene (*YHem1*)	*Saccharomyces cerevisiae*	*B. napus*	Salt (200 mmol L^{-1} NaCl)	Sun et al. (2015)
AtABI1	*A. thaliana*	*B. napus*	Drought (withholding water)	Babula-Skowrońska et al. (2015)
BnNAC485	*B. napus*	*B. napus*	Salt (200 mM NaCl), osmotic stress (300 mM mannitol), ABA treatment (1 μM ABA)	Ying et al. (2014)

(continued)

Table 10.2 (continued)

Gene name	Source of transgene	Transgenicplant developed	Type of abiotic stress tolerance conferred	References
Phosphatidylinositol-specific phospholipase C2 (*BnPLC2*)	*B. napus*	*B. napus*	Low temperature (−5 °C)	Nokhrina et al. (2014)
Glyoxalase II (*OsglyII*)	*O. sativa*	*B. juncea*	Salinity (150 and 200 mM NaCl)	Saxena et al. (2011)
Nicotianamine synthase (*OsNAS1*)	*O. sativa*	*B. napus*	Salinity(20 mMNa_2CO_3)	Kong et al. (2011)
Gamma-tocopherol methyl transferase (γ-*TMT*)	*A. thaliana*	*B. juncea*	Salt (400 mM NaCl), heavy metal (40 mM $CdCl_2$), osmotic stress (200 mM mannitol)	Yusuf et al. (2010), Kumar et al. (2013a, b, 2014)
Choline oxidase (*AgcodA*)	*Arthrobacter globiformis*	*B. campestris*	High temperature (45 °C), low temperature (1 °C), salinity (100, 200, and 300 mmol L^{-1} NaCl)	Wang et al. (2010)
Inositol polyphosphate kinase (*ThIPK2*)	*Thellungiella halophila*	*B. napus*	Salt (100 and 200 mM NaCl), drought (water witholding), oxidative stress (1 and 2% H_2O_2)	Zhu et al. (2009)
Na^+/H^+ antiporter (*PgNHX1*)	*Pennisetum glaucum*	*B. juncea*	Salinity (300 mM NaCl)	Rajagopal et al. (2007)
1-aminocyclopropane-1-carboxylate (ACC) deaminase	*Pseudomonas putida* UW4	*B. napus*	Flooding, Ni stress	Farwell et al. (2007)
Betaine aldehyde coding *Bet-a* gene	*E. coli*	*B. oleracea*	Salinity (150 and 300 mM NaCl)	Bhattacharya et al. (2004)
Pathogenesis-related (PR) protein gene (*PsPR10*)	*Pisum sativum*	*B. napus*	Salinity (75 mM NaCl)	Srivastava et al. (2004)
LEA protein gene (*CaLEA*)	*Capsicum annuum*	*B. campestris*	Salinity (250 mM NaCl), drought (10% PEG-6000), high temperature (45 °C)	Park et al. (2003)
Trehalose-6-phosphate synthase gene (*otsA*)	*E. coli*			
AgcodA	*A. globiformis*	*B. juncea*	Salinity (50, 100, 150, and 200 mM NaCl)	Prasad et al. (2000a, b)
Lectin	*Cicer arietinum*	*B. juncea*	Salinity (200 mM NaCl) and drought (200 mM mannitol) stress	Kumar et al. (2015)

domain proteins (BnNAC485) imparted enhanced salinity (200 mM NaCl) and drought (300 mM mannitol) tolerance in transgenic *B. napus*. Saha et al. (2016) reported that the overexpression of LEA protein genes isolated from *A. thaliana* conferred enhanced salt (up to 400 mM NaCl) and drought (40% PEG-6000) tolerance in transgenic *B. juncea* plants. Additionally, the study showed that the transgenic plants had higher survivability under stress conditions with improved chlorophyll and proline contents and increased antioxidant enzyme activities, including catalase, ascorbate peroxidase, guaiacol peroxidase, and glutathione reductase. Kim et al. (2017) showed the overexpression of dihydroflavonol-4-reductase (DFR) from *Arabidopsis* led to enhanced salinity tolerance in *B. napus*. The transgenic plants exhibited increased anthocyanin content, thereby reducing the accumulation of reactive oxygen species (ROS) under salinity stress. Luo et al. (2017) showed that the overexpression of *6b INTERACTING PROTEIN1 (BnSIP1-1)* in *Brassica* improved tolerance against salinity, osmotic pressure, and ABA treatment and enhanced seed germination in the transgenic plants. Yusuf et al. (2010) showed that transgenic *B. juncea* plants with increased α-tocopherol levels in plants because of overexpression of the gamma-tocopherol methyl transferase (*γ-TMT*) gene from *Arabidopsis* had better tolerance to high levels of NaCl (200 mM), $CdCl_2$ (20 mM), and mannitol (200 mM) stress as measured by different growth parameters. A higher photosynthetic efficiency was revealed by chlorophyll a fluorescence measurement. This lipid-soluble antioxidant was also found to modulate the other water-soluble components of plant antioxidant machinery under induced abiotic stress conditions in the transgenic *B. juncea* plants (Kumar et al. 2013a, b, 2014). In an another study, Kumar et al. (2015) showed that transgenic *B. juncea* plants harboring chickpea lectin gene showed enhancement in the tolerance to salinity (200 mM NaCl) and drought (200 mM mannitol) stress in comparison with untransformed control plants by enhancing the accumulation of osmolytes and reducing the cell membrane injury.

A recent study showed that the overexpression of *B. rapa* expansin-like B1 (*BrEXLB1*) conferred drought tolerance in transgenic *B. rapa* with sustainable seed germination, root development, and photosynthesis (Muthusamy et al. 2020).

10.6 Genetic Modification of *B. juncea* for Disease Resistance

Pathogenic infection and disease progression are greatly influenced by the environment. The study of disease occurrence in plants and its relation to factors, such as nutrition, humidity and wind, started during the eighteenth century (Colhoun 1973). Climate changes in the twenty-first century have been influencing host plant growth and plant–pathogen interaction in terms of susceptibility, pathogen reproduction, dispersal, and survival. The classic disease triangle proposed by Stevens (1960) illustrates interaction among susceptible host, virulent pathogen, and favorable environment for disease development. The climate change can have positive, negative, or neutral effects on reproductive fitness, virulence potential, and disease development pattern of pathogens. Simultaneously, the disease resistance pathways in plants, mediated by pattern-triggered immunity (PTI) and effector-triggered immunity (ETI), the RNA interference (RNAi), and networks of defense hormone are also affected by climate change. Hence, the changing pattern of disease and pathogens owing to climate change warrant an effort toward development of biotic stress-tolerant/resistant crops along with novel agricultural practices and sustainable agriculture. An indepth understanding of the mechanism of plant defense against pathogens is a prerequisite in designing new methods to improvise the current approaches of crop protection.

Improvement of the disease resistance potential of *B. juncea is* challenged by various factors such as environmental sensitivity, low heritability, complexity in disease resistance, costly and precise phenotyping, and limited availability of resistant germplasm. Moreover, contemporary protective measures rely on fungicides, resulting in skewed ecosystem and biomagnification. The advancements in "omics" technologies have provided a boost to the understanding of plant–pathogen interactions. Likewise, the transgenic technology and molecular-assisted breeding have emerged as a promising tool for the development of disease-resistant varieties.

10.6.1 Pests and Pathogens of *B. juncea*

There are various biotic constraints that limit and reduce the productivity of mustard crop. The most devastating diseases include alternaria blight (*Alternaria alternata*, *A. brassicae*, *A. brassicicola*, *A. raphani*), sclerotinia stem rot (*Sclerotinia sclerotiorum*), white rust (*Albugo candida*), downy mildew (*Hyaloperonospora brassicae*), blackleg (*Leptosphaeria maculans*, *L. biglobosa*), white leaf spot (*Neopseudocercosporella capsellae*), clubroot (*Plasmodiophora brassicae*), powdery mildew (*Erysiphe cruciferarum*), turnip yellow virus, cauliflower mosaic virus, turnip mosaic virus, and leaf blight (*Pseudomonas syringae*).

Of these, *Alternaria* blight, downy mildew, Sclerotinia stem rot, and white rust, account for major losses in pod formation (37–47%) and grain yield (17–54%). *Alternaria* species occur most commonly among the different *Brassica* crops. *A. brassicae* majorly affects the oil yielding *Brassica* (*B. juncea*, *B. rapa*). Other species of *Alternaria* (*A. alternata*, *A. brassicicola*, and *A. raphanin*) commonly infest vegetable brassicas. The symptoms of *Alternaria* infection could be observed on leaf, petiole, stem, inflorescence, siliquae, and seeds at the adult stage. *Alternaria* infection may lead to 11–100% loss in crop (quantitative as well as qualitative). The crop loss varies as per the time of infection, the prevailing environmental conditions, and the methods employed for control of infection, if any.

Sclerotinia sclerotiorum causes sclerotinial stem rot. The leaves and cotyledons are infested at the seedling stage. The fungus colonizes the intercellular spaces, causing necrosis of cotyledons and leaf lesions, before it enters a biotrophic phase while growing down the petiole and into the stem. The leaves and stems are infested at adult stage of plant. The symptoms include lesions filled with watery liquid and necrotic tissues harboring white fluffy mycelium. During the fruiting stage, the stem gets filled with numerous sclerotial bodies. As the number and size of the sclerotial bodies inside stem increases, the stem gets blackened and the nutrient flow in plants is obstructed. In severe situations, the plant dies.

Downy mildew, caused by the oomycete, *H. brassicae* infects the aerial parts of the plant. This pathogen infects commercially important *Brassica* crops such as broccoli, cabbage, radish, and rapeseed. The symptoms appear on cotyledons and leaves as yellowish green growth on the underside of leaves. In mature plants irregular angular yellow blotches are formed, which may have dark specks.

Among the insect pests, aphids pose a big havoc for *Brassica* crops. The most destructive pest of *Brassica* is mustard or turnip aphid, *Lipaphis erysimi pseudo brassicae*, which causes more than 50% reduction in yield. The aphids excrete honeydew, which promotes fungal growth (sooty mold). Furthermore, aphids of the *L. erysimi* group are carriers of more than 13 different viruses, including Turnip mosaic virus (TuMV). Among other destructive insect pests of *Brassica* crops are *Lepidopteran* pest, diamondback moth, and *Plutella xylostella* (Px).

Nematodes, as pathogens, feed on all parts of plants. There are specific parasitic nematodes that cause primarily soil-borne diseases and attack the root system of host plants. They produce the nutrient deficiency symptoms such as wilting or stunting.

10.6.2 Transgenic Technology for Disease Resistance in *B. juncea*

The germplasm of *B. juncea* lacks innate immunity against biotic stressors; hence, it is difficult to develop disease tolerant/resistant cultivars of *B. juncea* through conventional breeding approaches. Therefore, transgenic technologies are being used for the development of biotic stress-tolerant *B. juncea*. Some of the major biotic stress resistant lines developed so far, include those expressing chitinases, glucanases, and cry proteins. These transgenic plants show effective tolerance to pathogens, such as *Alternaria brassicae*, *L. maculans* or *S. sclerotiorum* or pests, such as *Lipaphiserysimi* or *Plutellaxylostella* (Poveda et al. 2020).

A large number of transgenic lines of different *Brassica* species have been generated by transformation with genes coding for chitinases. Chitinase is an enzyme (a pathogenesis-related protein) that catalyzes the breakdown of complex chitin, the main component of the fungal cell wall. Chitinase gene isolated from tobacco was transferred into *B. juncea* where the transgenic plants inhibited the *A. brassicae* fungal colony size by 12–56% in comparison with the untransformed control (Mondal et al. 2007). The *A. tumefaciens*-mediated transformation of *B. juncea* has been achieved using chitinase C (*ChiC*) encoding gene from *Streptomyces griseus* (Ahmad et al. 2015).

Brassica juncea has been transformed with synthetic chitinase gene (*NIC*). The leaf tissue extracts from transgenic plants exhibited considerable resistance and antifungal activity against *A. brassicae* (Munir et al. 2016). Transformation of *B. juncea* with two genes from barley, a class II chitinase gene and the gene coding for a ribosome inactivating protein (*RIP*), which inactivates foreign ribosomes in distantly related species and in other eukaryotes, including fungi, have been achieved (Chhikara et al. 2012). Transgeniic *B. juncea* expressing class I basic glucanase from tomato showed tolerance to *A. brassicae*. The lesions in transgenic plants were restricted in size, number, and spread as compared to untransformed control plants (Mondal et al. 2007).

During the past few decades, cationic antimicrobial peptides (CAPs) have emerged as very promising candidate gene for production of transgenic plants resistant to a broad spectrum of phytopathogens. The CAPs are structurally of two types: α-helical peptides, such as cecropins and magainins and β-sheet peptides, such as defensins, protegrins, and tachyplesins (Osusky et al. 2000; Phazang et al. 2020). Several transgenics of *Brassica* have been generated, which code for CAPs to produce antimicrobial proteins, and show enhanced disease tolerance and inhibition of pathogen growth. Wang and Fristensky (2001) generated rapeseed transgenic lines expressing α-helical peptides that were resistant against *L. maculans*. Rustagi et al. (2014, 2020) transformed *B. juncea* "Varuna" with *MsrA1* gene coding for a cecropin-melittin cationic peptide. The transgenic plants were able to counter the growth and disease progression caused by *A. brassicae* and *S. sclerotiorum*. In vitro assays showed that quantified protein from leaf tissues of transgenic plants inhibited the hyphal growth of *Alternaria* by 44–62%. The *Alternaria* infection was delayed and the number and size of lesions and time taken for complete leaf necrosis was less in transgenic plants as compared to untransformed control assessed. The disease protection varied from 69 to 85% in different transgenic lines. The transgenic plants showed tolerance to *S. sclerotiorum* infection as well. The sclerotial bodies formed in the stem of transgenic plants were less in number and smaller in size. The number of days for stem breaking was also more in the *MsrA1* gene expressing plants. A disease protection of 56–71.5% was obtained.

The recent studies on genetic transformation of *B. juncea* for tolerance to selected fungi and insect pests are summarized in Table 10.3.

10.7 Long Noncoding RNA: A Future Tool for Genetic Modification of *B. juncea*

One of the key mechanisms that cells use to regulate gene expression is through the noncoding RNAs (ncRNAs). Long noncoding RNAs (lncRNAs) are longer than 200 nucleotides (nt) unlike the 18–30 nt long small RNAs (miRNAs and siRNAs). The ncRNAs modulate the expression of genes at the transcriptional as

Table 10.3 Development of biotic stress resistant *Brassica juncea* plants through genetic engineering

S. No.	Name of gene	Gene source	Mechanism of action	Pathogen	References
1	*Lectin*	*Hevea brasiliensis*	Fungal immobilization by binding with carbohydrate in the cell wall	*A. brassicae*	Kanrar et al. (2002a, b)
2	*Class I chitinase*	*Nicotiana tabacum*	Degradation of fungal cell wall	*A. brassicae*	Mondal et al. (2003)
3	*Class I basic glucanase*	*Lycopersicon esculentum*	Degradation of fungal cell wall	*A. brassicae*	Mondal et al. (2007)
4	*Class II chitinase*	*Hordeum vulgare*	Degradation of fungal cell wall	*A. brassicae*	Chhikara et al. (2012)
5	*Ribosome inactivating protein (RIP)*	*Hordeum vulgare*	Inactivation of foreign ribosomes	*A. brassicae*	Chhikara et al. (2012)
6	*MsrA1*	Cecropin–melittin cationic peptide	Lethal membrane dysfunction by making pores through membrane of pathogen	*A. brassicae* and *S. sclerotiorum*	Rustagi et al. (2014)
7	*Lectin*	*Cicer arietinum*	Fungal immobilization by binding with the carbohydrate of cell wall	*A. brassicae*	Kumar et al. (2015)
8	Endochitinase 42 (*ech42*)	*Trichoderma virens*	Fungal cell wall degradation	*A. brassicae* and *A. brassisicola*	Kamble et al. (2016)
9	Non-expressor of pathogenesis-related (*NPR1*)	*B. juncea*	Activation of the salicylic acid (SA)-mediated plant defense	*A. brassicae* and *Erysiphe cruciferarum*	Ali et al. (2017)
10	Mitogen-activated protein kinase 3 (*MPK3*)	*B. juncea*	Activation of the SA-mediated plant defense	*A. brassicae*	Tasleem et al. (2017)
11	*NIC*	Synthetic chitinase	Fungal cell wall degradation	*A. brassisicola*	Munir et al. (2016)
12	Wheat germ agglutinin (*WGA*)	*Triticum aestivum*	Not mentioned	*Lipaphis erysimi*	Kanrar et al. (2002a, b)
13	Leaf agglutinin membrane (*ASAL*)	*Allium sativum*	Blockage of the insect gutepithelium	*L. erysimi*	Dutta et al. (2005)
14	*ACA Agglutinin*	*Allium cepa*	Blockage of the insect gut epithelial membrane	*L. erysimi*	Hossain et al. (2006)
15	*(E)-ß, farnesene Ebf*	*Myzus arvensis*	Volatile sesquiterpene as alarm pheromones	*L. erysimi*	Verma et al. (2015)
16	*LL Lentil lectin*	Lentil	Blockage of the insect gut epithelial membrane	*L. erysimi*	Rani et al. (2017a, b)

(continued)

Table 10.3 (continued)

S. No.	Name of gene	Gene source	Mechanism of action	Pathogen	References
17	*CPPI* chickpea protease inhibitor	*C. arietinum*	Disruption in assimilation of dietary protein	*L. erysimi*	Rani et al. (2017a, b)
18	*RiD* defensin	*Rorippa indica*	Inhibition of nutrient uptake	*L. erysimi*	Sarkar et al. (2017)
19	*HSPRO2* nematode resistance protein-like homolog	*R. indica*	Activation of basal plant resistance	*L. erysimi*	Bose et al. (2018, 2019)
20	*cry1Ac* protein	*Bacillus thuringiensis*	Lysis of the gut epithelial cells	*Helicoverpa armigera* and *Plutella xylostella*	Moon et al. (2007)
21	*cry1C* proteins	*B. thuringiensis*	Lysis of the gut epithelial cells	*H. armigera* and *P. Xylostella*	Kamble et al. (2013)
22	*BjuWRR1*, a CC-NB-LRR	*B. juncea*	Degradation of Avr proteins	White rust *Albugo candida*	Arora et al. (2019)

well as the post-transcriptional levels and can inhibit RNA translation into proteins or alter the RNA stability (Ponjavic et al. 2007; Peschansky and Wahlested 2014). Several groups have shown that these lncRNAs are involved in regulating gene expression under diverse biological functions in response to a variety of stresses such as salinity and drought, nitrogen/phosphate deficiency, excess light, extreme temperatures, hypoxia, and bacterial pathogens (Zhang et al. 2013; Hivrale et al. 2016; Zhao et al. 2016; Song and Zhang 2017).

10.7.1 lncRNAs Involved in Abiotic Stress Pathways in *Brassica*

Bhatia et al. (2020) identified 7613 putative lncRNAs through in silico genome analysis in *B. juncea*. Of these, 1614 lncRNAs were found to be differentially expressed in response to heat and drought stress conditions. They further characterized these lncRNAs by functional annotation through coexpression analysis and pathway enrichment analysis. The identified abiotic stress responsive lncRNAs were involved in both enzymatic and non-enzymatic antioxidants pathways such as glutathione metabolism and phenylpropanoid biosynthesis and cysteine metabolism. They also found that lncRNAs coexpress with transcription factors associated with abiotic stress response. The LncRNAs acting as putative targets and endogenous target mimics of miRNAs involved in response to heat and drought were also identified (Bhatia et al. 2020). A total of 2725, 1672, and 2810 lncRNAs were identified in three *Brassica* sp., *B. rapa*, *B. carinata*, and *B. hexaploid*, respectively, using next-generation sequencing (Wang et al. 2018). At least 725 lncRNAs were reported to be differentially expressed in the hexaploid compared with the parent species. Using bioinformatics tools, 109 lncRNAs were annotated and their target mRNAs were identified to be involved in diverse biological functions such as immune system processes, response to environmental stimulus, metabolic processes, biological regulation, reproduction, and pigmentation in *B. hexaploid* (Wang et al. 2018). The RNA-seq data also revealed 10,001 lncRNAs in *B. rapa* of which 9687 were novel and expressed upon heat and cold treatments. The coexpression network constructed for lncRNAs and miRNAs showed that lncRNAs regulated the expression of 192 and 67 target genes in response to heat and cold treatments, respectively. Also, lncRNA–pri-miRNA regulatory networks and their crosstalk under heat and cold stress were identified (Song et al. 2016). Under heat stress, interactions among the upregulated lncRNAs, mRNAs, and microRNAs showed that several phytohormones, including salicylic acid (SA) and brassinosteroid (BR), were linked with heat tolerance. At least 25 lncRNAs that were highly coexpressed with 10 heat responsive genes were reported, indicating special roles in heat tolerance. Thirty-nine lncRNAs were predicted as endogenous target mimics (eTMs) for 35 miRNAs, and five of them were validated to be involved in heat tolerance in Chinese cabbage (Wang et al. 2018). Zhang et al. (2018) further recognized 1299, 1885, and 1910 lncRNAs in *B. rapa*, *B. napus*, and *B. oleracea*, respectively, involved in lncRNA–miRNA mediated regulatory networks, proving that lncRNAs play key roles in stress responses, defense responses, multi-cellular organismal processes, immune system processes, developmental processes, and cell death (Zhang et al. 2018). In another study, Tan et al. (2020) compared transcripts from drought-tolerant and drought-sensitive genotypes of *B. napus* responding to drought stress and rehydration response at the seedling stage. They identified 34 transcription factors (TFs) corresponding to 126 differentially expressed lncRNAs in the drought-tolerant genotype and 45 TFs corresponding to 359 differentially expressed lncRNAs in the drought-sensitive genotype. Differential expression analysis of lncRNAs indicated that up and down regulated mRNAs coexpressed with lncRNAs participated in different metabolic pathways and were involved in diverse regulatory mechanisms in the drought-tolerant genotype. Especially, some lncRNAs were coexpressed with BnaC07g44670D (a homolog of ABF), which are associated with plant ABA hormone signal transduction. Additionally, some mRNAs colocalized with lncRNAs, XLOC_052298, XLOC_094954, and XLOC_012868, which were mainly found to be involved in the signal transport and defense against stress response (Tan et al. 2020).

10.7.2 lncRNAs Invoved in Biotic Stress Pathway in Brassica

lncRNAs have also been found to be involved in biotic stresses. Joshi et al. (2016) identified 3183 novel lncRNAs, which have been shown to be upregulated in response to *S. sclerotiorum* infection in *B. napus*. They also identified lncRNA–pri-miRNA regulatory networks and 41lncRNAs, which are precursors to miR394, miR156, and miR169. These miRNAs have earlier been reported to be upregulated upon infection with fungal pathogens, suggesting that lncRNAs have important functions in fungal phytopathogenesis (Joshi et al. 2016).

The long noncoding RNAs have so far not been used for genetic manipulations in *B. juncea*. Thus, understanding the lncRNA-mediated regulatory network in *Brassica* sp. can be extended to *B. juncea* and lncRNAs can be used in developing stress-tolerant plants.

10.8 Genetic Transformation of Brassica for Development of Edible Vaccines

Vaccination is the most effective way to prevent, control, and eradicate infectious diseases. Although, traditional mammalian cell expression system has been widely utilized for large-scale production of recombinant proteins or subunit vaccines, the plant-based recombinant expression system possesses significant advantages and represents a promising alternative to the current methods (Chan and Daniell 2015; Diamos et al. 2020). Oral delivery is immunologically more advantageous in generating effective mucosal and systemic immune response (Ko 2014). This dual immune response becomes pivotal, especially in case of infectious diseases where the pathogens make their entry into the system crossing the first line of defense (Saxena and Rawat 2013; Debnath et al. 2021).

The utilization of plants as protein production systems is wide and varied. The demonstration of expression of a vaccine antigen within plants started in 1990 with the efforts of Curtiss and Cardineau as they successfully expressed the *Streptococcus* mutant surface protein antigen A (SpaA) in tobacco (Curtiss and Cardineau 1990). Since then, several plant expression systems have been well established for a number of antigenic proteins such as surface protein antigen A (SpaA) of the *Streptococcus* mutans, hepatitis B surface antigen (HbsAg), *E. coli* heat–labile enterotoxin, Norwalk virus capsid protein and the rabies virus glycoprotein (Mason et al. 2005; Sahai et al. 2013). Several plant species like tobacco, potato, rice, wheat, maize, carrot, lettuce, celery cabbage, spinach, cauliflower, and brassica have been utilized for the production of pharmaceutical proteins and antibodies according to the desired mode of antigen delivery (Merlin et al. 2017; Kurup and Thomas 2020).

Brassica species have also been exploited for the expression of antigenic pathogenic proteins. Lee et al. (2011) expressed *Plasmodium vivax*-derived MLC chimeric recombinant gene in *B. napus* for the development of plant-based vaccine. First, they synthesized a synthetic chimeric recombinant gene consisting of merozoite surface protein-1 (MSP-1), which is the main antigenic protein of *P. vivax*, and a sporozoite stage surface antigen called circumsporozoite protein (CSP). Thereafter, cotyledonary explants of *B. napus* were transformed with the *MLC* gene construct via *Agrobacterium*-mediated transformation method. The results showed that BALB/c mice orally immunized with MLC chimeric recombinant protein produced antigen-specific IgG1 antibodies. Furthermore, they found a higher concentration of Th1-related cytokines, IL-12, TNF, and IFN-γ in the spleen of orally immunized mice. They concluded that the chimeric MLC recombinant protein produced in *B. napus* could potentially be used as potential as an excellent vaccine candidate for oral immunization against vivax malaria (Lee et al. 2011).

Gorantala et al. (2014) evaluated the expression of protective antigen (PA), an antigenic protein of *Bacillus anthracis*, in *B. juncea* for the development of oral anthrax vaccine. They transformed hypocotyl explants of *B. juncea* with a plant expression vector having PA gene by agroinfection. Functional activity of PA assessed by macrophage lysis assays confirmed the biological activity of plant-produced protective antigen protein. Furthermore, orally immunized BALB/c mice showed enhanced serum anti-PA IgG and IgA antibody titers. This study successfully demonstrated the generation of immune protective response against anthrax in orally immunized mice with transgenic *B. juncea*. In another study by the same group, an additional antigenic protein of *B. anthracis*, namely edema factor (EF), was expressed in *B. juncea* plant using *Agrobacterium*-mediated plant transformation method. The initial studies using the CHO cell lines confirmed that the antigenic protein expressed in plants is biologically active and can be used for oral immunization studies.

Recently, in the principal author's lab, one of the antigenic proteins of *B. anthracis,* namely edema factor (EF) and its catalytic mutant (EFM; H351AEF), has been expressed in *B. juncea* (communicated). Experiments on Chinese hamster ovary (CHO) cells confirmed that the plant-produced edema factor was biologically active and mutant of edema factor can be used along with protective antigen (PA) for the development of highly efficacious and safe vaccine against anthrax. The expression of mutant EF probably results in an otherwise structurally intact protein that could be used for the generation of immunological response.

For the production of vaccines, oil crops evidently provide production advantages along with formulation and delivery benefits. One major advantage of using oil crops as production hosts is that the oil bodies can be exploited to simplify the downstream processing and decrease the purification steps (Daniell et al. 2001). Being a major oil and edible crop plant, *B. juncea* has excellent potential for the production of biopharmaceuticals.

10.9 Conclusion

Genetic modification of *B. juncea* for biotic and abiotic stress tolerance has been achieved by genetic transformation. Among the various methods, *Agrobacterium*-mediated genetic transformation is most widely used and preferred mode. The novel technology of genome editing using the CRISPR/Cas9 system has opened ways of targeted gene modification to alleviate biotic and abiotic stresses. For an efficacious use of this technology, the protocol for gene editing needs to be optimized and suitable target genes need to be explored. The regulatory genes of *B. juncea* are crucial players in understanding the different molecular mechanisms involved in stress tolerance. A thorough knowledge about these genes is inevitable if novel genes are to be explored for stress tolerance mechanism. One of the key mechanisms of regulation of gene expression is through lncRNAs. The stress-specific lncRNAs may be identified and used for generating high-yielding, disease-resistant, and abiotic stress-tolerant Indian mustard lines. The study of pathways involved in the perception and response to stress and their regulatory molecules may unravel a set of key genes in *B. juncea*. We must have a thorough knowledge of the molecular mechanism of plant–pathogen interactions for the improvement of *Brassica* spp. Novel genes having regulatory role in disease tolerance need to be identified and characterized in this regard. Numerous advances have been achieved in molecular genetics of brassicas. Genetic engineering needs to be amalgamated with the crop breeding efforts to develop climate-resilient and high-yielding crops in future. Refined transgenic techniques for the transfer of DNA sequences will have immediate value as research tools. Every major discovery is accompanied with criticisms. The use of genetic modification (GM) techniques in edible foods involves several ethical, natural, and socioeconomic issues which will need consideration before the technique can be applied.

Acknowledgements AR, RR, and DK acknowledge DST-SERB project grant no. ECRA/2017/00563, EEQ/2016/000501 and EEQ/2016/000487, respectively. DK is also thankful to Department of Botany, Center for Advanced study in Botany, Institute of Science and Institute of Eminence (IoE) grant, Banaras Hindu University for providing necessary facilities and infrastructural support. We are thankful to Ms. Sahana Basu from Assam University, Silchar for her contribution in this chapter. The authors are grateful to Dr. Mukesh Jain, University of Florida, USA for his valuable inputs.

Conflict of Interest The authors declare no conflict of interest.

References

Abdelrahman AM, Al-Sadi A, Pour-Aboughadareh DJ, Burritt LS, Tran P (2018) Genome editing using CRISPR/Cas9–targeted mutagenesis: an opportunity for yield improvements of crop plants grown under environmental stresses. Plant PhysiolBiochem 131 (10):31–39

Ahmad B, Ambreen MS, Khan AH (2015) *Agrobacterium* mediated transformation of *Brassica juncea* (L.) Czern, with chitinase gene conferring resistance against fungal infections. Pak J Bot 47:211–216

Ahmed I, Yadav D, Shukla P, Vineeth TV, Sharma PC et al (2017) Constitutive expression of *Brassica juncea* annexin AnnBj2 confers salt tolerance and glucose and ABA insensitivity in mustard transgenic plants. Plant Sci 265:12–28

Ali S, Mir ZA, Tyagi A, Mehari H, Meena RP, Bhat JA, Yadav P, Papalou P, Rawat S, Grover A (2017) Overexpression of NPR1 in *Brassica juncea* confers broad spectrum resistance to fungal pathogens. Front Plant Sci 8:1693

Aminedi R, Dhatwalia D, Jain V, Ramcharan B (2019) High efficiency in planta transformation of Indian mustard (*Brassica juncea*) based on spraying of floral buds. Plant Cell Tiss Org Cult 138:229–237

Andres F, Coupland G (2012) The genetic basis of flowering responses to seasonal cues. Nat Rev Genet 13(9):627–639

Arora H, Padmaja KL, Paritosh K, Mukhi N, Tewari AK et al (2019) *BjuWRR1* a CC-NB-LRR gene identified in *Brassica juncea* confers resistance to white rust caused by *Albugo candida*. Theor App Genet 132:2223–2236

Augustine R, Majee M, Gershenzon J, Bisht NC (2013) Four genes encoding MYB28, a major transcriptional regulator of the aliphatic glucosinolate pathway, are differentially expressed in the allopolyploid *Brassica juncea*. J Exp Bot 64:4907–4921

Babula-Skowrońska D, Ludwików A, Cieśla A, Olejnik A, Cegielska-Taras T et al (2015) Involvement of genes encoding ABI1 protein phosphatases in the response of *Brassicanapus* L. to drought stress. Plant Mol Biol 88:445–457

Barfield DG, Pua EC (1991) Gene transfer in plants of *Brassica juncea* using *Agrobacterium tumefaciens* mediated transformation. Plant Cell Rep 10:308–314

Benedikt K, Leduc N, Sautter C, Potrykus I, Neuhaus G (1996) Transient marker-gene expression during zygotic in-vitro embryogenesis of *Brassica juncea* (Indian mustard) following particle bombardment. Planta 198:211–222

Bergman P, Glimelius K (1993) Electroporation of rapeseed protoplasts-transient and stable transformation. Physiol Plant 88:604–611

Bhatia G, Singh A, Verma D, Sharma S, Singh K (2020) Genome-wide investigation of regulatory roles of lncRNAs in response to heat and drought stress in *Brassica juncea* (Indian mustard). Environ Exp Bot 171:1–13

Bhattacharya RC, Maheswari M, Dineshkumar V, Kirti PB, Bhat SR et al (2004) Transformation of *Brassica oleracea* var, *capitata* with bacterial betA gene enhances tolerance to salt stress. Sci Hort 100:215–227

Bhuiyan MSU, Min SR, Jeong WJ, Sultana S, Choi KS et al (2011a) Overexpression of AtATM3 in *Brassica juncea* confers enhanced heavy metal tolerance and accumulation. Plant Cell Tiss Org Cult 107:69–77

Bhuiyan MSU, Min SR, Sultana S, Choi KS (2011b) An improved method for *Agrobacterium-mediated* genetic transformation from cotyledon explants of *Brassica juncea*. Plant Biotechnol 28:17–23

Bose S, Gangopadhyay G, Sikdar SR (2018) RiHSPRO2 a nematode resistance protein-like homolog from a wild crucifer *Rorippa indica* (L.) Hiern is a promising candidate to control mustard aphid *Lipaphiserysimi*. Arthropod Plant Interact 12:701–714

Bose S, Gangopadhyay G, Sikdar SR (2019) Rorippa indica HSPRO2 expression in transgenic *Brassica juncea* induces tolerance against mustard aphid *Lipaphiserysimi*. Plant Cell Tiss Org Cult 136:431–443

Byrne ME, Groover AT, Fontana JR, Martienssen RA (2003) Phyllotactic pattern and stem cell fate are determined by the Arabidopsis homeobox gene BELLRINGER. Development 130:3941–3950

Cartea ME, Velasco P (2008) Glucosinolates in *Brassica* foods: bioavailability in food and significance for human health. Phytochem Rev 7:213–229

Chan HT, Daniell H (2015) Plant-made oral vaccines against human infectious diseases—are we there yet? Plant Biotechnol J 13(8):1056–1070

Chauhan JS, Singh KH, Singh VV, Kumar S (2011) Hundred years of rapeseed-mustard breeding in India: accomplishments and future strategies. Indian J Agr Sci 81:1093–1109

Chen JL, Beversdorf WD (1992) Cryopreservation of isolated microspores of winter rapeseed (*Brassica napus* L.) for in vitro embryo production. Plant Cell Tissue Organ Cult 31:141–149

Cheng H, Desheng M, Wang H, Wang R, Jia Z, Chao LC, Hu Q (2021) Base editing with high efficiency in allotetraploid oilseed rape by A3A-PBE system. Plant Biotechnol J 19:87–97

Chhikara S, Chaudhury D, Dhankher OP, Jaiwal PK (2012) Combined expression of a barley class II chitinase and type I ribosome inactivating protein in transgenic *Brassica juncea* provides protection against *Alternaria brassicae*. Plant Cell Tissue Organ Cult 108:83–89

Cho L, Yoon J, An G (2017) The control of flowering time by environmental factors. Plant J 172:708–719

Clarke DB (2010) Glucosinolates, structures and analysis in food. Anal Method 2:310–325

Clough SJ, Bent AF (1998) Floral dip: a simplified method for *Agrobacterium*-mediated transformation of *Arabidopsis thaliana*. Plant J 16:735–743

Colhoun J (1973) Effect of environmental factors on plant disease. Annu Rev Plant Pathol 11:343–364

Curtiss RC, Cardineau GA (1990) World Intellectual Property Organization PCT/US89/03799, Washington University. St. Louis: 1990, Oral immunization by transgenic plants

Daniell H, Streatfield SJ, Wycoff K (2001) Medical molecular farming: production of antibodies biopharmaceuticals and edible vaccines in plants. Trends Plant Sci 6:219–226

Das BL, Goswami S, Ray S, Ghosh S, Bhattacharyya S et al (2006) *Agrobacterium*-mediated transformation of *Brassica juncea* with a cyanobacterial (*Synechocystis* PCC6803) delta-6 desaturase gene leads to production of gamma-linolenic acid. Plant Cell Tissue Organ Cult 86:219–231

Das M, Naiya H, Marik A, Mukherjee G, Seal A (2020) A protocol for functional study of genes in *Brassica juncea* by *Agrobacterium*-mediated transient expression: applicability in other *Brassicaceae*. J Plant Biochem Biotechnol 29:368–379

Debnath N, Thakur M, Khushboo, Negi NP, Gautam V, Yadav AK, Kumar D (2021) Insight of oral vaccines as an alternative approach to health and disease management: an innovative intuition and challenges. Biotechnol Bioeng 1–20. https://doi.org/10.1002/bit.27987. Epub ahead of print. PMID: 34755343

Diamos AG, Hunter JGL, Pardhe MD, Rosenthal SH, Sun H et al (2020) High level production of monoclonal antibodies using an optimized plant expression system. Front Bioeng Biotechnol 7:472

Dutta I, Majumder P, Saha P, Ray K, Das S (2005) Constitutive and phloem specific expression of Allium sativum leaf agglutinin (ASAL) to engineer aphid (*Lipaphiserysimi*) resistance in transgenic Indian mustard (*Brassica juncea*). Plant Sci 169:996–1007

Eapen S (2011) Pollen grains as a target for introduction of foreign genes into plants: an assessment. Physiol Mol Biol Plants 17:1–8

Fadhil AB, Saleh LA, Altamer DH (2020) Production of biodiesel from non-edible oil, wild mustard (*Brassica juncea* L.) seed oil through cleaner routes. Energy Sources, Part A: Recov Util Environ Effects 42:1831–1843

Fahey JW, Stephenson KK, Talalay P (1997) Glucosinolates, myrosinase, and isothiocyanates: three reasons for eating *Brassica* vegetables. In: Shibamoto T,

Terao J, Osawa T (eds) Functional food for disease prevention I. San Francisco, California, USA, pp 16–22

Fahey JW, Zalcmann AT, Talalay P (2001) The chemical diversity and distribution of glucosinolates and isothiocyanates among plants. Phytochemistry 56:5–51

Farooq N, Nawaz MA, Mukhtar Z, Ali I, Hundle P et al (2019) Investigating the in vitro regeneration potential of commercial cultivars of *Brassica*. Plants 8:558

Farwell AJ, Vesely S, Nero V, Rodriguez H, McCormack K et al (2007) Tolerance of transgenic canola plants (*Brassica napus*) amended with plant growth-promoting bacteria to flooding stress at a metal-contaminated field site. Environ Pollut 147:540–545

Feng YR, Sheliang C, Wang H, Yingpeng H, Xiangsheng LH et al (2020) Transcription Factor BnaA9, WRKY47 contributes to the adaptation of *Brassica napus* to low boron stress by up-regulating the boric acid channel gene BnaA3, NIP51. Plant Biotechnol J 18:1241–1254

Fornara F, Montaigu A, Coupland G (2010) SnapShot: control of flowering in *Arabidopsis*. Cell 141:3–5

Gan C (1989) Gene gun accelerates DNA-coated particles to transform intact cells. Scientist 3:25

Gasic K, Korban SC (2006) Indian mustard [*Brassica juncea* (L.) Czern.]. Meth Mol Biol 343:281–289

Gerber L, Zhang B, Roach M, Rende U, Gorzsás A et al (2014) Deficient sucrose synthase activity in developing wood does not specifically affect cellulose biosynthesis, but causes an overall decrease in cell wall polymers. New Phytol 203:1220–1230

Gerszberg A, Hnatuszko K, Kowalczyk T (2015) In vitro regeneration of eight cultivars of *Brassica oleracea* var capitata. In Vitro Cell Dev Biol-Plant 51:80–87

Gill SS, Gill R, Kumar G, Pareek A, Sharma PC et al (2012) Mustard: Approaches for crop improvement and abiotic stress tolerance. In: Tuteja N, Gill SS, Tiburcio AF, Tuteja R (eds) Improving crop resistance to abiotic stress, vol 2. Wiley Wiley-VCH Verlag GmbH & Co, Weinheim, pp 1351–1368

Gorantala J, Grover S, Rahi A, Chaudhary P, Rajwanshi R et al (2014) Generation of protective immune response against anthrax by oral immunization with protective antigen plant-based vaccine. J Biotechnol 176:1–10

Grover A, Pental D (2003) Breeding objectives and requirements for producing transgenics for major field crops of India. Curr Sci 84:310–320

Halkier BA, Gershenzon J (2006) Biology and biochemistry of glucosinolates. Ann Rev Plant Biol 57:303–333

Hardin SC, Huber SC (2004) Proteasome activity and the posttranslational control of sucrose synthase stability in maize leaves. Plant Physiol Biochem 42:197–208

Hirose T, Scofield GN, Terao T (2008) An expression analysis profile for the entire sucrose synthase gene family in rice. Plant Sci 174:534–543

Hivrale V, Zheng Y, Puli Jagadeeswaran COR, Gowdu K, Kakani VG et al (2016) Characterization of drought- and heat-responsive microRNAs in switchgrass. Plant Sci 242:214–223

Hong HN, Datla DW, Reed PS, Covello SL, Mac K et al (2002) High-level production of γ-linolenic acid in *Brassica juncea* using a Δ6 desaturase from *Pythium irregular*. Plant Physiol 129:354–362

Hossain MA, Maiti MK, Basu A, Sen S, Ghosh AK, Sen SK (2006) Transgenic expression of onion leaf lectin gene in Indian mustard offers protection against aphid colonization. Crop Sci 46:2022–2032

Huala E, Sussex IM (1992) LEAFY interacts with floral homeotic genes to regulate Arabidopsis floral development. Plant Cell 4:901–913

Huber SC, Huber JL, Liao PC, Gage DA, McMichael RW Jr et al (1996) Phosphorylation of serine-15 of maize leaf sucrose synthase: occurrence in vivo and possible regulatory significance. Plant Physiol 112:793–802

Jagannath AN, Arumugam V, Gupta A, Pradhan PK, Pental D (2002) Development of transgenic barstar lines and identification of a male sterile (barnase)/ restorer (barstar) combination for heterosis breeding in Indian oilseed mustard (*Brassica juncea*). Curr Sci 82:46–52

Ji MG, Park HJ, Cha JY, Kim JA, Shin GI et al (2020) Expression of *Arabidopsis thaliana* Thioredoxin-h2 in *Brassica napus* enhances antioxidant defenses and improves salt tolerance. Plant PhysiolBiochem 147:313–321

Jiang M et al (2020) *Brassica rapa* orphan genes largely affect soluble sugar metabolism. Hort Res 7. https://doi.org/10.1038/s41438-020-00403-z

Joshi RK, Megha S, Basu U, Rahman MH, Kav NN (2016) Genome wide identification and functional prediction of long non-coding RNAs responsive to *Sclerotinia sclerotiorum* infection in *Brassica napus*. PLoS One 11:e0158784

Juge N, Mithen RF, Traka M (2007) Molecular basis for chemoprevention by sulforaphane: a comprehensive review. Cell Mol Life Sci 64:1105–1127

Kamble S, Hadapad AB, Eapen S (2013) Evaluation of transgenic lines of Indian mustard (*Brassica juncea* L, Czern and Coss) expressing synthetic cry1Ac gene for resistance to *Plutellaxylostella*. Plant Cell Tissue Organ Cult 115:321–328

Kamble S, Mukherjee PK, Eapen S (2016) Expression of an endochitinase gene from *Trichoderma virens* confers enhanced tolerance to Alternaria blight in transgenic *Brassica juncea* (L.) czern and coss lines. Physiol Mol Biol Plants 22(1):69–76. https://doi.org/10.1007/s12298-016-0340-8

Kanrar S, Venkateswari J, Kirti P, Chopra V (2002a) Transgenic Indian mustard (*Brassica juncea*) with resistance to the mustard aphid (*LipaphiserysimiKalt.*). Plant Cell Rep 20:976–981

Kanrar S, Venkateswari JC, Kirti PB, Chopra V (2002b) Transgenic expression of hevein the rubber tree lectin in Indian mustard confers protection against *Alternaria brassicae*. Plant Sci 162:441–448

Kaur C, Kumar G, Kaur S, Ansari MW, Pareek A et al (2015) Molecular cloning and characterization of salt overly sensitive gene promoter from *Brassica juncea* (BjSOS2). Mol Biol Rep 42:1139–1148

Kim DH, Sung S (2014) Genetic and epigenetic mechanisms underlying vernalization. The Arabidopsis Book 12:e0171. https://doi.org/10.1199/tab.0171

Kim J, Lee WJ, Vu TT, Jeong CY, Hong SW, Lee H (2017) High accumulation of anthocyanins via the ectopic expression of AtDFR confers significant salt stress tolerance in *Brassica napus* L. Plant Cell Rep 36:1215–1224

Ko K (2014) Expression of Recombinant vaccines and antibodies in plants. Monoclonal Anti Immunodiagn 33:192–198

Kong F, Mao S, Du K, Wu M, Zhou XY et al (2011) Comparative proteomics analysis of OsNAS1 transgenic *Brassica napus* under salt stress. Chin Sci Bull 56:2343–2350

Koramutla MK, Ram C, Bhatt D, Annamalai M, Bhattacharya R (2019) Genome-wide identification and expression analysis of sucrose synthase genes in allotetraploid *Brassica juncea*. Gene 707:126–135

Kumar D, Yusuf MA, Singh P, Sardar M, Sarin NB (2013a) Modulation of antioxidant machinery in α-tocopherol-enriched transgenic *Brassica juncea* plants tolerant to abiotic stress conditions. Protoplasma 250:1079–1089

Kumar S, Sairam RK, Prabhu KV (2013b) Physiological traits for high temperature stress tolerance in *Brassica juncea*. Indian J Plant Physiol 18:89–93

Kumar D, Yusuf MA, Singh P, Sardar M, Sarin NB (2014) Histochemical detection of superoxide and H_2O_2 accumulation in *Brassica juncea* seedlings. Bio-protocol 4(8). https://doi.org/10.21769/BioProtoc.1108

Kumar D, Shekhar S, Bisht S, Kumar V, Varma A, Kumar M (2015) Ectopic overexpression of lectin in transgenic *Brassica juncea* plants exhibit resistance to fungal phytopathogen and showed alleviation to salt and drought stress. J Bioeng Biomed Sci 5:147

Kurup VM, Thomas J (2020) Edible vaccines: promises and challenges. Mol Biotechnol 62:79–90

Lee S, Kim J, Han JJ, Han MJ, An G (2004) Functional analyses of the flowering time gene *OsMADS50*, the putative suppressor of overexpression of *SOC1/Agamous-Like 20* (*SOC1/AGL20*) ortholog in rice. Plant J 38:754–764

Lee C, Kim HH, Choi KM, Chung KW, Choi YK, Jang MJ, Kim TS, Chung NJ, Rhie HG, Lee HS, Sohn Y, Kim H, Lee SJ, Lee WH (2011) Murine immune responses to a Plasmodium vivax-derived chimeric recombinant protein expressed in *Brassica napus*. Malar J 10:106

Lepiniec L, Debeaujon I, Routaboul J-M, Baudry A, Pourcel L, Nesi N, Caboche M (2006) Genetics and biochemistry of seed flavonoids. Annu Rev Plant Biol 576:405–430

Li J, Brader G, Kariola T, Palva ET (2006) WRKY70 modulates the selection of signaling pathways in plant defence. Plant J 46:477–491

Li F, Hao C, Yan L, Wu B, Qin X et al (2015) Gene structure, phylogeny and expression profile of the sucrose synthase gene family in cacao (*Theobroma cacao* L.). Genet 94:461–472

Li G, Xinyi W, Yaowen H, Maria MA, Jie L et al (2019) Orphan genes are involved in drought adaptations and ecoclimatic-oriented selections in domesticated cowpea. J Exp Bot 70:3101–3110

Li W, Chen F, Wang Y, Zheng H, Yi Q et al (2020) Genome-wide identification and functional analysis of ARF transcription factors in *Brassica juncea* var. tumida. PLoS One 15(4):e0232039

Liu C, Zhou J, Bracha-Drori K, Yalovsky S, Ito T, Yu H (2007) Specification of *Arabidopsis* floral meristem identity by repression of flowering time genes. Development 134:1901–1910

Liu LJ, Zhang YC, Li QH, Sang Y, Mao J et al (2008) COP1-mediated ubiquitination of CONSTANS is implicated in cryptochrome regulation of flowering in *Arabidopsis*. Plant Cell 20:292–306

Lohani N, Jain D, Singh MB, Bhalla PL (2020) Engineering multiple abiotic stress tolerance in canola *Brassica napus*. Front Plant Sci 11:3

Luo J, Tang S, Mei F, Peng X, Li J et al (2017) BnSIP1-1 a Trihelix family gene mediates abiotic stress tolerance and ABA signaling in *Brassica napus*. Front Plant Sci 8:44

Mason HS, Chikwamba R, Santi L, Mahoney RT, Arntzen CJ (2005) Transgenic plants for mucosal vaccines. Mucosal Immunol:1053–1060

Mayee P, Singh A (2016) Natural genetic variation in *Brassica* homologs of *FLOWERING LOCUS T* and characterization of its expression domains. J Plant Biochem Biot 25:270–277

Mehra SA, Pareek P, Bandyopadhyay P, Sharma PK, Pental D (2000) Development of transgenics in Indian oilseed mustard (*Brassica juncea*) resistant to herbicide phosphinothricin. Curr Sci 78:1358–1364

Melzer S, Lens F, Gennen J, Vanneste S, Rohde A et al (2008) Flowering-time genes modulate meristem determinacy and growth form in *Arabidopsis thaliana*. Nat Genet 40:1489–1492

Merlin M, Pezzotti M, Avesani L (2017) Edible plants for oral delivery of biopharmaceuticals. Br J Clin Pharmacol 83:71–81

Min BW, Cho YN, Song MJ, Noh TK, Kim BK et al (2007) Successful genetic transformation of Chinese cabbage using phosphomannose isomerase as a selection marker. Plant Cell Rep 26:337–344

Mondal KK, Chatterjee SC, Viswakarma N, Bhattacharya RC, Grover A (2003) Chitinase-mediated inhibitory activity of Brassica transgenic on growth of *Alternaria brassicae*. Curr Microbiol 47:171–173

Mondal KK, Bhattacharya RC, Koundal KR, Chatterjee SC (2007) Transgenic Indian mustard (*Brassica juncea*) expressing tomato glucanase leads to arrested growth of *Alternaria brassicae*. Plant Cell Rep 26:247–252

Moon HS, Halfhill MD, Good LL, Raymer PL, Stewart CN (2007) Characterization of directly transformed weedy *Brassica rapa* and introgressed *B. rapa* with *Bt cry1Ac* and *gfp* genes. Plant Cell Rep 26:1001–1010

Munir I, Hussan W, Kazi M, Farhatullah A, Iqbal A et al (2016) Production of transgenic *Brassica juncea* with the synthetic chitinase gene (*NIC*) Conferring resistance to *Alternaria brassicicola*. Pak J Bot 48:2063–2070

Muthusamy M, Kim JY, Yoon EK, Kim JA, Lee SI (2020) BrEXLB1 a *Brassica rapa* Expansin-Like B1 gene is associated with root development drought stress response and seed germination. Genes 11(4):404

Naeem I, Munir I, Durrett TP, Iqbal A, Aulakh KS et al (2020) Feasible regeneration and agrobacterium-mediated transformation of *Brassica juncea* with *Euonymus alatus* diacylglycerol acetyltransferase (EaDAcT) gene. Saudi J Biol Sci 27:1324–1332

Nakamura T, Song IJ, Fukuda T, Yokoyama J, Maki M et al (2005) Characterization of *TrcMADS1* gene of *Trillium camtschatcense* (Trilliaceae) reveals functional evolution of the *SOC1/TM3*-like gene family. J Plant Res 118:229–234

Nokhrina K, Ray H, Bock C, Georges F (2014) Metabolomic shifts in *Brassica napus* lines with enhanced BnPLC2 expression impact their response to low temperature stress and plant pathogens. GM Crops Food 5(2):120–131

Osusky M, Zhou G, Osuska L, Hancock RE, Kay WW et al (2000) Transgenic plants expressing cationic peptide chimeras exhibit broad-spectrum resistance to phytopathogens. Nat Biotechnol 18:1162–1166

Park SH, Jun SS, An G, Hong YN, Park MC (2003) A comparative study on the protective role of trehalose and LEA proteins against abiotic stresses in transgenic Chinese cabbage (*Brassica campestris*) overexpressing CaLEAorotsA. J Plant Biol 46:277–286

Paul Barratt DH, Derbyshire P, Findlay K, Pike M, Wellner N et al (2009) Normal growth of *Arabidopsis* requires cytosolic invertase but not sucrose synthase. Proc Natl Acad Sci USA 106:13124–13129

Pental D, Pradhan AK, Sodhi YS, Mukhopadhayaya A (1993) Variation amongst *Brassica juncea* cultivars for regeneration from hypocotyl explants and optimization of conditions for *Agrobacterium* mediated genetic transformation. Plant Cell Rep 12:462–467

Peschansky VJ, Wahlested C (2014) Non-coding RNAs as direct and indirect modulators of epigenetic regulation. Epigenetics 9:3–12

Phazang P, Negi NP, Raina M, Kumar D (2020) Plant antimicrobial peptides: next-generation bioactive molecules for plant protection. In: Kumar M, Kumar V, Prasad R (eds) Phyto-microbiome in stress regulation. Springer Nature, Singapore, pp 281–294. (ISBN-978-981-15-2576-6)

Ponjavic J, Ponting CP, Lunter G (2007) Functionality or transcriptional noise? Evidence for selection within long noncoding RNAs. Genome Res 17:556–565

Potts D, Rakow G, Males DR (1999) Canola quality *Brassica juncea*, a new oilseed crop for the Canadian prairies. In: Proceeding's 10th international rapeseed congress, Canberra, Australia, pp 26–29

Poveda J, Francisco M, Cartea ME, Velasco P (2020) Development of transgenic *Brassica* crops against biotic stresses caused by pathogens and arthropod pests. Plants 9:1664

Prasad K, Sharmila P, Kumar PA, Saradhi P (2000a) Transformation of *Brassica juncea* (L.) Czern with bacterial codA gene enhances its tolerance to salt stress. Mol Breed 6:489–499

Prasad K, Sharmila P, Saradhi P (2000b) Enhanced tolerance of transgenic *Brassica juncea* to choline confirms successful expression of the bacterial *codA* gene. Plant Sci 159:233–242

Prasad SC, Kisku AV, Sarin NB (2018) Understanding the Gamma-vacuolar processing enzyme gene regulation by promoter-gus fusion approach. Plant Archives 18(1):679–689

Probsting M, Schenke D, Hossain R, Thurau C, Wighardt L et al (2020) Loss-of-function of CRT1a (Calreticulin) reduces susceptibility to *Verticillium longisporum* in both *Arabidopsis thaliana* and oilseed rape (*Brassica napus*). Plant Biotechnol J 18:2328–2344

Rajagopal D, Agarwal P, Tyagi W, Pareek SL, Reddy MK et al (2007) *Pennisetum glaucum* Na^+/H^+ antiporter confers high level of salinity tolerance in transgenic *Brassica juncea*. Mol Breed 19:137–151

Rajwanshi R, Sarin NB (2013) Selection of genetically transformed *Brassica juncea* L, cv, Varuna (Indian mustard) based on Positech system. J Plant BiochemBiotechnol 22:214–221

Rajwanshi R, Roy SD, Pooggin M, Hohm T, Sarin NB (2007) Marker free approach for developing abiotic stress tolerant transgenic *Brassica juncea* (Indian mustard). In: Proceedings of the 2007 WSEAS international conference on cellular and molecular biology biophysics and bioengineering Athens Greece, pp 110–116

Rajwanshi R, Chakraborty S, Jayanandi K, Deb B, Lightfoot DA (2014) Orthologous plant micro RNAs: microregulators with great potential for improving stress tolerance in plants. Theor Appl Genet 127:2525–2543

Rajwanshi R, Kumar D, Yusuf MA, DebRoy S, Sarin NB (2016) Stress-inducible overexpression of *glyoxalase I* is preferable to its constitutive overexpression for abiotic stress tolerance in transgenic *Brassica juncea*. Mol Breed 36:76

Rani S, Sharma V, Hada A, Koundal KR (2017a) Efficient genetic transformation of *Brassica juncea* with lectin using cotyledons explants. Intl J Ad Biotechnol Res 7:1–12

Rani S, Sharma V, Hada A, Bhattacharya RC, Koundal KR (2017b) Fusion gene construct preparation with lectin and protease inhibitor genes against aphids and efficient genetic transformation of *Brassica juncea* using cotyledons explants. Acta Physiol Plant 39:115

Razzaq MK, Aleem M, Mansoor S, Khan MA, Rauf S et al (2021) Omics and CRISPR-Cas9 approaches for molecular insight, functional gene analysis, and stress tolerance development in crops. Intl J Mol Sci 22:1292

Roosjen M, Paque S, Weijers D (2018) Auxin response factors: output control in auxin biology. J Exp Bot 69:179–188

Roy SD, Saxena M, Bhomkar PS, Pooggin M, Hohn T et al (2008) Generation of marker free salt tolerant transgenic plants of *Arabidopsis thaliana* using the *gly I* gene and *cre* gene under inducible promoters. Plant Cell Tissie Organ Cult 95:1–11

Roy SD, Saxena M, Sarin NB (2009) Overexpression of *AtLEAFY* accelerates flowering in *Brassica juncea.* Crop Sci 49:930–936

Roy SD, Saxena M, Sarin NB (2011) Cloning and expression studies of a *LFY* cDNA from *Brassica juncea.* Adv Biosci Biotechnol 2:248–254

Ruan YL, Llewellyn DJ, Liu Q, Xu SM, Wu LM et al (2008) Expression of sucrose synthase in the developing endosperm is essential for early seed development in cotton. Funct Plant Biol 35:382–393

Rustagi A, Kumar D, Shekhar S, Yusuf MA, Misra S, Sarin NB (2014) Transgenic *Brassica juncea* plants expressing MsrA1 a synthetic cationic antimicrobial peptide exhibit resistance to fungal phytopathogens. Mol Biotechnol 56:535–545

Rustagi A, Negi NP, Choudhury HD, Mahajan A, Verma S, Kumar D, Rajwanshi R, Sarin NB (2020) Transgenic approaches for improvement of *Brassica* species. In: Wani SH, Thakur AK, Khan YJ (eds) *Brassica* Improvement. Springer, Cham Switzerland, pp 187–213

Saha B, Mishra S, Awasthi JP, Sahoo L, Panda SK (2016) Enhanced drought and salinity tolerance in transgenic mustard [*Brassica juncea* (L,) Czern & Coss,] overexpressing *Arabidops*is group 4 late embryogenesis abundant gene (AtLEA4-1). Environ Exp Bot 128:99–111

Sahai A, Shahzad A, Shahid M (2013) Plant edible vaccines: a revolution in vaccination. Recent Trends Biotechnol Therap Appl Med Plants 15:225–252

Sahni S, Prasad BD, Liu Q, Grbic V, Sharpe A, Singh SP, Krishna P (2016) Overexpression of the brassinosteroid biosynthetic gene DWF4 in *Brassica napus* simultaneously increases seed yield and stress tolerance. Sci Rep 6:28298

Samach A, Onouchi H, Gold SE, Ditta GS, Schwarz-Sommer Z et al (2000) Distinct roles of constans target genes in reproductive development of *Arabidopsis.* Science 288:1613–1616

Sarkar P, Jana K, Sikdar SR (2017) Overexpression of biologically safe Rorippaindica defensin enhances aphid tolerance in *Brassica juncea.* Planta 246:1029–1044

Saxena J, Rawat S (2013) Edible vaccines. Adv Biotechnol:207–226

Saxena M, Deb Roy S, Pareek S, Sopory SK, Sarin NB (2011) Overexpression of the glyoxalase II gene leads to enhanced salinity tolerance in *Brassica juncea.* Plant Sci J 5:23–28

Schultz EA, Haughn GW (1991) LEAFY, a homeotic gene that regulates inflorescence development in *Arabidopsis.* Plant Cell 3:771–781

Shekhawat K, Rathore SS, Premi OP, Kandpal BK, Chauhan JS (2012) Advances in agronomic management of Indian mustard (*Brassica juncea* (L.) Czernj. Cosson): an overview. Int J Agron 2012:1–14

Shen Q, Qin X, Xiang Y, Zhao Y, Du C (2013) Recurrent selection for improving quality characters of male sterile restorer lines of *Brassica napus.* Agric Sci Technol 14:763–766

Singh VV, Verma V, Pareek AK, Mathur M, Yadav R et al (2010) Optimization and development of regeneration and transformation protocol in Indian mustard using lectin gene from chickpea (*Cicer arietinum* L.). J Plant Breeding Crop Sci 1:306–310

Song Y, Zhang D (2017) The role of long noncoding RNAs in plant stress tolerance. Methods Mol Biol 1631:41–68

Song X, Liu G, Huang Z, Duan W, Tan H et al (2016) Temperature expression patterns of genes and their co-expression with LncRNAs revealed by RNA-Seq in non-heading Chinese cabbage. BMC Genomics 17:297

Sonntag K, Wang Y, Wallbrawn M (2004) A transformation method for obtaining marker-free plants based on phosphomannose isomerase. Acta Universitatis Latviensis 676:223–226

Srivastava S, Fristensky B, Kav NNV (2004) Constitutive expression of a PR10 protein enhances the germination of *Brassica napus* under saline conditions. Plant Cell Physiol 45(9):1320–1324

Stahl Y, Simon R (2009) Plant primary meristems: shared functions and regulatory mechanisms. Curr Opin Plant Biol 13:53–58

Stein O, Granot D (2019) An overview of sucrose synthases in plants. Front Plant Sci 10.https://doi.org/10.3389/fpls.2019.00095

Stevens RB (1960) Plant pathology an advanced treatise J.G. Horsfall and A.E. Dimond eds, Academic Press NY 3:357–429

Sun X, Feng X, Li C, Zhang Z, Wang L (2015) Study on salt tolerance with YHem1 transgenic canola (*Brassica napus*). Physiol Plant 154:223–242

Tan X, Li S, Hu L, Zhang C (2020) Genome-wide analysis of long non-coding RNAs (lncRNAs) in two contrasting rapeseed (*Brassica napus* L,) genotypes subjected to drought stress and rewatering. BMC Plant Biol 81:2–20

Tasleem M, Baunthiyal M, Taj G (2017) Induction of MPK3 MPK6 and MPK4 mediated defense signaling in response to Alternaria blight in transgenic *Brassica juncea.* Biosci Biotechnol Res Asia 14:1469–1474

Thakur AK, Parmar N, Singh KH, Nanjundan J (2020) Current achievements and future prospects of genetic engineering in Indian mustard (*Brassica juncea* L. Czern & Coss). Planta 2020:252–256

Tian Y, Deng F (2020) Phytochemistry and biological activity of mustard (*Brassica juncea*): a review. CyTA-J Food 18:704–718

Traka M, Mithen R (2009) Glucosinolate, isothiocyanates and human health. Phytochem Rev 8:269–282

Tyagi S, Mazumdar PA, Mayee P, Shivaraj SM, Anand S et al (2018) Natural variation in *Brassica FT* homeologs influences multiple agronomic traits including flowering time, silique shape, oil profile, stomatal morphology and plant height in *B. juncea*. Plant Sci 277:251–266

Ullah F, Banu A, Nosheen A (2012) Effects of plant growth regulators on growth and oil quality of canola (*Brassica napus* L.) under drought stress. Pak J Bot 44:1873–1880

Valverde F (2011) CONSTANS and the evolutionary origin of photoperiodic timing of flowering. J Exp Bot 62:2453–2463

van Dijk ADJ, Molenaar J (2017) Floral pathway integrator gene expression mediates gradual transmission of environmental and endogenous cues to flowering time. Peer J 5:e3197

Verma SS, Sinha RK, Jajoo A (2015) (E)-α-farnesene gene reduces *Lipaphiserysimi* colonization in transgenic *Brassica juncea* lines. Plant Signal Behav10: e1042636

von Wettstein D, Jende-Strid B, Ahrenst-Larsen B, Sørensen JA (1977) Biochemical mutant in barley renders chemical stabilization of beer superfluous. Carlsberg Res Commun 42:341–351

Wagner D (2016) Making flowers at the right time. Dev Cell 37:208–210

Wang Y, Fristensky B (2001) Transgenic canola lines expressing pea defense gene *DRR206* have resistance to aggressive blackleg isolates and to *Rhizoctonia solani*. Mol Breed 8:263–271

Wang J, Li Y, Liang C (2008) Recovery of transgenic plants by pollen mediated transformation in *Brassica juncea*. Transgenic Res 17:417–424

Wang Q, Xu W, Xue Q, Su W (2010) Transgenic *Brassica chinensis* plants expressing a bacterial *codA* gene exhibit enhanced tolerance to extreme temperature and high salinity. J Zhejiang Univ Sci 11:851–861

Wang Z, Wei P, Wu M, Xu Y, Li F et al (2015) Analysis of the sucrose synthase gene family in tobacco: structure, phylogeny, and expression patterns. Planta 242:153–166

Wang R, Zou J, Meng J, Wang J (2018) Integrative analysis of genome-wide lncrna and mRNA expression in newly synthesized *brassica* hexaploids. EcolEvol 12:6034–6052

Weigel D, Alvarez J, Smyth DR, Yanofsky MF, Meyerowitz EM (1992) LEAFY controls floral meristem identity in *Arabidopsis*. Cell 69:843–859

Wittstock U, Halkier BA (2002) Glucosinolate research in the *Arabidopsis* era. Trends Plant Sci 6:263–270

Wu J, Chen G, Xian L, Shengliang L, Qinfu Y, Fang SY, Zhang H, Wang Y (2020) Engineering herbicide-resistant oilseed rape by CRISPR/Cas9-mediated cytosine base-editing. Plant Biotechnol J 18:1857–1859

Xu F, Rong X, Huang X, Cheng S (2012) Recent advances of *Flowering Locus T* gene in higher plants. Int J Mol Sci 13:3773–3781

Yamaguchi K, Shinozaki K (1994) A novel cis-acting element in an Arabidopsis gene is involved in responsiveness to drought low temperature or high salt stress. Plant Cell 6:251–254

Yan MJ, Yamanaka N, Yamamoto F, Du S (2010) Responses of leaf gas exchange, water relations, and water consumption in seedlings of four semiarid tree species to soil drying. Acta Physiol Plant 32:183–189

Yang H, Deng L, Liu H, Fan S, Hua W, Liu J (2019) Overexpression of BnaAOX1b confers tolerance to osmotic and salt stress in rapeseed. G3 (Bethesda) 9:3501–3511

Ying L, Chen H, Cai W (2014) BnNAC485 is involved in abiotic stress responses and flowering time in *Brassica napus*. Plant PhysiolBiochem 79:77–87

Yu F, Lydiate DJ, Gugel RK, Sharpe AG, Rimmer SR (2012) Introgression of *Brassica rapa* subsp. sylvestris blackleg resistance into *B. napus*. Mol Breed 30:1495–1506

Yusuf MA, Sarin NB (2007) Antioxidant value addition in human diets: genetic transformation of *Brassica juncea* with gamma-TMT gene for increased alpha-tocopherol content. Transgenic Res 16:109–113

Yusuf MA, Kumar D, Rajwanshi R, Strasser RJ, Tsimilli-Michael M et al (2010) Overexpression of *γ-tocopherol methyl transferase* gene in transgenic *Brassica juncea* plants alleviates abiotic stress: physiological and chlorophyll a fluorescence measurement. BiochimBiophys Acta 1797:1428–1438

Zaman QU, Wen C, Hao M, Shi Y, Sun M (2019) CRISPR/Cas9-mediated multiplex genome editing of jagged gene in *Brassica napus*. Biomolecules 9:11

Zhang HX, Hodson JN, Williams JP, Blumwald E (2001) Engineering salt-tolerant Brassica plants: characterization of yield and seed oil quality in transgenic plants with increased vacuolar sodium accumulation. Proc Natl Acad Sci USA 98:12832–12836

Zhang Y, Zhang XO, Chen T, Xiang JF, Yin QF, Xing YH, Zhu S, Yang L, Chen LL (2013) Circular Intronic Long Noncoding RNAs. Mol Cell 51:792–806

Zhang J, Wei L, Jiang J, Mason AS, Li-Cui H et al (2018) Genome-wide identification putative functionality and interactions between lncRNAs and miRNAs in *Brassica* species. Sci Rep 8:4960

Zhao J, He Q, Chen G, Wang L, Jin B (2016) Regulation of non-coding RNAs in heat stress responses of plants. Front Plant Sci 97:1213

Zhao N, Liu C, Meng Y, Hu Z, Zhang M et al (2019) Identification of flowering regulatory genes in Allopolyploid *Brassica juncea*. Hort Plant J 5:109–119

Zhu YL, Pilon-Smits EAH, Jouanin L, Terry T (1999) Overexpression of glutathione synthetase in *Brassica junceae*nhances cadmium accumulation and tolerance. Plant Physiol 119:73–79

Zhu JQ, Zhang JT, Tang RJ, Lv QD, Wang QQ, Yang L, Zhang HX (2009) Molecular characterization of ThIPK2 an inositol polyphosphate kinase gene homolog from *Thellungiella halophila* and its heterologous expression to improve abiotic stress tolerance in *Brassica napus*. Physiol Plant 136:407–425

Zhu Z, Sun B, Xu X, Chen H, Zou L, Chen G, Cao B, Chen C, Lei J (2016) Overexpression of AtEDT1/HDG11 in Chinese Kale (*Brassica oleracea* var, alboglabra) enhances drought and osmotic stress tolerance. Front Plant Sci 7:1285

Molecular Linkage Mapping in *Brassica juncea*: Founding the Basis for Marker-Assisted Selection

11

Satish K. Yadava and Nirala Ramchiary

Abstract

Genetic maps are a prerequisite for understanding genome organization, map-based cloning, and marker-assisted selection of agronomically important traits in crops. The application of genetic markers to genetic mapping and quantitative trait locus analyses in *Brassica juncea* is now well-established. Extensive genetic mapping efforts have been made in *B. juncea* allowing inclusive genome-wide comparisons, discovery of evolutionary rearrangements, and identification of syntenic relationships with other *Brassica* species. The earliest genetic maps developed in *B. juncea* used different types of mapping populations and were based on DNA markers like restriction fragment length polymorphism, random amplified polymorphic DNA, and amplified fragment length polymorphisms which were gradually superseded by single-nucleotide polymorphisms due to their abundance in the genomes and their flexibility for high-throughput genotyping platforms. High-resolution genetic maps of *B. juncea* have provided the scaffolds on which whole-genome sequences have been assembled, thus representing a vital link between the breeding and sequencing research. Molecular breeding of several qualitative and quantitative traits, like white rust resistance, fatty acid contents, seed coat color, and glucosinolates content, have been successfully undertaken using a wide array of genetic markers and maps. Extensive research on white rust resistance utilizing diverse genotypes with varied levels of resistance/susceptibility to the disease by using different approaches, like bulked segregant analysis and comparative mapping with *Arabidopsis thaliana*, has led to the identification of key genetic loci imparting resistance against one of the most destructive diseases in *B. juncea*. Seed coat color is another trait of interest in breeding of *B. juncea* lines, as the character is known to be associated with protein and fiber content and has also been proposed as an important factor for increasing seed oil content. Different studies on the genetic improvement of the *B. juncea* cultivars leading to enhancement in the nutritional quality of the oil and meal have focused on composition and quantities of various fatty acids and glucosinolates. This chapter assembles the past and present research in classical and modern molecular breeding approaches, and describes the

S. K. Yadava
Centre for Genetic Manipulation of Crop Plants, University of Delhi South Campus, Benito Juarez Road, New Delhi 110021, India

N. Ramchiary (✉)
School of Life Sciences, Jawaharlal Nehru University, New Mehrauli Road, New Delhi 110067, India
e-mail: nramchiary@jnu.ac.in

C. Kole and T. Mohapatra (eds.), *The Brassica juncea Genome*, Compendium of Plant Genomes,
https://doi.org/10.1007/978-3-030-91507-0_11

advances with specific examples of crucial agricultural and economic traits in *B. juncea*.

11.1 Introduction

Brassica juncea is an important oilseed crop which is also cultivated as a condiment and vegetable in a few parts of the world. Diverse phenotypes of the crop include Indian oilseed mustard (*B. juncea* var. *oleifera*), root mustard (*B. juncea* var. *rapifera*), stem mustard (*B. juncea* var. *tsatsai*), leaf mustard (*B. juncea* var. *rugosa*), and cut leaf mustard (*B. juncea* var. *crispifolia*) (Labana and Gupta 1992). The vegetable types are mainly cultivated in China, while mustard is a principal edible oilseed crop in India and is grown to a lesser extent in Canada, Australia, Russia and Ukraine.

It has been reported that genetic variability in oilseed *B. juncea* is dispersed between two major genepools commonly described as the Indian and the east European genepools (Pradhan et al. 1993; Oram et al. 1999; Srivastava et al. 2001). Pradhan et al. (1993) explored genetic diversity in *B. juncea* based on pheno-morphological traits by utilizing Indian, exotic (Commonwealth of Independent States—former USSR) and some synthetic *B. juncea* accessions, and grouped them into different clusters, wherein the maximum genetic distance was observed between the lines belonging to the two germplasm types. Marker-based phylogenetic analyses undertaken in subsequent studies confirmed the existence of two genepools (Jain et al. 1994; Srivastava et al. 2001; Burton et al. 2004). These studies also revealed a very narrow genetic base of the Indian genepool which is also widely distant from the east European exotic genepool (Pradhan and Pental 2011).

Classical genetic analysis has been used to unravel the inheritance of various monogenic and polygenic traits. The development of several types of molecular markers has helped in the construction of genetic maps and dissection/identification of quantitative trait loci (QTLs), some of which have been used for marker-assisted selection (MAS) of the traits of interest. Furthermore, this information also leads to the identification and isolation of causative genes that increase our understanding of the trait variation. In this chapter, an attempt has been made to deliberate on the studies on the development of genetic maps and mapping of genes and QTLs in *B. juncea* using various molecular markers and mapping populations.

11.2 Construction of Genetic Maps

Several marker systems have been used to construct genetic maps in *B. juncea*. Classical mapping efforts in *B. juncea* included use of isozyme markers, the majority of which concentrated on peroxidase isozyme patterns (Edwards et al. 2007). In order to characterize 12 genotypes of Indian mustard, Kumar and Gupta (1985) performed a comparative study of the activities of the peroxidase and esterase isozymes based on the presence and absence of bands along with their intensity-based scores at five developmental stages of siliqua. The studies revealed the presence of similar profiles for the number of isozyme bands at all the five stages for peroxidase, and a variable number of isozyme bands for esterase indicating different isozymes were active at various stages of siliqua development. Thus, patterns based on the scores for the total peroxidase activity appeared to be more informative compared to the results on individual banding patterns and could therefore be successfully used to characterize different Indian mustard genotypes. Existence of separate morphotypes in *B. juncea* was shown by Chen and Tong (1985) based on polyacrylamide gel electrophoresis for studying peroxidase and acid phosphatase isozymes at three leaves seedling stage. Isozymes have also been used as genetic markers for discerning the hybridity in *B. juncea* compared to the parental lines and in breeding programs due to their relationships with agronomically important traits. However, these parameters are influenced by environmental factors and the developmental stage of the plant, and therefore, these markers were progressively substituted by DNA-based markers.

DNA markers have traditionally been classified into three main classes. These are as follows: (1) First-generation restriction-based DNA markers that include restriction fragment length polymorphism (RFLP), (2), second-generation amplification-based DNA markers like random amplified polymorphic DNA (RAPD), variable number of tandem repeats (VNTRs) that include both minisatellites and microsatellites (SSLP: simple sequence length polymorphism; SSRs: simple sequence repeats; STRs: short tandem repeats; STMS: sequence tagged microsatellite sites), and (3) third generation DNA markers such as single-nucleotide polymorphisms (SNPs) and indels (insertions and deletions).

11.2.1 Maps Developed Using First-Generation DNA Markers

The earliest report on genetic mapping presented the nature of intervarietal RFLPs and their inheritance, linkage relationship, and association with quantitative traits (Sharma et al. 1994) (Table 11.1). The study used *B. juncea* cv. Varuna and two exotic lines, BEC-144 and BEC-286 showing differences at both the morphological and molecular levels. The map covered a total length of 173.9 cM and included 15 loci distributed over six linkage groups (LGs) out of a total of 28 loci. Significant marker-quantitative trait associations for primary branches per plant, secondary branches per plant, and days to flowering were identified using single factor analysis of variance. Mohapatra et al. (1995) also assembled a partial RFLP genetic map of *B. juncea* from F_2 progenies of the same Varuna/BEC-144 cross and mapped 45 RFLP loci and one seed coat color marker on 14 LGs, covering 407.9 cM of the genome. Based on the same Varuna/BEC-144 cross, Upadhyay et al. (1996) carried out genetic mapping and QTL analysis using a F_2 population and showed linkage of RFLP markers with yellow seed coat color and six other traits—days to flowering, plant height, number of primary branches, secondary branches per primary branch, siliquae per secondary branch, and seeds per siliqua.

Cheung et al. (1997) developed a RFLP-based genetic map by using cDNA markers from *B. napus* for genotyping of a F_1-derived doubled haploid (DH) population of 119 individuals. The DH lines were developed from a cross between two lines—J90-4317 and J90-2733, which were contrasting for glucosinolates content, oil content, protein content, and white rust resistance. J90-4317 had both low oil and glucosinolates content, but had a high protein content and was susceptible to white rust, while J90-2733 had high levels of glucosinolates and oil content, but showed low protein content and was resistant to white rust. The map harbored 343 loci spanning a genetic length of 2073 cM. The markers were distributed over 18 major LGs and five small groups of two to five loci. A large percentage of marker loci was duplicated illustrating that complex duplications and subsequent rearrangements occurred after allopolyploidy. Cheung et al. (1998a, b) and Prabhu et al. (1998) later used this cross to identify markers and QTLs linked with white rust resistance in addition to seed quality traits.

In a mapping effort by Axelsson et al. (2000), an integrated map was developed by utilizing marker information from two individual RFLP-based maps using mapping populations that shared a common parent. The crosses were performed between a resynthesized *B. juncea* and *B. juncea* cultivars. Their results showed that the two individual maps were perfectly collinear with each other as demonstrated by the distribution and ordering of 137 common marker loci shared between the two maps. These maps comprised 183 and 296 loci and covered 1266 cM and 1041 cM units, respectively. The two maps allowed the first documentation of the A and B subgenome components of *B. juncea*. Disomic inheritance of parental alleles was observed at all loci, with chromosomes of

Table 11.1 Overview of the genetic mapping studies in *Brassica juncea*

Publication	Parental lines	Mapping population	Number of segregating progenies	Type of markers	Number of linkage groups	Total markers	Genetic length (cM)	Average interval size (cM)
Sharma et al. (1994)	Varuna, BEC-144	F2	50	RFLP	6	15	173.9	19.3
Upadhyay et al. (1996)	Varuna, BEC-144	F2	48	RFLP	9	25	243.3	16.2
Cheung et al. (1997)	J90-4317, J90-2733	DH	119	RFLP	18	326	1945.4	6.6
Axelsson et al. (2000)	J-o-3DH1, J-o-7DH1, resynthesized B. juncea line BjRS	BC1	120 (BjNAT), 60 (BjSYN)	RFLP	18 (BjNAT), 18 (BjSYN)	183 (BjNAT), 296 (BjSYN)	1,266 (BjNAT), 1,041 (BjSYN)	7.7 (BjNAT), 3.7 (BjSYN)
Sharma et al. (2002)	Varuna, BEC-144	RIL (F6)	94	RAPD	21	114	790.4	6.9
Lionneton et al. 2002	BJ-99, BJ-70	DH	131	AFLP, RAPD	18	273	1641	6.3
Mahmood et al. (2003a)	RLM-514 (HEP), LEP	DH	61 (S population), 51 (R population)	RFLP	18	280	1564	5.2
Pradhan et al. (2003)	Varuna, Heera	DH	123	RFLP, AFLP	18	1029	1628.7	3.5
Christianson et al. (2006)	AC Vulcan, UM 3132	F2	140	RFLP, SSR	18	323	1577.3	4.9
Ramchiary et al. (2007a)	Varuna, Heera	DH	123	RFLP, AFLP, SSR	18	1448	1840.1	1.3
Panjabi et al. (2008)	Varuna, Heera	DH	123	IP, RFLP and gene markers	18	533	1922.2	3.6
Yadava et al. (2012)	TM-4, Donskaja-IV	DH	100	IP, AFLP, SSR	18	911	1629.9	3.7
Paritosh et al. (2014)	Varuna, Heera	DH	123	SNP,IP	18	1708	1933.5	1.4
Zou et al. (2016)	Sichuan Yellow, purple mustard	F2	168	DArT-seq markers	18	3329	1578.5	0.5
Li et al. (2016)	Tumida, multiceps	F2	200	SSR	17	116	2061.0	17.9
Yang et al. (2016)	03A0106, T84-63	F2	100	SNP	18	5333	4710.3	0.9
Dhaka et al. (2017)	EH-2, Pusa Jaikisan	DH	182	SSR	18	860	2073.6	3.0

(continued)

Table 11.1 (continued)

Publication	Parental lines	Mapping population	Number of segregating progenies	Type of markers	Number of linkage groups	Total markers	Genetic length (cM)	Average interval size (cM)
Rout et al. (2018)	EH-2, J8, Donskaja-IV	DH	166 (EJ8), 163 (DE)	IP, SSR, SNP	18 (EJ8), 18 (DE)	388 (EJ8), 653 (DE)	1125.6 (EJ8), 1708.2 (DE)	3.9 (EJ8), 0.4 (DE)
Khattak et al. (2019)	G266, G302	F6 RILs	167	IP	18	304	1671.13	5.5
Paritosh et al. (2021)	Varuna, Heera, Tumida	DH	92 (VH), 119 (TuV)	IP, SSR, SNP	18 (VH), 18 (TuV)	3780 (VH), 9041 (TuV)	3232.3 (VH), 3417.4 (TuV)	0.9 (VH), 0.4 (TuV)

B. rapa and *B. nigra* pairing exclusively with their corresponding A or B subgenome homologues, respectively.

11.2.2 Maps Developed Using Second-Generation DNA Markers

Sharma et al. (1999) performed bulked segregant analysis (BSA) (Michelmore et al. 1991) of recombinant inbred lines (RILs) using RAPDs to identify markers linked to QTLs influencing seed oil content, a trait with quantitative mode of inheritance in *B. juncea*. The F_6 RILs were developed from an intervarietal cross (Varuna × BEC-144) that was previously used by Sharma et al. (1994). Sharma et al. (2002) used the same set of RILs for the construction of a linkage map and identification of QTLs influencing oleic acid level. The map included 114 markers distributed over 21 LGs spanning a genetic distance of 790.4 cM (Table 11.1).

Lionneton et al. (2002) constructed a molecular map of *B. juncea* using RAPD markers in conjunction with amplified fragment length polymorphisms (AFLPs) using 131 segregating F_1-derived DH progenies from a highly polymorphic BJ-99/BJ-70 cross. The parental lines were contrasting for plant height, flowering time, and seed coat color. BJ-99 was a yellow-seeded Russian genotype, taller in nature, and late-flowering, while BJ-70 was a brown-seeded Indian genotype with short stature and was early flowering. This genetic map harbored 273 loci (264 AFLPs, 9 RAPD) distributed over 18 LGs covering 1641 cM of the *B. juncea* genome. The average estimate of genome length was 1937.6 cM, and the estimated coverage of the *B. juncea* genome was 2251.5 cM. The framework map was presumed to cover 72.9% of this estimation.

Mahmood et al. (2003a) used reciprocal DH populations for studying the genes controlling the fatty acid profile in *B. juncea* using a genetic map with 316 RFLP loci. The DH populations were developed from a cross between two agronomically contrasting canola and non-canola quality *B. juncea* genotypes. The linked markers were grouped into 18 LGs and seven unlinked fragments, spanning a genetic length of 1564 cM. A total of 276 and 307 loci were mapped in the R and S populations, respectively, with 267 loci being common among the two populations. The nomenclature of the A subgenome LGs (A1-A10) of *B. juncea* was based on the A subgenome (N1-N10) LGs of *B. napus* (Cheung et al.1997; Butruille et al. 1999), while the B genome LGs were named as B1–B8. Gene order was shown to be preserved mostly, although frequent rearrangements were also identified in intra-genomic comparisons. The study was also the earliest to indicate that sex-based peculiarities were absent from recombination in *B. juncea*, and therefore, recombination did not depend on the direction of the cross. It was thus apparent that there were no restrictions on using a parent as the male or a female genotype in a crossing scheme. Moreover, integrated genetic maps of *B. juncea* could also be assembled from separate crosses as they were independent of male or female meiosis. The results also suggested that the direction of the cross in an experiment would not consequently impact fine-mapping for map-based cloning which requires accurate positioning of loci in a particular interval.

Pradhan et al. (2003) developed a high-density genetic linkage map of *B. juncea* using a F_1-derived DH population of 123 lines developed from a Varuna/Heera cross. The two parents are highly divergent as Varuna is a well-adapted, widely cultivated Indian genotype, while Heera is a canola quality exotic accession. These two parental lines were also highly contrasting for several agronomically important qualitative and quantitative traits. The 1,029 mapped loci (996 AFLPs and 33 RFLPs) were distributed over 18 LGs with a total map length of 1629 cM. The study also identified a set of 26 primer pairs for AFLP-based whole-genome selection (nine *Eco*RI/*Mse*I, six *Eco*RI/*Taq*I, six *Pst*I/*Mse*I, and five *Pst*I/*Taq*I) generating a total of 385 markers with a coverage of 96% of the genome, as mapped with the total set of 91 primer pairs. The whole-genome selection markers were spaced at an average interval of 6.8 cM between two consecutive markers. The selected set of AFLP primers could thus be used for whole-genome selection in backcross breeding programs in *B. juncea*. Ramchiary et al. (2007a) further saturated this genetic map by the addition of more AFLP, RFLP, SSR, and gene-based markers resulting in a revised linkage map of *B. juncea* consisting of 1448 markers (1297 AFLPs, 72 RFLPs, 69 SSRs, and 10 genic markers) with an overall genetic distance of 1840.1 cM.

Christianson et al. (2006) utilized a combination of microsatellite and RFLP markers to generate a linkage map based on a F_2 and a backcross population for genetic mapping of resistance to *Leptosphaeria maculans* in *B. juncea*. The parental lines included—'AC Vulcan', a *B. juncea* mustard cultivar, and UM3132, an inbred line of *B. juncea* described by Keri et al. (1997). Individual 'AC Vulcan' plants were pollinated by UM3132, and a single F_1 plant from this cross was self-pollinated for the development of the F_2 population segregating for resistance. The F_1 plant was also crossed with the susceptible parent UM3132 for the development of a first-generation backcross population. The total map with 323 loci covered 1577.3 Kosambi cM across 18 LGs. This map showed almost perfect collinearity with the genetic map developed by Axelsson et al. (2000) based on 75 common marker loci, except for four pairs of markers that were inverted between the two maps.

Panjabi et al. (2008) established the earliest and an exhaustive comparative linkage map of *B. juncea* showing a high level of synteny among the *A. thaliana* and *B. juncea* genomes. The map used single copy genes from Arabidopsis hosted on The Arabidopsis Information Resource (TAIR). Polymerase chain reaction (PCR) primers spanning intronic sequences were designed from genes which were present at approximate intervals of 100–200 kb. A total of 533 loci were mapped consisting of 486 intron polymorphism (IP), 34 RFLP, and 13 gene-based markers spanning 1992.2 cM. The 10 LGs of the A subgenome of *B. juncea* map were designated as A1–A10 corresponding to the nomenclature of LGs N1–N10 of the A subgenome of *B. napus* (Parkin et al. 2005). Homeologies among the A, B, and C subgenomes as described in this study were used for the nomenclature of the eight B subgenome LGs. *A. thaliana* genomic blocks were assigned to the LGs of the *B. juncea* linkage map on the basis of the distribution of 24 genomic blocks (A–X) described earlier for a hypothetical progenitor of the *A. thaliana* and Brassica ancestors by Schranz et al. (2006). This approach facilitated the identification of conserved blocks between *A. thaliana* and *B. juncea*. The study also identified the evolutionary rearrangements leading to karyotype diversification among the three genomes.

For QTL mapping of yield-associated traits in *B. juncea*, Yadava et al. (2012) constructed a genetic map consisting of AFLPs, IPs, and SSR markers. The TD map described in this study was derived from the F_1 of a cross of an Indian type (TM-4) with an exotic type (Donskaja-IV). The TD genetic map harbored 911 marker loci comprising of 585 AFLPs, 8 SSRs, and 318 IP loci with a total genetic length of 1629.9 cM. This study also reported the first detailed integrated map of *B. juncea* assembled by employing common markers as anchors among the two maps—Varuna × Heera map (described in Pradhan et al. 2003; Ramchiary et al. 2007a and Panjabi et al. 2008) and the TD map. This integrated map included a total of 2662 marker loci and spanned a total genetic length of 1927.1 cM. By projecting QTLs detected in the TD population on the QTLs identified in the VH population, and using VH map as the reference map, Yadava et al. (2012) optimized the number of consensus QTLs between the two populations, reduced the confidence intervals of these consensus or meta-QTLs, and consequently improved upon the accuracy and validity of the meta-QTLs substantially.

Khattak et al. (2019) also developed a genetic map based on IP markers using RILs for exploring the genetic basis of 2-propenyl and 3-butenyl glucosinolate biosynthesis by QTL mapping and RNA sequencing (RNA-seq) in *B. juncea*. The map spanned 1671.13 cM with an average distance of 5.5 cM between two consecutive markers and resulted in the identification of a novel major QTL for 2-propenyl and 3-butenyl glucosinolates on the LG A08.

11.2.3 Maps Developed Using Third-Generation DNA Markers

Paritosh et al. (2014) performed the RNA-seq of two *B. juncea* lines, Varuna and Heera, for the development of SNP markers with applications in both genome-wide and localized fine-mapping. Contigs obtained from the RNA-seq data were separated into gene models specific to the two subgenomes of *B. juncea* by comparisons with the A subgenome hosted on the Brassica database (BRAD) and to their work on the RNA-seq of *B. nigra* for the B subgenome (Paritosh et al. 2020). The contigs were also used to detect SNPs, which were placed on a genetic map developed by using F_1-derived DH lines based on the Varuna/Heera cross. The genotyping was based on a dataset of 999 SNPs identified in this study and 709 IP loci previously described in Panjabi et al. (2008). The total map length was 1933.5 cM, with an average interval size of 1.4 cM (Table 11.1). This map also provided a detailed understanding of the arrangement of the genomic blocks of *A. thaliana* (Schranz et al. 2006) in the A and B subgenomes of *B. juncea*. Different genes were also marked with codominant SNP markers in the vicinity of loci controlling glucosinolate (LGs A02, A03, A09 and B01) and erucic acid (LGs A08 and B07) traits, thus creating opportunities for the transfer of '00' traits from Heera into Indian lines. The study revealed the diverse nature of the constituent A and B subgenomes of *B. juncea* and strengthened the assumption that independent hexaploidization events have led to the evolution of the constituent A and B subgenomes of *B. juncea* based on comparative analysis of the arrangement of gene blocks and patterns of genome fragmentation.

Based on the GBS approach, DArT-seq, Zou et al. (2016) developed a dense *B. juncea* linkage map with high-throughput markers. An F_2 population of 168 segregating individuals derived from crossing Sichuan Yellow (SY) and purple mustard (PM) was used in the study. A total of 3329 of the markers were assigned into 18 large LGs with an average density of two loci per cM. These 3329 linked markers revealed 1570 distinct genetic loci on the 18 LGs and covered 1579 cM. Single markers consisted of 1132 unique loci, while 438 were defined by multiple markers and were classified as genetic 'bin loci'. By comparison of this map and the positioning of the marker sequences with respect to their organization in the physical genome of *A. thaliana*, the study reported both a highly consistent collinearity of the ancestral blocks and substantial block variation among the different A subgenomes. A total of 30 potential inversions distributed over large intervals and 20 potential translocation events among the three A subgenomes were identified by comparative analyses of genome sequences of *B. rapa* and *B. napus*, and marker sequences anchored on this genetic map.

Using 200 F_2 progenies developed from a cross between a tuber mustard and a leaf mustard, Li et al. (2016) developed a genetic map in *B. juncea* containing 116 SSR markers. A total of 520 pairs of SSR primer sequences designed from the transcriptome of the tuber mustard were used to amplify genomic regions with a two-stage screening being implemented for the identification of polymorphic markers between the two parental lines and the F_1 progeny. Seventeen LGs were identified in the linkage map with a total genetic distance of 2061.0 cM and an average distance of 17.92 cM between two consecutive markers. Based on this genetic map, QTL detection was performed for stem expansion traits pertaining to stem weight, stem diameter and stem length.

Yang et al. (2016) developed a high-resolution linkage map developed with 5333 bin markers distributed over the 18 pseudochromosomes while reporting the draft genome of the allopolyploid *B. juncea* var. tumida. The genome was constructed by a de novo assembly using shotgun reads, single-molecule long reads (PacBio sequencing), genomic (optical) mapping (BioNano sequencing) and genetic mapping.

Dhaka et al. (2017) identified genic SSRs in *B. juncea* and its ancestral species (*B. rapa* and *B. nigra*) by using unigene data of *B. rapa* harvested from NCBI Unigene database and transcriptomes of *B. nigra* and *B. juncea* developed by Paritosh et al. (2014). The study identified a total of 20,529 genic SSRs from the three *Brassica* species. The study also described the development of the first SSR-based genetic map of *B. juncea* using F_1-derived DH lines of EH-2/Pusa Jaikisan cross, as the segregating progenies. The map included 860 markers consisting of 462 unigene derived microsatellite (UGMs), 157 BAC-derived SSRs, and 241 IP loci, spanning 2073.6 cM.

Rout et al. (2018) analyzed oil content using eight bi-parental linkage maps of *B. juncea* developed using eight distinct mapping populations. The seven parental lines of these mapping populations were highly contrasting for quantities of oil and erucic acid in the seeds. These eight mapping populations were arranged into two groups of five and three populations, consisting of DH lines which were distinguished on the basis of whether they exhibited segregation for erucic acid (SE populations) or had zero erucic acid content (ZE populations). Three linkage maps ($EJ8^{A8B7}$, $EJ8^{Z}$, and DE^{Z}) were developed in this study–$EJ8^{A8B7}$ map consisted of 388 loci with a genetic length of 1125.6 cM, while the $EJ8^{Z}$ and DE^{Z} maps harbored 470 and 650 loci and spanned genetic distances of 1244.0 and 1680.2 cM, respectively. The other five genetic maps were reported earlier (Ramchiary et al. 2007a; Panjabi et al. 2008; Yadava et al. 2012; Paritosh et al. 2014; Rout et al. 2015; Dhaka et al. 2017), but were updated with SSR loci to extend the coverage of the *B. juncea* genome. An integrated genetic map was also reported in this study which was assembled by including the marker loci from the eight constituent maps using the common markers as anchors. This highly saturated map consisting of 3510 markers harbored a wide array of genetic markers including IP, SNP, genic SSR, genomic SSR, AFLP, and RFLP loci distributed over 2002.6 cM. A large fraction of about 70% of these mapped loci were unambiguously designated with their corresponding *A. thaliana* gene IDs and the genomic blocks of Arabidopsis (Schranz et al. 2006).

The latest linkage map to be developed in *B. juncea* was based on the novel GBS approach-based mapping performed by Paritosh et al. (2021) using *Sph*I-*Mluc*I and *Hinf*I-*Hpy*CH4IV restriction enzyme combinations. The study reported simultaneously but independently developed genetic maps from two F_1DH populations of *B. juncea*–VH and TuV. The VH genetic map which was firstly developed by

Pradhan et al. (2003) and later also included 833 markers comprising IP, genic SSR, and SNP loci (Panjabi et al. 2008; Paritosh et al. 2014) was further supplemented with GBS markers based on the *Sph*I-*Mluc*I enzyme combination. The VH genetic map harbored a total of 3780 markers among which a large proportion of 2947 were GBS-based SNP markers. The TuV genetic map was constructed from a population of 119 F_1-derived DH individuals of a Tumida/Varuna cross and consisted of 524, IP and SSR markers that were common with the VH population along with the 8517 GBS markers, based on the HinfI-*Hpy*CH4IV enzyme combination, which were added for a substantial genome coverage (Fig. 11.1). This map included a total of 9041 markers that were more than twice the number of markers in the VH map. The average distance of 0.4 cM between two consecutive loci in the TuV map was also less than half in comparison to 0.9 cM in the VH map. Both these maps were used to anchor scaffolds/contigs on to the 18 LGs of *B. juncea*, thus resulting in a chromosome-scale assembly of *B. juncea* line Varuna which appears to have a substantial improvement compared to the previous draft assembly of *B. juncea* var. tumida, a vegetable type of mustard (Yang et al. 2016).

11.3 Genetics and Mapping of Important Traits

An array of molecular markers has been used for mapping QTLs and cloning of genes influencing important agronomic, yield influencing, and seed oil quality traits in *B. juncea*. High-density genetic maps are an essential prerequisite for plant breeding applications such as the discovery of marker-trait associations and comparative genetic studies. In *B. juncea*, the principal objectives for genetic improvement of the crop apart from yield (described in Chap. 20) are white rust resistance, seed coat color, fatty acids, oil content, and glucosinolate content.

11.3.1 White Rust Resistance

The white rust disease caused by *Albugo candida* is one of the most destructive diseases of *B. juncea*. The accessions belonging to the Indian germplasm of *B. juncea* are extremely susceptible to *A. candida* (Li et al. 2008; Panjabi-Massand et al. 2010; Awasthi et al. 2012) and experience considerable damages in the crop yield (Lakra and Saharan 1989; Saharan and Verma 1992). Of the 13 pathotypes of *A. candida* studied in various *Brassica* species (Verma et al. 1999), the most prevalent race afflicting *B. juncea* has been recognized as race-2 (Petrie 1988; Rimmer et al. 2000). Extensive genetic research in *B. juncea* has shown that a single dominant gene governs resistance against race-2 (Tiwari et al. 1988; Sachan et al. 1995).

Rakow (1995) established a cross among two breeding lines of *B. juncea* (J90-4317 and J90-2733) with the objective of developing white rust resistant, high oil content, canola quality *B. juncea*. Among the two parental lines, J90-4317 (white rust sensitive) was used as the female parent, while J90-2733 (white rust resistant) was used as the male parent. Cheung et al. (1998a) identified a single gene responsible for conferring resistance to *A. candida* race-2 (*Acr*) using 119 F_1-derived DH progeny lines of the J90-4317/J90-2733 cross. The study identified a dominant RFLP marker, X140a, cosegregating with the *Acr* locus and two other RFLP markers, X42 (dominant) and X83 (codominant), closely linked with the *Acr* locus. Prabhu et al. (1998) also used the same DH population to identify RAPD loci associated with resistance/susceptibility to white rust using BSA. The resistance locus $Ac2_1$ was shown to be flanked by two markers, *WR2* and *WR3*, within an interval of 10 cM. These two markers were also observed to be highly effective in identifying the resistant/susceptible genotypes in the DH population.

RAPD markers were also used by Mukherjee et al. (2001) for genotyping of a set of 94 F_7 generation RILs obtained by selfing the F_2 progenies of a cross between Varuna and BEC-144,

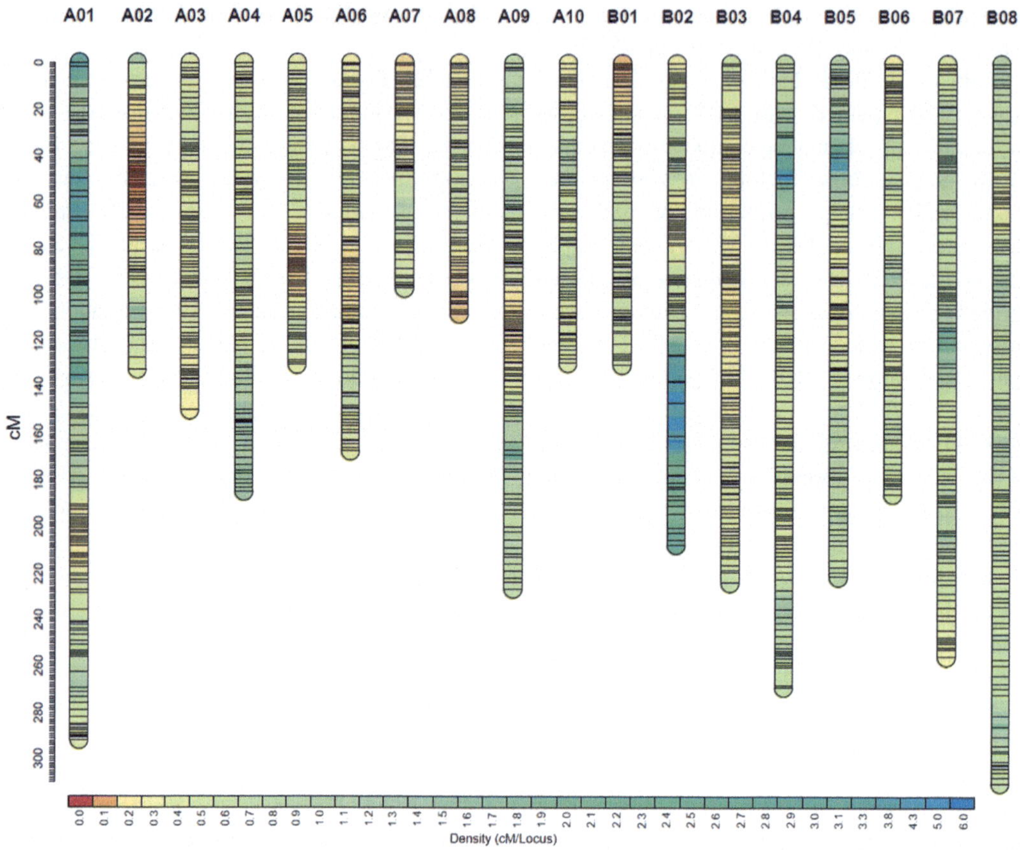

Fig. 11.1 Density gradient-based linkage map of *B. juncea* developed with intron polymorphism (IP), simple sequence repeats (SSRs), and genotyping by sequencing-based (GBS) single-nucleotide polymorphism (SNP) markers using DH lines of the Tumida/Varuna cross. The vertical ruler on the left represents the scale of measurement for the genetic lengths of the linkage groups. The density gradient at the bottom shows the color-coded density of markers in cM/locus. The figure was created using the R package LinkageMapView (Ouellette et al. 2018)

and reported the identification of flanking markers for a white rust resistance locus in BEC-144. A total of eight informative primers were able to distinguish the resistant bulk from the susceptible bulk. The rust resistance locus, designated as *Ac2* (t), was flanked by *OPB061000* and *OPN011000* at 5.5 cM and 9.9 cM, respectively. Absence of polymorphism between BEC-144 and Varuna for the markers which exhibited close linkage with the resistance locus $Ac2_1$ mapped by Prabhu et al. (1998), suggested that the locus mapped in this study might be different. Varshney et al. (2004) designed a PCR-based codominant cleaved amplified polymorphic sequence (CAPS) marker for the closely linked RAPD marker, *OPB061000* and a more tightly linked marker for the white rust resistance gene, using AFLP and BSA. They also validated the CAPS marker in two different F_2 populations developed from Varuna/BEC-144 and Varuna/BEC-286 crosses, and established its efficacy in MAS for white rust resistance.

Somers et al. (2002) focused on developing DNA markers for a novel *B. napus* locus ($Ac2V_1$) in a *B. juncea* genetic background using a BC_3F_2 population by identifying DNA markers derived from the introgressed *B. napus* chromosome segment. Two *B. juncea* cultivars were used as parental lines consisting of AC Vulcan (white rust race 2V-susceptible) and *B. napus* line S86-

69 (white rust resistant). BSA was performed with the two parents and two DNA bulks of eight white rust resistant and nine white rust-susceptible BC_3F_2 seedlings derived from a single BC_3F_1 plant. BSA using AFLP markers resulted in a total of 22 candidate markers being amplified. Eight of the 19 *B. napus* markers showed linkage to a single resistance gene.

Panjabi-Massand et al. (2010) used two east European germplasm lines, Heera and Donskaja-IV for understanding the genetics of resistance to *A. candida* race 2V. The study was based on two DH populations which were derived from the Varuna/Heera and TM-4/Donskaja-IV crosses. The Indian type parental lines–Varuna and TM-4 were susceptible, while the two east European lines–Heera and Donskaja-IV were partially and completely resistant, respectively. QTL analysis in the two DH populations revealed two major loci for resistance to white rust. In Heera, the locus *AcB1-A4.1* was located to LG A4, while in Donskaja-IV, the locus *AcB1-A5.1* was located to LG A5. *AcB1-A4.1* and *AcB1-A5.1* accounted for 76% and 70% of the total phenotypic variance, respectively. The study also described the use of comparative mapping with *A. thaliana* for generating marker loci closely linked with the white rust resistance loci. The comparative mapping showed that the *AcB1-A4.1* locus in *B. juncea* was homologous to the 'S block' of chromosome 5 of *A. thaliana* (Panjabi et al. 2008). A total of 54 IP primer sequences were developed from the S block of *A. thaliana* to densely map this QTL interval on the *B. juncea* map. Of the 54 primer pairs tested, eight markers mapped to the *AcB1-A4.1* region. *At5g41560* and *At5g41940*, within an interval of 1.2 cM on the VH map, showed perfect cosegregation with the phenotype except for the four individuals. Conversely, the QTL region of *AcB1-A5.1* in the TD map was observed to be homologous to the 'J' block of chromosome 2 of *A. thaliana* (Panjabi et al. 2008). A total of 147 IP primer sequences were constructed from the J block of *A. thaliana*, of which 38 pairs of sequences were polymorphic between the two parents. Cosegregation analysis with four of these markers which mapped to the QTL region, *AcB1-A5.1* was observed to lie between the markers *At2g34510* and *At2g36360*. The IP marker, *At2g36360* was found to be strongly linked with the resistance trait locus since no recombinants were observed. Based on these results, Arora et al. (2019) further showed the presence of a single CC-NB-LRR protein encoding R gene in Donskaja-IV (named as *BjuWRR1*) which conferred complete resistance to a range of isolates of *A. candida*. This was the first report of a gene conferring white rust resistance in the *Brassica* species (Arora et al. 2019). Identification of such resistance genes and markers will assist the introgression of resistance into otherwise susceptible varieties.

A new locus imparting resistance to *A. candida* isolate AcB1 was located by genetic mapping and QTL analysis to LG A6 of *B. juncea* by Bhayana et al. (2020). The study used a F_1-derived DH mapping population developed from a Tumida/Varuna cross. Tumida was observed to be resistant to the *A. candida* isolate AcB1, while Varuna was highly susceptible. The study used GBS markers to map the resistance locus to the interval of 63.0 cM–70.8 cM on LG A6 (Fig. 11.2). To identify the candidate gene(s) present within this region, the study used the currently available genome assemblies of *B. juncea* lines—Tumida (Yang et al. 2016) and Varuna (Paritosh et al. 2021). The study reported *BjuA046215*, a CC-NBS-LRR (CNL) type R gene with all the known subdomains of the proteins specified by such genes, to be the most probable candidate gene lying in the mapped interval that encodes a protein.

11.3.2 Seed Coat Color

Many studies have demonstrated the advantages of the yellow seed over brown seed characteristics of *Brassica* because of higher oil and protein content, and a lower fiber content in yellow-seeded *Brassica* genotypes (Stringam et al. 1974; Woods 1980; Hutcheson 1984; Shirzadegan and Röbbelen 1985; Rahman and McVetty 2011). It has been estimated that a shift to yellow-seeded cultivars can achieve an oil increase of up to 2% (Banga 1996).

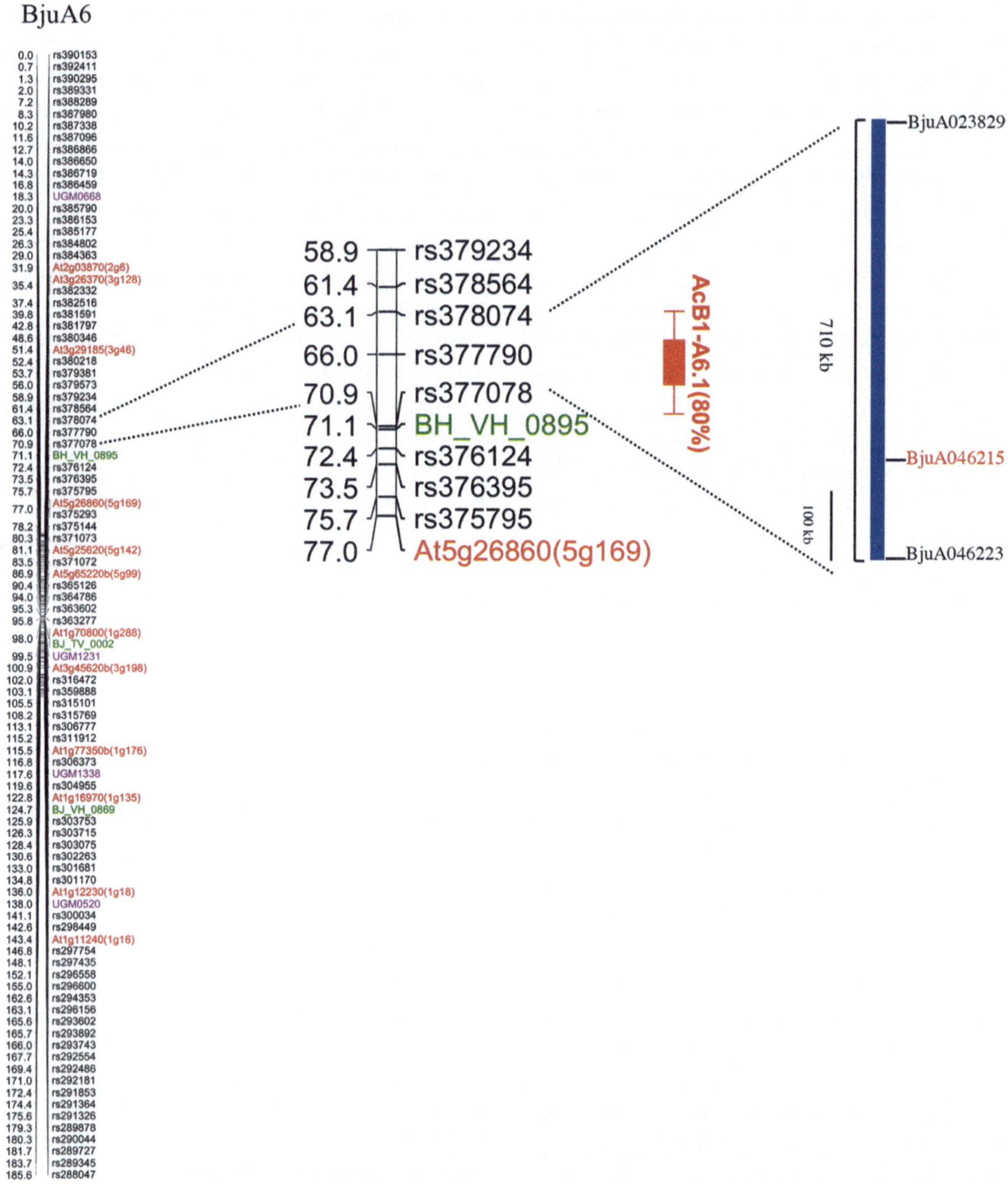

Fig. 11.2 Mapping of white rust resistance in the LG BjuA6. The LG contained 629 polymorphic markers (markers mapping at the same position were removed). The genic simple sequence repeats, intron polymorphism, genic single-nucleotide polymorphism, and GBS-based single-nucleotide polymorphism markers are represented with pink, red, green, and black color, respectively. A single major QTL (*AcB1*-A6.1; red bar) conferring resistance to white rust isolate AcB1 mapped to an interval of 63.0–70.8 cM. A CNL type R gene, *Bju046215*, was identified in the tumida genomic sequence spanning the QTL region (Adapted from Fig. 11.2 in Bhayana et al. 2020)

Different studies on seed coat color in *B. juncea* have shown that the trait is under the control of two loci segregating independently of each other, with either of these when harboring dominant alleles produce brown seeds. In addition, the maternal genotype also influences the manifestation of the trait (Vera et al. 1979; Vera and Woods 1982; Anand et al. 1985; Thiagarajah and Stringam 1993; Choudhary and Solanki 2007). While elucidating the genetics of seed coat color inheritance in *B. juncea*, Vera et al. (1979) concluded that the character is controlled by two duplicate pairs of genes (R_1 and R_2) for brown color, either of which can produce brown seed color when a single dominant allele is present, while yellow seed results when all alleles at both loci are recessive. By crossing a double heterozygote plant B-1092 (R1r1R2r2) of the cultivar Blaze with a yellow-seeded homozygous recessive (r1r1r2r2) plant from the same variety, Vera and Woods (1982) confirmed the regulation of seed coat color by two different dominant gene pairs by developing pure-breeding, brown-seeded strains carrying the gene for brown seed color at one locus, but not at the other, i.e., two different types of lines, homozygous dominant at one locus and homozygous recessive at the other (R1R1r2r2 and r1r1R2R2).

A combination of BSA and AFLP methodologies was employed by Negi et al. (2000) to detect marker loci linked to seed coat color in *B. juncea*. Three AFLP markers were specific for brown-seeded individuals and were therefore identified as being tightly linked to the seed coat color trait. For converting one AFLP marker into a sequence-characterized amplified region (SCAR) marker, the PCR walking methodology developed by Siebert et al. (1995) was employed to isolate genomic regions adjacent to the AFLP marker. The AFLP marker on conversion into the codominant SCAR marker (*SCM08*) generated amplification products which were polymorphic between yellow and brown-seeded individuals and easily differentiated between homozygous (BB) and heterozygous (Bb) brown-seeded individuals.

Upadhyay et al. (1996) used a F_2 population from an intervarietal cross of *B. juncea* cv. Varuna and exotic collection BEC-144 in a RFLP-based analysis for identifying association of markers with yellow seed coat color and six quantitative traits. The single factor analysis of variance revealed that the marker *BJG357c* showed tight linkage (3.9 cM) with yellow seed coat color locus (*r1*). The seed coat color trait was found to segregate into 15 (brown):1 (yellow) ratio confirming the digenic inheritance of the trait with duplicate gene action as reported earlier by Vera et al. (1979).

A population of 131DH lines produced from an F_1 plant of a cross between BJ-99 (with yellow seeds) and BJ-70 (with brown seeds) was used to develop an AFLP-based linkage map of *B. juncea* by Lionneton et al. (2002). The 131 DH progenies consisted of 100 lines with brown seed coats and 31 lines with yellow seed coats. This 3:1 ratio suggested a segregation of two Mendelian trait loci (MTL) determining seed coat color in this population which were assigned to two distinct LGs, with *Bjc1* and *Bjc2* cosegregating with the markers, E3M3_7 on LG3 and E8M7_4 on LG6, respectively.

Padmaja et al. (2005) employed microsatellite markers in *B. juncea* to map two independent loci controlling seed coat color using three populations consisting of segregating DH lines. Three SSR marker loci (*Ra2-A11*, *Na10-A08*, and *Ni4-F11*) were identified by BSA showing strong association with seed coat color. Subsequent genetic mapping located *Ra2-A11* and *Na10-A08* on LG1 at an interval of 0.6 cM from each other and marker *Ni4-F11* on LG2 of the linkage map of *B. juncea* (Pradhan et al. 2003).

Mahmood et al. (2005) also used a mapping population consisting of 112 DH lines from a cross of black/brown-seeded *B. juncea* 'RLM-514' of Indian origin, reciprocally crossed with a yellow-seeded *B. juncea* breeding line. Segregation analysis revealed that seed color was probably controlled by two unlinked loci with duplicate gene action. QTL analysis however identified three consistent QTLs across environments affecting seed color on LGs A6, A10 and B4. A major percentage of the phenotypic variances were described by these QTLs showing 43%, 31%, and 16% of the total variation, respectively, while these QTLs collectively exhibited a large fraction of 62% of the total available phenotypic variation in the population.

A F_2 population of 386 plants segregating for seed color was used by Huang et al. (2011) from a cross between two *B. juncea* landraces with light-yellow seed (Wuqi yellow mustard) and brown seed (Wugong mustard). Each F_2 individual was selfed to generate $F_{2:3}$ progenies, which were subsequently used to establish the genotype of each F_2 individual and to identify the markers linked to the yellow seed color allele. Four closely linked AFLP and SSR markers from a total of twenty AFLP and eight SSR markers were identified, which demonstrated good collinearity with those of *A. thaliana* chromosome 3, locating the homologue of the yellow seed color allele between *At3g14190* and *At3g32130*. The same cross was used for fine-mapping of the yellow seed trait in *B. juncea* by Huang et al. (2012). AFLP and SSR marker techniques were used to identify markers linked to the brown seed locus in a population of 1258 F_2 plants. The study identified 14 AFLP and 2 SSR markers that were closely linked to the brown seed locus from a total of 256 AFLP primer combinations and 456 pairs of SSR primers. The SSR marker *CB1022* showed codominant inheritance with its ability to distinguish between the different genotypes in the population. The study also reported the conversion of two AFLP markers (*EA06MC11* and *EA08MC13*) into SCAR markers, *SC1*, and *SC2*, which were subsequently mapped to the A9/N9 LG of *B. napus* based on synteny with the earlier reported genetic map of *B. napus* (Basunanda et al. 2010).

In continuation of their previous study in which two separate loci (*BjSc1* and *BjSc2*) were mapped (Padmaja et al. 2005), Padmaja et al. (2014) also identified three candidate flavonoid genes in *B. rapa*—*DFR*, *TT1*, and *TT8* on LG A9 based on a genome search using the Brassica database (BRAD, http://brassicadb.cn). Two homeologous *TT8* genes, *BjuA.TT8* and *BjuB.TT8* were mapped to LGs A9 and B3 of the *B. juncea* genome, respectively, and showed perfect cosegregation with *BjSc1* and *BjSc2* locus, respectively.

Huang et al. (2016) also used the Wuqi/Wugong mustard cross for the development of IP and SCAR markers linked to the yellow seed color gene in *B. juncea*. A total of twenty genes and eight marker sequences from the A09 chromosome were selected to design the IP and SCAR primers for screening a BC_8S_1 population consisting of 1256 individuals. Of the ten markers which were developed (five IP and five SCAR markers), two markers—*IP4* and *Y1* flanked the yellow seed color gene at distances of 0.1 and 0.3 cM, respectively. Another three IP markers, *IP1*, *IP2*, and *IP3* which were derived from *Bra036827*, *Bra036828*, and *Bra036829* independently, also cosegregated with the target gene. Detailed annotations of the genes in this region revealed that only *Bra036828* was associated with flavonoid biosynthesis and had high similarity with the *TRANSPARENT TESTA6* gene. The study concluded that *Bra036828* was possibly the gene responsible for yellow seed color.

11.3.3 Fatty Acids and Oil Content

In recent years, emphasis is being laid on the objectives of enhancement in essential nutritional quality of oil by development of low erucic acid cultivars. The oil quality of seed is determined by the underlying fatty acid composition, and high levels of erucic and eicosenoic acids are undesirable for oil quality compared to other oilseeds (Gupta and Pratap 2007). Therefore, the principal focus in breeding of *B. juncea* has been genetic improvement of seed quality by varying the proportion of fatty acids and to breed varieties suitable for nutritional and industrial purposes.

There have been several studies on genetic analysis of various fatty acids such as oleic and erucic acids performed by generating markers and identification of QTLs for these traits. The studies undertaken for understanding the genetics of erucic acid in *B. juncea* have revealed that the trait is under the control of two genes with additive effects, with zero-erucic acid content attributed to recessive genes (Kirk and Hurlstone 1983; Bhat et al. 2002; Gupta et al. 2004).

Earlier studies have used RFLP and AFLP markers for genetic mapping of QTLs associated with fatty acids including erucic acid trait in *B. juncea* (Cheung et al. 1998b; Lionneton et al. 2002; Mahmood et al. 2003a). Cheung et al.

(1998b) mapped the two *Fatty acid elongase* (*FAE*) genes as the loci controlling erucic acid (C22:1) in seeds using a segregating F_1DH population. The study also identified two QTLs associated with oil content which collocated with the *FAE* loci indicating that these two genes also influenced oil content in *B. juncea*. This conclusion was supported by the fact that the plants harboring low C22:1 alleles at both *FAE* loci also had low seed oil content. Mahmood et al. (2003a) confirmed the previous research on the inheritance of erucic acid content governed by a two-gene model with additive effects in *B. juncea* (Kirk and Hurlstone 1983) and in *B. napus* (Chen and Beversdorf 19901990).

Three further studies by Bhat et al. (2002), Lionneton et al. (2002) and Sharma et al. (2002) independently identified markers and QTLs linked with fatty acid (FA) content. According to Bhat et al. (2002), the gene located in the A subgenome displayed a greater influence on the total erucic acid content in *B. juncea* than the gene located in the B subgenome. However, Lionneton et al. (2002) observed a bimodal distribution for the erucic acid content supporting the hypothesis that only a single major gene controlled this trait in *B. juncea*, in contrast to the results reported by Kirk and Hurlstone (1983) and Cheung et al. (1998a) who suggested that two loci—one from *B. campestris* and the other one from *B. nigra*—controlled erucic acid synthesis. Sharma et al. (2002) used a linkage map constructed with RAPD markers for locating QTLs for oleic acid and identified two QTLs with genetic lengths of 14 cM and 10.6 cM for oleic acid levels on two different LGs of *B. juncea*. The two QTLs together described 32.2% of the total phenotypic variance, while the major QTL located on LG 9 explained 28.5% of the trait's phenotypic variance.

Mahmood et al. (2003a) performed the first detailed genetic analysis of markers and localization of QTLs associated with the oleic, linoleic, linolenic, and erucic acids content in *B. juncea*. The study revealed that two QTLs (*E1a and E1b*) were consistently associated with erucic, oleic, and linoleic acid content in multiple environments. The QTLs for erucic, oleic, and linoleic acids together accounted for 85.8%, 70.9%, and 68.3% of the total phenotypic variation, respectively. These QTLs also showed strong epistasis, and with inclusion of the epistatic term into the model explained approximately 98%, 97%, and 85% of the variation for erucic, oleic, and linoleic acids, respectively. The study also noted that the same QTLs which were associated with oleic, linoleic, and linolenic acids acted in an opposing manner, and there was also observed a significant negative relationship of erucic acid with oleic and linoleic acids. It was thus hypothesized that under the conditions where erucic acid pathway is functional, the majority of the oleic acid precursor is converted into erucic acid and only a small percentage is desaturated to linoleic acid leading to a strong negative correlation of erucic acid with oleic and linoleic acids.

Aggarwal et al. (2003) also carried out mapping of loci influencing the palmitic, stearic, oleic, linoleic, linolenic, and erucic acids content of mustard oil. The study reported similar findings as reported by Mahmood et al. (2003a) in that the content of erucic acid was negatively correlated with the contents of all other fatty acids, which were positively correlated. The results of QTL analysis using two different methods, single factor ANOVA ($P < 0.02$) and interval mapping ($LOD > 2.0$) were comparable and revealed the presence of two loci each on three different LGs for the linoleic, linolenic, and erucic acid contents. The loci for erucic and linoleic acid located on LG 9 appeared to be linked, while the position of the erucic acid locus exactly coincided with the oleic acid locus identified earlier (Sharma et al. 2002), thereby indicating a common locus for both the traits. A similar condition was reported for the LOD peaks for erucic and linoleic acid contents on the LG 17 in which the two markers *OPD06600* and *OPA11400* showed significant linkages between loci for linolenic acid and oleic acid. The study concluded that this marker interval affected four of the major fatty acids (oleic, linoleic, linolenic, and erucic acids) of mustard.

Further genetic analysis of the erucic acid trait influenced by the *FAE1* gene was carried out by

Gupta et al. (2004) by molecular tagging based on QTL analysis and identification of SNPs in the *FAE1* gene. The research utilized 123 DH lines derived from the high × low erucic acid (Varuna/Heera) cross previously described by Pradhan et al. (2003), and identified two QTLs associated with seed erucic acid content which were located to the LGs 3 and 17 of the *B. juncea* genome. As previously reported by Bhat et al. (2002), this study also revealed that the two QTLs contributed disproportionately to the levels of erucic acid in *B. juncea*. Two *FAE1* genes (*FAE1.1* and *FAE1.2*) were identified in both high and low erucic acid mustard lines in the study by PCR amplification and subsequent cloning and sequencing. The sequence comparisons among the corresponding *FAE1* genes between Varuna (high erucic acid) and Heera (low erucic acid) revealed a total of seven SNPs, four in *FAE1.1* and three in *FAE1.2*. Using the single-nucleotide primer extension method (SNuPE) of SNP genotyping (Chen et al. 1999), these two genes were mapped to two independent intervals on the genetic map of *B. juncea* which cosegregated with the two erucic acid QTLs. These results were further corroborated by genotyping of 40 segregating BC_5F_2 progenies derived from the same cross, with Varuna being used as the recurrent parent. Genotyping of these 40 BC_5F_2 progenies with the seven SNPs in the *FAE1* genes guided the identification of all nine genotypes as expected from the segregation of two independent loci. The results clearly showed that the zero-erucic types were characterized by *e1e1e2e2* genotypes, while the highest erucic-type individuals were identified by the *E1E1E2E2* genotypes. Also, the wild-type alleles showed an additive effect on the erucic acid phenotype with the heterozygote class (*E1e1E2e2*) showing an intermediate phenotype. Ramchiary et al. (2007a) also used the DH population from the same Varuna/Heera cross and identified QTLs for oil content using an AFLP-based genetic map of *B. juncea*. Two of the seven detected loci were major QTLs which mapped to the known positions of the two *FAE1* genes (*FAE1.1* and *FAE1.2*). It was also reported that the erucic acid content of the seed directly influenced seed oil content in *Brassica* species, as the rise in the molecular mass during the extension of oleic acid (C18:1) to erucic acid (C22:1) could explain the equivalent increase in the oil content. Conversely, a corresponding reduction in oil content has been observed with the first rapeseed varieties which were free of erucic acid (Ecke et al. 1995; Burns et al. 2003).

Jagannath et al. (2011) analyzed the phenotypic variation in different oil quality fractions among Varuna (with high erucic acid), Heera (with low erucic acid), and a zero-erucic Indian variety (ZE-Varuna). The study was based on DH lines derived from a Varuna/Heera cross and segregating for erucic acid content (SE), as described previously in Pradhan et al. (2003). A second population consisting exclusively of zero-erucic acid DH lines (ZE) was constituted from the original population of 1200 DH plants derived from the same Varuna/Heera cross. An AFLP-based skeletal linkage map was constructed in this study using these ZE DH lines for QTL analysis of oil quality fractions and oil content. The study reported novel loci which could potentially be used to improve oil quality and oil content in *B. juncea* under zero-erucic conditions. Furthermore, the results revealed that a constraint on the available substrate caused by a functional erucic acid pathway was responsible for reduced levels of linoleic and linolenic acid in Varuna, and not the effect of weaker alleles or enzyme limitation.

A comprehensive analysis of QTLs influencing oil content was undertaken by Rout et al. (2018) in eight different mapping populations. These DH populations were derived from seven different parents exhibiting extensive variation in the levels of oil and erucic acid in the seeds. The eight mapping populations as described in Sect. 11.2.3 were categorized as SE and ZE populations depending upon the segregation of the erucic acid trait. The five F_1DH source populations segregating for one or both the erucic acid genes located on the LGs A08 and B07 were randomly sampled for constituting the five SE populations ($EJ8^{A8B7}$, DE^{B7}, EPJ^{A8B7}, TD^{A8}, and VH^{A8B7}). QTL studies of the erucic acid trait in the three SE populations segregating for both the erucic acid genes ($EJ8^{A8B7}$, EPJ^{A8B7}, and VH^{A8B7}) consistently identified major QTLs on the LGs A08 and B07 comapping to the two *FAE1* loci in all the three environments. These QTLs explained large average phenotypic variances varying from 87.5% (in EPJ^{A8B7} population) to 94.0% (in VH^{A8B7}population). In contrast, the other two SE populations ($TD^{A8}$and DE^{B7}) detected major QTLs for the trait comapping with the respective *FAE1* locus in all the environments. Comparisons among the QTLs for seed oil content between the two types of populations (SE and ZE) showed that the 23 QTLs detected in the SE populations were exclusive and were entirely missing in the three ZE populations. These loci influencing oil content constituted 76% of the major QTLs detected in the five SE populations and merged into two consensus QTLs in SE populations upon meta-analysis (*Oil-C-A8-1* on LG A08 and *Oil-C-B7-1* on LG B07). In addition, results of multi-trait analysis to correlate structures among erucic acid and oil content in seeds revealed a strong pleiotropic effect between these two traits. As a corollary, the study also noted that the two 'Pleiotropic' consensus QTLs influencing oil content cannot be utilized in improving the oil content in ZE *B. juncea*.

Saini et al. (2019) explored the sequence variations in the promoter region of the gene encoding fatty acid elongase (*FAE1*) by evaluating promoter polymorphisms in the *FAE1.1* and *FAE1.2* genes associated with erucic acid content in *B. juncea*. A low erucic acid (Pusa Mustard 30) and a high erucic acid cultivar (Pusa Bold) were used for sequencing the upstream regions of *FAE1.1* and *FAE1.2* genes. Polymorphisms in the promoter regions of *FAE1.1* and *FAE1.2* were discovered in the form of a 28-bp deletion in *FAE1.1* and a 340-bp insertion of a transposon-like element in the *FAE1.2*, in the low erucic acid genotype. The markers based on sequence variability in the promoter regions clearly differentiated the genotypes between low erucic acid and high erucic acid groups of genotypes belonging to *B. juncea* and its two ancestral diploid species, *B. rapa* and *B. nigra*. Genotyping of these promoter-based markers on a BC_1F_2 population developed by crossing Pusa Mustard 30 with Pusa Bold perfectly distinguished the parents and illustrated unambiguous cosegregation with the biochemical phenotype in the backcross progenies without any recombination.

11.3.4 Glucosinolates Content

Several studies have been undertaken for understanding the genetics of seed glucosinolates in *B. juncea*. Love and coauthors in as early as 1990 studied the genetic control of glucosinolate composition in *B. juncea* in crosses between 2-propenyl type glucosinolate containing parental lines (Domo and ZEM 84–2293) and an Indian 3-butenyl type mustard line, 60143 (Love et al. 1990). Subsequent efforts by Stringam and Thiagarajah (1995) showed that glucosinolate biosynthesis was under the control of nuclear genes with five to nine genes influencing the contents of aliphatic glucosinolates in *B. juncea*.

An exhaustive genetic evaluation of the glucosinolate content and fractions in the varieties of *B. juncea* indicated a large contrast in the composition observed in the Indian cultivars compared to *B. juncea* germplasm from other countries (Sodhi et al. 2002). It was observed that the Indian varieties were rich with 2-propenyl (allyl) and 3-butenyl as the major fractions of glucosinolates, and 4-pentenyl as the minor fractions. Conversely, the exotic germplasm was typified by the existence of 2-propenyl as the major glucosinolate fraction, trace amounts of 3-butenyl, and a complete absence of the 4-pentenyl types. The inheritance of total glucosinolates was further investigated by using 752 F_1DH and 1263 BC_1 (BC_1DH) progenies of a cross between Varuna (with high glucosinolates) and Heera (a low glucosinolate line). The investigation revealed that the total glucosinolate was under the control of seven genes based on the segregation ratios calculated from the frequencies of the low glucosinolate individuals (Sodhi et al. 2002).

Molecular mapping of a number of seed quality traits *B. juncea* was undertaken by Cheung et al. in 1998 with a F_1DH mapping population derived from two parental lines, J90-4317 (low glucosinolates) and J90-2733 (high glucosinolates) (Cheung et al. 1998b). The study detected two QTLs associated with the trait of 2-propenyl and three QTLs for 3-butenyl glucosinolates which described large proportions of the total phenotypic variances. In a later study, Mahmood et al. (2003b) reported three QTLs for 2-propenyl glucosinolate which explained 78% of the total phenotypic variance, and five QTLs for total seed aliphatic glucosinolates which explained phenotypic variances between 30% and 45%. Lionneton et al. (2004) also analyzed sinigrin, gluconapin, and total GSL contents in a DH population which was derived from parents which contrasted for these fractions and total glucosinolates (Lionneton et al. 2004). Two QTLs each for sinigrin, gluconapin, and total glucosinolates were identified which explained between 4–18% of the trait variation.

Two different studies by Ramchiary et al. (2007b) and Bisht et al. (2009) have reported in details the genetic mapping of loci implicated in the biosynthesis of aliphatic glucosinolates. Ramchiary et al. (2007b) analyzed QTLs for various seed glucosinolates among early (F_1DH) and advanced generation (BC_4DH) populations and suggested that epistasis and additive effects of glucosinolate genes exist in different genetic backgrounds in *B. juncea*. The common QTLs detected in both F_1DH and BC_4DH populations exhibited a change in the landscape of the phenotypic variances explained by the QTLs from the former to the later generations. Notably, while some of the QTLs detected in the F_1DH population became insignificant in the BC_4DH population, new QTLs were identified in the BC_4DH population for various glucosinolates. This study, however, was not able to explain the homeologous effects of loci from the two subgenomes on the individual or total seed glucosinolate content. Bisht et al. (2009) followed these results and carried out fine-mapping of six seed glucosinolate QTLs (*J2Gsl1*, *J3Gsl2*, *J9Gsl3*, *J16Gsl4*, *J17Gsl5,* and *J3Gsl6*) reported by Ramchiary et al. (2007b) using candidate gene approach. Further, DNA sequences of the various genes implicated in the aliphatic glucosinolate biosynthesis pathway were identified in *A. thaliana* and *B. oleracea*, and the candidate genes were characterized in the two parental lines, Varuna (high glucosinolate) and Heera (low glucosinolate).

Rout et al. (2015) described the existing allelic variations for five different seed glucosinolate traits in *B. juncea* by performing QTL mapping with two DH mapping populations (DE and VH populations) derived from parental lines with contrasting seed glucosinolates profiles. The seed glucosinolate traits were evaluated in three environments and subjected to QTL analysis followed by QTL meta-analysis in Biomercator 2.1 (Arcade et al. 2004; Goffinet and Gerber 2000). A total of 60 QTLs were identified in the DE population from three environments which upon intra-population meta-analysis reduced to 17 S-QTLs (Stable QTL) and 15 E-QTLs (Environment QTL). The VH population detected a total of 58 QTLs from the three environments that were reduced to 15S-QTLs and 16E-QTLs upon meta-analysis. In both the DE and VH populations, majority of S-QTLs explained large proportions of phenotypic variances of the existing trait variation and were therefore regarded as major QTL. The study also identified three C-QTL (consensus QTL) following inter-population meta-analysis, *Bju.GSL.A3.1*, *Bju.GSL.A9.1* and *Bju.GSL.B4.1* all of which comapped with glucosinolate pathway candidate genes and were distributed over LGs A03, A09, and B04. The study also ranked the parental allelic variants of QTLs or of the comapped candidate gene(s) based on the phenotypic variances explained by the component QTLs. The results of the study appear significant for marker-assisted transfer of the low glucosinolate trait and for breeding of lines with lower seed glucosinolate content than are currently present in *B. juncea*.

Sharma et al. (2016) isolated and characterized an ortholog of *A. thaliana* gene *At1g16410*, described as *CYP79F1*, which is involved in the initial step of core glucosinolate biosynthesis. *CYP79F1* was mapped to the QTL *J16Gsl4* in LG B04 using the existing *B. juncea* genetic map of Panjabi et al. (2008). Functional validation of the candidate gene *CYP79F1* (named as *BjuB.*

CYP79F1) responsible for regulating sinigrin biosynthesis was undertaken using 95 F_2 progenies derived from a cross between Varuna and a QTL-NIL (QTL Near-isogenic line) *J16Gsl4* through cosegregation analysis of the genotype and the phenotype. All the F_2 progenies containing the Varuna allele of the gene (wild type) synthesized sinigrin while the progenies harboring the Heera allele were deficient in the synthesis of sinigrin.

In a recent study, Khattak et al. (2019) identified one novel major QTL for 2-propenyl (sinigrin) and 3-butenyl (gluconapin) glucosinolates on the LG A08 of the *B. juncea* genome using 167 F_6 RILs developed from a cross among two DH lines, G266, and G302. The seeds of the parental line G266 had low 2-propenyl and high 3-butenyl glucosinolates, while seeds of G302 had high 2-propenyl and low 3-butenyl GSL contents. The RILs were evaluated for the 2-propenyl and 3-butenyl GSL contents during 2016 and 2017 for undertaking QTL analysis. Two collocating major QTLs for sinigrin contents were identified on LG A08, and explained 42.3% and 42.6% of the total variation in 2016 and 2017, respectively. Another two QTLs for gluconapin were detected and colocalized with the QTLs for sinigrin content on the LG A08. The QTLs for gluconapin contents exhibited phenotypic variances of 31% and 38.4% of the total variation in 2016 and 2017, respectively. *At4g20150-1* was identified as the IP marker that was closest to the peak of the LOD profiles of sinigrin and gluconapin content.

11.3.5 Future Scope of Works

RFLP, RAPD, and AFLP markers were earlier used in the development of genetic maps and discovery of quantitative trait loci in *B. juncea*. Although there have been large-scale advancements in the discovery of genetic markers in *B. juncea* [(e.g., Illumina SNP chips, kompetitive allele-specific PCR (KASP) assays, GBS and sequencing technologies (e.g., Illumina, PacBio, Nanopore)], the genomes of only two accessions —Tumida and Varuna are currently available, representing only a small fraction of the species-wide genomic space (Yang et al. 2016; Paritosh et al. 2014, 2021). In view of the extreme diversity available among the different genotypes of *B. juncea*, concerted efforts are needed to detect further variations by including some additional reference genomes that will provide useful resources for genetic studies such as gene mapping and cloning. The way forward thus lies in undertaking pan-genomic studies in *B. juncea* which harvest the enormous variation available in different plant morphologies and entire sequence diversity, to provide useful resources for evolutionary studies, functional genomics, and breeding of cultivated Brassicas. Recently in *B. napus*, many presence and absence variations were identified using pan-genome comparative analysis based on de novo assembly and annotation of eight accessions (Song et al. 2020). A genome-wide association study based on these structural variants (SV) to explore their contribution to trait variation in three important yield-related traits (silique length, seed weight, and flowering time) led to discovery of causal associations between SVs and yield-related traits that were not captured by single-nucleotide polymorphisms-based genome-wide association study (GWAS) demonstrating the high worth of structural variants. With the availability of pan-genome in *B. juncea*, it is anticipated that breeders will be able to apply gene-editing technologies to rapidly manipulate and attain desirable agronomic traits in the germplasm lines, based on the information on causal genetic variants underlying these traits. The currently available genome assemblies and those that will be available in future will also permit greater utilization of the wide variability available within the different genepools of *B. juncea*.

References

Aggarwal RAK, Sharma R, Kumar R, Mohapatra T, Sharma RP (2003) Molecular mapping of loci affecting the contents of three major fatty acids in Indian mustard (*Brassica juncea* L). J Plant Biochem Biotechnol 12:131–137

Anand IJ, Reddy WR, Rawat DS (1985) Inheritance of seed coat colour in mustard. Indian J Genet 45:34–37

Arcade A, Labourdette A, Falque M, Mangin B, Chardon F, Charcosset A, Joets J (2004) BioMercator: integrating genetic maps and QTL towards discovery of candidate genes. Bioinformatics 20:2324–2326

Arora H, Padmaja KL, Paritosh K, Mukhi N, Tewari A, Mukhopadhyay A et al (2019) BjuWRR1, a CC-NB-LRR gene identified in *Brassica juncea*, confers resistance to white rust caused by *Albugo candida*. Theor Appl Genet 132:2223–2236

Awasthi RP, Nashaat NI, Kolte SJ, Tewari AK, Meena PD, Bhatt R (2012) Screening of putative resistant sources against Indian and exotic isolates of *Albugo candida* inciting white rust in rapeseed-mustard. J Oilseed Brassica 3(1):27–37

Axelsson T, Bowman CM, Sharpe AG, Lydiate DJ, Lagercrantz U (2000) Amphidiploid *Brassica juncea* contains conserved progenitor genomes. Genome 43:679–688

Banga SK (1996) Breeding for oil and meal quality. In: Chopra VL, Prakash S (eds) Oilseed and vegetable brassicas: Indian perspective. Oxford and IBH Publishing Company, New Delhi, pp 234–249

Basunanda P, Radoev M, Ecke W, Friedt W, Becker HC, Snowdon RJ (2010) Comparative mapping of quantitative trait loci involved in heterosis for seedling and yield traits in oilseed rape (*Brassica napus* L.). Theor Appl Genet 120(2):271–281

Bhat MA, Gupta ML, Banga SK, Raheja RK, Banga SS (2002) Erucic acid heredity in *Brassica juncea*—some additional information. Plant Breed 121:456–469

Bhayana L, Paritosh K, Arora H, Yadava SK, Singh P, Nandan D et al (2020) A mapped locus on LG A6 of *Brassica juncea* line Tumida conferring resistance to white rust contains a CNL type R gene. Front Plant Sci 10:1690

Bisht NC, Gupta V, Ramchiary N, Sodhi YS, Mukhopadhyay A, Arumugam N, Pental D, Pradhan AK (2009) Fine mapping of loci involved with glucosinolate biosynthesis in oilseed mustard (*Brassica juncea*) using genomic information from allied species. Theor Appl Genet 118:413–421

Burns MJ, Barnes SR, Bowman JG, Clarke MHE, Werner CP, Kearsey MJ (2003) QTL analysis of an intervarietal set of substitution lines in *Brassica napus*: (i) seed oil content and fatty acid composition. Heredity 90:39–48

Burton W, Ripley V, Potts D, Salisbury P (2004) Assessment of genetic diversity in selected breeding lines and cultivars of canola quality *Brassica juncea* and their implications for canola breeding. Euphytica 136:181–192

Butruille DV, Guries RP, Osborn TC (1999) Linkage analysis of molecular markers and quantitative trait loci in populations of inbred backcross lines of *Brassica napus* L. Genetics 153:949–964

Chen JL, Beversdorf WD (1990) Fatty acid inheritance in microspore-derived populations of spring rapeseed (*Brassica napus* L.). Theor Appl Genet 80:465–469

Chen S, Tong N (1985) A preliminary study of different morphotypes in mustard (*Brassica juncea*). Cruciferae Newsl 10:37

Chen X, Levine L, Kwok PY (1999) Fluorescence polarization in homogenous nucleic acid analysis. Genome Res 9:492–498

Cheung WY, Friesen LJ, Rakow G, Seguin-Swartz G, Landry BS (1997) An RFLP based linkage map of mustard *Brassica juncea* (L.) Czern. and Coss. Theor Appl Genet 94:841–851

Cheung WY, Gugel RK, Landry BS (1998a) Identification of RFLP markers linked to white rust resistance gene (Acr) in muatard (*Brassica juncea* (L.) Czern. & Coss.). Genome 41:626–628

Cheung WY, Landry BS, Raney P, Rakow GFW (1998b) Molecular mapping of seed quality traits in *Brassica juncea* L. Czern., and Coss. Acta Hort 459:139–147

Choudhary BR, Solanki ZS (2007) Inheritance of siliqua locule number and seed coat colour in *Brassica juncea*. Plant Breed 126:104–106

Christianson JA, Rimmer SR, Good AG, Lydiate DJ (2006) Mapping genes for resistance to *Leptosphaeria maculans* in *Brassica juncea*. Genome 49:30–41

Dhaka N, Mukhopadhyay A, Paritosh K, Gupta V, Pental D, Pradhan AK (2017) Identification of genic SSRs and construction of a SSR-based linkage map in *Brassica juncea*. Euphytica 213:15

Ecke W, Uzunova M, Weissleder K (1995) Mapping the genome of rapeseed (*Brassica napus* L.) II. Localization of genes controlling erucic acid synthesis and seed oil content. Theor Appl Genet 91:972–977

Edwards D, Salisbury PA, Burton WA, Hopkins CJ, Batley J (2007) Indian mustard. In: Kole C (ed) Genome mapping and molecular breeding in plants, vol Oilseeds. Springer. Berlin, Heidelberg, pp 179–210

Goffinet B, Gerber S (2000) Quantitative trait loci: a meta-analysis. Genetics 155:463–473

Gupta SK, Pratap A (2007) History, origin and evolution. In: Gupta SK (ed) Advances in botanical research-rapeseed breeding, vol 45. Academic Press. London, UK, pp 1–20

Gupta V, Mukhopadhyay A, Arumugam N, Sodhi YS, Pental D, Pradhan AK (2004) Molecular tagging of erucic acid trait in oilseed mustard (*Brassica juncea*) by QTL mapping and single nucleotide polymorphisms in *FAE1* gene. Theor Appl Genet 108:743–749

Huang Z, Zhang Y, Li HQ, Yang L, Ban YY, Xu AX, Xiao ES (2011) AFLP and SSR markers linked to the yellow seed colour gene in *Brassica juncea* L. Czech J Genet Plant Breed 47:149–155

Huang Z, Ban YY, Yang L, Zhang Y, Li HQ, Xiao ES, Xu AX, Zhang DH (2012) Fine mapping of the yellow seed locus in *Brassica juncea* L. Genome 55:8–14

Huang Z, Liu L, Lu H, Lang L, Zhao N, Ding J, Xu A (2016) Development of IP and SCAR markers linked to the yellow seed color gene in *Brassica juncea* L. Breed Sci 66(2):175–180

Hutcheson DS (1984) Performance of varietal hybrids and relationship of seed size and color with meal protein and crude fiber in *Brassica campestris* L. PhD Thesis. University of Saskatchewan, Saskatoon, Canada

Jagannath A, Sodhi YS, Gupta V, Mukhopadhyay A, Arumugam N, Singh I, Rohatgi S, Burma PK, Pradhan AK, Pental D (2011) Eliminating expression of erucic acid-encoding loci allows the identification of "hidden" QTL contributing to oil quality fractions and oil content in *Brassica juncea* (Indian mustard). Theor Appl Genet 122(6):1091–1103

Jain A, Bhatia S, Banga SS, Prakash S, Lakshmikumaran M (1994) Potential use of random amplified polymorphic DNA (RAPD) technique to study the genetic diversity in Indian mustard (*Brassica juncea*) and its relationship to heterosis. Theor Appl Genet 88:116–122

Keri M, van den Berg CJG, McVetty PBE, Rimmer SR (1997) Inheritance of resistance to *Leptosphaeria maculans* in *Brassica juncea*. Phytopathology 87:594–598

Khattak AN, Wang T, Yu K, Yang R, Wan W, Ye B, Tian E (2019) Exploring the basis of 2-propenyl and 3-butenyl glucosinolate synthesis by QTL mapping and RNA-sequencing in *Brassica juncea*. PLoS One 14(10):e0220597

Kirk JTO, Hurlstone CJ (1983) Variation and inheritance of erucic acid content in *Brassica juncea*. Z Pflanz 90:331–338

Kumar R, Gupta V (1985) Isozyme studies in Indian mustard (*Brassica juncea*). Theor Appl Genet 79:109–110

Labana KS, Gupta ML (1992) Importance and origin. In: Labana KS, Banga S, Banga SK (eds) Breeding oilseed brassicas. Springer-Verlag, Berlin, Heidelberg, pp 1–7

Lakra B, Saharan G (1989) Correlation of leaf and staghead infection intensities of white rust with yield and yield components of mustard. Indian J Mycol Plant Pathol 19:279–281

Li C, Sivasithamparam K, Walton G, Fels P, Barbetti M (2008) Both incidence and severity of white rust disease reflect host resistance in *Brassica juncea* germplasm from Australia, China and India. Field Crops Res 106:1–8

Li J, Zou X, Zhang L, Cao L, Chen L (2016) Linkage map construction using SSR markers and QTL analyses of stem expansion traits in *Brassica juncea*. Sci Hort 209:67–72

Lionneton E, Ravera S, Sanchez L, Aubert G, Delourme R, Ochatt S (2002) Development of an AFLP-based linkage map and localization of QTLs for seed fatty acid content in condiment mustard (*Brassica juncea*). Genome 45:1203–1215

Lionneton E, Aubert G, Ochatt S, Merah O (2004) Genetic analysis of agronomic and quality traits in mustard (*Brassica juncea*). Theor Appl Genet 109:792–799

Love HR, Rakow G, Raney JP, Downey RK (1990) Genetic control of 2-propenyl and 3-butenyl glucosinolate synthesis in mustard. Can J Plant Sci 70:425–429

Mahmood T, Ekuere U, Yeh F, Good AG, Stringam GR (2003a) RFLP linkage analysis and mapping genes controlling the fatty acid profile of *Brassica juncea* using reciprocal DH populations. Theor Appl Genet 107(2):283–290

Mahmood T, Ekuere U, Yeh F, Good AG, Stringam GR (2003b) Molecular mapping of seed aliphatic glucosinolates in *Brassica juncea*. Genome 46(5):753–760

Mahmood T, Rahman MH, Stringam GR, Raney JP, Good AG (2005) Molecular markers for seed colour in *Brassica juncea*. Genome 48:755–760

Michelmore RW, Paran I, Kesseli RV (1991) Identification of markers linked to disease resistance genes by bulked segregant analysis: a rapid method to detect markers in specific genomic regions by using segregating populations. Proc Natl Acad Sci USA 88:9828–9832

Mohapatra T, Sharma A, Upadhyay A, Chopra VL, Sharma RP (1995) RFLP mapping in Indian Mustard (*Brassica juncea*) Cruciferae Newsletter 17:22–23

Mukherjee AK, Mohapatra T, Varshney A, Sharma R, Sharma RP (2001) Molecular mapping of a locus controlling resistance to *Albugo candida* in Indian mustard. Plant Breed 120:483–487

Negi MS, Devic M, Delseny M, Lakshmikumaran M (2000) Identification of AFLP fragments linked to seed coat colour in *Brassica juncea* and conversion to a SCAR marker for rapid selection. Theor Appl Genet 101:146–152

Oram R, Salisbury P, Kirk J, Burton W (1999) *Brassica juncea* breeding. In: Salisbury PA, Potter TD, McDonald G, GreenAG (eds) Org Comm 10th Intl Rapeseed Congr, Canberra, ACT, Australia, pp 37–40

Ouellette LA, Reid RW, Blanchard SG, Brouwer CR (2018) Linkage map view-rendering high-resolution linkage and QTL maps. Bioinformatics 34:306–307

Padmaja KL, Arumugam N, Gupta V, Mukhopadhyay A, Sodhi YS, Pental D, Pradhan AK (2005) Mapping and tagging of seed coat colour and the identification of microsatellite markers for marker-assisted manipulation of the trait in *Brassica juncea*. Theor Appl Genet 111:8–14

Padmaja KL, Agarwal P, Gupta V, Mukhopadhyay A, Sodhi YS, Pental D, Pradhan AK (2014) Natural mutations in two homoeologous *TT8* genes control yellow seed coat trait in allotetraploid *Brassica juncea* (AABB). Theor Appl Genet 127(2):339–347

Panjabi P, Jagannath A, Bisht NC, Lakshmi Padmaja K, Sharma S, Gupta V, Pradhan AK, Pental D (2008) Comparative mapping of *Brassica juncea* and *Arabidopsis thaliana* using Intron Polymorphism (IP) markers: homeologous relationships, diversification and evolution of the A, B and C Brassica genomes. BMC Genomics 9:113

Panjabi-Massand P, Yadava SK, Sharma P, Kaur A, Kumar A, Arumugam N, Sodhi YS, Mukhopadhyay A, Gupta V, Pradhan AK, Pental D (2010) Molecular mapping reveals two independent loci conferring resistance to *Albugo candida* in the east European germplasm of oilseed mustard *Brassica juncea*. Theor Appl Genet 121(1):137–145

Paritosh K, Pradhan AK, Pental D (2020) A highly contiguous genome assembly of *Brassica nigra* (BB) and revised nomenclature for the pseudochromosomes. BMC Genomics 21:887

Paritosh K, Yadava SK, Singh P, Bhayana L, Mukhopadhyay A, Gupta V, Bisht NC, Zhang J, Kudrna DA, Copetti D, Wing RA, Reddy Lachagari VB, Pradhan AK, Pental D (2021) A chromosome-scale assembly of allotetraploid *Brassica juncea* (AABB) elucidates comparative architecture of the A and B genomes. Plant Biotechnol J 19:602–614

Paritosh K, Gupta V, Yadava SK, Singh P, Pradhan AK, Pental D (2014) RNA-seq based SNPs for mapping in *Brassica juncea* (AABB): Synteny analysis between the two constituent genomes A (from *B. rapa*) and B (from*B. nigra*) shows highly divergent gene block arrangement and unique block fragmentation patterns. BMC Genomics 15:396

Parkin IAP, Gulden SM, Sharpe AG, Lukens L, Trick M, Osborn TC, Lydiate DJ (2005) Segmental structure of the *Brassica napus* genome based on comparative analysis with *Arabidopsis thaliana*. Genetics 171:765–781

Petrie GA (1988) Races of *Albugo candida* (white rust and staghead) on cultivated Cruciferae in Saskatchewan. Can J Plant Pathol 10:142–150

Prabhu KV, Somers DJ, Rakow G, Gugel RK (1998) Molecular markers linked to white rust resistance in mustard *Brassica juncea*. Theor Appl Genet 97:865–870

Pradhan AK, Sodhi YS, Mukhopadhyay A, Pental D (1993) Heterosis breeding in Indian mustard (*Brassica juncea* L. Czern. & Coss.): analysis of component characters contributing to heterosis for yield. Euphytica 69:219–229

Pradhan AK, Gupta V, Mukhopadhyay A, Arumugam N, Sodhi YS, Pental D (2003) A high-density linkage map in *Brassica juncea* (Indian mustard) using AFLP and RFLP markers. Theor Appl Genet 106:607–614

Pradhan AK, Pental D (2011) Genetics of *Brassica juncea* L. In: Schmidt R, Bancroft I (eds) Genetics and Genomics of the Brassicaceae, vol 9. Springer, New York, pp 323–346

Rahman M, McVetty PBE (2011) A review of Brassica seed color. Can J Plant Sci 91:437–446

Rakow G (1995) Canola quality mustard. Proc Special Crop Opportunities Conf, Saskatoon. July 19–20, pp 220–225

Ramchairy N, Padmaja KL, Sharma S, Gupta V, Sodhi YS, Mukhopadhyay A, ArumugamN PD, Pradhan AK (2007a) Mapping of yield influencing QTL in *Brassica juncea*: implications for breeding of a major oilseed crop of dryland areas. Theor Appl Genet 115:807–817

Ramchiary N, Bisht NC, Gupta V, Mukhopadhyay A, Arumugam N, Sodhi YS, Pental D, Pradhan AK (2007b) QTL analysis reveals context-dependent loci for seed glucosinolate trait in the oilseed *Brassica juncea*: importance of recurrent selection backcross scheme for the identification of 'true' QTL. Theor Appl Genet 116:77–85

Rimmer S, Mathur S, Wu C (2000) Virulence of isolates of *Albugo candida* from western Canada to Brassica species. Can J Plant Pathol 22:229–235

Rout K, Sharma M, Gupta V, Mukhopadhyay A, Sodhi YS, Pental D, Pradhan AK (2015) Deciphering allelic variations for seed glucosinolate traits in oilseed mustard (*Brassica juncea*) using two bi-parental mapping populations. Theor Appl Genet 128(4):657–666

Rout K, Yadav BG, Yadava SK, Mukhopadhyay A, Gupta V, Pental D, Pradhan AK (2018) QTL landscape for oil content in *Brassica juncea*: analysis in multiple bi-parental populations in high and "0" erucic background. Front Plant Sci 9:1448

Sachan J, Kolte S, Singh B (1995) Genetics of resistance to white rust (*Albugo candida* race 2) in mustard (*Brassica juncea* (L.) Czern & Coss.). In: GCIRC 9th International Rapeseed Congr, Cambridge, UK, pp 1295–1297

Saharan G, Verma P (1992) White rusts: a review of economically important species. International Development Research Centre, Ottawa, pp 65–66

Saini N, Yashpal, Koramutla MK, Singh N, Singh S, Singh R, Yadav S, Bhattacharya R, Vasudev S, Yadava DK (2019) Promoter polymorphism in *FAE1.1* and *FAE1.2* genes associated with erucic acid content in *Brassica juncea*. Mol Breed 39:75

Schranz ME, Lysak MA, Mitchell-Olds T (2006) The ABC's of comparative genomics in the Brassicaceae: building blocks of crucifer genomes. Trends Plant Sci 11:535–542

Sharma A, Mohapatra T, Sharma R (1994) Molecular mapping and character tagging in *Brassica juncea*. I. Degree, nature and linkage relationship of RFLPs and their association with quantitative traits. J Plant Biochem Biotechnol 3:85–89

Sharma R, Mohapatra T, Mukherjee AK, Krishanpal, Sharma RP (1999) Molecular markers for seed oil content in Indian mustard. J Plant Biochem Biotechnol 8:99–102

Sharma R, Aggarwal RAK, Kumar R, Mohapatra T, Sharma RP (2002) Construction of an RAPD linkage map and localization of QTLs for oleic acid level using recombinant inbreds in mustard (*Brassica juncea*). Genome 45:467–472

Sharma M, Mukhopadhyay A, Gupta V, Pental D, Pradhan AK (2016) *BjuB.CYP79F1* regulates synthesis of propyl fraction of aliphatic glucosinolates in oilseed mustard *Brassica juncea*: functional validation through genetic and transgenic approaches. PLoS One 26:11(2):e0150060

Shirzadegan M, Röbbelen G (1985) Influence of seed colour and hull proportions on quality properties of seeds in *Brassica napus* L. Fette, Seifen, Anstrichmittel 87:235–237

Siebert PD, Chenchick A, Kellogg DE, Lukyanov KA, Lukyanov SA (1995) An improved PCR method for walking in uncloned genomic DNA. Nucleic Acids Res 23:1087–1088

Sodhi YS, Mukhopadhyay A, Arumugam N, Verma JK, Gupta V, Pental D, Pradhan AK (2002) Genetic analysis of total glucosinolate in crosses involving a high glucosinolate Indian variety and a low glucosinolate line of *Brassica juncea*. Plant Breed 121:508–511

Somers DJ, Rakow G, Rimmer SR (2002) *Brassica napus* DNA marker slinked to white rust resistance in *Brassica juncea*. Theor Appl Genet 104:1121–1124

Song JM, Guan Z, Hu J, Guo C, Yang Z, Wang S, Liu D, Wang B, Lu S, Zhou R, Xie WZ, Cheng Y, Zhang Y, Liu K, Yang QY, Chen LL, Guo L (2020) Eight high-quality genomes reveal pan-genome architecture and ecotype differentiation of *Brassica napus*. Nat Plants 6:34–45

Srivastava A, Gupta V, Pental D, Pradhan AK (2001) AFLP- based genetic diversity assessment amongst agronomically important natural and some newly synthesized lines of *Brassica juncea*. Theor Appl Genet 102:193–199

Stringam GR, Thiagarajah MR (1995) Inheritance of alkenyl glucosinolates in traditional and microspore-derived doubled haploid populations of *Brassica juncea* L. Czern. and Coss. In: Proceedings of 9th International Rapeseed Congress, Cambridge, pp 804–806

Stringam GR, McGregor DI, Pawlowski SH (1974) Chemical and morphological characteristics associated with seed coat colour in rapeseed. In: Proceedings of 4th International Rapeseed Conference Giessen, Germany, pp 99–108

Thiagarajah MR, Stringam GR (1993) A comparison of genetic segregation in traditional and microspore-derived populations of *Brassica juncea* L. Czern. And Coss. Plant Breed 111:330–334

Tiwari AS, Petrie GA, Downey RK (1988) Inheritance of resistance to *Albugo candida* race 2 in mustard (*Brassica juncea* (L.) Czern.). Can J Plant Sci 68:297–300

Upadhyay A, Mohapatra T, Pai R, Sharma R (1996) Molecular mapping and character tagging in Indian mustard (*Brassica juncea*) II. RFLP marker association with seed coat color and quantitative traits. J Plant Biochem Biotechnol 5:17–22

Varshney A, Mohapatra T, Sharma RP (2004) Development and validation of CAPS and AFLP markers for white rust resistance gene in *Brassica juncea*. Theor Appl Genet 109:153–159

Vera C, Woods D (1982) Isolation of independent gene pairs at two loci for seed coat color in *Brassica juncea*. CanJ Plant Sci 62:47–50

Vera C, Woods D, Downey R (1979) Inheritance of seed coat color in *Brassica juncea*. Can J Plant Sci 59:635–637

Verma PR, Saharan GS, Bartaria AM, Shivpuri A (1999) Biological races of *Albugo candida* on *Brassica juncea* and *B. rapa* var. toria in India. J Mycol Plant Pathol 29:75–82

Woods DL (1980) Association of yellow seed coat with other characteristics in mustard (*Brassica juncea* L.). Cruciferae Newsl 5:23–24

Yadava SK, Arumugam N, Mukhopadhyay A, Sodhi YS, Gupta V, Pental D et al (2012) QTL mapping of yield-associated traits in *Brassica juncea*: meta-analysis and epistatic interactions using two different crosses between east European and Indian gene pool lines. Theor Appl Genet 125:1553–1564

Yang J, Liu D, Wang X, Ji C, Cheng F, Liu B et al (2016) The genome sequence of allopolyploid *Brassica juncea* and analysis of differential homoeolog gene expression influencing selection. Nat Genet 48:1225–1232

Zou J, Hu D, Liu P, Raman H, Liu Z, Liu X, Parkin IAP, Chalhoub B, Meng J (2016) Co-linearity and divergence of the A subgenome of *Brassica juncea* compared with other *Brassica* species carrying different A subgenomes. BMC Genomics 17:18

Brassica juncea Genome Sequencing: Structural and Functional Insights

12

Thakku R. Ramkumar, Sagar S. Arya, Divyani D. Kumari, and Sangram K. Lenka

Abstract

Indian mustard (*Brassica juncea*) is one of the oldest cultivated amphidiploid crops by human civilization. Natural interspecific crossing between diploid progenitors *B. nigra* (BB) and *B. rapa* (AA) resulted in evolution of this allotetraploid plant species with a total genome size of 1068 Mb. For genetic improvement of the desired traits, it is a prerequisite to unravel the whole genome sequence of this allotetraploid crop and its diploid progenitors. There are several genome and transcriptome sequencing initiatives conducted in this regard to unravel structure and functional annotation of genes in the genome. Similarly, this genomic information was used to obtain species specific molecular insights into the important agronomic traits such as fatty acid biosynthesis, anti-nutritional factors, resilience to climatic perturbations and pathogen resistance. The nuclear and organellar genome sequencing efforts in *B. juncea*, therefore, helped in improving our understanding of the complex allotetraploid architecture and building a foundation to utilize the information for translational genomics and precession breeding in the future.

T. R. Ramkumar
Department of Microbiology and Cell Sciences, IFAS, University of Florida, Gainesville, FL 32611, USA

S. S. Arya · D. D. Kumari · S. K. Lenka (✉)
The Energy and Resources Institute, TERI-Deakin Nano Biotechnology Centre, Gurugram, Haryana 122003, India
e-mail: sagar.arya@teri.res.in

12.1 Introduction

Brassica juncea, commonly known as Indian mustard, is an allopolyploid (AABB) formed by the cross between *B. rapa* (AA) and *B. nigra* (BB). This *B. juncea* ($2n = 36$) allopolyploidy was discovered by Prakash and Chopra in 1991 by artificial resynthesizing (Prakash and Chopra 1991). Cytological investigations using flow cytometry revealed the genome size to be 922 Mb (Yang et al. 2016). Previously, the *B. juncea* genome size was believed to be 1068 Mb, of which, 529 Mb and 632 Mb has been acquired from *B. rapa* and *B. nigra*, respectively. The genome of *Brassica* spp. has been used to examine gene function divergence and genome evolution as a result of polyploidy, widespread duplication, and hybridization. These investigations and characterization are performed under the title of "Multinational Brassica Genome Project (MBGP)", established in 2002 (https://www.brassica.info/home/about_mbgp.html); the project's main goal was to distribute and organize Brassica genomic resources for the benefit of the global Brassica research community.

C. Kole and T. Mohapatra (eds.), *The Brassica juncea Genome*, Compendium of Plant Genomes,
https://doi.org/10.1007/978-3-030-91507-0_12

Table 12.1 Difference in the diploid and amphiploid genetic makeup of the *Brassica* species

Species	Genome type	n	IC	GS
B. rapa	AA	10	0.539	529
B. nigra	BB	8	0.647	632
B. oleracea	CC	9	0.71	696
B. juncea	AABB	18	1.092	1068
B. napus	AACC	19	1.154	1132
B. carinata	BBCC	17	1.308	1284

1C: refers to the size of haploid genome, GS: genome size in Mbps, n: number of homologous pairs of chromosomes

This helped greatly to study the qualitative and quantitative information of the diversity within *Brassica* spp. It allowed to generate a connection between phenotypic character and genomic loci responsible and to characterize the genes, and detail the substantial numbers of functional proteins within the organism. The database powered by these services has also enabled the identification and characterization of molecular genetic markers and the construction of high-resolution genetic maps. From this, gathering information of plant evolution complexity and its translation to the model plant has be come easy.

Genomic sequencing only provides information of one allelic form of each gene, but to find all variations in the heritable phenotype, three approaches are essential; gene expression information, comparison with known gene sequences, and genetic association studies. Further, functional genomics provides insights of the potential role of a gene by analyzing the pattern of gene expression. Since few genes are highly conserved, their core function can be predicted by comparing with annotated genes or available databases. Identifying the genetic markers can make it easy to pick certain alleles in the genome which are associated with important function. Genome sequence becomes applicable for crop improvement through marker-assisted breeding. Marker-assisted selection is likely to connect with a novel agronomic characteristic when an alternative allele of a gene is sequenced across different genotypes. These findings permit the identification of closely related genes with improved quality and quantities which target the improvement of biological pathways. The sequence of the *B. juncea* genome will identify orthologous genes, aim for germplasm enhancement. Table 12.1 provides with the basic understanding of the difference in the diploid and amphiploid genetic makeup of the *Brassica* species. The model plant *Arabidopsis* has high structural similarity with *Brassica* genes especially in the coding regions (~85% nucleotide conservation) (Cavell et al. 1998). Therefore, genome sequencing will serve as the foundation for translational genomics of this crop based on the functional insights of genes obtained from *Arabidopsis*. Here, we will contextualize the structural and functional details of *B. juncea* genome sequencing initiatives using various next-generation sequencing platforms.

12.2 Technological Advancements in Plant Genome Sequencing

There are several sequencing technologies that can be applied for genome sequencing in plants. These are primarily divided into two approaches, one of which is whole genome shotgun sequencing. In the shotgun sequencing approach, the complete genome is broken down into fragments ranging between 5 and 20 kb. Sequencing each fragment can be done by any suitable sequencing method and the sequence assembly can be done with computational tools to construct the whole genome sequence. Another approach is the application of bacterial artificial chromosomes (BACs) alongside genetic mapping, fingerprinting, and end sequencing. Till date, no technology exists that can read DNA from one end to the other of even moderately sized chromosome.

The initial microbial genome sequencing was primarily conducted by whole genome shotgun technology. In this approach genomic DNA is sheared to smaller insert libraries of 0.4–1.2 kb and 5–20 kb library. The 0.4–1.2 kb library is used for sequencing and the 5–20 kb library acted as template in assembly. This technology could not meet the demands of eukaryotic genome sequencing, which contains long range repeats. Several technologies such as BAC, yeast artificial chromosome (YAC), plasmic-based conventional clones with large-insert (PBC) and bacteriophage P1-derived artificial chromosome (PAC) were developed by several groups. Though the linear YAC vectors, with yeast centromere and arms, can carry inserts of 1000 kb, the limitations such as instability of insert, chimerism, less amenable in extraction and purification alongside similarly sized yeast chromosomes, led researchers to look for alternative technologies too (Anderson 1993). Meanwhile, in the early 1990s, the BAC library technology emerged as backbone technology to sequence the large complex eukaryotic genomes, while the microbial genome sequencing was dominated with whole genome shotgun technology (Zhang and Wu 2001). The BACs are vector derivatives of *E. coli* F factor with low copy number and autonomous replicative capability (Shizuya et al. 1992), and can host inserts of 120–200 kb efficiently, and even of 300 kb (Luo and Wing 2003). With larger insert sizes, the BAC-based sequencing gave an advantage during assembly over shotgun sequencing. The first BAC library for plants was developed for *Sorghum bicolor* (Woo et al. 1994), followed with rice (Tao et al. 1994), *Arabidopsis* (Choi et al. 1995), and other plants. Physical map of the candidate genomes was constructed using the corresponding species' library, which in turn were used to assist whole genome sequence assembly.

The chromosome mapping, often termed as physical mapping, is the sequential arrangement of several overlapping contigs into putative chromosome, on which DNA markers such as genes of agronomic traits, repeats, transposons, serve as landmarks. This gives the landscape of DNA markers with estimated distance, across the assembled chromosome. Genome assembly is the process of arranging nucleotide sequences into the correct manner and represent model for the actual genome. Physical mapping of chromosomes/genome is required to precisely place the assembled scaffolds at the right place of the genome, thus increasing the contiguity precision of genome construction. For instance, the 135 Mb *Arabidopsis* genome is comprised of five chromosomes. Hauge et al. (1991) utilized cosmid library of 17,000 cosmid clones to construct the physical map of *Arabidopsis thaliana* genome. These 17,000 cosmid clones were assembled into 750 contigs, spanning 90–95% of the entire *Arabidopsis* genome, however, lacking the landscape and DNA marker information. Later Schmidt et al. (1995) utilized YAC clones to construct the physical map of *Arabidopsis* chromosome 4 with much more resolution. This study resulted in the construction of four contigs with information about the distribution of 112 DNA markers, 20 unmapped genes, transposable elements and other random genomic DNA fragments, and covered 90–95% of 21 Mb chromosome 4 of Arabidopsis (Schmidt et al. 1995). Subsequently the physical maps of chromosome 5 (Schmidt et al. 1997; Kotani et al. 1997), chromosome 2 (Zachgo et al. 1996), and chromosome 3 (Camilleri et al. 1998) were also constructed using available YAC libraries. Following this, a complete *A. thaliana* physical map with BAC was constructed (Mozo et al. 1999). This model genome was further enriched with mutant information and gene expression data for functional annotation which subsequently became the cornerstone for Brassica genome annotation.

12.3 Sequencing Platforms and Data Generation for Brassica Genome

Multiple platforms were employed in *B. juncea* genome projects for sequencing data generation. Yang et al. (2016) used the Illumina HiSeq 2000,

Illumina HiSeq 2500 platforms for 13 paired-end (PE) and mate-paired (MP) Illumina libraries (175.8 ×) and PacBio RSII sequencing platform for 1 single-molecule reads library (12.03 ×), and 222 × of BioNano genome mapping data for *B. juncea* var. Tumida *T.* genome sequencing (Yang et al. 2016). Following this, Paritosh et al. (2021) sequenced *Brassica juncea* var. Varuna with dual library system, employing Illumina HiSeq 1000 sequencer platform for llumina PE libraries (200–350 bp) and PacBio RSII platform for PacBio long range (30–50 kb) DNA (PB) libraries. The combination of PE and PB gave higher resolution and better contiguity (Paritosh et al. 2021). The progenitor species with A genome, the *B. rapa* was sequenced with Illumina GA II platform with shotgun sequencing libraries ranging three different insert sizes as long (~2 kb, 5 kb and 10 kb), medium (~500 bp), and short (~200 bp), further supported with Sanger sequencing to fill the gaps in assembly (Wang et al. 2011). This reference genome was reannonated with sequencing data from Illumina 55 Gb (114 ×) PE, 8.7 Gb (18 ×) MP libraries with Illumina Hiseq2500 platform and 6.5 Gb (13.4 ×) reads of single-molecule sequencing PacBio PB libraries data with an average length of 12 Kb with PacBio Sequel II platform (Cai et al. 2017). The genome information is further improved with 19.40 Gb data generated obtained from PacBio library-based PacBio Sequel II platform (Zhang et al. 2018). Zhang et al. also utilized Illumina HiSeq 4000 platform, for Hi-C mapping with 2 × 125 bp reads, for improved chromosome. For the sequencing of the B genome progenitor *B. nigra*, Yang et al. (2016) generated 10 PE and MP Illumina libraries (95.99 ×) for *B. nigra* doubled haploid line (YZ12151). Perumal et al. (2020) used C2 and Ni100 cultivars. A total of 82 Gb (137 ×) for C2 cultivar and 115 Gb (192 ×) for Ni100 cultivar were generated with Illumina HiSeq 2500 and Oxford NanoporeMinION and GridION platforms were used for further assembly. Paritosh et al. (2020, 2021) constructed genome sequence for *B. nigra* var. Sangam with three different sized, PE libraries (200–350, 300–450, and 400–550 bp) and MP libraries (2–3, 4–6, and 10 kb) and Illumina HiSeq 1000 and Illumina MiSeq platforms.

12.4 k-mer Analysis and Error Correction

k-mers are short sequence substrings obtained from the insert reads. These are used to estimate the genome size, by dividing the total number of the k-mers with the depth of the major peaks. k-mers are also used to remove the errors during sequence assembly. A multiple appearance of specific k-mer in the read data reflects errorless sequencing, and error sequences were omitted with lower frequencies. K-mers of 16nt (Wang et al. 2011), 17nt (Yang et al. 2016; Perumal et al. 2020), 21nt (Paritosh et al. 2020), 24nt (Perumal et al. 2020) for short reads and even k-mers of 91 bp (Cai et al. 2017), and 100 bp (Paritosh et al. 2021) were used to cover the complex region with PacBio reads.

12.5 Tools and Technologies Used in *B. Juncea* and Its Progenitor Genomes Assembly

The de novo assembly with short reads of *B. juncea* var. Tumida reference genome was done by ALLPATHS-LG and the gaps were filled with Pacbio RS II reads (Yang et al. 2016). The Irys View RefAligner utility was utilized to draft assemble the genomic scaffolds, and the scaffolds were anchored to the chromosomal map obtained from BioNano data (Yang et al. 2016). The availability of *B. juncea* var. Tumida as reference made the *B. juncea* var. Varuna genome assembly easier. The assembly was carried out using Canu assembler (V1.4) with PacBio sequence reads and the resolution of the sequences was cross examined by mapping with Illumina short read sequences using BWA-MEM. This allowed the *B. juncea* var. Varuna genome to be more precise (Paritosh et al. 2021). The *B. rapa* reference genome (Wang et al. 2011) was de novo assembled using SOAPdenovo and was validated using NUCmer. The constructed chromosome assemblies were porous with gaps, in particular with the repetitive regions (Wang et al. 2011). The reference genome was reassembled to contigs by Cai et al. (2017) with new reads, as well as with the reads of Wang et al. (2011) using SOAPDenovo2. These assembled contigs were then linked into scaffolds using SSPACE (Cai et al. 2017). These scaffolds were then examined with BAC library sequences by BLAST as well as with Illumina short reads and anchored into super-scaffolds using GapCloser. The PacBio sequences were also used for gap filling using PBjelly_V15.2.20. The reassembled genome was then further validated with CEGMA analysis (Cai et al. 2017). Zhang et al. (2018) again constructed *B. rapa* genome by de novo assembly aided with Hi-C mapping data for better assembly. Using Canu (v1.5), the PacBio reads were assembled and the resolution was enriched by aligning with Illumina short reads using BWA (v0.7.15) and the assembly was then validated with Pilon (v1.22). The aligned and assembled scaffolds were further anchored onto chromosomes using BioNano optical mapping data with Lachesis. The Lachesis was also used to construct the spatial genome to visualize proximal interacting sequences with Hi-C data and plotting with ggplot2 package (Zhang et al. 2018). For *B. nigra* reference genome de novo assembly, the Illumina reads were assembled using ALLPATHS-LG and the gaps were then filled using GapCloser. The generated scaffolds were then aligned by BLAST with publicly available BAC library sequences of *B. nigra*. These scaffolds were then assembled into pseudo-chromosomes using available genetic map. The assembled genome sequence was further validated with CEGMA v.2.3 (Yang et al. 2016). Perumal et al. (2020) independently sequenced genomes of two other varieties of *B. nigra*. The generated Nanopore reads were assembled using CANU 1.6 and crosschecked with three other assemblers; SMARTDenovo, wtdbg, Miniasm (Perumal et al. 2020). The assembled contigs were iterated with high-quality Illumina reads using PILON. These iterated contigs were then scaffolded using Dovetail's HiRise pipeline and were anchored into pseudo-molecules using available genetic maps as references (Perumal et al. 2020). Paritosh et al. (2020, 2021) sequenced genome of *B. nigra* var. Sangam. The genome was assembled with Nanopore reads using CANU 1.6 and the contigs were mapped with high-quality Illumina reads using BWA-MEM (v0.7.12) for higher resolution and iteration with PILON (v1.23) (Paritosh et al. 2020). In parallel, the group also assembled the genome of *B. nigra* var. Sangam with Illumina reads using MaSuRcA, and the BWA-MEM was used to align and position these contigs with PacBio reads. The scaffolding was done with SSPACE-LongRead.pl script and SSPACE-STANDARD-3.0.pl script (Paritosh et al. 2021). The assembled contigs were anchored onto chromosomes using BioNano optical map data and the generated genetic data was polished with PILON (Paritosh et al. 2021).

12.6 Sequencing of Progenitor Diploid Genomes

Population analysis-based genetic linkage maps were earlier constructed for *B. oleracea* (Slocum et al. 1990), *B. rapa* (Song et al. 1991; Chyi et al. 1992), *B. napus* (Ferreira et al. 1994), *B. nigra* (Lagercrantz and Lydiate 1995), and other *Brassica* species as well (Thormann et al. 1994). *B. rapa* (AA) and *B. nigra* (BB) are the genomic progenitors of *B. juncea* (AABB). RNA-seq based single nucleotide polymorphism (SNP) marker mapping revealed collinearity between *B. nigra* genome and *B. juncea* B genome, and the *B. juncea* A genome with genome of *B. rapa* (Paritosh et al. 2014). The estimations by several groups put *B. rapa* (AA) genome to be 450–550 Mb, and confirmed to be of 442.9 MB with BioNano data (Zhang et al. 2018). The *B. rapa* genome underwent genome triplication, and localized rearrangements such as deletions, insertions, inversions, and substitutions (Hong et al. 2008). Wu et al. (2000) constructed binary vector-based library for *B. rapa* for cytoplasmic male sterility locus studies. Subsequently, a genome-wide BAC library was constructed for *B. rapa* jointly by multiple groups (Mun et al.

2008; Hong et al. 2006), and 12,017 BAC-end sequences were analyzed for TEs, SSRs, centromeric satellite repeats and genes, that could also act as markers. The coverage of genome with TEs (14%), 43,000 genes (16.8%) and 1392 different SSRs, with estimated 110,000 SSRs ($\sim$one SSR/$\sim$4.8 kb), were mapped with this and BAC-end sequence analyses (Hong et al. 2006, 2008). Meanwhile, multiple groups developed different *B. rapa* genetic linkage maps marked with different genetic markers (Kim et al. 2006; Choi et al. 2007). Wang et al. (2011) genome sequenced *B. rapa* by whole genome shotgun sequencing technology with short ($\sim$200 bp), medium ($\sim$500 bp), and long ($\sim$2, 5 and 10 kb) Illumina GA II libraries, and constructed an annotated draft genome. The paired short read sequences upon assembly covered 222 Mb deriving unique scaffolds. The long-read MP sequences ($\geq$2 kb) were not initially used to avoid chimeric reads and incorrect sequences that are common to such libraries. The obtained unique scaffolds were then matched with paired-end sequences to assemble the unique contigs and was matched and corrected with the available BAC libraries (Mun et al. 2008; Hong et al. 2006), and previously assembled chromosome sequences by BAC technology (Mun et al. 2010). Thus, the sequence assembly covered 283.8 Mb, with an estimated cover of >98% of the gene space (Wang et al. 2011). The gaps in draft genome were predominantly with repeat and centromeric regions. Search analyses with 214,425 expressed sequence tags (ESTs) and 52,712 unigenes from NCBI and other sources identified 41,174 putative coding genes. Analysis of *B. rapa* gene family revealed 1003 (5.9%) families out of 16,917 to be lineage specific (Wang et al. 2011), while the rest shared features closely with related plants such as *Arabidopsis, Carica papaya*, and *Vitis vinifera*. The genome data is further upgraded with more short reads and long PacBio reads with reannotation (Cai et al. 2017). The gaps in the previous versions are predominantly in the repeat regions and centromeric regions are due to usage of short read sequences. Zhang et al. (2018) overcome this drawback and further enriched the *B. rapa* genome sequence and developed a reference genome employing third generation sequencing technologies such as optical mapping, Hi-C, and single-molecule sequencing. This latest study with integrated approach put the *B. rapa* genome size to 442.9 Mb, shrinking the previous estimations of 450–550 Mb (Zhang et al. 2018). Optical mapping of *B. rapa* genome was performed alongside PacBio long-read sequencing with a focus on centromeric and repeat regions to fill the gaps from previous genome versions. Hi-C libraries were also generated to visualize the proximal and physically interacting DNA elements. The resequencing and reannotation provide the *B. rapa* genome with 45,985 protein-coding genes with 45,411 genes (98.75%) annotated on chromosomes, leaving behind 574 genes (1.25%) on scaffolds. The *B. nigra* (BB) genome has an estimated size of 570–607.8 Mb, while studies by Paritosh et al. (2020) suggested the *B. nigra* genome to 522 Mb. The *B. nigra* underwent whole genome duplication, and subsequent rearrangements, making it a complex genome for sequencing and annotation. Yang et al. (2016) genome sequenced the *B. nigra* genome, covering 396.9 Mb of the genome, and annotated 49,826 genes. Perumal et al. (2020) assembled *B. nigra* genome with whole genome shotgun Illumina PE (300–700 bp insert) sequencing, Roche/454and Illumina MP libraries (3–45 kb insert) and long reads from Nanopore sequencing. The short reads were assembled with the help of long reads and the genome was annotated with RNA seq data (Perumal et al. 2020). Perumal et al. (2020) predicted 59,877 and 67,030 genes in two different genotypes, of which 55,022 (92.0%) and 59,780 (89.2%) genes were annotated with Uniprot database. Following this, Paritosh et al. (2020) utilized Optical Mapping and Nanopore long-read sequencing to sequence and assemble the *B. nigra* genome with higher contiguity. The study identified 57,249 putative genes of which 42,444 genes were matched with transcriptome data. Table 12.2 enlists the sequencing and annotations efforts of *B. juncea* progenitors.

Table 12.2 Sequencing and annotation of *B. juncea* progenitors

NCBI accession	Organism	Description
PRJNA558855	*B. rapa*	Single-molecule real-time sequencing technology; Gene expression; multi-isolate
PRJNA445393	*B. rapa Subspecies Chinese*	Next-generation sequencing; transcriptome; multi-isolate
PRJNA244166	*B. rapa*	Transcriptome or gene expression
PRJNA298858	*B. rapa*	Global transcriptome profiling, transcriptome or gene expression; multi-isolate
PRJNA249065	*B. rapa*	Annotation; refSeqgenome; mono-isolate
PRJNA244166	*B. rapa*	Deep transcriptome sequencing; transcriptome or gene expression; mono-isolate
PRJNA636608	*B. nigra*	Nanopore sequencing; raw sequence reads; multispecies
PRJNA642332	*B. nigra* var. Sangam	Oxford nanoporeMinION; representative genome
PRJNA285130	*B. nigra* cultivar *inbred line YZ12151*	Illumina HiSeq; chromosome
PRJNA615316	*B. juncea*	Genome sequencing and assembly; multispecies
PRJNA516907	*B. nigra*	hybrid Illumina/Roche 454 sequencing; multispecies
PRJNA327666	*B. nigra*	RefSeqgenome; mono-isolate
PRJNA324621	*B. nigra* var. Sangam	Genome sequencing, assembly, raw sequence reads, transcriptome; mono-isolate
PRJNA320480	*B. nigra*	Genome sequencing, assembly, raw sequence reads, gene expression; multispecies
PRJNA311781	*B. nigra*	RefSeqgenome; mono-isolate

The data is deposited to NCBI database (date of access 20th Aug 2021)

12.7 Genome Sequencing of Cultivated *B. Juncea*

Sequencing of *B. juncea* has served to resolve the complicated allopolyploid genome using the tools and technologies described earlier. *B. juncea* carries an allopolyploid genome (AABB). The draft genome sequence for *B. juncea* was constructed (Yang et al. 2016) with multiple sequencing technologies primarily with PE and MP Illumina libraries, and certain portions with long-read Bionano and PacBio sequencing. Flow cytometric analysis suggested the genome size as 922 Mb. The first draft genome of *B. juncea* var. Tumida was done using shotgun reads by de novo assembly, single-molecule long reads (PacBio sequencing), genetic mapping and genomic (optical) mapping (BioNano sequencing). The assembled *B. Juncea* var. Tumida genome resolved the complication of allopolyploid genome, and it's constructed by de novo assembly. Subsequently, it was also demonstrated that the homeolog expression dominance in *B. juncea* to explore the A and B subgenomes for their transcriptional behavior. The draft genome covered 85% (784 Mb) of the genome and annotated with 80,050 genes, 21lncRNAs, 2,638 tRNAs, 511 rRNAs, 3,725 small RNAs, 1,402 microRNAs, and 15,418 small nuclear RNAs. The *B. juncea* genome appears to have lost identified 562 and 545 genes in A and B subgenomes (Yang et al. 2016). Genome-wide RNA-seq analysis of the homoeologous genes from different developmental stage, tissues and two newly resynthesized *B. juncea* revealed, in all the samples, on average 16.2% of genes displayed homeolog expression dominance, whereas only 8.2% of genes showed expression dominance toward to BjuB over BjuA excluding

B. juncea resynthesized lines. On the basis of this observation, its inferred that no significant global dominance for the subgenomes further strengthening recent polyplodization of *B. juncea* (Yang et al. 2016). The sequencing and assembly of the *B. juncea* genome facilitate the agricultural trait improvements in this important crop.

The constructed *B. juncea* genome was further enriched for improved contiguity using long read sequencing technologies (Paritosh et al. 2020). The chromosome-scale genome assembly of *B juncea* generates a complete architecture of the A and B genome. They subjected *B. juncea* var. Varuna to SMRT sequencing on the PacBio RSII platform. This study provided marked improvement in B genome assembly of Varuna variety, unlike earlier sequenced Tumida variety, exposing extensive gene block fragmentations and gene block associations than reported earlier. The quality of genome sequence was improved to the extent that, the longest contig covered 35.8 Mb and the largest scaffold covered 72.1 Mb. RNA-seq guided gene annotations revealed 101 959 putative genes, in A (46,381) and B (55,578) subgenomes, of which 93% genes were validated with uniprot entries (Paritosh et al. 2021). Transposable elements covered 385 Mb (45.8% of *B. juncea* genome) of the genome, with higher representation in B subgenome covering 259 Mb of 51% of the subgenome than the A subgenome, that covered 113 Mb of 33.9% of A subgenome (Paritosh et al. 2021). Further, we have also discussed organellar genome sequencing in details. The details of genome sequencing efforts in *B. juncea* are summarized in Table 12.3.

12.8 Genome Annotation

Genome annotation is a method of genome study to identify functional elements along the sequence. It is the description of individual genes and their proteins or predicting the meaning of the genome sequence. Such prediction can be structural annotation (including location of coding regions, open reading frames (ORFs), and their corresponding regulatory motifs, functional annotation (integration of additional data such as spatio-temporal expression profile at transcriptomic and proteomic level, interaction network, physiological and biochemical functions), and essential components (CDS, mRNAs, pseudogenes, promoters, and poly-A signals)). Annotation of *B. juncea* var. Varuna genome was carried out for transposable elements, centromeres, and genes. For transposable element *de-novo* prediction approach, centromeric region is identified by correlation plot and RNA-seq analysis is performed for gene annotation. There are 1590 consensus repeats generated when merged with *Arabidopsis thaliana* transposable element database, repeats were classified into retrotransposons, DNA transposons, and other repeats. Comparing to the A chromosome [(113 Mb (~33.9%)], the B chromosome [~259 Mb (~51%)] has more repeats. A genome of *B. juncea* has a centromeric region already identified in *B. rapa* (A genome) and *B. oleracea* (C genome), i.e., CentBr1 and CentBr2 but absent in the B genome centromeric region. However, seven new B genome-specific repeats along with three new A genome-specific centromeric repeats were identified from *B. juncea*. In gene annotation, a total of 40,208 full-length high-quality sequences were obtained and a total of 105,354 genes were predicted of which 48,270 in the A genome and 57,084 in the B genome (validated by the non-redundant proteins in the UniProt plant database). Other studies have found a total of 82,008 genes–38,232 (~79%), 43,776 (~76%) within the A and B genome, respectively (Paritosh et al. 2014; Yang et al. 2016). *B. juncea* var. Varuna, B genome has shown high similarity with *B. nigra* genome. Orthologous study has also identified a total of 19,404 orthologous groups of which 39,383 genes of A had orthologs in B, and 42,727 genes of B had orthologs in the A genome. Similarly, with using non-redundant nucleotide and NCBI protein sequences, SWISS-PROT, Gene ontology (GO), Cluster of Orthologous Groups (COG), and Kyoto Encyclopedia of Genes and Genomes (KEGG) databases, the Tumida variety was also annotated for 80,050 protein-coding genes. Additionally, Yang et al.(2016) identified 511

Table 12.3 Publicly available *B. juncea* genome sequences, assembly and associated initiatives useful for genome annotation

NCBI accession	Project data type	Scope	Institute	Remark
PRJNA719608	Transcriptome	Mono-isolate	Guizhou Academy of Agricultural Sciences, China	Transcriptome sequencings of *B. juncea* var. G170 (drought tolerant and sensitive) were performed for identifying differentially expressed genes related to drought stress. The identified genes were considered as potential candidates for enhancing drought tolerance in *B. juncea*
PRJEB41123	Genome assembly (Chromosome scale) of *B. juncea* (AABB) to compare the architecture of the A and B genomes	Mono-isolate	University of Delhi, India	A and B genome harbors high number of LTR/Gypsy retrotransposons, followed by centromeric repeats, and *B. nigra* specific gene clusters, that segregates the collinearity between A and B genomes
PRJNA673122	Raw sequence reads	Mono-isolate	University of York, UK	NA
PRJNA672814	Raw sequence reads	Multispecies	Huazhong Agricultural University, China	Anthocyanin studies of *B. juncea*
PRJNA670607	Transcriptome	Multispecies	Biozeron Shenzhen Inc., China	Gene expression studies of *B. juncea*
PRJNA662836	Raw sequence data	Multi-isolate	Punjab Agricultural University, India	*B. fruticulosa* is a wild species belonging to Brassicaceae, which carry resistance to various biotic stresses. Introgression lines were developed to transfer desirable variation from *B. fruticulosa* into *B. juncea*
PRJNA663021	Raw sequence data	Multispecies	Punjab Agricultural University, India	Molecular-genetic analysis of *B. juncea*: Development of *B. fruticulose* introgression lines to transfer desired variations into *B. juncea*
PRJNA639209	Raw sequence data from different inbred lines	Multi-isolate	Punjab Agricultural University, India	The project aims to SNP genotype of land races, advanced breeding materials, determinate mustard, resynthesized

(continued)

Table 12.3 (continued)

NCBI accession	Project data type	Scope	Institute	Remark
				genotypes and introgression lines of *B. juncea*. The data will be used to establish diversity conduits and association panels
PRJNA615316	Genome sequencing, assembly	Multispecies	Hunan Agricultural University, China	De novo assembly of a chromosome-scale genome of *B. juncea* landrace Sichuan Huangzi and resequencing of 480 *B. juncea* accessions
PRJNA562990	Transcriptome—raw sequence reads	Multispecies	Huazhong Agricultural University, China	Transcriptome profile revealed that cytokinin promoted callus sprouting in *B. juncea*
PRJNA550308	Genome sequencing and assembly	Mono-isolate	The Centre for Genetic Manipulation of Crop Plants, University of Delhi, India	Genome assembly (highly contiguous chromosome-scale) of an oleiferous type of *B. juncea* var. Varuna, with PacBio SMRT reads
PRJNA544908	Transcriptome—raw sequence reads	Multispecies	Xinyang Normal University, China	Transcriptome elucidation in the purple leaf of *B. juncea* and identified differentially expressed genes involved in anthocyanin biosynthesis
PRJNA507350	Transcriptome—raw sequence reads	Mono-isolate	University of York, UK	Transcriptomic validation in *B. juncea* by dissecting the genetic architecture of quality and agronomic traits
PRJNA499035	Raw sequence reads	Multispecies	Institute for Plant Breeding	Resequencing data for the parents of a population of interspecific *Brassica* hybrids. The parents are *B. napus* (Ag-Spectrum), *B. juncea* (Purple leaf mustard) and *B. carinata* (C2). The resulting population was allohexaploid
PRJNA497315	Mitochondrial genome and transcriptome analysis	Multispecies	Huazhong Agriculture University, China	Mitochondrial genome assembly of five alloplasmiccytoplasmic male sterile (CMS) lines in *B. juncea* and transcriptome analysis with the RNA editing profile

(continued)

Table 12.3 (continued)

NCBI accession	Project data type	Scope	Institute	Remark
PRJNA479948	Transcriptome—raw sequence reads	Multispecies	Central University of Punjab, India	Expression analysis to identify vital genes and pathways involved in the regulation of various metabolic and biological processes, particularly related to cadmium stress. A transcriptomic investigation to understand the response of *B. juncea* to cadmium stress could provide insights into improving its phytoremediation efficiency
PRJNA477240	Transcriptomics	Mono-isolate	Sichuan Agricultural University, China	Transcriptomicsto provides insight into stem development in *B. juncea* var. Tumida
PRJEB26751	SNP data	Mono-isolate	Sher-e-Kashmir University of Agricultural Sciences and Technology —Jammu, India	The project involves ddRAD-sequencing followed by identification of SNP identification
PRJNA471033	Raw sequence reads	Multi-isolate	University of York, UK	*B. juncea* (Varuna x Heera) developed from an DH population VHDH
PRJNA448707	Transcriptome or gene expression	Multi-isolate	University of Minnesota Duluth, Minnesota, US	RNA-Seq study of expression in *B. juncea* nectaries and leaves from different lines
PRJNA431509	Epigenomic data	Multispecies	ICAR—Institute of vegetable science, India	High through-put sequencing of small RNA
PRJNA430791	Raw sequence reads	Multispecies	School of Life Science, Lanzhou University, China	Investigated the molecular mechanisms of anthocyanin accumulation in *B. juncea* leaves
PRJNA395472	Transcriptome	Multi-isolate	Zhejiang A & F University, China	Transcriptional changes in *B. juncea* leaves after armyworm chewing
PRJNA383771	Raw sequence reads, assembly, transcriptome	Mono-isolate	ICAR-Indian Institute of Agricultural Biotechnology, India	RNA-seq of *B. juncea* under various Nitrate treatments
PRJNA339019	Transcriptome	Mono-isolate	Southeast Chongqing Academy of Agricultural Sciences, Chongqing, China	Transcriptomic profiling of arthrobacterium–induced systemic resistance in *B. juncea* var. Tumida infected by *Plasmodiophorabrassicae*

(continued)

Table 12.3 (continued)

NCBI accession	Project data type	Scope	Institute	Remark
PRJNA323808	Genome sequencing and assembly	Mono-isolate	SRM University, India	Whole genome sequencing and analysis of *B. juncea*
PRJNA321670	Transcriptome or gene expression	Multi-isolate	Department of Plant Molecular Biology, University of Delhi, India	The transgenic line BnCRY2aOE over-expressing a blue light photoreceptor (BnCRY2a) flowered earlier than wild-type plants. To identify the downstream candidate genes involved in regulating the early flowering transgenic phenotype, a genome-wide microarray analysis of the transgenic vs. wild type plants was performed. The microarray analysis unraveled the differential up-regulation of genes involved in flower development, cell differentiation and growth as well as hormone biosynthesis/signaling in comparison to wild-type
PRJNA319668	Transcriptome or gene expression	Multispecies	Zhejiang University, China	Analyzing and evaluating reversibility and heritability of DNA methylation resulting due to in vitro grafting between *B. oleracea* and *B. juncea*
PRJNA312980	Transcriptome or gene expression	Mono-isolate	Sichuan Normal University, China	Identified *B. juncea* transcripts using de novo assembly of the Illumina HiSeq 4000 sequencing
PRJNA301284	RefSeq genome	Mono-isolate	SRM University, India	*B. juncea* genome reference project
PRJNA298501	Raw sequence reads	Multispecies	Huazhong Agricultural University, China	Investigated the anthocyanin formation in five *Brassica* species, followed by transcriptome analysis between purple and green leaves. The response of potential key genes was examined for pigmentation as well as the physiological roles of anthyocyaninsin *Brassica* plant development

(continued)

Table 12.3 (continued)

NCBI accession	Project data type	Scope	Institute	Remark
PRJNA296365	Transcriptome or gene expression	Multi-isolate	Department of Botany, University of Delhi, India	Expression analysis of cold-stressed siliques in *B. juncea*
PRJNA290942	Transcriptome or gene expression	Mono-isolate	ICAR—National Research centre on plant biotechnology	Comparative transcriptome analysis between CMS (*Moricandiaarvensis*) *B. juncea* var. Pusa bold and its fertility restorer line
PRJEB9362	Transcriptome	Mono-isolate	Fondazione Edmund Mach, Italy	Identification of novel and conserved miRNAs involved in redox regulation of salt stress *B. juncea*
PRJNA289188	Transcriptome profiling	Mono-isolate	Zhejiang university, China	To obtain ESTs from *B. juncea* var. Tumida
PRJNA285130	Genome sequencing, assembly	Multispecies	Zhejiang university, China	*B. juncea* var. timuda genome sequencing and assembly
PRJNA277020	Transcriptome profiling	Multi-isolate	Banaras Hindu University, India	Microarray analysis of Indian mustard plants subjected to arsenate stress
PRJNA276704	Transcriptome or gene expression	Multi-isolate	Jawaharlal Nehru University, India	Assembly (de novo) and stress related transcriptomic profiling of a salinity-tolerant *B. juncea* var. CS52
PRJNA271638	Transcriptome or gene expression	Multi-isolate	ICAR-Directorate of Rapeseed Mustard Research, India	Elucidating salt (NaCl)-induced changes in whole transcriptome of *B. juncea* var. CS-52
PRJNA271633	Transcriptome or gene expression	–	ICAR-Directorate of Rapeseed Mustard Research, India	Salt stress induced changes in transcriptome of *B. juncea*
PRJNA270523	Transcriptome or gene expression	Multi-isolate	Plant Genomics and Stress Biology, Department of Botany, Delhi University, India	RNA-Seq of *B. juncea* in response to abiotic stresses
PRJNA245462	Transcriptome or gene expression	Multi-isolate	Centre for Genetic Manipulation of Crop Plants, Delhi University, India	Transcriptome profiling to identify differently expressed genes in various parts of *B. juncea* var. Heera
PRJNA244493	Transcriptome or gene expression	Mono-isolate	Bose Institute, India and Genotypic Technology Pvt. Ltd., India	Differentially expressed transcripts of susceptible *B. juncea* challenged with *Alternariabrassicicola*

(continued)

Table 12.3 (continued)

NCBI accession	Project data type	Scope	Institute	Remark
PRJNA231241	Transcriptome or gene expression	Multi-isolate	Plant Genomics and Stress Biology, Department of Botany, Delhi University, India	Genome-wide perspective of miRNAome in *B. juncea,* induced by abiotic stresses
PRJNA195539	Transcriptome or gene expression	Mono-isolate	Hunan Agricultural University, China	*B. juncea* transcriptome analysis
PRJNA185138	Transcriptome	Multi-isolate	Zhejiang University, China	Profiling of miRNA in *B. juncea* CMS and maintainer fertile lines by deep sequencing
PRJNA174360	Genetic map	Mono-isolate	–	Doubled haploid mapping population derived from a cross between Varuna and Heera was used to analyze yield influencing traits in *B. juncea.* SSR and RFLP markers were mapped onto an already existing AFLP map for the QTL analysis
PRJNA173786	Cheung_97 genetic map	Mono-isolate	–	Cheung_97 genetic map of *B. juncea* is based on RFLP detected by anonymous cDNA markers from *B. napus*
PRJNA173691	Panjabi_08 genetic map	Mono-isolate	Centre for Genetic Manipulation of Crop Plants, University of Delhi, India	Panjabi_08 genetic map of *B. juncea* was developed with PCR-based Intron Polymorphism markers to study segmental structure of the A and B genomes of *Brassica*
PRJNA169393	Transcriptome or gene expression	Multi-isolate	Agriculture and Food Systems, The University of Melbourne, Australia	Understanding the molecular mechanisms of *B. juncea* underpinning physiological tolerances to salinity and alkaline salinity
PRJNA136273	Transcriptome or gene expression	Multi-isolate	Prof. J.P. Khurana, Department of Plant Molecular Biology, University of Delhi, India	Elucidating the role of Cryptochrome 1, which modulates the development in *Brassica* by regulating the gene expression (involved in light, stress and phytohormone response)
PRJNA74717	RefSeq genome	Mono-isolate	Nanjing Agricultural University, State Key Laboratory of Crop Genetics and Germplasm Enhancement, P.R. China	Mitochondrial genome of *B. juncea*

(continued)

Table 12.3 (continued)

NCBI accession	Project data type	Scope	Institute	Remark
PRJNA72397	Transcriptome and gene expression	Multi-isolate	College of Bioinformation, Chongqing University of Posts and Telecommunications, China	Transcriptome analysis of the stem tumor mustard *B. juncea* var. Tumida
PRJNA43723	Pradhan_03 genetic map	Mono-isolate	University of Delhi, India	Pradhan_03 genetic map of *B. juncea* is derived by crossing of Varuna and Heera parental lines
PRJNA615316	RefSeq genome	–	Hunan Agricultural University, China	Representative genome of *B. juncea* var Sichuan Huangzi

Source NCBI-https://www.ncbi.nlm.nih.gov/bioproject/?term=brassica+juncea

rRNAs, 2,638 tRNAs,3,725 small RNAs, 21 long noncoding RNAs, 15,418 small nuclear RNAs and 1402 microRNAs, from the *B. juncea* genome and even extracted syntenic ortholog gene pairs (28,228 and 28,917) from the *B. juncea* subgenomes. Gene loss has been observed during speciation, a total of 562 (A subgenome) and 545 (B subgenome) genes were considered as lost, referring to their common ancestral genomes. Likewise, many structural and functional annotations have been performed to investigate the structure, function, and regulation of the genome which are enlisted in Table 12.4.

12.9 Organellar Genome Sequencing

12.9.1 Mitochondrial Genome Insights

Mitochondrial genomes in angiosperms are more complex compared to other organisms (Chang et al. 2011). Analyses of different angiosperm mitochondrial genome sequences have revealed several common characteristics; however, a number of diverse mitotypes have evolved within each genus/species. Sequencing and comparative investigations of six *Brassica* mitotypes indicates a mechanism for mitochondrial genome development in *Brassica*, which includes certain events of evolution like mutation, genome compaction, duplication, and rearrangements (Chang et al. 2011). Further, the analysis of the evolutionary relatedness between *Brassica* mitotypes provides information of mitochondrial genome of *B. juncea*. The size of whole single circular mitochondrial genome of *B. juncea* is 219,766 bp [GenBank: JF920288], which is reported to be similar to that of *B. rapa*, i.e., 219,747 bp. Whereas, the mitochondrial genome sizes of *B. carinata* and *B. oleracea* are greater than both *B. juncea* and *B. rapa*, i.e., 232,241 bp and 360,271 bp, respectively. The report suggests that the *B. juncea* mitotype is an evolutionary outcome of an inherited mitotype of (*B. rapa*) cam with certain modifications (Chang et al. 2011). The percentage of the total functional genes in the mitochondrial genome of *B. juncea* are almost the same compared to *B. rapa, B. oleracea* and *B. napus* mitotypes, expect *B. carinata*, where the similarity is 27.98% (Chang et al. 2011).

One of the main aspects of mitochondria genomes is producing cytoplasmic male sterility (CMS), which is essential to explore the heterosis in crop plants and is routinely utilized as a model to investigate the interactions between nuclear and cytoplasmic. These interactions are either spontaneous or formed through interspecific

Table 12.4 Structural and functional annotations performed in *B. juncea*

Annotation	Description	References
Nucleotide-binding site leucine-rich repeat genes	Identified 289 NLR genes with a ratio of 0.61:1 of toll/interleukin-1 receptor-NLR to non-toll/interleukin-1 receptor-NLR genes	(Inturrisi et al. 2020)
Heat shock transcription factor	A total of 60 HSF transcription factors were identified; establish a phylogenetic relationship	(Li et al. 2020a)
Superoxide dismutase proteins (SOD), belongs to the family of metalloenzyme	Identified 29 genes; and cis-regulatory elements in the promoter region, of which 10 SOD genes were abiotic stress-responsive	(Verma et al. 2019)
Receptor-like protein (RLP) and Receptor-like kinase (RLK) genes	A total of 228 RLPS and 493 RLKS were identified	(Yang et al. 2021)
Jasmonate ZIM domain proteins	38 genes were identified in Tumida variety	(Cai et al. 2020)
Auxin response factors	65 *B. Juncea* genes that encode ARF proteins were identified, further promoter *cis*-element also demonstrated in all the 65 genes	(Li et al. 2020b)
Teosinte branched1/*Cycloideal* proliferating cell factors	Identified 62 genes from the *B. Juncea* var. Tumida	(He et al. 2020)
Chitinase gene family	Identified 47 genes	(Mir et al. 2020)
Nonexpressor of pathogenesis-related genes	Identified 19 genes, which cluster into six separate groups in the genome of *B. juncea* var. Tumida	(Wang et al. 2021)
GRAS transcription factors	A total of 88 GRAS genes were identified	(Li et al. 2019)
Regulatory roles of lncRNAs	Identified 7613 lncRNAs, of which 1614 lncRNAs are involved in heat and drought related stress response	(Bhatia et al. 2020)

crossing, which results into an alloplasmic condition/cytoplasmic substitution (Gaikwad et al. 2006). CMS characterized, particularly by non-Mendelian inheritance and the suppression of the production of viable pollen. The CMS could result due to mutations, recombination, or rearrangements in the genome of mitochondria. Knowledge of the *B. juncea* mitochondrial genome is crucial to develop superior phenotypes. In this quest, numerous attempts are being undertaken to sequence and assemble the mitochondrial genomes of *B. juncea* lines. For instance, a study performed on five alloplasmic *B. juncea* lines revealed that the mitochondrial genomes from 221 to 256 kb (Wu et al. 2019), which is somewhat greater compared to the typical *B. juncea* size reported earlier (Chang et al. 2011). Another comparative study between the mitochondrial genomes of the *hau* CMS line and its iso-nuclear maintainer line in *B. juncea* showed a difference in the genome sizes and GC content (Heng et al. 2014). Mitochondrial genome of *B. junceahau* CMS is reported to be of 247,903 bp in size with a GC content of 45.08%, whereas that of another normal line (J163-4) and a maintainer line are 219,863 bp in size with a GC content of 45.23%. Further, the mitochondrial genome has numerous genes, of which, 35 are protein encoding, 25 are tRNA, 3 are rRNAs, and 29 are ORFs of unknown function. Whereas, genes in maintainer line are 36 for protein encoding, 22 for tRNA, 3 for rRNAs, and 31 are unidentified ORFs. In addition, sub-stoichiometrical coexistence of distinct mitotypes is confirmed in *hau* CMS lines as well as its maintainer lines in *B. juncea* (Heng et al. 2014). Further, it was demonstrated that a cytotoxic protein ORF288 associated with male sterility in *B. juncea* causes aborted pollen development (Jing et al. 2012). The toxicity generating region of ORF288 is reported to be located near the N-terminus, which repressed the growth when heterologously expressed in *E. coli* (Heng et al. 2018). However, heterologous expression of ORF288 portions indicates that the region which induces CMS is present between amino acids 73 and 288, whose heterologous expression did not inhibit *E. coli* growth. It is reported that the transcript levels of *orf288* are associated with altered nuclear gene expression and the *hau* CMS system. Apart from *orf288*, it also reported that *orf220* gene causes male sterility in *B. juncea* (Yang et al. 2010). Both positive and negative correlations have been reported between the occurrence of CMS and CMS-associated orfs, when CMS-associated orfs were targeted and expressed in mitochondria. Certain orfs can cause male- or semi-sterility, while some do not. Understanding of nuclear-mitochondrial compatibility is crucial as it can result in differential expression of mitochondrial genes, e.g., mitochondrial *aptα* gene (Gaikwad et al. 2006). Furthermore, it is essential to identify the molecular basis of mitochondrial recombination in the male sterile cytoplasmic hybrids to understand the underlying mechanism as well as the environmental factors affecting fertility reversion. For instance, analysis of floral bud transcriptome under both pollination and non-pollination state showed a variability in the expression of Muts HOMOLOG1 (*MSH1*) (a nuclear gene which regulates illegitimate recombination in plant mitochondria) in response to different sugars, which shows that physiological changes are involved in the pollination signaling and fertility reversion in CMS plants (Zhao et al. 2021). This is indicative that the gynodioecy is a reproductive plan of action that might incorporate ecologically responsive genes such as *MSH1* as a switch for fertility-sterility transition under reproductive isolation. It was found that the mitochondrial genome in revertant lines of *B. juncea* CMS cytoplasm undergoes substoichiometric shift to suppress *orf220* copy number, whereas *MSH1-RNAi* with increased *orf220* copy number are male sterile (Zhao et al. 2016). This provides a valuable insight into substoichiometric shift in CMS induction, fertility reversion as well as interplay of *MSH1*.

12.9.2 Chloroplast Genome Sequencing

Chloroplast is a vital organelle in plants, which harbors genetic and enzymatic resource required for photosynthesis, as well as various important biosynthesis pathways such as pigments, fatty acids, vitamins, and amino acids. Understanding of the chloroplast genome is therefore essential to enhance the qualitative and quantitative traits of *B. juncea*. As chloroplast genome engineering of *B. jucnea* through chloroplast can increase the oil content, improve oil quality, as well as make it resistant to most of the abiotic and biotic factors. Prabhudas et al. (2016) are the first to sequence the complete chloroplast genome of *B. juncea* (Indian mustard). The sequencing carried out on Illumina Hiseq 2500 platform generated 100 bp paired raw reads, which were qualitatively trimmed followed by read quality assessment.

The chloroplast genome size is reported to be 153,483 bp, with 36.36% of GC content (Prabhudas et al. 2016). The assembled chloroplast genome exhibited a quadripartite structure consisting of an 83,286 bp large single copy region and an 17,775 bp single copy region, which are separated by a pair of IRa and IRb (inverted repeats a and b) of 26,211 bp each. The GC content was 34.12, 29.20, and 42.34% for large single copy region, single copy region, and inverted repeats, respectively. In the entire chloroplast genome, a total number of 113 genes were annotated, of which, 79 genes are protein encoding, 30 are tRNA, and 4 are rRNA. Total 15 genes including 9 protein-encoding genes and 6 tRNA genes had either a single of couple of introns. Gene duplications in inverted repeat regions were extended to 4 rRNA, 7 tRNA, as well as 6 protein-encoding genes.

B. juncea share the chloroplast genome with one of the hybridization donor *B. rapa*, which fits the U model (Li et al. 2017). The phylogenetic analysis indicated that the branch length of *B. juncea* chloroplast genome is close to *B. oleracea*, *B. napus*, and *B. rapa* more distant from *B. nigra* and *B. carinata*, suggesting a divergence from the two Brassica's.

12.10 Conclusion

Rapid development in next-generation sequencing technologies has undoubtedly accelerated the sequencing programs of *B. juncea* genomes along with its progenitors to generate metadata at less cost and time. The *B. juncea* bio-projects have sequenced, assembled, analyzed and have annotated the genome to highlight the impact of recombination between progenitor-genomes of *Brassica* spp. Apart from providing fundamental knowledge about the evolution of *B. juncea*, the bio-projects also helped in the identification of candidate genes involved in various physiological and biochemical processes. Importantly, the information can be utilized form trait mapping to targeted tinkering of the genetic makeup of *B. juncea* and to further improve resistance to biotic and abiotic stresses, enhance climate resilience, increase yield as well as quality and nutritional traits. The availability of the genomic data will further advance the knowledge of cultivated *B. juncea* lines and help to produce better tool kits for molecular breeding and crop improvement. Considering the level of diversity and wealth of the species in the *Brassica* genus, there is a lot yet to explore about *B. juncea* to clarify the relationship between genotypes/phenotypes. Similarly, progress in proteome and transcriptome research has the capacity to reveal the structural and functional role of genes in *B. juncea*, altered in response to physiological and environmental processes and pathogen attacks. Overall, the information of nuclear and organellar genomes, as well as the increasing amount of "omics" resources will certainly contribute toward the effective use of *B. juncea*.

References

Anderson C (1993) Genome shortcut leads to problems. Science 259:1684–1688

Bhatia G, Singh A, Verma D, Sharma S, Singh K (2020) Genome-wide investigation of regulatory roles of lncRNAs in response to heat and drought stress in *Brassica juncea* (Indian mustard). Environ Exp Bot 171:103922

Cai C, Wang X, Liu B, Wu J, Liang J, Cui Y, Cheng F, Wang X (2017) *Brassica rapa* genome 2.0: a reference upgrade through sequence re-assembly and gene re-annotation. Mol Plant 10:649–651

Cai Z, Chen Y, Liao J, Wang D (2020) Genome-wide identification and expression analysis of jasmonate ZIM domain gene family in tuber mustard (*Brassica juncea* var. tumida). PLoS One 15:e0234738

Camilleri C, Lafleuriel J, Macadre C, Varoquaux F, Parmentier Y, Picard G, Caboche M, Bouchez D (1998) A YAC contig map of Arabidopsis thaliana chromosome 3. Plant J 14:633–642

Cavell A, Lydiate D, Parkin I, Dean C, Trick M (1998) A 30 centimorgan segment of Arabidopsis thaliana chromosome 4 has six collinear homologues within the *Brassica napus* genome. Genome 41:62–69

Chang S, Yang T, Du T, Huang Y, Chen J, Yan J, He J, Guan R (2011) Mitochondrial genome sequencing helps show the evolutionary mechanism of mitochondrial genome formation in Brassica. BMC Genomics 12:1–12

Choi S, Creelman RA, Mullet JE, Wing RA (1995) Construction and characterization of a bacterial artificial chromosome library of *Arabidopsis thaliana*. Plant Mol Biol Rep 13(2):124–128

Choi SR, Teakle GR, Plaha P, Kim JH, Allender CJ, Beynon E, Piao ZY, Soengas P, Han TH, King GJ (2007) The reference genetic linkage map for the multinational *Brassica rapa* genome sequencing project. Theor Appl Genet 115(6):777–792

Chyi Y-S, Hoenecke M, Sernyk J (1992) A genetic linkage map of restriction fragment length polymorphism loci for *Brassica rapa* (syn. campestris). Genome 35(5):746–757

Ferreira M, Williams P, Osborn T (1994) RFLP mapping of *Brassica napus* using doubled haploid lines. Theor Appl Genet 89(5):615–621

Gaikwad K, Baldev A, Kirti P, Mohapatra T, Bhat S, Chopra V, Prakash S (2006) Organization and expression of the mitochondrial genome in CMS (Moricandia) *Brassica juncea*: nuclear-mitochondrial incompatibility results in differential expression of the mitochondrial atpα gene. Plant Breed 125(6):623–628

Hauge BM, Giraudat J, Hanley S, Hwang I, Kohchi T, Goodman HM (1991) Physical mapping of the Arabidopsis genome and its applications. In: Plant Molecular Biology vol 2. Springer, pp 239–248

He J, He X, Chang P, Jiang H, Gong D, Sun Q (2020) Genome-wide identification and characterization of

TCP family genes in *Brassica juncea* var. tumida. Peer J 8:e9130

Heng S, Wei C, Jing B, Wan Z, Wen J, Yi B, Ma C, Tu J, Fu T, Shen J (2014) Comparative analysis of mitochondrial genomes between the hau cytoplasmic male sterility (CMS) line and its iso-nuclear maintainer line in *Brassica juncea* to reveal the origin of the CMS-associated gene orf288. BMC Genomics 15 (1):1–12

Heng S, Gao J, Wei C, Chen F, Li X, Wen J, Yi B, Ma C, Tu J, Fu T (2018) Transcript levels of orf288 are associated with the hau cytoplasmic male sterility system and altered nuclear gene expression in *Brassica juncea*. J Exp Bot 69(3):455–466

Hong CP, Plaha P, Koo D-H, Yang T-J, Choi SR, Lee YK, Uhm T, Bang J-W, Edwards D, Bancroft I (2006) A Survey of the *Brassica rapa* genome by BAC-end sequence analysis and comparison with Arabidopsis thaliana. Molecules Cells (Springer Science and Business Media BV) 22(3)

Hong CP, Kwon S-J, Kim JS, Yang T-J, Park B-S, Lim YP (2008) Progress in understanding and sequencing the genome of *Brassica rapa*. Int J Plant Genomics 2008

Inturrisi F, Bayer PE, Yang H, Tirnaz S, Edwards D, Batley J (2020) Genome-wide identification and comparative analysis of resistance genes in *Brassica juncea*. Mol Breed 40(8):1–14

Jing B, Heng S, Tong D, Wan Z, Fu T, Tu J, Ma C, Yi B, Wen J, Shen J (2012) A male sterility-associated cytotoxic protein ORF288 in *Brassica juncea* causes aborted pollen development. Jexp Bot 63(3):1285–1295

Kim JS, Chung TY, King GJ, Jin M, Yang T-J, Jin Y-M, Kim H-I, Park B-S (2006) A sequence-tagged linkage map of *Brassica rapa*. Genet 174(1):29–39

Kotani H, Nakamura Y, Sato S, Kaneko T, Asamizu E, Miyajima N, Tabata S (1997) Structural analysis of *Arabidopsis thaliana* chromosome 5. II. Sequence features of the regions of 1,044,062 bp covered by thirteen physically assigned P1 clones. DNA Res 4 (4):291–293

Lagercrantz U, Lydiate DJ (1995) RFLP mapping in *Brassica nigra* indicates differing recombination rates in male and female meioses. Genome 38(2):255–264

Li P, Zhang S, Li F, Zhang S, Zhang H, Wang X, Sun R, Bonnema G, Borm TJ (2017) Front Plant Sci 8:111

Li M, Sun B, Xie F, Gong R, Luo Y, Zhang F, Yan Z, Tang H (2019) Identification of the GRAS gene family in the *Brassica juncea* genome provides insight into its role in stem swelling in stem mustard. Peer J 7: e6682

Li M, Xie F, Li Y, Gong L, Luo Y, Zhang Y, Chen Q, Wang Y, Lin Y, Zhang Y (2020a) Genome-wide analysis of the heat shock transcription factor gene family in *Brassica juncea*: structure, evolution, and expression profiles. DNA Cell Biol 39(11):1990–2004

Li W, Chen F, Wang Y, Zheng H, Yi Q, Ren Y, Gao J (2020b) Genome-wide identification and functional analysis of ARF transcription factors in *Brassica juncea* var. tumida. PloS One 15(4):e0232039

Luo M, Wing RA (2003) An improved method for plant BAC library construction. In: Plant functional genomics. Springer, pp 3–19

Mir ZA, Ali S, Shivaraj S, Bhat JA, Singh A, Yadav P, Rawat S, Paplao PK, Grover A (2020) Genome-wide identification and characterization of Chitinase gene family in *Brassica juncea* and Camelina sativa in response to Alternaria brassicae. Genomics 112 (1):749–763

Mozo T, Dewar K, Dunn P, Ecker JR, Fischer S, Kloska S, Lehrach H, Marra M, Martienssen R, Meier-Ewert S (1999) A complete BAC-based physical map of the *Arabidopsis thaliana* genome. Nat Genet 22 (3):271–275

Mun J-H, Kwon S-J, Yang T-J, Kim H-S, Choi B-S, Baek S, Kim JS, Jin M, Kim JA, Lim M-H (2008) The first generation of a BAC-based physical map of *Brassica rapa*. BMC Genomics 9(1):1–11

Mun J-H, Kwon S-J, Seol Y-J, Kim JA, Jin M, Kim JS, Lim M-H, Lee S-I, Hong JK, Park T-H (2010) Sequence and structure of *Brassica rapa* chromosome A3. Genome Biol 11(9):1–12

Paritosh K, Pradhan AK, Pental D (2020) A highly contiguous genome assembly of *Brassica nigra* (BB) and revised nomenclature for the pseudochromosomes. BMC Genomics 21(1):1–12

Paritosh K, Yadava SK, Singh P, Bhayana L, Mukhopadhyay A, Gupta V, Bisht NC, Zhang J, Kudrna DA, Copetti D (2021) A chromosome-scale assembly of allotetraploid *Brassica juncea* (AABB) elucidates comparative architecture of the A and B genomes. Plant Biotechnol J 19(3):602

Paritosh K, Gupta V, Yadava SK, Singh P, Pradhan AK, Pental D (2014) RNA-seq based SNPs for mapping in *Brassica juncea* (AABB): synteny analysis between the two constituent genomes A (from *B. rapa*) and B (from *B. nigra*) shows highly divergent gene block arrangement and unique block fragmentation patterns. BMC Genomics 15(1):1–14

Perumal S, Koh CS, Jin L, Buchwaldt M, Higgins EE, Zheng C, Sankoff D, Robinson SJ, Kagale S, Navabi Z-K (2020) A high-contiguity *Brassica nigra* genome localizes active centromeres and defines the ancestral Brassica genome. Nat Plants 6(8):929–941

Prabhudas SK, Raju B, Kannan Thodi S, Parani M, Natarajan P (2016) The complete chloroplast genome sequence of Indian mustard (*Brassica juncea* L.). Mitochondrial DNA A 27(6):4622–4623

Prakash S, Chopra V (1991) Cytogenetics of crop Brassicas and their allies. In: Developments in Plant Genetics and Breeding, vol 2. Elsevier, pp 161–180

Schmidt R, West J, Love K, Lenehan Z, Lister C, Thompson H, Bouchez D, Dean C (1995) Physical map and organization of *Arabidopsis thaliana* chromosome 4. Science 270(5235):480–483

Schmidt R, Love K, West J, Lenehan Z, Dean C (1997) Description of 31 YAC contigs spanning the majority of *Arabidopsis thaliana* chromosome 5. Plant J 11 (3):563–572

Shizuya H, Birren B, Kim U-J, Mancino V, Slepak T, Tachiiri Y, Simon M (1992) Cloning and stable maintenance of 300-kilobase-pair fragments of human DNA in *Escherichia coli* using an F-factor-based vector. Proc Natl Acad Sci 89(18):8794–8797

Slocum M, Figdore S, Kennard W, Suzuki J, Osborn T (1990) Linkage arrangement of restriction fragment length polymorphism loci in *Brassica oleracea*. TheorAppl Genet 80(1):57–64

Song K, Suzuki J, Slocum M, Williams P, Osborn T (1991) A linkage map of*Brassica rapa* (syn. campestris) based on restriction fragment length polymorphism loci. Theor Appl Genet 82(3):296–304

Tao Q, Zhao H, Qiu L, Hong G (1994) Construction of a full bacterial artificial chromosome (BAC) library of *Oryza sativa* genome. Cell Res 4(2):127–133

Thormann C, Ferreira M, Camargo L, Tivang J, Osborn T (1994) Comparison of RFLP and RAPD markers to estimating genetic relationships within and among cruciferous species. Theor Appl Genet 88(8):973–980

Verma D, Lakhanpal N, Singh K (2019) Genome-wide identification and characterization of abiotic-stress responsive SOD (superoxide dismutase) gene family in *Brassica juncea* and *B. rapa*. BMC Genomics 20 (1):1–18

Wang X, Wang H, Wang J, Sun R, Wu J, Liu S, Bai Y, Mun J-H, Bancroft I, Cheng F (2011) The genome of the mesopolyploid crop species *Brassica rapa*. Nat Genet 43(10):1035–1039

Wang P, Zhao Z, Zhang Z, Cai Z, Liao J, Tan Q, Xiang M, Chang L, Xu D, Tian Q (2021) Genome-wide identification and analysis of NPR family genes in *Brassica juncea* var. tumida. Gene 769:145210

Woo S-S, Jiang J, Gill BS, Paterson AH, Wing RA (1994) Construction and characterization of bacterial artificial chromosome library of Sorghum bicolor. Nucleic Acids Res 22(23):4922–4931

Wu Y, Tulsieram L, Tao Q, Zhang H-B, Rothstein SJ (2000) A binary vector-based large insert library for *Brassica napus* and identification of clones linked to a fertility restorer locus for Ogura cytoplasmic male sterility (CMS). Genome 43(1):102–109

Wu Z, Hu K, Yan M, Song L, Wen J, Ma C, Shen J, Fu T, Yi B, Tu J (2019) Mitochondrial genome and transcriptome analysis of five alloplasmic male-sterile lines in *Brassica juncea*. BMC Genomics 20 (1):1–15

Yang J, Liu X, Yang X, Zhang M (2010) Mitochondrially-targeted expression of a cytoplasmic male sterility-associated orf220 gene causes male sterility in *Brassica juncea*. BMC Plant Biol 10 (1):1–10

Yang J, Liu D, Wang X, Ji C, Cheng F, Liu B, Hu Z, Chen S, Pental D, Ju Y (2016) The genome sequence of allopolyploid *Brassica juncea* and analysis of differential homoeolog gene expression influencing selection. Nat Genet 48(10):1225–1232

Yang H, Bayer PE, Tirnaz S, Edwards D, Batley J (2021) Genome-wide identification and evolution of receptor-like kinases (RLKs) and receptor like proteins (RLPs) in *Brassica juncea*. Biology 10(1):17

Zachgo EA, Wang ML, Dewdney J, Bouchez D, Camilleri C, Belmonte S, Huang L, Dolan M, Goodman HM (1996) A physical map of chromosome 2 of *Arabidopsis thaliana*. Genome Res 6(1):19–25

Zhang H-B, Wu C (2001) BAC as tools for genome sequencing. Plant Physiol Biochem 39(3–4):195–209

Zhang L, Cai X, Wu J, Liu M, Grob S, Cheng F, Liang J, Cai C, Liu Z, Liu B (2018) Improved *Brassica rapa* reference genome by single-molecule sequencing and chromosome conformation capture technologies. Hortic Res 5(1):1–11

Zhao N, Xu X, Wamboldt Y, Mackenzie SA, Yang X, Hu Z, Yang J, Zhang M (2016) MutS HOMOLOG1 silencing mediates ORF220 substoichiometric shifting and causes male sterility in *Brassica juncea*. J Exp Bot 67(1):435–444

Zhao N, Li Z, Zhang L, Yang X, Mackenzie SA, Hu Z, Zhang M, Yang J (2021) MutS HOMOLOG1 mediates fertility reversion from cytoplasmic male sterile *Brassica juncea* in response to environment. PlantCell Environ 44(1):234–246

Brassica juncea Genome Assemblies—Characteristics and Utilization

13

Kumar Paritosh and Deepak Pental

Abstract

Brassica juncea (AABB, $n = 18$, ~922 Mb) is a natural allopolyploid of *B. rapa* (AA, $n = 10$) and *B. nigra* (BB, $n = 8$). Oleiferous types of *B. juncea* (mustard) are extensively cultivated in the Indian subcontinent. We reported a highly contiguous genome assembly of *B. juncea* variety Varuna, a mega variety belonging to the Indian gene pool of mustard using PacBio SMRT sequencing technology and optical mapping. Two genetic maps using F1DH populations—Varuna × Heera (VH) and Tumida × Varuna (TuV) were used to assign the scaffolds and contigs to 18 LGs of *B. juncea*. We also assembled the genome of *B. nigra* variety Sangam using long-read ONT sequencing and optical mapping. The two assemblies were annotated for transposons, centromeric repeats, and protein-encoding genes. Highly contiguous genome assemblies of *B. juncea*, *B. rapa*, *B. nigra,* and *B. oleracea* (CC, $n = 9$) by our group and others have allowed structural analysis of the genomes in terms of gene block arrangements and gene content. *B. juncea* is a strict allopolyploid of *B. rapa* and *B. nigra* with no apparent gene loss or genome reshuffling or homeologous exchanges. Comparison of the A, B, and C genomes shows a common inter-paleogenome non-contiguous gene block association pointing to a common origin for the three diploid genomes in contradiction to A/C and the B genomes belonging to two different plastid lineages. We propose a new nomenclature for the B genome LGs/pseudochromosomes based on the homoeology between the B genome and the A (*B. rapa*) genome, using the A genome nomenclature as the baseline being the first sequenced Brassica genome. Genome assemblies of *B. juncea* Varuna and more recently line YS and short-read sequences of extensive germplasm are proving very useful for breeding of both oleiferous and vegetable types of mustard.

13.1 Introduction

This review describes our efforts on the genome assembly of an Indian gene pool line of oilseed mustard—*Brassica juncea* variety Varuna and puts it in the context of the genome assemblies of other *B. juncea* lines, Brassica species belonging to U's triangle (Nagaharu 1935) and the model crucifer species *Arabidopsis thaliana*. We also discuss our work since publishing of the genome assembly of *B. juncea* Varuna on bringing genetics and genomics studies together for precision breeding of oilseed mustard for higher yields.

K. Paritosh · D. Pental (✉)
Centre for Genetic Manipulation of Crop Plants, University of Delhi South Campus, Benito Juarez Road, New Delhi 110021, India

C. Kole and T. Mohapatra (eds.), *The Brassica juncea Genome*, Compendium of Plant Genomes,
https://doi.org/10.1007/978-3-030-91507-0_13

B. juncea belongs to the family Brassicaceae. Model species *A. thaliana* (At), also belonging to the family Brassicaceae, was the first plant genome that was sequenced. A high-quality reference genome of At was assembled using a minimum tiling path of bacterial artificial chromosomes (BACs) that were sequenced by Sanger technology (Arabidopsis Genome Initiative 2000). Even though, At is one of the smaller genomes ($\sim$125 Mb) known among the flowering plants, the genome assembly took considerable effort and resources. However, the effort was worthwhile as it led to much deeper insights into all the facets of plant biology using At as a model species.

It was but natural that At genome assembly and its impact on plant biology would spur interest in the genomes of other members of the family Brassicaceae; particularly, the Brassica species (tribe Brassiceae) belonging to the U's triangle (Nagaharu 1935) that constitute some of the most extensively grown oilseed and vegetable crops, worldwide. U's triangle describes the relationship of three "diploid" species—*B. rapa* (Bra, AA, $n = 10$, $\sim$485 Mb), *B. nigra* (Bni, BB, $n = 8$, $\sim$522 Mb), and *B. oleracea* (Bol, CC, $n = 9$, $\sim$630 Mb) and their three allopolyploids –*B. juncea* (Bju, AABB, $n = 18$, $\sim$922 Mb), *B. napus* (Bna, AACC, $n = 19$, $\sim$1130 Mb), and *B. carinata* (Bca, BBCC, $n = 17$, $\sim$1100 Mb).

However, the genome size of the diploid Brassica species is too large, 4 to 5 times the size of the At genome; the genome assemblies of the Brassica species had to wait for the short-read sequencing technologies. The first draft genome assembly published was that of *B. rapa* through an international effort (Wang et al. 2011). Between 2000, the year At genome assembly was published and 2011, the year *B. rapa* draft assembly was reported, many interesting insights into the Brassica genomes were published. Martin Lysak's group, using labelled At BACs for in-situ hybridizations with the chromosome preparations of the species of U's triangle and some other species belonging to the tribe Brassiceae, showed that each probe hybridized to a minimum of three chromosomal regions in the diploids and six in the allotetraploids (Lysak et al. 2005). This work clearly showed three orthologous regions with gene collinearity in diploid Brassica genomes vis-a-vis At genomic regions. Isobel Parkin's group showed by genetic mapping in *B. napus*, using genic restriction fragment length polymorphism (RFLP) probes from the At genome, in general – three mapping positions for each probe in the A genome linkage groups (LGs) and a similar number in the C genome LGs. Linked RFLP markers in the At genome were linked over long genetic intervals in the A and C genomes of *B. napus*. Both the pioneering studies showed genome triplication and gene collinearity defining syntenic regions between At and the diploid Brassica genomes.

Our laboratory also showed, using intron length polymorphism (ILP) markers that were developed from the conserved exonic regions enclosing an intron of different At genes, that both the A and the B genomes of *B. juncea* had up to three mapping locations for most of the genic markers (Panjabi et al. 2008). Syntenic regions showing gene collinearity could also be mapped. Thus, it was clear that the three diploid genomes A, B, and C are paleohexaploids with extensive stretches of gene collinearity with the At genome but a reduced chromosome number due to genome reshuffling. Comparative architecture of the At and the A genome of *B. rapa* was well summarized by Schranz et al. (Schranz et al. 2006), marking gene blocks in At that defined both the gene collinear regions and the reshuffling of At gene blocks in the A (*B. rapa*) genome.

Publication of a draft genome assembly of *B. rapa* Chiifu (Wang et al. 2011) was a watershed moment for the genomics of Brassica species. The assembly confirmed that *B. rapa* genome was the result of a genome triplication event. The three constituent genomes had undergone extensive chromosomal rearrangements leading to gene block reshuffling vis-à-vis the gene block order in At. The most significant observation was that the three constituent genomes had undergone differential gene losses and were termed—least fractionated (LF), medium fractionated (MF1), and most fractionated (MF2). A similar situation was shown to exist in the *B. oleracea* draft genome assembly by Liu et al. (2014). A draft genome assembly of *B. napus* variety Darmor-bzh (a winter type oleiferous rapeseed) using short-read sequencing was published by Chalhoub et al. (2014).This was followed by a draft genome assembly of *B. napus* ZS-81, a semi-winter morphotype grown in China, by Sun et al. (2017).

13.2 Third-Generation Sequencing Technologies and Brassica A and C Genomes

Long read sequencing technologies–Pacific Bioscience's (PacBio) single-molecule real-time (SMRT) technology and Oxford Nanopore's sequencing technology (ONT) changed the scope of plant genome assemblies dramatically—allowing chromosome-scale assemblies (reviewed by Todd and Van Busen 2020). The long read sequence data were supported by new assembler software–Falcon (Chin et al. 2016) and Canu (Koren et al. 2017). Two other technologies—chromatin confirmation capture (Hi-C) and optical mapping facilitated higher-order scaffolding of long-read contigs providing chromosome-scale assemblies. Long-read technologies, however, have one major flaw–sequences have a significantly higher error rate as compared to the short-read Illumina sequences. Assembler software such as Canu and Falcon achieve self-correction in the hierarchical genome assembly process (HGAP). Programs like Pilon (Walker et al. 2014) use short-read Illumina sequences to polish long-read sequencing based contigs.

The third-generation technologies have been extensively used in the recent years for Brassica genome assemblies. *B. rapa* Chiifu assembly was improved by SMRT sequencing and Hi-C scaffolding (Zhang et al. 2018). Genome assemblies of *B. rapa* and *B. oleracea* were published using ONT long reads and optical mapping (Belser et al. 2018). Genome assemblies of eight different *B. napus* varieties, grown in different agro-ecologies globally, have been assembled by ONT sequencing and Hi-C scaffolding and extensively compared to study the pan-genome structure in this important oilseed crop grown in the sub-temperate regions of the world (Song et al. 2020).Genome assembly of *B. napus* variety Darmor-bzh has also been improved by third-generation technologies (Rousseau-Gueutin et al. 2020). While the short-read-based assembly of Darmor-bzh reported abundant homeologous exchanges (Chalhoub et al. 2014), the long read assembly has not reported any.

13.3 Genome Assemblies of *B. juncea*

B. juncea (mustard) is an important oilseed crop of the Indian subcontinent, well adapted to dry-land cultivation (Chauhan et al. 2011). Although there is interest in the crop in other parts of the world due to its productivity under low moisture availability, oleiferous types of *B. juncea* are extensively cultivated in South Asia only. We reported two divergent gene pools in the oleiferous mustards—the Indian gene pool and the East European gene pool (Pradhan et al. 1993). Significant variation has also been reported in *B. juncea* from China with distinct vegetable and oleiferous types (Kang et al. 2021).

It is quite creditable that despite very little area under *B. juncea*, Ming Fang Zhang's laboratory in China undertook the task of putting together a draft genome assembly of a vegetable type line of *B. juncea* type Tumida (Yang et al. 2016). The assembly was based on Illumina short reads providing a $\sim 176 \times$ coverage, gap filling with $\sim 12 \times$ SMRT sequences, and long-range scaffolding with BioNano optical mapping.

In a quest to search for new single nucleotide polymorphism (SNP)-based markers for mapping different agronomic traits in *B. juncea* doubled haploid (DH) populations, we identified transcriptome-based markers between the Indian and East European gene pool lines Varuna and Heera (Paritosh et al. 2014). Later, after the genome of line Tumida was published, we found many discrepancies between our SNP-based genetic map and the published genome assembly. Our laboratory felt the need for a genome assembly of an oleiferous type of mustard,

Table 13.1 Assembly statistics of *B. juncea* variety Varuna and *B. nigra* variety Sangam genomes

	B. juncea	*B. juncea* A	*B. juncea* B	*B. nigra* B
PacBio	✓	✓	✓	
Nanopore				✓
BioNano	✓	✓	✓	✓
Linkage Map	✓	✓	✓	✓
Total assembly size	869,539,390	–	–	515,400,203
Number of contigs	1253	–	–	1549
Longest contig	35,843,917	–	–	17,509,570
N50 contig length	5,734,093	–	–	1,488,221
Number of scaffolds	61	–	–	15
Longest scaffold	72,092,522	–	–	115,616,497
N50 scaffold length	33,612,814	–	–	68,578,869
Unscaffolded contigs	714 (partial)	–	–	1,051(partial)
Number of pseudochromosomes/LGs	18	10	8	8
Scaffolds assigned to LGs	56	37	19	14
Length of assigned sequences to LGs	840,173,505	333,021,048	507,152,457	505,183,631
Length of unassigned sequences	35,468,516	–	–	30,296,383

particularly of an Indian gene pool line, as the crop is so important for meeting the edible oil requirements of a growing population in South Asia. We, therefore, undertook the sequencing of *B. juncea* variety Varuna as it is one of the most extensively grown mega variety of mustard released by the public-funded breeding programs in India (Chauhan et al. 2011).

To assemble the genome of *B. juncea* variety Varuna (genome size ~922 Mb), high molecular weight DNA was subjected to SMRT sequencing on a PacBio RSII platform at the University of Arizona Genome Sequencing facility. No PacBio machine was available in India at that time. A total of 9,735,857 reads were obtained providing ~100 × coverage of the genome. The reads were assembled into 1253 contigs with an N50 value of ~5.7 Mb using Canu assembler. The contigs were subjected to hierarchical mapping using three different optical maps. Two maps were developed with nicks, labels, repairs, and stains (NLRS)-based labeling using *I BssS*I and *I* BspQI enzymes and one with Direct labeling and stain (DLS) labeling using the DLE-1 enzyme. Corrections were made at each mapping stage, and corrected sequences were mapped again on the consensus maps. A total of 45 contigs were identified with misassemblies which were corrected by breaking the misassembled regions and aligning the edited contigs again to the consensus maps. The assembly after corrections and multiple rounds of scaffolding consisted of 61 scaffolds and 714 unscaffolded contigs/fragments (Table 13.1). Scaffolds were error-corrected using Illumina short reads (~40 × coverage) in three iterated rounds of the Pilon program.

As paralogs within the A and B genomes are evolutionarily more distant than the homoeologs between the A and B genomes, there is always a possibility of misincorporation of conserved homoeologous regions of one of the constituent genomes into the other. We used two strategies to identify such misassemblies, independent of the assembly, and corrections carried out using optical mapping. We sequenced a DH line of *B. nigra* (BB) Sangam, genome size ~522 Mb, using the Illumina HiSeq platform obtaining contigs with an N50 value of ~85.9 kb. The assembled draft genome covered ~470.8 Mb of the *B. nigra* genome (for details see Paritosh et al. 2021).

We compared the 1253 SMRT-based *B. juncea* contigs with the *B. rapa* (A) genome sequences (Wang et al. 2011) and of *B. nigra* (B) to identify the A and B subgenome specific contigs in the *B. juncea* assembly. Predicted protein-encoding genes in the *B. juncea* contigs and scaffolds of *B. nigra* were subjected to a reciprocal protein blast to identify the best matching homologous regions. *B. juncea* SMRT contigs with 95% genes showing the best matching orthologous genes to either of the two parental genomes were grouped as the A or B genome-specific contigs. Of the 1253 contigs, 691 contigs were found to belong to the B genome and 505 contigs were specific to the A genome; 57 contigs remained unassigned.

We further used two genetic maps of *B. juncea* to assign the scaffolds and contigs obtained with SMRT sequencing and optical mapping to 18 linkage groups (LGs) of *B. juncea*. Genetic maps have also been useful in discerning misassemblies. We used a map of *B. juncea* developed from a F1DH population derived from a cross between Varuna x Heera (an East European gene pool line); this VH population map besides the earlier mapped 833 ILP and 999 SNP markers was further enriched with additional 2947 GBS (genotyping by sequencing) markers. Fifty-six of the 61 scaffolds and nine of the 714 contigs could be assigned to the 18 *B. juncea* LGs to constitute 18 pseudochromosomes. The GBS markers were used to compare their position on the genetic map with their physical location on the pseudochromosomes—a very high correlation was observed. However, an involved comparison between genetic marker order and their physical location identified 18 misassemblies. All these 18 misassemblies had been observed in the optical maps also.

To make sure that homoeology between the A and the B genomes is not confounding the genome assembly of *B. juncea,* we used another FlDH population (name TuV) developed from a cross between *B. juncea* lines Tumida (earlier sequenced vegetable type mustard) and Varuna. The TuV genetic map containing 524 ILP and simple sequence repeat (SSR) markers common with the VH population and additional 8517 GBS markers showed correlation with the physical location of the markers on the optical map corrected SMRT scaffolds. The TuV genetic map helped in assigning two more scaffolds to the pseudochromosomes.

13.4 The Final *B. juncea* Varuna Genome Assembly

We found that optical mapping was able to capture all the misassemblies and provided accurate scaffolding of the SMRT contigs. Further three rounds of Pilon using Illumina short-read sequences were sufficient to remove errors resulting from the long read sequencing chemistry.

The final genome assembly of *B. juncea* Varuna consisted of 58 scaffolds and nine contigs assigned to the 18 pseudochromosomes (Fig. 13.1); three scaffolds and 705 contigs remained unassigned. The size of the assigned sequences was calculated to be around 840.2 Mb (~91.2% of the estimated genome size of 922 Mb) around 35.4 Mb of the sequenced genome (3.7%of the total genome) remained unassigned. Three of the pseudochromosomes were covered with only one scaffold, four with two scaffolds, six with three, and the rest with four or more scaffolds and a few contigs.

A comparison of the chromosomes level assembly of *B. juncea* Varuna genome with the assembly earlier reported for *B. juncea* Tumida (V1.1) (Yang et al. 2016), and its improved version V1.5 (Yang et al. 2018) showed a high level of contiguity and lesser number of gaps in the Varuna assembly (for detail see Paritosh et al. 2021). Misassemblies were found to be more pronounced in the assembly of the B genome of *B. juncea* Tumida. The coverage achieved for the B genome in the Varuna genome assembly was significantly higher −~507.1 Mb, as compared to the earlier reported coverage of ~395.9 Mb in V1.1 (Yang et al. 2016) and ~377.6 Mb in V1.5 (Yang et al. 2018) in *B. juncea*Tumida.

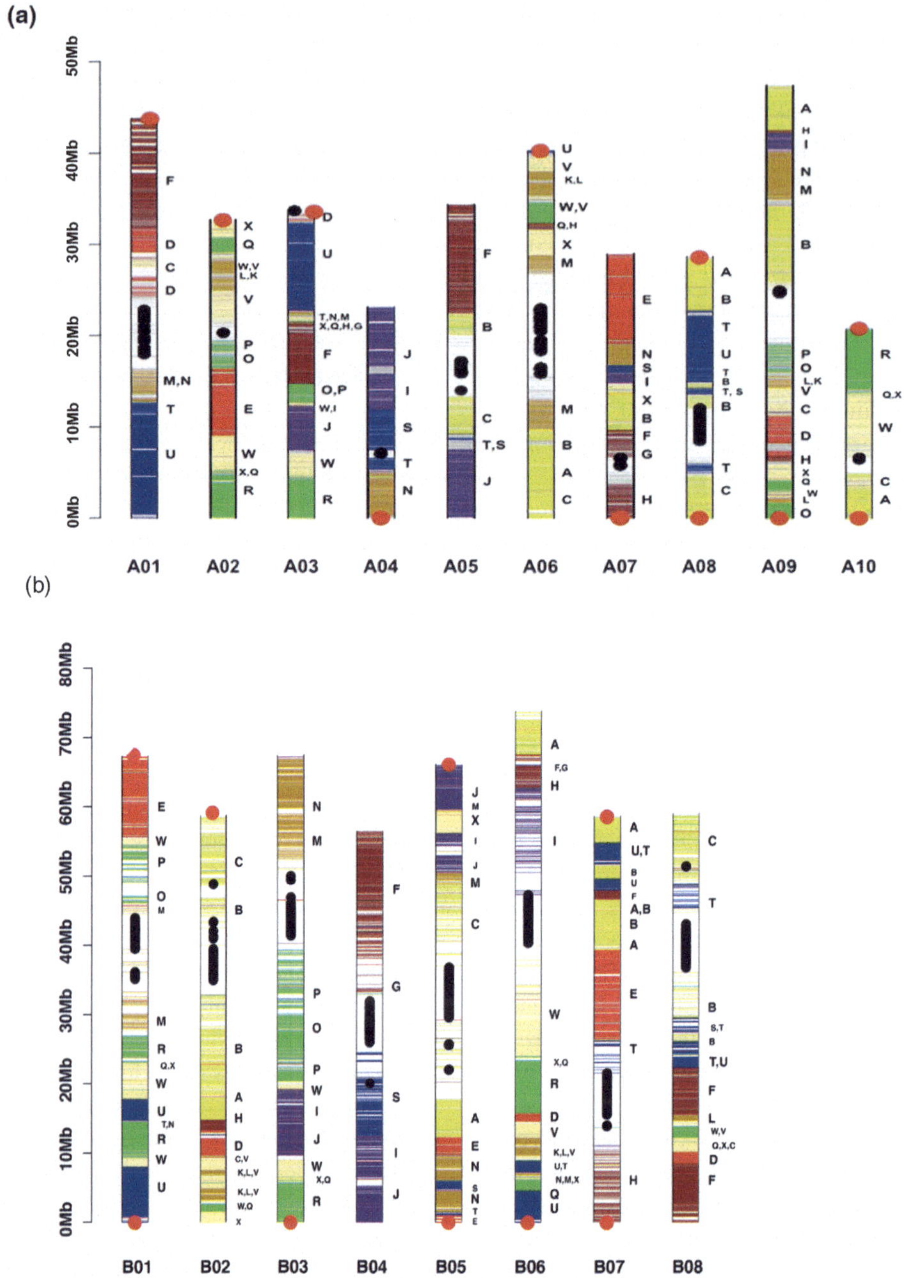

Fig. 13.1 Graphical representation of the *Brassica juncea* (AABB) pseudochromosomes; **a** 10 chromosomes of the A genome, **b** 8 chromosomes of the B genome. A new nomenclature has been adopted for the B genome chromosome following Paritosh et al. (2020). Horizontal lines represent the predicted genes; color of the lines was assigned based on the synteny with the *A. thaliana* gene blocks (A–X) as defined by Schranz et al. (2006); black and red dots represent the centromeric and telomeric repeats, respectively. Figure modified from Paritosh et al. (2021) incorporating changes in the nomenclature of the pseudochromosomes of the B genome following Paritosh et al. (2020)

B. juncea Varuna genome assembly was annotated for transposable elements, centromeres, and protein-encoding genes. Around 385 Mb (~45.8%) of the assembled genome of Varuna was found to be constituted of transposable elements (TEs) – the B genome had a higher repeat content ~259 Mb (~51%) as compared to the A genome – ~113 Mb (~33.9%). Retrotransposons—LTR/Copia and LTR/Gypsy were the predominant TEs present in the Varuna genome.

Candidate centromeric regions were identified from the correlation plots between GBS markers on the genetic maps and the physical positions of their respective tags on the pseudochromosomes; regions with significantly low recombination frequencies were marked as potential centromeric regions and analyzed for repeats that were absent from other parts of the pseudochromosomes. CentBr1 and CentBr2, identified as centromere-specific sequences in *B. rapa* (A genome) and *B. oleracea* (C genome) in earlier studies (Wang et al. 2011; Liu et al. 2014), were found to be present in the A genome of *B. juncea* but absent from the centromeric regions of all the pseudochromosomes of the B genome. Three new A genome-specific centromeric repeats and seven B genome-specific repeats were identified. Besides centromeric regions, highly contiguous genome assembly could allow us to identify telomeres in 15 out of 18 assembled pseudochromosomes.

Genome sequences assigned to the 18 pseudochromosomes of Varuna were repeat-masked and analyzed for gene content with Augusts software (Stanke and Morgenstern 2005), using 543 randomly selected full-length CDS obtained from the PacBio-based RNA-seq analysis as the training set. A total of 101,959 genes were predicted to which 46,381 in the A genome, and 55,578 in the B genome were assigned to the individual chromosomes. The earlier *B. juncea* Tumida genome assembly, predominantly based on short-read sequencing, had reported only 80,050 genes.

It can be concluded that the third-generation genome assembly technologies have provided a superior genome assembly in *B. juncea* Varuna as has been reported in other crop species including *B. napus*.

13.5 Evolutionary Implications of the *B. juncea* Genome Assemblies

While the accuracy of the genome assembly is of prime importance, all high impact journals would like to have some new biological insights. For us, the most interesting biological aspects were to compare the structure of the A and B genome of *B. juncea* as the Varuna assembly was the first contiguous assembly for the B genome. We took the A and B genomes of *B. juncea* Varuna as representative of the A and B genomes of *B. rapa* and *B. nigra*. The logic of this was based on a comparison of the assembly of the A genome of *B. juncea* Varuna with the most recent *B. rapa* Chiifu assembly (V3.0) (Zhang et al. 2018). A gene-to-gene correlation analysis based on the least Ks values showed very high collinearity between the BjuA and BraA genomes. The *B. rapa* V3.0 reported a gene content of 45,985 genes, whereas our analysis had predicted 46,381 genes in the BjuA genome. This analysis confirmed that there have been no significant structural changes in the BjuA genome after the allopolyploidy event and that the two A genomes —BjuA and BraA can be taken as equivalent. The B genome of *B. juncea* Varuna contained 57,084 genes, and the genome assembly was highly contiguous.

While genome triplication (the ***b*** event) and subsequent gene fractionation and gene blocks rearrangements are a characteristic feature of the genera belonging to the tribe Brassicaceae (Lysak and Koch 2011), some intriguing questions remain—were structurally similar or dissimilar genomes involved in the ***b*** event and were there a single or more than one independent hybridization events? Two different models have been proposed for the evolution of the Brassica genomes. One model suggested genome triplication with three near-similar proto-Calepineae karyotype (PCK; $n = 7$) genomes that precipitated differential gene loss, reduction in chromosome number, and new gene block associations followed by changes resulting from homoploid hybrid speciation (Cheng et al. 2017). Another model on genome triplication suggested that two of the three paleogenomes—MF1 and MF2, formed an allotetraploid with MF1 establishing dominance in gene retention, followed by crosses with the LF genome (the actual ***b*** event) leading to chromosomal reshuffling and further gene fractionation (Tang et al. 2012). However, neither a single nor a two-step ***b*** event resolves the issue of the presence of two distinct plastid lineages (the rapa lineage and the nigra lineage) in the tribe Brassiceae (Warwick and Black 1991; Pradhan et al. 1992; Arias and Pires 2012; Li et al. 2017). The two-step model was developed further to accommodate the presence of the two distinct plastid genome lineages by suggesting reciprocal crosses between the tetraploid (MF1 + MF2) lineage and the LF lineage (Sharma et al. 2014).

We assigned all the At gene blocks present in the A and B genomes of *B. juncea* and the C genome of *B. oleracea* (Belser et al. 2018) to the three constituent paleogenomes—LF, MF1, and MF2. The A and C genomes besides being less divergent showed extensive similarity in their gene block arrangements as compared to the B genome homoeologs. Contiguous genome assemblies have shown the A, B and C genomes to be even more fragmented than reported earlier (Wang et al. 2011; Liu et al. 2014; Belser et al. 2018; Zhang et al. 2018) with extensive intra-block fragmentation. Gene block associations reported in the selected taxa of different tribes identified in the family Brassicaceae have been used by Lysak et al. (2016) to understand inter-tribe and intra-tribe evolutionary relationships. We scanned the gene block associations in the BjuA, BjuB, and BolC genomes and classified these under four categories—(a) intra-paleogenome contiguous gene block associations, (b) intra-paleogenome non-contigous gene block associations, (c) inter-paleogenome contiguous gene block associations, and (d) inter-paleogenome non-contiguous gene block associations (Fig. 13.2).

The ancestral intra-paleogenome contiguous gene block associations—A-B, F-G, G-H, T-U, and W-X, reported to be absent in the A and C genomes of *B. rapa, B. oleracea,* and *B. napus* were found in one or the other constituent paleogenomes. The earlier assemblies missed these associations, and therefore, it was concluded that their absence was a characteristic feature of the tribe Brassiceae (Lysak et al. 2016). Intra-paleogenome non-contiguous gene block associations are key to understand the pre-***b*** event structure of the three paleogenomes. A previously reported association V-K-L-W-Q-X does not exist in any of the Brassicaceae paleogenomes; only V-K-L is a shared association. We report some new associations in this category—three of these Q-X, W-X-Q-R, and V-K-L are present in all the three paleogenomes pointing to a shared ancestry; W-E, M–N-T-U, and O-P-W are present only in $MF1_A$, $MF1_B$, and $MF1_C$. Novel associations in the MF1 genome could predate the ***b*** event and could have contributed to gene block fragmentation upon hybridization.

The most intriguing structural aspect was the presence of a common inter-paleogenome non-contiguous gene block association J_{MF1}-I_{MF1}-S_{MF2}-S_{LF} lending support to a monophyletic origin of the A and B genomes (Fig. 13.2). A gene-to-gene collinearity analysis showed similar junctions in the J_{MF1}-I_{MF1}-S_{MF2}-S_{LF} of the A, B, and C genomes (Paritosh et al. 2021).

13.6 Post *B. juncea* Varuna Assembly Developments

While the *B. juncea* genome assembly manuscript was under review, we put together a highly contiguous genome assembly of a DH line of *B. nigra* Sangam using Nanopore MinION platform and optical mapping (Paritosh et al. 2020). The obtained long-reads provided $\sim 100\times$ coverage of the *B. nigra* genome if we consider the genome size to be ~ 522 Mb. The raw reads were assembled into 1549 contigs with an N50 value of ~ 1.48 Mb using the Canu assembler (Table 13.1). The total size of the assembled contigs was ~ 515.4 Mb, covering $\sim 98\%$ of the

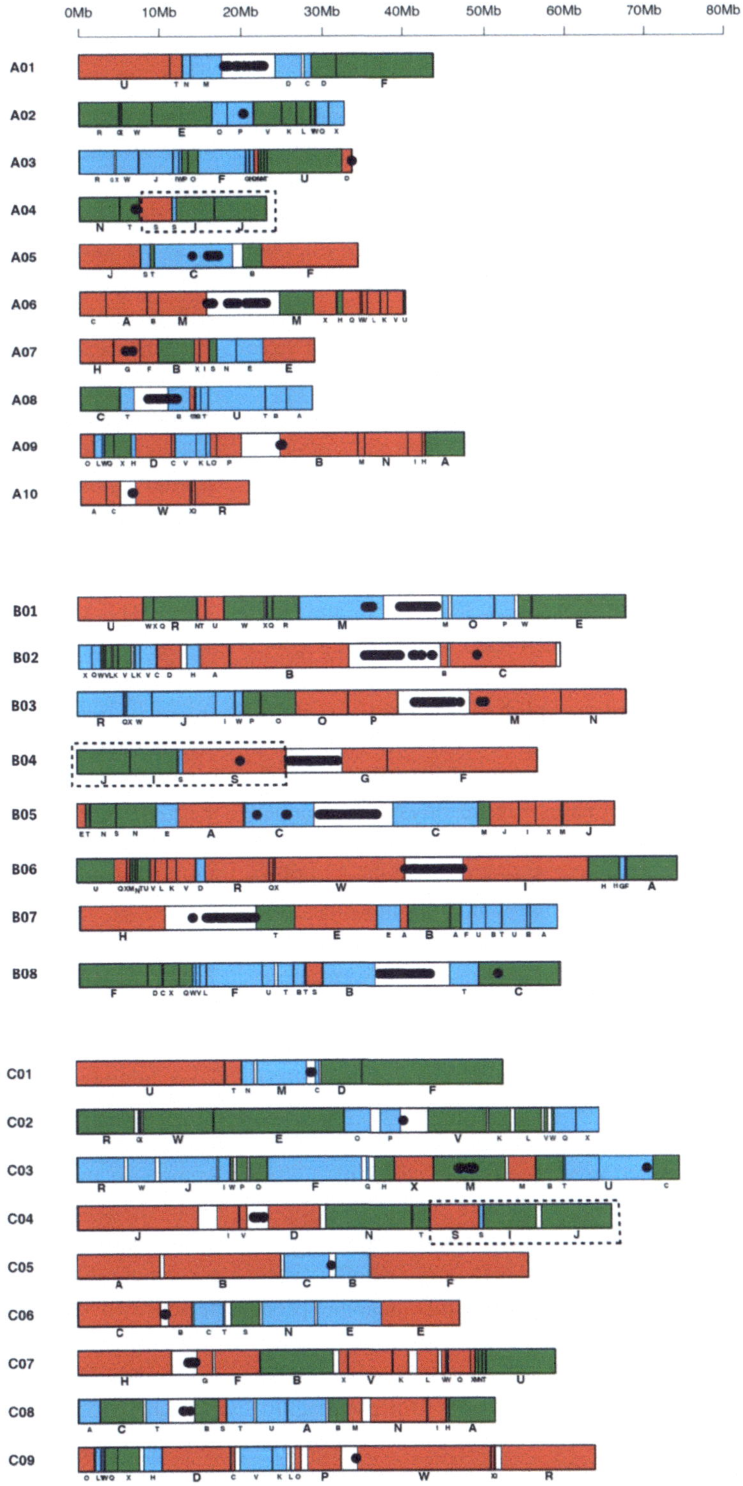

Fig. 13.2 Gene block arrangements in the A and B genome pseudochromosomes of *Brassica juncea* and C genome of *B. oleracea* (Belser et al. 2018). LF, MF1, and MF2 paleogenomes are represented by red, green, and blue colors, respectively. Gene block arrangements in the A and C genomes were found to be more similar as compared to the B genome. The inter-paleogenome arrangement J_{MF1}-I_{MF1}-S_{MF2}-S_{LF} is conserved in all the three diploid genomes—A, B, and C, suggesting a common origin for the A/C and B genomes. Modified from Paritosh et al. (2021), pseudochromosomes nomenclature for BjuB genome is as suggested for the BniB genome by Paritosh et al. (2020)

B. nigra genome. Nanopore contigs were error-corrected with ~100 × Illumina PE reads using the Pilon program for five iterative cycles.

Optical mapping was used for finding the misassemblies in the contigs and for assembling the contigs into scaffolds. Two different optical maps, one with direct label and stain (DLS) technology using the DLE-I enzyme and the other with nick, label, repair, and stain (NLRS) technology using *BssSI* enzyme, were developed. A hybrid assembly protocol was used, which generated 15 scaffolds with an N50 value of ~70.4 Mb covering ~506.4 Mb of the genome. One hundred forty-eight contigs were found to contain misassemblies, mostly due to the merger of some of the highly conserved syntenic regions. A total of 1051 unmapped sequence fragments with an N50 value of ~36.7 kb, covering ~30.4 Mb of the genome, remained unscaffolded (Table 13.1).

A genetic map of *B. nigra*, with 2723 markers, was used to validate the integrity of the scaffolds and to assign these to the eight pseudochromosomes—BniB01—BniB08. The GBS-based genetic markers were physically mapped on the scaffolds; no misassemblies were observed. Fourteen out of 15 scaffolds could be assembled into eight pseudochromosomes; fifteenth scaffold contained the plastid genome. Five out of the eight chromosomes were represented by a single scaffold each; the remaining three chromosomes consisted of two, three, and four scaffolds. The size of the final *B. nigra* genome that could be assigned to the pseudochromosomes was ~505.18 Mb (~96.7% of the estimated genome size). A total of 57,249 protein-encoding genes were predicted in the *B. nigra* genome using Augustus program.

A comparison of the *B. nigra* (BniB) pseudochromosomes with the BjuB genome pseudochromosomes showed very high gene collinearity between the two B genomes except for the three inversions in the Bni pseudochromosome. The gene content of the two B genomes —57,249 in BniB vs 57,084 in BjuB was also found to be similar. Further no exchanges (≥ 5 consecutive genes) were observed between the homoeologus regions of the BjuA and BjuB genomes. Therefore, the comparative structural analysis of the BjuA and BjuB genomes with BolC in Paritosh et al. (2021) was perfectly valid.

In 2020, two more *B. nigra* genome assemblies of lines Ni100 and C2 were published (Perumal et al. 2020). The two *B. nigra* assemblies do not record any inversions observed in the *B. nigra* Sangam genome. The *B. nigra* Ni100 and C2 genomes show total gene collinearity with one another and the B genome of *B. juncea*. Incidentally, the two genome assemblies used a *B. juncea* VH genetic map developed in collaboration with our lab using transcriptome-based genetic markers (Harper et al. 2020).

More recently, a genome assembly of *B. juncea* variety Sichuan Yellow (SY) has been published using PacBio SMRT sequencing and optical mapping and Hi-C analysis (Kang et al. 2021). The VH genetic map has been used also in this assembly to assign contigs to the pseudochromosomes. The statistics of this assembly are very similar to the *B. juncea* Varuna assembly stats reported by us. While the SY assembly is comparable to the Varuna assembly in stats, in the gene block arrangement on the pseudochromosomes of SY, many small gene blocks are amiss.

13.7 Evolutionary Conclusions on the B Genome

To accommodate two plastid lineages in Brassica coenospecies—the *B. rapa* lineage and *B. nigra* lineage, we expected that the A/C and the B genome, if they had resulted from two independent crosses with divergent female parents, should not have any common inter-paleogenome non-contiguous gene block associations. The presence of such an association in all the three genomes points to a monophyletic origin of the A/C and B genomes (Fig. 13.2). Genome assemblies of some more taxa belonging to the tribe Brassicecae could reveal a triplicated genome with no common inter-paleogenome non-contiguous gene block association.

We carried out an analysis of the extent of homoeology between the BraA (Zhang et al.

2018) pseudo chromosomes and the *B. nigra* Sangam pseudochromosomes (Paritosh et al. 2020). We proposed a new nomenclature for the pseudochromosomes of the B genome which reflects homoeologous relationships with the BraA genome—the first diploid species that was sequenced from the tribe Brassiceae. We had earlier given a nomenclature based on the synteny between the A and B genomes of *B. juncea* using ILP and SNP markers (Panjabi et al. 2008; Paritosh et al. 2014). However, the presence of a common inter-paleogenome non-contiguous gene block association in the three genomes A/C and B requires a further change in the nomenclature to acknowledge this new evidence. Figures 13.1 and 13.2 are, therefore, based on the new nomenclature for the pseudochromosomes of the B genome.

All nomenclatures should reflect relatedness; the new BniB and BjuB nomenclature reflect the structural relationship between the pseudochromosomes of the A, B and C genomes more accurately.

We hope that this new nomenclature will be accepted by the Brassica research community.

13.8 Implications of *B. juncea* Genome Assemblies for Plant Breeding

We had reported earlier that the Indian and the East European gene pool lines are heterotic for yield (Pradhan et al. 1993). The lines of the two gene pools are highly contrasting in many yield-influencing traits. Our laboratory has mapped several qualitative and quantitative trait loci using different mapping populations developed from crosses between the Indian and East European gene pool lines (Table 13.2).

Along with the genome assembly of *B. juncea* variety Varuna, we carried out short-read sequencing of six parental lines that have been used to develop the F1DH populations mentioned in Table 13.2. We sequenced three lines of the Indian gene pool—Varuna, Pusa Jaikisan, and Kranti, five lines of the East European gene pool—Heera, EH2, Donskaja-IV, Skorospieka, Cutlass, and two intermediate lines—TM4 and J8 (unpublished results). Short-read sequences of ten different Chinese gene pool lines reported by Yang et al. (2016) were also used for the analysis. Short-read sequences of all the lines were mapped on the *B. juncea* Varuna assembly, and SNPs were identified between the sequenced lines using GATK-based realignment pipeline (Van der Auwera et al. 2013).

PCA analysis between all the lines using total identified SNPs showed that the Indian gene pool has least genetic variability, whereas the East European gene pool lines are more variable. The Chinese gene pools lines grouped into two different clusters, one for the vegetable types and the other for the oilseed types. Intermediate type lines grouped apart from the Indian, East European, and the Chinese gene pools (Fig. 13.3).

All the genetic intervals mapped for different yield related quantitative trait loci (QTLs) in different mapping populations (Table 13.2) have been converted to their physical positions on the Varuna genome assembly (unpublished results). SNPs, both synonymous and non-synonymous, present in the genic regions within the physical intervals have been marked and can be used for fine mapping and discovery of the candidate gene(s). The Varuna genome assembly and short-

Table 13.2 Summary of different populations developed by CGMCP and used for mapping yield-influencing quantitative traits

Mapping population	Parental lines	Size of mapping population	References
VH	Varuna × Heera	166	Ramchiary et al. (2007), Rout et al. (2018)
EPJ	EH2 × PusaJaikisan	182	Dhaka et al. (2017), Rout et al. (2018)
TD	TM4 × Donskaja-IV	166	Yadava et al. (2012), Rout et al. (2018)

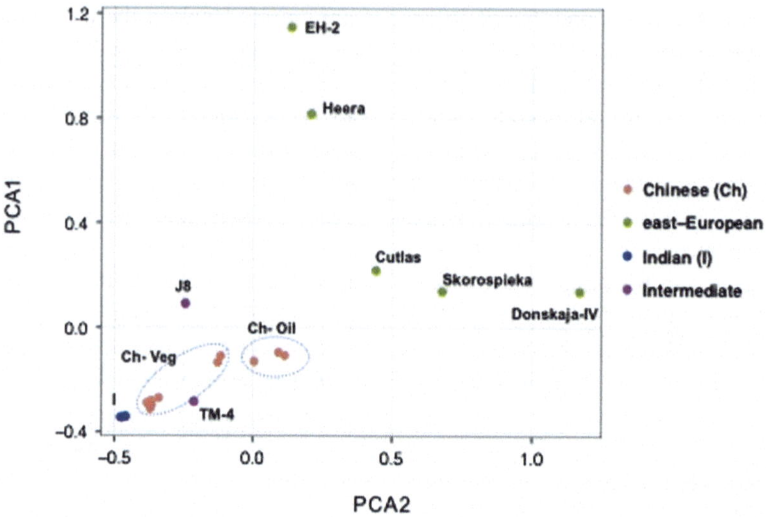

Fig. 13.3 Genetic variability among some select lines of *Brassica juncea* based on principal component analysis (PCA). SNPs present in three lines of the Indian gene pool—Varuna, PusaJaikisan, and Kranti, five lines of the East European gene pool—Heera, EH2, Donskaja-IV, Skorospieka, Cutlass, two Intermediate lines—TM4, J8, and Chinese lines with accession numbers—SRR2050307, SRR2057910, SRR2057917, SRR2057922, SRR2057923, SRR2057924, SRR2057925, SRR2057927, SRR2057935, SRR2057939 (Yang et al. 2016) were analyzed

read sequences of the germplasm available with us have been used to identify allelic diversity for the white rust resistance conferring CC-NBS-LRR (CNL type R gene *BjuWRR1*) mapped in Donskaja-IV in the TD population and *BjuWRR2* in line Tumida in TuV population (Arora et al. 2019; Bhayana et al. 2019).The recent publication on the genome assembly of a Chinese type of *B. juncea* has also reported short-read sequences of 480 lines collected from different parts of the world and mapped many traits by association mapping (Kang et al. 2021). Allelic diversity can now be studied in a much wider germplasm.

The highly contiguous genome assemblies of Varuna and a Chinese yellow seeded line (YS) along with the short-read sequences of extensive variability available in *B. juncea* will support future research on trait mapping, gene identification, allelic diversity analysis, and trait diversification in both oilseed and vegetable types of mustard. However, answering some of the lingering evolutionary questions and the possibility of mobilizing genes form other taxa of the tribe Brassiceae to crop Brassicas, including *B. juncea,* would require some more genomes to be sequenced.

References

Arias T, Pires JC (2012) A fully resolved chloroplast phylogeny of the brassica crops and wild relatives (Brassicaceae: Brassiceae): novel clades and potential taxonomic implications. Taxon 61:980–988

Arora H, Padmaja KL, Paritosh K, Mukhi N, Tewari AK, Mukhopadhyay A, Gupta V, Pradhan AK, Pental D (2019) BjuWRR1, a CC-NB-LRR gene identified in *Brassica juncea*, confers resistance to white rust caused by *Albugo candida*. Theor Appl Genet 132 (8):2223–2236

Belser C, Istace B, Denis E, Dubarry M, Baurens FC, Falentin C, Genete M, Berrabah W, Chevre AM, Delourme R, Deniot G, Denoeud F, Duffe P, Engelen S, Lemainque A, Manzanares-Dauleux M, Martin G, Morice J, Noel B, Vekemans X, D'Hont A, Rousseau-Gueutin M, Barbe V, Cruaud C, Wincker P, Aury JM (2018) Chromosome-scale assemblies of plant genomes using nanopore long reads and optical maps. Nat Plants 4:879–887

Bhayana L, Paritosh K, Arora H, Yadava SK, Singh P, Nandan D, Mukhopadhyay A, Gupta V, Pradhan AK, Pental D (2019) A mapped locus on LG A6 of *Brassica juncea* line tumida conferring resistance to white rust contains a CNL type R gene. Front Plant Sci 10:1690

Chalhoub B, Denoeud F, Liu S, Parkin IA, Tang H, Wang X, Chiquet J, Belcram H, Tong C, Samans B, Correa M, Da Silva C, Just J, Falentin C, Koh CS, Le Clainche I, Bernard M, Bento P, Noel B, Labadie K, Alberti A, Charles M, Arnaud D, Guo H, Daviaud C, Alamery S, Jabbari K, Zhao M, Edger PP, Chelaifa H, Tack D, Lassalle G, Mestiri I, Schnel N, Le Paslier MC, Fan G, Renault V, Bayer PE, Golicz AA, Manoli S, Lee TH, Thi VH, Chalabi S, Hu Q, Fan C, Tollenaere R, Lu Y, Battail C, Shen J, Sidebottom CH, Wang X, Canaguier A, Chauveau A, Berard A, Deniot G, Guan M, Liu Z, Sun F, Lim YP, Lyons E, Town CD, Bancroft I, Wang X, Meng J, Ma J, Pires JC, King GJ, Brunel D, Delourme R, Renard M, Aury JM, Adams KL, Batley J, Snowdon RJ, Tost J, Edwards D, Zhou Y, Hua W, Sharpe AG, Paterson AH, Guan C, Wincker P (2014) Plant genetics. Early allopolyploid evolution in the post-Neolithic *Brassica napus* oilseed genome. Science 345:950–953

Chauhan J, Singh K, Kumar S (2011) Hundred years of rapeseed-mustard breeding in India: accomplishments and future strategies. Indian J Agri Sci 81:1093–1109

Cheng F, Liang J, Cai C, Cai X, Wu J, Wang X (2017) Genome sequencing supports a multi-vertex model for Brassiceae species. Curr Opin Plant Biol 36: 79–87

Chin CS, Peluso P, Sedlazeck FJ, Nattestad M, Concepcion GT, Clum A, Dunn C, O'Malley R, Figueroa-Balderas R, Morales-Cruz A, Cramer GR, Delledonne M, Luo C, Ecker JR, Cantu D, Rank DR, Schatz MC (2016) Phased diploid genome assembly with single-molecule real-time sequencing. Nat Methods 13:1050–1054

Dhaka N, Rout K, Yadava SK, Sodhi YS, Gupta V, Pental D, Pradhan AK (2017) Genetic dissection of seed weight by QTL analysis and detection of allelic variation in Indian and east European gene pool lines of *Brassica juncea*. Theor Appl Genet 130:293–307

Harper AL, He Z, Langer S, Havlickova L, Wang L, Fellgett A, Gupta V, Kumar Pradhan A, Bancroft I (2020) Validation of an associative transcriptomics platform in the polyploid crop species *Brassica juncea* by dissection of the genetic architecture of agronomic and quality traits. Plant J 103:1885–1893

Kang L, Qian L, Zheng M, Chen L, Chen H, Yang L, You L, Yang B, Yan M, Gu Y, Wang T, Schiessl SV, An H, Blischak P, Liu X, Lu H, Zhang D, Rao Y, Jia D, Zhou D, Xiao H, Wang Y, Xiong X, Mason AS, Chris Pires J, Snowdon RJ, Hua W, Liu Z (2021) Genomic insights into the origin, domestication and diversification of *Brassica juncea.* Nat Genet 53:1392–1402

Koren S, Walenz BP, Berlin K, Miller JR, Bergman NH, Phillippy AM (2017) Canu: scalable and accurate long-read assembly via adaptive k-mer weighting and repeat separation. Genome Res 27:722–736

Li P, Zhang S, Li F, Zhang S, Zhang H, Wang X, Sun R, Bonnema G, Borm TJ (2017) A phylogenetic analysis of chloroplast genomes elucidates the relationships of the six economically important Brassica species comprising the triangle of U. Front Plant Sci 8:111

Liu S, Liu Y, Yang X, Tong C, Edwards D, Parkin IA, Zhao M, Ma J, Yu J, Huang S, Wang X, Wang J, Lu K, Fang Z, Bancroft I, Yang TJ, Hu Q, Wang X, Yue Z, Li H, Yang L, Wu J, Zhou Q, Wang W, King GJ, Pires JC, Lu C, Wu Z, Sampath P, Wang Z, Guo H, Pan S, Yang L, Min J, Zhang D, Jin D, Li W, Belcram H, Tu J, Guan M, Qi C, Du D, Li J, Jiang L, Batley J, Sharpe AG, Park BS, Ruperao P, Cheng F, Waminal NE, Huang Y, Dong C, Wang L, Li J, Hu Z, Zhuang M, Huang Y, Huang J, Shi J, Mei D, Liu J, Lee TH, Wang J, Jin H, Li Z, Li X, Zhang J, Xiao L, Zhou Y, Liu Z, Liu X, Qin R, Tang X, Liu W, Wang Y, Zhang Y, Lee J, Kim HH, Denoeud F, Xu X, Liang X, Hua W, Wang X, Wang J, Chalhoub B, Paterson AH (2014) The *Brassica oleracea* genome reveals the asymmetrical evolution of polyploid genomes. Nat Commun 5:3930

Lysak MA, Koch MA (2011) Phylogeny, genome, and karyotype evolution of Crucifers (Brassicaceae). In: Schmidt R, Bancroft I (eds) Genetics and genomics of the brassicaceae. Springer, New York, NY, pp 1–31

Lysak MA, Koch MA, Pecinka A, Schubert I (2005) Chromosome triplication found across the tribe Brassiceae. Genome Res 15:516–525

Lysak MA, Mandakova T, Schranz ME (2016) Comparative paleogenomics of crucifers: ancestral genomic blocks revisited. Curr Opin Plant Biol 30:108–115

Nagaharu U (1935) Genome analysis of Brassica with special reference to the experimental formation of B. napus and peculiar mode of fertilization. Jpn J Bot 7:389–452

Panjabi P, Jagannath A, Bisht NC, Padmaja KL, Sharma S, Gupta V, Pradhan AK, Pental D (2008) Comparative mapping of *Brassica juncea* and *Arabidopsis thaliana* using Intron Polymorphism (IP) markers: homoeologous relationships, diversification and evolution of the A, B and C Brassica genomes. BMC Genomics 9:113

Paritosh K, Gupta V, Yadava SK, Singh P, Pradhan AK, Pental D (2014) RNA-seq based SNPs for mapping in *Brassica juncea* (AABB): synteny analysis between the two constituent genomes A (from *B. rapa*) and B (from *B. nigra*) shows highly divergent gene block arrangement and unique block fragmentation patterns. BMC Genomics 15:396

Paritosh K, Pradhan AK, Pental D (2020) A highly contiguous genome assembly of Brassica nigra (BB) and revised nomenclature for the pseudochromosomes. BMC Genomics 21:887

Perumal S, Koh CS, Jin L, Buchwaldt M, Higgins EE, Zheng C, Sankoff D, Robinson SJ, Kagale S, Navabi ZK, Tang L, Horner KN, He Z, Bancroft I, Chalhoub B, Sharpe AG, Parkin IAP (2020) A high-contiguity *Brassica nigra* genome localizes active centromeres and defines the ancestral Brassica genome. Nat Plants 6:929–941

Pradhan AK, Prakash S, Mukhopadhyay A, Pental D (1992) Phytogeny of Brassica and allied genera based on variation in chloroplast and mitochondrial DNA patterns: molecular and taxonomic classifications are incongruous. Theor Appl Genet 85:331–340

Pradhan AK, Sodhi YS, Mukhopadhyay A, Pental D (1993) Heterosis breeding in Indian mustard (*Brassica-Juncea* L Czern and Coss)—analysis of component characters contributing to heterosis for yield. Euphytica 69:219–229

Ramchiary N, Padmaja KL, Sharma S, Gupta V, Sodhi YS, Mukhopadhyay A, Arumugam N, Pental D, Pradhan AK (2007) Mapping of yield influencing QTL in *Brassica juncea*: implications for breeding of a major oilseed crop of dryland areas. Theor Appl Genet 115:807–817

Rousseau-Gueutin M, Belser C, Da Silva C, Richard G, Istace B, Cruaud C, Falentin C, Boideau F, Boutte J, Delourme R, Deniot G, Engelen S, de Carvalho JF, Lemainque A, Maillet L, Morice J, Wincker P, Denoeud F, Chevre AM, Aury JM (2020) Long-read assembly of the *Brassica napus* reference genome Darmor-bzh. Gigascience 9

Rout K, Yadav BG, Yadava SK, Mukhopadhyay A, Gupta V, Pental D, Pradhan AK (2018) QTL landscape for oil content in *Brassica juncea*: analysis in multiple bi-parental populations in high and '0' erucic background. Front Plant Sci 9:1448–1463

Schranz ME, Lysak MA, Mitchell-Olds T (2006) The ABC's of comparative genomics in the Brassicaceae: building blocks of crucifer genomes. Trends Plant Sci 11:535–542

Sharma S, Padmaja KL, Gupta V, Paritosh K, Pradhan AK, Pental D (2014) Two plastid DNA lineages–Rapa/Oleracea and Nigra–within the tribe Brassiceae can be best explained by reciprocal crosses at hexaploidy: evidence from divergence times of the plastid genomes and R-block genes of the A and B genomes of Brassica juncea. PLoS One 9:e93260

Song JM, Guan Z, Hu J, Guo C, Yang Z, Wang S, Liu D, Wang B, Lu S, Zhou R, Xie WZ, Cheng Y, Zhang Y, Liu K, Yang QY, Chen LL, Guo L (2020) Eight high-quality genomes reveal pan-genome architecture and ecotype differentiation of Brassica napus. Nat Plants 6:34–45

Sun F, Fan G, Hu Q, Zhou Y, Guan M, Tong C, Li J, Du D, Qi C, Jiang L, Liu W, Huang S, Chen W, Yu J, Mei D, Meng J, Zeng P, Shi J, Liu K, Wang X, Wang X, Long Y, Liang X, Hu Z, Huang G, Dong C, Zhang H, Li J, Zhang Y, Li L, Shi C, Wang J, Lee SM, Guan C, Xu X, Liu S, Liu X, Chalhoub B, Hua W, Wang H (2017) The high-quality genome of *Brassica napus* cultivar 'ZS11' reveals the introgression history in semi-winter morphotype. Plant J 92:452–468

Tang H, Woodhouse MR, Cheng F, Schnable JC, Pedersen BS, Conant G, Wang X, Freeling M, Pires JC (2012) Altered patterns of fractionation and exon deletions in *Brassica rapa* support a two-step model of paleohexaploidy. Genetics 190:1563–1574

Van der Auwera GA, Carneiro MO, Hartl C, Poplin R, Del Angel G, Levy-Moonshine A, Jordan T, Shakir K, Roazen D, Thibault J, Banks E, Garimella KV, Altshuler D, Gabriel S, DePristo MA (2013) From FastQ data to high confidence variant calls: the Genome Analysis Toolkit best practices pipeline. Curr Protoc Bioinformat 43:11 10 11–11 10 33

Walker BJ, Abeel T, Shea T, Priest M, Abouelliel A, Sakthikumar S, Cuomo CA, Zeng Q, Wortman J, Young SK, Earl AM (2014) Pilon: an integrated tool for comprehensive microbial variant detection and genome assembly improvement. PLoS One 9:e112963

Wang X, Wang H, Wang J, Sun R, Wu J, Liu S, Bai Y, Mun JH, Bancroft I, Cheng F, Huang S, Li X, Hua W, Wang J, Wang X, Freeling M, Pires JC, Paterson AH, Chalhoub B, Wang B, Hayward A, Sharpe AG, Park BS, Weisshaar B, Liu B, Li B, Liu B, Tong C, Song C, Duran C, Peng C, Geng C, Koh C, Lin C, Edwards D, Mu D, Shen D, Soumpourou E, Li F, Fraser F, Conant G, Lassalle G, King GJ, Bonnema G, Tang H, Wang H, Belcram H, Zhou H, Hirakawa H, Abe H, Guo H, Wang H, Jin H, Parkin IA, Batley J, Kim JS, Just J, Li J, Xu J, Deng J, Kim JA, Li J, Yu J, Meng J, Wang J, Min J, Poulain J, Wang J, Hatakeyama K, Wu K, Wang L, Fang L, Trick M, Links MG, Zhao M, Jin M, Ramchiary N, Drou N, Berkman PJ, Cai Q, Huang Q, Li R, Tabata S, Cheng S, Zhang S, Zhang S, Huang S, Sato S, Sun S, Kwon SJ, Choi SR, Lee TH, Fan W, Zhao X, Tan X, Xu X, Wang Y, Qiu Y, Yin Y, Li Y, Du Y, Liao Y, Lim Y, Narusaka Y, Wang Y, Wang Z, Li Z, Wang Z, Xiong Z, Zhang Z (2011) Brassica rapa Genome Sequencing Project C. The genome of the mesopolyploid crop species *Brassica rapa*. Nat Genet 43:1035–1039

Warwick SI, Black LD (1991) Molecular systematics of Brassica and allied genera (Subtribe Brassicinae, Brassiceae)—chloroplast genome and cytodeme congruence. Theor Appl Genet 82:81–92

Yadava SK, Arumugam N, Mukhopadhyay A, Sodhi YS, Gupta V, Pental D, Pradhan AK (2012) QTL mapping of yield associated traits in *Brassica juncea*: meta-analysis and epistatic interactions using two different crosses between east European and Indian gene pool lines. Theor Appl Genet 125:1553–1564

Yang J, Liu D, Wang X, Ji C, Cheng F, Liu B, Hu Z, Chen S, Pental D, Ju Y, Yao P, Li X, Xie K, Zhang J, Wang J, Liu F, Ma W, Shopan J, Zheng H, Mackenzie SA, Zhang M (2016) The genome sequence of allopolyploid *Brassica juncea* and analysis of differential homoeolog gene expression influencing selection. Nat Genet 48:1225–1232

Yang J, Ji C, Liu D, Wang X, Zhang M (2018) Reply to: 'Organization of the genome sequence of the polyploid crop species *Brassica juncea*'. Nat Genet 50:1497–1498

Zhang L, Cai X, Wu J, Liu M, Grob S, Cheng F, Liang J, Cai C, Liu Z, Liu B, Wang F, Li S, Liu F, Li X, Cheng L, Yang W, Li MH, Grossniklaus U, Zheng H, Wang X (2018) Improved *Brassica rapa* reference genome by single-molecule sequencing and chromosome conformation capture technologies. Hortic Res 5:50

14 Resequencing in *Brassica juncea* for Elucidation of Origin and Diversity

Lei Kang, Lunwen Qian, Hao Chen, Liu Yang, and Zhongsong Liu

Abstract

Brassica juncea (L.) Czern. et Coss. was formed 8000–14,000 years ago by hybridization between *B. rapa* and *B. nigra*. Domestication and improvement of this species have resulted in diverse morphotypes, which produce seed oil, condiment, swollen root and stem vegetables and leafy vegetables. Various studies had been done in the past to investigate the genetic diversity and origin of *B. juncea* germplasm and varieties using morphological traits and biochemical markers. However, the origin and domestication of the ancient allotetraploid species *B. juncea* remain uncertain. The genome sequence of three types *B. juncea* has been reported, which provides an opportunity to resolve the origin, domestication and diversification of *B. juncea*. Genetic variation in mustard was identified by resequencing of 480 global accessions. Population genetic and phylogenetic analyses revealed evidence for three clade and six distinct genetic groups. Nuclear and organelle genomic analyses supported a monophyletic origin of *B. juncea* in West Asia. During mustard eastward spread following three independent migrations, new forms were evolved by gene mutations and introgressions. Sweep scan, genome-wide association study and RNA seq identified causal variants and genes for flowering time and morphological variations associated with domestication and diversification of the versatile mustard. In this chapter, we provide comprehensive insight into the diversity, origin and domestication of *B. juncea*.

14.1 Introduction

The allotetraploid *Brassica juncea* (AABB, $2n = 36$) is originated from interspecific crosses between the parental species *B. rapa* (AA, $2n = 20$) and *B. nigra* (BB, $2n = 16$) (UN 1935). *B. juncea* is an economically important species, grown worldwide as oilseed, condiment and vegetable crops. Taxonomically, four subspecies had been recognized in *B. juncea* that were domesticated for a variety of uses (Gladis and Hammer 1992). These include *juncea* (seed mustard) that is used as oilseed and a condiment, *integrifolia* (leaf mustard) with a diverse variation of leaf, *napiformis* (root mustard) with a tuberous root and *tumida* (stem mustard) with an enlarged edible stem (Fig. 14.1).

Brassica juncea had a long history of domestication and cultivation. Carbonized mustard seeds were unearthed at Banpo site in China, gonging back to New Stone Age (about 4800

L. Kang · L. Qian · H. Chen · L. Yang · Z. Liu (✉)
College of Agronomy, Hunan Agricultural University, Changsha 410128, China
e-mail: zsliu48@hunau.net

C. Kole and T. Mohapatra (eds.), *The Brassica juncea Genome*, Compendium of Plant Genomes,
https://doi.org/10.1007/978-3-030-91507-0_14

Fig. 14.1 Morphotype of *B. juncea* varieties. **a** *B. juncea* var. *napiformis*. **b** *B. juncea* var. *tumida*. **c**, **d** *B. juncea* var. *multisect* Bailey. **e** *B. juncea* var. *rugosa*. **f** *B. juncea* var. *lapita*. **g** *B. juncea* cultivar Sichuan Yellow. **h** *B. juncea* accession PI 426178 (Afghanistan cultivar)

BC) by ^{14}C analysis (Wu et al. 2009). There are earliest historical records of mustard as a condiment (spice) in Sanskrit and Sumerian literatures dating back to 3000 BC (Hemingway 1995). Vavilov (1951) proposed Afghanistan and its contiguous regions were original place of *B. juncea* origin, with Asia Minor, central/western China and eastern India were diversity centers. However, many investigators (Bailey1930; Mizushima and Tsunoda 1967; Sun 1970; Hinata and Prakash 1984) argued that *B. juncea* originated in the Middle East where the distribution of progenitors *B. rapa* and *B. nigra* are overlapped. Morphological studies proposed a China origin of *B. juncea* (Prain 1898; Sinskaja 1928), and the view has been supported by recently genome resequencing analysis (Yang et al. 2016, 2018). By contrast, the biochemical evidence (Vaughan et al. 1963), molecular markers (Song et al. 1988; Chen et al. 2013) and chloroplast-specific simple sequence repeat (SSR) markers (Kaur et al. 2014) suggested a polyphyletic origin. To date, there is no generally accepted hypothesis for the center of origin and domestication events of *B. juncea*, due to the limited number and diversity of accessions analyzed or narrow breadth of the genetic data in these studies. Recently, the production of vast genomic data enabled researchers to understand the origin and domestication of crop plants (Larson et al. 2014). In this chapter, we employed the population genomics strategy to insight into the origin and diversity of *B. juncea*.

14.2 A Chromosome-Level Assembly of a Yellow-Seeded *Brassica juncea*

The first draft genome of *B. juncea* var. *tumida* (assembled 784 Mb) was assembled by Illumina reads, PacBio single-molecule reads, BioNano optical mapping and genetic mapping (Yang et al. 2016). Recently, an oleiferous type of

B. juncea var. Varuna (assigned 841.2 Mb) was assembled by PacBio single-molecule real-time reads, BioNano optical mapping and genetic mapping (Paritosh et al. 2021). Another oilseed type with yellow coat, *B. juncea* var. Sichuan Yellow (SY, Fig. 14.1g), was assembled by PacBio long-read integrated with BioNano optical mapping and Hi-C data (Kang et al. 2021). 98.4 Gb subreads were generated with an N50 length of 13.6 Kb on the PacBio Sequel platform, and 137.85 Gb Illumina reads from 350 bp library. First, PacBio reads were assembled into contigs, and later polished using the high-quality Illumina reads, further extended the scaffolds by BioNano optical map. The pseudo-chromosomes were constructed by a high-density linkage map and Hi-C data. The total size of the assembled SY was 933.5 Mb, with 867.3 Mb (~92.9%) assigned to the 18 chromosomes, which is superior to previously published genomes of stem (Yang et al. 2016) and Indian (Paritosh et al. 2021) mustards in terms of the assembled scaffolds size, accuracy, contiguity and anchorage. The contiguity and completeness of the SY genome were supported by BUSCO and CEGMA assessments, by alignment of randomly selected bacterial artificial chromosomes (BACs) and paired BAC-end sequence from the Purple-leaf Mustard BAC library (Liu et al. 2016), by evaluation of long-terminal repeat (LTR) continuity by the LTR Assembly Index (Ou et al. 2018), by genome-ordered graphical genotypes (He and Bancroft 2018), and by the consistency of SY with those of *B. rapa* (Z1) (Belser et al. 2018) and *B. nigra* (Ni100-LR) (Perumal et al. 2020) and other reported *Brassica* genomes (Song et al. 2020; Yang et al. 2016; Paritoshet al. 2021).

14.3 Genetic Diversity, Structure and Relationship of *Brassica Juncea*

Genetic diversity evaluation is a prerequisite for germplasm utilization, genetic improvement and utilization of heterosis. The genetic diversity and structure of *B. juncea* have been studied using ecogeographic, phenotypic and molecular markers. Two geographical races of oilseed mustard have been described, the Indian race and the Oriental race (Vaughan et al. 1963; Vaughan 1977). Based on amplification fragment length polymorphism (AFLP) markers, Srivastava et al. (2001) assessed that the Indian and Chinese oilseed mustard formed one cluster, while the Australia, Canada, Eastern Europe and Russia accessions formed another cluster. The genetic diversity of 119 global oilseed mustards were investigated using 99 SSR markers, and two distinct groups were also identified (Chen et al. 2013). Group 1 consisted of germplasm from central and western India and eastern China, and group 2 comprised central and western China, European and Australian accessions, as well as northern and eastern India lines. Wu et al. (2009) assessed genetic diversity of 95 oil and vegetable mustards from China, France, India, Pakistan and Japan, using sequence related amplified polymorphism (SRAP) markers. Three distinct groups were clearly identified, vegetable, spring and winter oilseed mustard, which reflected their growth habit. The winter oil mustard accessions from southwestern China were closely related to Indian mustard, in agreement with previous AFLP studies (Srivastava et al. 2001). The genetic diversities among 101 oilseed mustard landraces from western China were analyzed using AFLP, SRAP and SSR markers (Xu et al. 2008). The winter types showed a higher genetic diversity than spring types, and the genetic differences were mainly due to geographical and ecological conditions. Genetic diversity of 73 mustard accessions from China Tibet and 35 from southwest of China, northwest of China and India was analyzed using SRAP markers and phenotypic traits (Song et al. 2013). China Tibet accessions were clustered into a distinct group, showing higher genetic diversity than others. Sun et al. (2018) divided 25 wild mustard populations from China into Northern and Southern lineages, and proposed two migration routes, eastward along the Yellow River or Yangtze River, respectively.

China has the richest vegetable mustard resources and has been cultivated in the last sixth century AD (Jia 2009). In total, 16 varieties with typical characters were identified (Yang et al. 1989). Among these, 11 varieties of leaf mustard were classified based on diverse leaf shapes, showing most highly diversified (Yang et al. 1989; Yao et al. 2012). Using 69 microsatellite loci, 34 Chinese vegetable mustards are divided into seven groups, inconsistent with the morphological classifications, which may be due to lack of sufficient genetic markers (Yao et al. 2012). Song et al. (1988) distinguished root mustard from other type *B. juncea* accessions using specific probe and enzyme digest. Tuber (stem) mustard is an agriculturally and economically important vegetable, cultivated along the Yangtze River. 133 stem mustard cultivars were divided into up-, mid- and down-Yangtze River groups, according to their geographical distributions (Fang et al. 2013). The mid-Yangtze River group had highest genetic diversity, indicating that it was the diversity center of tuber mustard, and then this crop was spread along the Yangtze River in both directions.

The genome of *B. juncea* var. *tumida* (T84-66) has been assembled (Yang et al. 2016), which provides an opportunity for detecting genomic validation in mustard accessions. A total of 891,289 single nucleotide polymorphisms (SNPs) were identified in a panel of 109 *B. juncea* accessions, using specific-locus amplified fragment sequencing (Yang et al. 2018). Four groups were recovered, including root mustard, leaf/stem mustard, Oil (IN) (India mustard) and Oil (A, C, E) (accessions from Australia, Northwest China and Europe). It proposed that the root mustard diversified first, the Oil (A, C, E) group diversified second, and the Indian mustard domesticated from Leaf/Stem mustard (Yang et al. 2018).

To construct a comprehensive map of genetic variations of *B. juncea*, Kang et al. (2021) resequenced 480 accessions, represented the global diversity of mustard. These accessions were sequenced with a median depth 15-fold and scored 4.53 million high-quality SNPs. The genetic structures of the global mustard accessions were characterized for clusters (K) from 2 to 10, and the marginal likelihood plateaued at $K = 6$. A total of 90 admixed lines with their main genetic components less than 0.6 were excluded, for better clarify the relationships of mustard population. Based on phylogenetic and principal component analyses, the remaining accessions indicated three distinct clades, which exhibited strong geographic separation and distinctive botanical features. Clade I comprised root mustard from Northeast Asia and was recovered as sister to all other *B. juncea* subspecies. Clade II corresponded to seed mustard distributed in West Asia, Central Asia, and Northwest China along the ancient Steppe Route (Christian 2000). Clade III comprised Indian mustard and South China vegetable mustard along the South Silk Road.

Clades II and III contained a large number of versatile accessions and were further classified into two and three groups, respectively (Kang et al. 2021). Finally, the global *B. juncea* accessions were classed into six genetic groups (G1–G6). G1 is the root mustard group. G2, consistent of yellow-seeded or Oriental mustard, were mainly from Northwest China, as well as some lines collected from the former Soviet Union, Canada and Europe with recorded introductions from China (Musil 1948; Pustovoit 1973; Oram et al. 2005). G3 comprised seed mustard from Tibet, central and western Asia and Eastern Europe. G4, also yellow-seeded mustard, comprised mainly accessions from Southwest China. G5 comprised leaf, stem and several seed mustards, which were sourced from southern China, Japan and the USA (Bailey 1930; Musil 1948). G6, brown or Indian mustard, were almost from South Asia.

The genetic divergence among the six diverse groups was assessed by calculating their pairwise fixation statistics (*Fst*) (Kang et al. 2021). G1 showed very strong differentiation from the other five groups, while G2 was found to have minimum divergence from G3. The nucleotide diversity (π) and linkage disequilibrium (LD) analyses showed that G5 had the highest nucleotide diversity and the greatest linkage disequilibrium (LD) decay. In contrast, G6 showed the lowest nucleotide diversity, consistent with earlier reports (Srivastava et al. 2001; Khan et al. 2008).

14.4 Cytoplasmic Variation in *B. juncea*

The amphidiploid *B. juncea* is originated from hybridization between *B. rapa* and *B. nigra*, with *B. rapa* as the mother parent (Uchimiya and Wildman 1978; Erickson et al. 1983; Song et al. 1988; Chang et al. 2011; Li et al. 2017). However, Kaur et al. (2014) reported that the East European and northwestern India *B. juncea* accessions carried cytoplasm from *B. nigra*, genotyped using chloroplast-specific SSRs. Recent studies also have found *B. juncea* lines from north and eastern India owned the cytoplasm of *B. nigra* (Banga and Banga 2016). To better evaluate the cytoplasmic donor and variation in a global panel of *B. juncea*, Kang et al. (2021) have assembled 478 chloroplast (CP) and 10 mitochondrial (MT) genomes. The cpDNA-based phylogeny revealed that 467 mustards were clustered in the *B. rapa* lineage, accorded with the previous conclusions that *B. rapa* was the cytoplasm donor. The other 11 mustard lines were clustered in the *B. rapa* group, suggesting that an introgressive hybridization occurred between *B. rapa* and *B. juncea*. Based on SNP and InDel (insertion/deletion) variants, the *B. juncea* CP and MT genomes were concurrently divided into three types and were named plasmotype I-III. All root mustard (G1) harbored plasmotype I, while all Indian mustard (G6) and most (94.2%) leaf/stem mustard (G5) carried plasmotype III. The others three oilseed mustard groups owned all three plasmotypes, with plasmotype II being predominant (Kang et al. 2021).

14.5 The Origin of *Brassicas juncea*

Previous studies suggested that the *B. juncea* had evolved from multiple interspecific hybridizations events between wild *B. rapa* and *B. nigra* (Vaughan et al. 1963; Vaughan 1977; Prakash and Hinata 1980; Song et al. 1988; Chen et al. 2013; Kaur et al. 2014). In a recent study, Yang et al. (2016) revealed a monophyletic origin for *B. juncea* by phylogenetic analyses of A-subgenomes of *Brassica* species. His team (2018) further demonstrated that the root mustard diversified first, followed by oilseed mustard, leaf mustard and stem mustard. Subgenome phylogenies confirmed that the G1 root mustard was closest to its progenitor species, supported the hypothesis that *B. juncea* is a monophyletic origin (Kang et al. 2021).

Organellar DNAs inherited from a maternal parent are often used to investigate origin and phylogenetic relationship of a species. *B. rapa* is the maternal ancestor of *B. juncea* (Uchimiya and Wildman 1978; Erickson et al. 1983; Song et al. 1988; Chang et al. 2011; Li et al. 2017; Kang et al. 2021), instead of earlier report that the cytoplasmic donor of East European and north India *B. juncea* landrace was *B. nigra* (Kaur et al. 2014; Banga and Banga 2016). CP and MT phylogenies and PCR validations led to the identification of the three plasmotypes of *B. juncea*, and the cytoplasmic donor is the *B. rapa* (Kang et al. 2021). Plasmotype I of mustard was inherited from *B. rapa* and then evolved into plasmotype II and III via InDels and base substitution. It also revealed that the *B. juncea* is a monophyletic origin based on variation of cytoplasmic DNA.

Where *B. juncea* originated geographically has been disputed for almost a century. Historically, three geographic regions, i.e., the Middle East (Burkill 1930; Mizushima and Tsunoda 1967; Sun 1970; Hinata and Prakash 1984), Central Asia (Vavilov 1997), and China (Yang et al. 2018), were proposed to be the putative center of origin for *B. juncea*. However, the Middle East is the only region where wild *B. rapa* and *B. nigra* overlapped (Tsunoda1980; OECD 2016), and wild *B. juncea* forms are growing (Olsson 1960; Mizushima and Tsunoda1967; Tsunoda and Nishi 1968; Kayaçetin 2019; Dönmeza et al. 2021). As above, G3 had all three plasmotypes and higher nucleotide diversity, which means that the source region of the G3 accessions is a plausible origin center of

B. juncea. Taken together, *B. juncea* was supported to originated in West Asia (Middle East) (Kang et al. 2021).

14.6 The Spread and Domestication of *Brassicas juncea*

B. juncea was originated from the natural cross between the two diploid progenitors ~8000–14,000 years ago (Kang et al. 2021), which was slightly latter than the previous estimate of ~0.039–0.055 million years ago derived by Bayesian Markov chain Monte Carlo simulation (Yang et al. 2016). There are at least three independent evolutionary routes of the *B. juncea* by demographic history model.

Root mustard (G1) was the first group to differentiate from the ancient population, approximately ~2500 years ago. Root mustard is mainly cultivated in Mongolia and northern China and is morphologically (lobed leaves and enlarged taproot: Sinskaja 1928; Bailey 1930) and physiologically (vernalization requirement, Kumazawa and Abe 1955) distinct from other subspecies. Early taxonomists treated root mustard (an introduction from China) as a separate species (*B. napiformis*) (Bailey 1930). In contrast with the hypothesis by Yang et al (2018), we think there is little possibility that root mustard originated in China because of the absence of the diploid progenitor species *B. nigra*. We speculated that it was likely domesticated in Mongolia and northeast China, on the basis of historical records of the Ming Dynasty (Wang 1587) that documented the spread of root mustard from the north to south China.

The G3 seed mustard was evolved from wild mustard. When it spread eastward into Afghanistan, diverse accessions were developed, while accessions from its different parts can be clearly distinguished in nuclear or cytoplasmic genomes. Subsequently, the G3 mustard migrated to northwest China through the ancient Steppe Route, finally arrived Tibet via the Hexi corridor. A new yellow-seeded mustard (G2) was domesticated from G3 by spontaneous gene mutations about 500 years ago (Liu et al. 2009, 2015), probably in Xinjiang (Vavilov 1992).

The G3 mustard spread via the passes of Hindu Kush, directly into Punjab (the heart of past India, present Pakistan). In Pakistan's Harappa, the ancient mustard (2400–1700 BC) was excavated (Hutchinson et al. 1976). The *B. juncea* accessions from southern Afghanistan, bordering Pakistan and India, are clustered into the same G6 group and have plasmotype III as Indian mustard, which implies that Indian mustard (G6) was domesticated from G3. Indian mustard then spread eastward (Pokhariaet al. 2017; Rahman et al. 2020), a new form with broad-leaf mustard, var. *rugosa* (Fig. 14.1e) (Prain 1898) formed probably at Kautilya Dynasty (around 300 BC) (Prakash et al. 2012). These broad-leaf mustards spread through ancient trade routes into Southwest China and harvested as vegetable before the sixth century AD. Later, in the eighteenth century, a mustard form with swollen stem, var. *tumida* was developed from var. *rugosa* in the Sichuan Basin (Chen 1982). The G4 was proved that evolved from hybridization of G2 from northwestern China with G5 in southwestern China (Kang et al. 2021).

14.7 Adaptability of *Brassica juncea* to Flowering Time in Different Geographic Regions

Flowering time is an essential feature associated with crop domestication. As a cumulative long-day plant species, *B. juncea* does not require strict vernalization for flowering, except for root mustard and a few leaf mustard varieties (e.g., *var. multiceps*). We used the cross-population composite likelihood ratio test (XP-CLR) to detect signatures associated with flowering time during *B. juncea* domestication (Kang et al. 2021). By comparing with G6 (early flowering), 43 and 38 candidate selection regions were identified in G1 (need vernalization) and G2 (flowering under long day), respectively, including 63 flowering time candidate genes.

Meanwhile, our genome-wide association study (GWAS) has identified 56 candidate genes, of which 12 were detected by both. Two genes, BjuA10g14550S (*SRR1*) and BjuB05g31990S (*VIN3*) involved in photoperiod and vernalization pathways (Staiger et al. 2003; Sung and Amasino 2004) were significantly associated with flowering time in four environments. The strong LD was found between *SRR1* and *VIN3*, suggesting that selection for environmental adaptability has led to the coevolution of important flowering genes in various flowering pathways (Kang et al. 2021). Hap1 of *SRR1* and *VIN3* were present in the G1 root mustard accessions with the latest flowering times, which was domesticated in cold and high latitude area. Hap2 of *SRR1* and *VIN3* related to the G2 and G3 seed mustards with moderate flowering times, which were developed under long day and great diurnal temperature range. Hap3 of *SRR1* and *VIN3* mainly corresponded to G4, G5 and G6 accessions with earliest-flowering, which were domesticated in warm, short-day environments.

14.8 The Genetic Basis of Morphological Diversification in *Brassica juncea*

14.8.1 Seed Size

Brassica juncea harbors significant variation in thousand seed weight (TSW) varies from around 0.29 g in leaf mustard to more than 7.0 g in oilseed mustard (Dhaka et al. 2017; Kang et al. 2021). Dhakaet al. (2017) reported six 'consensus' QTLs in four mapping populations derived from three Indian and three Eastern European accessions in different environments and predicted 11 candidate genes. Allelic diversity of 10 functional genes for seed size were examined in 123 *B. juncea* accessions from Australia, China, Eastern Europe and India (Sra et al. 2019). Two genes, *IKU1* and *GRF2* were identified as targets of artificial selection based on sequence diversity.

To understand the genetic base of seed weight among the diverse groups of mustard, a total of 65 and 22 candidate genes were identified as responsible for TSW by XP-CLR and GWAS, respectively (Kang et al. 2021). Of which, seven were detected by both. For instance, BjuA04g00760S (*CYP78A9*) was significantly associated with TSW, which was reported to regulate seed weight in *B. napus* (Shi et al. 2019). Four haplotypes were detected in *CYP78A9*, Hap1 related to ten G5 vegetable accessions with the lightest TSW, whereas Hap4 correspond to seven G3 oilseed accessions with the highest TSW. Another candidate gene, BjuB05g28000S (*CaM7*), regulating the seed weight in *Gossypium hirsutum* (Cheng et al. 2016a, b), also had four haplotypes. *CaM7*-Hap1 related to the lowest TSW accession from G1, while *CaM7*-Hap4 was present in ten oilseed mustards from G2 with the largest TSW. However, Hap2 and Hap3 of *CaM7* related to accessions corresponded to these with *CYP78A9*-Hap2 and -Hap3 in four environments, produced bigger seeds in long-day than short-day environments. Moreover, these two gene were up-regulated in the big-seeded accession 7981, contrasted with the small-seeded accession SY. The results revealed that selection of these haplotypes adapted to local photoperiod during domestication, leaded to diversification of mustard seed size (Kang et al. 2021).

14.8.2 Root Expansion

Artificial selection of *B. juncea* resulted in the edible storage organs' formation, including root expansion, and stem swelling (Fig. 14.1). Vegetative storage organs are also observed in other *Brassica* species, including turnip (*B. rapa* ssp. *rapa*), kohlrabi (*B. oleracea*) and rutabaga (*B. napus*). In *Brassica*, the swollen root and stem mainly store glucose and fructose, instead of starch as in potato tubers (Liu et al. 2019). Several reports have revealed that possible genomic selective regions during domestication and regulatory genes involved in swollen organs formation

(Cheng et al. 2016a, b; An et al. 2019; Liu et al. 2019). However, the molecular mechanism underlying storage organs' formation and its regulation are poorly understood. The selection signatures in root mustard were investigated using XP-CLR. Fourteen candidate genes involved in auxin signaling, sugar transport, cell division and cell wall metabolism were identified that potentially control tuberous roots forming in root mustard (Kang et al. 2021). For example, *CDC48A4* (*BjuA03g27650S*) regulated of cell division and growth (Rancour et al. 2004) showing significant genotype diversities between root and non-root mustard was up-regulated during root enlarging. The expression of *EXPB1* (*BjuB02g61740S*) was down-regulated in root mustard, consistent with the corresponding *EXPB1* orthologs action during storage root development in radish (Xie et al. 2018) and sweet potato (Noh et al. 2013).

14.8.3 Stem Swelling

Stem mustard is an important vegetable crop, and its enlarged edible stem used to produce pickles (e.g., Fuling hot pickled tuber mustard). As mentioned above, it's domesticated from leaf mustard. The stem of stem mustard can enlarge to more than 20 cm in diameter, whereas the diameter of leaf mustard's stem is usually less than 5 cm (Shi et al. 2012). The enlargement of the swollen stem is a complex biological process and has been reported that regulated by temperature, photoperiod, cell division and cell wall remodeling (Shi et al. 2012; Sun et al. 2012; Li et al. 2020; Zhang et al. 2020). Twelve candidate genes were identified in stem mustard, related to cell division, cell expansion, auxin signaling and glucose transport (Kang et al. 2021). The functions of these genes are further verified by expression data. For example, BjuA05g02460S which is orthologous to the gene *GRF7* regulating leaf and stem development (Wang et al. 2014) was up-regulated during stem swelling. Two genes, BjuA10g12920S (*IAA33*) and BjuB03g51870S (*ARF5*) related to auxin were down-regulated after stem swelling.

14.9 Conclusions

In summary, the *B. juncea* is a monophyletic origin in West Asia, raised 8000–14,000 years ago. Six genetic clusters were domesticated through gene mutations, introgressions and artificial selection during its eastward spread.

References

An H, Qi X, Gaynor ML, Hao Y, Gebken SC et al (2019) Transcriptome and organellar sequencing highlights the complex origin and diversification of allotetraploid *Brassica napus*. Nat Commun 10(1):2878

Bailey LH (1930) The cultivated *Brassicas*. Second Paper. Gentes Herb 2:211–267

Banga SS, Banga S (2016) Genetic diversity and germplasm patterns in *Brassica juncea*. In: Rajpal V, Rao S, Raina S (eds) Gene pool diversity and crop improvement. Springer International Publishing, Switzerland, pp 163–186

Belser C, Istace B, Denis E, Dubarry M, Baurens FC et al (2018) Chromosome-scale assemblies of plant genomes using nanopore long reads and optical maps. Nat Plants 4:879–887

Burkill IH (1930) The Chinese mustards in the Malay Peninsula. Gad's Bull 5:99–117

Chang S, Yang T, Du T, Huang Y, Chen J et al (2011) Mitochondrial genome sequencing helps show the evolutionary mechanism of mitochondrial genome formation in *Brassica*. BMC Genomics 12:497

Chen SR (1982) The origin and differentiation of mustard varieties in China. Cruciferae Newsl 7:7–10

Chen S, Wan Z, Nelson MN, Chauhan JS, Redden R et al (2013) Evidence from genome-wide simple sequence repeat markers for a polyphyletic origin and secondary centers of genetic diversity of *Brassica juncea* in China and India. J Hered 104:416–427

Cheng F, Sun R, Hou X, Zheng H, Zhang F et al (2016a) Subgenome parallel selection is associated with morphotype diversification and convergent crop domestication in *Brassica rapa* and *Brassica oleracea*. Nat Genet 48:1218–1224

Cheng Y, Lu L, Yang Z, Wu Z, Qin W et al (2016b) GhCaM7-like, a calcium sensor gene, influences cotton fiber elongation and biomass production. Plant Physiol Biochem 109:128–136

Christian D (2000) Silk roads or steppe roads? the silk roads in world history. J World Hist 11:1–26

Dhaka N, Rout K, Yadava SK, Sodhi YS, Gupta V et al (2017) Genetic dissection of seed weight by QTL analysis and detection of allelic variation in Indian and east European gene pool lines of *Brassica juncea*. Theor Appl Genet 130(2):293–307

Dönmeza AA, Aydına ZU, Wang XW (2021) Wild Brassica and its close relatives in Turkey, the genetic treasures. Hortic Plant J 7(2):97–107

Erickson LR, Straus NA, Beversdorf WD (1983) Restriction patterns reveal origins of chloroplastgenomes in *Brassica* amphidiploids. Theor Appl Genet 65:201–206

Fang P, Chen FB, Yao QL, Yang KC, Zhou GF et al (2013) Analysis of genetic diversity in the tuber mustard (*Brassica juncea* var. *tumida* Tsen et Lee) in the Yangtze river basin of China. Genet Resour Crop Evol 60:129–143

Gladis T, Hammer K (1992) The *Brassica* collection in Gatersleben: *Brassica juncea*, *Brassica napus*, *Brassica nigra*, and *Brassica rapa*. Feddes Rep 103:469–507

He Z, Bancroft I (2018) Organization of the genome sequence of the polyploid crop species *Brassica juncea*. Nat Genet 50:1496–1497

Hemingway J (1995) The mustard species: condiment and food ingredients use and potential as oilseed crops. In: Kimber DS, McGregor DI (eds) Brassica oilseeds: production and utilization. CAB Press, pp 373–383

Hinata K, Prakash S (1984) Ethnobotany and evolutionary origin of Indian oleiferous *Brassicae*. Indian J Genet 44:102–112

Hutchinson J, Allchin FR, Vishnu-Mittre (1976) India; local and introduced crops. Phil Trans R Soc Lond B 275:129–141

Jia SX (2009) Qi Min Yao Shu (Important Arts for the Peoples Welfare written in the sixth century AD. Shanghai Classics Publishing House

Kang L, QianLW ZM, Chen LY, Chen H et al (2021) Genomic insights into the origin, domestication and diversification of *Brassica juncea*. Nat Genet 53:1392–1402

Kaur P, Banga S, Kumar N, Gupta S, Akhatar J et al (2014) Polyphyletic origin of *Brassica juncea* with *B. rapa* and *B. nigra* (*Brassicaceae*) participating as cytoplasm donor parents in independent hybridization events. Amer J Bot 101:1157–1166

Kayaçetin F (2019) Morphological characterization and relationships among some important wild and domestic Turkish mustard genotypes (*Brassica spp.*). Turk J Bot 43:499–515

Khan MA, Rabbani MA, Munir M, Ajmal SK, Malik MA (2008) Assessment of genetic variation within Indian mustard (*Brassica juncea*) germplasm using random amplified polymorphic DNA markers. J Integr Plant Biol 50:385–392

Kumazawa S, Abe S (1955) Studies on varieties of mustard. J Jap Soc Hort Sci 24:69–84

Larson G, Piperno DR, Allaby RG, Purugganan MD, Andersson L et al (2014) Current perspectives and the future of domestication studies. Proc Natl Acad Sci USA 111:6139–6146

Li P, Zhang S, Li F, Zhang S, Zhang H et al (2017) A phylogenetic analysis of chloroplast genomes elucidates the relationships of the six economically important *Brassica* species comprising the triangle of U. Front Plant Sci 8:111

Li M, Xie F, He Q, Li J, Liu J et al (2020) Expression Analysis of XTH in Stem Swelling of Stem Mustard and Selection of Reference Genes. Genes (basel) 11 (1):113

Liu XJ, Yuan MZ, Guan CY, Chen SY, Liu SY et al (2009) Inheritance, mapping, and origin of yellow-seeded trait in *Brassica juncea*. Acta Agron Sinica 35:839–847

Liu Z, Liu X, Lu Y, Guan C, Chen S et al (2015) Domestication and molecular mechanism underlying yellow seed in *Brassica juncea* (*L.*) czern & coss. In: 14th International rapeseed congress, Saskatoon, Canada

Liu X, Lu Y, Yan M, Sun D, Hu X et al (2016) Genome-wide identification, localization, and expression analysis of proanthocyanidin-associated genes in *Brassica*. Front Plant Sci 7:1831

Liu M, Bassetti N, Petrasch S, Zhang N, Bucher J et al (2019) What makes turnips: anatomy, physiology and transcriptome during early stages of its hypocotyl-tuber development. Hortic Res 6:38

Mizushima U, Tsunoda S (1967) A plant exploration in *Brassica* and allied genera. Tohoku J Agric Res 17:249–277

Musil AF (1948) Distinguishing the species of *Brassica* by their seed. US Dep Agric Misc Publ No. 643

Noh SA, Lee HS, Kim YS, Paek KH, Shin JS et al (2013) Down-regulation of the *IbEXPB1* gene enhanced storage root development in sweet potato. J Exp Bot 64:129–142

OECD (2016) Brassica crops. In: Safety assessment of transgenic organisms in the environment. OECD Publishing, pp 151–291

Olsson G (1960) Species crosses within the genus *Brassica* I Artificial *Brassica juncea* Coss. . Hereditas 46:171–222

Oram RN, Kirk JTO, Veness PE, Hurlstone CJ, Edlington JP et al (2005) Breeding Indian mustard [*Brassica juncea* (L.) Czern.] for cold-pressed, edible oil production: a review. Aust J Agric Res 56:581–596

Ou S, Chen J, Jiang N (2018) Assessing genome assembly quality using the LTR assembly index (LAI). Nucleic Acids Res 46:e126

Paritosh K, Yadava SK, Singh P, Bhayana L, Mukhopadhyay A et al (2021) A chromosome-scale assembly of allotetraploid *Brassica juncea* (AABB) elucidates comparative architecture of the A and B genomes. Plant Biotechnol J 19:602–614

Perumal S, Koh CS, Jin L, Buchwaldt M, Higgins EE et al (2020) A high-contiguity *Brassica nigra* genome localizes active centromeres and defines the ancestral *Brassica* genome. Nat Plants 6:929–941

Pokharia AK, Sharma S, Tripathi D, Mishra N, Pal JN et al (2017) Neolithic early historic (2500–200 BC) plant use: the archaeobotany of Ganga Plain, India. Quatern Intl 443:223–237

Prain D (1898) The mustards cultivated in Bengal. Agric Ledger 5:1–80

Prakash S, Hinata K (1980) Taxonomy, cytogenetics and origin of crop *Brassica*, a review. Opera Bot 55:1–57
Prakash S, Wu X, Bhat SR (2012) History, evolution and domestication of *Brassica* crops. Plant Breed Rev 35:19–84
Pustovoit VS (1973) Handbook of selection and seed growing of oil plants. Israel Program for Scientific Translations
Rahman M, Castillo CC, Murphy C, Rahman SM, Fuller DQ (2020) Agricultural systems in Bangladesh: the first archaeobotanical results from early historic Wari-Bateshwar and early medieval Vikrampura. Archaeol Anthropol Sc 12:37
Rancour DM, Park S, Knight SD, Bednarek SY (2004) Plant UBX domain-containing protein 1, PUX1, regulates the oligomeric structure and activity of *Arabidopsis* CDC48. J Biol Chem 279:54264–54274
Shi H, Wang LL, Sun LT, Dong LL, Liu B et al (2012) Cell division and endoreduplication play important roles in stem swelling of tuber mustard (*Brassica juncea* Coss. var. *tumida* Tsen et Lee). Plant Biol 14:956–963
Shi L, Song J, Guo C, Wang B, Guan Z et al (2019) A CACTA-like transposable element in the upstream region of *BnaA9.CYP78A9* acts as an enhancer to increase silique length and seed weight in rapeseed. Plant J 98:524–539
Sinskaja EN (1928) The oleiferous plants and root crops of the family *Cruciferae*. Bull Appl Bot Genet Plant Breed 19:3–646
Song KM, Osborn TC, Williams PH (1988) *Brassica* taxonomy based on nuclear restriction fragment length polymorphisms (RFLPs). 1. Genome evolution of diploid and amphidiploid species. Theor Appl Genet 75:784–794
Song W, Xu K, Li F, Chen B, Cai M et al (2013) Genetic diversity of *Brassica juncea* from Tibet autonomous region of China. Chin J Oil Crop Sci 35:153–161
Song JM, Guan Z, Hu J, Guo C, Yang Z et al (2020) Eight high-quality genomes reveal pan-genome architecture and ecotype differentiation of *Brassica napus*. Nat Plants 6:34–45
Sra SK, Sharma M, Kaur G, Sharma S, Akhatar J et al (2019) Evolutionary aspects of direct or indirect selection for seed size and seed metabolites in *Brassica juncea* and diploid progenitor species. Mol Biol Rep 46:1227–1238
Srivastava A, Gupta V, Pental D, Pradhan AK (2001) AFLP-based genetic diversity assessment amongst agronomically important natural and some newly synthesized lines of *Brassica juncea*. Theor Appl Genet 102:193–199
Staiger D, Allenbach L, Salathia N, Fiechter V, Davis SJ et al (2003) The *ArabidopsisSRR1* gene mediates phyB signaling and is required for normal circadian clock function. Genes Dev 17:256–268
Sun VG (1970) Breeding plants of *Brassica*. J Agron Assoc China 71:141–152
Sun Q, Zhou GF, Cai YF, Fan YH, Zhu XT et al (2012) Transcriptome analysis of stem development in the tumourous stem mustard *Brassica juncea* var. *tumida* Tsen et Lee by RNA sequencing. BMC Plant Biol 12:53
Sun XQ, Qu YQ, Li MM, Song XL, Hang YY (2018) Genetic diversity, genetic structure and migration routes of wild *Brassica juncea* in China assessed by SSR markers. Genet Resour Crop Evol 65:1581–1590
Sung S, Amasino RM (2004) Vernalization in *Arabidopsis thaliana* is mediated by the PHD finger protein VIN3. Nature 427:159–164
Tsunoda S (1980) Eco-physiology of wild and cultivated forms in *Brassica* and allied genera. In: Tsunoda S, Hinata K, Gomez-Campo C (eds) Brassica crops and wild allies. Japan Scientific Societies Press, Tokyo, pp 109–120
Tsunoda S, Nishi S (1968) Origin, differentiation and breeding of cultivated *Brassica*. Pro XII Intl Congr Genet 2:77–88
Uchimiya H, Wildman SG (1978) Evolution of fraction 1 protein in relation to the origin of amphidiploid *Brassica* species and other members of the Cruciferae. J Hered 69:299–303
UN (1935) Genomic analysis in *Brassica* with special reference to the experimental formation of *B. napus* and peculiar mode of fertilization. Jpn J Bot 7:389–452
Vaughan JG (1977) A multidisciplinary study of the taxonomy and origin of *Brassica* crops. Bioscience 27:35–40
Vaughan JG, Hemingway JS, Schofield HJ (1963) Contributions to a study of variation in *Brassica juncea* Coss. & Czern. J Linn Soc (bot) 58:435–447
Vavilov NI (1951) Phytogeographic basis of plant breeding. In: Chester KS (ed) The origin, variation, immunity and breeding of cultivated plants. Chronica Bot 13:14–56
Vavilov NI (1992) Origin and geography of cultivated plants (translated by Love D) Cambridge University Press
Vavilov NI (1997) Five continents. International Plant Genetic Resources Institute, Rome, Italy
Wang SM (1587) Xue Pu Za Shu (Explanations of horticultural plants, written in 1587)
Wang F, Qiu N, Ding Q, Li J, Zhang Y et al (2014) Genome-wide identification and analysis of the growth-regulating factor family in Chinese cabbage (*Brassica rapa L.* ssp. *pekinensis*). BMC Genomics 15:807
Wu X, Chen B, Lu G, Wang H, Xu K et al (2009) Genetic diversity in oil and vegetable mustard (*Brassica juncea*) landraces revealed by SRAP markers. Genet Resour Crop Evol 56:1011–1022
Xie Y, Xu L, Wang Y, Fan L, Chen Y et al (2018) Comparative proteomic analysis provides insight into a complex regulatory network of taproot formation in radish (*Raphanus sativus* L.). Hortic Res 5:51
Xu A, Ma C, Xiao E, Quan J, Ma C et al (2008) Genetic diversity of *Brassica juncea* from western China. Acta Agron Sin 34:754–763
Yang YG, Liu NC, Chen XQ, Chen CL, Zhou GF et al (1989) A study on classification of mustard. Acta Hort Sinica 16:114–121

Yang J, Liu D, Wang X, Ji C, Cheng F et al (2016) The genome sequence of allopolyploid *Brassica juncea* and analysis of differential homoeolog gene expression influencing selection. Nat Genet 48:1225–1232

Yang J, Zhang C, Zhao N, Zhang L, Hu Z et al (2018) Chinese root-type mustard provides phylogenomic insights into the evolution of the multi-use diversified allopolyploid *Brassica juncea*. Mol Plant 11:512–514

Yao QL, Chen FB, Fang P, Zhou GF, Fan YH et al (2012) Genetic diversity of Chinese vegetable mustard (*Brassica juncea* Coss) landraces based on SSR data. Biochem Syst Ecol 45:41–48

Zhang L, Li Z, Garraway J, Cai Q, Zhou Y et al (2020) The casein kinase 2 β subunit CK2B1 is required for swollen stem formation via cell cycle control in vegetable *Brassica juncea*. Plant J 104:706–717

Large Scale Genome Analysis: Genome Sequences, Chromosomal Reorganization, and Repetitive DNA in *Brassica juncea* and Relatives

15

Manosh Kumar Biswas, Trude Schwarzacher, and J. S. Pat Heslop-Harrison

Abstract

Sequencing technology and bioinformatics tools enable us to generate and analyses whole genome sequences of Brassica crops and wild relatives. High quality assembled references genome for the Brassica species are available, aiding downstream genome scale analysis such as variant calling, chromosomal rearrangement, translocations, repetitive DNA diversity, genome wide association mapping and transcriptomics. In this chapter, we are illustrating the current state of large scale genome analysis in *Brassica juncea* and its relatives.

15.1 Introduction

Brassica juncea (Brassicaceae or Cruciferae), as discussed in other chapters, is the most important oilseed in India and Bangladesh, and extensively grown (along with *B. napus*) in China and Canada (Bañuelos et al. 2013; Agrawal et al. 2020, 2021). One of the species in the small genus *Brassica* (37 species), *B. juncea* is globally distributed except in polar regions and it has been cultivated for centuries in many parts of the world including Asia, Europe, central Africa and southern Russia. Apart from oil production, *B. juncea* is also cultivated for vegetables where leaves, seeds, stems and roots are used in Asian, European and African cuisine, emphasizing the morphological and genetic diversity in the species. Four main cultivar groups of *B. juncea* cultivars are recognized: *integrifolia*, *juncea*, *napiformis* and *tsatsai* (Hanelt et al. 2001). Integrifolia produces leafy vegetables popular in China, Japan and Korea. *Napiformis* are root mustard, and *tsatsai*, also known as multishoot mustard, and very popular in China and Korea where it is used to make pickles (Ghawi et al. 2014; Wiersema and León 1999). *Juncea*, by far the most important group, is the oilseed used for oil production (with seed meal as co-product), animal fodder, green manure or mulch, medicinal purposes and some industrial uses. *B. juncea* oil may be sharp (hot or spicy) from mustard oil with erucic acid (restrictions on import and use in many countries), although low erucic acid, canola, types are widely grown.

Within *Brassica* species, genomes with distinct genetic identity are designated by letters. The three major diploid species are *B. rapa* (AA genomes, $2n = 20$), *B. nigra* (BB genomes, $2n = 16$) and *B. oleracea* (CC genomes, $2n = 18$). Three allotetraploid species evolved from pair-wise natural hybridization between the

M. K. Biswas · T. Schwarzacher · J. S.P. Heslop-Harrison (✉)
Department of Genetics and Genome Biology, University of Leicester, Leicester L1 7RH, UK
e-mail: phh4@le.ac.uk; phh@molcyt.com

M. K. Biswas
e-mail: mkb35@leicester.ac.uk

C. Kole and T. Mohapatra (eds.), *The Brassica juncea Genome*, Compendium of Plant Genomes,
https://doi.org/10.1007/978-3-030-91507-0_15

three basic diploid species: *B. juncea* (AABB, $2n = 36$), *B. napus* (AACC, $2n = 38$) and *B. carinata* (BBCC, $2n = 34$). Thus, *B. juncea* is an allopolyploid of two diploid species *B. rapa* (AA) and *B. nigra* (BB) and *B. juncea* genome contains the largely conserved genomes of both progenitor species (Axelsson et al. 2000; Qiao et al. 2020). Although known from the wild, resynthesized hybrids have also been made by crossing the ancestral species to increase the genetic diversity available (Bansal et al. 2009). *Brassica juncea* is self-pollinated, though 20–30% cross-pollination has been recorded. Bees are the main pollinator of this species. *Brassica juncea* is suitable for cultivation in moisture-limited (drought) and hot conditions. The genome sizes—number of base pairs in an unreplicated haploid chromosome set—of diploid species are estimated to be 529 Mbp (*B. rapa*) and 639 Mbp (*B. oleracea)* (Belser et al. 2018) and that of *B. nigra* averaging 580 Mpb (Perumal et al. 2020), 3–4 times larger than the model *Brassicaceae* species *Arabidopsis thaliana*, but only 15% of diploid cereals crops such as barley or rye. *B. juncea* has an estimated genome size of 910–920 Mb (Yang et al. 2016; Yang et al. 2018a; Paritosh et al. 2021), somewhat smaller than *B. napus*, 1130–1280 Mbp (Song et al. 2021).

Multidisciplinary approaches including genetics, genomics, cytogenetics and bioinformatics assist with meeting breeding challenges and improvement of *Brassica* crop species. Over the last few decades, a large number of breeding lines, reference genome assemblies, transcriptome sequences and phenotypic data have been generated for *B. juncea* and its relatives for addressing questions about diversity and genetic improvement. In this chapter, we discuss the recent progress and future path of the large scale of genome or genomic data analysis, and the nature of sequences included in the genome, for accelerating genetic improvement of *B. juncea* and its relatives.

15.2 Genome Sequences

Complete genomic DNA sequences have been generated for the diploid and tetraploid *Brassica* species, and these have been of increasingly high-quality since publication of the first plant genome sequence, the Brassicaceae species *A. thaliana*, in 2000 (https://doi.org/10.1038/35048692). By 2020, extensive and detailed reports of genome sequences of crop Brassicas are appearing regularly, using increasingly high-quality sequencing methods supported by improved assembly algorithms, bioinformatics and databases (including NCBI Genbank, Swis-sProt, EBI and the China National GeneBank Database). Web-searching, now dominated by Google, allows genome browsers and databases to be found and accessed: individual databases often become outdated or superseded within a few months of publication, because of the increased amount of data to be included, improvements in algorithms and presentation, the high cost of database maintenance to keep Websites available and compatible with newer Internet browser versions, changing Website availability restrictions and often dependency on short-term grant funding for maintenance. For example, a state-of-the-art database in 2010 of restriction fragment polymorphisms or amplified fragment length polymorphisms (AFLPs), written with a Flash-based graphical interface, would neither have useful information nor work in a browser by 2021.

15.2.1 Whole Genome Assemblies of *Brassica* Species

Following the sequencing of the *A. thaliana* genome (Arabidopsis Genome Initiative 2000), many comparisons of *Brassica* gene sequences became possible, before complete genome sequencing of diploid and tetraploid Brassicas. After a period involving sequencing of large-insert clones (bacterial artificial chromosomes, BACs), genome sequencing relied on the availability of massive shotgun (random) approaches, sequencing the genome tens to hundreds of times (the Genome Coverage, Table 15.1) as random fragments less than 100–300 bp length, before assembly into contigs (contiguous stretches of DNA) using computational approaches. Using these draft assemblies, the genes of *B. juncea* (AB genomes) were studied in detail in the early 2010s (Wang et al. 2011), often through reference to the diploid A genome (*B. rapa*) progenitor and Arabidopsis, since DNA sequencing and assembly methods were not suitable for a tetraploid species at that time. A range of sequencing technologies with different capabilities has been used (Table 15.2) since, leading to the high-quality *B. juncea* sequences (Yang et al. 2016; Paritosh et al. 2021) now published (Table 15.1). In the diploid species sequenced first, genetic recombination maps from segregating hybrid-derived populations analysed with polymorphic DNA markers were important for linking fragmentary assemblies (with the contigs joined by inserting a row of NNNN indicating unknown nucleotides). More recently, long-molecule sequencing technology (PacBio Sequel or Oxford Nanopore Technology), along with optical mapping (BioNano) means that genetic maps, although useful, are not essential for sequence assembly.

Sequence assembly quality is measured using different parameters (measured with various tools and algorithms), and these parameters need to be evaluated in the context of the use of the assembly. The parameter N50 is widely quoted, and gives the length of the shortest continuous stretch of DNA (or contig) that, along with longer contigs, represents half the total genome length. Thus, a complete, end-to-end, assembly would have an N50 of the median chromosome length, and the total assembly include each individual chromosome; such complete assemblies are becoming available in the 2020s for the first time. N50 is the minimum number of contigs which include half of all the DNA sequence: a perfect genome sequence will have an N50 of half the number of chromosomes of the organism. A highly fragmented (often called 'draft') assembly of a genome sequence may have an N50 of 10s of thousands of contigs (so N50 will be a few kb), and there will be gaps and unassembled regions. A fragmented or partial genome sequence is useful for gene sequence identification and marker development. Study of genome and chromosomal rearrangements, copy number variation (CNV) or linkage require more complete assembly (lower N50 and larger N50) than an assembly used for identification and analysis of genes.

It is normal for genome assemblies to be improved by their original authors in the years following the original publication. This leads to Version numbers, V1 being followed by V2, and sometimes minor changes (e.g. V1.4) are indicated after a decimal point. These revisions, often only reported in on-line databases, may include additional sequencing (sometimes using a different, long-read, technology), or genetic linkage maps, or may be software based, with improved calling of bases from the original sequencing, or, very often, using improved assembly algorithms and manual optimization.

15.2.1.1 Diploid Assemblies

The whole genome of the diploid *B. rapa* ssp. *pekinensis* line Chiifu-401–42 was sequenced and assembled using 72 × coverage of paired short read sequences generated by Illumina GA II technology with stringent assembly parameters (Wang et al. 2011). This draft genome was 283.8 Mb long and calculated to cover >98% of the gene space. This assembly integrated ~200,000 sequences from the ends of large DNA clones (BAC-end sequences) into 159 super-scaffolds that represented 90% of the assembled sequences, with an N50 scaffold size of 1.97 Mb. After that ten pseudo-chromosomes were assembled using 1400 genetic markers mapped on the scaffolds. This assembly was the first sequenced genome among the *Brassica* species and it was used as references genome of other species. The first version of the assembly updates with additional short read data was released as v2.0. The scaffold order was improved in the version v2.5 (http://brassicadb.

Table 15.1 Status of the whole genome sequence of *Brassica juncea* and its relatives in 2021

Species	No of Genotypes sequenced	Genome	No. of chromosomes	Assembly level and N50 (number of Scaffold)	Genome coverage	Assembly	Sequencing platform	Assembly methods	References
B. rapa	6	AA	10	Chromosome N50 Scaffold: 39	57x–442x	GCA_000309985.3; GCA_900412535.3; GCA_016163755.1; GCA_003434825.1; GCA_008629595.1; GCA_017639395.1	PacBio RSII; MinION sequencing; Oxford nanopore;PacBio Sequel; Illumina MiSeq; Illumina HiSeq;	Canu v. 1.5; Canu v. 1.6; Meraculous v. May_2013; AllPaths v. FEB-2015; MaSuRCA v. 3.2.2; ALLMAPS v. 2019-March	NCBI; BRAD Zhang et al. (2018) Wang et al (2011)
B. nigra	2	BB	8	Chromosome N50 Scaffold: 38	96x–100x	GCA_016432835.1; GCA_001682895.1	Oxford NanoporeMinION; Illumina HiSeq	Canu v. 1.4; AllPaths v. 44,540	NCBI Perumal et al. (2020) Paritosh et al. (2020)
B. oleracea	5	CC	9	Chromosome N50 Scaffold:224	38x–243x	GCA_000695525.1; GCA_900416815.2; GCA_018177695.1; GCA_902726615.1; GCA_000604025.1;	Illumina GAII; Illumina HiSeq; 454; PacBio Sequel; Hi-C	SOAPdenovo v. 1.05; FALCON v. 1.8.7	NCBI Sun et al. (2019)
B. juncea	**3**	**AABB**	**18**	**Chromosome N50 Scaffold**	**110x–410x**	**GCA_015484525.1; GCA_001687265.1 NB Version 1 for Yang et al. Currently V1.5 only available on BRAD GCA_018703725.1**	**Illumina HiSeq; PacBio; RSII; Bionano**	**ALLPATHS-LG; PBjelly_V15.2.20 Jellyfish(v.2.2.9) FALCON Quiver Pilon**	**NCBI BRAD** http://cgmcp.du.ac.in/ Yang et al. (2016) Paritosh et al. (2021) Kang et al (2021)

(continued)

Table 15.1 (continued)

Species	No of Genotypes sequenced	Genome	No. of chromosomes	Assembly level and N50 (number of Scaffold)	Genome coverage	Assembly	Sequencing platform	Assembly methods	References
B. napus	3	AACC	19	Chromosome/scaffold N50 Scaffold:160	22x–200x	GCA_000686985.2; GCA_000751015.1; GCA_014170575.1	Illumina HiSeq 2000; Illumina MiSeq	SOAPdenovo v. DEC-2018	NCBI Chen et al. (2021) Sun et al. (2017)
B. carinata	1	BBCC	17	Chromosome N50 Scaffold: 8	301x	GCA_016771965.1	Illumina; Nanopore; Hi-C	canu v. 1.8	NCBI Song et al. (2021)

Table 15.2 Key sequencing technologies used for *Brassica* species genome assemblies current in 2021

Technology	Features	Error rate
Nanopore MinIon, PromethIONOxford Nanopore Technology ONT	Single-molecule sequencing with read length from short to to >1 Mb	error rate >10%
Illumina HiSeq, MiSeq	150–300 bp reads, usually from both ends of a fragment (paired ends) from 300 bp to 30 kb long (mate pairs)	<0.01%
Sanger	Reads 500–1000 bp of single DNA clones or PCR products. Very expensive per base but used to fill gaps of a few kb, and confirm polymorphisms	1–0.1%
Hi-C	Chromatin conformation and capture. Cross-couples DNA molecules and sequences both ends, identifying those together in nuclei. The hybrid DNA molecules are then sequenced at each end with Illumina technology	n/a
Sequel/Pac BioPacBioSMRTreads	Reads of 5–20 kb	5–15% error rate
BioNano	Optical mapping of patterns of hybridization to stretched single molecules	
BGI DNBSeq	Alternative technology to Illumina. 50–150 bp reads, usually paired ends from 100 to 300 bp fragments	<0.01% error rate;

n/a not applicable

org/brad/datasets/pub/Genomes/Brassica_rapa/V2.0/V2.5/). The *B. rapa* genome, like most other species, includes highly repeated sequences and complex centromeric regions which were under-represented in the early assemblies. It is not possible to assemble the genome completely using short read technologies only, even when supplemented with large-insert BAC clone data, and therefore the genome v2.5 is quite fragmented, allowing comprehensive analysis of the gene space but limiting the utility of this assembly for extensive genetic and genomic analyses. In order to expand the utility of the reference genome of *B. rapa*, at the structural level, a massively improved assembly v3.0 (Zhang et al. 2018) has been created using single-molecule sequencing (PacBio), optical mapping (BioNano), and chromosome conformation capture technologies (Hi-C). In the earlier versions, the genetic map derived from the crosses between distantly related cultivars of *B. rapa* was used to assign scaffolds to chromosomes, which may have resulted in errors in the assembly including possible overlapping sequences. v3.0 serves as an important community resource for genetic and genomic studies in *B. rapa*, facilitating breeding, and comparative genomic analysis with other *Brassica* species.

A high-quality continuous DNA sequence of the second diploid progenitor of *B. juncea*, the B genome *B. nigra*, was published in 2020 (Perumal et al. 2020). Although survey sequences and many shorter sequences were published earlier, the lower economic value of *B. nigra* (black mustard) and more limited interest in its genetic diversity for introgression compared to the A genome, lowered the incentive for sequencing. Now, *B. nigra* (genotypes Ni100 and CN115125) genome sequencing and assembly was performed using a combination of Nanopore sequencing, with Illumina error correction, Hi-C sequencing and genetic mapping technology. Reads were assembled to the contigs then the assembly was polished with short reads (Illumina). The final contigs were developed from SMARTdenovo using 30–64× coverage of CANU-corrected reads. The assembly genome was 447 Mb, representing 78% of the predicted genome size of genotype Ni100 (Perumal et al. 2020). Genotyping-by-sequencing data was applied to anchor assembled scaffolds to eight pseudo-chromosomes.

For the remaining diploid species, the C genome *B. oleracea*, five genotypes have been sequenced and assembled (Table 15.1), although only two were assembled at chromosomal level. The draft assembly of *B. oleracea* var. *capitata* line 02–12 genome was made by including Illumina, Roche 454 and Sanger sequence data (Liu et al. 2014). The assembly represent 85% of the estimated 630 Mb genome with more than 98% of the gene space. A popular inbreed line *B. oleracea* L. var. *botrytis* (C-8) was sequenced and assembled using 69.06 Gb of high-quality PacBio long reads and 45.99 Gb of Illumina reads (Sun et al. 2019) represents ~200 fold coverage of the cauliflower genome. All these reads assembled in to 1.4 k contigs and size of the genome was 584 Mb. Quality of the assembly verified by mapping the Illumina reads to the contigs by BWA-MEM tools. Almost 99% of the reads mapped to the assembly, which covered 99.15% of the assembled sequence.

Additional diploid *Brassica* sequences are being generated, particularly for genomes where introgression of agronomic and quality traits, including biotic and abiotic stress resistances, are valuable, with the assembly quality depending on the aims of the experiment. For example, a de novo draft assembly of *B. fruticulosa* with 2,710,391 scaffolds a maximum of 61 kb long (N50 483 bp) was valuable to identify introgression (Agrawal et al. 2021).

15.2.1.2 *Brassica juncea* Genome Assemblies

By 2021, three complete genome sequences of the tetraploid *B. juncea* (AABB genome) had been published. Compared to a diploid species, the tetraploid has about two loci, one in each genome, for each sequence, giving an additional challenge to making an assembly. The diploid species may also have two alleles at each locus from the immediate parents (reduced by using inbred lines) as well as different chromosome rearrangements, including deletions and duplications within and between chromosomes. In Brassica, remnants of ancestral whole genome duplication events are a particular challenge because the diploid species are themselves mesohexaploid so each diploid genome is triplicated with three copies.

Yang et al. (2016) used an inbred (homozygous) line of *B. juncea* var. *tumida* for whole genome sequencing and assembly. In combination of 12× coverage with PacBio single-molecule long reads and 176× coverage Illumina shotgun reads, a 784 Mb genome assembly representing 85% of the estimated *B. juncea* genome was completed. The first version of assembly was refined using a million BioNano data reads, each over 150 kb length (222 genome coverage), with the final assembly giving a 955 Mb genome length with scaffold N50 of 1.5 Mb. Then a high-resolution genetic map with 5,333 bin markers were used to construct 18 pseudo-chromosomes of the *B. junea* genome. A total 91.5% from A-subgenome and 72.3% from B-subgenome sequences were anchored onto the 10 and 8 pseudo-chromosomes of the assembled *B. juncea* genome, respectively. Finally, *B. juncea* chromosomes sorted into the 402.1 Mb *B. juncea* A and 547.5 Mb *B. juncea* B subgenomes. The assembly further validate by mapping PacBio data, published BACs from *B. nigra*, 458 core eukaryotic genes and expressed sequence tags (ESTs) from the US National Centre for Biotechnology Information (NCBI) database.

In the first version of the *B. juncea* genome Yang et al. (2016), He and Bancroft (2018) showed clearly the that there were problems with the contiguity of the two genomes present, due to inter-homeolog polymorphisms (IHPs), that were more abundant than the allelic single nucleotide polymorphisms (SNPs) used for genetic mapping. A careful transcriptome analysis showed the scaffold chaemerism in the first version of the assembly. Subsequent versions of the assembly have improved and updated the genome sequence of *B. juncea* (Yang et al. 2018b and BRAD v. 1.5, not published in Genbank). He and Bancroft (2018) exemplify the challenges of assembly of a tetraploid genome with their criticism of its chaemeric nature and potential jumps between the diploid genomes.

Paritosh et al. (2021) sequenced and assembled the genome of the *B. juncea* variety Varuna, giving a second de novo *B. juncea* genome assembly. SMRT (single-molecule real-time) sequencing on the PacBio RSII and the Illumina HiSeq platform were used to generate raw reads. PacBio reads were assembled with Canu assembler and obtained 1253 contigs then Illumina HiSeq reads were mapped on the PacBio contigs; 98% of the short reads mapped to the PacBio-contigs demonstrating that the long-read sequences had covered most of the genome. Contigs were then corrected for the mis-assemblies and scaffolded with BioNano optical mapping. To generate the final chromosome level assembly of the *B. juncea* genome, three linkage maps were used (two for *B. juncea* and one for *B. nigra*), giving the highly contiguous genome assembly with an N50 value of >5 Mb.

Complementing the *B. juncea* (AB) sequence, Paritosh et al. (2021) also sequenced a doubled haploid (DH) line of *B. nigra* (B) variety Sangam using the Illumina HiSeq platform and assembled in to 57 941 contigs with an N50 value of ~ 34.6 kb. These *B. nigra* contigs were scaffolded with the Oxford Nanopore long-read sequence and finally assembled into eight pseudo-chromosomes and it represent around 346.9 Mb of B. nigra out of 470.8 Mb sequenced *B. nigra* genome.

Recently, third *B. juncea* genome has been sequenced and chromosome-scale de novo assembly reported by the Kang et al (2021). The whole genome was assembled using PacBio long-read, Illumina short read sequencing, BioNano optical mapping and Hi-C seq data. First PacBio (~93×) reads were assembled into contigs then Illumina reads (~130×) were applied to contig correction for generation of the first version of the assembly (V1). An optical consensus map generated using BioNano (202× coverage) data and this map implemented to assemble 1,897 super-scaffolds with an N50 of 5.87 Mb (assembly V2). All these super- scaffolds assembled into pseudo-chromosomes were checked against a high-density genetic linkage map consisting of 15,543-markers. Finally, Hi-C data was used to confirm the pseudo-chromosomes. The assembly size of this yellow-seeded *B. juncea* genome is 933,496,244 bp containing 82,723 protein coding genes and ~50% of the genome annotated as TE elements.

15.2.1.3 Other Tetraploid Sequence Assemblies

The *B. carinata* (BC) genome was de novo assembled using 83.09 Gb (76×) ONT Nanopore, 180.53 Gb (166×) Hi-C sequencing and 64.24 Gb (59 ×) Illumina reads. The assembly was 1086.8 Mb, representing 94.44% of the estimated genome. The *B. carinata* genome is larger than two other tetraploid species *B. juncea* (~955 Mb) and *B. napus* "Darmor-bzh" (849.7 Mb) (Chalhoub et al. 2014; Yang et al. 2016; Yang et al. 2018a; Song et al. 2021). Around 94% (1,019.1 Mb) of the assembled genome were anchored to 17 chromosomes. Among these 52% (530.6 Mb) and 48% (488.5 Mb) reads were anchored onto the eight (*B. carinata* B) and nine (*B. carinata* C) pseudo-chromosomes, respectively.

The first draft genome of *B. napus* was developed from the European winter-type cultivar 'Darmor-bzh' (Chalhoub et al. 2014), and has been largely used for DNA marker development and mapping. A second European winter-type cultivar 'Tapidor' was also assembled with Illumina short reads (Bayer et al. 2017), and a Chinese semi-winter-type cultivar 'Ningyou7' was assembled from short reads and gaps were filled in by PacBio reads (Zou et al. 2019). The genome of *B. napus* cultivar 'ZS11', with high oil content and high seed production, was assembled based on a hybrid strategy using BAC large-insert clones and short reads (Sun et al. 2017). Recently an improved assembly has been released for the cultivar 'ZS11' (Chen et al. 2021) WITH long PacBioSMRT reads, genetic maps and Hi-C technologies.

15.2.2 Transcriptome Sequences and Assembly of *Brassica* Species

Large scale RNA-sequence data (EST) is a powerful and cost-efficient method for transcription profiling, gene annotation (gene modelling) for the assembled genome (Ozsolak and Milos 2011), and expression measurement of the genes in different tissues or under different conditions (Mortazavi et al. 2008). RNA-seq data also enables detection of alternative splicing (AS) events at single nucleotide resolution (Sultan et al. 2008), increasingly recognized as a major aspect of regulation of cell development. A total of ~27 million Illumina RNA-seq paired-end reads were generated for the eight different tissues of two *B. rapa* cultivars (Chiifu-401-42 and Caixin accession, L58). These RNA-seq along with public EST of *Brassica* species were used for gene modelling of the sequenced *B. rapa* genome. Similarly, *B. nigra* was annotated using a combination of sequenced 39 Gb RNA-seq data from *B. nigra* cv. Ni100 and C2 and public Brassica EST sequence data. *B. oleracea* genome was annotated using only Brassica ESTs from NCBI. For the *B. juncea* and *B. napus* genome annotation publicly available transcripts from Brassica and its relative species were used. Large number of high-quality full-length transcript sequences were generated from the six different tissue samples (root, stem, leaf, flower, silique and seed) of *B. carinata* applying the PacBio single-molecule real-time sequencing technology. Further Illumina RNA-seq data from six tissues and drought-treatment leaf samples of *B. carinata* were generated for *B. carinata* genome annotation as well as predict the differential expressed gene in this species.

Comparative transcriptomics approaches is popular to identify putative genes underlying (responsible) for the particular traits. It is one kinds of snapshot methods to detect genes and their putative role/function. In this approach, total RNA from the samples were extracted and cDNA library prepared and subsequent the library sequenced to get raw RNA sequences reads, after that a series of bioinformatics tools and methods use to analysis and compare these transcriptome data. In Brassica comparative transcriptomics approaches applied for characterization many genes including disease resistance, aphid resistant, stress tolerance (salt, biotic, abiotic), anthocyanin etc.

Chao et al. (2020) developed a massive number of RNA-seq data from 103 tissues from rapeseed cultivar Zhong Shuang11 (ZS11) at seven developmental stages (seed germination, seedling, bolting, initial flowering, full-bloom, podding and maturation). In this analysis, the expression patters of 101,040 genes were determined by estimate FPKM value. All the analysis are deposited in a searchable database called Brassica EDB (https://brassica.biodb.org/index).

15.3 Genome Composition and Genome Size

The DNA sequence includes genes and regulatory sequences, but the majority of the sequence of most plants and animals are represented by repetitive elements (Biscotti et al. 2015). A typical plant genome includes both tandemly repeated satellite DNA motifs, located at a small number of chromosomal location, and transposable elements which may be dispersed throughout the genome or amplified at particular locations; together, 60% or more of the nuclear DNA may be represented by these repetitive sequences. Genome size variations among species occurs by duplication of parts of the genome including genes, but more often through gain or loss of repetitive DNA: transposable elements insertion and deletion events occur in the Brassicaceae during the evolution of the species (Nouroz et al. 2017). Large scale genome sequencing and comparative genomics analysis of *Brassica* species enable us to explore the genomic composition and evolution events of the species.

Based on assemble and annotation information, we compared the gene number, genome size and gene density among the six *Brassica* species. In terms of assemble genome size *B. carinata* genome is the largest and it is 3-2fold bigger than *B. rapa*, *B. nigra* and *B. olerecia* genomes (Fig. 15.1). Similarly, maximum number of the gene counts in the *B. carinata* genome, *B. nigra* and *B. oleracea* genomes contains similar number of the genes. The highest gene density was found in the *B. rapa* genome (1 gene per 8 kb genome).

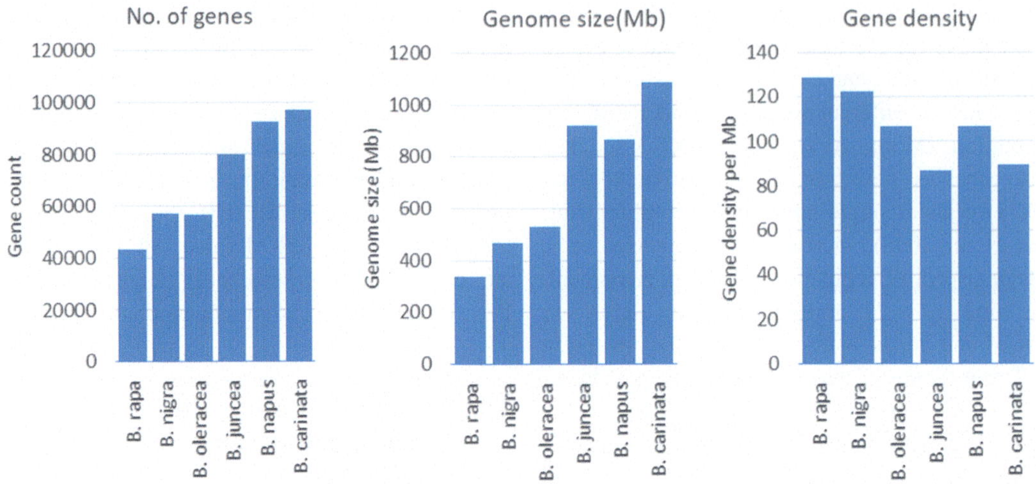

Fig. 15.1 Comparative illustration of the gene number, genome size and gene density of six *Brassica* species

Around 47% of the *B. juncea* genome consists TEs; A genome (113 Mb; ~33.9%) has less repeat content than the B genome (~259 Mb; ~51%).LTR/Copia and LTR/Gypsy, types retrotransposons were the predominant in *B. juncea* genome (Paritosh et al. 2021). Among the six *Brassica* species *B. oleracea* contains highest proportion of TEs in its genome and it was 70% (Fig. 15.2).

Simple sequences repeats—arrays of tandem repeats of short nucleotide motifs typically between 1 and 5 bp long—also are found abundantly in all plant genomes. Being typically dispersed in the genome, and showing rapid and

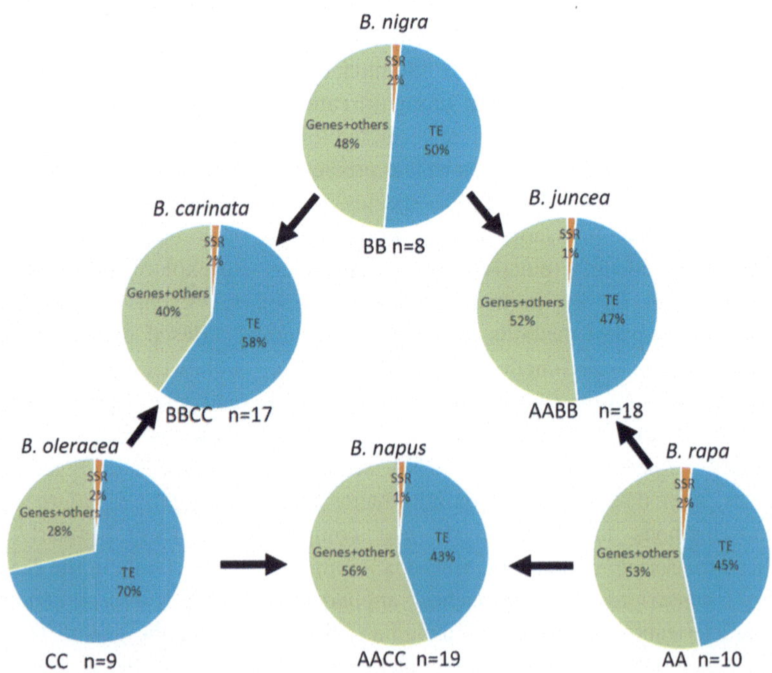

Fig. 15.2 Comparative features of the genome composition of six *Brassica* species

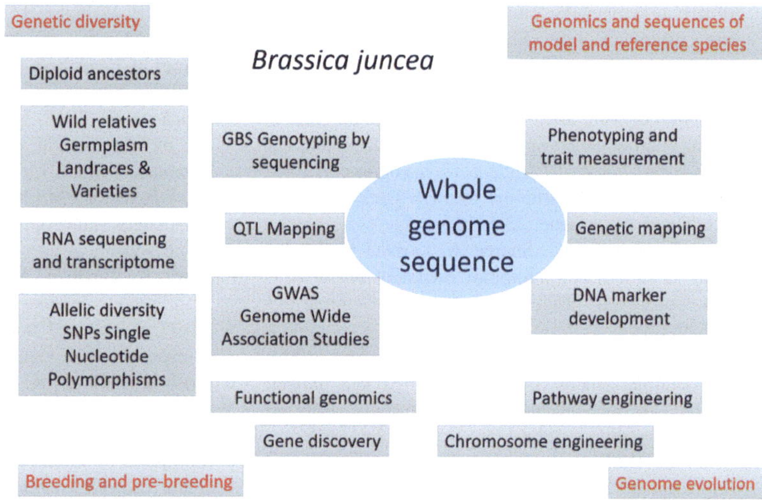

Fig. 15.3 Application of whole genome sequencing

heritable evolutionary changes in the number of copies of the repeats, flanking PCR primers can be used to generate molecular markers. Thousands of potential SSR markers have been identified in *Brassica* species using whole genome sequencing data and transcriptome sequence data or BAC-end sequences. Now, hundreds of the SSR markers have been optimized and mapping in the Brassica genomes, and are now used extensively for linkage map development, studying genetic diversity and for analysis of population structures (Fig. 15.3).

15.4 Conclusions and Applications

As for many other important crops, the availability of whole genome sequence in *B. juncea*, as well as sequences for diploid relatives and the wider pool of germplasm, is proving critical for identifying key genes and their controlling elements related to agronomy, biotic and abiotic stress resistance, yield and quality. Whole genome sequencing enables development of DNA markers at known genomic locations and enables mapping of genes. Quantitative trait loci (QTLs) are identified in DNA sequences where polymorphisms have been identified, by measuring phenotypes in genetically segregating populations of a few hundred plants derived from contrasting inbred parents. Genome wide association studies (GWAS) identify phenotypic traits that correlate with particular DNA polymorphisms in measurements of the trait in large numbers—ideally hundreds, but often tens—of plants which are not closely related. With a genetic map, segregation of markers can be examined, along with character traits. The genome sequence also underpins the generation of a pangenome for the species, more broadly allowing the complete allelic diversity of genes to be identified. Importantly for the future, copy number variation (CNV) and structural variation will be increasingly easy to analyse with long-molecule DNA sequencing. The sequencing information can be combined with epigenomics and modification of expression. The availability of high-quality genomic DNA sequences from *Brassica juncea* is underpinning future improvement of the crop.

Acknowledgements The work was supported under the Newton-Bhabha Fund UK-India Pulses and Oilseeds Research Initiative, with funding from UK's Official Development Assistance Newton Fund awarded by UK Biotechnology and Biological Sciences Research Council (BB/R019819/1).

References

Agrawal N, Gupta M, Banga SS, Heslop-Harrison JP (2020) Identification of chromosomes and chromosome rearrangements in crop brassicas and *Raphanus sativus*: a cytogenetic toolkit using synthesized massive oligonucleotide libraries. Front Plant Sci 11:598039

Agrawal N, Gupta M, Atri C, Akhatar J, Kumar S et al (2021) Anchoring alien chromosome segment substitutions bearing gene(s) for resistance to mustard aphid in *Brassica juncea-B. fruticulosa* introgression lines and their possible disruption through gamma irradiation. Theor Appl Genet 23:1–16

Axelsson T, Bowman C, Sharpe A, Lydiate D, Lagercrantz U (2000) Amphidiploid *Brassica juncea* contains conserved progenitor genomes. Genome 43 (4):679–688

Bansal P, Kaur P, Banga SK, Banga SS (2009) Augmenting genetic diversity in *Brassica juncea* through its resynthesis using purposely selected diploid progenitors. Int J Plant Breed 3(1):41–45

Bañuelos G, Dhillon K, Banga S (2013) 16 Oilseed Brassicas. In: Biofuel crops: production, physiology and genetics, pp 339

Bayer PE, Hurgobin B, Golicz AA, Chan CKK, Yuan Y et al (2017) Assembly and comparison of two closely related *Brassica napus* genomes. Plant Biotech J 15 (12):1602–1610

Belser C, Istace B, Denis E, Dubarry M, Baurens FC et al (2018) Chromosome-scale assemblies of plant genomes using nanopore long reads and optical maps. Nat Plants 4(11):879–887

Biscotti MA, Olmo E, Heslop-Harrison JS (2015) Repetitive DNA in eukaryotic genomes. Chr Res 23:415–420

Chalhoub B, Denoeud F, Liu S, Parkin IA, Tang H et al (2014) Early allopolyploid evolution in the post-neolithic *Brassica napus* oilseed genome. Science 345 (6199):950–953

Chao H, Li T, Luo C, Huang H, Ruan Y et al (2020) BrassicaEDB: a gene expression database for Brassica crops. Int J Mol Sci 21(16):5831

Chen X, Tong C, Zhang X, Song A, Hu M et al (2021) A high-quality *Brassica napus* genome reveals expansion of transposable elements, subgenome evolution and disease resistance. Plant Biotech J 19(3):615–630

Ghawi SK, Shen Y, Niranjan K, Methven L (2014) Consumer acceptability and sensory profile of cooked broccoli with mustard seeds added to improve chemoprotective properties. J Food Sci 79(9):S1756–S1762

Hanelt P, Buttner R, Mansfeld R (2001) Mansfeld's encyclopedia of agricultural and horticultural crops (except Ornamentals)

He Z, Bancroft I (2018) Organization of the genome sequence of the polyploid crop species *Brassica juncea*. Nat Genet 50:1496–1497

Kang L, Qian L, Zheng M, Chen L, Chen H et al (2021) Genomic insights into the origin, domestication and diversification of *Brassica juncea*. Nat Genet 53:1392–1402

Liu S, Liu Y, Yang X, Tong C, Edwards D et al (2014) The *Brassica oleracea* genome reveals the asymmetrical evolution of polyploid genomes. Nat Com 5 (1):1–11

Mortazavi A, Williams BA, McCue K, Schaeffer L, Wold B (2008) Mapping and quantifying mammalian transcriptomes by RNA-Seq. Nat Methods 5(7):621–628

Nouroz F, Noreen S, Khan MF, Ahmed S, Heslop-Harrison JP (2017) Identification and characterization of mobile genetic elements LINEs from *Brassica genome*. Gene 627:94–105

Ozsolak F, Milos PM (2011) RNA sequencing: advances, challenges and opportunities. Nat Rev Genet 12 (2):87–98

Paritosh K, Pradhan AK, Pental D (2020) A highly contiguous genome assembly of *Brassica nigra* (BB) and revised nomenclature for the pseudochromosomes. BMC Genomics 21(1):1–2

Paritosh K, Yadava SK, Singh P, Bhayana L, Mukhopadhyay A et al (2021) A chromosome-scale assembly of allotetraploid *Brassica juncea* (AABB) elucidates comparative architecture of the A and B genomes. Plant Biotech J 19(3):602

Perumal S, Koh CS, Jin L, Buchwaldt M, Higgins EE et al (2020) A high-contiguity *Brassica nigra* genome localizes active centromeres and defines the ancestral Brassica genome. Nat Plants 6(8):929–941

Qiao J, Zhang X, Chen B, Huang F, Xu K et al (2020) Comparison of the cytoplastic genomes by resequencing: insights into the genetic diversity and the phylogeny of the agriculturally important genus Brassica. BMC Genomics 21(1):1–15

Song X, Wei Y, Xiao D, Gong K, Sun P et al (2021) *Brassica carinata* genome characterization clarifies U's triangle model of evolution and polyploidy in Brassica. Plant Physiol 186(1):388–406

Sultan M, Schulz MH, Richard H, Magen A, Klingenhoff A et al (2008) A global view of gene activity and alternative splicing by deep sequencing of the human transcriptome. Science 321(5891):956–960

Sun F, Fan G, Hu Q, Zhou Y, Guan M et al (2017) The high-quality genome of *Brassica napus* cultivar 'ZS 11'reveals the introgression history in semi-winter morphotype. The Plant J 92(3):452–468

Sun D, Wang C, Zhang X, Zhang W, Jiang H et al (2019) Draft genome sequence of cauliflower (*Brassica oleracea* L. var. botrytis) provides new insights into the C genome in Brassica species. Hort Res 6:1–11

Wang X, Wang H, Wang J, Sun R, Wu J et al (2011) The genome of the mesopolyploid crop species *Brassica rapa*. Nat Genet 43(10):1035–1039

Wiersema JH, León B (1999) World economic plants: a standard reference. CRC Press

Yang J, Liu D, Wang X, Ji C, Cheng F et al (2016) The genome sequence of allopolyploid *Brassica juncea* and analysis of differential homoeolog gene expression influencing selection. Nat Genet 48(10):1225–1232

Yang J, Liu D, Wang X, Ji C, Cheng F et al (2018a) Author Correction: The genome sequence of allopolyploid *Brassica juncea* and analysis of differential homoeolog gene expression influencing selection. Nat Genet 50:1616–1616. https://doi.org/10.1038/s41588-018-0227-4

Yang J, Zhang C, Zhao N, Zhang L, Hu Z et al (2018b) Chinese root-type mustard provides phylogenomic insights into the evolution of the multi-use diversified allopolyploid Brassica juncea. Mol plant 11(3):512–514

Zhang L, Cai X, Wu J, Liu M, Grob S et al (2018) Improved *Brassica rapa* reference genome by single-molecule sequencing and chromosome conformation capture technologies. Hort Res 5(1):1–11

Zou J, Mao L, Qiu J, Wang M, Jia L et al (2019) Genome-wide selection footprints and deleterious variations in young Asian allotetraploid rapeseed. Plant Biotech J 17(10):1998–2010

Brassica juncea L.: Chloroplast Genome

16

Sudheesh K. Prabhudas, Raju Balaji, and Purushothaman Natarajan

Abstract

Brassica juncea L. is one of the major edible oilseed crops of India. It is an amphidiploid species with *Brassica rapa* L. and *Brassica nigra* L. as its progenitor species. The *B. juncea* L. chloroplast genome sequence is similar to its maternal progenitor species *B. rapa* L. The chloroplast genome encodes 113 genes consisting of 4 rRNA, 30 tRNA, and 79 protein-coding genes. Phylogenetic analysis of multiple accessions of chloroplast genomes of six major *Brassica* species of the triangle of U reconfirmed their divergence into two major lineages, the rapa/oleracea and the nigra lineages. The rapa/oleracea lineage consisted of three major groups consisting of four of the six major species of *Brassica* and the oleracea lineage consisted of the remaining two species. The usage of multiple accessions of chloroplast genomes further resolved the phylogenetic analysis, and in the future, using more accessions might resolve the actual divergence time among the species.

16.1 Introduction

Chloroplast (cp) is the plant organelle belonging to the plastid group. It evolved by endosymbiogenesis of an ancient cyanobacterium into a non-photosynthetic eukaryotic ancestor cell, and the evolution of photosynthetic eukaryotes dates back to 1.25 billion years ago (Gray 1992; Gibson et al. 2018). Chloroplasts are the metabolic centers of plants playing a pivotal role in photosynthesis, de novo fatty acid synthesis, amino acid synthesis, and innate immunity of the plant (Padmanabhan and Dinesh-Kumar 2010; Reyes-Prieto and Moustafa 2012; Li-Beisson et al. 2017). The cp population in a single cell varies from 20 to 120 based on tissue type and ploidy of the organism (Mochizuki and Sueoka 1955). During the course of evolution, most of the cyanobacterium genome migrated to the host nuclear genome, and the resulting cp genome has been reduced to harbor 120–130 genes, mostly coding for transcription, translation, and photosynthesis. The organization of the cp genome is essentially circular and quadripartite, with two single-copy sections (long single-copy region, LSC, and short single-copy region, SSC) separated by a pair of inverted repeat regions (IR). The cp genomes having linear, stem-

P. Natarajan
Department of Biology, West Virginia State University, Institute, WV 25112, USA

R. Balaji · P. Natarajan (✉)
Department of Genetic Engineering, SRM Institute of Science and Technology, Kattankulathur, Chennai 603203, India
e-mail: purushon@srmist.edu.in

S. K. Prabhudas
Nutrition, Genetics and Biotechnology Division, ICAR—Central Institute of Brackishwater Aquaculture, Chennai 600028, India

C. Kole and T. Mohapatra (eds.), *The Brassica juncea Genome*, Compendium of Plant Genomes,
https://doi.org/10.1007/978-3-030-91507-0_16

loop structure and multiple mini circular fragmented genomes have also been reported in Cladophorales and Dinoflagellate plastids (Bendich 2004; De Vries and Archibald 2018). The cp genome is usually maternally inherited in angiosperms, with a few exceptions of paternal and bi-parental inheritance (Birky 1995; Corriveau and Coleman 1988).

Brassica juncea L. is one of the six major crops belonging to the genus *Brassica*. Three are the parental diploid crops belonging to the triangle of U consisting of *Brassica nigra* L. (BB, $2n = 16$), *Brassica rapa* L. (AA, $2n = 20$), and *Brassica oleracea* L. (CC, $2n = 18$). Three are the amphidiploid offsprings of the triangle of U comprising *Brassica juncea* L. (AABB, $2n = 36$), *Brassica napus* L. (AACC, $2n = 38$), and *Brassica carinata* L. (BBCC, $2n = 34$) (Nagaharu 1935). *B. juncea* L. is a hybrid between *B. rapa* L. and *B. nigra* L. and its cp genome is identical to that of its maternal ancestor *B. rapa* L.

16.2 *Brassica juncea* Chloroplast Genome

The first report of complete cp genome of *B. juncea* L. utilized the shallow whole-genome sequence data from the Illumina sequencing technology (Prabhudas et al. 2016a). The cp genome exhibited circular quadripartite structure with a GC content of 36.36% and total length of 153,483 bp. The LSC and SSC were 83,826 bp and 17,775 bp with GC content of 34.12% and 29.20% respectively. The IR pair were 26,211 bp long each with a GC content of 42.34%. The presence of rRNA genes in IRs accounts for its high GC content. The total 113 annotated genes include 4 rRNA, 30 tRNA, and 79 protein-coding genes (Fig. 16.1). Currently, NCBI hosts five different accessions of *B. juncea* L. cp genomes four of which have the same sequence length of 153,483 bp, and one accession has total length of 153,490 bp.

16.3 Chloroplast Genes

The cp genome of *B. juncea* L. consists of a total of 85 protein-coding genes which includes six duplicated genes present in each of the inverted repeats. Eight rRNA genes are present in both the inverted repeats and 37 tRNA genes. Despite having a prokaryotic origin some of the cp genes developed introns in their sequence. Genes like ATP synthase subunit F (*atpF*), NADH dehydrogenase subunit A (*ndhA*), large ribosomal subunit 2 (*rpl2*), RNA polymerase subunit C1 (*rpoC1*), certain tRNA genes (*trnV*-UAC, *trnL*-UAA, *trnK*-UUU, *trnI*-GAU, and *trnA*-UGC) have developed a single intron whereas genes like protease (*clpP*) and conserved open reading frame (*ycf3*) has developed two introns in their sequence. The small ribosomal subunit 12 exhibits trans-splicing with its 5′ end in the LSC and two copies of its 3′ region present in each of the IR regions. Table 16.1 furnishes the genes in the cp genome of *B. juncea* L. classified based on their functions.

In the *B. juncea* cp genome, the relative synonymous codon usage (RSCU) and amino acid frequency of the protein-coding sequences were analyzed and presented in Table 16.2. Leucine (30.9%) and cysteine (3.5%) were the most and least common amino acids in the cp-encoded proteins, respectively. The codons AUU (12.3%; encoding isoleucine) and UGA (0.1 percent; encoding translational stop) were the most and least utilized, respectively. We discovered 32 non-preferred (RSCU < 1) and 32 preferred (RSCU > 1) synonymous codons. In general, synonymous codons with a C-ending were preferred, while the non-preferred codons more likely ended in U. In the *B. juncea* cp genome, no codons were characterized as rare (RSCU < 0.3).

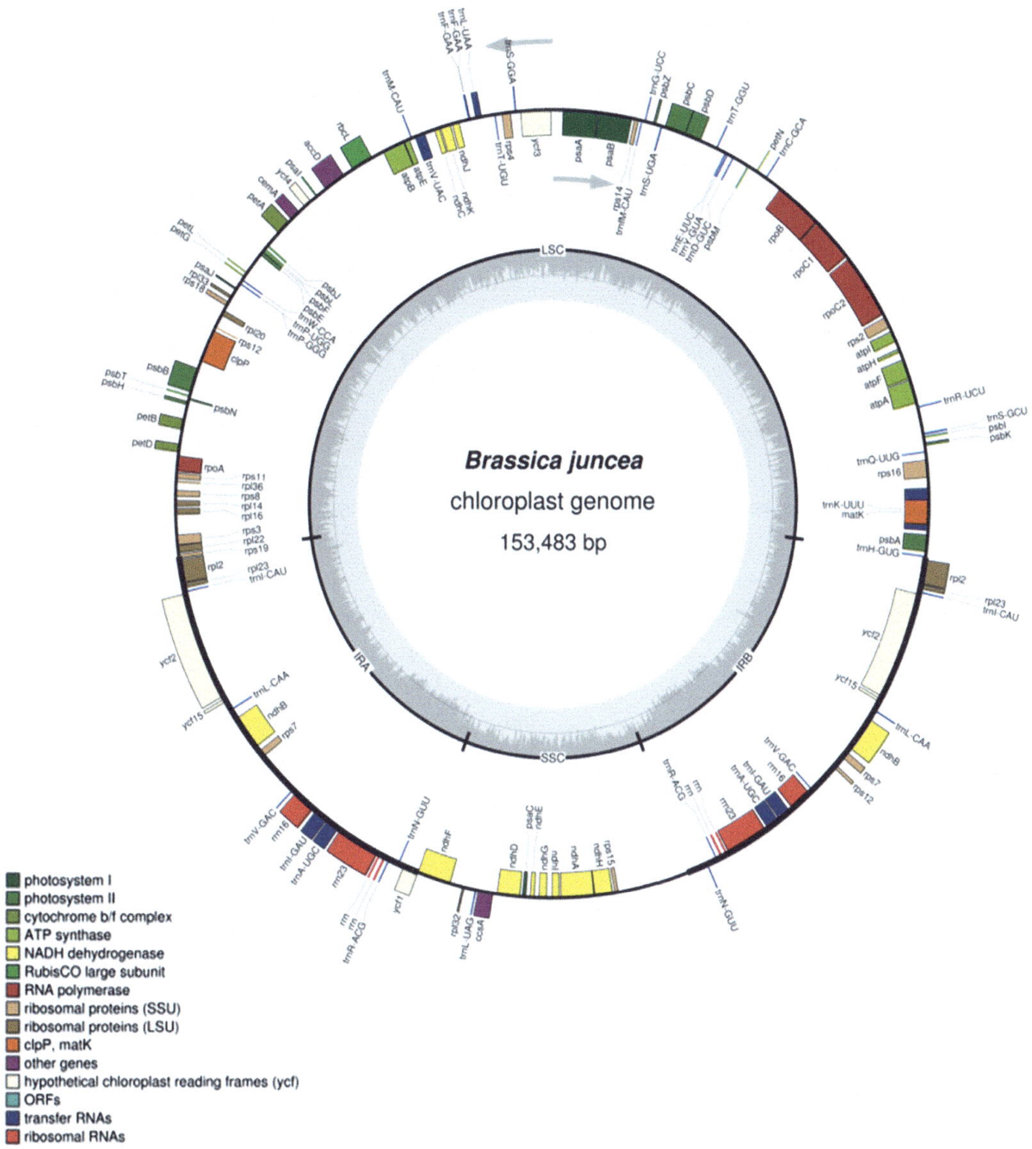

Fig. 16.1 Chloroplast genome map of *Brassica juncea* L.

16.4 Simple Sequence Repeat (SSR) and Inverted Repeat (IR) Analysis

SSRs are found across the genome at different loci (Asaf et al. 2017). The cp genome exhibits uniparental inheritance and SSRs have diversity within the same species. As a result, cp SSRs have been widely employed as molecular markers in target gene calibration, genetic map generation, and mapping research (Flannery et al. 2006; Yin et al. 2017). There are a total of 67 SSRs in the *B. juncea* L. cp genome, with 62 mononucleotides (mono) and 5 dinucleotides (di) SSRs. Mono SSR is the most common type of SSR; T type SSR accounts for 53.73 percent of total SSRs.

Assessment of the SSRs in the seven cp genomes of *Brassica* (Fig. 16.2) showed that all

Table 16.1 List of genes in *B. juncea* chloroplast genome

Function	Gene group	Gene names					
Photosynthesis	ATP synthase	*atpA*	*atpB*	*atpE*	*atpF*[a]	*atpH*	*atpI*
	Cytochrome b/f complex	*petA*	*petB*	*petD*	*petG*	*petL*	*petN*
	NADH dehydrogenase	*ndhA*[a]	*ndhB*	*ndhC*	*ndhD*	*ndhE*	*ndhF*
		ndhG	*ndhH*	*ndhI*	*ndhJ*	*ndhK*	
	Photosystem I	*psaA*	*psaB*	*psaC*	*psaI*	*psaJ*	
	Photosystem II	*psbA*	*psbB*	*psbC*	*psbD*	*psbE*	*psbF*
		psbH	*psbI*	*psbJ*	*psbK*	*psbL*	*psbM*
		psbN	*psbT*	*psbZ*			
	Large subunit of Rubisco	*rbcL*					
Self-replication	Ribosomal proteins (SSU)	*rps2*	*rps3*	*rps4*	*rps7*	*rps8*	*rps11*
		rps12[c]	*rps14*	*rps15*	*rps16*	*rps18*	*rps19*
	Ribosomal proteins (LSU)	*rpl2*[a]	*rpl14*	*rpl16*	*rpl20*	*rpl22*	*rpl23*
		rpl32	*rpl33*	*rpl36*			
	RNA polymerase	*rpoA*	*rpoB*	*rpoC1*[a]	*rpoC2*		
	Ribosomal RNAs	*rrn4.5*	*rrn5*	*rrn16*	*rrn23*		
	Transfer RNA's	*trnA*-UGC[a]	*trnC*-GCA	*trnD*-GUC	*trnE*-UUC		
		trnK-UUU[a]	*trnS*-UGA	*trnG*-UCC	*trnH*-GUG		
		trnL-UAA[a]	*trnT*-GGU	*trnF*-GAA	*trnI*-CAU		
		trnI-GAU[a]	*trnL*-CAA	*trnL*-UAG	*trnN*-GUU		
		trnfM-CAU	*trnM*-CAU	*trnP*-UGG	*trnP*-GGG		
		trnT-UGU	*trnQ*-UUG	*trnR*-UCU	*trnS*-GCU		
		trnV-GAC	*trnR*-ACG	*trnS*-*GGA*	*trnS*-GGA		
		trnV-UAC[a]	*trnW*-CCA	*trnY*-GUA			
Other genes	Subunit of acetyl-CoA-carboxylase, c-type cytochrome synthase, envelope membrane protein, protease, maturase K	*accD, ccsA, cemA, clpP*[b], *matK*					
Proteins of unknown function	Conserved open reading frames (ORF/ycf)	*ycf1*	*ycf2*	*ycf3*	*ycf4*	*ycf15*	

[a] Consists of one intron
[b] Consists of two introns
[c] Exhibits trans-splicing

Table 16.2 The codon frequency and relative synonymous codon usage (RSCU) in the protein-coding sequences of the *B. juncea* cp genome

Amino acid	Codon	Amino acid residue (%)	RSCU	Amino acid	Codon	Amino acid residue (%)	RSCU
Ala	GCU(A)	7.2	1.85	Leu	CUC(L)	2.1	0.41
Ala	GCC(A)	2.3	0.6	Leu	CUA(L)	4.2	0.82
Ala	GCA(A)	4.4	1.12	Leu	CUG(L)	1.9	0.37
Ala	GCG(A)	1.7	0.42	Lys	AAA(K)	11.1	1.49
Arg	CGU(R)	3.8	1.34	Lys	AAG(K)	3.8	0.51
Arg	CGC(R)	1.2	0.44	Met	AUG(M)	6.6	1
Arg	CGA(R)	3.8	1.35	Phe	UUU	11.4	1.34
Arg	CGG(R)	1.4	0.5	Phe	UUC	5.6	0.66
Arg	AGA(R)	5	1.75	Pro	CCU(P)	4.7	1.59
Arg	AGG(R)	1.8	0.63	Pro	CCC(P)	2.2	0.73
Asn	AAU(N)	10.6	1.54	Pro	CCA(P)	3.3	1.14
Asn	AAC(N)	3.1	0.46	Pro	CCG(P)	1.6	0.54
Asp	GAU(D)	9.2	1.61	Ser	UCU(S)	6.5	1.76
Asp	GAC(D)	2.2	0.39	Ser	UCC(S)	3.3	0.89
Cys	UGU(C)	2.7	1.53	Ser	UCA(S)	4.4	1.19
Cys	UGC(C)	0.8	0.47	Ser	UCG(S)	2.2	0.59
Gln	CAA(Q)	7.9	1.54	Ser	AGU(S)	4.5	1.21
Gln	CAG(Q)	2.3	0.46	Ser	AGC(S)	1.4	0.37
Glu	GAA(E)	11.1	1.5	Thr	ACU	6.1	1.64
Glu	GAG(E)	3.7	0.5	Thr	ACC	2.6	0.71
Gly	GGU(G)	6.5	1.31	Thr	ACA	4.5	1.22
Gly	GGC(G)	1.9	0.38	Thr	ACG	1.6	0.44
Gly	GGA(G)	8.2	1.65	Trp	UGG(W)	4.9	1
Gly	GGG(G)	3.2	0.65	Tyr	UAU(Y)	8.6	1.62
His	CAU(H)	5	1.5	Tyr	UAC(Y)	2	0.38
His	CAC(H)	1.7	0.5	Val	GUU(V)	5.8	1.47
Ile	AUU(I)	12.3	1.49	Val	GUC(V)	1.9	0.5
Ile	AUC(I)	4.8	0.58	Val	GUA(V)	5.7	1.45
Ile	AUA(I)	7.7	0.93	Val	GUG(V)	2.3	0.58
Leu	UUA	10.4	2.02	STOP	UAA(*)	0.6	1.8
Leu	UUG(L)	5.9	1.15	STOP	UAG(*)	0.3	0.78
Leu	CUU(L)	6.4	1.25	STOP	UGA(*)	0.1	0.42

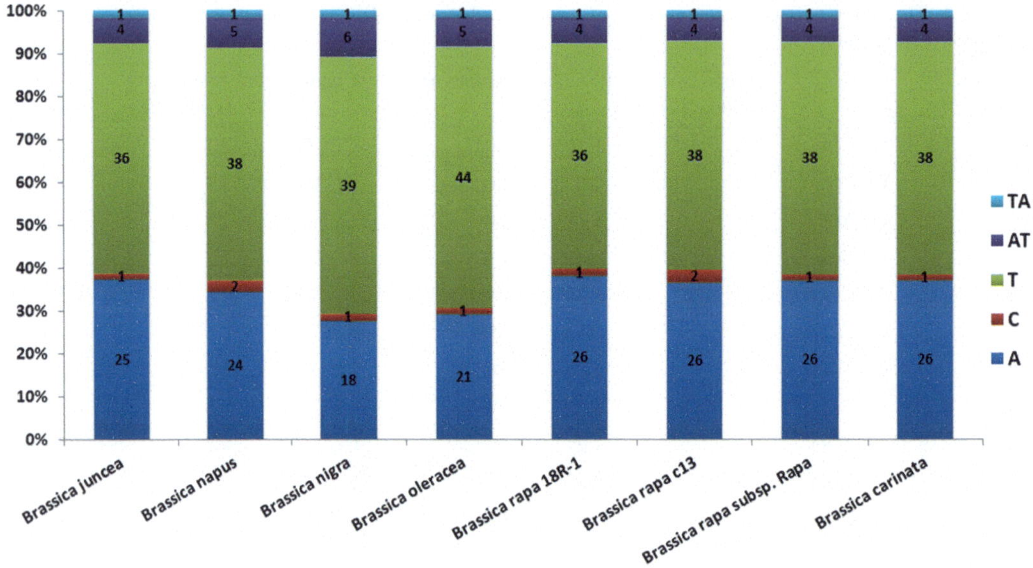

Fig. 16.2 Simple sequence repeat (SSR) motifs of seven chloroplast genomes of *Brassica* species

seven cp genomes are almost similar. Most of the SSRs in the seven cp genomes consist of mono and di repeat motifs. The mono repeats vary from 62 (*B. juncea*) to 66 (*B. oleracea* and *B. rapa* c13), and the di repeats vary from 5 (*B. juncea, B. rapa* c13, *B. rapa* 18R-1, and *B. rapa* subsp. Rapa) to 7 (*B. nigra*). However, there is only one trinucleotide repeat present in the *B. nigra* species of cp genome. The comparison is given in Table 16.3.

The IR scope program was used to analyze the expansion and contraction of typical cp genomes from *Brassica* species from the organellar genome database and nucleotide database. (Amiry-ousefi et al. 2018). The LSC regions ranged between 83,030 and 83,558 bp, the SSC regions ranged from 17,683 to 17,834 bp and the IR regions ranged from 26,035 to 26,219 bp as shown in Fig. 16.3. The IR regions were almost similar in size among most of the species except for *B. napus* L. cp genome. The IR and LSC of *B. napus* L. exhibited a slight contraction in comparison to the other cp genomes, which is also reflected in its cp genome size. The remaining cp genomes had only a few base pairs difference at the IR junctions.

16.5 Phylogenetic Study in Brassica

The cp gene sequences obtained using Sanger sequencing technique have played a crucial role in phylogenetic studies and have resolved many evolutionary relationships between closely related species (Palmer et al. 1983). The next-generation sequencing techniques make it easy to get entire cp genomes by employing shallow whole-genome sequencing without the need to extract cp genomes individually (Prabhudas et al. 2016b). The evolutionary study based on cp genes and nuclear genome blocks revealed that there were two lineages Rapa/Oleracea lineage and the nigra lineage that diverged approximately 12 Mya and underwent two polyploidization events to give rise to the three diploid species of triangle of U (Sharma et al. 2014). In a more recent study, about 60 different accessions of cp genome sequences belonging to all the six species of triangle of U were used to evaluate their evolutionary relationship. The phylogenetic tree revealed four major clades with *Brassica rapa* L., the progenitor species of *Brassica juncea* L., occurring in clades I and III

Table 16.3 Comparison table of SSR's in chloroplast genomes of major *Brassica* species

Species	Repeat/motif length	Number of repeats											Total	%
		6	7	8	9	10	11	12	13	14	15	> 15		
Brassica juncea	A	0	0	0	0	13	5	1	3	2	1	0	25	37.31
	C	0	0	0	0	0	0	0	1	0	0	0	1	1.49
	T	0	0	0	0	9	13	5	5	2	0	2	36	53.73
	AT	2	1	1	0	0	0	0	0	0	0	0	4	5.97
	TA	0	1	0	0	0	0	0	0	0	0	0	1	1.49
		Total number of repeats											67	
Brassica napus	A	0	0	0	0	0	9	8	0	3	1	3	24	34.29
	C	0	0	0	0	0	2	0	0	0	0	0	2	2.86
	T	0	0	0	0	0	15	10	5	4	1	3	38	54.29
	AT	0	2	1	2	0	0	0	0	0	0	0	5	7.14
	TA	0	1	0	0	0	0	0	0	0	0	0	1	1.43
		Total number of repeats											70	
Brassica nigra	A	0	0	0	0	0	8	3	0	3	2	2	18	26.87
	C	0	0	0	0	0	1	0	0	0	0	0	1	1.49
	T	0	0	0	0	0	14	7	6	9	0	3	39	58.21
	AT	0	5	1	0	0	0	0	0	0	0	0	6	8.96
	TA	0	0	0	0	1	0	0	0	0	0	0	1	1.49
	TC	0	1	0	0	0	0	0	0	0	0	0	1	1.49
	ATA	1	0	0	0	0	0	0	0	0	0	0	1	1.49
		Total number of repeats											67	
Brassica oleracea	A	0	0	0	0	8	7	1	2	2	1	0	21	29.17
	C	0	0	0	0	0	1	0	0	0	0	0	1	1.39
	T	0	0	0	0	18	11	4	3	4	4	0	44	61.11
	AT	2	3	0	0	0	0	0	0	0	0	0	5	6.94
	TA	0	0	0	1	0	0	0	0	0	0	0	1	1.39
		Total number of repeats											72	
Brassica rapa 18R-1	A	0	0	0	0	14	5	2	1	3	1	0	26	38.24
	C	0	0	0	0	0	1	0	0	0	0	0	1	1.47
	T	0	0	0	0	10	12	6	4	2	2	0	36	52.94
	AT	2	1	1	0	0	0	0	0	0	0	0	4	5.88
	TA	0	1	0	0	0	0	0	0	0	0	0	1	1.47
		Total number of repeats											68	
Brassica rapa c13	A	0	0	0	0	14	5	2	1	3	1	0	26	36.62
	C	0	0	0	0	1	1	0	0	0	0	0	2	2.82
	T	0	0	0	0	11	13	5	4	3	0	2	38	53.52
	AT	2	1	1	0	0	0	0	0	0	0	0	4	5.63
	TA	0	1	0	0	0	0	0	0	0	0	0	1	1.41
		Total number of repeats											71	

(continued)

Table 16.3 (continued)

Species	Repeat/motif length	Number of repeats											Total	%
		6	7	8	9	10	11	12	13	14	15	> 15		
Brassica rapa subsp. Rapa	A	0	0	0	0	14	5	1	1	3	1	1	26	37.14
	C	0	0	0	0	0	1	0	0	0	0	0	1	1.43
	T	0	0	0	0	11	13	5	4	3	0	2	38	54.29
	AT	2	1	1	0	0	0	0	0	0	0	0	4	5.71
	TA	0	1	0	0	0	0	0	0	0	0	0	1	1.43
		Total number of repeats											70	
Brassica carinata	A	0	0	0	0	14	5	1	1	3	1	1	26	37.14
	C	0	0	0	0	0	1	0	0	0	0	0	1	1.43
	T	0	0	0	0	11	13	5	4	3	0	2	38	54.29
	AT	2	1	1	0	0	0	0	0	0	0	0	4	5.71
	TA	0	1	0	0	0	0	0	0	0	0	0	1	1.43
		Total number of repeats											70	

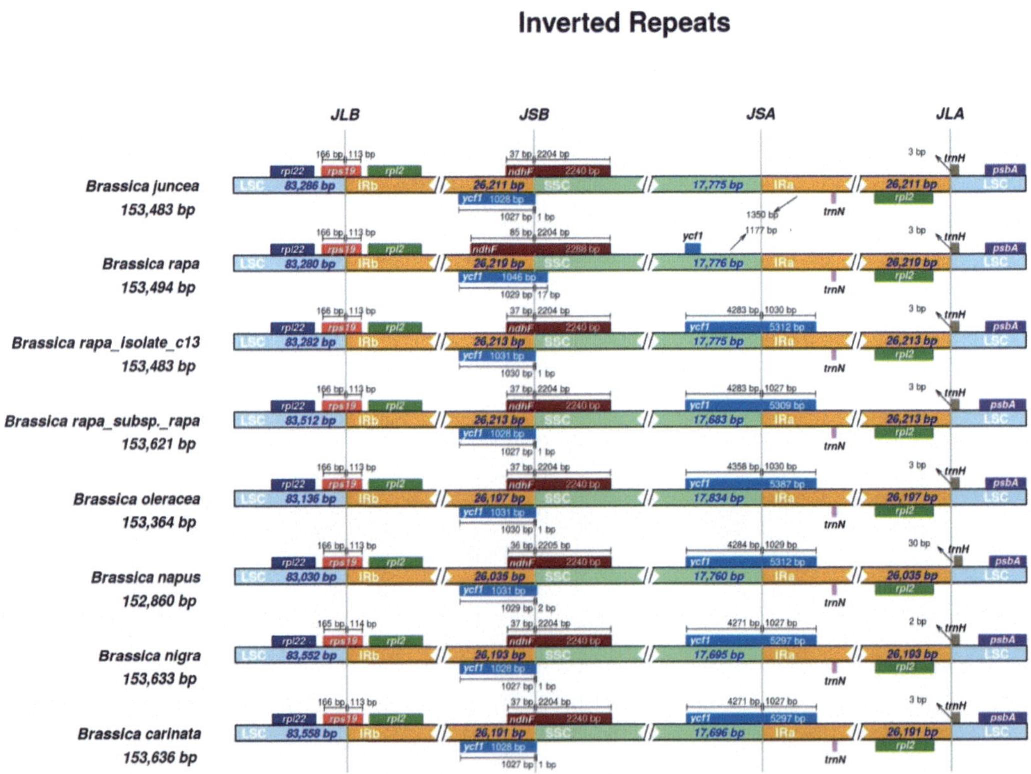

Fig. 16.3 Inverted repeat junction sites. The genes transcribed on the sense and antisense strand are depicted on the top and bottom of each track respectively for the respective species. The junction sites between each of the two corresponding sites are denoted as JLB (LSC/IRb), JSB (IRb/SSC), JSA (SSC/IRa), and JLA (IRa/LSC). The scale bar above and below the genes represent the number of bases present in each corresponding region at the junctions and the arrows point to the distance towards the start or stop site of the nearest gene to the junction site

pointing to having two different cp lineages. All of *Brassica juncea* L cp accessions grouped with the clade I, whereas the *Brassica napus* L. accessions grouped with either of the lineages of its progenitor *Brassica rapa* L. *Brassica oleracea* L. occupied the clade II and *Brassica nigra* L. and *Brassica carinata* L. shared the clade IV, since there was only single accession for each of them (Li et al. 2017). In an attempt to replicate the phylogenetic study, we mined the 47 cp genome sequences available in NCBI for *B. rapa* L., *B. oleracea* L., *B. napus* L., *B. juncea* L., *B. carinata* L. and *B. nigra* L. each having 11, 11, 10, 6, 5 and 4 accessions respectively. *Arabidopsis thaliana* L. cp genome was used as an outgroup and the sequence alignment was performed using the MUSCLE module of MEGAX tool and the overhang sequences in the alignment were trimmed. The best fit model for generating a maximum likelihood tree was suggested as GTR + G + I by the MEGAX tool (Kumar et al. 2018). The phylogenetic tree was generated using MrBayes v3.2.7 tool with the alignment file from MEGAX tool as input (Ronquist et al. 2012). Tree generated by MrBayes was visualized in FigTree v1.4.4 (Rambaut 2009).

The phylogenetic tree in Fig. 16.4 concurs with the findings of Sharma et al. (2014), and show the divergence of Brassica into the nigra lineage containing *B. nigra* L. and *B. carinata*

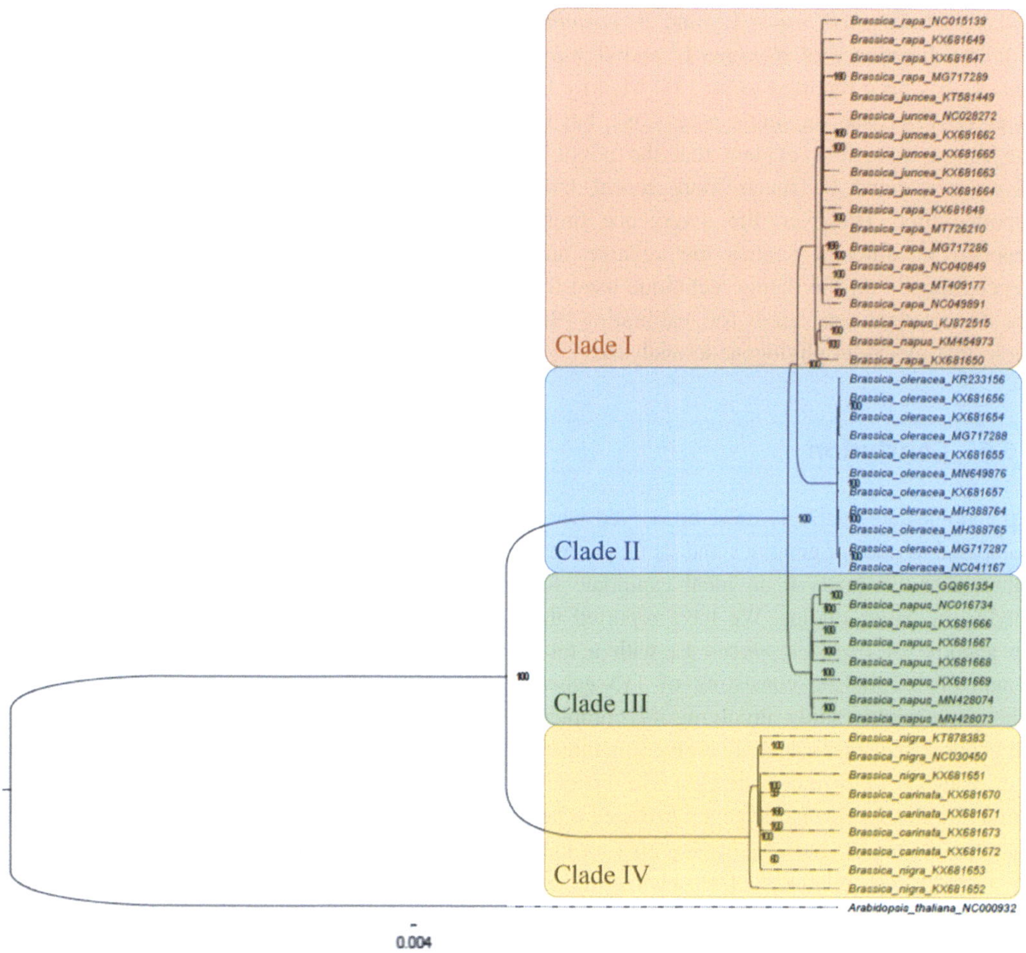

Fig. 16.4 Phylogenetic tree constructed using MrBayes using the GTR + G + I nucleotide model for the alignment of complete chloroplast genomes of *Brassica* genus. The probability percentage is displayed at the nodes of the tree

L. and Rapa/Oleracea lineage containing *B. rapa* L., *B. juncea* L., *B. oleracea* L. and *B. napus* L. The tree also had four clades similar to that of Li et al. (2017). Clade I was represented by *B. rapa* L., *B. juncea* L., and two accessions of *B. napus* L. The clustering of all *B. juncea* L. accessions with *B. rapa* L. proves its monophyletic origin. Clade II was completely represented by *B. oleracea* L. Clade III in our case represented completely *B. napus* L. and none of the *B. rapa* L. accessions, one of the reasons could be the absence of the broccoletto accession of *B. rapa* L. used by Li et al., in NCBI. However, the grouping of *B. napus* L. into clade III in itself hints at the presence of the second type of *B. rapa* L. ancestral species other than the one present in Clade I. Clade IV was represented by the nigra lineage (*B. nigra* L., and *B. carinata* L.). The divergence of *B. nigra* L. and *B. carinata* L. has been reported to be 1.07 Mya by Li et al. (2017) based on single accessions, but in the current tree, it is evident that the possible divergence could be much more recent than reported time. However, the divergence times reported in different studies are relative and depend mostly on the dating technique used for the fossils that are used for calibrating the molecular clocks in phylogenetic analysis.

16.6 Conclusion

Cp genomes are relatively more conserved when compared to nuclear genomes and its uniparental inheritance has made it an ideal candidate for phylogeny-based studies. We have reported the cp genome of *Brassica juncea* L. with a total length of 153,483 bp consisting of 113 genes. The phylogenetic study involving 47 complete cp genome sequences gave a resolved picture of the evolution of *Brassica* species. The cp genomes of other species in the genus *Brassica* other than the six major species can also give an idea of inter-species relations. The availability of multiple accessions of complete cp genomes also helps resolve the intra- and inter-species relationship of Brassicaceae family.

References

Amiryousefi A, Hyvönen J, Poczai P (2018) IRscope: an online program to visualize the junction sites of chloroplast genomes. Bioinformatics 34(17):3030–3031

Asaf S, Khan AL, Khan MA, Waqas M, Kang SM et al (2017) Chloroplast genomes of *Arabidopsis halleri* ssp. *gemmifera* and *Arabidopsis lyrata* ssp. *petraea*: structures and comparative analysis. Plant Breed Biotechnol 5(3):221–227

Bendich AJ (2004) Circular chloroplast chromosomes: the grand illusion. Plant Cell 16(7):1661–1666

Birky CW (1995) Uniparental inheritance of mitochondrial and chloroplast genes: mechanisms and evolution. Proc Natl Acad Sci USA 92(25):11331–11338

Corriveau JL, Coleman AW (1988) Rapid screening method to detect potential biparental inheritance of plastid DNA and results for over 200 angiosperm species. Amer J Bot 75(10):1443–1458

de Vries J, Archibald JM (2018) Plastid genomes. Curr Biol 28(8):R336–R337

Flannery ML, Mitchell FJ, Coyne S, Kavanagh TA, Burke JI et al (2006) Plastid genome characterisation in Brassica and Brassicaceae using a new set of nine SSRs. Theor Appl Genet 113(7):1221–1231

Gibson TM, Shih PM, Cumming VM, Fischer WW, Crockford PW et al (2018) Precise age of *Bangiomorpha pubescens* dates the origin of eukaryotic photosynthesis. Geology 46(2):135–138

Gray MW (1992) The endosymbiont hypothesis revisited. Int Rev Cytol 141:233–357

Kumar S, Stecher G, Li M, Knyaz C, Tamura K (2018) MEGA X: molecular evolutionary genetics analysis across computing platforms. Mol Biol Evol 35 (6):1547–1549

Li-Beisson Y, Neunzig J, Lee Y, Philippar K (2017) Plant membrane-protein mediated intracellular traffic of fatty acids and acyl lipids. Curr Opin Plant Biol 40:138–146

Li P, Zhang S, Li F, Zhang S, Zhang H et al (2017) A phylogenetic analysis of chloroplast genomes elucidates the relationships of the six economically important *Brassica* species comprising the triangle of U. Front Plant Sci 8:111

Mochizuki A, Sueoka N (1955) Genetic studies on the number of plastid in stomata. Cytologia 20(4):358–366

Nagaharu U (1935) Genome analysis in Brassica with special reference to the experimental formation of *B. napus* and peculiar mode of fertilization. Jpn J Bot 7:389–452

Padmanabhan MS, Dinesh-Kumar SP (2010) All hands on deck—the role of chloroplasts, endoplasmic reticulum, and the nucleus in driving plant innate immunity. Mol Plant-Microbe Interact 23(11):1368–1380

Prabhudas SK, Raju B, Kannan Thodi S, Parani M, Natarajan P (2016a) The complete chloroplast genome sequence of Indian mustard (*Brassica juncea* L.). Mitochon DNA Part A 27(6):4622–4623

Prabhudas SK, Prayaga S, Madasamy P, Natarajan P (2016b) Shallow whole genome sequencing for the assembly of complete chloroplast genome sequence of *Arachis hypogaea* L. Front Plant Sci 7:1106

Rambaut A (2009) FigTree v1.3.1. Institute of Evolutionary Biology, University of Edinburgh, Edinburgh. http://tree.bio.ed.ac.uk/software/figtree/

Reyes-Prieto A, Moustafa A (2012) Plastid-localized amino acid biosynthetic pathways of Plantae are predominantly composed of non-cyanobacterial enzymes. Sci Rep 2(1):1–12

Ronquist F, Teslenko M, Van Der Mark P, Ayres DL, Darling A et al (2012) MrBayes 3.2: efficient Bayesian phylogenetic inference and model choice across a large model space. Syst Biol 61(3):539–542

Sharma S, Padmaja KL, Gupta V, Paritosh K, Pradhan AK et al (2014) Two plastid DNA lineages—Rapa/Oleracea and Nigra—within the tribe Brassiceae can be best explained by reciprocal crosses at hexaploidy: evidence from divergence times of the plastid genomes and R-block genes of the A and B genomes of *Brassica juncea*. PLoS One 9(4):e93260

Yin D, Wang Y, Zhang X, Ma X, He X et al (2017) Development of chloroplast genome resources for peanut (*Arachis hypogea* L.) and other species of Arachis. Sci Rep 7(1):1–11

Transcriptomics Research and Resources in *Brassica* spp.

17

A. N. Rai, M. Rao, N. C. Gupta, Anamika Kashyap, Sapna Thakur, P. Bharadwaj, Sampath Perumal, and P. Suprasanna

Abstract

Brassica species present maximum diversity and play an important role in agri and horticulture sectors. Brassica transcriptome landscape has facilitated the identification of agronomically important genes relevant for biotic and abiotic stress tolerance and other traits. Genomes of five important Brassica family members including *B. rapa*, *B. oleracea*, *B. nigra*, *B. napus*, and *B. juncea* have been assembled to provide valuable genomics information on agronomic traits for use in molecular breeding. The whole genome transcriptomic (RNA Seq) analysis tools have become significant for further investigation and analysis of crop diversity and loci governing important traits. Over the years, RNA-seq in Brassicas has expanded rapidly providing gainful insights into differential gene expression, genome structure, diversity, evolutionary analysis, and marker development. The sequencing tools for Brassica crops and the resultant genomic databases are definitely making strides in unraveling genomics detailing of glucosinolates, anthocyanins, disease resistance, flowering, and hormones. In this article, we present an overview on the transcriptomics research in Brassica species and discuss the advances in genomics tools such as RNAinterference and genome editing. Blend of genomics and breeding efforts should foster the development of climate smart Brassicas to achieve sustainability in the times of changing environment.

A. N. Rai · P. Suprasanna (✉)
Ex-Head, NABTD, Nuclear Agriculture & Biotechnology Division, Bhabha Atomic Research Centre, Mumbai 400085, India
e-mail: archurai@barc.gov.in

M. Rao · N. C. Gupta · A. Kashyap
ICAR National Institute of Plant Biotechnology, New Delhi 110012, India
e-mail: mahesh.rao@icar.gov.in

S. Thakur · P. Bharadwaj
Molecular Genetics Laboratory, Department of Botany, Central University of Punjab, VPO Ghudda, Bathinda, Punjab 151401, India
e-mail: pankajihbt@gmail.com

S. Perumal
Global Institute for Food Security (GIFS), University of Saskatchewan, Saskatoon, SK S7N 0W9, Canada

17.1 Introduction to Crop and Evolution

Brassica crops provide the maximum diversity of products from a single genus *Brassicaceae* includes 372 plant genera and almost 4060 are accepted species names (*Brassicaceae*—The Plant List) and 3660 species are classified within the 321 genera (Kiefer et al. 2014). Brassica species play an essential role in agriculture and horticulture (Rakow 2004; El-Esawi 2016). Annual coverage of cultivation of Brassica

C. Kole and T. Mohapatra (eds.), *The Brassica juncea Genome*, Compendium of Plant Genomes,
https://doi.org/10.1007/978-3-030-91507-0_17

oilseed crops is ~ 34 million hectares of the world's agricultural land (FAO 2013). India stands third in rapeseed-mustard production with a total of 12–15% of cultivated oilseeds' area (Venkattakumar and Padmaiah 2010). Members of Brassica are mostly adaptive to lower temperatures and hence are well adapted to cultivation at high elevations and as winter crops in the subtropical areas. In temperate zones, oilseed rape (*Brassica napus*) and turnip rape (*Brassica rapa*) are predominately cultivated, while Indian mustard or rai (*Brassica juncea*) is cultivated as major oil source in the subtropics of Asia. The three allotetraploids (*B. juncea*, *B. carinata*, and *B. napus*) account for 12% of edible oil production of the world (http://www.fao.org/faostat). Besides this, Brassicas serve as leaf, flower, and root vegetables that are eaten fresh, cooked and processed and also being used as fodder and forage. There is a wide variation among the Brassicas for morphological and adaptive traits which has been useful for breeding for improved cultivars (Jambhulkar 2015; Rai et al. 2021).

Wild diploid Brassica and their related hybrid amphidiploids have evolved naturally and are confirmed by extensive experimental crosses between diploid and/or tetraploid followed by karyotyping and microscopic observations at the synapsis stage of meiosis in these crosses (Cheng et al. 2014). Based upon the studies, the genetic relationships of these species were identified by the Korean botanist Nagaharu (1935) that three basic diploid Brassica forms were probably the parents of subsequent amphidiploid crops. *Brassica nigra* (black mustard), the ancestor of culinary mustards, is found as annual herb growing in the rocky Mediterranean coasts. Natural populations of *B. oleracea* and associated types have been identified as potential progenitors of many European cole vegetables which are capable of conserving water and nutrients. The putative ancestor of *B. rapa* may have originated from the high plateau regions in today's Iran–Iraq–Turkey which had the ability to grow rapidly in the hot, dry conditions, forming copious seed (Dixon 2007; El-Esawi 2016). *Brassica carinata* ($n = 17$) hybrid might have originated from the hybridization of *B. oleracea* ($n = 9$) with wild or semi-domesticated forms of *B. nigra* ($n = 8$). Another amphidiploid, *Brassica juncea* ($n = 18$) is a hybridization product of *B. rapa* ($n = 10$) and *B. nigra* ($n = 8$) (Frandsen 1943). The third amphidiploid, *B. napus* ($n = 19$) developed from a cross between *B. rapa* ($n = 10$) and *B. oleracea* ($n = 9$). Besides these, an additional gene pool involves genera and species related to Brassica crops in 36 cytodemes such as Diplotaxis, Enarthrocarpus, Eruca, Erucastrum, Hirschfeldia, Rhynchosinapis, Sinapis, Sinapodendron, and Trachystoma genera (Harbered 1976; Branca and Cartea 2011). The nuclear DNA content among the different species in Brassicaceae has a very narrow range (0.16 pg < 1C < 1.95 pg) much lower than Poaceae, and Fabaceae suggesting a dynamic, genome size divergence during evolution in the Brassica members. Genetic relationship of the Brassica species and genome size are presented in Fig. 17.1. Despite such conservative DNA content, a great deal of structural evolution of genomes has taken place during the evolution (Lagercrantz and Lydiate 1996; Lan et al. 2000). According to Song et al. (1995) genome instability was the basis for all the genomic changes observed in allopolyploids.

The evolution of *Brassica* and allied genera from a common ancestor with $n = 6$ was explained through the phylogenetic studies suggesting an increase in the number of chromosomes and partial homology of A, B, and C genomes (Branca and Cartea 2011). Whole genome sequencing and comparative genomic analysis based on the genome sequences of *B. rapa* and *A. thaliana* further suggested the whole genome triplication (WGT) phenomenon in the speciation and morphotype diversification of *Brassica* spp. After WGT, extensive genome fractionation, block reshuffling and chromosome reduction produced the stable diploid species (Cheng et al. 2017a, b). Further rearrangement of these species and their hybridization has led to Brassica speciation (Cheng et al. 2014). Genome sequencing of *B. juncea* and *B. napus* revealed that A subgenomes of these species had independent origins. Homoeolog expression dominance has been observed between subgenomes of

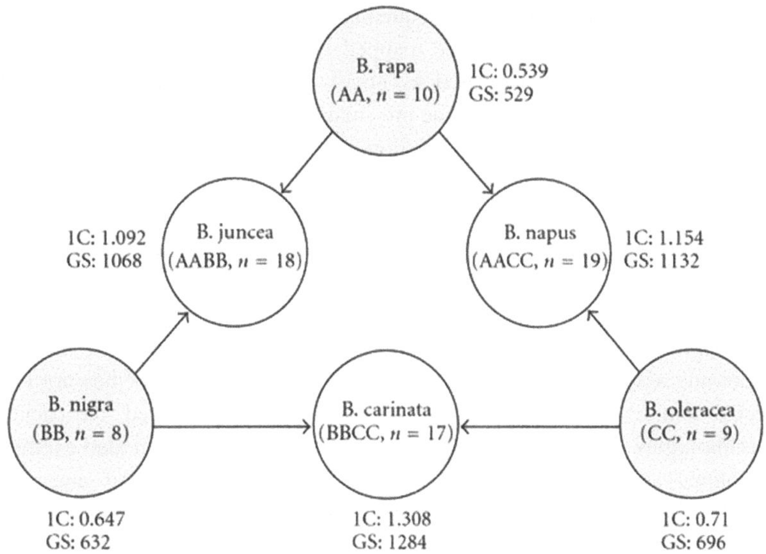

Fig. 17.1 Genetic relationship of the Brassica species [1C, 1C nuclear DNA content (pg); GS, genome size (Mbp)] (Johnston et al. 2005; Chang et al. 2008)

allopolyploid *B. juncea* and differentially expressed genes for glucosinolates and lipid metabolism showed more selection potential over neutral genes (Yang et al. 2016). In *B. napus*, transcriptomic shock was found to be dominated, and variation in the expression level dominance biasness was observed from tissue to tissue along with more transgressive upregulation, rather than down regulation (Li et al. 2020).

17.2 Transcriptome Studies

In the initial era of genomics, gene expression studies were initially restricted to few/specific genes using techniques like expressed sequence tags (ESTs) (Marra et al. 1998), Northern hybridization (Alwine et al. 1977), PCR analyses of specific genes (Becker-André and Hahlbrock 1989). This was followed by genome scale approaches to transcript characterization, namely serial analysis of gene expression (SAGE) (Velculescu et al. 1995) and DNA microarrays (Lockhart et al. 1996) which allowed a direct transcript quantification and discovery of new genes. With the advancement of sequencing techniques, i.e., next generation sequencing (Margulies et al. 2005), the whole genome transcriptomics (RNA Seq) has become a significant tool for transcriptome analysis of non-model organisms (Ellegren et al. 2012; Lamichhaney et al. 2012).

17.2.1 Transcriptome Sequencing

RNA-Seq combines the high-throughput sequencing methodology with computational methods to capture and quantify transcripts (Ozsolak and Milos 2011) in a tissue, organ, or organism (Martin et al. 2013; Conesa et al. 2016). This technique enables comparative quantification of total gene expression in different tissues, developmental stages, or environmental conditions and has been used to identify genes responsible for specific biological or regulatory functions. Moreover, a comprehensive "snapshot" of the total transcripts present in a sample can be developed to determine the presence or absence of specific transcripts and quantify transcript abundance. RNA-seq can also provide valuable information on unusual transcriptional events, such as alternate splicing, gene fusion, and novel transcripts (Mutz et al. 2013). There are three basic strategies for RNA-seq analysis: genome mapping, transcript mapping, and reference-free assembly (Conesa et al. 2016). In case of genome mapping, all the resultant RNA-seq reads are mapped against the organism's reference genome for transcript identification which can be subsequently quantified. Transcripts not able to be mapped to the reference genome are identified as novel transcripts and all the relevant genome information is used to predict novel transcript function enabling further genome annotation (Conesa et al. 2016; Yang et al. 2016). Finally, reference-free assembly uses an RNA-seq derived transcript profile to de novo assemble a complete transcriptome in the absence of a reference genome; this approach is also known as de novo transcriptome assembly (Grabherr et al. 2011). Several next generation sequencing (NGS) ctechnologies have been developed for transcriptome analysis, including Illumina, Solexa, SOLID, and Roche 454 (Conesa et al. 2016). Of these, Illumina has become the predominant transcriptome platform for NGS research, due its cost-effectiveness and high-throughput nature. In the "short-read sequencing," total transcript can be sequenced in short (< 500 bp) fragments, which are then bioinformatically assembled with or without a reference genome to obtain full-length transcripts and isoforms. These total transcripts may then be annotated using reference databases for functional characterization and comparative analyses (Garg and Jain 2013).

17.2.2 Long-Read-Based Transcriptome Sequencing

Recent improvements in long-read sequencing (LS) technologies, such as Oxford Nanopore Technologies (ONT) and PacBio (PB), have enabled the direct RNA and cDNA sequencing of full-length transcriptomes (Cui et al. 2020). With the ability to sequence polynucleotide molecules which are hundreds of thousands of nucleotides in length, long-read transcriptome sequencing has greatly improved the ability to obtain full-length transcript information (Wang et al. 2016a, b). Furthermore, LS-based transcriptomics provided support for alternate splicing analysis and complete isoform characterization, which paved the ways for existing genome annotations and gene models. Recently, LS-based maize transcriptome analysis helped to identify the most comprehensive mRNA profile to date, including identification of 57% novel transcripts and isoforms. In *B. napus*, single molecule long-read sequence analysis provided a highly accurate and comprehensive transcriptome, in which approximately 15,000 genes (18%) were identified as multi-exonic and showed complex alternative splicing (Yao et al. 2020). These data facilitate a critical new understanding of *B. napus* transcriptomics for functional genomics research. Such work has not only revealed the previously unexplored intricacies of *B. napus* transcriptomes, but also exemplifies the importance of LS in exploring and understanding transcriptome complexities (Wang et al. 2016a, b).

The PacBio single-molecule real time (SMRT) sequencing approach has been employed for transcriptome sequencing of many different plant species, including maize, rice, coffee bean, *Amborella trichopoda*, *Rhododendron lapponicum*, and *B. napus* (Cheng et al. 2017a, b; Yao et al. 2020). Using the SMRT approach, Sun et al. (2019) reported the genome assembly of cauliflower of 584.60-Mb size constituting 47,772, 56.65% repetitive sequences. The study also found larger genome size of cauliflower than A genome of *B. rapa*, the B genome of *B. nigra,* and the A or B subgenome of *B. napus* and *B. juncea*. Interestingly, cauliflower had the same number of genes as that in C genome *Brassica* species, and higher abundance of repetitive sequences and other noncoding sequences. In another study, SMRT sequencing was employed to generate transcriptome of Xinjiang green and purple turnips, (*Brassica rapa* var. *rapa*) at five developmental stages. The results have yielded a novel resource of alternative splicing, simple sequence repeats, long-noncoding RNAs for use in future genomics research of turnips (Zhuang et al. 2020). In contrast, transcriptomic study using Oxford Nanopore Technologies (ONT) has been severely limited, owing primarily to the low-throughput and high read-error rates associated with the platform. However, it is likely that the continued improvements in the long-read RNA-seq technologies will make these studies attractive and affordable in the near future (Cui et al.

2020). As both ONT and PB LS-based transcriptome analyses have been minimally explored in Brassica genomes, these platforms are expected to play an important role in developing a comprehensive transcriptome atlas of *Brassica* species.

17.2.3 Single-Cell Transcriptomics

Single-cell transcriptomics or single-cell RNA sequencing (scRNA-seq) has been used to study cell-to-cell gene expression variation within a cell population, which in turn helps to identify the developmental trajectory of individual cell types (Tang et al. 2011; Shulse et al. 2019). Drop-seq is a recently developed high-throughput scRNA-seq method which encapsulates and separates cells in emulsified droplets, enabling the user to transcriptionally profile hundreds of thousands of cells in a single experiment (Macosko et al. 2015). Recently, Drop-seq profiling of > 12,000 *Arabidopsis* root cells revealed distinct cell types involved in different root stages and developmental activities (Shulse et al. 2019). In this study, the authors demonstrated the rapid identification of rare and novel cell types from plant tissue and simultaneous characterization of multiple and different cell types. This analysis also demonstrated the ability to determine the cell-specific transcriptional response of environmental stimuli such as exogenous sucrose treatment. Such approaches will greatly enhance our understanding of the functional role of tissues, cells, and genes in plant developmental processes and environmental responses. The full potential of this recently evolving technology in plant research is just now being realized and scRNA-seq is expected to be used extensively in future for many plant species, including *Brassica* (Shaw et al. 2021).

17.2.4 Considerations Regarding RNA Seq

RNA-seq is an efficient technique, showing high resolution and cost advantages for profiling of gene expression between samples or differential expression (DE). However, there are several sources of sequencing bias and systematic noise because of wrong base calls, sequence quality biases (Dohm et al. 2008; Hansen et al. 2010), variability in sequence depth (Sendler et al. 2011) and differences in the composition and coverage of raw sequence data generated from technical and biological replicate samples (Lü et al. 2009).

Thus, the guidelines and standards have been defined by ENCODE to emphasize upon the best practices designed to get quality transcriptome measurements. RNA seq experiments should be performed with two or more biological replicates. A typical R^2 (Pearson) correlation of gene expression (RPKM) between two biological replicates, for RNAs should be between 0.92 and 0.98 and the experiments with biological correlations below 0.9 should either be explained or repeated. Experiments related to global view of gene expression typically require 30–60 million reads per sample, whereas 100–200 million reads required to get an in-depth view of the transcriptome or new transcripts assembly. For RNA-seq, sequencing platforms giving reads of $\geq$ 75 bp length is optimal to minimize the sequencing cost. Other recommendations needs to be taken care as suggested by ENCODE to design the transcriptome experiments (ENCODE 2011, 2016) for significant finding.

17.3 Transcriptome Research in the Brassica Genome

The use of RNA-seq in Brassica research has expanded rapidly in the areas of de novo transcriptome assembly and analyses, differential expression triggered by various biotic and abiotic stresses, noncoding RNA analyses, investigations of genome structure, diversity and genome origin, evolutionary analysis, and marker development (Bancroft et al. 2011; Izzah et al. 2014; Kim et al. 2014; Parkin et al. 2014; Wang et al. 2015). So far, complete sequencing has been reported in five important Brassica family members which include diploids [*B. rapa* Wang et al. 2011a, b, c; *B. oleracea*, Liu et al. 2014;

B. nigra, Perumal et al. 2020] and allotetraploids [*B. napus*, Chalhoub et al. 2014; *B. juncea*, Yang et al. 2016]. In all cases, de novo transcriptome assembly played an important role in decoding the final whole-genome transcripts. Critically, the *Brassica* transcriptome landscape has facilitated the identification of agronomically important genes, such as those relevant to biotic and abiotic stress tolerance (Mohd Saad et al. 2021). For example, transcriptome analysis in *B. napus* was used to elucidate the genes involved lipid and glucosinolate biosynthesis (Chalhoub et al. 2014) which could greatly accelerate Brassica breeding programs. Besides there are several other agronomic traits which are targeted for Brassica improvement, and genomic approaches are now sought to aid the breeding efforts (Fig. 17.2).

RNA-seq-based genome analysis also provided a valuable foundation for the understanding the phenomena of biased gene fractionation and genome dominance of the mesohexaploid *B. oleracea* genome, whereby one subgenome exhibits transcriptional dominance over the two other subgenomes (Parkin et al. 2014). In addition, a transcriptomic approach employed to dissect the complexity of the origin and diversification of the *B. napus* genome found that over 8000 differentially expressed genes are associated with diversification in this species (An et al. 2019). Furthermore, RNA-seq has been used in *Brassica* species to identify the roles of non-coding RNAs (ncRNAs), particularly micro-RNAs and long ncRNAs, in important biological process such as abiotic stress (Ahmed et al. 2020). Harper et al. (2020) provided confirmatory results on the Associative Transcriptomics platform in *Brassica juncea*. Using a diverse panel of *B. juncea* accessions, transcriptome data was mapped to pan-transcriptome. The authors identified several single nucleotide polymorphism variants and measured the quantity of thousands of transcripts. The study identified potential candidate gene *BjA.TTL* for seed weight trait and other markers for seed color and vitamin E content.

17.3.1 Transcriptomic Studies Related to Biotic Stresses

Plant disease and pests cause significant yield loss in *Brassica* spp. Major 16 disease and 37 insect pests have been reported in mustard or oilseed rape growing regions (Zheng et al. 2020a, b). The development of host resistance is one of the most desirable and cost-effective method for disease control. Plant-pathogen interaction is a

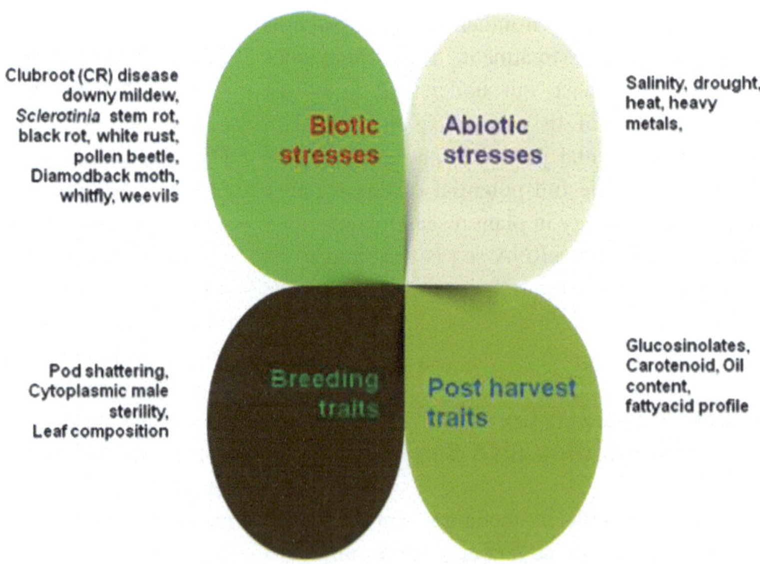

Fig. 17.2 General breeding considerations for the improvement of Brassica family members

broad process and starts with the detection of microbial elicitors, pathogen-associated molecular patterns (PAMPs) by the membrane-localized receptor proteins with PRRs motif of plants (Dodds and Rathjen 2010; Zipfel 2014). Plant immunity is mainly effector-triggered immunity (ETI) constituting the hypersensitive response (HR), however mostly, the effective resistance against pathogen is imposed through PAMP-triggered immunity (PTI) (Neik et al. 2017). Plants also develop broad-spectrum immunity through various hormonal signaling pathways (Kazan and Lyons 2014).

The differentially expressed genes, QTLs, and the corresponding pathways play important role in host–pathogen interaction and other biotic stresses have become more apparent with the transcriptome profiling in several *Brassica* species. The RNA-Seq analysis has strengthened the basic understanding of the defense mechanism and the factors imparting tolerance toward the diseases like clubroot disease caused by *Plasmodiophora brassicae* in *B. rapa* (Chu et al. 2014; Fu et al. 2019), *B. napus* (Hejna et al. 2019) and *B. juncea* (Luo et al. 2018). Similarly, the RNA-Seq studies have also unraveled the defense mechanism for the disease like Fusarium wilt (*F. oxysporum*), Sclerotinia stem rot (*S. sclerotiorum*), Blackleg (*Leptosphaeria maculans*), Downy mildew (*Hyaloperonospora brassicae*), etc. (Table 17.1A). Most of the studies suggested upregulation of genes related to salicylic acid (SA), jasmonic acid (JA)/ethylene (ET) and brassinosteroid (BR) signaling pathways induced after the pathogen infection. The other components and the pathways providing a shield of host defense against the invading pathogens include secondary metabolites, phenolics, signal transduction, phytohormones, Studies have thrown light on the enrichment of genes in metabolic processes, plant-pathogen interactions, plant hormone signal transduction, glucosinolate biosynthesis, cell wall thickening, chitin metabolism and pathogenesis-related (PR) genes and pathways (Jia et al. 2017). Transcriptomic studies have also revealed insights on the host-defense mechanism (s) for insect-pest attack (Table 17.1B) which includes pathways of cell wall synthesis, secondary metabolite production, redox homeostasis, phytohormones signaling, glucosinolate biosynthesis and degradation (Gruber et al. 2018).

17.3.2 Transcriptomic Studies Related to Abiotic Stress

Abiotic stresses have become one of the major threats which restrict crop production and productivity. These influence plant growth at all the phenological stages and induce yield losses depending on stress intensity and durability. Comprehensive studies regarding abiotic stress impact and indices used to assess the impact of these stress have been compiled by Rai et al (2021). Abiotic stress tolerance is a quantitative trait and involves cross talk between various signaling, metabolic, and defense pathways (Fig. 17.3).

Transcriptomic studies have been performed to understand the plant stress responses to different abiotic stresses and the tolerance mechanisms. Genome-wide gene expression analysis under drought, salinity, heat, cold, Cadmium metal stress and combined stresses have been performed using RNA seq. These studies have led to generation of enormous datasets which are now being utilized to understand the abiotic stress responses. For example, the major upregulated transcripts identified belong to classes like transcription factors, kinases, heat shock factors (HSFs), calcium signaling pathways, ROS detoxification. Yue et al. (2021) identified candidate heat stress tolerance genes by comparative transcriptomics study on contrasting *B. rapa* accessions subjected to long-term heat stress treatment. There were notable alterations in functional gene expression, especially of processes related to ER protein processing, hormones and signal transduction pathways. Transcriptomic studies related to abiotic stresses in various *Brassica* species are summarized in Table 17.2.

Table 17.1 Brassica transcriptomics related to biotic stresses

S. No.	Trait	Brassica species	Outcome	References
A	*Disease resistance*			
1	Clubroot disease (*Plasmodiophora brassicae*)	*Brassica rapa*	Upregulation of the genes related to salicylic acid (SA), jasmonic acid (JA)/ethylene (ET), and brassinosteroid (BR) signaling pathways and are important at the late stage of the infection	Fu et al. 2019
2	Clubroot disease (*P. brassicae*)	*B. rapa* ssp. chinensis	Upregulation of the genes of defense-response, biological processes (jasmonate and ethylene signaling and metabolism), defensive deposition of callose and biosynthesis of indole-containing compounds	Chu et al. 2014
3	Clubroot disease (*P. brassicae*)	*Brassica rapa* ssp. pekinesis	Upregulation of DEGs enriched in metabolic process, biological regulation, response to stimulus, plant–pathogen interaction, plant hormone signal transduction, genes related to disease-resistance, calcium ion influx, glucosinolate biosynthesis, cell wall thickening, salicylic acid (SA) homeostasis, chitin metabolism, pathogenesis-related (PR) pathway, etc. in the resistant line and, upregulation of the indole acetic acid (IAA) and cytokinin-related genes in the susceptible line	Jia et al. 2017
4	Clubroot disease (*P. brassicae*)	*B. napus*	Upregulation of the genes related to indole-3-acetic acid (IAA) signal transduction, cytokinin synthesis, and myrosinase synthesis	Chen et al. 2015
5	Clubroot disease (*P. brassicae*)	*B. napus*	• Identification of genes for defense pathway, phytohormone pathway • Transcription factors with plant defense domains ERF, bZIP, WRKY, MYB, plant defense *cis*-regulatory ET/JA motifs, G-box, GCC-box, W-box and pathogen-related proteins	Hejna et al. 2019
6	Clubroot disease (*P. brassicae*)	*B. napus*	• Receptor-like protein (*RLP*) genes, resistance (*R*) genes, and genes involved in salicylic acid (SA) signaling, Ca^{2+} signaling, leucine-rich repeat (LRR) receptor kinases (*RLKs*) genes, the respiratory burst oxidase homolog (RBOH) proteins, and transcription factors [*WRKYs*, ethylene-responsive factors, and basic leucine zippers (*bZIPs*)]	Zhou et al. 2020
7	Clubroot disease (*P. brassicae*)	*B. nigra*	• Identification of *Rcr6* for clubroot resistance, located on B7 chromosome	Chang et al. 2019
8	Clubroot disease (*P. brassicae*)	*B. juncea* cv. *tumida* Tsen	• Accumulation of defense proteins/genes–pathogenesis-related proteins, pathogen-associated molecular pattern-triggered immunity, and effector-triggered immunity signaling pathways, signaling and ROS and cell wall modification	Luo et al. 2018

(continued)

Table 17.1 (continued)

S. No.	Trait	Brassica species	Outcome	References
9	Fusarium wilt (*F. oxysporum*)	*B. oleracea*	• Activation of the early defense systems, MAPK signaling pathway, calcium signaling and salicylic acid-mediated hypersensitive response (SA-mediated HR), Ethylene (ET)- and jasmonic (JA)-mediated pathways and the lignin biosynthesis pathway • Expression of the defense-related genes encoding pathogenesis-related (PR) proteins, UDP-glycosyltransferase (UDPG), pleiotropic drug resistance, ATP-binding cassette transporters (PDR-ABC transporters), myrosinase, transcription factors and kinases	Xing et al. 2016
10	Fusarium wilt (*F. oxysporum*)	*B. oleracea*	• Identification of differentially expressed NBS-LRR and *WRKY* transcription factors genes, one potential effector, two elicitors and six virulence factors	Liu et al 2020
11	Fusarium wilt (*F.oxysporum*)	*B. rapa* var. pekinensis	• Identification of candidate genes *Bra012688* and *Bra012689* for fusarium yellows resistance	Shimizu et al. 2014
12	Fusarium wilt (*F. oxysporum*)	*B. rapa*	• The antagonistic transcriptional response between SA and JA/ET has observed in Fusarium yellows resistant lines	Miyaji et al. 2021
13	Sclerotinia stem rot (*Sclerotinia sclerotiorum*)	*B. napus*	• Identification of genes for detoxification, secondary metabolites, effectors, signaling, development, oxalic acid and ROS production	Seifbarghi et al. 2017
14	Sclerotinia stem rot (*S. sclerotiorum*)	*B. napus*	• The peroxisome related pathways along with cell wall degradation and detoxification of host metabolites as the key mechanisms underlying pathogenesis of *S. sclerotiorum* on *B. napus*	Chittem et al. 2020
15	Sclerotinia stem rot (*S. sclerotiorum*)	*B. napus*	• Rapid induction of key pathogen responsive genes including glucanases, chitinases, peroxidases and WRKY Transcription factors • Induction of genes involved in plant-pathogen interactions; Identification of many novel disease responsive genes including TFs associated with jasmonate (JA) and ethylene (ET) signaling	Joshi et al. 2016
16	Sclerotinia stem rot (*S. sclerotiorum*)	*B. napus*	• Activation of the immune response, sulfur metabolism, especially glutathione (GSH) and glucosinolates in both R and S genotypes • R-genotype-specific genes related to the jasmonic acid pathway, lignin biosynthesis, defense response, signal transduction and encoding transcription factors • Identification of SNP-trait association including a tau class glutathione S-transferase (*GSTU*) gene cluster	Wei et al. 2015

(continued)

Table 17.1 (continued)

S. No.	Trait	Brassica species	Outcome	References
17	Sclerotinia stem rot (*S. sclerotiorum*)	*B. oleracea*	• Downregulation of the virulence genes of *S. sclerotiorum* including polygalacturonase, chitin synthase, secretory proteins, and oxalic acid biosynthesis in R line of *B. oleracea* after 12hpi • The R line of *B. oleracea* mediated suppression the pathogen establishment by a quick accumulation of ROS via activating Ca^{2+} signaling and repressing the oxalic acid generation in the pathogen	Ding et al. 2019
18	Sclerotinia stem rot (*S. sclerotiorum*)	*B. napus*	• Stage-specific DEGs: DNA binding, ATP binding, ion binding and oxidoreductase activity after 6 and 24 hpi. Most of the hydrolase activity after 48 hpi	Peng et al. 2017
19	Sclerotinia stem rot (*S. sclerotiorum*)	*B. napus*	• Identification of 17 QTLs and 36 putative candidate genes for SSR resistance upregulated in resistant lines	Qasim et al. 2020
20	Sclerotinia stem rot (*S. sclerotiorum*)	*B. napus*	• Identification of 7 QTLs and revealed activation of JA- and ethylene-mediated responses for SSR resistance	Bergmann et al. 2021
21	Blackleg (*Leptosphaeria maculans*)	*B. napus*	• Role of *LepR1–AvrLepR1* gene interaction in resistance reaction and their association with the spatial transcriptional gradients during ETD linked with pathogen detection, IGS production and hormone signaling	Becker et al. 2017
22	Blackleg (*Leptosphaeria maculans*)	*B. napus*	• Upregulation of the membrane targeting proteins, ribosome and suppression of programmed cell death, as a resistant reaction, whereas downregulation of the SA and JA pathways as a susceptible reaction • A threshold level of SA and JA signaling is required for the activation of *Rlm1*-mediated resistance	Zhai et al. 2021
23	Blackleg (*Leptosphaeria maculans*)	*B. napus*	• Identification of plant defense related genes (*LepR3 and Rlm2*) and proteins involved in host-plant and pathogen interactions	Zhou et al. 2019
24	Blackleg (*Leptosphaeria* species)	*B. napus*	• *L. biglobosa* "*canadensis*," induced more cell wall degrading genes • *L. maculans* "*brassicae*" induced genes in the Carbohydrate-Binding Module class (CAZy, CBM50) that evade the basal innate immunity of the host plant	Lowe et al. 2014
25	Downy mildew (*Hyaloperonospora brassicae*)	*B. rapa* L. ssp. *pekinensis*	• Identified and characterized the long noncoding RNAs involved in resistance to downy mildew • The long noncoding RNA, MSTRG.19915, a natural antisense transcript of a MAPK gene, *BrMAPK15*	Zhang et al. 2021

(continued)

Table 17.1 (continued)

S. No.	Trait	Brassica species	Outcome	References
26	Downy mildew (*Hyaloperonospora brassicae*)	*B. rapa* L. ssp. *pekinensis*	• Identification of 54 DEGs involved in plant-pathogen interaction, and 33 transcription factors	Zheng et al. 2020a, b
27	Leaf rust (*Alternaria brassicicola*)	*B. juncea*	• A mutually shared function of *NAC*s in abiotic and biotic stresses was revealed in *B. juncea* and *S. alba*	Mondal et al. 2020
28	Black rot (*Xanthomonas campestris*)	*B. oleracea*	• Upregulation of the genes of glucosinolate biosynthetic and catabolic pathways, ROS scavenging, hormonal, receptor-kinase-related and nucleotide-binding site (NBS)-encoding resistance genes during the early infection stage	Sun et al. 2020
29	Soft rot (*Erwinia carotovora*)	*B. rapa* ssp. *pekinensis*	• Strong activation of the downstream defense-related genes of PTI (*CPK, CML, RBOH MPK3*, and *MPK4*) during infection • Identification of immunity inducing role of endogenous hormones (auxins, JAs, and SA), exogenous auxins (MEJA and BTH), and increased expression of the genes for glucosinolate and lignin biosynthesis	Liu et al. 2019a, b
30	Turnip mosaic virus (TuMV)	*B. rapa* ssp. *pekinensis*	• Candidate genes of calcium signaling pathways, heat shock proteins, WRKY transcription factors, and non-specific lipid transfer proteins	Lyu et al. 2020
B	*Insects-pests*			
31	Silverleaf whitefly (SLWF)	Arabidopsis	• Repression of JA, and induction of SA-regulated genes; Accumulation of callose synthase gene RNAs and callose deposition in SLWF-infected tissues	Louisa et al. 2007
32	Flea beetle	*B. napus*	• Up-regulation of the genes involved in cell wall synthesis, secondary metabolite production, redox, stress and hormone-related responses, glucosinolate biosynthesis and degradation	Gruber et al. 2018
33	*Plutella xylostella*	*Arabidopsis thaliana*	• Overexpression of the genes associated with plant signal molecules or phytohormones, octadecanoid signaling	Ehlting et al. 2008

17.3.3 Transcriptomic Studies Related to Other Traits

Hybrid lethality is an important criterion especially in view of problems in gene exchange and stabilization of a breeding population. Xiao et al. (2021) observed that hybrid lethality in cabbage is the result of program cell death, and hence studied the transcriptome which showed the activation of defense pathways, hormonal and MAPK signaling pathway, related to Ca^{2+} and hydrogen peroxide. Transcriptomic studies to decipher the heterosis event in *B. oleracea* suggested the involvement of regulatory processes involving light and hydrogen peroxide-mediated signaling pathways (Li et al. 2018). In *B. napus*, biomass and yield traits, and harvest index traits related genes were identified using RNA seq

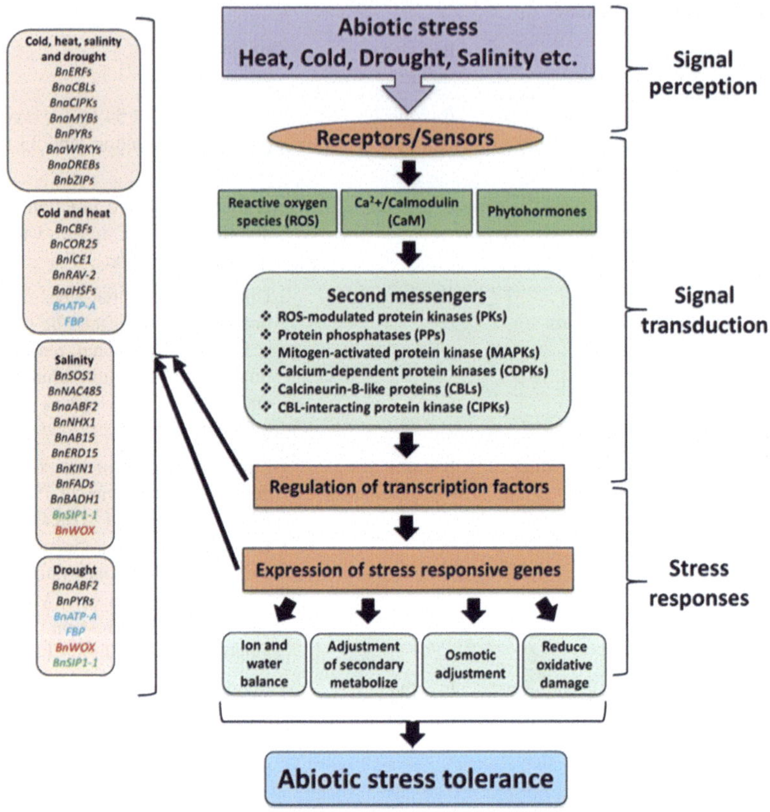

Fig. 17.3 Abiotic stress responsive pathways in plants, from signal perception to downstream stress responses. *Source* Ali Raza et al. (2021)

(Lu et al. 2017; Lu et al. 2017). Flowering time is an important agronomic trait. Natural variation in the expression levels of floral repressor *FLOWERING LOCUS C* (*FLC*) leads to differences in vernalization. In *Brassica napus,* nine copies of *FLC* have been found which control time of vernalization and the transcriptome study suggested the dynamic shift in the expression of multiple paralogs of *BnaFLC* (Calderwood et al. 2021). The RNA seq-based studies have also helped in deciphering the mechanism involved in bolting, flowering, leaf color, petal color and size, seed color, embryo development, and oil accumulation (Table 17.3). Dynamic gene expression changes of acyl-CoA-binding proteins, BnACBP2 and BnACBP6 were found to regulate the distribution of lipids in embryos and seed coats of *B. napus* suggesting their importance in fatty acid and triacylglycerol biosynthesis and oil accumulation (Pan et al. 2019). In *B. rapa*, which is important as a vegetable and oil crop, seed related traits like size, color, and oil content assume great relevance. Niu et al. (2020) studied transcriptomes of seed samples and developed transcriptional networks to identify key regulatory genes governing the above traits. This study has further highlighted regulatory networks through transcription factors like *TT8*, *WRI1*, *FUS3*, and *CYCB1;* genes underlying the trait variation in the seeds for use in biotechnological efforts to breed high yield and improved oil content in Brassica crops.

17.4 RNA-seq-Based Marker Development for Genotype Analysis

RNA-seq analyses have become an important resource for developing polymorphic genetic markers, such as expressed sequence tag (EST)-derived simple sequence repeat (SSR) markers

Table 17.2 Brassica transcriptomics related to abiotic stress

S. No.	Trait	Brassica species	Outcome	References
1	High temperature and drought stress	*B. juncea*	• 886 and 2834 transcripts, respectively coding for transcription factors, kinases, heat shock factors (HSFs) and dehydration responsive element-binding (DREB) families	Bhardwaj et al. 2015
2	Salinity tolerance	*B. juncea*	• Upregulation of the genes associated with ROS detoxification, sulfur assimilation and calcium signaling pathways	Sharma et al. 2015
3	Salt stress	*B. napus*	• Revelation of the genes involved in proline metabolism, inositol metabolism, carbohydrate metabolic processes and oxidation–reduction processes in the salt-stress response at the germination stage	Long et al. 2015
4	Salt stress	*B. napus*	• DEGs encoding transcription factors (582) and transporter genes (438)	Yong et al. 2014
5	Drought and salt stresses	*B. napus*	• Responsive transcription factors, *BnMYB44* and *BnVIP1*	Shamloo-Dashtpagerdi et al. 2018
6	Drought stress	*B. rapa* ssp. *pekinensis*	• Activation of transcription factor genes containing domain of AP2/ERFs, bHLHs, NACs and bZIPs. Differential acclimation responses in glucosinolate metabolism in leaves and roots	Eom et al. 2018
7	Dehydration stress	*B. rapa* L. ssp. *pekinensis*	• Identification of 37 transcription factors, 28 signal transduction, and 61 water- and osmosensing-responsive genes	Yu et al. 2012
8	Heat stress	*B. rapa* ssp. *chinensis*	• Upregulation of DEGs of transcription factors (TFs), kinases/phosphatases, related to photosynthesis and effectors of homeostasis. *NAC069* TF in all the heat treatment stages	Wang et al. 2016a, b
9	Low-temperature Stress	*B. napus*	• Upregulation of ABA and IP3/Ca^{2+} signal transduction; Protein serine/threonine kinases, myo-inositol-1-phosphate synthases and calmodulins	Xian et al. 2017
10	Cold and freezing stress	*B. napus*	• 47,328 DEGs, Snf1-related protein kinase 2 (SnRK2), ABA receptors (PYR/PYL/RCAR)	Xin et al. 2019
11	Cold stress	*B. rapa*	• DEGs of phenylpropanoid biosynthesis, phytohormone signal transduction, ribosome biogenesis, MAPK signaling pathway, basal transcription factors, and photosynthesis	Ma et al. 2019a, b
12	Cold stress	*B. juncea*	• Identification of core cold-inducible transcripts (283), expression patterns of gene families for transcription factors (TFs), transcription regulators (TRs) and kinases, and induction of cold stress-responsive protein kinases only during the early silique developmental stage	Sinha et al. 2015
13	Freezing stress	*B. napus*	• Identification of DEGs for carbohydrates and energy metabolism, signal transduction, amino acid metabolism and translation. Up-regulation of DEGs enriched in plant hormone signal transduction, starch and sucrose metabolism pathways	Pu et al. 2019

(continued)

Table 17.2 (continued)

S. No.	Trait	Brassica species	Outcome	References
14	Cadmium stress	*B. juncea*	• Altered gene expression related to plant hormones, calcium signaling, and MAP kinases altered Cd stress	Thakur et al. 2019
15	Chromium stress	*Brassica napus*	• Up-regulation of the several number of stress-responsive DEGs, related metabolic pathways like the tryptophan, vitaminB6 sulfur and nitrogen in cultivar ZS 758 and zeatin biosynthesis in cultivar Zheda 622. Cr also highlighted the numerous TFs and proteins	Gill et al. 2016

and single nucleotide polymorphisms (SNPs). Such markers enable high-throughput and cost-effective genotyping analysis (Paritosh et al. 2013; Izzah et al. 2014), and have various applications in plant breeding, including genetic diversity and population structure analysis, linkage mapping, mapping quantitative trait loci (QTLs) and association analysis, marker-assisted selection, and evolutionary analysis (Izzah et al. 2014; Ding et al. 2015; Chen et al. 2017). RNA-seq-based EST-SSR or SNP markers are developed using expressed transcripts or unigenes, and are therefore expected to have a higher correlation with functional traits than traditional genome-wide SSR and SNP markers (Chen et al. 2017). Furthermore, RNA-seq-based EST-SSR marker development requires minimal labor compared to the conventional approach of EST library-based SSR marker development (Tóth et al. 2000).

RNA-seq-based EST-SSR and SNP markers have been developed for many plant species, including *Brassica* spp. SNP markers developed from a complete transcriptome assembly of 40 *B. napus* lines helped to elucidate the impact of polyploidy on breeding and evolution of the *B. napus* genome (Bancroft et al. 2011). In this study, over 23,000 SNP markers were used to create multiple linkage maps without a reference genome, and elucidated the genome rearrangements and genomic inheritance of the allotetraploid *B. napus* genome (Bancroft et al. 2011).

Gene expression and transcriptome diversity are contributed by a central mechanism known as alternative splicing which is responsible for plant development, evolution, complexity, and adaptation (Mastrangelo et al. 2012; Ganie and Reddy 2021). Typical codominant markers InDel and SNP are highly polymorphic and are used in marker-assisted selection, genetic mapping, identification, and characterization of brassica germplasm. Three available transcriptome datasets of cabbage were collected to study alternative splicing events and markers like InDel, SNP, SSR markers. Novel mRNA transcripts among these three cabbage transcriptomes were identified via alignment of short reads to the cabbage genome dataset (Xu et al. 2019). InDel genetic markers were used for studying genetic diversity in 36 cabbage genotypes and the transcriptomic analysis showed 20.8% alternate splicing events in the total cabbage genome.

17.5 Genomic and Computational Databases for *Brassica* spp.

Genomic tools and resources are important in revolutionizing the field of Brassica improvement. With the advancement in sequencing technology, mass sequencing of genomes of various crops have become possible. The custom computational tools and databases play important role in proper utilization of the huge genomic data being produced. Some of the genomic databases for important oilseed crop Brassicas are being outlined in this section.

Table 17.3 Brassica transcriptomics for yield and other attributes

S. No.	Trait and technique	Brassica species	Outcome	References
Yield				
1	Heterosis	*B. oleracea* L var. italic	• Identification of 53 candidate genes for curd yield heterosis and regulatory processes involving light and hydrogen peroxide-mediated signaling pathways proposed to be functionally important in yield or biomass heterosis	Li et al. 2018
2	Yield and yield attributing traits	*B. napus*	• Identification of 14 candidate genes important for the developmental processes and biomass accumulation,	Lu et al. 2017
3	Flowering time regulation	*B. napus* L	• Identification of 36 genes associated with flowering time, seed yield, or both, and novel indications for neofunctionalization in homologs of known flowering time regulators like *VIN3* and *FUL*	Shah et al. 2018
4	Harvest index-related traits	*B. napus*	• Identification of 33 candidate genes functionally associated with photosynthesis, inflorescence, and silique development	Lu et al. 2017
5	Diversity	*B. napus*	• Genetic diversity analysis • Detection of 8,187 differentially expressed genes with implications for *B. napus* diversification	An et al. 2019
6	Purple leaf color	*B. juncea*	• Identification of 2,286 differentially expressed genes between the purple and green leaves • 213 differently expressed transcription factors, and role of *MYB* and *bHLH* transcription factors in anthocyanin biosynthesis • Up regulation of *BjTT8* and *BjMYC2* and anthocyanin biosynthetic genes (*BjC4H*, *BjDFR*, and *BjANS*) involved in the activation of the purple leaf formation in *B. juncea*	Heng et al. 2020
7	White petal color	*B. napus*	• Identification of lower levels of lutein and zeaxanthin responsible for white petal color • *BnNCED4b* involved in carotenoid degradation and abnormally high expression in WP petals • Identification of transcription factor *BNWRKY22* upstream of *BnNCED* promoting carotenoid degradation	Jia et al. 2021
8	Petal size	*B. rapa*	• Identification of 52 differentially expressed genes (DEGs) involved in control of petal size variation in rapeseed • Identification of *BnaA05.RAP2.2* in the negative control of petal size via ethylene signaling pathway	Qian et al. 2021
9	Flowering diversity	*B. napus*	• Variation of *FLC* expression during cold treatment between paralogues • Total *FLC* expression dynamics between cultivars rather than specific *FLC* paralogues expression	Calderwood et al. 2021

(continued)

Table 17.3 (continued)

S. No.	Trait and technique	Brassica species	Outcome	References
10	Early and late bolting	*B. rapa*	• Identification of six unigenes encoding the indole-3-acetic acid-induced protein ARG7 (*BraA02g009130*), auxin-responsive protein SAUR41 (*BraA09g058230*), serine/threonine-protein kinase BSK11 (*BraA07g032960*), auxin-induced protein 15A (*BraA10g019860*), and abscisic acid receptor PYR1 (*BraA08g012630* and *BraA01g009450*), putative candidates for the late bolting trait	Wei et al. 2021
11	Yellow seed coat color	*B. rapa*	• Identification of 19 unigenes associated with the phenylpropanoid, flavonoid, flavone and flavonol biosynthetic pathways as involved in seed coat color formation • Down regulation of *BrTT8* and *BrMYB5* in yellow seed	Ren et al. 2021
12	Embryo development genes	*B. rapa*	• Predominant expression of fatty acid biosynthesis, biosynthesis of secondary metabolites, and photosynthesis-related genes in embryos • Upregulation of genes for lipid metabolism and storage proteins in the middle and late stages of embryo development	Zhang et al. 2014
13	Oil content	*B. napus*	• Identification of 64 lipid metabolism-related DEGs, 14 of which are involved in triacylglycerols (TAGs) biosynthesis and assembly	Xiao et al. 2019
14	Marker development	*B. oleracea*	• Identification of InDel, SNP based markers	Xu et al. 2019

17.5.1 Brassica Database (BRAD)

The Brassica database, BRAD is a decade old database and was built after the whole genome sequencing of *Brassica rapa* (Chiifu-401-42) (Cheng et al. 2011). It is a web-based genomic database which can be accessed through http://brassicadb.org and alternative domain (http://brassicadb.cn/). Major sections of the database include Browse, Search, Tools, Download, and Links.

Browse: It contains information on genetic markers, gene families, various genes (glucosinolate gene, anthocyanin genes, resistance genes, flower genes, and auxin genes) and some basic phenotype and species information. Markers and map, subsection of Browse section gives information of a reference genetic linkage map and covers all ten chromosomes. The genetic map was constructed using a population (RCZ16_DH) of 119 doubled haploid (DH) lines obtained from F1 cross between DH line (Z16) and rapid cycling inbred line (L144) (Wang et al. 2011a, b, c). A total of 182 gene families in *B. rapa* corresponding to that in *A. thaliana* are given under the subsection gene families. Another subsection under *Browse* is *Glucosinolate genes* which describes 102 putative genes and corresponding *A. thaliana* orthologs (Wang et al. 2011a, b, c). Similarly, under *Anthocyanin genes* 73 genes of *B. rapa* as orthologs of 41 anthocyanin biosynthetic genes are given (Guo et al. 2014). Also, the other subsections consist of 244 resistance genes, 136 flowering genes, 342 auxin genes, and 3561 transcription factors of genes of *B. rapa*.

Search: This section provides the option of keyword search for annotations, syntenic genes, non-syntenic ortholog and gene sequence, and flanking regions. Searching a gene ID under *annotations* provides result in five databases (Gene Ontology, InterPro domain, KEGG, Swissprot, and Trembl) and orthologous genes as well as BLASTX (best hit) to *A. thaliana*.

Syntenic genes and non-syntenic orthologs between Brassicaceae and *A. thaliana*, a well-studied model plant can be accessed using a simple keyword search in BRAD. Insyntenic genes three abbreviations, viz. LF, MF1, and MF2, are used for least fractionized, moderate fractionized, and most fractionized, respectively, to denote subgenomes. Non-syntenic genes in BRAD are determined using two rules that the BLASTP alignment identity should be more than 70% and the genes should not be syntenic orthologs (Cheng et al. 2012). By using the flanking region search in BRAD, users can find the genomic elements such as genes, miRNA, tRNA, rRNA, snRNA, transposons, and genetic markers that flank the region of interest.

Tools: BRAD provides with two embedded tools, viz. BLAST and Genome browse (Gbrowse). BLAST can be used for sequence analysis while Gbrowse can be used to visualize *B. rapa* genome. Under the alternative domain of BRAD (http://brassicadb.cn/) JBrowse is integrated to visualize the genome of 35 species.

17.5.2 Brassica Genome

This database contains repeat information related to Brassica at http://www.Brassicagenome.net (Wang et al. 2011a, b, c; Golicz et al. 2016; Hurgobin et al. 2018). The database *Brassica genome* is maintained through grants from the University of Western Australia and the Australian Research Council. The pangenome of *B. oleracea*, *B. rapa*, and *B. napus* can be downloaded from this database. It contains an integrated analysis tool Blast Gbrowse by which a query sequence can be blast against available *Brassica* genomes and resulting hits can be viewed using Genome Browser. Furthermore, pangenome of *B. oleracea*, *B. rapa*, and *B. napus* can be viewed and searched using embedded tool JBrowse genome browse.

17.5.3 brassica.Info

"brassica.Info" was established under Multinational Brassica Genome Project (MBGP) in 2002 and since then it collates and shares the open source information regarding Brassica genetics and genomics. Information regarding Brassicales Map Alignment Project (BMAP) can also be retrieved through this platform. The major sections of "brassica.info" include genome, phenome, tools, infome, crop use, and outreach. The section genome contains download links to reference annotated Brassica genomes, pan-genomes of *B. oleracea* and *B. napus*, 52 *B. napus* re-sequenced genomes, 4.3 million SNPs and other *Brassica* genome resources. The section phenome contains link to important research articles related to *Brassica* ionome, metabolome, proteome, and transcriptome. Under tools section, information regarding clone libraries, genetic markers, research populations (mapping population, TILLING population, mutant population, and *Brassica rapa* Fast plants) is provided. Another important section of "brassica.info" is infome under which links to a range of databases and web portals relating to Brassica genetics and genomics are given.

17.5.4 BnPIR: *Brassica napus* Pan-Genome Information Resource

More whole genomes have been sequenced owing to the advancement in sequencing technology. Moreover, for the better understanding of genome complexity and genetic difference analysis pan-genomes has been proposed. So, based on the genome sequence of eight representative rapeseed cultivars and 1688 rapeseed re-sequencing data, BnPIR database (http://cbi.hzau.edu.cn/bnapus) was constructed (Song et al. 2020). It is a comprehensive functional genomic database and its important sections include pan browser, search (gene, species, gene expression, transposable elements, population variation and NLR genes), Gbrowse, tools (blast, KEGG/GO

Table 17.4 Genomic databases of *Brassica*

Sr. No.	Name of the database	Link
1	Brassica Database (BRAD)	http://brassicadb.org/ http://brassicadb.cn/
2	Brassica genome	http://www.Brassicagenome.net/
3	brassica.Info	https://www.brassica.info/
4	Brassica napus pan-genome information resource (BnPIR)	http://cbi.hzau.edu.cn/bnapus/
5	BrassicaDB	http://brassica.nbi.ac.uk/BrassicaDB/
6	Bolbase	http://ocri-genomics.org/bolbase/
7	BrassicaEDB	https://biodb.swu.edu.cn/brassica/

enrichment, homologous region, orthologous, phylogenetic tree, seq_fetch), and KEGG pathway for all the eight representative rapeseed cultivars, viz. Gangan, Zheyou7, Shengli, Tapidor, Quinta, Westar, No2127, and ZS11. The pan-genome is displayed using JBrowse and details of a query gene can be visualized using Gbrowse. Also Gbrowsesynteny can be used to identify gene structural differences. Overall the database BnPIR contains gene classification and annotation, (presence/absence variations) PAV and phylogenetic information, sequence and expression data, and common tools for multi-omics analysis.

17.5.5 BrassicaDB

The database BrassicaDB (http://brassica.nbi.ac.uk/BrassicaDB/index.html) contains information on genetic maps, markers, sequence accessions, "BBSRC set" of Brassica SSR markers and bibliographic information related to *B. napus* and *B. oleracea*. *Brassica* BLAST server is embedded in the database. This database was funded by BBSRC UK CropNet until 2003. However, newly deposited data is still automatically updated periodically in the database. Chao et al. (2020) developed the Brassica Expression Database (BrassicaEDB, https://biodb.swu.edu.cn/brassica/) for the brassica research community to retrieve the expression level data for target genes in different tissues and in response to different treatments to elucidate gene functions and explore the biology of rapeseed at the transcriptome level.

17.5.6 Bolbase

The database Bolbase (http://ocri-genomics.org/bolbase) contains genome data of *B. oleracea* and provides comparative genomics information including syntenic regions (Yu et al. 2013). The database Bolbase contains two important sections: (1) genomic data and genomic component data (2) analysis on syntenic regions. The information on genomic data includes genome sequence, scaffold and psuedochromosome sequences while genomic component data mainly includes gene structure, location, functional annotation, orthologs, syntenic regions, repeats elements, and predicted noncoding RNAs. Major sections of the database include browse, synteny, search, and document. Bolbase contains important tools including keyword and similarity search, and an embedded generic genome browser (GBrowse) for visualization (Table 17.4).

17.6 Functional Genomics and Its Role in Brassica Improvement

17.6.1 Functional Genomics

Functional genomics research in Brassica has enabled the understanding the function and regulation of several genes associated with productivity related traits. Loss of function or knockout mutants can be created using techniques such as mutagenesis, RNA interference, and CRISPR/Cas9. Mehmood et al. (2021) analyzed cold-stress responses in tolerant and sensitive rapeseed lines using RNA-Seq and found involvement of pathways of photosynthesis, antioxidant defense, and energy metabolism. Further authors validated the function of three genes (*nir*, *cml*, and *cat*) by analyzing the T-DNA insertion lines mutant lines of Arabidopsis and suggested varied freezing response. Function of a gene can be assigned using mutant analysis which further can provide important information on its regulation and metabolic activity. In mutagenesis, mutation in a specific gene is produced to disrupt its function and phenotype of the mutant is then observed for assigning function to the particular gene. One of the most important objectives of mutagenesis is to produce maximum genetic variation (Sikora et al. 2011). Ethyl methanesulfonate or EMS is the most commonly used chemical mutagen while other chemical mutagens such as sodium azide and methylnitrosourea are also in use (Sikora et al. 2011).

17.6.2 TILLING for Identification of Genes Related to Erucic Acid and Abiotic Stress Tolerance

Targeting induced local lesions in genomes (TILLING) is an efficient technique to detect mautagenesis (McCallum et al. 2000). TILLING as a reverse genetics tool provide numerous advantages in functional genetics. It can be applied to any species irrespective of its genome size and ploidy level. This technique combines the advantage of classical mutagenesis for producing high frequency of mutation and high throughput screening for nucleotide polymorphism (Kurowska et al. 2011). TILLING has been applied for important crops including *B. oleracea* (Himelblau et al. 2009), *B. rapa* (Stephenson et al. 2010), and *A. thaliana* (Greene et al. 2003). Briefly, TILLING includes three major steps, i.e., (1) mutant population generation, (2) detection of mutation, and (3) analysis of mutant phenotype. Sequencing of target gene can be done to confirm the mutation, and phenotyping of M3 individuals is done for the analysis (Kurowska et al. 2011). The seeds and DNA samples from M2 population are archived and form TILLING platform. RevGenUK (http://revgenuk.jic.ac.uk/about.html) and CAN-TILL (http://www.botany.ubc.ca/can-till/) are the TILLING platforms related to Brassica (Himelblau et al. 2009; Stephenson et al. 2010).

A TILLING platform in *B. napus* was constructed using EMS for functional genomics and generated two mutated populations derived from cv. Ningyou7. Furthermore, these populations were used for forward genetic screen for gene discovery. The TIILING platform was tested for mutations in fatty acid elongase1 (*FAE1*) gene, an important gene in erucic acid biosynthesis. Using reverse genetics screening, 19 mutations for *FAE1* in 1344 M2 plants could be identified out of which three mutations were associated with reduction in erucic acid content (Wang et al. 2008). Another TILLING platform in diploid Brassica (*B. rapa*) was also created using EMS and is available publicly through RevGenUK platform (Stephenson et al. 2010).

Phytoremediation potential of various species of genus *Brassica* is well reported in literature (Rizwan et al. 2018; Thakur et al. 2019; Raj et al. 2020). Function of a vacuolar transporter, i.e., calcium exchanger 1 (*CAX1*), was examined in *B. rapa* using TILLING. The mutants for the gene *CAX1* were created through TILLING. It was revealed that *BraA.cax1a* mutation enhances cadmium uptake capacity but *BraA.cax1a*-12 mutants were found suitable for phytoremediation as it accumulated threefold more cadmium than parental line as well as greater cadmium tolerance (Navarro-León et al. 2019). A mutant (*BraA.hma4a-3*) detected through TILLING, having mutation for HMA4 transporter in *B. rapa,* was found to be a better zinc accumulator than parental line (R-o-18). Moreover, *BraA.hma4a*-3 plants showed better tolerance toward zinc toxicity (Blasco et al. 2019). Another study found that *BraA.hma4a-3* mutants can accumulate greater amount of cadmium in leaves and showed better tolerance to cadmium toxicity than parental line (Navarro-León et al. 2019).

17.6.3 RNA Interference

RNA interference (RNAi) is an important tool of functional genomics. RNAi has been used successfully to find out the function and biological role of genes in crops including wheat, cotton and *B. napus* (Travella et al. 2006; Abdurakhmonov et al. 2016). It is a universal eukaryotic process of sequence-specific gene silencing (Hannon 2002). Dicer enzymes recognize and cleave dsRNA into siRNA (21–25 bp long double stranded fragments) which is further processed into single stranded “passenger” and “guide” RNAs. While the “passenger” RNA is degraded “guide” RNA recognize and digest the target RNA through RNA-induced silencing complex (Hannon 2002).

For its use as functional genomics tool, knock out lines are generated and phenotype is tested to characterize the function of knock out gene. RNAi as a functional genomics tool has many advantages such as multiple target genes silencing (McGinnis et al. 2007). Using RNAi, a loss-of-function analysis for *BnaNPR1* was performed and it was found that *BnaNPR1* repression is associated with reduction in *S. sclerotiorum* resistance in *B. napus* (Diepenbrock 2000). Another study demonstrated the function of *BnGPAT19* and *BnGPAT21* in *B. napus* using RNAi. Suppression of *BnGPAT19* and *BnGPAT21* resulted in thinner cuticle and necrotic lesions on fungal inoculation, indicating the possible role of these genes in cuticular wax biosynthesis (Wang et al. 2020a, b).

Glucoraphanin is a glucosinolate found in Brassicales and its breakdown product sulphoraphane is known to have anti-cancerous properties (Fahey et al. 1997; Variyar et al. 2014). It is known that GSL-ALK enzyme catalyze conversion of glucoraphanin to undesirable products; a total of 29 transgenic lines (knock-down of gene *GSL-ALK*) of *B. juncea* were created using RNAi. Silencing of GSL-ALK enzyme led to reduction in undesirable glucosinolates while the growth and seed quality was not hampered as compared to untransformed control (Augustine and Bisht 2015). Similarly, in another study the transgenic *B. juncea* lines (*BjMYB28* gene suppressed) were created using RNAi, which leads to reduction in glucosinolate content without affecting its growth and development (Augustine et al. 2013).

17.7 Genome Editing Tools

Advancements in genome editing techniques, especially the Clustered regularly interspaced short palindromic repeat /CRISPR-associated protein 9 (CRISPER/Cas9) has become a powerful tool for plant functional genomics research (Feng et al. 2013; Shan et al. 2013; Liu et al. 2016). Using CRISPER/Cas9, the target DNA is cut and which then is repaired by non-homologous end-joining giving rise to indel mutations. Knockout mutants created using the CRISPER/Cas9 technology can be used for loss of function analysis (Puchta 2017; Liu et al. 2019a, b). Further, high throughput functional screening can be done as it is programmable and highly precise (Liu et al. 2019a, b). This technology has been successfully used in different plant species. However, in Brassica, there are few successful examples of genome editing. In *B. napus,* the modification of the metabolic pathway for fatty acid synthesis was done using a CRISPR/Cas9-based editing of target gene, fatty acid desaturase 2 gene (*FAD2*), responsible for the catalysis of the desaturation of oleic acid. Seeds of one of the mutants having *fad2*_Aa allele with a 4-bp deletion was found to have significantly high oleic acid over the wild-type seeds (Okuzaki et al. 2018). Pod shattering is a problem for achieving higher yield in rapeseed cultivation. Zaman et al. (2019) successfully reported multiplex editing of five homeologs *BnJAG.A02*, *BnJAG.C02*, *BnJAG.C06*, *BnJAG.A07*, and *BnJAG.A08*. The knockout mutants showed altered pod shape and size phenotypes. One mutant, (*BnJAG.A08*-NUB-Like paralog of the *JAG* gene) had significant change in the pod dehiscence and resistance to pod shattering by ~ twofold. Ma et al. (2019a, b) synthesized a tandemly arrayed tRNA-sgRNA sequence to simultaneously generate several sgRNAs by employing the plant endogenous tRNA processing system in cabbage. Target genes included, phytoene desaturase gene (*BoPDS*), self-incompatibility determinant gene (*BoSRK3*), and the male sterility associated gene (*BoMS1*). The application of CRISPR/Cas9 system in *B. campestris* was studied by targeting the pectin-methylesterase genes *Bra003491*, *Bra007665*, and *Bra014410*. Results have shown the introduction of mutations at the rate, ranging from 20 to 56%. The study has highlighted the potential of CRISPR/Cas9 system for single and multiplex genome editing in a stable and inheritable manner (Xiong et al. 2019). Jeong et al. (2019) have successfully used CRISPER/cas9 system to modify the early-flowering trait in *B. rapa* by designing seven guide RNAs to target the FLOWERING LOCUS C (*FLC*). The double knockouts, *BraFLC2* and *BraFLC3* showing indel efficiency of 97.7 and 100%, were found to have early-flowering phenotype without depending on vernalization. Yellow seed color is a desirable trait for seed quality. By using CRISPR/Cas9 editing, yellow-seeded mutants were generated in rapeseed having mutations in the target gene, *BnTT8* gene. The mutants were found genetically stable with high seed oil, protein content and modified fatty acid (FA) profile with no compromise on yield (Zhai et al. 2020).

17.8 Conclusions and Future Perspective

The Brassica family has a wide spectrum of phenotypic and genomic plasticity. Breeding aimed at improvement in traits for biotic and abiotic stress tolerance, and nutritional quality besides yield associated characters is a continued priority. Advances in genomics tools have opened up new avenues in the detection of genetic basis of trait variation and development of molecular markers for accelerating introgression of useful traits (Hu et al. 2021). Transcriptomics advances including RNA-Seq technologies are now increasingly used for profiling gene expression of thousands of genes in spatial and temporal mode. The availability of assembled genomes has enabled molecular marker development, marker-aided selection and functional genomics of important agronomic traits for designing better crops. In this context, functional gene characterization through approaches like loss of function mutants has become valuable for information on regulatory, developmental, biochemical and metabolic networks. Besides other tools like TILLING for fatty acid biosynthesis, insertional mutagenesis and RNA interference for disease resistance and glucosinolates synthesis have been useful in Brassica breeding and improvement.

The study of transcriptomes in *Brassica* crops has provided significant resource on genome structure, diversity and genome origin, evolutionary analysis, differential gene expression and marker development. This has become possible because of the advances in genome sequencing of important *Brassica* species (*B. rapa*, *B. oleracea* and *B. nigra*, *B. napus*, *B. juncea*) for investigating the whole-genome transcripts, identification of agronomically important genes for stress tolerance, and lipid and glucosinolate biosynthesis. Brassica genome databases are an information gateway for unraveling pathways of biological processes regulated by noncoding RNAs (ncRNAs), particularly microRNAs and long ncRNAs. Further to these developments, the genome editing based on CRISPR/Cas9 system for single and multiplex genome editing has opened up means for designing Brassicas with useful targeted and precise trait modifications. Genomics of the Brassicaceous crops along with other omics technologies offer immense scope for designing highly productive new crop varieties in the Brassica family.

References

Abdurakhmonov IY, Ayubov MS, Ubaydullaeva KA, Buriev ZT, Shermatov SE et al (2016) RNA interference for functional genomics and improvement of cotton (*Gossypium sp*). Front Plant Sci 7:202

Ahmed W, Xia Y, Li R, Bai G, Siddique KHM et al (2020) Non-coding RNAs: functional roles in the regulation of stress response in Brassica crops. Genomics 112:1419–1424

Alwine JC, Kemp DJ, Stark GR (1977) Method for detection of specific RNAs in agarose gels by transfer to diazobenzyloxymethyl-paper and hybridization with DNA probes. Proc Natl Acad Sci USA 74:5350–5354

An H, Qi X, Gaynor ML, Hao Y, Gebken SC et al (2019) Transcriptome and organellar sequencing highlights the complex origin and diversification of allotetraploid *Brassica napus*. Nat Commun 10:2878

Augustine R, Bisht NC (2015) Biofortification of oilseed *Brassica juncea* with the anti-cancer compound glucoraphanin by suppressing GSL-ALK gene family. Sci Rep 5:1–12

Augustine R, Mukhopadhyay A, Bisht NC (2013) Targeted silencing of BjMYB28 transcription factor gene directs development of low glucosinolate lines in oilseed *B. rassica juncea*. Plant Biotechnol J 11:855–866

Bancroft I, Morgan C, Fraser F, Higgins J, Wells R et al (2011) Dissecting the genome of the polyploid crop oilseed rape by transcriptome sequencing. Nat Biotechnol 29:762–766

Becker MG, Zhang X, Walker PL, Wan JC, Millar JL et al (2017) Transcriptome analysis of the *Brassica napus–Leptosphaeria maculans* pathosystem identifies receptor, signaling and structural genes underlying plant resistance. Plant J 90(3):573–586

Becker-André M, Hahlbrock K (1989) Absolute mRNA quantification using the polymerase chain reaction (PCR). A novel approach by a PCR aided transcript titration assay (PATTY). Nucleic Acids Res 17:9437–9446

Bergmann T, Menkhaus J, Schemmel M, Ye W, Hasler M et al (2021) QTL mapping and transcriptome analysis of Sclerotinia-resistance in the wild cabbage species *Brassica oleracea* var. villosa. ResearchSquare. https://doi.org/10.21203/rs.3.rs-309600/v1

Bhardwaj AR, Joshi G, Kukreja B, Malik V, Arora P et al (2015) Global insights into high temperature and drought stress regulated genes by RNA-Seq in economically important oilseed crop *Brassica juncea*. BMC Plant Biol 15:9

Blasco B, Navarro-León E, Ruiz JM (2019) Study of Zn accumulation and tolerance of HMA4 TILLING mutants of *Brassica rapa* grown under Zn deficiency and Zn toxicity. Plant Sci 287:110201

Branca F, Cartea E (2011) Brassica. In: Kole C (ed) Wild crop relatives: genomic and breeding resources, oilseeds. Springer, Berlin, Heidelberg, pp 17–36

Calderwood A, Lloyd A, Hepworth J, Tudor EH, Jones DM et al (2021) Total FLC transcript dynamics from divergent paralogue expression explains flowering diversity in *Brassica napus*. New Phytol 229:3534–3548

Chalhoub B, Denoeud F, Liu S, Parkin IAP, Tang H et al (2014) Early allopolyploid evolution in the post-Neolithic *Brassica napus* oilseed genome. Science 345:950–953

Chang A, Lamara M, Wei Y, Hu H, Parkin IAP et al (2019) Clubroot resistance gene Rcr6 in *Brassica nigra* resides in a genomic region homologous to chromosome A08 in *B. rapa*. BMC Plant Biol 19:224

Chang PH, Soo-Jin K, Jung SK, Tae-Jin Y, Beom-Seok P, Yong PL (2008) Progress in understanding and sequencing the genome of *Brassica rapa*. Int J Plant Genomics 582837:9pp. https://doi.org/10.1155/2008/582837

Chao H, Li T, Luo C, Huang H, Ruan Y et al (2020) Brassica EDB: a gene expression database for Brassica crops. Int J Mol Sci 21(16):5831

Chen J, Li R, Xia Y, Bai G, Guo P et al (2017) Development of EST-SSR markers in flowering Chinese cabbage (*Brassica campestris* L. ssp. *chinensis* var. utilis Tsen et Lee) based on de novo transcriptomic assemblies. PLoS One 12:e0184736

Chen SW, Liu T, Gao Y, Zhang C, Peng SD et al (2015) Discovery of clubroot-resistant genes in *Brassica napus* by transcriptome sequencing. Genet Mol Res 15(3):gmr.15038243

Cheng B, Furtado A, Henry RJ (2017a) Long-read sequencing of the coffee bean transcriptome reveals the diversity of full-length transcripts. Giga Sci 6

Cheng F, Liang J, Cai C, Cai X, Wu J et al (2017b) Genome sequencing supports a multi-vertex model for Brassiceae species. Curr Opin Plant Biol 36:79–87

Cheng F, Liu S, Wu J, Fang L, Sun S et al (2011) BRAD, the genetics and genomics database for Brassica plants. BMC Plant Biol 11:136

Cheng F, Wu J, Fang L, Wang X (2012) Syntenic gene analysis between *Brassica rapa* and other Brassicaceae species. Front Plant Sci 3:198

Cheng F, Wu J, Wang X (2014) Genome triplication drove the diversification of Brassica plants. Hortic Res 1:14024

Chittem K, Yajima WR, Goswami RS, Mendoza LEdR et al (2020) Transcriptome analysis of the plant pathogen *Sclerotinia sclerotiorum* interaction with resistant and susceptible canola (*Brassica napus*) lines. PLoS One 15(3):e0229844

Chu M, Song T, Falk KC, Zhang X, Liu X et al (2014) Fine mapping of Rcr1 and analyses of its effect on transcriptome patterns during infection by *Plasmodiophora brassicae*. BMC Genomics 15:1166

Conesa A, Madrigal P, Tarazona S, Gomez-Cabrero D, Cervera A et al (2016) A survey of best practices for RNA-seq data analysis. Genome Biol 17:13

Cui J, Lu Z, Xu G, Xu G, Wang Y et al (2020) Analysis and comprehensive comparison of PacBio and nanopore-based RNA sequencing of the Arabidopsis transcriptome. Plant Methods 16:1–13

Diepenbrock W (2000) Yield analysis of winter oilseed rape (*Brassica napus* L.): a review. Field Crops Res 67:35–49

Ding Q, Li J, Wang F, Zhang Y, Li H et al (2015) Characterization and development of EST-SSRs by deep transcriptome sequencing in Chinese cabbage (*Brassica rapa* L. ssp. pekinensis). Int J Genom 2015:1–11. https://doi.org/10.1155/2015/473028

Ding Y, Mei J Cha Y, Yu Y, Shao C et al (2019) Simultaneous transcriptome analysis of host and pathogen highlights the interaction between *Brassica oleracea* and *Sclerotinia sclerotiorum*. Phytopathology 109:542–550

Dixon GR (2007) Origin and diversity of Brassica and its relatives. In: Dixon GR (ed) Vegetable Brassicas and related Crucifers. CABI, Wallingford, UK

Dohm JC, Lottaz C, Borodina T, Himmelbauer H (2008) Substantial biases in ultra-short read data sets from high-throughput DNA sequencing. Nucleic Acids Res 36:e105. https://doi.org/10.1093/nar/gkn425

Dodds PN, Rathjen JP (2010) Plant immunity: towards an integrated view of plant–pathogen interactions. Nat Rev Genet 11:539–548

Ehlting J, Chowrira SG, Mattheus N, Aeschliman DS, Arimura GI et al (2008) Comparative transcriptome analysis of *Arabidopsis thaliana* infested by diamond back moth (*Plutella xylostella*) larvae reveals signatures of stress response, secondary metabolism, and signalling. BMC Genomics 9:154

El-Esawi MA (2016) Genetic diversity and evolution of Brassica genetic resources: from morphology to novel genomic technologies—a review. Plant Genet Resour 1(5):1–12

Ellegren H, Smeds L, Burri R, Olason PI, Backström N et al (2012) The genomic landscape of species divergence in *Ficedula flycatchers*. Nature 491:756–760

ENCODE V1.0 (2016) Revised guidelines and best practices for RNA-Seq.

Eom SH, Baek SA, Kim JK, Hyun TK (2018) Transcriptome analysis in Chinese cabbage (*Brassica rapa* ssp. *pekinensis*) provides the role of glucosinolate metabolism in response to drought stress. Molecules 23:1186

Fahey JW, Zhang Y, Talalay P (1997) Broccoli sprouts: an exceptionally rich source of inducers of enzymes that protect against chemical carcinogens. Proc Natl Acad Sci USA 94:10367–10372

FAOSTAT (2013) FAO Statistics online database, "Production / Crops, – rapeseed, Year 2013". Food andAgriculture Organization of the United Nations. http://faostat3.fao.org/home/E. Accessed 5 Jan 2016

Feng Z, Zhang B, Ding W, Liu X, Yang D-L et al (2013) Efficient genome editing in plants using a CRISPR/Cas system. Cell Res 23:1229–1232

Frandsen KJ (1943) The experimental formation of *Brassica juncea* CZERN et Coss. Ilansk Botanisk Arkiv 11:1–17

Fu P, Piao Y, Zhan Z, Zhao Y, Pang W et al (2019) Transcriptome profile of *Brassica rapa* L. reveals the involvement of jasmonic acid, ethylene, and brassinosteroid signaling pathways in clubroot resistance. Agronomy 9(10):589

Ganie SA, Reddy ASN (2021) Stress-induced changes in alternative splicing landscape in rice: functional significance of splice isoforms in stress tolerance. Biology (basel) 10(4):309

Garg R, Jain M (2013) RNA-Seq for transcriptome analysis in non-model plants. In: Legume genomics. Methods in molecular biology, vol 1069, pp 43–58

Gill RA, Ali B, Cui P, Shen E, Farooq MA et al (2016) Comparative transcriptome profiling of two *Brassica napus* cultivars under chromium toxicity and its alleviation by reduced glutathione. BMC Genomics 17(1):885

Golicz AA, Bayer PE, Barker GC, Edger PP, Kim HR et al (2016) The pangenome of an agronomically important crop plant *Brassica oleracea*. Nat Commun 7:1–8

Grabherr MG, Haas BJ, Yassour M, Levin JZ, Thompson DA et al (2011) Full-length transcriptome assembly from RNA-Seq data without a reference genome. Nat Biotechnol 29:644

Greene EA, Codomo CA, Taylor NE, Henikoff JG, Till BJ et al (2003) Spectrum of chemically induced mutations from a large-scale reverse-genetic screen in Arabidopsis. Genetics 164:731–740

Gruber M, Alahakoon U, Taheri A, Nagubushana N, Zhou R et al (2018) The biochemical composition and transcriptome of cotyledons from *Brassica napus* lines expressing the AtGL3 transcription factor and exhibiting reduced flea beetle feeding. BMC Plant Biol 18:64

Guo N, Cheng F, Wu J, Liu B, Zheng S et al (2014) Anthocyanin biosynthetic genes in *Brassica rapa*. BMC Genomics 15:426

Hannon GJ (2002) RNA interference. Nature 418:244–251

Hansen KD, Brenner SE, Dudoit S (2010) Biases in Illumina transcriptome sequencing caused by random hexamer priming. Nucleic Acids Res 38:e131. https://doi.org/10.1093/nar/gkq224

Harbered DJ (1976) Cytotaxonomic studies of Brassica and related genera. In: Vaughan JG, MacLeod AJ, Jones MG (eds) The biology and chemistry of the Cruciferae. Academic, London, pp 47–68

Harper AL, He Z, Langer S, Havlickova L, Wang L et al (2020) Validation of an associative transcriptomics platform in the polyploid crop species *Brassica juncea* by dissection of the genetic architecture of agronomic and quality traits. Plant J 103:1885–1893

Hejna O, Havlickova L, He Z, Bancroft I, Curn V (2019) Analysing the genetic architecture of clubroot resistance variation in *Brassica napus* by associative transcriptomics. Mol Breed 39:112

Heng S, Wang L, Yang X, Huang H, Chen G et al (2020) Genetic and comparative transcriptome analysis revealed DEGs involved in the purple leaf formation in *Brassica juncea*. Front Genet 11:322

Himelblau E, Gilchrist EJ, Buono K, Bizzell C, Mentzer L et al (2009) Forward and reverse genetics of rapid-cycling *Brassica oleracea*. Theor Appl Genet 118:953–961

Hu D, Jing J, Snowdon RJ, Mason AS, Shen J, Meng J, Zou J (2021) Exploring the gene pool of Brassica napus by genomics-based approaches. Plant Biotechnol J. https://doi.org/10.1111/pbi.13636

Hurgobin B, Golicz AA, Bayer PE, Chan C-KK, Tirnaz S et al (2018) Homoeologous exchange is a major cause of gene presence/absence variation in the amphidiploid *Brassica napus*. Plant Biotechnol J 16:1265–1274

Izzah NK, Lee J, Jayakodi M, Perumal S, Jin M et al (2014) Transcriptome sequencing of two parental lines of cabbage (*Brassica oleracea* L. var. capitata L.) and construction of an EST-based genetic map. BMC Genomics 15:149

Jeong SY, Ahn H, Ryu J, Oh Y, Sivanandhan G et al (2019) Generation of early-flowering Chinese cabbage (*Brassica rapa* spp. *pekinensis*) through CRISPR/Cas9-mediated genome editing. Plant Biotechnol Rep 13:491–499

Jambhulkar SJ (2015) Induced mutagenesis and allele mining. In: Kumar A, Banga SS, Meena PD, Kumar PR (eds) Brassica oilseeds: breeding and management. CABI, Wallingford, UK, pp 53–67

Johnston JS, Pepper AE, Hall AE (2005) Evolution of genome size in Brassicaceae. Ann Bot 95:229–235

Jia H, Wei X, Yang Y, Yuan Y, Wei F et al (2017) Root RNA-seq analysis reveals a distinct transcriptome landscape between clubroot-susceptible and clubroot-resistant Chinese cabbage lines after *Plasmodiophora brassicae* infection. Plant Soil 421:93–105

Jia L, Wang J, Wang R, Duan M, Qiao C et al (2021) Comparative transcriptomic and metabolomic analyses of carotenoid biosynthesis reveal the basis of white petal color in *Brassica napus*. Planta 253:8

Joshi RK, Megha S, Rahman MH (2016) A global study of transcriptome dynamics in canola (*Brassica napus* L.) responsive to *Sclerotinia sclerotiorum* infection using RNA-Seq. Gene 590(1):57–67

Kazan K, Lyons R (2014) Intervention of phytohormone pathways by pathogen effectors. Plant Cell 26:2285–2309

Kiefer M, Schmickl R, German DA, Mandáková T, Lysak MA et al (2014) BrassiBase: introduction to a novel knowledge database on Brassicaceae evolution. Plant Cell Physiol 55(1):e3

Kim HA, Lim CJ, Kim S, Choe JK, Jo S-H et al (2014) High-throughput sequencing and de novo assembly of *Brassica oleracea* var. Capitata L. for transcriptome analysis. PLoS One 9:e92087

Kurowska M, Daszkowska-Golec A, Gruszka D, Marzec M, Szurman M et al (2011) TILLING-a shortcut in functional genomics. J Appl Genet 52:371–390

Lagercrantz U, Lydiate D (1996) Comparative genome mapping in Brassica. Genetics 144:1903–1910

Lamichhaney S, Barrio AM, Rafati N, Sundström G, Rubin C-J et al (2012) Population-scale sequencing reveals genetic differentiation due to local adaptation in Atlantic herring. Proc Natl Acad Sci USA 109:19345–19350

Lan T-H, DelMonte TA, Reischmann K, Hyman J, Kowalski SP et al (2000) An EST-enriched comparative map of *Brassica oleraceae* and *Arabidopsis thaliana*. Genome Res 10:776–788

Li H, Yuan J, Wu M, Han Z, Li L et al (2018) Transcriptome and DNA methylome reveal insights into yield heterosis in the curds of broccoli (*Brassica oleracea* L var. italic). BMC Plant Biol 18:168

Li M, Wang R, Wu X, Wang J (2020) Homoeolog expression bias and expression level dominance (ELD) in four tissues of natural allotetraploid *Brassica napus*. BMC Genomics 21:330

Liu D, Chen M, Mendoza B, Cheng H, Hu R et al (2019a) CRISPR/Cas9-mediated targeted mutagenesis for functional genomics research of crassulacean acid metabolism plants. J Exp Bot 70:6621–6629

Liu D, Hu R, Palla KJ, Tuskan GA, Yang X (2016) Advances and perspectives on the use of CRISPR/Cas9 systems in plant genomics research. Curr Opin Plant Biol 30:70–77

Liu M, Wu F, Wang S, Lu Y, Chen X et al (2019b) Comparative transcriptome analysis reveals defense responses against soft rot in Chinese cabbage. Hortic Res 6:68

Liu S, Liu Y, Yang X, Tong C, Edwards D et al (2014) The Brassica oleracea genome reveals the asymmetrical evolution of polyploid genomes. Nat Commun 5:3930

Liu X, Zhao C, Yang L, Zhuang M, Zhang Y et al (2020) A time-resolved dual transcriptome analysis reveals the molecular regulating network underlying the compatible/incompatible interactions between cabbage (*Brassica oleracea*) and *Fusarium oxysporum* f. sp. *conglutinans*. Plant Soil 448:455–478

Lockhart DJ, Dong H, Byrne MC, Follettie MT, Gallo MV et al (1996) Expression monitoring by hybridization to high-density oligonucleotide arrays. Nat Biotechnol 14:1675–1680

Long W, Zou X, Zhang X (2015) Transcriptome analysis of canola (*Brassica napus*) under salt stress at the germination stage. PLoS One 10(2):e0116217

Louisa AK, Xinping C, Frances M, Walling LL (2007) Arabidopsis transcriptome changes in response to phloem-feeding silver leaf whitefly nymphs. Similarities and distinctions in responses to aphids. Plant Physiol 143(2):849–865

Lowe RGT, Cassin A, Grandaubert J, Clark BL, Wouw APVd et al (2014) Genomes and transcriptomes of partners in plant-fungal-interactions between canola (*Brassica napus*) and two *Leptosphaeria* species. PLoS One 9(7):e103098

Lü B, Yu J, Xu J, Chen J, Laib M (2009) A novel approach to detect differentially expressed genes from count-based digital databases by normalizing with housekeeping genes. Genomics 94:211–216

Lu K, Peng L, Zhang C, Lu J, Yang B et al (2017) Genome-wide association and transcriptome analyses reveal candidate genes underlying yield-determining traits in *Brassica napus*. Front Plant Sci 8:206

Luo Y, Dong D, Su Y, Wang X, Peng Y et al (2018) Transcriptome analysis of *Brassica juncea* var. tumida Tsen responses to *Plasmodiophora brassicae* primed by the biocontrol strain *Zhihengliuella aestuarii*. Funct Integr Genom 18:301–314

Lyu S, Gao L, Zhang R, Zhang C, Hou X (2020) Correlation analysis of expression profile and quantitative iTRAQ-LC-MS/MS proteomics reveals resistance mechanism against TuMV in Chinese cabbage (*Brassica rapa* ssp. *pekinensis*). Front Plant Sci 11:963

Ma C, Zhu C, Zheng M, Liu M, Zhang D et al (2019a) CRISPR/Cas9-mediated multiple gene editing in *Brassica oleracea* var. capitata using the endogenous tRNA-processing system. Hortic Res 6:20

Ma L, Coulter JA, Liu L, Zhao Y, Chang Y et al (2019b) Transcriptome analysis reveals key cold-stress-responsive genes in winter rapeseed (*Brassica rapa* L.). Int J Mol Sci 20:1071

Macosko EZ, Basu A, Satija R, Nemesh J, Shekhar K et al (2015) Highly parallel genome-wide expression profiling of individual cells using nanoliter droplets. Cell 161:1202–1214

Margulies M, Egholm M, Altman W, Attiya S, Bader JS et al (2005) Genome sequencing in microfabricated high-density picolitre reactors. Nature 437:376–380

Marra MA, Hillier L, Waterston RH (1998) Expressed sequence tags: establishing bridges between genomes. Trends Genet 14:4–7

Martin L, Fei Z, Giovannoni J, Rose JKC (2013) Catalyzing plant science research with RNA-seq. Front Plant Sci 4:66

Mastrangelo AM, Marone D, Laidò G, Leonardis AMD, Vita PD (2012) Alternative splicing: enhancing ability to cope with stress via transcriptome plasticity. Plant Sci 185–186:40–49

McCallum CM, Comai L, Greene EA, Henikoff S (2000) Targeting induced locallesions in genomes (TILLING) for plant functional genomics. Plant Physiol 123:439–442

McGinnis K, Murphy N, Carlson AR, Akula A, Akula C et al (2007) Assessing the efficiency of RNA interference for maize functional genomics. Plant Physiol 143:1441–1451

Mehmood SS, Lu G, Luo D, Hussain MA, Raza A et al (2021) Integrated analysis of transcriptomics and proteomics provides insights into the molecular regulation of cold response in *Brassica napus*. Environ Exp Bot 187. https://doi.org/10.1016/j.envexpbot.2021.104480

Miyaji N, Shimizu M, Takasaki-Yasuda T, Dennis ES, Fujimoto R (2021) The transcriptional response to salicylic acid plays a role in Fusarium yellows resistance in *Brassica rapa* L. Plant Cell Rep 40:605–619

Mohd Saad NS, Severn-Ellis AA, Pradhan A, Edwards D, Batley J (2021) Genomics armed with diversity leads the way in Brassica improvement in a changing global environment. Front Genet 12:600789. https://doi.org/10.3389/fgene.2021.600789

Mondal B, Mazumder M, Mukherjee A, Ghosh S, De A et al (2020) Association of *Alternaria brassicicola* induced NAC transcription factors with desiccation and wound responses in Indian mustard. Physiol Mol Plant P 112:101540

Mutz K-O, Heilkenbrinker A, Lönne M, Walter J-G, Stahl F (2013) Transcriptome analysis using next-generation sequencing. Curr Opin Biotechnol 24:22–30

Nagaharu U (1935) Genome analysis in Brassica with special reference to the experimental formation of *B. napus* and peculiar mode of fertilization. Jpn J Bot 7:389–452

Navarro-León E, Oviedo-Silva J, Ruiz JM, Blasco B (2019) Possible role of HMA4a TILLING mutants of *Brassica rapa* in cadmium phytoremediation programs. Ecotoxicol Environ Saf 180:88–94

Neik TX, Barbetti MJ, Batley J (2017) Current status and challenges in identifying disease resistance genes in *Brassica napus*. Front Plant Sci 8:1788. https://doi.org/10.3389/fpls.2017.01788

Niu Y, Wu L, Li Y, Huang H, Qian M et al (2020) Deciphering the transcriptional regulatory networks that control size, color, and oil content in *Brassica rapa* seeds. Biotechnol Biofuels 13:90

Okuzaki A, Ogawa T, Koizuka C, Kaneko K, Inaba M et al (2018) CRISPR/Cas9-mediated genome editing of the fatty acid desaturase 2 gene in *Brassica napus*. Plant Physiol Biochem 131:63–69

Ozsolak F, Milos PM (2011) RNA sequencing: advances, challenges and opportunities. Nat Rev Genet 12:87–98

Pan L, Helen KW, John LH, Chye M-L, Scofield S (2019) Comparative transcriptomics analysis of *Brassica napus* L. during seed maturation reveals dynamic changes in gene expression between embryos and seed coats and distinct expression profiles of acyl-CoA-binding proteins for lipid accumulation. Plant Cell Physiol 60(12):2812–2825

Paritosh K, Yadava SK, Gupta V, Panjabi-Massand P, Sodhi YS et al (2013) RNA-seq based SNPs in some agronomically important oleiferous lines of *Brassica rapa* and their use for genome-wide linkage mapping and specific-region fine mapping. BMC Genomics 14:1–13

Parkin IA, Koh C, Tang H, Robinson SJ, Kagale S et al (2014) Transcriptome and methylome profiling reveals relics of genome dominance in the mesopolyploid *Brassica oleracea*. Genome Biol 15:R77

Peng Q, Xie Q, Chen F, Zhou X, Zhang W et al (2017) Transcriptome analysis of sclerotinia sclerotiorum at different infection stages on *Brassica napus*. Curr Microbiol 74:1237–1245

Perumal S, Koh CS, Jin L, Buchwaldt M, Higgins EE et al (2020) A high-contiguity *Brassica nigra* genome localizes active centromeres and defines the ancestral Brassica genome. Nat Plants 6:929–941

Pu Y, Liu L, Wu J, Zhao Y, Bai J et al (2019) Transcriptome profile analysis of winter rapeseed (*Brassica napus* L.) in response to freezing stress, reveal potentially connected events to freezing stress. Int J Mol Sci 20:2771

Puchta H (2017) Applying CRISPR/Cas for genome engineering in plants: the best is yet to come. Curr Opin Plant Biol 36:1–8

Qasim MU, Zhao Q, Shahid M, Samad RA, Ahmar S et al (2020) Identification of QTLs containing resistance genes for sclerotinia stem rot in *Brassica napus* using comparative transcriptomic studies. Front Plant Sci 11:776

Qian M, Fan Y, Li Y, Liu M, Sun W et al (2021) Genome-wide association study and transcriptome comparison reveal novel QTL and candidate genes that control petal size in rapeseed. J Exp Bot 72 (10):3597–3610

Rai AN, Bhujbal S, Jambhulkar SJ (2021) Development of abiotic stress tolerant mustard genotype through induced mutagenesis. In: Singh S, Singh P, Rangabhashiyam S, Srivastava KK (eds) Global climate change. Elsevier, Netherlands, pp 213–233

Raj D, Kumar A, Maiti SK (2020) *Brassica juncea* (L.) *Czern.* (Indian mustard): a putative plant species to facilitate the phytoremediation of mercury contaminated soils. Int J Phytoremediation 22:733–744

Rakow G (2004) Species origin and economic importance of Brassica. In: Pua EC, Douglas CJ (eds) Biotechnology in agriculture and forestry, vol 54. Springer, New York, Berlin, Heidelberg, pp 3–11

Raza A, Razzaq A, Mehmood SS, Hussain MA, Wei S et al (2021) Omics: the way forward to enhance abiotic stress tolerance in *Brassica napus* L. GM Crops Food 12(1):251–281

Ren Y, Zhang N, Li R, Ma X, Zhang L (2021) Comparative transcriptome and flavonoids components analysis reveal the structural genes responsible for the yellow seed coat color of *Brassica rapa* L. Peer J 9:e10770

Rizwan M, Ali S, ur Rehman MZ, Rinklebe J, Tsang DCW et al (2018) Cadmium phytoremediation potential of Brassica crop species: a review. Sci Total Environ 631:1175–1191

Seifbarghi S, Borhan MH, Wei Y, Coutu C, Robinson SJ et al (2017) Changes in the *Sclerotinia sclerotiorum* transcriptome during infection of *Brassica napus*. BMC Genomics 18:266

Sendler E, Johnson GD, Krawetz SA (2011) Local and global factors affecting RNA sequencing analysis. Anal Biochem 419:317–322

Shah S, Weinholdt C, Jedrusik N, Molina C, Zouet J et al (2018) Whole-transcriptome analysis reveals genetic factors underlying flowering time regulation in rapeseed (*Brassica napus* L.). Plant Cell Environ 41 (8):1935–1947

Shamloo-Dashtpagerdi R, Razi H, Ebrahimie E, Niazi A (2018) Molecular characterization of *Brassica napus* stress related transcription factors, BnMYB44 and BnVIP1, selected based on comparative analysis of *Arabidopsis thaliana* and *Eutrema salsugineum* transcriptomes. Mol Biol Rep 45(5):1111–1124

Shan Q, Wang Y, Li J, Zhang Y, Chen K et al (2013) Targeted genome modification of crop plants using a CRISPR-Cas system. Nat Biotechnol 31:686–688

Sharma R, Mishra M, Gupta B, Parsania C, Singla-Pareek SL et al (2015) De novo assembly and characterization of stress transcriptome in a salinity-tolerant variety CS52 of *Brassica juncea*. PLoS One 10(5):e0126783

Shaw R, Tian X, Xu J (2021) Single-cell transcriptome analysis in plants: advances and challenges. Mol Plant 14:115–126

Shimizu M, Fujimoto R, Ying H, Pu ZJ, Ebe Y et al (2014) Identification of candidate genes for fusarium yellows resistance in Chinese cabbage by differential expression analysis. Plant Mol Biol 85:247–257

Shulse CN, Cole BJ, Ciobanu D, Lin J, Yoshinaga Y et al (2019) High-throughput single-cell transcriptome profiling of plant cell types. Cell Rep 27:2241–2247

Sikora P, Chawade A, Larsson M, Olsson J, Olsson O (2011) Mutagenesis as a tool in plant genetics, functional genomics, and breeding. Intl J Plant Genom 2011:314829

Sinha S, Raxwal VK, Joshi B, Jagannath A Katiyar-Agarwal S et al (2015) De novo transcriptome profiling of cold-stressed siliques during pod filling stages in Indian mustard (*Brassica juncea* L.). Front Plant Sci 6:932

Song JM, Liu DX, Xie WZ, Yang Z, Guo L et al (2020) BnPIR: *Brassica napus* pan-genome information resource for 1,689 accessions. Plant Biotechnol J 19 (3):412–414

Song K, Lu P, Tang K, Osborn TC (1995) Rapid genome change in synthetic polyploids of Brassica and its implications for polyploid evolution. Proc Natl Acad Sci USA 92:7719–7723

Stephenson P, Baker D, Girin T, Perez A, Amoah S et al (2010) A rich TILLING resource for studying gene function in *Brassica rapa*. BMC Plant Biol 10:1–10

Sun D, Wang C, Zhang X, Zhang W, Jiang H et al (2019) Draft genome sequence of cauliflower (*Brassica oleracea* L. var. botrytis) provides new insights into the C genome in Brassica species. Hortic Res 6:82

Sun Q, Zhang E, Liu Y, Xu Z, Hul M et al (2020) Transcriptome analysis of two lines of *Brassica oleracea* in response to early infection with *Xanthomonas Campestris* pv. *campestris*. Can J Plant Pathol 43:127–139

Tang F, Lao K, Surani MA (2011) Development and applications of single-cell transcriptome analysis. Nat Methods 8:S6–S11

Thakur S, Choudhary S, Bhardwaj P (2019) Comparative transcriptome profiling under cadmium stress reveals the uptake and tolerance mechanism in *Brassica juncea*. J Plant Growth Regul 38:1141–1152

The ENCODE Consortium (2011) Standards, guidelines and best practices for RNA-Seq

Tóth G, Gáspári Z, Jurka JJGR (2000) Microsatellites in different eukaryotic genomes: survey and analysis. Genome Res 10:967–981

Travella S, Klimm TE, Keller B (2006) RNA interference-based gene silencing as an efficient tool for functional genomics in hexaploid bread wheat. Plant Physiol 142:6–20

Variyar PS, Banerjee A, Jincy C, Suprasanna P (2014) Role of glucosinolates in plant stress tolerance. In: Ahmad P, Rasool S (eds) Emerging technologies and management of crop stress tolerance: volume 1—biological techniques. Academic, pp 271–291

Velculescu VE, Zhang L, Vogelstein B, Kinzler KW (1995) Serial analysis of gene expression. Science 270:484–487

Venkattakumar R, Padmaiah M (2010) Adoption behaviour of oilseed growers in India. Indian J Ext Educ 10:75–83

Wang A, Hu J, Huang X, Li X, Zhou G et al (2016a) Comparative transcriptome analysis reveals heat-responsive genes in Chinese cabbage (*Brassica rapa* ssp. chinensis). Front Plant Sci 7:93

Wang B, Tseng E, Regulski M, Clark TA, Hon T et al (2016b) Unveiling the complexity of the maize transcriptome by single-molecule long-read sequencing. Nat Commun 7:11708

Wang D, Yang C, Dong L, Zhu J, Wang J et al (2015) Comparative transcriptome analyses of drought-resistant and-susceptible *Brassica napus* L. and development of EST-SSR markers by RNA-Seq. J Plant Biol 58:259–269

Wang H, Wu J, Sun S, Liu B, Cheng F et al (2011a) Glucosinolate biosynthetic genes in *Brassica rapa*. Gene 487:135–142

Wang J, Singh SK, Geng S, Zhang S, Yuan L (2020a) Genome-wide analysis of glycerol-3-phosphate O-acyltransferase gene family and functional characterization of two cutin group GPATs in *Brassica napus*. Planta 251:1–16

Wang N, Wang Y, Tian F, King GJ, Zhang C et al (2008) A functional genomics resource for *Brassica napus*: development of an EMS mutagenized population and discovery of FAE1 point mutations by TILLING. New Phytol 180:751–765

Wang T, Guo S, Jiang Y, Zou J, Yu K et al (2020b) Comparative transcriptome and iTRAQ-based proteome analysis in mature leaves of *Brassica carinata* provides insights into the purple leaf color diversity. J Hortic Sci Biotechnol 96(4):444–455

Wang X, Wang H, Wang J, Sun R, Wu J et al (2011b) The genome of the mesopolyploid crop species *Brassica rapa*. Nat Genet 43:1035–1039

Wang Y, Sun S, Liu B, Wang H, Deng J et al (2011c) A sequence-based genetic linkage map as a reference for *Brassica rapa* pseudochromosome assembly. BMC Genomics 12:239

Wei L, Jian H, Lu K, Filardo F, Yin N et al (2015) Genome-wide association analysis and differential expression analysis of resistance to Sclerotinia stem rot in *Brassica napus*. Plant Biotechnol J 14(6):1368–1380

Wei X, Rahim M A, Zhao Y, Yang S, Wang Z et al (2021) Comparative transcriptome analysis of early- and late-bolting traits in Chinese cabbage (*Brassica rapa*). Front Genet 12:590830

Xian M, Luo T, Khan MN (2017) Identifying differentially expressed genes associated with tolerance against low temperature stress in *Brassica napus* through transcriptome analysis. Intl J Agric Biol 19:273–281

Xiao Z, Zhang C, Tang F, Yang B, Zhang L et al (2019) Identification of candidate genes controlling oil content by combination of genome-wide association and transcriptome analysis in the oilseed crop *Brassica napus*. Biotechnol Biofuels 12:216

Xiao Z, Liu X, Fang Z, Yang L, Zhang Y et al (2021) Transcriptome and plant hormone analyses provide new insight into the molecular regulatory networks underlying hybrid lethality in cabbage (*Brassica oleracea*). Planta 253:96

Xin H, Xianchao N, Pan X, Wei L, Min Y et al (2019) Comparative transcriptome analyses revealed conserved and novel responses to cold and freezing stress in *Brassica napus* L. G3 Genes|Genomes|Genetics 9 (8):2723–2737

Xing M, Lv H, Ma J, Xu D, Li H et al (2016) Transcriptome profiling of resistance to *Fusarium oxysporum* f. sp. conglutinans in cabbage (*Brassica oleracea*) roots. PLoS One 11(2):e0148048

Xiong X, Liu W, Jiang J, Xu L, Huang L et al (2019) Efficient genome editing of *Brassica campestris* based on the CRISPR/Cas9 system. Mol Genet Genom 294:1251–1261

Xu Y, Zeng A, Song L, Li J, Yan J (2019) Comparative transcriptomics analysis uncovers alternative splicing events and molecular markers in cabbage (*Brassica oleracea* L.). Planta 249(5):1599–1615

Yang J, Liu D, Wang X, Ji C, Cheng F et al (2016) The genome sequence of allopolyploid *Brassica juncea* and analysis of differential homoeolog gene expression influencing selection. Nat Genet 48:1225–1232

Yao S, Liang F, Gill RA, Huang J, Cheng X et al (2020) A global survey of the transcriptome of allopolyploid *Brassica napus* based on single-molecule long-read isoform sequencing and Illumina-based RNA sequencing data. Plant J 103:843–857

Yong H-Y, Zou Z, Kok E-P, Kwan B-H, Chow K et al (2014) Comparative transcriptome analysis of leaves and roots in response to sudden increase in salinity in *Brassica napus* by RNA-seq. BioMed Res Int 467395

Yu J, Zhao M, Wang X, Tong C, Huang S et al (2013) Bolbase: a comprehensive genomics database for *Brassica oleracea*. BMC Genomics 14:1–7

Yu S, Zhang F, Yu Y, Zhang D, Zhao X et al (2012) Transcriptome profiling of dehydration stress in the Chinese cabbage (*Brassica rapa* L. ssp. pekinensis) by tag sequencing. Plant Mol Biol Rep 30:17–28

Yue L, Li G, Dai Y, Sun X, Li F et al (2021) Gene co-expression network analysis of the heat-responsive core transcriptome identifies hub genes in *Brassica rapa*. Planta 253:111

Zaman QU, Chu W, Hao M, Shi Y, Sun M et al (2019) CRISPR/Cas9-mediated multiplex genome editing of JAGGED gene in *Brassica napus* L. Biomolecules 9 (11):725

Zhai C, Liu X, Song T, Yu F, Peng G (2021) Genome-wide transcriptome reveals mechanisms underlying

Rlm1-mediated blackleg resistance on canola. Sci Rep 11:4407

Zhai Y, Yu K, Cai S, Hu L, Amoo O et al (2020) Targeted mutagenesis of BnTT8 homologs controls yellow seed coat development for effective oil production in *Brassica napus* L. Plant Biotechnol J 18(5):1153–1168

Zhang B, Su T, Li P, Xin X, Cao Y et al (2021) Identification of long noncoding RNAs involved in resistance to downy mildew in Chinese cabbage. Hortic Res 8:44

Zhang Y, Peng L, Wu Y, Xin X, Cao Y et al (2014) Analysis of global gene expression profiles to identify differentially expressed genes critical for embryo development in *Brassica rapa*. Plant Mol Biol 86:425–442

Zheng H, Zhang Y, Li J, He L, Wang F et al (2020a) Comparative transcriptome analysis between a resistant and a susceptible Chinese cabbage in response to *Hyaloperonospora brassicae*. Plant Signal Behav 15:1777373

Zheng X, Koopmann B, Ulber B, von Tiedemann A (2020b) A global survey on diseases and pests in oilseed rape—current challenges and innovative strategies of control. Front Agron.https://doi.org/10.3389/fagro.2020.590908

Zhou Q, Galindo-González L, Manolii V, Hwang S-F, Strelkov SE (2020) Comparative transcriptome analysis of Rutabaga (*Brassica napus*) cultivars indicates activation of salicylic acid and ethylene-mediated defenses in response to *Plasmodiophora brassicae*. Intl J Mol Sci 21(21):8381

Zhou T, Xu W, Hirani AH, Tuan PA, Ayele BT et al (2019) Transcriptional insight into *Brassica napus* resistance genes LepR3 and Rlm2-mediated defense response against the *Leptosphaeria maculans* infection. Front Plant Sci 10:823

Zhuang H, Wang Q, Han H, Liu H, Wang H (2020) Single-molecule real-time transcript sequencing of turnips unveiling the complexity of the turnip transcriptome. G3 (Bethesda) 10(10):3505–3514

Metabolomics of *Brassica juncea* 18

Rimaljeet Kaur and Anju Sharma

Abstract

Plant metabolomics is an extensive, unbiased, and highly efficient phytochemical analysis of vast array of metabolites from a single extract. It has contributed significantly in understanding the basic metabolic mechanisms of plants, identification of chemical compounds that participate and respond to various environmental changes, recognition of defensive metabolites against pathogens and insects' infestation, chemical footprints at different phases of growth and development, deployment of gene mutations, discovery of agronomically important metabolites, accumulation and regulation of phytonutrients, etc. Indian mustard, *Brassica juncea* is an economical oilseed crop known for its edible oil, vegetable, condiment, and animal feed. It serves as a good candidate for studying metabolic profile of primary and secondary metabolites, responsive chemical compounds to various abiotic and biotic stresses, metabolic networks, and understanding of metabolic mechanisms and pathways. Indian mustard is the best representative for phytoremediation as it possesses capability of storing heavy metals directly as a crop and indirectly as an excellent gene source for improving other targeted crops. These species exhibit the capacity of sequestering metals by thiol-peptide compounds, contents of leaf pigments, enzymatic and non-enzymatic antioxidants. *B. juncea* as a vegetable crop is also known to produce several bioactive phytochemicals and nutrients of pharmaceutically relevant properties. Taken altogether, the chapter encompasses various metabolomic tools responsible for generating fast and wide information of metabolites, detailed metabolic profiling under heavy metal stress, therapeutic potential, and nutritional quality attributes in *B. juncea*. The information being assembled would aid in gaining better insights to understand the complexity of *B. juncea* metabolic networks, responses of metabolites to environmental and genetic alterations and basic knowledge of *B. juncea* genotypes *w.r.t* biodiversity, development, physiology, tissue identity, resistance, etc.

R. Kaur (✉) · A. Sharma
Department of Biochemistry, College of Basic Sciences & Humanities, Punjab Agricultural University, Ludhiana 141004, India
e-mail: rimal-pbg@pau.edu

A. Sharma
e-mail: anjusharma3@pau.edu

18.1 Introduction

With the advent of leading-edge metabolomic technologies following genomics era, researchers gained access to characterize a cell or organism at functional level. The information being

C. Kole and T. Mohapatra (eds.), *The Brassica juncea Genome*, Compendium of Plant Genomes,
https://doi.org/10.1007/978-3-030-91507-0_18

generated possesses enriched precision and speed in generating nutritionally improved germplasm, which is a prime concern for providing food safety and abundance (Parry and Hawkesford 2012). Metabolomics is the most complex and major breakthrough in science which allows speedy, accurate, and broad-spectrum detection of metabolites from single isolation process (Heyman and Dubery 2016; van Dam and Bouwmeester 2016; Wuolikainen et al. 2016). Specifically, metabolomics provides comprehensive investigation of cellular metabolites participating in diverse cellular events and metabolic pathways, thereby constituting the physiological state of a cell (Yonekura-Sakakibara and Saito 2006; Hong et al. 2016). Remarkably, metabolomics uncovers the impact of genes upon metabolic pathways, metabolic regulation, and integrated networking of various metabolites (Wen et al. 2015). Metabolomic platform encompassing large scale annotation of known and unknown metabolites is an integrated approach of improving attributes of interest (Matsuda et al. 2010; Lei et al. 2011; Pandey et al. 2016). Recent efforts in plant metabolomics have directed the improvement of yield and stress-related traits. Integration of various omics technologies guided researchers to functionally characterize massive genes and metabolites, identification of commercially important trait-specific markers. Successively, it improves the capability of a plant breeder for making favorable decisions while crop improvement.

Metabolomics expedites the selection of superior attributes and augments breeding materials in various crop species (Oikawa et al. 2008; Fernie and Schauer 2009; Daygon and Fitzgerald 2013; Simo et al. 2014). Metabolic profiling of *Brassica juncea* revealed the existence of various bio-active compounds such as carbohydrates, flavonoids, glycosides, phenolics, proteins, sterols, and triterpene alcohols (Allen et al. 1976; Li et al. 2000; Yokozawa et al. 2002; Das et al. 2009; Jung et al. 2009). Interestingly, several researchers have investigated regulation of key metabolites in *B. juncea*, especially under heavy metal stress conditions (Bauddh and Singh 2012; Mohamed et al. 2012; Ahmad and Gupta 2013; Kanwar et al. 2015). Given this, the chapter includes significant lead gained in the past decade in the field of phytoremediation and therapeutic properties of *B. juncea* metabolomics along with a brief discussion on the current contribution and the future scope of metabolomics in accelerating crop improvement.

18.2 Classification of Metabolomic Approaches and Metabolome Technologies

Metabolomics approaches are classified into targeted or non-targeted. Targeted metabolomics requires prior knowledge of metabolites and involves quantitative analysis of selected groups of metabolites (e.g., amino acids, lipids, sugars, and/or fatty acids). Thus, this approach investigates specific metabolic pathways, validates identified biomarkers, and does not achieve global coverage (Roberts et al. 2012). On the contrary, non-targeted metabolomics approaches are involved in global profiling of the metabolome such as biomarker discovery (Vinayavekhin et al. 2010). Thus, non-targeted metabolomics often provides more information than targeted metabolomics, but targeted metabolomics typically is more quantitative. A comprehensive view of metabolome is not provided by a single analytical technique and thus, multiple technologies are required for detailed and complete information on metabolome (Hall et al. 2002; Sumner et al. 2002). Metabolome profiling requires coupling of various biochemical techniques such as infrared spectroscopy (IR), nuclear magnetic resonance (NMR), thin layer chromatography (TLC), HPLC coupled with ultraviolet and photodiode array detection (LC/UV/PDA), coupling of capillary electrophoresis to ultraviolet absorbance detection (CE/UV), laser-induced fluorescence detection (CE/LIF), mass spectrometry (CE/MS), coupling of chromatographic techniques such as gas chromatography, liquid chromatography, high performance liquid chromatography with mass spectrometry (GC/MS,

LC/MS, HPLC/MS) and NMR (LC/NMR/MS/MS), Fourier transform ion cyclotron mass spectrometry (FTMS) (Oliver et al. 1998; Tweeddale et al. 1998; Bailey et al. 2000a, b; Fraser et al. 2000; Arlt et al. 2001; Bligny and Douce 2001; Ratcliffe and Shachar-Hill 2001; Baggett et al. 2002; Soga et al. 2002; Huhman and Sumner 2002; Aharoni et al. 2002) have all been used.

The most suitable technology is employed by considering speed, selectivity, and sensitivity features. For instance, NMR is rapid and selective but possesses low sensitivity; CE/LIF, CE/UV is highly sensitive but lacks selectivity; GC/MS and LC/MS provide good sensitivity and selectivity but require longer time for analysis.

18.3 Metabolic Profiling of *B. juncea* Under Heavy Metal Stress

Heavy metals are regarded as toxic environmental pollutants due to environmental and ecological concerns (Nagajyoti et al. 2010). Non-biodegradable nature of heavy metals causes long lasting persistence in the environment (Singh et al. 2011). They get frequently adsorbed on the soil surface, followed by gradual distribution in the soil. Plants accumulate heavy metals upon exposure and adversely affect its growth and development (Ramasubbu and Prabha 2012). Due to their long-duration persistence, these metals accumulate in the human body through food chain, which results in serious ill effects on health (Flora et al. 2012). The first line of plant defense to heavy metals stress is the production of reactive oxygen species (ROS) and hence elicits oxidative stress in plants (Mithofer et al. 2004). Plants exhibit various defensive mechanisms in response to stress like chelation, detoxification, exclusion of metal ions through phytoremediation, and activation of various stress protective proteins and osmolytes (Vamerali et al. 2010; Zhang et al. 2013; Zhou et al. 2013; Feng et al. 2021).

18.3.1 Cadmium Stress

Cadmium (Cd) being a phytotoxic metal affects plant growth, development, and vital metabolic processes. It causes browning of root tips, root length inhibition, biomass reduction, chlorosis, and even death and affects water and nutrient uptake, photosynthesis, and respiration (Singh et al. 2011). Cd toxicity causes blockage of essential functional groups in biomolecules, overproduction of ROS which triggers DNA and proteins damage, removal of cofactors from enzymes and makes them inactive, etc. (Mithofer et al. 2004; Ramasubbu and Prabha 2012; Flora et al. 2012). In order to alleviate Cd toxicity, plants exhibit various resistance mechanisms such as immobilization of cell walls, compartmentalization of vacuoles, exclusion of plasma membrane, etc. Compounds with higher percentage of cysteine (Cys) sulfhydryl residues such as thiol peptides play a critical role in Cd detoxification. Non-enzymatic antioxidant, reduced form of glutathione (γ-Glu-Cys-Gly, GSH) provides tolerance and sequestration of heavy metals by chelation and protect cells from oxidative stress damage (Vamerali et al. 2010; Jiang et al. 2012). GSH also acts as a precursor of phytochelatins (PCs) which are family of peptides with the general structure $(\gamma\text{-Glu-Cys})_n$-Gly, where n = number of repetitions of the γ-Glu-Cys unit, that can vary from 2 to 11 (Zhang et al. 2013). PCs possess high antioxidant capacity and reduce free metal concentration from the cytosol by binding and transporting the metal to specific compartments such as vacuoles (Zhou et al. 2013). Synthesis of PCs has been reported in Cd stress Indian mustard plants (Shekhawat et al. 2012).

Plants have acquired several resistance mechanisms for ROS detoxification viz., activation of enzymatic and non-enzymatic antioxidants. Enzymatic antioxidants include superoxide dismutases (SOD), peroxidases (POD), catalases (CAT), polyphenyl oxidase (PPO), and glutathione peroxidase (GPOX), and antioxidant metabolites such as ascorbate, glutathione, and α-tocopherol, scavenges free radicals and provide an important line of defense against Cd-induced oxidative stress (Jiang et al. 2012). Metabolic pathways such as ascorbate–glutathione cycle reduce H_2O_2 by utilizing ascorbate in an ascorbate peroxidase (APX)-dependent reaction and recycles ascorbate in the reduced form by dehydroascorbate reductase by using GSH as a substrate. Oxidized glutathione is, further, reduced by glutathione reductase in the presence of NAD(P)H (Jiang et al. 2012). Stress protective proteins, e.g., heat shock proteins also protect plants against oxidative damage (Donnelly and Robinson 1991; Chen and Yang 2012). Another resistance mechanism in plants involves the action of plant growth regulators (PGRs) and secondary metabolites. Plant hormones including auxins, abscisic acid, brassinosteroids, and polyamines regulate growth and development metabolic processes. PGRs act as stress protectants by activating antioxidative defense systems in plants and scavenges ROS (Haubrick and Assmann 2006; Sharma and Bhardwaj 2007; Verma et al. 2012).

Different families such as Brassicaceae, Fabaceae, and Caryophyllaceae exhibit the potential of tolerating high concentrations of metal(loid)s in the soil (Arnon 1949). Brassicaceae is the best-represented family among metal-accumulator families, and Indian mustard is a strong tool for developing effective Cd phytoremediation strategies due to its ability to store high concentrations of heavy metals in its organs (Velikova et al. 2000). Therefore, Indian mustard is involved directly as a phytoremediation crop and indirectly as an enriched source of genes for improving other phytoremediation-targeted plants (Maclachlan and Zalik 1963). In *B. juncea*, an increased level of Cd toxicity enhanced H_2O_2 content which might be due to cell membrane destabilization (Srivastava et al. 2005). Additionally, activities of SOD, PPO, GST, and GPOX enzymes in *B. juncea* were upregulated on exposure to Cd stress which causes detoxification and scavenging of free radicals like DPPH, superoxide radical, singlet oxygen, etc. (Fig. 18.1). Bauddh and Singh (2012) studied Cd tolerance, its phytoremediation, and the role of antioxidants from the contaminated soil of two oil-yielding plants *Ricinus communis* (L.) and *B. juncea* (L.). Plants were treated with 25, 50, 75, 100, and 150 mg Cd/kg soil for 60 days and observed that the contents of two non-enzymatic antioxidants viz. proline and malondialdehyde increased in the leaves of both the species with increase in Cd level whereas soluble protein decreased. The total metal accumulation, bioaccumulation, and translocation of Cd from roots to shoot were also found in both *B. juncea* and *R. communis* at all Cd concentrations. Further, Mohamed et al. (2012) observed that with an increase in Cd concentrations, a progressive accumulation of Cd occurs in roots and shoots, along with an organ-dependent alteration in mineral uptake. Root/shoot length, fresh/dry weight, chlorophyll, and carotenoid contents were decreased due to Cd treatment. Cd exposure unregulated xanthophyll cycle, thereby indicating the requirement of protecting photosynthetic apparatus from photoinhibition. Cd stress also revealed that shoots are less efficient than roots in scavenging ROS due to differential activities of POD and CAT and solely with respect to the highest Cd concentration, by ascorbate level. Such variability in antioxidant capacity might explain differences in lipid peroxidation of roots and shoots. Furthermore, glutathione and phytochelatin contents are significantly increased due to Cd stress in both shoots and roots, regardless of the metal concentration involved.

Tan et al. (2021) compared the effects of Cd stress in roots of *B. juncea* after 48 h and 7 d by metabolite profiling. Significant differences were observed in amino acids, organic acids, carbohydrates, lipids, flavonoids, alkaloids, and indoles and these metabolites exhibit important participation in adapting *B. juncea* roots to Cd stress. *B. juncea* roots regulate the synthesis of amino acids such as arginine, valine, leucine, isoleucine; lipid metabolism such as linoleic acid, glycerophospholipid, and alpha-linolenic acid metabolism and ABC transporters and thus, resist 48 h Cd stress. However, 7 d Cd stress was

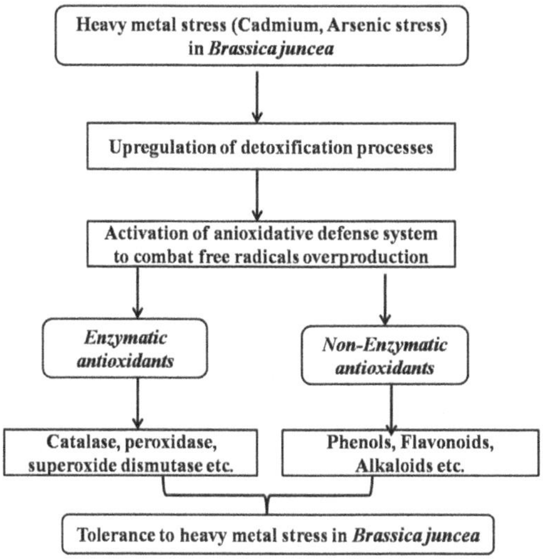

Fig. 18.1 Regulation of heavy metal detoxifying metabolites in *B. juncea*

resisted by regulating linoleic acid, alpha-linolenic acid, glycerophospholipid metabolism, and synchronizing the activity of ABC transporters in *B. juncea* (Sect. 18.3.1).

18.3.2 Arsenic Stress

Arsenic (As) is non-essential and pervasive metalloid that possesses high level of toxicity and imposes environmental and health concerns. Naturally, As occurs through dissolution and the adsorption of As compounds onto pyrite ores into the water by geochemical factors as well as anthropogenic through insecticide and herbicide uses, phosphate fertilization, and also from the semi-conductor industry (Mondal et al. 2006). The oxidation states of As are −3, 0, +3, and +5 and it exists in the form of arsenious acids (H_3AsO_4, $H_3AsO_4^-$, $H_3AsO_4^{2-}$) and arsenic acids (H_3AsO_4, $H_2AsO_4^-$, $HAsO_4^{2-}$), arsenates, arsenites, methyl arsenic acid, dimethyl arsenic acid, and arsine, etc. in nature (Mohan and Pittman 2007). Irrigation strategies involving contaminated groundwater causes entry of As into the food chain, e.g., in rice (Meharg 2004). The harmful effects of As are counteracted by adopting two main practices viz., phytoremediation which involves the removal of As from the environment, and safe crops development which can be raised in contaminated environments (Tripathi et al. 2008).

Various cellular, morphological, physiological, and biochemical processes such as shoot and root growth, germination, and biomass production are interrupted upon exposure of plants to As (Ahsan et al. 2008). Generation of ROS is also induced by As exposure, despite the fact that As is a non-redox-active metalloid (Srivastava et al. 2007; Mishra et al. 2008), owing to its intra-conversion from one ionic form to other (Mylona et al. 1998). The formation of free radicals leads to oxidative stress that results in lipid peroxidation, cell damage or even, cell death, biological macromolecule decay, deconstruction of membrane DNA, and proteins damage (Chardi et al. 2009; Flora 2011). To alleviate As stress, plants have developed various defensive metabolic pathways which not only maintain free metalloid ion concentration to minimum (primary detoxification) but also prohibit/repair the damage caused due to free ions (secondary detoxification) (Bleeker et al. 2006; Srivastava et al. 2007; Mishra et al. 2008). Free metal ions are detoxified with the activation of enzymatic and non-enzymatic antioxidants, hormones, etc. (Peleg and Blumwald 2011; Table 18.1) For instance, Brassinosteroids (BRs), a group of polyhydroxylated steroidal hormones, combat various stresses like heat, cold, drought, salt, heavy metals, pest infestation and pathogen infection (Bajguz and Hayat 2009; Kanwar et al. 2013) by crosstalking with other PGRs like salicylic acid, jasmonic acid, cytokinins, auxins, etc. PGRs trigger defensive mechanisms and protect the plant from various stresses (Bajguz and Hayat 2009).

Studies by Garg and Singla (2011) have reported that As (V) toxicity declined transpiration rate, water and essential nutrients transportation, and root activity. As also causes leaf abscission, plant growth inhibition, and biomass reduction by reacting with proteins' sulphydryl groups (Shao et al. 2011). Indian mustard is grown hydroponically displayed reduction in plant growth parameters due to As stress (Khan et al. 2009). Lipid peroxidation occurs as a result of oxidative stress caused by As stress leads to formation of Malondialdehyde (MDA) which serves as an indicator of oxidative stress and tissue damage in plants and animals. Several studies in *B. juncea* have reported that stress marker parameters like MDA and proline contents were upregulated due to As (V) toxicity at various growth and development stages. MDA enhancement due to As (V) leads to the production of superoxide radicals which increases lipid peroxidation (Khan et al. 2009; Ahmad and Gupta 2013; Kanwar et al. 2015). In order to alleviate oxidative stress, plants activate their antioxidative defense system (Kanwar et al. 2012). *B. juncea* plants treated with As (V) results in enhanced synthesis of stress protective proteins and antioxidative enzymes activity such as SOD, CAT, and GPX (Table 18.1; Gupta et al. 2009; Kanwar et al. 2015).

The potential of a plant to resist As stress depends upon its early perception. Studies have concluded that phosphate starvation via As (V) stress downregulates various genes in plants. On the contrary, plants have evolved As (V) sensing system which works antagonistic to the phosphate-sensing mechanism (Catarecha et al. 2007; Abercrombie et al. 2008). Important steps involved in As tolerance have been implicated as reduction of arsenate [As(V)] to arsenite [As(III)], organ-and tissue specific and subcellular distribution of As, complexation with sulfur-containing ligands, and vacuolar sequestration (Srivastava et al. 2012; Kumar et al. 2015). These mechanisms to avoid As toxicity are coordinated through transcriptome and proteome changes (Chakrabarty et al. 2009; Yu et al. 2012).

The underlying principle of As detoxification is chelation with sulfur-containing compounds such as GSH, PCs, and sequestration of complexes in the vacuoles (Table 18.1; Raab et al. 2005; Bleeker et al. 2006). The upregulation of PCs due to As stress is observed in hypertolerant, hyperaccumulator and non-hyperaccumulator plants (Hartley-Whitaker et al. 2001; Cai et al. 2004; Srivastava et al. 2007; Mishra et al. 2008). Abrupt biosynthesis of PCs via de novo pathway due to As stress requires increased GSH biosynthesis which further depends on cysteine biosynthesis due to upliftment of sulfur assimilation pathway (Rother et al. 2008). Hence, exposure of plants to heavy metal stress might lead to sulfur deficiency.

Several studies have depicted the significant role of thiols in As stress and detoxification by plants (Cai et al. 2004; Mishra et al. 2008; Castrillo et al. 2013). Transcriptomic profiling of rice plants treated with As (V) stress revealed the involvement of various genes of sulfur assimilation pathways such as *CS* and *cECS*, GSH biosynthetic, and transportation pathways (Norton et al. 2008). Enzyme activity, phytochelatin synthase was upregulated to maintain the rich supply of GSH. Thus, As stress triggers sulfate assimilation pathway (Harada et al. 2001) in plants. PGRs not only regulates growth and development but also respond to various abiotic and biotic stresses (Wang et al. 2002; Maruyama-Nakashita et al. 2004). For instance, jasmonic acid triggers GSH biosynthesis genes and regulates plant responses to sulfur depletion (Xiang and Oliver 1998; Nikiforova et al. 2003). Hence, As stressed plants activate jasmonate synthesis which might be involved in perception and signal transduction; thereby targets various defensive genes to combat As stress (Grsic et al. 1999). Further, signaling of As in *B. juncea* indicates the pivotal role of jasmonates, and microRNA specific microarray analysis in *B. juncea* also indicated the involvement of As-specific microRNAs in regulating the metabolism and function of PGRs like auxins, ABA, jasmonates, and sulfur metabolism (Srivastava et al. 2009, 2013). Therefore, two main strategies are responsible for As detoxification:

(i) Modification of signal transduction pathways on perceiving As stress which involves PGRs such as jasmonic acid, ethylene, auxins, and cytokinins to induce abrupt biochemical alterations as well as long-term adaptation responses.
(ii) Upregulation of As detoxification processes through activation of sulfate assimilation pathways to combat As stress.

Table 18.1 Key responsive metabolites in *B. juncea* under heavy metal stress

Type of heavy metal stress	Metabolite involved	Status	References
Cd stress	H_2O_2, SOD, PPO, GST, GPOX, POD, CAT, ascorbate, glutathione, phytochelatin, proline, MDA, flavonoids, indoles, alkaloids	Increase	Srivastava et al. (2005), Mohamed et al. (2012), Bauddh and Singh (2012), and Tan et al. (2021)
	Soluble protein, chlorophyll, carotenoid	Decrease	Mohamed et al. (2012)
	Xanthophyll cycle	Activation	
As stress	SOD, CAT, GPX, ROS, brassinosteroids, glutathione, phytochelatin, phytochelatin synthase, malondialdehyde, proline, jasmonate, abscisic acid, auxin, superoxide radicals	Increase	Grsic et al. (1999), Bleeker et al. (2006), Srivastava et al. (2007), Mishra et al. (2008), Ahsan et al. (2008), Norton et al. (2008), Gupta et al. (2009), Khan et al. (2009), Srivastava et al. (2009), Ahmad and Gupta (2013), and Kanwar et al. (2015)

18.4 Therapeutic Potential of *B. juncea*

B. juncea possess diverse medicinal properties from seed and oil products. Seeds are an excellent source of protein and edible oil. An essential oil, popularly known as mustard oil is used in cosmetics for hair control (Yu et al. 2003). Commercialized oils of *B. juncea* exhibit chief pungent chemical constituent, Allyl isothiocyanate. This isothiocyanate is most important cancer chemo-preventive phytochemical and also acts as antimicrobial agent against various organisms (Luciano and Holley 2009; Okulicz 2010; Zhang et al. 2010). Metabolic profiling of *B. juncea* revealed the biosynthesis of various biochemical constituents/phytochemicals such as carbohydrates, protein, flavonols, glucosinolates, phenolics, sterols, and triterpene alcohols (Fabre et al. 1997; Li et al. 2000; Yokozawa et al. 2002; Das et al. 2009; Jung et al. 2009).

18.4.1 Glucosinolates

Glucosinolates are glucose derivatives attached to sulphonated oxime through aliphatic, aromatic, or heterocyclic side chains (Chew 1988). Brassicaceae plants constitute 200 diverse glucosinolate compounds which are mainly categorized into two main classes: aliphatic and aromatic glucosinolates. Aliphatic glucosinolates, commonly found in Brassica species, possess but-3-enyl and prop-2-enyl (Fenwick et al. 1983) compounds. Seeds *of B. juncea* exhibit three different types of glucosinolates such as Sinalbin, Glucoarabin, and Glucohirsutin (Fabre et al. 1997). Sinigrin and glucoraphanin are mainly present in seeds and leaves of *B. juncea* which act as a precursor of anti-cancerous compounds, allyl isothiocyanate, and sulphoraphane, respectively. Sinigrin has goitrogenic activity while allyl isothiocyanate possesses fungicidal, antitumor, antimicrobial, and antioxidant activities.

18.4.2 Flavonoids and Other Phenolic Compounds

Flavonoids and hydroxycinnamic acids are enriched polyphenols in Brassica. Among flavonoids, flavonols and anthocyanins exist largely in *Brassica* species. Chemically, flavonoids are comprised of fifteen carbons with two aromatic rings connected by a three carbon bridge. Quercetin, Kaempferol, and Isorhamnetin are the main flavonols and are commonly found as O-glycosides. Colored flavonoids and anthocyanins have antioxidative activities and the most common are pelargonidin, cyanidin, delphinidin, peonidin, petunidin, and malvidin with cyanidin (Sadilova et al. 2006; McDougall et al. 2007; Moreno et al. 2010) A novel and rare Kaempferol-3-Glucoside-2″-Rhamnoside-7-Rhamnoside, was isolated from *B. juncea*. Earlier known compounds from the leaves of *B. juncea* are kaempferol-3-O-(2-O-feruloyl-β-D-glucopyranosyl-(1 → 2)-β-D-glucopyranoside)-7-O-β-D-glucopyranoside, kaempferol-3-O-β-D-glucopyranosyl-(1 → 2)-O-β-D-glucopyranoside-7-O-β-D-glucopyranoside, 1-O-sinapoyl-glucopyranoside, kaempferol-3-O-(2-O-sinapoyl)-β-D-glucopyranosyl-(1 → 2)-β-D-glucopyranoside-7-O-β-D-glucopyranoside, kaempferol-3-O-β-D-glucopyranosyl-(1 → 2)-β-D-glucopyranoside-7-O-β-D-glucopyranosyl-(1 → 6)-β-D-glucopyranoside, and kaempferol-3-O-(2-O-sinapoyl)-β-D-glucopyranosyl-(1 → 2)-β-D-gluco-pyranoside-7-O-β-D-glucopyranosyl-(1 → 6)-β-D-glucopyranoside (Kim et al. 2002; Jung et al. 2009). The major flavonoid in leaves of *B. juncea* is Isorhamnetin 3,7-di-O-β-D-glucopyranoside (Isorhamnetin diglucoside) (Yokozawa et al. 2002).

Mustard meal is a good source of phenolic compounds which are bitter and astringent in nature. These compounds are now emerging as value-added products due to their antioxidative properties. The major phenolic compound in mustard meal is sinapic acid (SA) (MW—224.2 Da) and sinapine (MW—275 Da) (Das et al. 2009).

18.4.3 Proteins, Phytosterols, and Alcohols

Brassica species exhibit two major storage proteins viz., napin and cruciferin which constitute 45–50% and 25%, respectively (Appelqvist et al. 1973). Napin of *B. juncea* consists of two polypeptides, a small subunit of 29 amino acids and a large subunit of 86 amino acids which are held together by disulfide bridges (Dasgupta et al. 1995). These bridges provide stability and compactness to napin. Trypsin inhibitor is the precursor of napin which is antifungal and allergenic in nature (Jyothi et al. 2007). Another antifungal protein, juncin (18.9 kDa) was extracted from *B. juncea* seeds (Ye and Ng 2009).

Phytosterols are highest in concentration (64 mg/g) among various edible fats and oils in mustard oil (Sabir et al. 2003). Long chain alcohols and phytosterols are active components of "healthy vegetables" in mustard seeds and are useful in impeding hypercholesterolemia (Table 18.2).

18.5 Concluding Remarks and Future Perspectives

Following the plant metabolomics development, an advanced shift from single metabolite analysis to high throughput assays of profiling various metabolites in one go has paved the way for generating metabolic networks, identification of biomarkers, and antioxidative constituents under various abiotic and biotic stress conditions. Metabolomics in *B. juncea* has allowed the identification of metabolic changes and their impact during stress conditions. Additionally, seeds and oil metabolite profiling in *B. juncea* revealed the existence of therapeutic potential due to glucosinolates, flavonoids, phytosterols, etc. In the last decade, integrating layers of metabolomics with other omics approaches has uncovered a plethora of known and novel metabolites which contribute towards *B. juncea* improvements in terms of resistance, yield, oil quality, etc. To this end, high throughput avenues such as GWAS, GBS, transcriptomics, proteomics has allowed efficient integration of metabolite profiling in designing and developing agronomically superior plants that adequately would meet the challenges of twenty-first century agriculture.

Table 18.2 Pharmacological properties of metabolites in *B. juncea*

S. No.	Metabolite	Therapeutic property	References
1	Glucosinolates Sinigrin (allyl glucosinolate)	Goitrogenic	Yu et al. (2003)
2	Isothiocyanate I. Allyl isothiocyanate II. Phenyl isothiocyanate	Fungicidal activity, antitumor activity, antimicrobial activity, antioxidant activity	Luciano and Holley (2009) and Kumar et al. (2009)
3	Phenolic constituents (sinapic acid, sinapine)	Anxiolytic activity, antioxidant, cognition improving activity	Karakida et al. (2007)
4	Fatty acids (α-linolenic acid)	Astrocyte developing activity and other health benefits	Joardar and Das (2007)
5	Kaempferol glycosides	Antioxidant activity	Kim et al. (2002) and Jung et al. (2009)
6	Flavonoid (Isorhamnetin diglucoside)	Antioxidant effect	Yokozawa et al. (2003)
7	Proteins I. Napins	Antifungal Allergenicity	Jyothi et al. (2007)
	II. Juncin	Antifungal	Ye and Ng (2009)

References

Abercrombie JM, Halfhill MD, Ranjan P, Rao MR, Saxton AM, Yuan JS, Stewart CN Jr (2008) Transcriptional responses of *Arabidopsis thaliana* plants to As(V) stress. BMC Plant Biol 8:87

Aharoni A, de Vos CH, Verhoeven HA, Maliepaard CA, Kruppa G, Bino RJ, Goodenowe DB (2002) Non-targeted metabolome analysis by use of Fourier Transform Ion Cyclotron Mass Spectrometry. OMICS 6(3):217–234

Ahmad MA, Gupta M (2013) Exposure of *Brassica juncea* (L.) to arsenic species in hydroponic medium: comparative analysis in accumulation and biochemical and transcriptional alterations. Environ Sci Pollut Res Int 20(11):8141–8150

Ahsan N, Lee D-G, Alam I, Kim PJ, Lee JJ, Yo A, Kwak SS, Lee IJ, Bahk JD, Kang KY, Renaut J, Komatsu S, Lee BH (2008) Comparative proteomic study of arsenic-induced differentially expressed proteins in rice roots reveals glutathione plays a central role during As stress. Proteomics 8(17):3561–3576

Allen SE, Grimshaw HM, Rowland AP (1976) Chemical analysis. In: Chapman SB (ed) Methods in plant ecology. Blackwell Scientific Publications, Oxford, UK, pp 311–314

Appelqvist LA, Ohlson R, Sprague MA (1973) Rapeseed: cultivation, composition, processing and utilization. Soil Sci 116(6):453

Arlt K, Brandt S, Kehr J (2001) Amino acid analysis in five pooled single plant cell samples using capillary electrophoresis coupled to laser-induced fluorescence detection. J Chromatogr A 926(2):319–325

Arnon DI (1949) Copper enzymes in isolated chloroplasts. Polyphenoloxidase in *Beta vulgaris*. Plant Physiol 24(1):1–15

Baggett BR, Cooper JD, Hogan ET, Carper J, Paiva NL, Smith JT (2002) Profiling isoflavonoids found in legume root extracts using capillary electrophoresis. Electrophoresis 23(11):1642–1651

Bailey NJ, Cooper P, Hadfield ST, Lenz EM, Lindon JC, Nicholson JK, Stanley PD, Wilson ID, Wright B, Taylor SD (2000) Application of directly coupled HPLC-NMR-MS/MS to the identification of metabolites of 5-trifluoromethylpyridone (2-hydroxy-5-trifluoromethylpyridine) in hydroponically grown plants. J Agric Food Chem 48(1):42–46

Bailey NJ, Stanley PD, Hadfield ST, Lindon JC, Nicholson JK (2000) Mass spectrometrically detected directly coupled high performance liquid chromatography/nuclear magnetic resonance spectroscopy/mass spectrometry for the identification of xenobiotic metabolites in maize plants. Rapid Commun Mass Spectrom 14(8):679–684

Bajguz A, Hayat S (2009) Effects of brassinosteroids on the plant responses to environmental stresses. Plant Physiol Biochem 47(1):1–8

Bauddh K, Singh RP (2012) Cadmium tolerance and its phytoremediation by two oil yielding plants *Ricinus communis* (L.) and *Brassica juncea* (L.) from the contaminated soil. Int J Phytoremed 14(8):772–785

Bleeker PM, Hakvoort HWJ, Bliek M, Souer E, Schat H (2006) Enhanced arsenate reduction by a CDC25-like tyrosine phosphatase explains increased phytochelatin accumulation in arsenate-tolerant *Holcus lanatus*. Plant J 45(6):917–929

Bligny R, Douce R (2001) NMR and plant metabolism. Curr Opin Plant Biol 4(3):191–196

Cai Y, Su J, Ma LQ (2004) Low molecular weight thiols in arsenic hyperaccumulator *Pteris vittata* upon exposure to arsenic and other trace elements. Environ Pollut 129(1):69–78

Castrillo G, Sanchez-Bermejo E, de Lorenzo L, Crevillen P, FraileEscanciano A, Mohan TC, Mouriz A, Catarecha P, Sobrino-Plata J, Olsson S, Leo Del Puerto Y, Mateos I, Rojo E, Hernández LE, Jarillo JA, Piñeiro M, Paz-Ares J, Leyva A (2013) WRKY6 transcription factor restricts arsenate uptake and transposon activation in Arabidopsis. Plant Cell 25 (8):2944–2957

Catarecha P, Segura MD, Franco-Zorrilla JM, Garcia-Ponce B, Lanza M, Solano R, Paz-Ares J, Leyva A (2007) A mutant of the Arabidopsis phosphate transporter PHT1;1 displays enhanced arsenic accumulation. Plant Cell 19(3):1123–1133

Chakrabarty D, Trivedi PK, Misra P, Tiwari M, Shri M, Shukla D, Kumar S, Rai A, Pandey A, Nigam D, Tripathi RD, Tuli R (2009) Comparative transcriptomic analysis of arsenate and arsenite stresses in rice seedlings. Chemosphere 74(5):688–702

Chardi AS, Matutano CP, Borras M, Nadal J (2009) Bioaccumulation of metals and effects of a landfill in small mammals. Part III: structural alterations. Environ Res 109(8):960–967

Chen J, Yang ZM (2012) Mercury toxicity, molecular response and tolerancein higher plants. Biometals 25:847–857

Chew FS (1988) Biological effects of glucosinolates. In: Cutler HG (ed) Biologically active natural products: potential use in agriculture. American Chemical Society, Washington, USA, pp 155–181

Das R, Bhattacherjee C, Ghosh S (2009) Preparation of mustard (*Brassica juncea* L.) protein isolate and recovery of phenolic compounds by ultrafiltration. Ind Eng Chem Res 48(10):4939–4947

Dasgupta J, Dasgupta S, Ghosh S, Roy B, Mandal RK (1995) Deduced amino acid sequence of 2S storage protein from Brassica species and their structural features. Indian J Biochem Biophys 32(6):378–384

Daygon V, Fitzgerald M (2013) Application of metabolomics for providing a new generation of selection tools for crop improvement. In: Daygon V, Fitzgerald M (eds) Hot topics in metabolomics: food and nutrition. Future science book series. Future Science Ltd., London, pp 6–16

Donnelly JK, Robinson DS (1991) Superoxide dismutase. In: Robinson DS, Eskin NAM (eds) Oxidative enzymes in foods. Elsevier Applied Science, London, UK, pp 49–91

Fabre N, Bon M, Moulis C, Fourastfazn I, Stanislas E (1997) Three glucosinolates from seeds of *Brassica juncea*. Phytochemistry 45(3):525–527

Feng Z, Ji S, Ping J, Cui D (2021) Recent advances in metabolomics for studying heavy metal stress in plants. Trends Analyt Chem 143:116402

Fenwick GR, Heaney RK, Mullin WJ (1983) Glucosinolate and their breakdown products in food and food plants. Crit Rev Food Sci Nutr 18(2):123–201

Fernie AR, Schauer N (2009) Metabolomics-assisted breeding: a viable option for crop improvement? Trends Genet 25(1):39–48

Flora G, Gupta D, Tiwari A (2012) Toxicity of lead: a review with recent updates. Interdiscip Toxicol 5 (2):47–58

Flora SJS (2011) Arsenic-induced oxidative stress and its reversibility. Free Radic Biol Med 51(2):257–281

Fraser PD, Pinto ME, Holloway DE, Bramley PM (2000) Application of high-performance liquid chromatography with photodiode array detection to the metabolic profiling of plant isoprenoids. Plant J 24(4):551–558

Garg N, Singla P (2011) Arsenic toxicity in crop plants: physiological effects and tolerance mechanisms. Environ Chem Lett 9(3):303–321

Grsic S, Kirchheim B, Pieper K, Fritsch M, Hilgenberg W, Ludwig-Miuller J (1999) Induction of auxin biosynthetic enzymes by jasmonic acid and in club root diseased Chinese cabbage plants. Physiol Planta 105 (3):521–531

Gupta M, Sharma P, Sarin NB, Sinha AK (2009) Differential response of arsenic stress in two varieties of *Brassica juncea* L. Chemosphere 74(9):1201–1208

Hall R, Beale M, Fiehn O, Hardy N, Sumner L, Bino R (2002) Plant metabolomics as the missing link in functional genomics strategies. Plant Cell 14(7):1437–1440

Harada E, Coi Y-E, Tsuchisaka A, Obata H, Sano H (2001) Transgenic tobacco plants expressing a rice cysteine synthase gene are tolerant to toxic levels of cadmium. J Plant Physiol 158(5):655–661

Hartley-Whitaker J, Ainsworth G, Vooijs R, Ten Bookum WM, Schat H, Meharg AA (2001) Phytochelatins are involved in differential arsenate tolerance in *Holcus lanatus* L. Plant Physiol 126(1):299–306

Haubrick LL, Assmann SM (2006) Brassinosteroids and plant function: some clues, more puzzles. Plant Cell Environ 29(3):446–457

Heyman HM, Dubery IA (2016) The potential of mass spectrometry imaging in plant metabolomics: a review. Phytochem Rev 15:297–316

Hong J, Yang L, Zhang D, Shi J (2016) Plant metabolomics: an indispensable system biology tool for plant science. Int J Mol Sci 17(6):E767

Huhman DV, Sumner LW (2002) Metabolic profiling of saponins in *Medicago sativa* and *Medicago truncatula* using HPLC coupled to an electrospray ion-trap mass spectrometer. Phytochemistry 59(3):347–360

Jiang HLM, Che LL, Nie L, Yang ZM (2012) BjHO-1 is involved in the detoxification of heavy metal in India mustard (*Brassica juncea*). Biometals 25(6):1269–1279

Joardar A, Das S (2007) Effect of fatty acids isolated from edible oilslike mustard, linseed or coconut on astrocytes maturation. CellMol Neurobiol 27(8):973–983

Jung HA, Woo JJ, Jung MJ, Hwang GS, Choi JS (2009) Kaempferol glycosides with antioxidant activity from *Brassica juncea*. Arch Pharm Res 32(10):1379–1384

Jyothi TC, Sinha S, Singh SA, Surolia A, Appu Rao AG (2007) Napin from *Brassica juncea*: thermodynamic and structural analysis of stability. Biochim Biophys Acta 1774(7):907–919

Shekhawat K, Rathore SS, Premi OP, Kandpal BK, Chauhan JS (2012) Advances in agronomic management of Indian mustard (*Brassica juncea* (L.) Czernj. Cosson): an overview. Int J Agron 2012:1–14

Kanwar MK, Bhardwaj R, Arora P, Chowdhary SP, Sharma P, Kumar S (2012) Plant steroid hormones produced under Ni stress are involved in the regulation of metal uptake and oxidative stress in *Brassica juncea* L. Chemosphere 86(1):41–49

Kanwar MK, Bhardwaj R, Chowdhary SP, Arora P, Sharma P, Kumar S (2013) Isolation and characterization of 24-epibrassinolide from *Brassica juncea* L. and its effects on growth, Ni uptake and antioxidant defense of Brassica plants and *in vitro* cytotoxicity. Acta Physiol Plant 35(4):1351–1362

Kanwar MK, Poonam BR, Bhardwaj R (2015) Arsenic induced modulation of antioxidative defense system and brassinosteroids in *Brassica juncea* L. Ecotoxicol Environ Safe 115:119–125

Karakida F, Ikeya Y, Tsunakawa M, Yamaguchi T, Ikarashi Y, Takeda S, Aburada M (2007) Cerebral protective and cognition-improving effects of sinapic acid in rodents. Biol Pharm Bull 30(3):514–519

Khan I, Ahmad A, Iqbal M (2009) Modulation of antioxidant defence system for arsenic detoxification in Indian mustard. Ecotoxicol Environ Saf 72(2):626–634

Kim JE, Jung MJ, Jung HA, Woo JJ, Cheigh HS, Chung HY, Choi JS (2002) A new kaempferol 7-O-triglucoside from the leaves of *Brassica juncea* L. Arch Pharm Res 25(5):621–624

Kumar A, D'Souza SS, Tickoo S, Salimath BP, Singh HB (2009) Antiangiogenic and proapoptotic activities of allyl isothiocyanate inhibit ascites tumor growth *in vivo*. Integr Cancer Ther 8(1):75–87

Kumar S, Dubey RS, Tripathi RD, Chakrabarty D, Trivedi PK (2015) Omics and biotechnology of arsenic stress and detoxification in plants: current updates and prospective. Environ Int 74:221–230

Lei Z, Huhman DV, Sumner LW (2011) Mass spectrometry strategies in metabolomics. J Biol Chem 286 (29):25435–25442

Li Q, Eigenbrode SD, Stringam GR, Thiagarajah MR (2000) Feeding and growth of *Plutella xylostella* and *Spodoptera eridania* on *Brassica juncea* with varying glucosinolate concentrations and myrosinase activities. J Chem Ecol 26:2401–2419

Luciano FB, Holley RA (2009) Enzymatic inhibition by allyl isothiocyanate and factors affecting its antimicrobial action against *Escherichia coli* O157: H7. Int J Food Microbiol 131(2–3):240–245

Maclachlan S, Zalik S (1963) Plastid structure, chlorophyll concentration and free amino acid composition of a chlorophyll mutant of barley. Can J Bot 41 (7):1053–1062

Maruyama-Nakashita A, Nakamura Y, Yamaya T, Takahashi H (2004) A novel regulatory pathway of sulfate uptake in Arabidopsis roots: implication of CRE1/WOL/AHK4-mediated cytokinin-dependent regulation. Plant J 38(5):779–789

Matsuda F, Hirai MY, Sasaki E, Akiyama K, Yonekura-Sakakibara K, Provart NJ, Sakurai T, Shimada Y, Saito K (2010) AtMetExpress development: a phytochemical atlas of Arabidopsis development. Plant Physiol 152(2):566–578

McDougall GJ, Fyffe S, Dobson P, Stewart D (2007) Anthocyanins from red cabbage—stability to simulated gastrointestinal digestion. Phytochemistry 68 (9):1285–1294

Meharg AA (2004) Arsenic in rice—understanding a new disaster for South-East Asia. Trends Plant Sci 9 (9):415–417

Mishra S, Srivastava S, Tripathi RD, Trivedi PK (2008) Thiol metabolism and antioxidant system complement each other during arsenate detoxification in *Ceratophyllum demersum* L. Aquat Toxicol 86(2):205–215

Mithofer A, Schulze B, Boland W (2004) Biotic and heavy metal stress response in plants: evidence for common signals. FEBS Lett 566(1–3):1–5

Mohamed AA, Castagna A, Ranieri A, Toppi LS (2012) Cadmium tolerance in *Brassica juncea* roots and shoots is affected by antioxidant status and phytochelatin biosynthesis. Plant Physiol Biochem 57:15–22

Mohan D, Pittman CU Jr (2007) Arsenic removal from water/wastewater using adsorbents—a critical review. J Hazard Mater 142(1–2):1–53

Mondal P, Majumdar CB, Mohanty B (2006) Laboratory based approaches for arsenic remediation from contaminated water: recent developments. J Hazard Mater 137(1):464–479

Moreno DA, Perez-Balibrea S, Ferreres F, Gil-Izquierdo A, Garcia-Viguera C (2010) Acylated anthocyanins in broccoli sprouts. Food Chem 123(2):358–363

Mylona PV, Polidoros AN, Scandalios JG (1998) Modulation of antioxidant responses by arsenic in maize. Free Radic Biol Med 25(4–5):576–585

Nagajyoti PC, Lee KD, Sreekanth TVM (2010) Heavymetals, occurrence and toxicity for plants: a review. Environ Chem Lett 8(3):199–216

Nikiforova V, Freitag J, Kempa S, Adamik M, Hesse H, Hoefgen R (2003) Transcriptome analysis of sulfur depletion in *Arabidopsis thaliana*: interlacing of biosynthetic pathways provides response specificity. Plant J 33(4):633–650

Norton GJ, Lou-Hing DE, Meharg AA, Price AH (2008) Rice arsenate interactions in hydroponics: whole genome transcriptional analysis. J Expt Bot 59 (8):2267–2276

Oikawa A, Matsuda F, Kusano M, Okazaki Y, Saito K (2008) Rice metabolomics. Rice 1:63–71

Okulicz M (2010) Multidirectional time-dependent effect of sinigrin and allyl isothiocyanate on metabolic parameters in rats. Plant Foods Hum Nutr 65 (3):217–224

Oliver SG, Winson MK, Kell DB, Baganz F (1998) Systematic functional analysis of the yeast genome. Trends Biotechnol 16(9):373–378

Pandey MK, Roorkiwal M, Singh VK, Ramalingam A, Kudapa H, Thudi M, Chitikineni A, Rathore A, Varshney RK (2016) Emerging genomic tools for legume breeding: current status and future perspectives. Front Plant Sci 7:455

Parry MA, Hawkesford MJ (2012) An integrated approach to crop genetic improvement. J Integr Plant Biol 54:250–259

Peleg Z, Blumwald E (2011) Hormone balance and abiotic stress tolerance in crop plants. Curr Opin Plant Biol 14(3):290–295

Raab A, Schat H, Meharg AA, Feldmann J (2005) Uptake, translocation and transformation of arsenate and arsenite in sunflower (*Helianthus annuus*): formation of arsenic–phytochelatin complexes during exposure to high arsenic concentrations. New Phytol 168(3):551–558

Ramasubbu R, Prabha C (2012) Effect of heavy metals on growth and biochemical components of plants. LAP Lambert Academic Publishing, Mauritus

Ratcliffe RG, Shachar-Hill Y (2001) Probing plant metabolism with NMR. Annu Rev Plant Physiol Plant Mol Biol 52:499–526

Roberts LD, Souza AL, Gerszten RE, Clish CB (2012) Targeted metabolomics. Curr Protoc Mol Biol PMC

Rother M, Krauss G-J, Grass G, Wesenberg D (2008) Sulphate assimilation under Cd^{2+} stress in *Physcomitrella patens*—combined transcript, enzyme and metabolite profiling. Plant Cell Environ 29(9):1801–1811

Sabir SM, Hayat I, Gardezi SDA (2003) Estimation of sterols in edible fats and oils. Pak J Nutr 2(3):178–181

Sadilova E, Stintzing FC, Carle R (2006) Anthocyanins, colour and antioxidant properties of eggplant (*Solanum melongena* L.) and violet pepper (*Capsicum annuum* L.) peel extracts. Z Naturforsch C 61(7–8):527–535

Shao Y, Jiang LN, Zhang DJ, Ma LJ, Li CX (2011) Effects of arsenic, cadmium and lead on growth and respiratory enzymes activity in wheat seedlings. Afr J Agric Res 6(19):4505–4512

Sharma P, Bhardwaj R (2007) Effect of 24-epibrssinolide on seed germination, seedling growth and heavy metal uptake in *Brassica juncea* L. Gen Appl Plant Physio 33(1–2):59–73

Simo C, Ibáñez C, Valdés A, Cifuentes A, García-Cañas V (2014) Metabolomics of genetically modified crops. Int J Mol Sci 15(10):18941–18966

Singh R, Gautam N, Mishra A, Gupta R (2011) Heavy metals and living systems: an overview. Indian J Pharmacol 43(3):246–253

Soga Y, Ueno Y, Naraoka H, Ohashi Y, Tomita M, Nishioka T (2002) Simultaneous determination of anionic intermediates for *Bacillus subtilis* metabolic pathways by capillary electrophoresis electrospray ionization mass spectrometry. Anal Chem 74 (10):2233–2239

Srivastava M, Ma LQ, Singh N, Singh S (2005) Antioxidant responses of hyper-accumulator and sensitive fern species to arsenic. J Exp Bot 56(415):1335–1342

Srivastava S, Mishra S, Tripathi RD, Dwivedi S, Trivedi PK, Tandon PK (2007) Phytochelatins and antioxidant systems respond differentially during arsenite and arsenate stress in *Hydrilla verticillata* (L.f.). Environ Sci Technol 41(8):2930–2936

Srivastava S, Srivastava A K, Suprasanna P, D'Souza SF (2009) Comparative biochemical and transcriptional profiling of two contrasting varieties of *Brassica juncea* L. in response to arsenic exposure reveals mechanisms of stress perception and tolerance. J Exp Bot 60(12):3419–3431

Srivastava S, Srivastava AK, Suprasanna P, D'Souza SF (2013) Identification and profiling of arsenic stress-induced microRNAs in *Brassica juncea*. J Exp Bot 64 (1):303–315

Srivastava S, Suprasanna P, D'Souza SF (2012) Mechanisms of arsenic tolerance and detoxification in plants and their application in transgenic technology: a critical appraisal. Int J Phytoremed 14(5):506–517

Sumner LW, Duran AL, Huhman DH, Smith JT (2002) Metabolomics: a developing and integral component in functional genomic studies of *Medicago truncatula*. In: Romeo JT, Dixon RA (eds) Recent advances in phytochemistry. Elsevier, Pergamon, Oxford, UK, pp 31–61

Tan P, Zeng C, Wan C, Liu Z, Dong X, Peng J, Lin H, Li M, Liu Z, Yan M (2021) Metabolic profiles of *Brassica juncea* roots in response to cadmium stress. Metabolites 11:383

Tripathi RD, Srivastava S, Mishra S, Singh N, Tuli R, Gupta DK, Maathuis FJM (2008) Arsenic hazards: strategies for tolerance and remediation by plants. Trends Biotechnol 25:158–165

Tweeddale H, Notley-McRobb L, Ferenci T (1998) Effect of slow growth on metabolism of *Escherichia coli*, as revealed by global metabolite pool ("Metabolome") analysis. J Bacteriol 180(19):5109–5116

Vamerali T, Bandiera M, Mosca G (2010) Field crops for phytoremediation of metal-contaminated land. A review. Environ Chem Lett 8(1):1–17

van Dam NM, Bouwmeester HJ (2016) Metabolomics in the rhizosphere: tapping into belowground chemical communication. Trends Plant Sci 21:256–265

Velikova V, Yordanov I, Edreva A (2000) Oxidative stress and some antioxidant systems in acid rain-treated bean plants protective role of exogenous polyamines. Plant Sci 151(1):59–66

Verma A, Malik CP, Gupta VK (2012) In vitro effects of brassinosteroids on the growth and antioxidant enzyme activities in groundnut. ISRN Agron 1–8

Vinayavekhin N, Homan EA, Saghatelian A (2010) Exploring disease through metabolomics. ACS Chem Biol 5(1):91–103

Wang KLC, Li H, Ecker JR (2002) Ethylene biosynthesis and signaling networks. Plant Cell 14(1):S131–S151

Wen W, Li K, Alseekh S, Omranian N, Zhao L, Zhou Y, Xiao Y, Jin M, Yang N, Liu H, Florian A, Li W, Pan Q, Nikoloski Z, Yan J, Fernie AR (2015) Genetic determinants of the network of primary metabolism and their relationships to plant performance in a maize recombinant inbred line population. Plant Cell 27 (7):1839–1856

Wuolikainen A, Jonsson P, Ahnlund M, Antti H, Marklund SL, Moritz T, Forsgren L, Andersen PM, Trupp M (2016) Multi-platform mass spectrometry analysis of the CSF and plasma metabolomes of rigorously matched amyotrophic lateral sclerosis, Parkinson's disease and control subjects. Mol Biol Syst 12(4):1287–1298

Xiang C, Oliver DJ (1998) Glutathione metabolic genes coordinately respond to heavy metals and jasmonic acid in Arabidopsis. Plant Cell 10(9):1539–1550

Ye X, Ng TB (2009) Isolation and characterization of juncin, an antifungal protein from seeds of Japanese takana (*Brassica juncea* var. integrifolia). J Agric Food Chem 57(10):4366–4371

Yokozawa T, Kim HY, Cho EJ, Choi JS, Chung HY (2002) Antioxidant effects of Isorhamnetin 3, 7-di-O-beta-D-glucopyranoside isolated from mustard leaf (*Brassica juncea*) in rats with streptozotocin-induced diabetes. J Agric Food Chem 50(19):5490–5495

Yokozawa T, Kim HY, Cho EJ, Yambi N, Choi JS (2003) Protective effects of mustard leaf (*Brassica juncea*) against diabetic oxidative stress. J Nutr Sci Vitaminol 49:87–93

Yonekura-Sakakibara K, Saito K (2006) Review: genetically modified plants for the promotion of human health. Biotechnol Lett 28(24):1983–1991

Yu JC, Jiang ZT, Li R, Chan SM (2003) Chemical composition of the essential oils of *Brassica juncea* (L.) Coss. grown in different regions, Hebei, Shaanxi and Shandong of China. J Food Drug Anal 11(1):22–26

Yu LJ, Luo YF, Liao B, Xie LJ, Chen L, Xiao S, Li JT, Hu SN, Shu WS (2012) Comparative transcriptome analysis of transporters, phytohormone and lipid metabolism pathways in response to arsenic stress in rice (*Oryza sativa*). New Phytol 195(1):97–112

Zhang X, Fan X, Li C, Nan Z (2010) Effects of cadmium stress on seed germination, seedling growth and antioxidative enzymes in *Achnatherum* inebrians plants infected with a *Neotyphodium* endophyte. Plant Growth Regul 60:91–97

Zhang LW, Song JB, Shu XX, Zhang Y, Yang ZM (2013) miR395 is involved in detoxification of cadmium in *Brassica napus*. J Hazard Mater 250–251:204–211

Zhou SZ, Yang SN, Li H, Zhu CC, Liu ZP, Yang ZM (2013) Molecular dissection of mercury-responsive transcriptome and sense/antisense genes in *Medicago truncatula*. J Hazard Mater 252–253:123–131

19 Proteomics Approach to Uncover Key Signalling Pathways in *Brassica juncea* in Abiotic and Biotic Stress

Ankita Sehrawat and Renu Deswal

Abstract

Brassica juncea (Indian mustard) is an important oil yielding crop. Total and sub-organelle proteome of *B. juncea* is analyzed by SDS-PAGE, two-dimensional gel electrophoresis (2-DE), 2-D DIGE, iTRAQ, gel free/in-gel trypsin digestion, MALDI-ToF/ToF and nLC-MS/MS. For sub-organelle proteome (cuticle, apoplast and nucleus) purity of samples was assessed by checking the presence of a cytosolic marker, glucose-6-phosphate dehydrogenase and a chloroplastic marker, RuBisCO. Enzymes associated with "mustard oil bomb" (myrosinase and its associating proteins) are common in apoplast and cuticle. S-nitrosylation (a post-translational modification) analyzed in the apoplast and nucleus showed the importance of sub-organelle proteomics in identifying regulation of Brassicaceae specific, defence (glucosinolate-hydrolysis pathway), stress and signalling pathway. Proteome analysis of Cadmium (Cd) treated Brassica plants showed the importance of peptide methionine sulfoxide reductase, 2-nitropropane dioxygenase, O-acetyl serine sulfhydrylase and glutathione S-transferase in the Cd hyperaccumulation and tolerance. Functional categorization of the mass spectrometry (MS) identified proteins suggested that during both abiotic and biotic stress, pathways of cellular detoxification (including enzymes of ascorbate glutathione cycle, superoxide dismutase and glutathione S-transferase), redox homeostasis, defence and photosynthesis are modulated. Interestingly, no detailed study has been reported on other PTMs except S-nitrosylation in *B. juncea*. To have an overview of *B. juncea* proteomics, collective information on the proteome based studies are presented.

19.1 Introduction

Brassica juncea is a Brassicaceae (Cruciferae) member, commonly known as Indian mustard, yellow sarson, brown mustard, Chinese mustard or oriental mustard. The leaves are simple and petioled; ovate or obovate shaped; the flowers show raceme inflorescences; flower bisexual with four free sepals and four yellow coloured petals, along with two longer and two shorter stamens.

A. Sehrawat (✉)
Maitreyi College, University of Delhi, New Delhi 110021, India
e-mail: asehrawat@maitreyi.du.ac.in

R. Deswal
Department of Botany, University of Delhi, New Delhi 110007, India
e-mail: rdeswal@botany.du.ac.in

C. Kole and T. Mohapatra (eds.), *The Brassica juncea Genome*, Compendium of Plant Genomes,
https://doi.org/10.1007/978-3-030-91507-0_19

B. juncea is an important oilseed crop cultivated in central and south Asia, China, northern Africa, Europe and North America. It is a globally important agronomic crop, as it is the second-largest cultivated oilseed around the world, supplying around 15% of the global consumption of edible vegetable oils (Rahman et al. 2021). In India, it is mainly cultivated in the north-western agroclimatic zone, where productivity of the crop is affected by abiotic stress. Abiotic stress affects its growth and development, and therefore, its oil yield. A total of 17% yield reduction was shown in Brassica plants in high temperature stress (Lobell and Asner 2003). Similarly, high salinity levels in the north-western regions also reduce its crop productivity. In addition to abiotic stress, *Albugo candida*(an oomycete) causes white rust disease in *B. juncea* and reduces its oil quality and yield(up to 20–60%)(Arora et al. 2019). As Indian mustard is grown in more than five million hectares of land in India, the impact of both biotic and abiotic factors is high.

Plants sense environment and adapt at cellular, physiological and developmental level by evolving diverse mechanisms to achieve an optimized growth and reproductive success. Stress induces various changes by modifying genome, proteome as well as metabolome. Plethora of information on gene and protein databases along with the recent development in protein identification techniques such as MALDI-ToF and mass spectrometry (MS) has greatly strengthened the proteomics approach for understanding the effect of stress on plants (Hakeem et al. 2012). Proteome analysis of *B. juncea* will be important in understanding molecular regulatory mechanisms, manipulations and stress-induced acclimatization occurring at the cellular level.

Here information on the proteome based studies is presented to give an overview of *B. juncea* proteomics and to provide the current status of *B. juncea* research. This will help in understanding various regulatory pathways involved in providing stress tolerance in this important oilseed crop.

19.2 Proteomics of Brassica

Protein extraction is one of the most crucial steps for proteome analysis both during gel based-[two-dimensional gel electrophoresis (2-DE)] and gel free proteomics. This is because the quality of the results depends upon the removal of interfering substances such as phenolics, pigments, terpenes, lipids, oxidative enzymes and inhibitory ions present in the plant tissue. These interfering substances if not removed will result in poor resolution of the protein reduce spot number and finally lead to poor results. Proteome analysis can be done at whole plant level or sub-cellular level. Total and sub-organelle proteome of *B. juncea* is analyzed by sodium dedocyl sulfate–polyacrylamide gel electrophoresis (SDS-PAGE), 2-DE, 2-D DIGE (Two-dimensional difference gel electrophoresis), iTRAQ (Isobaric tags for relative and absolute quantitation), gel free/in-gel trypsin digestion, MALDI-ToF/ToF, and nLC-MS/MS.

19.2.1 Total Proteome

19.2.1.1 Seedlings

To get the optimum yield, protocol should be standardized for different samples. As *B. juncea* seedlings are rich in secondary metabolites, therefore, either trichloroacetic acid (TCA)—acetone precipitation or phenol extraction methanol—ammonium acetate precipitation methods are most commonly used for removing interfering substances. 2-D gel comparisons of phenol and TCA/acetone-based extraction buffers showed both the methods are suitable for protein extraction from *Brassica* seedlings (unpublished work, Dr. Renu Deswal). However, compared with TCA/acetone method, the protein yield is comparatively higher in the phenol-based method.

19.2.1.2 Seed

Recently, a simple and robust protocol for single rapeseed seed protein extraction (SRPE) was established by Rahman et al. (2021). Quality and yield of proteins extracted by SRPE method were checked by downstream gel-based approach (1DSDS-PAGE), which showed clear and well-resolved polypeptides of the characteristic two triplets for cruciferin and a single doublet for napin. The predicted polypeptide of napin and cruciferin on SDS-PAGE was confirmed by western blotting using anti-2S napin and anti-12S cruciferin antibody. This method was shown to be useful for seed proteome extraction from various *Brassica* species, including *B. juncea, B. rapa, B. fruticulosa* and *B. nigra* and also highlighted the usefulness of the method across a range of species differing in pigmentation and oil content. This study provided a useful methodology (SRPE method) for identifying and characterizing very high amount of seed storage proteins (SSPs) in an economical and rapid way. Moreover, the samples can further be used for MS.

Proteomics of a cell is too complex to understand only by total proteome analysis. Organelles provide a step up because they compartmentalize a set of proteins in the cell, therefore, help in the analysis of the physiology and metabolism in time and space (Plöscher et al. 2009). Moreover, low abundant targets are masked by abundant proteins such as RuBisCO (ribulose-1, 5-bisphosphate carboxylase/oxygenase), RuBisCO activase, glyceraldehyde-3-phosphate dehydrogenase (GAPDH) and heat shock proteins, asthese compete with the low abundant proteins either during column chromatography or 2-DE or MS identification during total proteome analysis, leading to waste of effort and time. In contrast, organelle proteomics will selectively remove these abundant proteins and help in providing deeper insight into the role of regulatory and low abundant proteins in stress and signalling.

19.2.2 Organelle Proteomics

Due to multicompartmental and dynamic nature of protein localization, the revelation of the sub-cellular distribution of proteins is one of the major challenges in cell biology. Characterization of organelle proteome will help in understanding the protein functions and dynamics. Unfortunately, there is no report on sub-organelle proteome of *B. juncea* except cuticle. For reports which are available pertains to distribution and identification of S-nitrosylated proteins in different cellular compartments such as apoplast and nucleus. The reason for the selection of apoplast and nucleus for analysis was their crucial role in signalling perception response. Purity of sub-organelle fractions was assessed by either performing the enzyme assays for the cytosolic marker enzyme (glucose-6-phosphate dehydrogenase, G6PDH) or by performing the western blotting using anti-RuBisCO antibodies for RuBisCO (a chloroplastic protein) contamination. The isolated sub-organelle proteins were resolved on the 2-D SDS-PAGE and the proteins were identified by MS.

19.2.2.1 Cuticle

Cuticle is the outermost layer of the aerial parts of the plant with a protective role against herbivores, pathogens and UV radiation. It is the perception site for biotic and abiotic stress signalling. However, only a few reports are available on the cuticle proteome. Recently, leaf cuticle proteome of *B. juncea* was isolated using organic solvents [chloroform–methanol, 2:1(v/v)], and identified by gel-based and quantitative shotgun proteomics (Arya et al. 2021). A total of 615 proteins were identified, out of which 169 (27%) had a signal peptide indicating their extracellular localization. Identified targets include xyloglucan endotransglucosylase/hydrolase protein, succinate-semialdehyde dehydrogenase, GDSL esterse/Lipase ESM1, myrosinase, MADS-box transcription Factor 32 and PsbP domain-containing protein. Functional categorization using QuickGO predicted the involvement of identified targets in catabolism (21%), peptidase activity (13%), oxidation–reduction (12%), defence (9%), fatty acid binding (9%), nutrient reservoir activity (8%), chitin-binding (7%) and lipid transport (2%). Interestingly, antifreeze activity was shown in the cuticle extracts using nanoliter osmometer-phase contrast microscopy. However, further work is required to understand the role of antifreeze activity in *B. juncea* cuticle.

19.2.2.2 Apoplast

B. juncea apoplastic proteins were isolated from seedlings using sodium acetate buffer (pH 3) by vacuum infiltration method (Sehrawat and Deswal 2014a). The quality of the apoplastic fluid assessed with G6PDH assay and western blotting showed negligible cytoplasmic and chloroplastic contamination. Contamination less apoplastic fractions were resolved on the 2-D gel strips, pH 3–10 which showed that 93% spots focused on an acidic pH. This is because of the acidic nature of plant apoplast (pH < 6.5). Resolution of the samples on the Zoom gels (pH 4–7) further confirmed the acidic nature of the extracted apoplastic proteins. Around 109 spots were resolved on the 2-D gel and identified by MALDI-TOF/TOF and nLC − MS/MS. The subcellular location of the identified proteins was predicted by TargetP program (www.cbs.dtu.dk/services/TargetP). Proteins with signal peptide were predicted by SignalP-4.0 server (www.cbs.dtu.dk/services/SignalP/), while proteins without typical signal peptide sequences, but following non-classical secretion mechanism was predicted by SecretomeP program (http://www.cbs.dtu.dk/services/SecretomeP-1.0). Using these prediction tools and literature studies, 75% targets were localized in the apoplast. Functional categorization after MS identification of S-nitrosylated proteins showed 41% and 38% targets to bemetabolic/cell wall modifying and stress-related, respectively (detailed in Sect. 14.3.3). Apoplast being the first line of defence, majority of the identified targets were expected to be stress, defence and redox responsive proteins. Identification of non-secretory S-nitrosylatedproteins [acyl-binding/lipid transfer protein, cysteine proteases, Cu/Zn superoxide dismutase (SOD) and dehydroascorbate reductase (DHAR)] suggests cold stress and nitric oxide induced transport of these proteins to the apoplast (detailed in the Sect. 14.4.3). Interestingly, enzymes associated with "mustard oil bomb", i.e. myrosinase and its associating protein are common in apoplast and cuticle. This study was crucial in understanding the role of sub-organelle proteomics for identification of spatially regulated targets of cold stress signalling.

19.2.2.3 Nucleus

Nucleus is regarded as CPU (central processing unit) of the cell, as it regulates all the important cellular events. It is also regarded as principal information and administrative centre of the cell (Pandey et al. 2018). Nuclei from *B. juncea* seedlings were isolated using sucrose density gradient method (Sehrawat et al. 2019). The integrity of the isolated nuclei was checked using differential interference phase contrast microscopy and fluorescence microscopy with the help of DAPI (4′,6-diamidino-2-phenylindole) staining. Samples with intact, bright and homogenously glowing nuclei were used for the proteome analysis. For the isolated nuclei, nuclear proteins were isolated using TriPure reagent. This step helps in removing the nucleic acids which interfere in isoelectric focusing of the proteins. Chloroplastic contamination was checked by measuring chlorophyll content using a spectrophotometer and western blotting using anti-RuBisCO antibodies. Cytoplasmic contamination was measured in the final nuclear fractions by assessing a cytoplasmic marker (G6PDH enzymes). Highly pure nuclear fractions were used for the proteome analysis. Nuclear proteins were resolved as 178 reproducible spots on the 2-DE gel. Nuclear localization signal (NLS) in the MS identified targets such as myrosinase binding protein 2-like, PYK 10 binding protein 1, PYK10-binding protein 1-like and PYK 10 binding protein 2-like was predicted by BaCelLo (balanced sub-cellular localisation predictor, http://gpcr.biocomp.unibo.it/bacello/), which confirmed targets nuclear localization. Plastid sub-organelle proteome is not yet analyzed in *B. juncea*. Nevertheless, in *Sinapis alba* (white mustard, a Brassicaceae member), a specific purification strategy for the analysis of low abundant plastid proteome was shown (Schröter et al. 2014). Additionally, using plastid sub-proteome analysis of *Sinapis alba*, plastid-encoded RNA polymerase (PEP), plastid transcriptionally active chromosome proteins, RNA-binding proteins, a novel thioredoxin as a subunit of PEP, a 2-Cys-peroxiredoxin complex and ferredoxin: NADP-oxidoreductase, were identified as potential redox regulators (Schröter et al.

2010).Plants are constantly affected by both biotic and abiotic stress, which not only affect their yield but also their growth and survival. In response to stress, plants have evolved various mechanisms to perceive and acclimate. The role of proteins in understanding plant stress response is important, as they regulate physiological characteristics and help in adapting to changing environment. Proteins are "critical executors" of regulatory pathways (Liu et al. 2019). Therefore, it is important to identify and characterize these proteins and their functions in stress.

19.3 Proteome Analysis Under Abiotic and Biotic Stresses

19.3.1 Salinity Stress

B. juncea shows severe yield loss due to salinity stress mainly in the north-western region of India (Yousuf et al. 2016a). Comparative proteome analysis of a *B. juncea* salt-tolerant genotype (CS-52) with a salt-sensitive genotype (*Pusa-Varuna*) using 2-DE showed differential accumulation of 21 proteins. The identified proteins were involved in osmoregulation, ion homeostasis, energy metabolism, antioxidant defence system, photosynthesis, protein synthesis and stabilization. Interestingly, both Fe-SOD and Cu/Zn-SOD showed two-fold expression in stress highlighting the importance of these redox enzymes in maintaining the redox homeostasis in salinity stress. CS-52 showed higher expression of proteins involved in maintenance of turgor and salt levels in comparison with *PusaVaruna* (Yousuf et al. 2016a).

Glucose is a pivotal signalling molecule in mediating metabolic and stress responses. In a recent study, effect of glucose (2–8%) supplementation in providing salinity tolerance (50–100 mM) in *B. juncea* plants was shown (Sami et al. 2021). In this study, seeds were given salinity treatment by soaking them in three concentrations of NaCl (0, 50 or 100 mM for 8 h), followed by foliar spray treatment with four concentrations of glucose (0, 2, 4, or 8%) for 5 days. Salt stress reduced chlorophyll fluorescence parameters (ϕPSII, Fv/Fm, qP and ETR). Glucose-mediated recovery in chlorophyll fluorescence parameters was observed in NaCl-treated plants, probably due to enhanced thylakoid stabilization and organization, which further increased photosynthetic efficiency. Scanning electron micrographs showed enhanced stomatal aperture after glucose foliar spray to NaCl-treated plants, indicating the role of sugars as an osmolyte and in regulating stomatal aperture movements, which further enhanced gaseous exchange. Moreover, glucose enhanced photosynthetic efficiency by up-regulating activity of photosynthetic enzymes (RuBisCO, carbonic anhydrase), glycolytic enzymes (hexokinase) and Calvin cycle enzymes (succinate dehydrogenase and fumarase). Glucose is not only used as a source of energy by stressed plant but also it enhances osmotic pressure under salt stress probably due to the increased expression of sugar metabolism enzymes such as soluble sugar invertase (SAI) and neutral invertase (NI). To sum up, glucose application as foliar spray increased plant growth, photosynthetic efficiency, activities of antioxidative enzymes, reduced ROS (superoxide anion and hydrogen peroxide) and malondialdehyde content in salinity exposed plants. Spray of 4% glucose proved best in alleviating the harmful effects of salt stress. This study clearly indicated positive effect of glucose during salt stress in Indian mustard and demonstrated it as a significant signalling molecule in mediating stress tolerance mechanism.

19.3.2 Drought Stress

Drought stress is one of the most harmful stresses which causes the retardation of plant growth and development, and finally yield loss. A few attempts have been made in the past to analyze the effects of drought stress on transcriptome of *B. juncea*. However, to the best of our knowledge, till now there is no report on analysis of the proteome of drought stress treated *B. juncea* plants. Nevertheless, proteome analysis in a related species, *Brassica napus* leaves in drought stress using iTRAQ LC–MS/MS, identified 1,976 proteins with 417 proteins showing significant changes under drought stress (Koh et al. 2015). Moreover, effect of drought stress on protein modification was analyzed by examining the post-translational modifications (PTMs) of the identified proteins through MS/MS peptide sequencing, which showed 9.2% proteins (181 of the 1976 proteins) with oxidized peptides, while only 0.3% proteins (5 of the 1976 proteins) were identified with phosphorylated peptides. Proteins associated with photosynthesis, protein synthesis, stress and defence showed increased accumulation, while metabolism, protein folding, degradation and signalling related proteins showed decreased accumulation. Interestingly, a positive correlation between protein expression and gene transcription was observed.

19.3.3 Cold Stress

Proteomics of crude, apoplast, nuclear fractions in the cold stress treated *B. juncea* seedlings showed the modulation of proteins of Brassicaceae specific glucosinolate-hydrolysis pathways such as myrosinase, myrosinase binding protein, myrosinase binding protein-related protein and myrosinase associated proteins, therefore, highlighting the novel role of glucosinolate-hydrolysis pathway in alleviating cold stress (Abat and Deswal 2009, Sehrawat and Deswal 2014a; Sehrawat et al. 2019). Additionally, β-glucosidase was also identified as a cold responsive target in the nucleus. α-hydroxynitrile glucoside/β-glucosidase system/cyanide bomb is one of the most well characterized plant defence systems in the herbivory attack. Interestingly, β-glucosidases (in white mustard, maize and almonds) resist digestion by *Spodoptera littoralis* larvae (a lepidopteran) and remain intact in the gut lumen, where they catalyze the activation of plant défense after ingestion (Vassão et al. 2018). Cold stress studies on *B. juncea* also indicated stress and nitric oxide (NO) mediated cross-talk (detailed in the following section). Interestingly, C-repeat binding factor (CBF) dependent cold stress signalling cascade was analyzed in *B. juncea* and two novel isoforms of an upstream regulator of CBF, inducer of CBF expression (ICE, BjICE46 and BjICE53) were isolated (Kashyap and Deswal 2019). This study showed ICE-CBF pathway is conserved in *B. juncea* along with the differential regulation of the ICE isoforms in the cold stress.

19.3.4 Cadmium Stress

B. juncea is known as potential heavy metal-phytoextractor for the bioremediation. It can accumulate high levels of heavy metals from the polluted soils (Alvarez et al. 2009). In the past twenty-five years, *B. juncea* has been used as a model plant to investigate the mechanisms of cadmium (Cd) accumulation. Effect of Cd (250 μM) on *B. juncea* roots using a proteomic approach was first shown by fluorescence two-dimensional difference gel electrophoresis (2-D DIGE) and iTRAQ (Alvarez et al. 2009). Proteins associated with sulphur assimilation, xenobiotics, redox homeostasis showed up-regulation, while proteins associated protein synthesis and processing were down-regulated. This study indicated the involvement of peptide methionine sulfoxide reductase and 2-nitropropane dioxygenase in redox homeostasis and glutathione S-transferase, O-acetyl serine sulfhydrylase and glutathione-conjugate membrane transporter in the Cd hyperaccumulation and tolerance. In another study, two weeks old *B. juncea* seedlings were treated with $CdCl_2$ (25, 50 and 100 μM) and the leaves were used for gel-based proteomics (2-DE) and quadrupole time-of flight mass spectrometry. Comparison of 2-D gels of control and Cd treated plants showed lower abundance of 23 spots, while 13 spots showed higher abundance (D'Alessandro et al. 2013). Proteomics analysis indicated Cd-dependent alteration of beta carbonic anhydrase and phosphoglycolate accumulation which resulted in the increased photorespiration. It was observed that in response to exposure to $CdCl_2$, Cd accumulated in roots and shoots, altered photosystem efficiency (especially at 50–100 μM), reduced transpiration and ATP synthesis, resulting in the reduction in the plant growth.

19.3.5 Oxidative Stress

Hydrogen peroxide (H_2O_2) and other reactive oxygen species (ROS) act as a signalling molecules in response to the oxidative stress. Redox status of *Brassica juncea* roots' proteome was analyzed using 2-DE proteomics in response to exogenous H_2O_2 and buthionine sulfoximine (BSO, depletes glutathione and accumulates endogenous H_2O_2, Alvarez et al. 2011). A total of 103 spots were identified, of which 29 proteins were associated with amino acid biosynthesis, redox homeostasis and glycolysis, which showed a change in the redox status in response to H_2O_2 and BSO treatments. Identified targets involved in the redox homeostasis were DHAR, glutathione S-transferases and H-type thioredoxins. Proteins involved in the amino acid synthesis and proteolysis were cobalamin-independent methionine synthase, 3-phosphoshikimate 1-carboxyvinyltransferase, PAA2 20S proteasome subunit, CLP protease proteolytic subunit 2 and 20S proteasome a-subunit C1. Interestingly, 14-3-3 proteins, general regulatory factor 10 (GRF10 or GFe) and GF14l, involved in brassinosteroid signalling, were identified in the oxidative state, indicating that redox modification of 14-3-3 proteins may change protein conformation, therefore, modifying protein–protein interactions and signalling pathways. This study identified oxidative stress redox-regulated proteins using a proteomics (2-DE) approach and indicated the role of redox-regulated pathways including glutathione-ascorbate cycle in the redox homeostasis.

In another study, H_2O_2 modified proteins were identified using two-dimensional diagonal redox SDS-PAGE to understand thiol-disulphide exchange in *B. juncea* seedlings following H_2O_2 (10 mM) treatment (Prakash and Deswal 2017). Redox responsive polypeptides (11) were identified as RuBisCO large subunit, myrosinase, cruciferin and NLI [Nuclear LIM (Lin11, Isl-1 & Mec-3 domains)] interacting protein phosphatase. Redox modulation of RuBisCO large subunit was further proved by western blotting. Interestingly, small subunit of RuBisCO did not show any redox changes, indicating differential regulation of the two subunit in response to oxidative stress. In silico analysis showed more than 90% of the identified targets (except NLI interacting protein) have oxidation sensitive cysteines, which further strengthened the importance of proteomics.

Interestingly, although the system (*B. juncea*) is same, targets in the above mentioned studies were not similar, probably because of the different plant parts (roots and seedlings) used for the analysis and also the lack of exhaustive MS identification of proteins targets.

19.3.6 Elevated CO_2

It is well known that carbon (C) and nitrogen (N) are two essential elements involved in normal plant growth and development. Proteome analysis of N-efficient and N-inefficient *B. juncea*, grown under low-Nand elevated CO_2 using 2-DE and MALDI-TOF/TOF showed 158 candidate protein spots, of which, 72 spots were identified (Yousuf et al. 2016b). Identified targets were involved in energy metabolism, protein synthesis, photosynthesis, transport, nitrogen metabolism, signal transduction and defence. Identification of proteins such as oxygen-evolving enhancer protein, elongation factor-TU, RuBisCOactivase, PII-like protein and cyclophilin gave clue about how N-efficient cultivar of *B. juncea* adapt to low N supply under elevated CO_2 conditions.

19.3.7 Albugo candida

Albugo candida (an oomycete) causes white rust disease in *B. juncea* and reduces its oil quality. Proteome study by Q-TOF–MS/MS of *B. juncea* plants infected with white rust disease showed modulation of 19 proteins including thaumatin-like protein (PR-5), peptidyl-prolyl *cis/trans* isomerase showing the importance of defence-related proteins in pathogenesis by suppressing the host cell's immune response (Kaur et al. 2011).

PTMs are involved in the regulation of many regulatory and metabolic processes. Recent methodological progress has enabled identification of numerous PTMs including phosphorylation, glutathionylation, acetylation, Tyr nitration, methylation, S-nitrosylation, glycosylation and sumoylation. These modifications regulate metabolism, photosynthesis, DNA replication, transcriptional- and translational-machinery. In the following section, post-translational modifications analyzed in *B. juncea* are described.

19.4 Post-translational Modifications (PTMs) Analysis in *Brassica juncea*

19.4.1 Phosphorylation

Protein phosphorylation plays an important role in coordinating the signalling events involved in the development of plant. Brassinosteroids (BRs, a type of phytohormone) are shown to regulate phosphorylation in Brassicaceae members. In a study based on MS-based phosphoproteomics on BR treated plants, 1104 unique phosphorylated peptides from 739 unique phosphoproteins were identified (Lin et al. 2015). The protein–protein interaction showed that majority of the identified phosphoproteins have strong connections with BR signalling components. This study suggested that many downstream proteins of BR signalling were induced by phosphorylation through kinases.

19.4.2 Glycosylation

An exhaustive identification of glycosylated proteins in *B. juncea* has not been done yet. In the *B. juncea* cuticle proteome, PTMs were analyzed using PTM viewer which predicted 66% N-glycosylation, 40% N-terminal proteolysis and 32% phosphorylation (Arya et al. 2021). Con A affinoblotting showed N-glycosylation of myrosinase and GDSL esterase/lipase, ESM1.

19.4.3 S-nitrosylation

NO is a remarkably important signalling molecule, involved in regulating various defence and stress responses. NO regulates proteins through PTMs such as S-nitrosylation and tyrosine nitration (Kovacs and Lindermayr 2013). S-nitrosylation (a reversible modification) involves the formation of a covalent bond between NO and reactive cysteine residues (Romero-Puertas et al. 2013). It is associated with the regulation of more than 1000 physiologically relevant targets indicating its broad-spectrum regulation. Physiological significance of S-nitrosylation in cold stress was first analyzed in *B. juncea* seedlings (Abat and Deswal 2009). This study was important in showing NO based modulation of RuBisCO in cold stress. S-nitrosylation analysis in apoplastic and nuclear fractions of *B. juncea* showed that stress-induced NO regulates cellular detoxification by modulating redox enzymes (superoxide dismutase, glyoxalase I, ascorbate glutathione cycle and glutathione S-transferase, Sehrawat et al. 2014a; Sehrawat et al. 2019). Moreover, specifically in the apoplast, cell wall modifying and proteolytic enzymes with signal peptides were identified to be S-nitrosylated showing spatial regulation by NO (Sehrawat and Deswal 2014a). Interestingly, enzymes associated with glucosinolate hydrolysis pathway were shown to be specifically S-nitrosylated in cold stress, indicating a novel regulation of Brassicaceae specific pathway by NO. Moreover, for the first time in *B. juncea*, S-nitrosylation of enzymes of both "mustard oil bomb" and "cyanide bomb" were identified which indicated NO mediated regulation of both these pathways. In addition, S-nitrosylation analyzed in RuBisCO depleted fractions (to remove abundant proteins) identified Glycolytic and Calvin cycle enzymes as targets of NO, which is an indicator that not only stress and redox related pathways but metabolic enzymes are also a target for S-nitrosylation (Sehrawat and Deswal 2013). 2-D proteome map of apoplastic and RuBisCO depleted S-nitrosylated *B. juncea* samples was published on WORLD2-D repository with database number 0080 (https://world-2dpage.expasy.

org/repository/database=0080) and 0084 (https://world-2dpage.expasy.org/repository/database=0084) respectively.

These studies indicated the significance of S-nitrosylation in regulating not only metabolic and photosynthetic targets but also stress, redox and defence related proteins. However, tyrosine nitration (addition of a nitro group to a biomolecule), a marker of the nitrosative stress and S-Glutathionylation (GSH binding to a reactive protein Cys thiol) is not yet explored in any *Brassica* spp. Nevertheless, literature survey of *B. juncea*S-nitrosylated targets for other NO based PTMs (tyrosine nitration and S-glutathionylation) indicated 37% of the S-nitrosylated targets to be tyrosine nitrated and/or S-glutathionylated (Sehrawat and Deswal 2014a). This suggests a probable cross-talk between these NO based PTMs for the pathways of cellular detoxification, protein synthesis and folding, signalling, glycolysis and

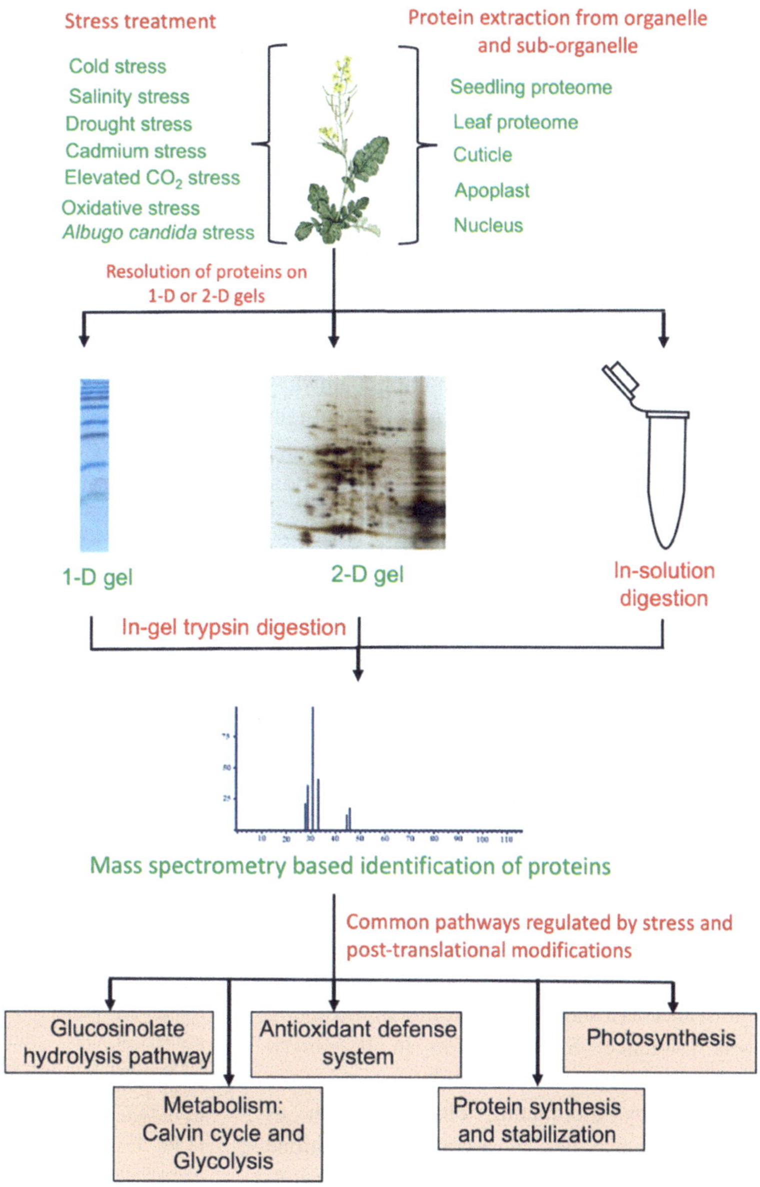

Fig. 19.1 Proteome analysis in *Brassica juncea*, organelle explored for stress responsive signaling and common pathways regulated in stress

photosynthesis. However, this story is incomplete till all the major PTMs including phosphorylation, oxidation–reduction, glycosylation, S-nitrosylation, nitration and glutathionylation are analyzed in same tissue type and stress conditions. Interestingly, NO production increases in the cold stress and this cold enhanced NO not only regulates at protein level, but also at the gene level. Transcriptional and post-translational regulation of CBF1 by NO through S-nitrosylation was shown for the first time in *Lycopersicon esculentum* (Kashyap et al. 2015). However, effect of NO on Brassica CBF is yet to be analyzed in any abiotic stress.

19.5 Conclusion

Brassica juncea (Indian mustard) is a major oil yielding crop of Brassicaceae. *B. juncea* faces yield loss due to both abiotic- (cold, salinity and heavy metal) and biotic stress, due to which its annual oil requirements are met by import. Therefore, to comprehend signalling pathways involved in providing stress tolerance is important for generating stress tolerant crops. For analyzing the stress modulated signalling pathways, proteomics approach (both gel-based and gel-free) is most widely used. Current status of proteome analysis in *B. juncea* presented in this chapter showed that both gel-based and gel free approaches have been used for proteome identification at cellular and sub-cellular level. Organelle proteomics via apoplastic, cuticle fractions and nuclei isolation showed the importance of sub-organelle proteomics in understanding the role of Brassicaceae specific glucosinolate-hydrolysis pathway, pathways of cellular detoxification, redox homeostasis and photosynthetic enzymes(Glycolysis and Calvin cycle), in abiotic and biotic stress signalling (Fig. 19.1). Ameliorative effect of glucose on salinity stress in *B. juncea* indicated its significance as a signalling molecule in mediating stress tolerance. S-nitrosylation in *B. juncea* showed the effect of NO in regulating cold stress signalling by modulating enzymes such as myrosinase, enzymes of ascorbate glutathione cycle, superoxide dismutase and glutathione S-transferase, cell wall modifying and proteolytic enzymes. As only S-nitrosylation is analyzed in details in *B. juncea*, further work is required to understand PTMs cross-talk in the Indian mustard particularly phosphorylation, glycosylation, tyrosine nitration and S-glutathionylation. Interestingly, NPs-protein corona (nanoparticle adsorbed proteins) may influence *B. juncea* yield, as around 27% of the corona protein fraction represented energy-yielding pathways, while 10% of the corona proteins are proteases (Prakash and Deswal 2020). This is first report of corona characterization in plant system and can act as a base study for future work.

References

Abat JK, Deswal R (2009) Differential modulation of S-nitrosoproteome of *Brassica juncea* by low temperature: change in S-nitrosylation of Rubisco is responsible for the inactivation of its carboxylase activity. Proteomics 9:4368–4380

Alvarez S, Berla BM, Sheffield J, Cahoon RE, Jez JM et al (2009) Comprehensive analysis of the *Brassica juncea* root proteome in response to cadmium exposure by complementary proteomic approaches. Proteomics 9:2419–2431

Alvarez S, Galant A, Jez JM, Hicks LM (2011) Redox-regulatory mechanisms induced by oxidative stress in *Brassica juncea* roots monitored by 2-DE proteomics. Proteomics 11:1346–1350

Arora H, Padmaja KL, Paritosh K, Mukhi N, Tewari AK et al (2019) BjuWRR1, a CC-NB-LRR gene identified in *Brassica juncea*, confers resistance to white rust caused by *Albugo candida*. TheorAppl Genet 132:2223–2236

Arya M, Prakash S, Sougrakpam Y, Deswal R (2021) *Brassica juncea* leaf cuticle proteome analysis shows myrosinase protein, antifreeze activity, and post-translationally modified secretory proteins. Plant Physiol Biochem 161:234–247

D'Alessandro A, Taamalli M, Gevi F, Timperio AM, Zolla L et al (2013) Cadmium stress responses in *Brassica juncea*: hints from proteomics and metabolomics. J Proteome Res 12:4979–4997

Hakeem KR, Chandna R, Ahmad A, Qureshi MI, Iqbal M (2012) Proteomic analysis for low and high nitrogen-responsive proteins in the leaves of rice genotypes grown at three nitrogen levels. Appl Biochem Biotechnol 168:834–850

Kashyap P, Sehrawat A, Deswal R (2015) Nitric oxide modulates *Lycopersicon esculentum* C-repeat binding factor 1 (LeCBF1) transcriptionally as well as post-translationally by nitrosylation. Plant Physiol 96:115–123

Kashyap P, Deswal R (2019) Two ICE isoforms showing differential transcriptional regulation by cold and hormones participate in *Brassica juncea* cold stress signaling. Gene 695:32–41

Kaur P, Jost R, Sivasithamparam K, Barbetti MJ (2011) Proteome analysis of the *Albugo candida–Brassica juncea* pathosystem reveals that the timing of the expression of defence-related genes is a crucial determinant of pathogenesis. J Exp Bot 62:1285–1298

Koh J, Chen G, Yoo MJ, Zhu N, Dufresne D et al (2015) Comparative proteomic analysis of *Brassica napus* in response to drought stress. J Proteome Res 14:3068–3081

Kovacs I, Lindermayr C (2013) Nitric oxide-based protein modification: formation and site-specificity of protein S-nitrosylation. Front Plant Sci 4:37

Lin LL, Hsu CH, Hu CW, Ko SY, Hsieh HY et al (2015) Integrating phosphoproteomics and bioinformatics to study brassinosteroid-regulated phosphorylation dynamics in Arabidopsis. BMC Genomics 16:533

Liu Y, Lu S, Liu K, Wang S, Huang L et al (2019) Proteomics: a powerful tool to study plant responses to biotic stress. Plant Methods 15:135

Lobell DB, Asner GP (2003) Climate and management contributions to recent trends in U.S. agricultural yields. Science 299:1032

Pandey A, Chakraborty S, Chakraborty N (2018) Nuclear proteome: isolation of intact nuclei, extraction of nuclear proteins, and 2-DE analysis. Methods Mol Biol 1696:41–55

Prakash S, Deswal R (2017) Identification and in silico analysis of major redox modulated proteins from *Brassica juncea* seedlings using 2D redox SDS PAGE (2-dimensional diagonal redox sodium dodecyl sulphate polyacrylamide gel electrophoresis). Protein J 3:64–76

Prakash S, Deswal R (2020) Analysis of temporally evolved nanoparticle-protein corona highlighted the potential ability of gold nanoparticles to stably interact with proteins and influence the major biochemical pathways in *Brassica juncea*. Plant Physiol 146:143–156

Plöscher M, Granvog B, Reisinger V, Masanek A, Eichacker LA (2009) Organelle proteomics. Methods Mol Biol 519:65–82

Rahman M, Liu L, Barkla BJ (2021) A single seed protein extraction protocol for characterizing Brassica seed storage proteins. Agronomy 11:107

Romero-Puertas MC, Rodríguez-Serrano M, M. Sandalio L (2013) Protein S-nitrosylation in plants under abiotic stress: an overview. Front Plant Sci 4:373

Sami F, Husna Siddiqui H, Pravej A, Shamsul H (2021) Glucose-induced response on photosynthetic efficiency, ROS homeostasis, and antioxidative defense system in maintaining carbohydrate and ion metabolism in Indian mustard (*Brassica juncea* L.) under salt-mediated oxidative stress. Protoplasma258:1–20

Schröter Y, Steiner S, Matthäi K, Pfannschmidt T (2010) Analysis of oligomeric protein complexes in the chloroplast sub-proteome of nucleic acid-binding proteins from mustard reveals potential redox regulators of plastid gene expression. Proteomics 10:2191–2204

Schröter Y, Steiner S, Weisheit W, Mittag M, Pfannschmidt T (2014) A purification strategy for analysis of the DNA/RNA-associated sub-proteome from chloroplasts of mustard cotyledons. Front Plant Sci 5:55

Sehrawat A, Deswal R (2013) Nitric oxide-cold stress signalling crosstalk-evolution of a novel regulatory mechanism. Proteomics 13:12–13

Sehrawat A, Deswal R (2014a) S-nitrosylation analysis in *Brassica juncea* apoplast highlights the importance of nitric oxide in cold-stress signaling. J Proteome Res 13:2599–2619

Sehrawat A, Sougrakpam Y, Deswal R (2019) Cold modulated nuclear S-nitrosoproteome analysis indicates redox modulation of novel Brassicaceae specific, myrosinase and napin in *Brassica juncea*. Environ Exp Bot 161:312–333

Vassão DG, Wielsch N, Gomes AMMM, Gebauer-Jung S, Hupfer Y, Svatoš A, Gershenzon J (2018) Plant defensive beta-glucosidases resist digestion and sustain activity in the gut of a lepidopteran herbivore. Front Plant Sci 9:1389

Yousuf PY, Ahmad A, Arshid HG, Muhammad I (2016a) Salt stress-induced modulations in the shoot proteome of *Brassica juncea* genotypes. Environ Sci Pollut Res 23:2391–2401

Yousuf PY, Ganie AH, Khan I, Qureshi MI, Ibrahim MM et al (2016b) Nitrogen-efficient and nitrogen-inefficient Indian Mustard showed differential expression pattern of proteins in response to elevated CO_2 and low nitrogen. Front Plant Sci 7:1074

Genomics Approaches to Understand the Evolution of *Brassica juncea* 20

Snehdeep Kaur, Kawalpreet Kaur, Kusum Rana, and Simarjeet Kaur Sra

Abstract

Brassica juncea is an important allopolyploid species in six major cultivated species of Brassica genus which is used for edible and industrial oil, condiments, medicines and animal feed. This allopolyploid species originated by spontaneous hybridization between *B. rapa* and *B. nigra* followed by chromosome doubling. Several cytogenetic and molecular studies have been exploited to understand the evolutionary history of this allopolyploid crop in past decades. However, the genomic era has revolutionized this research by explaining the molecular basis of crop evolution and domestication. Various genomic resources have been employed to study the genetic diversity, construction of linkage maps and comparative mapping in *B. juncea*. Comparative sequence studies across the species in Brassicaceae have already been used to understand genome structure, evolutionary history and detection of conserved genomic regions. Moreover, the accessibility of *B. juncea* genome sequence has transformed this research in the area of mustard genomics and proposes a basis for evolutionary and comparative genomics. The molecular basis of alterations in the gene expression because of neofunctionalization, subfunctionalization and gene silencing during polyploidization has also been studied using transcriptomic studies. Next- and third-generation sequencing technologies provide cost-effective and time efficient platforms to generate more sequencing data in less time. These tools have been widely used to develop single nucleotide polymorphism (SNP) marker, transcriptomic data and genotyping of diverse germplasm assemblage. The comprehensive studies of Brassica chloroplast DNA and mitochondrial DNA also play an essential role in phylogenetic analysis and delineating the evolutionary history of *B. juncea*. Consequently, the availability of chloroplast and mitochondrial genome sequence of different lines of *B. juncea* assists in the study of cytoplasmic variations among genotypes of allopolyploid and its diploid progenitors. In this chapter, we will discuss about the evolution of *B. juncea* and its interspecific relationships with other Brassica species using comparative analysis, high-throughput genome sequencing and gene expression studies in Indian mustard.

S. Kaur · S. K. Sra (✉)
Department of Plant Breeding and Genetics, Punjab Agricultural University, Ludhiana, Punjab, India
e-mail: snehdeepkaur@pau.edu

K. Kaur
Department of Agricultural, Food and Nutritional Science, University of Alberta, Edmonton, AB, Canada
e-mail: kawalpre@ualberta.ca

K. Rana
Department of Biotechnology, Punjab University, Chandigarh, India

C. Kole and T. Mohapatra (eds.), *The Brassica juncea Genome*, Compendium of Plant Genomes,
https://doi.org/10.1007/978-3-030-91507-0_20

20.1 Introduction

Allopolyploid Indian mustard (*Brassica juncea* L. Czern & Coss, 2n = 4x = 36, genome AABB) is a natural amphidiploid species that originated due to spontaneous hybridization between *B. rapa* and *B. nigra* followed by chromosome doubling, at least 10,000 years ago (Prakash et al. 2009). Cytogenetic and hybridization studies have revealed that Indian mustard is an important allopolyploid species of the U's triangle which consists of vegetable, oilseed and condiment crops (U 1935; Vaughan and Hemingway 1959; Woods et al. 1991; Burton et al. 2004). The common ancestor of the tribe Brassicaceae is known to endure polyploid events, i.e., complete genome duplication. The isozyme analysis suggested that aneuploidy plays an imperative role comparative to polyploidy in the evolution of higher base chromosome numbers in this tribe (Anderson and Warwick 1999). Brassica genomics offers a better understanding of the genomic changes that occurred following the formation of allotetraploid and explains the rapid phenotypic evolution of *B. juncea* subspecies. Evidently, a research explains the dynamics of genomic changes and the consequences of nuclear-cytoplasmic interaction in synthetic allotetraploid species (Lukens et al. 2006). China is one of the prime ancestries of *B. juncea* and artificial selection exhibited a significant role in the evolution of Chinese mustard. Phylogenetic studies have been performed using chloroplast and nuclear simple sequence repeat (SSR) markers for trait differentiation and domestication. These results show polyphyletic origin of Indian mustard with both diploid parents as cytoplasmic donors in independent hybridization events (Kaur et al. 2014). The Chalcone synthase gene has been used for the evolutionary study of *B. juncea* which depicts that cultivated mustard evolved from wild mustard, and the greater nucleotide variation in the B genome observed as compared to A genome (Chen et al. 2016). Gene sequence variations for seed traits have been studied in natural and resynthesized forms of Indian mustard along with diploid progenitors which exhibit superior directional selection for seed size comparative to seed metabolites (Sra et al. 2019).

Recent advances in the development of genomic tools have facilitated comparative mapping, population genetic and genomic studies to explain the molecular basis of crop evolution and domestication. The evolutionary history of any crop is best studied using three approaches: high-quality reference genome assembly of the crop species and their wild relatives, comparative mapping and genomic characterization of a diverse germplasm panel. The Arabidopsis genome sequence has revolutionized our understanding in terms of molecular biology and offers clues to its evolutionary history and comparative biology (The Arabidopsis Genome Initiative 2000). The genome size of diploid Brassica species varied from 529 to 696 Mb and for polyploids it ranged from 1068 to 1284 Mb which is bigger than that of *A. thaliana* (125 Mb). This is suggestive of new gene interactions via subfunctionalization and/or neofunctionalization of paralogs in Brassica crops (Lynch and Force 2000; He and Zhang 2005; Roth et al. 2007). Yang et al. (2016) reported a draft genome sequence of a vegetable type *B. juncea* var. *tumida* which was assembled by de novo assembly using shotgun reads, PacBio sequencing, optical mapping, aiming to resolve complex allopolyploid genomes. The outcomes show the monophyletic origin of A subgenome of *B. juncea* and its evolution into vegetable and oil types. The homoeolog expression dominance exists between both the subgenomes of *B. juncea* which contribute to selection response and predict the directional impacts of selection in the genome of polyploid species. A significantly improved and highly contiguous genome assembly was later on reported of an *oleiferous* type of *B. juncea* variety Varuna with ~100x PacBio single-molecule real-time (SMRT) long reads forming contigs which were improved for the misassembles and scaffold with BioNano optical mapping (Paritosh et al. 2020).

Comparative sequence analysis has been widely utilized in Brassicaceae to understand the genome structure, detection of conserved regions and evolutionary history across the species. Earlier attempts on comparative mapping between different Brassica genomes have relied on the usage of restriction fragment length polymorphism (RFLP) markers (Lagercrantz 1998; Lan et al. 2000; Babula et al. 2003; Lukens et al. 2003). In later years, intron polymorphism (IP) markers based analysis discovered a high degree of collinearity between A and B genomes of *B. juncea* with the A and B genomes of *B. napus* and *B. nigra* respectively (Panjabi et al. 2008). Second- and third-generation sequencing tools have been widely exploited to develop cost-efficient and time-saving approaches for single nucleotide polymorphism (SNP) marker development, genotyping and expression profiling in Brassica species. In spite of wide research, the evolution of Indian mustard and the interspecific relationships within Brassica genus remains unsettled and confused. In this chapter, we will resolve these questions with the aid of the advancement made in comparative genomics, high-throughput genome sequencing and gene expression studies in Indian mustard.

20.2 Linkage Map

Linkage mapping is a prerequisite for explaining the origin and evolution of Indian mustard and dissecting the genetic relationships between diploid and amphidiploid Brassica species. The detailed molecular linkage maps are the valuable tools for understanding the genome architecture and evolution, comparing gene order as well as organization in various Brassica species (For review, see Cheung et al. 1997). Although many genetic maps have been constructed in various commercially important *Brassica* species over the past 20 years, very little information is available for Indian mustard. Therefore, continuous efforts are being made to elucidate molecular markers for comparative or evolutionary studies. Several molecular maps using different types of genetic markers and mapping populations have already been developed in *B. juncea* (Table 20.1).

DNA-based molecular markers like RFLP and amplified fragment length polymorphism (AFLP) have been extensively exploited for developing linkage maps in Brassicas (Quiros 1999), including Indian mustard (Cheung et al. 1997; Axelsson et al. 2000; Pradhan et al. 2003). The RFLP map consisted of 343 markers spanning a distance of 2073 cM whereas the AFLP map comprised 1029 markers with the coverage of 1629 cM. So, the AFLP map included more markers but spanned less distance on the map. SSR markers are much more desirable as compared to RFLP or AFLP markers, owing to their abundance, high polymorphism and codominance (Schlötterer 2004). A high-density linkage map has been constructed in the F1 derived DH population of *B. juncea* using 996 AFLP and 33 RFLP markers that can be used for transferring agronomically important traits from wild germplasm into the well-adapted cultivars (Pradhan et al. 2003). The first SSR based linkage map consisting of 860 markers has been developed in the F1 derived DH population of

Table 20.1 Molecular maps using different types of genetic markers and mapping populations in Indian mustard

Type of marker	Population	References
RFLP	F_1 derived DH population of J90-4317 × J90-2733	Cheung et al. (1997)
RFLP	BC_1 population of (J-o-7DH1 × J-o-3DH1) × J-o-3DH1	Axelsson et al. (2000)
AFLP and RFLP	F_1 derived DH population of Varuna × Heera	Pradhan et al. (2003)
DArT	89 genotypes of *B. rapa*, *B. napus*, *B. juncea*, and *B. carinata*	Raman et al. (2012)
SNP	DH population of Varuna × Heera	Paritosh et al. (2014)
DArT	F_2 population derived from Sichuan yellow/purple mustard	Zou et al. (2016)
SSR	F_1 derived DH population of EH-2 and Pusa Jai Kisan	Dhaka et al. (2017)

lines EH-2 and Pusajaikisan (Dhaka et al. 2017). Presence of the non-parental bands in the mapping population was an interesting finding in this study. Some of the orthologous Arabidopsis genes were found that were involved in abiotic stress responses. This study developed 2118 genic SSRs which could be used in comparative mapping.

Genetic basis of adaptation is observed using quantitative trait locus (QTL) mapping in evolutionary studies which elucidate the divergence between discriminated populations, moreover it determines the QTL effects and their associations in the population. Various gene-specific markers have been used for developing high-density linkage maps providing excellent platforms to detect molecular markers linked with specific regions. Genetic mapping in Indian mustard has identified random amplified polymorphic DNA (RAPD), RFLP, AFLP and SSR markers associated with various disease resistances (Mukherjee et al. 2001; Saal et al. 2004; Panjabi-Massand et al. 2010; Singh et al. 2015; Rana et al. 2017; Atri et al. 2019). Other QTLs controlling some important agronomic and quality traits have also been mapped in *B. juncea* (Cheung et al. 1998; Negi et al. 2000; Lionneton et al. 2002; Mahmood et al. 2003a, b; Ramchiary et al. 2007). SSR marker set has also been used to study the genetic variations among all the six *Brassica* species of U's triangle and 98.15% of cross-transferability across the species was observed (Thakur et al. 2018). This set would enable evaluation of genetic variability in Brassica germplasm and evolutionary relationship among *Brassica* species.

Unnecessary and overlapping bands because of coamplification of loci reduce the efficiency of SSR markers in Indian mustard. SNP markers are superior to SSRs owing to their proficiency to monitor genotypes at comparatively low cost as well as abundance and distribution of these markers in the genomes. High-density molecular map of *B. juncea* has been developed with transcriptome based SNP markers using double haploid lines derived from the cross of two lines, one Indian type (Heera) and other east European type (Varuna; Paritosh et al. 2014). In this study, block architecture of A and B genomes has been compared which demonstrated wide variances in gene block associations and block fragmentation patterns. These variations between A and B genomes exhibit that both the genomes evolved from independent hexaploid events. Diversity array technology (DArT) is DNA hybridization-based technology and acceptable in various polyploid species. DArT polymorphisms are outcomes of nucleotide polymorphisms within the recognition sites of a restriction enzyme. DArT markers have been developed in *Brassica* species comprising A or C genomes. Total 1547 polymorphic DArT markers of superior quality have been documented and exploited to evaluate molecular diversity among 89 genotypes of *B. rapa, B. napus, B. juncea* and *B. carinata* (Raman et al. 2012). A dense linkage map has been constructed using the F_2 population derived from Sichuan Yellow/Purple Mustard to elucidate subgenome differentiation in *B. juncea* in comparison to *B. rapa* and *B. napus* (Zou et al. 2016). This map includes 3329 DArT-sequence markers on 18 linkage groups covering 1579 cM. A strong collinearity of ancestral blocks among different A subgenomes has been observed along with considerable block variation supporting the idea of distinct genomic diversity. This population genetic study on A genome comprising Brassica species proves that *B. juncea* has discrete genomic diversity and evolved from a diverse A genome progenitor of *B. napus*.

Polyploidy is a key force in the evolution of crops where the addition of the genomes alters the genome organization and expressions in polyploid species. Recently, kompetitive allele-specific PCR (KASP) markers have been developed in polyploid crop wheat by successful conversion of Illumina Infinium assays from 90K SNP array into robust locus-specific KASP markers (Makhoul et al. 2020). This approach can also be applied in amphidiploid species *B. juncea* for conversion of trait associated SNP markers from SNP array into KASP markers. These KASP markers can be further used to evaluate the population structure, genetic diversity and genetic relationship within Indian mustard germplasm as well as in comparison with diploid progenitors.

Molecular markers are tremendous tools to observe the genetic relationships along with genomic evolution of polyploid species like Brassicas. Various evolutionary relationships have been established among Brassica species using molecular markers that have assisted in understanding the history of A genome in Indian mustard. Likewise, these studies can also be exploited in understanding the history of the B genome of *B. juncea*. The genetic map, which has been constructed from different markers, can also be used for comparative mapping in Brassica species.

20.3 Comparative Mapping

Comparative genetic mapping is an important tool for studying plant genome organization allowing for the understanding of chromosomal relocations resulting due to speciation. It has also been used to transfer important characters from model plant species to related crops. It has been used in agronomically important Brassica species and their relative *Arabidopsis thaliana*. The generation of the linkage maps in the allopolyploid Brassica species provided the data for first comparative mapping in Brassicaceae (Parkin et al. 1995; Axelsson et al. 2000). Prior to the sequencing of Arabidopsis genome, comparative mapping mainly focused on gene and chromosome segment duplications (Slocum et al. 1990; McGrath et al. 1990; Song et al. 1991). Very early findings were related to the degree of collinearity between Arabidopsis and Brassica species representing some conserved regions among the chromosomal differences (Kowalski et al. 1994; Lagercrantz 1998). On the other hand, after the sequencing of the Arabidopsis genome, a new era of comparative mapping began which enabled the study of genome evolution and origin (Lynch and Conery 2000; Vision et al. 2000).

The conserved map locations of various candidate genes for flowering time, inflorescence morphology and seed glucosinolate biosynthesis, already characterized in *A. thaliana*, and correlated with QTL loci in *B. rapa, B. napus, B. oleracea* and *B. juncea* (Osborn et al. 1997; Lan and Paterson 2000; Long et al. 2007; Bisht et al. 2009). This technique helps in identifying candidate genes of QTLs as well as conserved QTLs involved in agronomically important traits like length of silique, seed size, silique number etc. across the different lines and species (Pradhan and Pental 2011; Sharma et al. 2014). Similarly, candidate genes from QTL region for aliphatic glucosinolate biosynthetic pathway have been characterized using comparative mapping (Bisht et al. 2009). In this report, the sequence information was retrieved from *A. thaliana* and *B. oleracea* for various aliphatic glucosinolate biosynthetic pathway genes. Both high and low glucosinolate *B. juncea* varieties Varuna and Heera have been used for amplification and sequence analysis of the candidate genes from *A. thaliana* and *B. oleracea*, respectively. Li et al. (2013) conducted a comparative mapping study in which they identified conserved functional genomic regions having QTLs related to morphological and yield traits in *B. rapa, B. juncea* and *B. nigra*.

The initial studies have suggested that the Brassica diploid genomes have evolved from an ancient polyploid. The subsequent comparative mapping studies with Arabidopsis genome cemented the fact that a member of Brassicaceae had departed from the Brassica species 14–24 million years ago (Koch et al. 2000). The comparative map has been developed between *B. juncea* and Arabidopsis using PCR-based intron polymorphism (IP) markers to evaluate the genome wide synteny between these two species (Panjabi et al. 2008). This map was used to study homeologous relationship, genomic diversity and evolution of A, B and C genomes of Brassica species. The comparison represents a highly conserved collinearity between A genome of *B. juncea* and *B. napus* and between B genome of *B. juncea* and *B. nigra* concluding the absence of genomic perturbation during the development of allopolyploid Brassica species. After the assembly of the first *B. juncea* reference genome (Yang et al. 2016), it became much easier to understand the genomic basis of selection in this species. The evolutionary history of nucleotide-binding site leucine-rich repeat (NLR) gene family in *B. juncea* has been elucidated by comparative analysis with *B. rapa* and *B. nigra* (Inturrisi et al. 2020). The study reveals that NLR genes are comparable among the three species in terms of number, chromosomal distribution and domain. Total 45% of these genes have been found to be clustered in the genome of Indian mustard. These comparative genomic findings will enable the researchers to identify and characterize orthologous candidate genes of key agronomic traits in Brassicas as well as assist in constructing an integrated linkage map of Brassica crops.

20.4 Next-Generation and Third-Generation Sequencing Techniques for Evolutionary Genomics

The presence of two subgenomes (A and B) hinders genomics in *B. juncea* due to poor discrimination between homeologous segments during formation of sequence assembly. Nevertheless, innovative next-generation sequencing (NGS) technologies and bioinformatics analyses have a potential for dissection of complex genomes and studying evolutionary mechanisms of polyploidization (Aversano et al. 2012; Edwards et al. 2013). NGS, a cost-effective and time efficient approach to produce numerous short DNA sequences, is being progressively applied in plant genome de novo sequencing (Al-Dous et al. 2011; Xu et al. 2011), methylation sequencing (Cokus et al. 2008; Lister et al. 2008, 2009; Dowen et al. 2012), resequencing (Lam et al. 2010; Ashelford et al. 2011), small RNA and transcriptome sequencing (Buggs et al. 2012; Strickler et al. 2012) for functional and comparative genomics studies. Nowadays, novel sequencing techniques are long-read third-generation sequencing compared to short-read second-generation techniques and these approaches generate more data in a short time. These long reads significantly enhance the precision of genome assemblies by spanning the highly repetitive fragments. Third-generation sequencing technologies include Pacific Biosciences (PacBio) with their single-molecule real-time SMRT technology and Oxford Nanopore Technologies (ONT; Li et al. 2017; Van Dijk et al. 2018). These technologies have also been applied in various *Brassica* species (Belser et al. 2018; Lv et al. 2020; Perumal et al. 2020; Mathieu et al. 2020; Paritosh et al. 2020, 2021). Second- and third-generation sequencing approaches are most effective for genotyping, transcriptomic studies and organelle genome sequencing in evolutionary studies of Brassicas.

20.4.1 Genotyping

The discovery of SNPs is lagging particularly in Indian mustard due to its complex allotetraploid genome with highly repetitive regions and narrow genetic base. The duplication and triplication events of A and B genomes due to polyploidization complicate the process (Yang et al. 2006). The advent of several NGS platforms like 454 sequencing (Roche Applied Science), Solexa (now Illumina) technology (Illumina inc.), SOLiD (Applied Biosystems) and Pacific Biosciences (PacBio) have revolutionized genetics by generating robust genotypic data for advanced molecular research. NGS mediated genotyping has become a highly cost-efficient agri-genomics tool in crop species especially for those having large and complex genomes.

SNP genotyping arrays and KASP assays are universal marker tools which only evaluate known alleles at specific loci and very likely miss out rare variants along with common alleles missing from the samples used to develop the assays. So, the reduced representation sequencing (RRS) approach of NGS technologies has the ability to genotype novel and existing markers using either whole genome sequence or a smaller, reproducible fraction thereof as a query. Various RRS approaches are available for genotyping in crop species, such as diversity arrays technology (DArT, Jaccoud et al. 2001), complexity reduction of polymorphic sequences (CRoPS; Van Orsouw et al. 2007), restriction associated DNA sequencing (RADseq; Baird et al. 2008), genotyping by sequencing (GBS,

Table 20.2 Different NGS based genotyping techniques used in *B. juncea*

Sequencing platform	Traits studied	Sequencing method	References
ddRAD	Days to flowering, days to maturity, plant height, siliqua length, seeds per siliqua and thousand seed weight	Illumina HiSeq 2000 platform	Sudan et al. (2019)
GBS	Disease resistance against *Sclerotinia sclerotiorum*	Illumina® HiSeq platform	Rana et al. (2019)
GBS	Pod shattering, pod length and number of seeds per pod	Illumina® HiSeq platform	Kaur et al. (2020)
GBS	Oil, protein, and glucosinolates	Illumina® HiSeq platform	Akhatar et al. (2020)
GBS	Days to flower initiation, days to fifty percent flowering, days to complete flowering, days to maturity, growing degree days, helio-thermal unit, photo-thermal units and plant height	Illumina® HiSeq platform	Akhatar et al. (2021)

Elshire et al. 2011) and double digest restriction associated DNA sequencing (ddRAD-seq; Peterson et al. 2012). A number of these genotyping techniques have already been explored in *B. juncea* (Table 20.2).

The genotypic data is used to examine the genetic architecture of several quantitative traits by conducting genome wide association study (GWAS), using an association panel comprising diverse genotypes. SNPs have been generated by using ddRAD approach and NGS technology, which were further exploited for association analysis in *B. juncea* (Sudan et al. 2019). Common SNPs have been identified in this study which are associated with various morphological traits. These identified common genomic regions emphasize the potential function of genes linked with common chromosomal regions controlling specific phenotypic traits in Indian mustard. Moreover, GWAS studies have been performed in *B. juncea* to explore the genetic architecture of several quality traits. Association studies for seed quality traits (oil, protein and glucosinolates) and flowering with plant height traits have been performed independently based on the same genotyping by sequencing (GBS) approach using the Illumina® HiSeq platform (Akhatar et al. 2020, 2021). The Annotation results predicted 21 orthologs of the functional candidate genes regulating the seed quality traits and 30 candidate genes for flowering and plant height traits. Indian mustard, being a polyploid crop, exhibits more than one copy of predicted candidate genes on A and B genome chromosomes based on in silico study of the reference genome sequence. GWAS has also been carried out in *B. juncea* and its progenitor species using GBS and DArT tools on a diverse set of germplasm (Kaur et al. 2020). Few common genes along with some species-specific genes identified in this study, which predicts the role of these genes in three pod traits, i.e., regulating rupture energy, pod length and seeds per pod. Therefore, this study reveals the evolutionary history of the genes related to shattering traits in comparison of *B. juncea* with its progenitors. Total 84 *B. juncea–Erucastrum-cardaminoides* introgression lines have been used for GWAS analysis using the GBS approach, and many SNPs identified on chromosomes A03, A06, and B03. These SNPs were associated with Sclerotinia stem rot resistance (Rana et al. 2019). Consequently, evolutionary history of the wild species *E. cardaminoides* can be further studied which has resistance to various biotic stresses, and this study will assist in marker assisted introgression of these significant resistances into novel varieties of Indian mustard.

High-quality reference genome sequences with maximum coverage are a base of the foundation of omics investigations especially in polyploid species. The upgraded NGS assembly algorithms and data types allow the construction of chromosome-level genome assemblies (Mascher et al. 2017). Roche 454 GS FLX platform has been efficiently used in Brassica genome sequencing as it produces comparatively long length reads and enhances specificity of mapping homeologous fragments (Wang et al. 2011). Yang et al. (2016), reported the draft genome of *B. juncea* var. *tumida*, assembled using de novo assembly via shotgun reads and single-molecule long reads (PacBio sequencing) as well as through optical and genetic mapping. It serves as a model to investigate the genomic basis of evolution in polyploidy species. The comparison of the A subgenome following resequencing of eighteen accessions indicates a closer relationship of *B. juncea* subvarieties with *B. rapa* ssp. *tricolaris* group. The findings also exhibit the monophyletic origin of A subgenome of *B. juncea* and its evolution into vegetable and oilseed type. A highly contiguous genome assembly has also been reported of an *oleiferous* type B. *juncea* variety Varuna using ~100x PacBio SMRT reads (Paritosh et al. 2021). In this chromosome-scale assembly, BioNano optical mapping has been used to correct and scaffold the assembled contigs. The assembled genome depicts that the B genome has a considerably superior content of LTR/Gypsy retrotransposons, diverse centromeric repeats and loads of *B. nigra* specific gene clusters comparative to A genome which split the gene colinearity between both the genomes.

20.4.2 Gene Expression Studies

Over the evolutionary history, polyploid undergoes enormous changes in the gene expression following neofunctionalization, subfunctionalization, gene silencing etc. These gene conversion episodes were often selected by humans/breeders during the process of domestication and yield enhancement. Eventually, the newly developed polyploid is stable and adaptive to a wide range of environmental conditions. The molecular basis of these changes during polyploidization has been studied previously using resynthesized *B. napus* with respect to flowering time differences (Pires et al. 2004). This study was carried out using flowering time differences and the results exhibited changes at transcript levels of *BnFLC2* and *BnFLC3* between the early and late flowering lines. To study the adaptation and evolutionary genomics in Brassicaceae, a robust platform comprising 135,201 genes has been developed for whole-transcript profiling of Brassica species (Love et al. 2010).

Transcriptomics is a tool to acquire information of gene expression using less number of test sequences. In the new sequencing era, Higgins et al. (2012) reported mRNA-Seq as an efficacious approach for both qualitative and quantitative studies of gene expression in polyploids. They used Illumina mRNA-Seq data to allocate the expression between the A and C genome homoeologous in both natural and resynthesised *B. napus*. Such studies are seemingly effective for studying the molecular and expressional basis associated with evolution, stability and adaptation of polyploid species. The molecular mechanisms and regulatory pathway associated with development of stem swelling in *B. juncea* var. *tumida* were unraveled using the gene expression studies through the first large scale de novo transcriptome assembly (Sun et al. 2012). The documentation of the genes involved in tumorous stem mustard swelling was completed using sequencing of transcriptome at different stem developmental stages and subsequently comparing it with that of a mutant variety missing swollen stems. Total 146,265 assembled unigenes and 105,555 annotated unigenes were predicted in this study. The research results further enhance our knowledge on functional genomics in Indian mustard and the evolutionary relationships of the family Brassicaceae.

The annotated genes in *B. juncea* genome assembly have been validated using the transcriptomics data and these results depict that differentially expressed genes contributed largely to selection response (Yang et al. 2016). Homoeolog expression dominance was found to be prevalent between two subgenomes where the differentially expressed genes exhibited more selection pressure as compared to neutral genes. Subsequently, this leads to selection of the genes associated with glucosinolate and lipid metabolism in subspecies of *B. juncea*. The syntenic and orthologous genes identified from two subgenomes and their respective progenitor genomes provide deeper insights into allopolyploid speciation and its correlation with domestication (Chalhoub et al. 2014). A novel Associative Transcriptomics platform has been established in *B. juncea* and total 355,050 SNP variants were detected (Harper et al. 2020). Further, association study analyzed candidate genes for vitamin E content, seed weight and seed colour. This new approach facilitates rapid advancement to study the gene sequence and expression variations in the diversity panel.

Two different comparative transcriptomics studies have identified the differentially expressed genes related to aphid defense and seed coat color in *B. juncea* (Liu et al. 2013; Duhlian et al. 2020). Recently, transcriptome analysis has been performed to characterize the molecular mechanism involved in anthocyanin biosynthesis in *B. juncea*. The backcrossed BC3 segregating population for purple and green leaves was used to identify the differentially expressed genes. A total of 2286 genes exhibited differential expression. Out of these 1593 genes were upregulated and 693 genes were downregulated. The study assists in understanding the critical genes and pathways involved in anthocyanin biosynthesis in *B. juncea* (Heng et al. 2020). Different regulation pathways of various development stages can be studied at gene expression level using innovative transcriptomics tools. This may help for further exploration of comparative and evolutionary functional genomics of Indian mustard.

20.4.3 Pangenome

In this genomic era, multiple reference genomes are available in crop species with high-quality assemblies which can be utilized to find out the structural variations. Although more than one reference genomes were available in Brassica species, it cannot cover the whole gene content of a species because of structural variants, like the copy number variants (CNVs) and presence/absence variants (PAVs). So, the lack of agronomically important genes in the reference genome leads to the construction of pangenomes in *Brassica* species. Pangenome has been constructed in *B. oleracea* utilizing diverse varieties and their wild relative (Golicz et al. 2016). The genes linked to key agronomic traits were annotated and many of the genes influenced by PAVs and CNVs. This study of the pangenome of *B. oleracea* depicts that about 20% of the genes are impacted by PAVs. Resistance genes have also been characterized by pangenome study in *B. oleracea* and *B. napus* species and total 59 and 1749 resistance gene analogs (RGAs) were observed respectively (Bayer et al. 2019; Dolatabadian et al. 2020). Another recent finding on pangenome of *B. napus* has identified the clusters of specific genes, core gene and dispensable gene (Song et al. 2020). Total 8 genotypes of *B. napus* were assembled utilizing de novo approach in this study, and genome wide comparative study was performed to identify the causal PAVs. A pan transcriptome reference has also been generated in *B. juncea* using diverse genotypes and specific markers have been detected which are useful to study the genetic variations in this crop (Harper et al. 2020). Consequently, use of pangenome to study the structural variations among diverse germplasm of Indian mustard, its diploid progenitors and wild relatives can provide better understanding of the evolution of this amphidiploid crop.

20.5 Organellar Genome Organization and Sequence

The comprehensive studies of Brassica chloroplast DNA and mitochondrial DNA play a crucial role in phylogenetic analysis and delineating the evolutionary history of *B. juncea*. The nuclear genome is inherited from both parents whereas the organelle genomes including chloroplastic and mitochondrial genomes are maternally inherited across generations (Mogensen 1996; Jansen and Ruhlman 2012). Therefore, the study of organelle genomes is essential to find out the maternal inheritance in amphiploid species *B. juncea*. Moreover, the evolutionary relationships are easily figured out using cytoplasmic genome studies because it is much smaller than the nuclear genome.

20.5.1 Chloroplast Genome

The chloroplast genome (cpDNA) of *B. juncea* L. was sequenced using the next-generation sequencing method by Prabhudas et al. (2015). The complete size was around 153,483 bp and consists of large single copy (LSC) and short single copy (SSC) region of 83,286 bp and 17,775 bp respectively and two inverted repeats regions that were separated by 26,211 bp each. In *B. juncea*, cpDNA is small, circular, double stranded (100–300 kb) and highly conserved structure as compared with mitochondrial genome (Birky 2001; Rubeson and Jansen 2005). Whereas in the related allotetraploid species *B. napus,* one cpDNA haplotype was identified using genotyping and TILLING analysis (Allender and King 2010; Qiao et al. 2016). The *B. juncea*, cpDNA have GC content of 36.36% and the gene content of cpDNA comprises 79 protein-coding genes, 30 tRNA genes and four rRNA genes. Moreover, phylogenetic study using cpDNA with other related species of Brassica genus revealed that *B. juncea* showed more similarity to *B. rapa* (Prabhudas et al. 2015). The phylogenetic analysis based on 42 chloroplast genomes of Brassica genus revealed that 13, 14 and 24 accessions of *B. rapa*, *B. juncea* and *B. napus* respectively were separately clustered into a species-specific group (Qiao et al. 2020). Also, they demonstrated that Brassica genus was divided into three different clades, of which *B. rapa, B. juncea* and few *B. nap*us accessions constitute Clade I and most importantly Indian mustard accessions didn't diverge into any secondary branches thus confirming their monophyletic maternal origin from cam-type *B. rapa*.

20.5.2 Mitochondrial Genome

The mitochondrial genome (mtDNA) of *B. juncea* is a circular, multipartite structure and larger (219,766 bp) than cpDNA (153,483 bp). The mtDNA of *B. juncea* was comparatively smaller than *B. carinata* and *B. oleracea* and the G + C content was approximately 45.33% (Chang et al. 2011; Heng et al. 2014). The mtDNA gene content of *B. juncea* was 54%, protein coding genes constitute 34%, rRNA (3%), and tRNA (18%). Heng et al. (2014) studied the mechanism of cytoplasmic male sterility (CMS) in *B. juncea*, using the comparative analysis of mitochondrial DNA of the CMS line and its maintainer line. This comparative study has validated the cytoplasmic male sterility-linked gene, *orf288*, in *B. juncea*. Furthermore, it helps in understanding of evolutionary and structural variances between the diverse mitotypes. The genome of five alloplasmic male sterile lines has been sequenced and the transcriptome has also been explored in *B. juncea* (Wu et al. 2019). Comparative alignment of these sequences showed high homology and similar segment arrangement; also these five genotypes conserved the maximum part of their ancestral mitochondrial genome. This genome indicated about 21% promiscuous sequences and the sequences derived from the chloroplast are conserved among Cruciferae. Moreover, the phylogenetic analysis placed these genomes close to their wild relatives. Total 12 mtDNA genomes from six Brassica species were used for phylogenomics to detect the maternal parents of the polyploid Brassica species (Xue et al. 2020). This finding revealed that *B. juncea* was comparatively less divergent than *B. napus* and confirmed its maternal origin from *B. rapa*. Therefore, the structural differences during evolution of Brassica mt genomes are primarily because of the repeat mediated recombinations.

20.6 Conclusion

Science has progressed beyond the age of genomics and into the age of comparative genomics, where the pieces of the evolutionary puzzle will begin to fall into place. Advancement has been made in the diverse areas of Brassica genomics, including evolutionary genomics, throughout the last two decades. But, the progress in *B. juncea* genomics has been delayed because of the unavailability of genome sequence of this species. However, an excellent advancement has been observed in *B. juncea* genomics from the last five years after the accessibility of the whole genome sequence. Different sequence based molecular markers were already available in Indian mustard which were generated using sequence information from progenitor or other Brassica species. These markers were exploited for evolutionary studies in *B. juncea*. Nowadays, pangenome comparative analysis has been used in various Brassica species, which assist in better understanding of the genome architecture and can be employed in Indian mustard. A pan transcriptome reference has already been established in *B. juncea* using transcriptome data of a diverse germplasm panel which is used to study the genetic variations. The field of paleogenomics has advanced because of the current improvements in high-throughput DNA sequencing tools as well as in bioinformatics which will make it possible to retrieve precious information regarding the evolutionary history of Indian mustard.

References

Akhatar J, Singh MP, Sharma A, Kaur H, Kaur N et al (2020) Association mapping of seed quality traits under varying conditions of nitrogen application in *Brassica juncea* L. Czern&coss. Front Genet 11:744

Akhatar J, Goyal A, Kaur N, Atri C, Mittal M et al (2021) Genome wide association analyses to understand genetic basis of flowering and plant height under three levels of nitrogen application in *Brassica juncea* (L.) Czern&Coss. Sci Rep 11:4278

Al-Dous EK, George B, Al-Mahmoud ME, Al-Jaber MY, Wang H et al (2011) De novo genome sequencing and comparative genomics of date palm (*Phoenix dactylifera*). Nat Biotechnol 29(6):521–527

Allender CJ, King GJ (2010) Origins of the amphiploid species *Brassica napus* L investigated by chloroplast and nuclear molecular markers. BMC Plant Biol 10:54

Anderson JK, Warwick SI (1999) Chromosome number evolution in the tribe Brassiceae (Brassicaceae): evidence from isozyme number. Plant Syst Evol 215:255–285

Ashelford K, Eriksson ME, Allen CM, D'Amore R, Johansson M et al (2011) Full genome re-sequencing reveals a novel circadian clock mutation in Arabidopsis. Genome Biol 12(3):R28

Atri C, Akhatar J, Gupta M, Gupta N, Goyal A et al (2019) Molecular-genetic analysis of defensive responses of *Brassica juncea-B. fruticulosa* introgression lines to *Sclerotinia* infestation. Sci Rep 9:17089

Aversano R, Ercolano MR, Caruso I, Fasano C, Rosellini D et al (2012) Molecular tools for exploring polyploid genomes in plants. Intl J Mol Sci 13 (8):10316–10335

Axelsson T, Bowmann CM, Sharpe AG, Lydiate DJ, Lagercrantz U (2000) Amphidiploid *Brassica juncea* contains conserved progenitor genomes. Genome 43:679–688

Babula D, Kaczmarek M, Barakat A, Delseny M, Quiros CF et al (2003) Chromosomal mapping of *Brassica oleracea* based on ESTs from *Arabidopsis thaliana*: complexity of the comparative map. Mol Gen Genom 268:656–665

Baird NA, Etter PD, Atwood TS, Currey MC, Shiver AL et al (2008) Rapid SNP discovery and genetic mapping using sequenced RAD markers. PLoS One 3(10):e3376

Bayer PE, Golicz AA, Tirnaz S, Chan CK, Edwards D et al (2019) Variation in abundance of predicted resistance genes in the *Brassica oleracea* pangenome. Plant Biotechnol J 17(4):789–800

Belser C, Istace B, Denis E, Dubarry M, Baurens FC et al (2018) Chromosome-scale assemblies of plant genomes using nanopore long reads and optical maps. Nat Plants 4(11):879–887

Birky CW Jr (2001) The inheritance of genes in mitochondria and chloroplasts: Laws, mechanisms, and models. Annu Rev Genet 35:125–148

Bisht NC, Gupta V, Ramchiary N, Sodhi YS, Mukhopadhyay A et al (2009) Fine mapping of loci involved with glucosinolate biosynthesis in oilseed mustard (*Brassica juncea*) using genomic information from allied species. Theor Appl Genet 118:413–421

Buggs RJA, Renny-Byfield S, Chester M, Jordon-Thaden IE, Viccini LF et al (2012) Next-generation sequencing and genome evolution in allopolyploids. Am J Bot 99(2):372–382

Burton W, Ripley V, Potts D, Salisbury PA (2004) Assessment of genetic diversity in selected breeding lines and cultivars of canola quality *Brassica juncea* and their implications for canola breeding. Euphytica 136:181–192

Chalhoub B, Denoeud F, Liu S, Parkin IAP, Tang H et al (2014) Early allopolyploid evolution in the post-Neolithic *Brassicanapus* oilseed genome. Science 345 (6199):950–953

Chang S, Yang T, Du T, Huang Y, Chen J et al (2011) Mitochondrial genome sequencing helps show the evolutionary mechanism of mitochondrial genome formation in Brassica. BMC Genomics 12:497

Chen FB, Liu HF, Yao QL, Fang P (2016) Evolution of mustard (*Brassica juncea* Coss) subspecies in China: evidence from the chalcone synthase gene. Genet Mol Res 6:15(2)

Cheung WY, Friesen L, Rakow GFW, Seguin-Swartz G, Landry BS (1997) A RFLP-based linkage map of mustard [*Brassica juncea* (L.) Czern. and Coss.]. Theor Appl Genet 94:841–851

Cheung WY, Landry BS, Raney P, Rakow GFW (1998) Molecular mapping of seed quality traits in *Brassica juncea* L. Czern. and Coss. Actahortic 459:139–148

Cokus SJ, Feng S, Zhang X, Chen Z, Merriman B et al (2008) Shotgun bisulphite sequencing of the Arabidopsis genome reveals DNA methylation patterning. Nature 452(7184):215–219

Dhaka N, Mukhopadhyay A, Paritosh K, Gupta V, Pental D et al (2017) Identification of genic SSRs and construction of a SSR-based linkage map in *Brassica juncea*. Euphytica 213:15

Dolatabadian A, Bayer PE, Tirnaz S, Hurgobin B, Edwards D et al (2020) Characterization of disease resistance genes in the *Brassica napus* pangenome reveals significant structural variation. Plant Biotechnol J 18(4):969–982

Dowen RH, Pelizzola M, Schmitz RJ, Lister R, Dowen JM et al (2012) Widespread dynamic DNA methylation in response to biotic stress. Proc Natl Acad Sci USA 109(32):E2183-2191

Duhlian L, Koramutla MK, Subramanian S, Chamola R, Bhattacharya R (2020) Comparative transcriptomics revealed differential regulation of defense related genes in *Brassica juncea* leading to successful and unsuccessful infestation by aphid species. Sci Rep 10:10583

Edwards D, Batley J, Snowdon RJ (2013) Accessing complex crop genomes with next-generation sequencing. Theor Appl Genet 126(1):1–11

Elshire RJ, Glaubitz JC, Sun Q, Poland JA, Kawamoto K et al (2011) A robust, simple genotyping by-sequencing (GBS) approach for high diversity species. PLoS One 6(5):e19379

Golicz A, Bayer P, Barker G, Edger PP, Kim H et al (2016) The pangenome of an agronomically important crop plant *Brassica oleracea*. Nat Commun 7:13390

Harper AL, He Z, Langer S, Havlickova L, Wang L et al (2020) Validation of an associative transcriptomics platform in the polyploid crop species *Brassica juncea* by dissection of the genetic architecture of agronomic and quality traits. Plant J 103:1885–1893

He X, Zhang J (2005) Rapid subfunctionalization accompanied by prolonged and substantial neofunctionalization in duplicate gene evolution. Genetics 169:1157–1164

Heng S, Wei C, Jing B, Wan Z, Wen J et al (2014) Comparative analysis of mitochondrial genomes between the hau cytoplasmic male sterility (CMS) line and its iso-nuclear maintainer line in *Brassica juncea* to reveal the origin of the CMS-associated gene orf288. BMC Genomics 15(1):322

Heng S, Wang L, Yang X, Huang H, Chen G et al (2020) Genetic and comparative transcriptome analysis revealed degs involved in the purple leaf formation in *Brassica juncea*. Front Genet 11:322

Higgins JA, Magusin A, Trick M, Fraser F, Bancroft I (2012) Use of mRNA-Seq to discriminate contributions to the transcriptome from the constituent genomes of the polyploid crop species *Brassica napus*. BMC Genomics 13(1):247

Inturrisi F, Bayer PE, Yang H, Tirnaz S, Edwards D et al (2020) Genome-wide identification and comparative analysis of resistance genes in *Brassica juncea*. Mol Breed 40(8):78

Jaccoud D, Peng K, Feinstein D, Kilian A (2001) Diversity arrays: a solid state technology for sequence information independent genotyping. Nucleic Acids Res 29(4):e25

Jansen RK, Ruhlman TA (2012) Plastid genomes of seed plants. In: Bock R, Knoop V (eds) Genomics of chloroplasts and mitochondria. Springer, Netherlands, pp 103–126

Kaur P, Banga S, Kumar N, Gupta S, Akhatar J et al (2014) Polyphyletic origin of *Brassica juncea* with *B. rapa* and *B. nigra* (Brassicaceae) participating as cytoplasm donor parents in independent hybridization events. Am J Bot 101:1157–1166

Kaur J, Akhatar J, Goyal A, Kaur N, Kaur S et al (2020) Genome wide association mapping and candidate gene analysis for pod shatter resistance in *Brassica juncea* and its progenitor species. Mol Biol Rep 47(4):2963–2974

Koch MA, Haubold B, Mitchell-Olds T (2000) Comparative evolutionary analysis of chalcone synthase and alcohol dehydrogenase loci in Arabidopsis, Arabis, and related genera (Brassicaceae). Mol Biol Evol 17:1483–1498

Kowalski SP, Lan TH, Feldmann KA, Paterson AH (1994) Comparative mapping of *Arabidopsis thaliana* and *Brassica oleracea* chromosomes reveals islands of conserved organization. Genetics 138:499–510

Lagercrantz U (1998) Comparative mapping between *Arabidopsis thaliana* and *Brassica nigra* indicates that Brassica genomes have evolved through extensive genome replication accompanied by chromosome fusions and frequent rearrangements. Genetics 150:1217–1228

Lam HM, Xu X, Liu X, Chen W, Yang G et al (2010) Resequencing of 31 wild and cultivated soybean genomes identifies patterns of genetic diversity and selection. Nat Genet 42:1053–1059

Lan T-H, Paterson AH (2000) Comparative mapping of quantitative trait loci sculpting the curd of *Brassica oleracea*. Genetics 155:1927–1954

Lan T-H, DelMonte TA, Reischmann KP, Hyman J, Kowalski SP et al (2000) An EST-enriched comparative map of *Brassica oleracea* and *Arabidopsis thaliana*. Genome Res 10:776–788

Li X, Ramchiary N, Dhandapani V, Choi SR, Hur Y et al (2013) Quantitative trait loci mapping in *Brassica rapa* revealed the structural and functional conservation of genetic loci governing morphological and yield component traits in the A, B, and C subgenomes of Brassica species. DNA Res 20:1–16

Li C, Lin F, An D, Wang W, Huang R (2017) Genome sequencing and assembly by long reads in plants. Genes 9(1):6

Lionneton E, Ravera S, Sanchez L, Aubert G, Delourme R et al (2002) Development of an AFLP-based linkage map and localisation of QTLs for seed fatty acid content in condiment mustard (*Brassica juncea*). Genome 45:1203–1215

Lister R, O'Malley RC, Tonti-Filippini J, Gregory BD, Berry CC et al (2008) Highly integrated singlebase resolution maps of the epigenome in Arabidopsis. Cell 133(3):523–536

Lister R, Pelizzola M, Dowen RH, Hawkins RD, Hon G et al (2009) Human DNA methylomes at base resolution show widespread epigenomic differences. Nature 462(7271):315–322

Liu X, Lu Y, Yuan Y, Liu S, Guan C et al (2013) De novo transcriptome of *Brassica juncea* seed coat and identification of genes for the biosynthesis of flavonoids. PLoS One 8(8):e71110

Long Y, Shi J, Qiu D, Li R, Zhang C et al (2007) Flowering time quantitative trait loci analysis of oilseed Brassica in multiple environments and genome wide alignment with Arabidopsis. Genetics 177:2433–2444

Love CG, Graham NS, Lochlainn SO, Bowen HC, May ST et al (2010) A Brassica exon array for whole-transcript gene expression profiling. PLoS One 5(9):e12812

Lukens L, Zou F, Lydiate D, Parkin I, Osborn T (2003) Comparison of a *Brassica oleracea* genetic map with the genome of *Arabidopsis thaliana*. Genetics 164:359–372

Lukens LN, Pires JC, Leon E, Vogelzang R, Oslach L et al (2006) Patterns of sequence loss and cytosine methylation within a population of newly resynthesized *Brassica napus* allopolyploids. Plant Physiol 140(1):336–348

Lv H, Wang Y, Han F, Ji J, Fang Z et al (2020) A high-quality reference genome for cabbage obtained with SMRT reveals novel genomic features and evolutionary characteristics. Sci Rep 10(1):12394

Lynch M, Conery JS (2000) The evolutionary fate and consequences of duplicate genes. Science 290:1151–1155

Lynch M, Force A (2000) The probability of duplicate gene preservation by subfunctionalization. Genetics 154:459–473

Mahmood T, Ekuere U, Yeh F, Good AG, Stringam GR (2003a) RFLP linkage analysis and mapping genes controlling the fatty acid profile of *Brassica juncea* using reciprocal DH populations. Theor Appl Genet 107:283–290

Mahmood T, Ekuere U, Yeh F, Good AG, Stringam GR (2003b) Molecular mapping of seed aliphatic glucosinolates in *Brassica juncea*. Genome 46:753–760

Makhoul M, Rambla C, Voss-Fels KP, Hickey LT, Snowdon RJ et al (2020) Overcoming polyploidy pitfalls: a user guide for effective SNP conversion into KASP markers in wheat. Theor Appl Genet 133:2413–2430

Mascher M, Gundlach H, Himmelbach A, Beier S, Twardziok SO et al (2017) A chromosome conformation capture ordered sequence of the barley genome. Nature 544:427–433

Mathieu R-G, Belser C, Da Silva C, Richard G, Istace B et al (2020) Long-read assembly of the *Brassica napus* reference genome Darmor-bzh. Gigascience 9 (12):giaa137

McGrath JM, Quiros CF, Harada JJ, Landry BS (1990) Identification of *Brassica oleracea* monosomic alien chromosome addition lines with molecular markers reveals extensive gene duplication. Mol Gen Genet 223(2):198–204

Mogensen H (1996) The hows and whys of cytoplasmic inheritance in seed plants. Am J Bot 3:383–404

Mukherjee AK, Mohapatra T, Varshney A, Sharma R, Sharma RP (2001) Molecular mapping of a locus controlling resistance to *Albugo candida* in Indian mustard. Plant Breed 120:483–497

Negi MS, Devic M, Delseny M, Lakshimikumaran M (2000) Identification of AFLP fragments linked to seed coat colour in *Brassica juncea* and conversion to a SCAR marker for rapid selection. Theor Appl Genet 101:146–152

Osborn TC, Kole C, Parkin IA, Sharpe AG, Kuiper M et al (1997) Comparison of flowering time genes in *Brassicarapa, B. napus* and *Arabidopsis Thaliana*. Genetics 146:1123–1129

Panjabi P, Jagannath A, Bisht NC, Padmaja KL, Sharma S et al (2008) Comparative mapping of *Brassica juncea* and *Arabidopsis thaliana* using Intron Polymorphism (IP) markers: homoeologous relationships, diversification and evolution of the A, B and C Brassica genomes. BMC Genomics 9:113

Panjabi-Massand P, Yadava SK, Sharma P, Kaur A, Kumar A et al (2010) Molecular mapping reveals two independent loci conferring resistance to Albugo candida in the east European germplasm of oilseed mustard *Brassica juncea*. Theor Appl Genet 121 (1):137–145

Paritosh K, Gupta V, Yadava SK, Singh P, Pradhan AK et al (2014) RNA-seq based SNPs for mapping in *Brassica juncea* (AABB): synteny analysis between the two constituent genomes A (from *B. rapa*) and B (from *B. nigra*) shows highly divergent gene block arrangement and unique block fragmentation patterns. BMC Genomics 15:396

Paritosh K, Pradhan AK, Pental D (2020) A highly contiguous genome assembly of *Brassica nigra* (BB) and revised nomenclature for the pseudochromosomes. BMC Genomics 21:887

Paritosh K, Yadava SK, Singh P, Bhayana L, Mukhopadhyay A et al (2021) A chromosome-scale assembly of allotetraploid *Brassica juncea* (AABB) elucidates comparative architecture of the A and B genomes. Plant Biotechnol J 19(3):602–614

Parkin IA, Sharpe AG, Keith DJ, Lydiate DJ (1995) Identification of the A and C genomes of amphidiploid *Brassica napus* (oilseed rape). Genome 38:1122–1131

Perumal S, Koh CS, Jin L, Buchwaldt M, Higgins EE et al (2020) A high-contiguity *Brassica nigra* genome localizes active centromeres and defines the ancestral Brassica genome. Nat Plants 6:929–941

Peterson BK, Weber JN, Kay EH, Fisher HS, Hoekstra HE (2012) Double digest RADseq: an inexpensive method for de novo SNP discovery and genotyping in model and non-model species. PLoS One 7(5): e37135

Pires JC, Zhao J, Schranz ME, Leon EJ, Quijada PA et al (2004) Flowering time divergence and genomic rearrangements in resynthesized Brassica polyploids (Brassicaceae). Biol J Linn Soc 82:675–688

Prabhudas SK, Raju B, Thodi SK, Parani M, Natarajan P (2015) The complete chloroplast genome sequence of Indian mustard (*Brassica juncea* L.). Mitochondrial DNA Part A 27(6):4622–4623

Pradhan AK, Pental D (2011) Genetics of *Brassica juncea*. In: Schmidt R, Bancroft I (eds) Genetics and genomics of the brassicaceae. Plant Genetics and genomics: crops and models, vol 9, Springer, New York, NY

Pradhan AK, Gupta V, Mukhopadhyay A, Arumugam N, Sodhi YS et al (2003) A high-density linkage map in *Brassica juncea* (Indian mustard) using AFLP and RFLP markers. Theor Appl Genet 106:607–614

Prakash S, Bhat SR, Quirós CF, Kirti PB, Chopra VL (2009) Brassica and its close allies: cytogenetics and evolution. Plant Breed Rev 31:21–187

Qiao J, Cai M, Yan G, Wang N, Li F et al (2016) High-throughput multiplex cpDNA re-sequencing clarifies the genetic diversity and genetic relationships among *Brassica napus, Brassica rapa* and *Brassica oleracea*. Plant Biotechnol J 14:409–418

Qiao J, Zhang X, Chen B, Huang F, Xu K et al (2020) Comparison of the cytoplastic genomes by re-sequencing: insights into the genetic diversity and the phylogeny of the agriculturally important genus Brassica. BMC Genomics 21:480

Quiros CF (1999) Genome structure and mapping. In: Gomez Campo C (ed) Biology of Brassica coenospecies. Elsevier, Amsterdam, pp 217–245

Raman H, Raman R, Nelson MN, Aslam MN, Rajasekaran R et al (2012) Diversity array technology markers: genetic diversity analyses and linkage map construction in rapeseed (*Brassica napus* L.). DNA Res 19(1):51–65

Ramchiary N, Padmaja KL, Sharma S, Gupta V, Sodhi YS et al (2007) Mapping of yield influencing QTL in *Brassica juncea*: implications for breeding of a major oilseed crop of dryland areas. Theor Appl Genet 115(6):807–817

Rana K, Atri C, Gupta M, Akhatar J, Sandhu PS et al (2017) Mapping resistance responses to *Sclerotinia* infestation in introgression lines of *Brassica juncea* carrying genomic segments from wild Brassicaceae *B. fruticulosa*. Sci Rep 7:5904

Rana K, Atri C, Akhatar J, Kaur R, Goyal A et al (2019) Detection of first marker trait associations for resistance against *Sclerotinia sclerotiorum* in *Brassica juncea–Erucastrumcardaminoides* introgression lines. Front Plant Sci 10:1015

Roth C, Rastogi S, Arvestad L, Dittmar K, Light S et al (2007) Evolution after gene duplication: models, mechanisms, sequences, systems and organisms. J Exp Zool 306B:58–73

Rubeson LA, Jansen RK (2005) Chloroplast genomes of plants. In: Henry RJ (ed) Plant diversity and evolution: genotypic and phenotypic variation in higher plants. Wallingford, CABI

Saal B, Brun H, Glais I, Struss D (2004) Identification of a *Brassica juncea*-derived recessive gene conferring resistance to *Leptosphaeriamaculans* in oilseed rape. Plant Breed 123:505–511

Schlötterer C (2004) The evolution of molecular markers—just a matter of fashion? Nat Rev Genet 5(1):63–69

Sharma A, Li X, Lim YP (2014) Comparative genomics of Brassicaceae crops. Breed Sci 64:3–13

Singh BK, Nandan D, Ambawat S, Ram B, Kumar A (2015) Validation of molecular markers for marker-assisted pyramiding of white rust resistance loci in Indian mustard (*Brassica juncea* L.). Can J Plant Sci 95:939–945

Slocum MK, Figdore SS, Kennard WC, Suzuki JY, Osborn TC (1990) Linkage arrangement of restriction fragment length polymorphism loci in *Brassica oleracea*.Theor Appl Genet 80:57–64

Song KM, Suzuki JY, Slocum MK, Williams PH, Osborn TC (1991) A linkage map of *Brassica rapa* (syn. campestris) based on restriction fragment length polymorphism loci. Theor Appl Genet 82:296–304

Song JM, Guan Z, Hu J, Guo C, Yang Z et al (2020) Eight high-quality genomes reveal pan-genome architecture and ecotype differentiation of *Brassica napus*. Nat Plants 6:34–45

Sra SK, Sharma M, Kaur G, Sharma S, Akhatar J et al (2019) Evolutionary aspects of direct or indirect selection for seed size and seed metabolites in *Brassica juncea* and diploid progenitor species. Mol Biol Rep 46:1227–1238

Strickler SR, Bombarely A, Mueller LA (2012) Designing a transcriptome next-generation sequencing project for a nonmodel plant species. Am J Bot 99(2):257–266

Sudan J, Singh R, Sharma S, Salgotra RK, Sharma V et al (2019) ddRAD sequencing-based identification of inter-genepool SNPs and association analysis in *Brassica juncea*. BMC Plant Biol 19:594

Sun Q, Zhou G, Cai, Fan Y, Zhu X et al (2012) Transcriptome analysis of stem development in the tumourous stem mustard *Brassica juncea* var. *tumida* Tsen et Lee by RNA sequencing. BMC Plant Biol 12:53

Thakur AK, Singh KH, Singh L, Nanjundan J, Khan YJ et al (2018) SSR marker variations in Brassica species provide insight into the origin and evolution of Brassica amphidiploids. Hereditas 155:6

The Arabidopsis Genome Initiative (2000) Analysis of the genome sequence of the flowering plant *Arabidopsis thaliana*. Nature 408:796–815

U N (1935) Genome analysis in Brassica with special reference to the experimental formation of *B. napus* and peculiar mode of fertilization. Jpn J Bot 7:389–452

Van Dijk EL, Jaszczyszyn Y, Naquin D, Thermes C (2018) The third revolution in sequencing technology. Trends Genet 34:666–681

Van Orsouw NJ, Hogers RCJ, Janssen A, Yalcin F, Snoeijers S et al (2007) Complexity reduction of polymorphic sequences (CRoPSTM): a novel approach for large-scale polymorphism discovery in complex genomes. PLoS One 2:1172–1182

Vaughan JG, Hemingway JS (1959) The utilization of mustards. Econ Bot 13:196–204

Vision TJ, Brown DG, Tanksley SD (2000) The origin of genomic duplications in Arabidopsis. Science 290:2114–2117

Wang X, Wang H, Wang J, Sun R, Wu J et al (2011) The genome of the mesopolyploid crop species *Brassica rapa*. Nat Genet 43(10):1035–1039

Woods DL, Capcara JJ, Downey RK (1991) The potential of mustard (*Brassica juncea* (L) Coss) as an edible oil crop on the Canadian prairies. Can J Plant Sci 71:195–198

Wu Z, Hu K, Yan M, Song L, Wen J et al (2019) Mitochondrial genome and transcriptome analysis of five alloplasmic male-sterile lines in *Brassica juncea*. BMC Genomics 20:348

Xu X, Pan S, Cheng S, Zhang B, Mu D et al (2011) Genome sequence and analysis of the tuber crop potato. Nature 475(7355):189–195

Xue JY, Wang Y, Chen M, Dong S, Shao ZQ et al (2020) Maternal inheritance of U's triangle and evolutionary process of Brassica mitochondrial genomes. Front Plant Sci 11:805

Yang TJ, Kim JS, Kwon SJ, Lim KB, Choi BS et al (2006) Sequence-level analysis of the diploidization process in the triplicated Flowering Locus C region of *Brassica rapa*. Plant Cell 18(6):1339–1347

Yang J, Liu D, Wang X, Ji C, Cheng F et al (2016) The genome sequence of allopolyploid *Brassica juncea* and analysis of differential homoeolog gene expression influencing selection. Nat Genet 48:1225–1232

Zou J, Hu D, Liu P, Raman H, Liu Z et al (2016) Co-linearity and divergence of the A subgenome of *Brassica juncea* compared with other Brassica species carrying different A subgenomes. BMC Genomics 17:18

21 Regulatory Genes in Development and Adaptation, and Their Utilization in Trait Improvement in *Brassica juncea*: Challenges and Opportunities

Sandip Das and Anandita Singh

Abstract

The transformational agenda of the United Nations (UN) on Sustainable Development recognizes the crucial role of agriculture in mitigation and adaptation to climate change for safeguarding food security and alleviating hunger. Widespread crop losses are foretold in climate-change-induced disrupted agro-ecosystems meriting urgent intervention. The discovery of novel genes and alleles governing key agronomic traits underpins generation of climate-resilient crops capable of sustaining optimal performance despite environmental onslaughts. Regulatory genes and associated cis-elements are implicated in shaping development and adaptations thereby qualifying these as targets for enhancing productivity and building resilience in crops. Given the regulatory role of transcription factors, activators, non-coding RNAs, and associated *cis*-regulatory elements in trait manifestation, characterization of these is essential for generating crop ideotypes. *Brassica juncea* (Indian Mustard), is an important oil-seed crop of India that faces significant yield gaps owing to environmental stresses, and high consumer demand. It is an interesting crop model to understand the consequences of polyploidy in context to natural variation and phenotypic plasticity, and to evolve insights on how regulatory networks underlying trait manifestation are established in polyploids. This chapter presents recent advances in functional characterization of regulatory genes governing developmental and adaptive traits of agronomic relevance in *B. juncea*. In the initial sections, identification and functional characterization of regulatory genes involved in such adaptive life-history traits as environmental stress response, flowering, development of pollen, ovule, seed, and fruit are discussed. Characterization of genes governing oil-quality traits, biosynthesis of fatty acids, and glucosinolates are presented. Special case studies describing, characterization of non-coding RNAs in stress response and cis-regulatory elements are mentioned. Finally, applications emerging from the basic theoretical framework as representative examples of deployment of regulatory genes for generation of climate-smart cis- and transgenic *B. juncea* are discussed. Major research gaps, open questions, and challenges constraining improvement programs in *B. juncea* are identified to highlight the course of and future studies.

S. Das (✉)
Department of Botany, University of Delhi, New Delhi 110007, India
e-mail: sdas@botany.du.ac.in

A. Singh (✉)
Department of Biotechnology, Institutional Area, TERI School of Advanced Studies, Plot no. 10, Vasant Kunj, New Delhi 110070, India
e-mail: asingh@terisas.ac.in

C. Kole and T. Mohapatra (eds.), *The Brassica juncea Genome*, Compendium of Plant Genomes,
https://doi.org/10.1007/978-3-030-91507-0_21

21.1 Introduction

Accelerating crop improvement in Anthropocene is the greatest global challenge of present times. With an estimated rise of 1–3.7 °C in mean surface temperature by the turn of century (IPCC 2014, 2018), climate-change modelers predict severely compromised agricultural productivity (Bita and Gerats 2013; Cohn et al. 2017). Altered climates with modified ecological dynamics threaten widespread crop loss owing to undesirable overlaps between life cycles of novel pathogens and plants. Generation of climate-resilient crops is thus of paramount importance. Sustaining food, nutrition, and energy security within finite natural resources is only possible by innovative integration of modern agri-genomic technologies. For unlocking the true potential of genomics-assisted breeding methodologies towards design of future crops (Peleman and van der Voort 2003; Varshney et al. 2021), unraveling the molecular bases of agronomic traits is an essential pre-requisite. Transcription factor genes, non-coding RNAs, and associated *cis*-regulators are key components of gene regulatory networks underlying development and adaptive traits, many of which are agronomic traits of interest. Identification and functional analysis of regulatory genes and elements are thus crucial for crop improvement schemes.

The family Brassicaceae comprises 3709 species and several members are of immense value as crops of economic trade, and as model plants to understand the linkages between natural variation, evolutionary developmental pathways, and quantitative agronomic traits in context of polyploidy (Warwick and Hall 2009; Huang et al. 2016; Francis et al. 2021). Among the various species, Brassicas are valued as source of seed oil, vegetable and horticultural varieties, spices, and condiments and are particularly interesting for understanding polyploidy driven origin of morphological novelties and adaptive diversity (Warwick 2011; Cheng et al. 2014; Koenig and Weigel 2015). The amphidiploid *Brassica juncea* ($2n = 4x = 36$; *AABB* genome), a natural hybrid between *B. rapa* ($2n = 2x = 20$; *AA* genome) and *B. nigra* ($2n = 2x = 16$; *BB* genome), is of special relevance to food culture and oil-seed economy of India.

Although, developing molecular insights on structure and function of genomes is challenging in plants with moderate to large polyploid genomes (Fu et al. 2016), availability of whole-genome sequence information constitutes a rich genomic resource to be leveraged for genetic dissection of key agronomic traits. Fortuitously, complete or near-complete draft sequence of a number of Brassicaceae members is now available at Brassica Database (BRAD, www.brassicadb.org, http://brassicadb.cn/#/SpeciesInfo/), including an annotated genome of allotetraploid *B. juncea* var. Tumida (*AABB*) (Yang et al. 2016). The genome sequence of constituent ancestral base-genome karyotypes *AA, BB, CC* (*B. rapa, B. nigra*, and *B. oleracea*, respectively) can also be accessed at BRAD for large-scale structural and functional studies. The characteristic evolutionary history of genus *Brassica* and genetic relationships shared with other members of Brassicaceae is well dissected at molecular genetic level and sequence analysis revealed that rounds of whole-genome duplication followed by gene fractionation have modeled the architecture of extant *Brassica* species (Bowers et al. 2003; Yu et al. 2017; Nikolov et al. 2019). Duplicated genes present in partially redundant notional sub-genomes tend to diversify and are suggested to have opted into novel biological roles via large-scale rewiring of regulatory networks. While comparative genomics approaches permit extension of mechanistic insights from close relatives like *A. thaliana* into *B. juncea*, experimental validation is nevertheless an imperative given the possibility of polyploidy induced homolog and homeolog divergence. Deciphering functional consequences of polyploidy-induced gene diversification in *B. juncea* is an exciting area of research with promising implications in crop improvement.

The high sensitivity of mustard to climatic variables with adverse impacts on productivity is well known. Elevated temperatures coinciding with onset of flowering and pod filling reduce yield potential of *B. juncea* and as low as 1 °C rise in ambient temperature results in a 17% yield loss in Brassicas (Lobell and Asner 2003; Boomiraj et al. 2010). The threshold temperature for flowering ranges from 4 to 25 °C in Indian mustard and even incremental temperature rise compromises seed yield (Shekhawat et al. 2012). *B. juncea* is also severely affected by various abiotic stresses including, but not limited to, high temperature, drought, salinity, frost; and biotic stresses because of sucking insects such as *Lipaphiserysimi* (aphid); fungi such as *Sclerotinia sclerotiorum* (stem rot), *Albugo candida* (white rust), *Alternaria brassicicola* (black spot), *Leptosphaeria maculans* (blackleg), *Alternaria brassicae*, *A. brassicicola*, and *A. raphanin* (alternaria blight), *Hyaloperonospora brassicae* (downy mildew), *Neopseudocercosporella capsellae* (white leaf spot), *Plasmodiophora brassicae* (clubroot), *Erysiphe cruciferarum* (powdery mildew); and viruses such as Turnip yellow virus (formerly Beet western yellow virus), Cauliflower mosaic virus, Turnip mosaic virus; and bacteria-*Pseudomonas syringae* (leaf blight). Further, there is a need to improve fatty acid profile as an increased proportion of PUFAs such as oleic acid and γ-linoleic acid (Sinha et al. 2007), low viscosity (Naeem et al. 2020), and enhanced nutritional value by biofortification (Yusuf and Bhalla-Sarin 2007). Developmental traits such as flowering and maturity time, root system architecture for efficient nutrient and water acquisition, lateral branching, pod shape, and size, are target traits of agro-economic importance. Multidisciplinary interventions are therefore required to improve seed yield, oil quality, growth, development, and adaptations against biotic and abiotic stresses (Pradhan and Pental 2011).

With this perspective, we present a survey of developments in functional characterization of key regulatory genes and cis-elements underpinning developmental and adaptive traits in *B. juncea*. With an overall goal of crop improvement, following agronomic traits requiring biotechnological intervention will be discussed:

1. Seed germination and seedling establishment
2. Plant architecture
3. Leaf development
4. Root development and architecture including mineral and water uptake
5. Phase transition, floral induction
6. Flower development
   (a) Sepal
   (b) Petal
   (c) Stamen
   (d) Gynoecium
7. Fertilization, and seed/silique development (including dehiscence/pod shattering)
8. Oil quality and quantity.

21.2 Regulatory Genes During Vegetative Phase of Growth

This section provides a summary of research findings on regulatory genes and elements involved in seed germination, establishment of seedling, leaf and root development, plasticity in root system architecture, and overall plant growth and architecture.

Heat stress adversely impact percentage germination and successful establishment of seedlings in *B. juncea*. Studies related to seed germination in *B. juncea* have therefore been performed with the objective of understanding stress response to heat and heavy metal stress during seedling stage (Sharma et al. 2013; Thakur and Sharma 2016; Pandey and Penna 2017; Rai et al. 2020). Comparison of transcript levels of *HEAT SHOCK TRANSCRIPTION FACTOR 21 (HSF21)* and *HSFA7A* across seedlings of 34 genotypes of *B. juncea* have led to categorization of varieties into five distinct classes viz. highly heat tolerant, heat tolerant, moderately heat tolerant, heat sensitive, and highly heat sensitive (Rai et al. 2020).

Although functional analysis of genes involved in leaf development has not been reported from *B. juncea* hitherto, a large number of molecular regulators have been identified that are involved in regulation of leaf size and related trait biomass in *Brassica* species. Some of these include homologs of *AINTEGUMENTA*

(*BrANT*), *AINTEGUMENTA-Like* (*BrAIL*), *AUXIN-REGULATED GENE INVOLVED IN ORGANSIZE (BrARGOS), patatin-related phospholipase A (pPLA)* family members (*pPLAI, pPLAIIα, pPLAIIβ, pPLAIIγ, pPLAIIδ, pPLAIIε, pPLAIIIp, PLAIIIα, pPLAIIIβ, pPLAIIIγ,* and, *pPLAIIIδ*), *AUXIN-REPRESSED PROTEIN 1* (*BrARP1*), *DORMANCY-ASSOCIATED PROTEIN 1* (*BrDRM1*), *ETHYLENE RESPONSE FACTORS* (*BrERF*), *Repressor of ga1* (*BrRGA1*), TCP family members (eg. *BrTCP2*), member of the B3 superfamily (*NGATHA/NGA/BrNGA1*), *BrPHYB,* and, *MIR319, MIR394* (Karamat et al. 2021). A combination of bulked segregant analysis and transcriptome sequencing of the bulked pools has been used to identify and clone the gene *REDUCE COMPLEXITY* (*RCO; BjRCO*). Characterization of *BjRCO* gene revealed its product to be an HD-Zip class of transcription factor and was shown to be responsible for leaf lobing in *B. juncea* (Heng et al. 2020b).

Anthocyanin pigmentation has been reported in leaves of *B. juncea* purple mustard. The locus responsible for purple coloration was identified through genetic mapping, bulked segregant analysis, and annotated as *BjPLI1* (Zhao et al. 2017) and later as *BjPur* (Heng et al. 2020a). The locus was subsequently cloned and shown to encode an *R2R3-MYB* gene. An insertion in the first intron of the *R2R3-MYB* gene was found to be responsible for green and purple coloration in leaves of *B. juncea* (Heng et al. 2020a). The authors also propose that *BjTT8*, *BjMYC2*, *BjC4H*, *BjDFR*, and *BjANS* (Heng et al. 2020a, b, c) or homologs of *AtPAP1*, *AtPAP2*, *AtMYB113*, and *AtMYB114* (Zhao et al. 2017), that are involved in anthocyanin biosynthesis, are linked to purple coloration in *B. juncea*. The studies are considered significant as most genes implicated in defense response against pathogens are also upregulated in purple leaf *B. juncea* (Heng et al. 2020a, b, c).

Among growth habit and plant architecture traits, the gene for determinacy, *Sdt1* (homolog of *TFL1* from *A. thaliana*) was mapped in a segregating population of *B. juncea*. Genotypes with determinate and indeterminate growth habits were generated artificially by crossing *B. napus* (*AACC*) and *B. carinata* (*BBCC*) and identifying plants with *AABB* chromosome constitution (Kaur and Banga 2015). In another study, the Gα subunit of the G-protein complex comprising of Gα, Gβ, and Gγ has been shown to be a key regulator of overall plant architecture, including organ size and seed weight. Downregulation of Gα1 homologs (*BjuA.Gα1* and *BjuB.Gα1*) led to reduction in plant height (between 32 and 58%), hypocotyl, root, internode length, and seed weight (between 11–13%) whereas constitutive overexpression increases plant height, hypocotyl, internode length, and seed weight (between 7 and 25%) in *B. juncea* (Kumar and Bisht 2020).

Previous studies had led to identification, isolation and characterization of members of auxin efflux carrier proteins- *PIN1*, *PIN2* and *PIN3*, members of PIN family (Ni et al. 2002a, b). In a recent study, as many as 65 members of the *Auxin Response Factor (ARF)* gene families were identified from *B. juncea* var. *Tumida* genome (Li et al. 2020a, b, c). Although all the ARFs exhibited spatio-temporal variation in transcript abundance during growth and developmental phases, detailed analysis indicated that *BjARF2b_A, BjARF3b_A, BjARF6b_A, and BjARF17a_B* are involved in auxin signaling and are highly expressed in tumor stem. Homologs of *A. thaliana ARFs*, *BjARF10*, *BjARF16*, and *BjARF17* genes were found to be regulated by *miR160*, and, *BjARF6* and *BjARF8* by *miR167* (Li et al. 2020a, b, c).

Several other families of regulatory genes that are involved in development and adaptive processes during vegetative phase of growth and have been characterized include 62 members of *TCP* gene family (He et al. 2020); 88 members of the *GRAS* gene family (Li et al. 2019a, b); 29 members of the *SOD* gene family (Verma et al. 2019); 14 members of the *SUCROSE SYNTHASE* (*SUS*) gene family (Li et al. 2021); 60 genes of *HEAT SHOCK TRANSCRIPTION FACTOR (HSF)* family (Li et al. 2020a, b, c); 57 members of *SQUAMOSA promoter binding protein* (SBP)-box gene family (Li et al. 2020a, b, c); 19 genes of *NON-EXPRESSOR*

PATOGENESIS-RELATED (*NPR*) gene family (Wang et al. 2021); and, 300 genes of the *NAC* gene family (Phukan et al. 2016; Jiang et al. 2021), to cite a few examples.

21.3 Regulatory Genes During Phase Change and Reproduction

21.3.1 Flowering Time and Flower Development

Understanding flowering time control and flower development, two key life-history traits, has received widespread attention in crops, including *Brassica* species. Accelerated floral transition is a target trait promising mitigation of yield penalties caused by environmental stresses. Homologs of several regulatory genes involved in photoperiod and vernalization mediated floral induction pathways in *A. thaliana* such as *CONSTANS (CO; BjuACO4) FLOWERING LOCUS T (FT; BjuAFT1, BjuBFT4), SUPPRESOR OF OVEREXPRESSION OF CONSTANS (SOC1; BjuASOC1 and BjuASOC4), and FLOWERING LOCUS C (FLC; BjuFLC)* have been identified and characterized (Zhao et al. 2019). Other than these, the study also identified homologs of *GI, FRI, PRC2, VRN1, VRN2, VIN3, FCA, FPA, FY, LD, FLD, FLK, FVE, REF6, FD, LFY*, and *AGL24* from the genome of *B. juncea* and analyzed copy number variation (Zhao et al. 2019). Functional characterization identified a *DNA J* homolog (*J3*) from *B. juncea* that was found to be an upstream regulator of *SOC1* as J3 protein binds to promoter of *SOC1* and regulates flowering time (Zhou et al. 2018). In a parallel study, the histone deacetylase protein *HDA9* (*RPD3*-like histone deacetylase) was strongly induced under short day conditions and was found to physically interact with promoters of *BjSOC1* and *BjAGL24* to control flowering time (Jiang et al. 2018). *AGL24* has furthermore been shown to interact with *AGL18* and *AGL19* proteins to regulate flowering time in *B. juncea* (Li et al. 2019a, b; Wang et al. 2020). Other key regulators of flowering time that have been characterized include homologs of *LFY, SOC1*, and *FT*. Overexpression of *LFY* homolog isolated from *B. juncea* triggered extreme earliness in *B. juncea* under field conditions (Dhakate et al. 2017). Similarly, analysis of gain-of-function as well as artificial miRNA based mutant phenotypes of *BjFT* and *BjSOC1* in native *B. juncea* background under net-house conditions showed multiple roles beyond flowering (Tyagi et al. 2018, 2019). Overexpression of *B. juncea FT* homolog influenced agronomic traits such as silique shape, oil profile, stomatal morphology, and plant height (Tyagi et al. 2018). *SUPPRESSOR OF OVEREXPRESSION OF CONSTANS1* was found to influence other agronomic traits such as lateral branching, oil quality, and seed yield, apart from flowering time, in *B. juncea* (Tyagi et al. 2019).

21.3.2 Pollen Development and Male Sterility

Hybrid seed production through induction of cytoplasmic male sterility (CMS) and restoration system has been a major research goal for *Brassica* yield improvement. Most CMS systems developed in *B. juncea* have relied on use of alien cytoplasm through generation of interspecific or intergeneric hybrids; a few have also been obtained as spontaneous mutants such as the *hau* CMS system (Rao et al. 1994; Prakash et al. 1998; Pathania et al. 2003; Wan et al. 2008, 2014; Yamagishi and Bhat 2014; Atri et al. 2016). Molecular regulators and genetic basis of CMS systems have been well dissected and shown to involve rearrangements in the organellar genome. For example, *ORF108* derived from the mitochondrial genomes of *Diplotaxis erucoides, Diplotaxis berthautii, Diplotaxis catholica* and *Moricandia arvensis* has been implicated in CMS systems developed through intergeneric crosses with *B. juncea* (Kumar et al. 2012); as has been duplication of *COX1* gene in mitochondrial genome of *B. juncea* in a cross with *Diplotaxis catholica* (Bhat et al. 2005, 2006, 2008; Pathania et al. 2007; Ashutosh et al. 2008). Similarly, *ORF263* derived from mitochondrial genome of *Brassica fruticulosa,* and *ORF220,* and *ORF288* from mitochondrial genome of *Brassica rapa*, have also been shown to induce CMS in *B. juncea* (Landgren et al 1996; Jing-Hua et al. 2005; Yang et al. 2005, 2009, 2010; Jing et al. 2012; Zhao et al. 2016). Several genes such as *MutS Homolog1* have been shown to restore fertility by "sub-stoichiometric shifting (SSS)" of *ORF220* (Zhao et al. 2021). Transcript of *atp6* has been associated with *Ogura* CMS system that causes conversion of stamens to petaloid structures; whereas key regulators are yet to be identified in systems as exemplified by *oxa* CMS (Meur et al. 2006; Heng et al. 2018a, b). The sequence of *MYB80* homologs was found to be conserved between *A. thaliana*, *B. juncea*, *B. napus,* and *Gossypium hirsutum*; and its role was demonstrated in tapetum and pollen development as the *A. thaliana myb80* mutants display male-sterile phenotypes, with fertility restored by complementation with functional *MYB80* even from *G. hirsutum* (Xu et al. 2014). The product of the regulatory gene *RUB* (*related to ubiquitin 2*) *conjugating enzyme* (*RCE*; *BjRCE1*) has been demonstrated to be responsible for altered auxin homeostasis in CMS lines of *B. juncea* (Yang et al. 2012).

In a study based on comparative analysis, mitochondrial genome and transcriptome of five *B. juncea* CMS systems-*Diplotaxis catholica*-based, *Diplotaxis berthautii*-based, *Diplotaxis erucoides*-based, *Brassica oxyrrhina*-based, and, *Moricandia arvensis*-based male-sterile lines were investigated (Wu et al. 2019). The mitochondrial genomes range from 221 kb (*D. catholica*) to 256 kb (*M. arvensis*) with 32 protein coding genes in all species, except 31 in *D. catholica*. A comparative analysis of the mitochondrial genome and transcriptome data revealed that *ORF108* may be a common regulator of CMS system in *Diplotaxis berthautii*-based, *Diplotaxis erucoides*-based, and, *Moricandia arvensis*-based systems, but not in *Brassica oxyrrhina* and *Diplotaxis catholica* based CMS systems (Wu et al. 2019). Similar comparative analysis of mitochondrial genomes of hau CMS and its maintainer line revealed highly rearranged nature of mitochondrial genome of *hau* CMS and the origin of *orf288* that is associated with *hau* CMS (Heng et al. 2014).

21.3.3 Female Gametophyte, Embryo, Seed, and Fruit Development

Investigations into development of female gametophyte of *B. juncea* are rather limited and include reports on histological details (Smith and Sterner 1994), and role of *CLAVATA 1 (CLV1)* gene homologs, *BjMc1/Bjln1* (*BjuA07.CLV1*) in multi-locular phenotypes that have been described (Xiao et al 2013; Xu et al. 2014a, b; Xu et al. 2017; Chen et al. 2018; Xiao et al. 2018).

Nectaries, both floral and extra-floral (EFNs), are responsible for attracting pollinators, and animal visitors. The structure, quantitative, and qualitative composition of nectar in terms of sugar, amino acid from both floral and extra-floral nectaries in *B. juncea* have been investigated; yet there is paucity of studies relating to identification and characterization of regulatory genes. A nectary functional genomics endeavor is nevertheless ongoing for a number of plants including *B. juncea* (http://www.nectarygenomics.org/node/558; Masierowska 2003; Mathur et al. 2013; Nagpal et al. 2020).

Embryo development is known to be auxin regulated. In order to investigate the mechanism of action of auxin, embryo development in *B. juncea* was studied under *in-vitro* conditions using indole-3-acetic acid (IAA; natural auxin), *N*-(1-naphthyl)-thalamic acid (NPA; auxin transport inhibitor), and *p*-chlorophenoxyisobutyric acid (PCIB; anti-auxin) (Hadfi et al. 1998). Dramatic defects such as complete inhibition of morphogenesis, or formation of embryos with only shortened hypocotyl were observed in the presence of IAA. Similarly, inhibition of auxin transport led to formation of "twin embryos", "split-collar", and "collar-cotyledons", broadening of embryo axis, and formation of "twin radicles"(Hadfi et al. 1998). Defects in cotyledon number, hypocotyl, and radicle development were also observed when embryos of *B. juncea*, at different stages of development, were subjected to anti-auxin (Hadfi et al. 1998). Some of the developmental defects obtained in this study were found to mimic mutants observed in *A. thaliana*. Such studies have laid the groundwork for understanding role of auxin as a regulatory molecule affecting embryo development, and auxin signaling network in *B. juncea* (Hadfi et al. 1998; Li et al. 2020a, b, c).

Enzyme kinetics and activity of acid and alkaline phosphatases, and peroxidases have been shown to be associated with seed development in *B. juncea* (Saroop et al. 1999, 2002). Distinct patterns of cytoplasmic and wall-bound acid phosphatase were observed with significantly higher activity of wall-bound acid phosphatase. The activity of both increases as the seed development proceeds from 10 to 50-days post-anthesis and fertilization (Saroop et al. 1999).

Seed related traits are of paramount importance as seed is the source of mustard oil. Understanding of developmental process and seed characteristics viz. seed size, number, weight, and seed coat color is crucial for crop improvement programs. In a study, the micro-sculpting patterns and sub-cellular changes during seed coat development of *B. juncea* have been documented (Zeng et al. 2004, 2006). Although molecular regulators have not been identified, such studies form a basis for future investigations aimed at deciphering developmental patterns and analysis of seed coats in *B. juncea*. In addition to seed coat patterns, loci that are responsible for seed coat color, have been identified, rapid tests for screening of seed coat color developed, and yellow-seeded varieties have been developed (Vera and Woods 1982; Abraham and Bhatia 1986; Lu et al. 2012).

Comparative transcriptome of seed coat from near-isogenic brown seeded and inbred yellow-seeded lines revealed close to 800 transcripts that are upregulated and 500 are downregulated in brown testa; transcripts of *DFR* (*dihydroflavonol reductase*), *LDOX* (*leucoantho-cyanidin dioxygenase*), and *ANR* (*anthocyanidin reductase*) for late flavonoid biosynthesis are either absent or detected at very low abundance in the yellow-seeded testa (Liu et al. 2013). Sequence data, expression analysis and mapping of *F3H* (*Flavanone-3-hydroxylase*), *DFR* (*dihydroflavonol reductase*), *BAN* (*BANYULS; Anthocyanidin reductase -ANR*), *TT1* (*Transparent Testa 1;* C2H2 zinc finger protein), *TT2* (*Transparent Testa 2;* R2R3-MYBprotein), *TT8* (*Transparent testa 8*; bHLH protein) and *TTG1* (*Transparent Testa Glabra1;* WD40 regulatory protein) across brown and yellow-seeded *B. juncea* varieties and mapping population identified alleles of *TT8* controlling seed coat color (Yan et al. 2010; Padmaja et al. 2014).

In order to identify genomic loci regulating locule number as bilocular or trilocular, which is directly related to seed yield, genetic mapping studies identified *BjMC1* locus as being tightly linked to trilocular condition. Subsequently, map-based cloning and sequence characterization revealed *BjMC1* locus to be encoding a homolog of *CLAVATA1* (*CLV1*), or *Bj-BARELY ANY MERISTEM* (*BjBAM*), the product of which is a transmembrane protein. Insertion of a *copia-like* retrotransposon in the coding region of the *BjCLV1* was found to be responsible for trilocular condition in *B. juncea* (Xu et al. 2017; Chen et al. 2018). Genetic mapping and functional genomics have demonstrated that *SHATTERPROOF1/2 (SHP1/2)*, *FRUITFULL (FUL)*, *ENDO-BETA-MANNASE7 (MAN7)*, and *NAC SECONDARY WALL THICKENING PROMOTING FACTOR 2* (*NST2*) are key regulatory genes controlling pod shattering in *B. juncea* (Ostergaard et al. 2005; Kaur et al. 2020).

21.4 Regulatory Genes and Oil-Quality Traits

Brassica juncea is commercially valued as a source of seed oil, and efforts have been made to enhance not only oil quantity but also improve the quality parameters including increasing beneficial fatty acids and reducing harmful components.

21.4.1 Biosynthesis of Fatty Acids

The most important product of commerce from *B. juncea* is the seed oil, and biotechnological interventions have constantly aimed at increasing the oil content and modification of fatty acid profile. Towards this goal, a number of genes and QTLs involved in regulation of fatty acid biosynthesis and storage have been identified from *B. juncea* (Rout et al. 2018).

Among the various regulators of fatty acid biosynthesis and storage that have been identified, cloned, and analyzed from *B. juncea*, *Oleoyl-Acyl Carrier Protein (ACP) Thioesterase* (chain length determinant in Oleic acid biosynthesis; Mandal et al. 2002); *Stearoyl-ACP Desaturase* (responsible for catalyzing conversion of stearoyl-ACP to oleoyl-ACP; Vageeshbabu et al. 1996); four copies of long chain saturated *acyl-ACP thioesterases* (*BjFatB*; Mandal et al. 2000; Sinha-Jha et al. 2010); *Fatty acid elongase* (*FAE1*) responsible for very long chain fatty acid synthesis (C22; erucic acid; Venkateswari et al. 1999; Yadav et al. 2003; Gupta et al. 2004; Saini et al. 2016; 2019); *plastidial omega-3 desaturase* (*BjFAD7*; Garg et al. 2000, 2001); endoplasmic reticulum-associated *oleate desaturase* involved in conversion of C18:1 oleic acid to C18:2 linolenic acid (*FAD2*; Suresha et al. 2012, 2013; Suresha and Santha 2013). A total of 57 *Fatty acid desaturases* (*FAD*) genes have also been identified through in-silico analysis of the *B. juncea* genome (Xue et al. 2020).

21.4.2 Biosynthesis of Glucosinolates

Glucosinolates, a Sulphur- rich secondary metabolite with a large number of structural variants found in members of Brassicaceae are both beneficial, as well as known to contain anti-nutritional properties (Fahey et al. 1997; Halkier and Gershenzon 2006; Juge et al. 2007; Cartea and Velasco 2008; Clarke 2010). Based on leads obtained in *A. thaliana* and *Brassica* by previous researchers, homologs and homeologs of *MYB28* were identified from *B. juncea* as key regulators of aliphatic glucosinolate biosynthesis (Bisht et al. 2009; Augustine et al. 2013a; Augustine and Bisht 2015). Functional complementation using reference strain and *myb28* mutants of *A. thaliana* using *BjuMYB28* homeologs, and silencing of the endogenous copies using RNAi strategy have confirmed the role of *MYB28* in biosynthesis of aliphatic glucosinolate (Augustine et al. 2013a, b).

Other regulatory genes governing glucosinolate biosynthesis *in B. juncea* that have been identified and characterized include *CYP79F1* (*CYP79F1; Di-homo-methionine N-hydroxylase*), *CYP83A1* (*Cytochrome P450 83A1*), *SUR1* (*SUPERROOT1*), *UGT74B1* (*UDP-glucose:thiohydroximate S-glucosyltransferase*), *SOT16* (sulfotransferase), *CYP79A2 (CYTOCHROME P450 79A2; Phenylalanine N-monooxygenase)*, *CYP83B1 (ALTERED TRYPTOPHAN REGULATION 4; ATR4, CYTOCHROME P450, FAMILY 83SUBFAMILY B, POLYPEPTIDE 1)*, *CYP79B2* (*CYTOCHROME P450, FAMILY 79, SUBFAMILY B, POLYPEPTIDE* 2) and *CYP79B3* (*CYTOCHROME P450, FAMILY 79, SUBFAMILY B, POLYPEPTIDE 3*; *Tryptophan N-monooxygenase* 2) (Yang et al. 2014; Meenu et al. 2015; Sharma et al. 2016).

21.5 Regulatory Genes in Nutrient Uptake

Molecular and genetic analysis of association panel of 92 inbred lines at S7 generation of *B. juncea* diversity fixed foundation set (*BjDFFS*) were subjected to genome-wide association studies (GWAS) under varying nitrogen application (N-0 and N-100 kg/ha). Several strong marker-trait associations (MTAs) on chromosomes *A01, A06, B02, B04, B05*, and *B08* were identified and analysis of the target genomic locations on these chromosomes identified genes such as *NITRATE REDUCTASES* [*AFB-3* (31.3); *HRS-1 HOMOLOGUE 1* (12.3–12.4)], *NITRITE REDUCTASE* [*BAT-1* (13.2); *GLN1*.3]; *myb-like* transcription factor family protein, high-affinity nitrate transporters [*ATNRT2.6*; *ATNRT2.3*], *GLUTAMINE SYNTHETASE 1.3, BIDIRECTIONAL AMINO ACID TRANSPORTER 1, AUXIN SIGNALING F-BOX 3*, and, *OXIDOREDUCTASES* key regulators of nitrogen uptake (Gupta et al. 2019a). Mapping studies using an association panel of 92 diverse germplasm lines of *B. juncea*, under varying doses of nitrogen application (low dose of 75 kg/ha; standard dose of 100 kg/ha; and, high dose of 125 kg/ha), permitted identification of close to 30 regulatory genes that are involved in plant growth and height (*CDF1, CO, FLC, AGL24, GNC, FAF2,* etc.) and reproductive traits (*AP1, FVE, FRI, FLC, and CO*) in *B. juncea* that are affected by N-application. These genes are known to be involved in light perception, circadian rhythm, floral meristem identity, regulation of flowering time, gibberellic acid pathways, and overall plant development (Akhatar et al. 2021).

Genes for sulfate uptake and transport such as *low-affinity sulfate transporter* (*LAST*), *ATP sulfurylase* (*ATPS*), *APS reductase* (*APSR*), and *Myrosinase* (*MYR*; which may be involved in glucosinolate breakdown to provide additional sources of Sulphur under stress) have been isolated and cloned from *B. juncea* (Heiss et al. 1999). The study revealed that the expression state of these genes is also modulated under Cadmium stress (Heiss et al. 1999). Similar interactions between nitrogen and Carbon were observed when *B. juncea* was grown in combinations of varying N and C concentrations and expression patterns of several genes viz. *BjNRT1.1, BjNRT1.8, BjNR1* and *BjNR2, BjAMT1.2, BjAMT2* and BjPK, *BjAMT2* and *BjGS1.1* involved in N-uptake, assimilation, and metabolism were analyzed (Goel et al. 2016). A separate study identified genes involved in nitrogen use efficiency (NUE) by comparing the transcript profile of roots of *B. juncea* var. Pusa Bold that has high NUE and *B. juncea* var. Pusa Jaikisan with low NUE (Goel et al. 2018). The study identified genes responsible for N-uptake such as Nitrate transporters-*NRT1.1, NRT1.8,* and *NRT2.1*; N-assimilation such as Nitrate reductase (*NR1, NR2), Nitrite reductase (NiR), Glutamine synthetase* (*GS1.3*), and *Ferredoxin-dependent glutamate synthase* (*Fd-GOGAT*); N-remobilization such as *Glutamate dehydrogenase* (*GDH2*), *Asparagine synthetase (ASN2–3), Alanine aminotransferase* (*ALaT)*; and, transcription factors like *mitochondrial Transcription tERmination Factor* (*mTERF*), *FHA, Orphan, bZip,* and *FAR1* as key regulators of NUE in *B. juncea* (Goel et al. 2018). These genes involved in N-uptake, transport, assimilation, and metabolism were also previously been found to be modulated upon various abiotic stresses including salt (150 mM NaCl), osmotic (250 mM Mannitol), cold (4 °C), and heat (42 °C) stress at different time points of 1-h, or 24-h post-stress treatment (Goel and Singh 2015).

21.6 Regulatory Genes Involved in Biotic and Abiotic Stress

21.6.1 Biotic Stress Tolerance

Several fungal, bacterial, viral and insect pests pose serious threats to *B. juncea* at various stages of its life cycle. Prominent among these include blackleg disease caused by *Leptosphaeria maculans;* stem rot by *Sclerotinia sclerotiorum*; white rust by *Albugo candida*; alternaria blight by *Alternaria brassicae, A. brassicicola,* and *A. raphanin*; downy mildew by *Hyaloperonospora brassicae;* white leaf spot by *Neopseudocercosporella capsellae*; clubroot by *Plasmodiophora brassicae*; powdery mildew by *Erysiphe cruciferarum*; viral diseases by Turnip yellow virus, Cauliflower mosaic virus, and Turnip mosaic virus; leaf blight by *Pseudomonas syringae*; and sucking aphid insect *Lipaphis erysimi* (Kalt.) (Inturrisi et al. 2021). Over the years, several pathotypes of the various causative agents have been identified with each strain exhibiting variable extent of disease severity attributable to molecular diversity and ploidy level variation in the pathogen, and the host microbiome (Jouet et al. 2019). In addition, co-occurrence of multiple pathogens such as *Hyaloperonospora brassicae* (downy mildew) and *Albugo candida* (white rust) also impacts disease severity (Kaur et al. 2011b). Complexities of such interactions between pathogens and host have made identification of conserved, universal molecular regulators for several pathogen and pests an uphill task.

Aphid infestation is a major cause of yield loss in *B. juncea* and identification of regulatory genes involved in conferring aphid tolerance is therefore critical. A transcriptome-based comparison of mustard leaves infested with *Lipaphis erysimi* (mustard aphid; compatible interaction with mustard) and *Aphis craccivora* (cowpea aphid; incompatible interaction with mustard) was performed. Structural grouping of the differentially expressed genes revealed members of seven transcription factor gene families, namely *AP2, LBD, e2f-dp, ERF, G2*-like, *HB*-other, and *GATA* were specific to the transcriptome of *L. erysimi* infested tissue; and the major class of differentially expressed transcripts belonged to *bHLH* (11 members), *GeBP* (9 members), *NAC* (8 members), *B3* (5 members) and *CO*-like (5 members) (Duhlian et al. 2020).

Among several candidates, PR5, a thaumatin-like protein was identified through comparative proteomics analysis to be associated with resistance against *Albugo candida* in a tolerant *B. juncea* var. CBJ 001, but not in the susceptible variety RH819 (Kaur et al. 2011a). Of the four genes gene that encodes *Toll-like (interleukin-1) receptor-nucleotide-binding-leucine-rich repeat (TIR-NB-LRR)* proteins- *WRR4A, WRR4B, WRR8,* and *WRR9*, only *WRR4B* was identified in *A. thaliana* and found to confer resistance to *A. candida* race 2 (Ac2V) isolate in *B. juncea* (Borhan et al. 2010; Cevik et al. 2019). *WRR4* was found to confer resistance against race 2 (pathogen of *B. juncea*) and race 7 (pathogen of *B. napus*) of *A. candida* and is thus considered a broad range molecular regulator conferring tolerance (Borhan et al. 2010). Several other loci linked to white rust resistance have been identified in *B. juncea*. These include *Ac2(t)* (Mukherjee et al. 2001); *AcB1-A4.1* (Varshney et al. 2004); *Acr/Ac2a1* (Prabhu et al. 1998); *Ac2V1* (Somers et al. 2002); *AcB1-A4.1* (on *LG4/A4*) and *AcB1-A5.1* (on *LG5/A5*) (Panjabi-Massand et al. 2010; Singh et al. 2015); and *AcB1* on *LG6* (Bhayana et al. 2020). Although multiple loci conferring resistance to white rust has been identified and molecular markers generated for marker-assisted breeding and selection, molecular regulators that confer resistance/tolerance associated with majority of these loci are yet to be identified and isolated. However as a rare exception, characterization of the genomic loci encompassing *AcB1-A5.1* revealed the presence of *coiled-coil nucleotide-binding-leucine-rich repeat (CC-NB-LRR)* protein encoding *R* gene which was annotated as *BjuWRR1*. Functional analysis using *cis*-genics revealed the ability of *BjuWRR1* to confer tolerance against a range of *A. candida* isolates in *B. juncea* (Panjabi-Massand et al. 2010; Arora et al. 2019). Similarly, characterization of *AcB1* on *LG6* also identified a *CC-NBS-LRR* encoding *R* gene, responsible for white rust resistance (Bhayana et al. 2020).

Molecular regulators including candidate genes and/or loci that have been identified in *B. juncea* for other diseases include *Rlm5* and *Rlm6* genes encoding *nucleotide-binding site leucine-rich-repeat (NLR) leucine-rich repeat RLK (LRR-RLK)* and *LRR-RLP* genes, against blackleg disease (*Leptosphaeria maculans*; Balesdent et al. 2002; Plissonneau et al. 2018; Yang et al. 2021b); members of chitinase gene family against *Alternaria brassicae* (*Alternaria* blight/black spot disease; Mir et al. 2020; Yadav et al. 2020), and, members of pathogenesis related (PR) gene family that are part of salicyclic acid and jasmonic acid dependent signalling pathway (Ali et al. 2017a). The transcript of *jasmonic acid carboxylmethyl transferase (JMT)* was found to be induced 48 h post-infection with *Alternaria brassicicola* in *B. juncea* and has been proposed to be a candidate regulator against necrotrophic fungal pathogens (Meur et al. 2015).

Based on analysis of introgression lines of *B. juncea* containing genetic segments from *Erucastrum cardaminoides* and *B. fruticulosa*, several candidate genes that confer tolerance to *Sclerotinia sclerotiorum* that causes stem rot disease, have been identified. These include members of *TIR-NBS-LRR* gene family, *chitinase* family, *malectin/receptor-like protein kinases*, *defensin-like* (*DEFL*) genes, *desulphoglucosinolate sulphotransferase* protein, *lipoxygenases (LOX)*, and regulatory components of effector-triggered immunity (ETI) and pathogen-associated molecular patterns (PAMPs) pathways (Rana et al. 2017, 2019; Atri et al. 2019).

Brassica juncea suffers significant yield losses due to attack by *Plasmodiophora brassicae* (clubroot disease). Molecular regulator as a source for clubroot resistance has not yet been detected in *B. juncea*, although candidate genes such as *Rcr6* have been found on B-genome of *Brassica nigra* that is homologous to a chromosomal segment on *A08* of *B. rapa* (Hasan et al. 2012; Hasan and Rahman 2018; Chang et al. 2019). In a report, one accession of *B. juncea* (accession 1012) was however been found to be resistant to clubroot disease (Liu et al. 2018). Similarly, two loci- *Rcr3* and *Rcr9*^{wa}, with three candidate clubroot resistance genes each, *Bra020951*, *Bra020974*, and *Bra020979* for *Rcr3*; and *Bra020827, Bra020828, Bra020814* for*Rcr9*^{wa}, has been identified in *B. rapa* (Karim et al. 2020). Another study also identified two QTLs—*qBrCR38*-1 and *qBrCR38*-2, with strong association with clubroot resistance in *B. rapa* (Zhu et al. 2019). With *B. juncea* being an allopolyploid of AB genomes, it is intriguing and would be of immense interest to understand the lack of clubroot resistance in *B. juncea* though "resistance genes" have been identified in both the parental A and B genomes (Cheng et al. 2018; Hasan and Rahman 2018; Zhu et al. 2019; Karim et al. 2020). Efforts, though only partially successful, have therefore been made to introduce clubroot resistance by resynthesizing *B. juncea* using *B. rapa* and *B. nigra* as parents (Hasan and Rahman 2018). Jasmonic acid (JA) is known as a crucial component of defense response, and *JAZ* domain protein (*Jasmonate ZIM domain proteins*) are critical regulators of JA signaling during both abiotic and biotic stress. The entire repertoire of 18 regulatory genes belonging to the *JAZ* family is encoded by the *B. juncea* var. Tumida (tuber mustard) genome was identified and characterized. Out of these 18, two *JAZ* genes, *BjuA030800* and *BjuA007483* were found to be exclusively expressed in root and leaf, respectively. In addition, the expression of 13 *JAZ* regulatory genes—*BjuB031487, BjuB021388, BjuA027037, BjuA029428, BjuB029798, BjuA005572, BjuA045157, BjuB007213, BjuB025543, BjuA022138, BjuB030035, BjuA022588* and *BjuA047148* were upregulated after infection with *Plasmodiophora brassicae*, and expression levels *of BjuB026559, BjuB032915, BjuB029529, BjuA034780, BjuA001107,* and *BjuA001950* were downregulated or repressed. The transcriptional regulation of nearly the entire JAZ gene family in *B. juncea* var. Tumida was are thus affected by *Plasmodiophora brassicae* infection (Cai et al. 2020). The steady-state expression levels of some, not all of the *B. juncea JAZ* genes (*BjuB032431, BjuB007213, BjuB043343, BjuB031487, BjuB011370, BjuA030507, BjuB010656, BjuA007483, BjuB032915 BjuB035964, BjuA029428, BjuB029798, BjuA005572, BjuB025543, BjuB029529, BjuA034780,* and *BjuA027422*) were also found to be upregulated after 200 mM salt (NaCl) treatment indicating a cross-talk and dual role of *JAZ* domain proteins in management of biotic and abiotic stress (Cai et al. 2020).

Although one *B. juncea* variety RDV29 that shows resistance against *Erysiphe cruciferarum* (causal organism of powdery mildew) has been identified, no gene/s or molecular regulator has yet been isolated and characterized (Nanjundan et al. 2020). Similarly, studies have identified tolerant varieties of *B. juncea* against downy mildew (*Hyaloperonospora brassicae/Peronospora parasitica*) which cause yield losses of between 40 and 66% in susceptible varieties (Meena et al. 2014). Several researchers have hypothesized the presence of multiple genes/alleles conferring resistance against various strains of the fungal pathogen without actually identifying the molecular regulator (Nashaat et al. 2004; Mohammed et al. 2019).

Black rot (causal organism *Xanthomonas campestris*) is another bacterial disease that causes significant yield losses in *B. juncea* (Vicente and Holub 2013); and three resistance genes—*R1* (*Xca1* locus), *R4* (*Xca4* locus), and *R5* are possibly present in *B. juncea* (Vicente et al. 2002; Fargier and Manceau 2007).

Nucleotide-binding site leucine-rich repeat genes (NLRs) are known as major regulators of disease resistance, and identification of homologs of *NLRs* are key to formulating strategies for imparting disease tolerance. A total of 289 homologs of *NLRs* were identified from *B. juncea* genome and analysis revealed that these can be categorized into ten classes and two subfamilies—*toll/interleukin-1 receptor-NLR (TNL)* and *coiled-coil-NLR (CNL)* proteins (Inturrisi et al. 2020). Of these 289 genes, 145 were derived from B-genome and 119 from the A-genome. In a separate study, a total of 493 *Receptor-like kinases (RLKs)* including *LysM-RLKs* and *LRR-RLKs*, and 228 *Receptor-like proteins (RLPs)* including *LysM-RLPs* and *LRR-RLPs*, were identified from the genome of *B. juncea* (Yang et al. 2021a).

In a genome-wide association study (GWAS), 167 *B. juncea* varieties that differed with respect to tolerance/susceptibility to *Leptosphaeria maculans* were genotyped using the 90 K *Brassica* SNP array (Illumina). GWAS analysis identified seven SNPs showing strong association to *Rlm6* genes/loci present on chromosomes *A07* and *B04*. Detailed characterization of the chromosomal regions flanked by the SNPs showed 16 *nucleotide-binding site leucine-rich-repeat (NBS-LRR), leucine-rich repeat RLK (LRR-RLK),* and *Leucine-rich repeat receptor-like proteins (LRR-RLP)* genes that correspond to the *Rlm6* genes/*loci* on chromosomes A07 and B04 (Yang et al. 2021b).

Another study identified six candidate NAC transcription factors-*NAC19, NAC36, NAC55, NAC62, NAC72,* and *NAC81* from *B. juncea*, based on their expression pattern in *Sinapis alba*, a species exhibiting tolerance to *Alternaria brassicicola* causing black spot disease. Expression analysis showed that the selected NACs were not only induced by *Alternaria* brassicicola but also by desiccation stress, wounding, as well as methyl-jasmonate in *B. juncea* (Mondal et al. 2020). A clear correlation was observed between expression pattern and disease tolerance in *B. juncea* tolerant variety (PAB9511) and *B. juncea* susceptible variety (B85) (Mondal et al. 2020).

Several viral diseases are known to attack *Brassica* crops and cause enormous yield losses such as the Turnip mosaic virus (TuMV; Walsh and Jenner 2002; Guo et al. 2005). The first gene reported to confer resistance to a number of TuMV pathotypes is *TuRBJU01* (TuMV Resistance in *B. juncea* No. 1; Nyalugwe et al. 2015, 2016). The gene *TuRBJU01* is yet to be isolated and characterized and the exact molecular identity of the gene still remains unknown. In contrast, bulked segregant analysis of susceptible and resistant *B. juncea* identified a locus termed as *retr03* (*recessive TuMV resistance 03*). Sequence and functional analysis of *retr03* revealed it to be allelic to *eukaryotic translation initiation factor 2B-beta* (*eIF2Bβ*; Shopan et al. 2017). Based on subsequent analysis of *eIFs* from *B. juncea*, the authors propose that other sub-groups of eIFs, namely *eIF2β, eIF2α, eIF2Bβ, EF1A*, and *PABP* could potentially be utilized as sources for TuMV resistance (Shopan et al. 2020; Palukaitis and Kim 2021). The *alternative oxidase* (*AOX*) gene, from *B. juncea* (*BjAOX1a*), has also been proposed to increase resistance to TuMV as the inhibition of *BjAOX1a* protein by treatment of plants with salicyl hydroxamic acid (SHAM) caused an increase in systemic movement of the TuMV coat protein (Zhu et al. 2012).

21.6.2 Abiotic Stress Including Heavy Metal Tolerance

Annexin gene family proteins are Ca^{2+}-dependent plasma membrane-localized proteins and known to play role in various developmental and stress responses (Konopka-Postupolska et al. 2011; Jami et al. 2012). The annexin gene family was found to have six members in *B. juncea*-*AnnBj1-4*, *AnnBj6-7* (Jami et al. 2008, 2009, 2010). Although the members of the *B. juncea* Annexin family were found to be differentially expressed in various tissues, *AnnBj2* was found to be highly expressed when seedlings were subjected to salt stress using 200 mM NaCl, and, ABA (Ahmed et al. 2017). Functional characterization of *AnnBj2* using constitutive overexpression transgenic lines demonstrated its potential to tolerate salt, and ABA stress whereas no difference in plant performance was found when non-transgenic and *AnnBj2* transgenic plants were grown on 200 mM Mannitol (Ahmed et al. 2017). Molecular and physiological analysis of the transgenic plants revealed an increase in relative water content, proline, Ca^{2+}, and K^{+}, and a decrease in ROS and Na^{+} levels. The tolerance was hypothesized to be mediated by both ABA-dependent and ABA-independent pathways based on upregulation of *RAB18* (ABA-dependent) and *DREB2B* (ABA-independent) genes (Ahmed et al. 2017).

A major pathway that has previously been shown to be involved in ion homeostasis during salinity stress is the *Salt Overly Sensitive (SOS)* pathway (Ji et al. 2013). Twelve homologs of this regulatory gene family were identified from the genome of *B. juncea* var. *Tumida*, and these included paralogs and homeologs of *SOS1, SOS4, SOS5, SOS6* (two copies each), *SOS2* (one copy), and *SOS3* (three copies) (Cheng et al. 2019b). These homologs exhibited differential spatio-temporal expression patterns during development and under various stresses. For example, the two homologs of *SOS1* are highly expressed in both root and inflorescence. However, of the two copies of *SOS1*, *SOS1*-2 is expressed in leaves whereas *SOS1-1* is not. Similarly, the highest expression of *SOS4*-1 is seen in stem whereas that of *SOS4*-2 is seen in root and inflorescence. When seedlings were exposed to various abiotic stresses (200 mM NaCl, 50μMABA, low temperature at 4 °C) and biotic stress (*Plasmodiophora brassicae*), transcripts of several of the *SOS* genes were found to be up- or downregulated indicating their involvement in stress adaptive responses. For example, all homologs of *SOS1, SOS2,* and *SOS3* were upregulated, whereas those of *SOS4, SOS5,* and *SOS6* were downregulated under salinity stress. In comparison to this, *SOS3-1*(upregulated) and *SOS3-2*(downregulated) showed contrasting expression patterns when exposed to 50μM ABA. Similar expression dynamics were also observed under cold temperature and biotic stress indicating the complex regulatory network of adaptive response mediated by members of *SOS* gene family in *B. juncea* (Cheng et al. 2019b). Conserved function was demonstrated when *BjSOS3* homolog was used to complement, and rescue the *A. thaliana SOS3* mutant (*Atsos3*) and confer salinity tolerance (Nutan et al. 2018).

Co-repeat binding factor (CBF) and *Inducer of CBF Expression (ICE)* are key regulatory genes in cold temperature stress signaling. Two isoforms of *ICE-BjICE46* and *BjICE53* were isolated and functionally characterized. While *BjICE46* was found to be induced by salinity stress, MeJA, and ABA, *BjICE53* was found to be induced by cold stress and non-responsive to hormonal induction. *Co-repeat binding factor (CBF)*, the downstream target of ICE, was also identified and isolated from *Brassica juncea*. The authors also expressed *CBF* and validated the binding of CBF protein to Drought Responsive Elements (*DRE*; TACCGACAT) that are present in promoter region of cold regulated genes through EMSA (Kashyap and Deswal 2019). This study thus demonstrated the putative regulatory role played by *B. juncea* homologs of CBF and ICE in management of abiotic stress response (Kashyap and Deswal 2019). The transcript profile of developing siliques subjected to cold stress has also been cataloged using NGS technology, and 283 transcripts that are upregulated during all stages of embryo development have been identified (Sinha et al. 2015). Some of the genes that are cold-induced include homologs of *Nodulin MtN3 (SWEET13)*, *β-amyrin-synthase*, *ORG3-like*, *Oxy-phytodienoate reductase3 (OPR3)*, *WRKY48*, *COLD REGULATED (COR) 14 (COR14)*, *COR27*, *β-amylase*, and, *EARLY RESPONSIVE TO DEHYDRATION (ERD)*. Transcription factors and regulatory genes that are cold specific, early developments stage specific, or, late development stage specific belong to diverse classes as exemplified by *ABI3VP1*, *AP2-EREBP*, *bZIP*, *C2H2*, *HSF*, *MADS-box, Myb, Plant AT-rich sequence, and zinc-binding proteins (PLATZ), Sigma70-like, Trihelix, WRKY, AT-Rich Interaction Domain (ARID), Gcn5-related N-acetyltransferases (GNAT), Tumor necrosis factor receptor-associated factor (TRAF), DOMAIN OF UNKNOWN FUNCTION 26 (DUF26) kinases, Ca-dependent protein kinases (CDPK),* and *SNF1-related protein kinases* (SnRKs) were identified (Sinha et al. 2015).

Other reports on identification, and characterization of regulators from *B. juncea* that are involved in various abiotic and biotic stress tolerances include *Dehydrins* (Yao et al. 2005); *Expansin* (*BjEXPA1*; Sun et al. 2011); *Dehydration responsive element binding transcription factor* gene (*BjDREB1B*; Cong et al. 2008); 25 members of the *pyrabactin resistance (PYR)/ PYR1-like (PYL)* gene family exemplified by *BjuPYL3, BjuPYL4s, BjuPYL5s, BjuPYL6s, BjuPYL7s, BjuPYL8s, BjuPYL10s, BjuPYL11s,* and*BjuPYL13* that are regulatory components of ABA receptor (*RCAR*) and proposed to regulate stress response and stem development (Cheng et al. 2019a); 60 members of *Heat shock factor (HSF)* gene family (Li et al. 2020a, b, c); 47 members of the *chitinase* gene family (Zhao and Chye 1999; Mir et al. 2020); *BjPR1* gene, that plays a role in systemic acquired resistance (SAR; Ali et al. 2018); *BjMYB1* and its interaction with promoter of the *chitinase* gene, *BjCHI1* (Gao et al. 2016).

Metals and heavy metals such as Cd, Pb, As, Se, and Ni are known to significantly alter plant growth and reduce plant productivity. It is therefore important to investigate the molecular basis of heavy metal toxicity and mechanisms to ameliorate stress (Singh et al. 2016). The effect of heavy metal pollutants on *B. juncea* is two-fold. On one hand, heavy metal causes sub-optimal growth and yield loss, and on the other hand, *B. juncea* has been shown to be a hyperaccumulator of heavy metals and therefore valued as a plant useful for phytoremediation (Singh and Fulekar 2012; Rathore et al. 2019). It is therefore important to understand and identify molecular regulators that not only allow *B. juncea* to grow on metal contaminated soil but also allow *B. juncea* to act as a hyperaccumulator for phytoremediation. At least 73 Cd-responsive genes that are differentially regulated were identified using cDNA-AFLP (Fusco et al. 2005). Sequence analysis of these cDNA-AFLP fragments and expression analysis revealed their identities with a few examples being *MYB59*; Transcription factor *GBF5*; *bZIP* family transcription factor *TGA3*; *Glutamine synthetase-chloroplast precursor glutamate–ammonialigase* (*GS2*); *Cysteine synthase (O-acetylserinesulfhydrylase) (O-acetylserine(thiol)-lyase); histone deacetylase*; *β-Hydroxyacyl-ACP dehydratase; PIP1*, to name a few (Fusco et al 2005). In another study, RNA-sequencing was employed to identify genes involved in uptake, and detoxification of Cd (Thakur et al. 2019). A total of 616 membrane transporters exemplified by *NRAMP1* (metal transporter), *MTPC2* and *MTP11* (metal tolerance proteins), *PCR2, PCR6* (plant cadmium resistance protein), and *cadmium-transporting ATPase*, were found to be upregulated. Transcripts of *HMA2, HMA3, HMA4* (cadmium/zinc-transporting ATPase), *CAX1* (high-affinity calcium antiporter), and *IRT1* (iron transport protein) were downregulated. Molecular regulators annotated as part of various pathways were also identified and included 87 transcripts from glutathione (GSH) metabolism, 112 transcripts from phenylpropanoid biosynthesis, 125 transcripts of MAPK signalling [eg., *MKS1; WRKY33; WRKY22; MKK4/5; MKK9*; aminocyclopropane-1-carboxylate synthase (*ACS*); serine/threonine-protein kinase *OXI1* (*OXI1*); calmodulin (*CALM*); *RBOH*], 14 transcripts of calcium signaling [e.g., *PLCD; MCU; VDAC2*], and,184 transcripts of plant hormone signal transduction [e.g., *AUX/IAA*; ARFs; auxin-responsive GH3 gene family; *SAUR*; *CRE1* (cytokinin response 1); *A-ARR; B-ARR*; *gibberellin receptor* (*GID1*); *PIF4*; *adenylatedimethylallyltransferase* (*IPT*); *PP2C*; brassinosteroid (*BR*), jasmonic acid (JA) and salicylic acid signaling; ethylene receptor / *ETHYLENE RESPONSE 1* (*ETR*); *ETHYLENE INSENSITIVE 3* (*EIN3*); *ETHYLENE RESPONSE FACTOR 1 (ERF1); ETHYLENE INSENSITIVE 2 (EIN2); BRASSINOSTEROID INSENSITIVE 1 (BRI1); BRASSINOSTEROID-INSENSITIVE 2 (BIN2)* (Thakur et al. 2019).

Comparative proteome analysis of *B. juncea* roots when subjected to Cd stressusing 2-D gel electrophoresis and multiplexed isobaric tagging technology (iTRAQ) revealed several classes of proteins that are involved in alternative redox regulation such as peptide methionine sulfoxide reductase, and, 2-nitropropane dioxygenase; and proteins that are involved in Cd hyper-accumulation and tolerance such as O-acetylserinesulfhydrylase, glutathione-S-transferase, and glutathione-conjugate membrane transporters. Other than these, proteins such as Hydroxy-methyltransferase, Aspartate aminotransferase, Stromal ascorbate peroxidase, WD-40 repeat protein, Glutathione S-transferase, Germin-like protein, Putative Myrosinase-binding protein, Selenium-binding protein-like, and proteins that are part of an antioxidant mechanism, defense response, thiol metabolism, xenobiotic detoxification, and several other categories were identified (Alvarez et al. 2009).

Regulatory genes and networks involved in Arsenic tolerance have been identified through transcriptome analysis of shoot and root of *B. juncea* in a time-dependent manner under Arsenate stress (Srivastava et al. 2015). Some of these regulatory genes identified in the study are *PIP1;2, PIP2;2, SULTR2;1, APS1, FSD2, CAT3, OPR1, CTR1, DOF6.8, WRKY33, WRKY6, ACA13* (Srivastava et al. 2015). A total of 50 heavy metal transporter genes (partial sequences) from *natural resistance-associated macrophage proteins* (*NRAMP*; 23 genes) and *yellow stripe-like proteins* (*YSL*; 27 genes) were isolated based on a combination of RT-PCR and Amplified rDNA Restriction Analysis (ARDRA) from *B. juncea*. Out of these 50 genes, *BjYSL6*.1 was found to be upregulated in shoots of Cd-treated plants, and *BjYSL5*.8 was found to be upregulated in roots of Pb-treated plants of *B. juncea* (Das et al. 2011).

Various other studies that have reported identification of regulatory genes implied in heavy metal tolerance are genes for Pb and Cd responses (Dalyan et al. 2017); heavy metal cation efflux transporter genes—*BjCET1, BjCET2, BjCET3,* and *BjCET4* involved in Zn and Cd assimilation (Xu et al. 2009; Lang et al. 2005); regulators of Cd-uptake, transport and assimilation such as *NRAMP1*; *MTPC2*, and *MTP11* (both encoding metal tolerance protein), *cadmium-transporting ATPase*, *PCR2*, and *PCR6* (both encoding plant cadmium resistance proteins), *HMA2*, *HMA3*, and *HMA4* (all encoding cadmium/zinc-transporting ATPases), *CAX1* (high-affinity calcium antiporter), glutathione transporter (BjGT1) upon Cd stress (Bogs et al. 2003) and, *IRT1* (iron transport protein) (Thakur et al. 2019); *HMA4A* involved in Zn and Cd-stress response (Wang et al. 2019); *BjYSL7* (yellow stripe-like 7), a metal-nicotinamide (NA) transporter that is responsible for Ni and Cd transport (Wang et al 2013); a total of 19 genes that are differentially expressed upon Cd-stress include auxin responsive *GH3, ARF-like small GTPases/ARFs, APSR2 (APS reductase), ARD/ARD'* (*acid-reductone dioxygenase*), members of *Nop* family that are involved in pre-RNA processing ribonucleoproteins, C3HC4-type RING finger (zinc finger) proteins, *DAG* (diacylglycerol) kinases, members of halo acid dehalogenase-like hydrolase, predicted membrane proteins (*KOG3491*), and ribosome-associated membrane protein families (*RAMP4*; Minglin et al. 2005); *mitochondrial γ-glutamyl cysteine synthetase* involved in Cd-stress (Schafer et al. 1998); *hemeoxygenase* (*BjHO1*) gene that is regulated by Zn, Cd, Hg and Pb stress (Li et al. 2012); *SK2-type dehydrin* genes-*DHN2, DHN3,* involved in Cd and Zn stress (Xu et al. 2008).

21.7 Non-coding RNAs Regulating Development and Stress Response

The power of NGS technology has propelled the discovery and profiling of non-coding RNAs involved in development and stress management in plants.

Development of CMS and restorer lines in *B. juncea* for hybrid seed production has been a crucial strategy to increase yield. Identification of key regulators such as small RNAs, involved in floral organ development and to induce CMS is, therefore, a critical research goal. Based on small RNA profiling, a total of 275 miRNAs of which 78 were novel and 197 previously reported, were identified from different stages of reproductive organ development in *B. juncea* (Yang et al. 2013). In addition to these, a total of 47 miRNA that is differentially expressed between CMS and the fertility maintainer (MF) lines were also identified. Among the novel miRNA identified during various stages of organ development, several were mapped onto chloroplast (*PC-5p-13, PC-3p-14, PC-3p-39, PC-5p-40, PC-3p-54, PC-3p-72, PC-5p-75, PC-5p-88* and *PC-3p-90*) and or onto mitochondrial (*PC-5p-17, PC-3p-18, PC-3p-25* and *PC-5p-26*) genomes. In addition, miRNAs annotated as *pre-miRNA22/23, pre-miRNA24/25, pre-miRNA73/74,* and *pre-miRNA75/76* mapped to both nuclear and mitochondrial genomes (Yang et al. 2013). Comparison of CMS and MF lines revealed that such miRNAs among several others, *miR156a, miR156e, miR158a, miR159a, miR164c, miR165a, miR167a, miR319b, miR390a, miR395d, miR396b, miR408, miR845a* are differentially regulated (Yang et al. 2013).

A genome-wide "*miRNAOme*" of *Brassica juncea* was generated by growing the seedlings under salinity (150 mM and 250 mM NaCl), drought (20% PEG and 300 mM Mannitol), and heat (35 °C and 42 °C) stresses for different time periods, followed by small RNA profiling (Bhardwaj et al. 2014). The authors identified at least 126 novel miRNAs that had not been reported until then in any other plant species, and 51 conserved miRNAs. Validation of novel miRNAs revealed that the levels of *Bju-N30* were significantly downregulated in heat, drought, and salinity stresses. In contrast, an abundance of *miRNA Bju-N29* and *Bju-N21* increases upon both salt and drought stresses but decreases when subjected to high-temperature stress. Similar differential expression was observed for several other novel miRNAs such as *Bju-N31* (increased under salinity stress), *Bju-N35* (reduced under high temperature and salinity stress). Of the six conserved miRNAs investigated, viz. *miR168_1, miR169_3, miR172_2, miR390_1, miR394_1, miR395_2,* and *miR828_1*, transcript levels of *miR395_2* was found to be significantly increased under salinity, drought, and heat stresses. In contrast, transcript abundance of *miR390_1* and *miR172_2* was reduced across all the three stresses investigated (Bhardwaj et al. 2014). In a separate study, functional characterization of *MIR172* genes from *B. juncea* revealed conserved role in regulating floral timing, as well as altered function in regulating floral organ development when homeologs and paralogs of *BjuMIR172* was ectopically expressed in *B. juncea* background (Shivraj et al. 2018). All the homeologs of *MIR172b (miR172b, miR172b'), MIR172d (miR172d, miR172d'),* and, *MIR172e (miR172e, miR172e'),* caused earliness in flowering; in addition, transgenic lines over-expressing *miR172b, miR172b', miR172d,* and *miR172d'* exhibited defects in floral organs whereas *miR172e, miR172e'* did not (Shivaraj et al. 2018). Other researchers have employed *in-silico* analysis and comparative genomics to identify networks of miRNAs and their putative targets that are enriched in various pathways under abiotic stress responses. Furthermore, SSRs, and SNPs have been identified in miRNAs that are part of the abiotic stress regulatory network for the purpose of generating markers, and for marker-assisted breeding (Singh et al. 2017).

Custom microarray was employed to identity regulatory miRNAs that are significantly differentially expressed under Arsenic (As) stress in *B. juncea*. Several miRNAs such as *miR156, miR169, miR172, miR395, miR838, and miR854, miR319, miR167, miR164,* and *miR159* that are otherwise known to be involved in development, mineral nutrition, and hormonal responses were identified but their precise role in As stress response is yet to be deciphered (Srivastava et al. 2013). Studies have also identified close to 7600 long non-coding (lnc) RNAs from the *B. juncea* genome; of these ~7600 lncRNAs, ca. 1600 lincRNAs may play a role during heat and drought stress responses as these were found to be co-expressed with various genes belonging to transcription factor classes such as *MIKC-MADS, NAC, Myb, bHLH, HSF, HD-Zip, CO-Like, ERF,* and *NF-γC*, that were previously known to be involved in abiotic stress management (Bhatia et al. 2020).

21.8 cis-Regulatory Elements in *Brassica Juncea*

Identification and functional analysis of cis-regulatory elements such as promoters are key to generation of cis-genic and transgenic plants that would allow highly specific regulation of transgene expression. Towards this goal, the promoter associated with *B. juncea SOS2* (*Salt*

Overly Sensitive 2), a key regulator of salinity stress was functionally characterized (Liu et al. 2000; Kaur et al. 2015). Functional analysis of the 713 bp promoter fragment and nested deletions using promoter::GUS reporter fusion constructs demonstrated upregulation of transcriptional activity under salt, desiccation, cold, and heat stress, and ABA, thus indicating the involvement of *B. juncea SOS2* in mediating multiple stress responses (Kaur et al. 2015). In another study, promoters of *KCS6* and *KCS5*, key enzymes in cuticular wax biosynthesis were isolated from *B. juncea* and functionally characterized. Analysis of GUS reporter activity revealed that the promoter can regulate transcription of GUS reporter gene in a wide range of tissues including leaves, inflorescence, all floral organs, and siliques (Singh et al. 2020). The authors also demonstrate the role of part of first intron from *AtKCS6* as negative regulator of transcription, and, that the transcription is under epigenetic control. A comparative analysis of homologs from *B. juncea* on the epigenetic regulation of promoter activity and role of intron is pending and would throw light on the transcriptional regulation of cuticular wax biosynthesis (Singh et al. 2020). Among other genes, promoter analysis for transcriptional regulation of genes such as *Metal Transport Protein (BjMTP1)* under heavy metal stress (Ni, Cd, Zn) (Muthukumar et al. 2007); promoter of *BjPR1* gene (Ali et al. 2018); promoter of *Chitinase* gene, *BjCHI1* that responds to wounding stress, salt, PEG, and MeJA treatment (Wu et al. 2009); promoter of *Oleosin* gene (Hanur et al. 2004); promoter of a *sugar transporter protein* gene, *BjSTP4,* that is responsive to infection by the mustard aphid-*Lipaphis erysimi* (Ram et al. 2020); phloem-specific promoters from *GLP13 (Germin-like proteins), GS3A (Glutamine synthase), TGG1 (Myrosinase 1/β-thioglucosideglucohydrolase 1), GAS1 (GA signal-transduction), SUC2 (Sucrose synthase), SULTR2 (Sulfate transporter),* and *PP2 (Phloem Protein 2)* genes (Koramutla et al. 2016), to name a few, have been studied. Characterization of promoter homeologs *SOC1* derived from three sub-genomes (LF, MF1, and MF2) of *B. juncea* using expression studies have revealed regulatory diversification via differential partitioning of transcription factor binding sites (Sri et al. 2020).

Promoter associated with several microRNA genes of *B. juncea* has also been characterized and reported. In a study published in 2018, Jain et al. (2018) performed functional analysis of the promoter homologs and homeologs of *MIR164* from *B. juncea*. Given the polyploid nature of genome of *B. juncea*, homeologs of *MIR164A* (two copies from *A* genome), *MIR164B*, and *MIR164C* (three copies from *A* genome) were identified. Characterization revealed functional and regulatory diversification amongst homeologs when transcriptional regulation of reporter gene activity in developing leaf, and floral organs were compared. Comparative analysis with promoter homologs from *A. thaliana* also indicated functional diversification because of sequence and transcription factor binding site (TFBS) variation (Jain et al. 2018). Similar functional diversification of transcriptional regulation has also been observed amongst promoter homeologs of *MIR159* and *MIR319* from *B. juncea* at seedling stage, and developmental stage of leaves, in floral organs, during cold and heat stress, and under regimes of various hormones (Gibberellic acid-3; ABA; Ethylene; 1-naphthaleneacetic acid/ NAA; 6-Benzylaminopurine/6-BAP; Salicylic acid; and MethylJasmonate) (Chauhan et al. 2020; Arora nee Joshi et al. 2021).

21.9 Deploying Regulatory Genes for Generation of cis-Genics and Transgenics for Crop Improvement

Genetic modification and transgenic technology in *Brassica juncea* using regulatory genes and elements from *B. juncea* genome, as well as heterologous system, have been attempted to modify several traits (Thakur et al. 2020). In the following section, a few representative examples that illustrate the significance of transgenesis for B. *juneca* improvement are presented.

21.9.1 Engineering Male Sterility for Heterotic Breeding

Inducing male sterility is necessary for hybrid seed production. Apart from a number of naturally occurring sources of CMS (e.g., *pol, ogu, hau, tour* systems), strategies have been developed wherein heterologous genes, such as *Cysteine Protease* from *Arachis diogoi* (*AdCP*) have been utilized. This gene-*AdCP* was expressed under TA-29, a tapetum specific promoter to induce male sterility in *Brassica juncea* var. Pusa Jaikisan (Gautam et al. 2019).

21.9.2 Modulating Flowering Time and Flower Development

Modification of flowering time to achieve early and delayed flowering has been a major breeding objective in *Brassica*. *AGL18*, a key regulator of flowering time was isolated from *B. juncea* and used to generate a cis-genic in *B. juncea*. Comparative analysis of gain-of-function and loss-of-function cis-genic lines showed that *BjAGL18* delays flowering time by inducing expression of *BjAGL15*, a floral repressor and, inhibiting expression of *BjSOC1*, a floral integrator (Yan et al 2018).

In a separate study, reduction in levels of *BjSOC1* by gene silencing employing artificial miRNA caused delayed flowering in *B. juncea* (Tyagi et al. 2019). The authors also propose that increasing the levels of *SOC1* in *B. juncea* can lead to achieving earliness in flowering time. In addition to altering flowering time, altered expression of *SOC1* in transgenic plants also altered fatty acid profiles especially those of Palmitic acid (C16:0), Oleic acid (C18:1), Linoleic acid (C18:2), Linolenic acid (C18:3), Eicosenoic acid (C20:1), and Erucic acid (C22:1) (Tyagi et al. 2019).

21.9.3 Fruit Development Traits

Preventing pod shattering is a key trait that that has been selected during the process of domestication as it allows seeds to be harvested upon maturity, and reduces seed loss. *FRUITFULL (FUL)*, along with *SHATTERPROOF 1* (*SHP1*) and *SHATTERPROOF 2* (*SHP2*) were identified in *A. thaliana* as one of the key regulators that control valve margin development (Gu et al. 1998; Liljegren et al. 2000). Taking advantage of the conserved pathway of fruit development between *A. thaliana* and *Brassica* species, constitutive overexpression of *A. thaliana FRUITFULL* in *B. juncea* was shown to prevent pod shattering, a trait desirable by canola farmers (Østergaard et al. 2006).

21.9.4 Modification of Fatty Acid and Glucosinolates for Seed, Oil Quality, and Quantity

The seed oil of *B. juncea* is known to contain <10% saturated fatty acid (C16:0; C18:0), ca. 15% each of MUFA (Mono-unsaturated fatty acid; C18:1) and PUFA (Polyunsaturated fatty acids; C18:2, C18:3), and 40–50% of Very-long-chain-unsaturated fatty acids (VLCUFA; C22:1) and is not considered desirable for human consumption (Kaushik and Agnihotri 2000; Sinha et al. 2007; Singh and Singh 2008).

In order to alter the fatty acid profile, and to reduce the proportion of saturated fatty acid fraction, the *ADS1* gene from *A. thaliana* was overexpressed in *B. juncea*. *ADS1* encodes the mammal/yeast/cyanobacteria homolog of acyl-CoA-Δ9 desaturases/acyl-lipid-Δ9 desaturase. Desaturases are responsible for introduction of double bonds and a key enzyme for synthesis of MUFA and PUFA (Nakamura and Nara 2004; Dar et al. 2017). Overexpression of *ADS1* in *B. juncea* led to significant reduction in levels of palmitic acid (16:0), stearic acid (18:0), arachidic acid (20:0), behenic acid (22:0), and lignoceric acid (24:0) with slight increase in oleic acid (18:1) (Yao et al. 2003).

The gene for diacyl-glycerol acyltransferase, *DGAT*, which is responsible for acetylation of diacyl-glycerol (DAG) to triacyl-glycerol (TAG) is a key enzyme in seed oil biosynthesis. The oil biosynthesis pathways being conserved across species permitted seed-specific overexpression of *DGAT* from *A. thaliana* into *B. juncea* RLM198 with nearly 8% increase in oil content, and up to 14% increase in fatty acid content in transgenic *B. juncea* plants (Savadi et al. 2016).

In another report, *B. juncea* was transformed with a gene for FatB thioesterase (*MbFatB*) from the Indian butter tree, *Diploknema butyracea* (Sapotaceae). The *MbFatB* gene was expressed under the control of napin promoter for seed-specific expression which led to significantly higher levels of C16:0 (between 2 and 3 fold), C18:1 (between 15 and 44% increase), and C18:2 (24–110% increase), and, between 64 and 82% decrease in levels of C22:1. This caused an increase in the ratio of C18:1 (desirable) /C22:1 (undesirable) from 0.5 in control to 2.3 (in transgenic plants). In parallel, *B. juncea* was also transformed with a hairpin dsRNA construct against *BjFAE1* (Fatty acid elongase 1), the key enzyme for C22:1 biosynthesis. In these transgenic plants, a significant drop of between 80 and 82% in C22:1 levels along with increase in C18:1 (between 66 and 75%) and C18:2 (between 36 and 59%) was observed (Sinha et al. 2007).

Other studies that have attempted to alter fatty acid profile using regulators of fatty acid biosynthesis included use of antisense technology designed against *B. rapa fad2* to engineer existing low-erucic *B. juncea* lines with elevated levels of oleic acid, and reduced levels of linoleic acid (Sivaraman et al. 2004); overexpression of *FAE1*, and antisense mediated downregulation of *BjFAE1* gene to increase (by up to 36%) and reduce (by up to 86%) erucic acid levels, respectively (Kanrar et al. 2006); use of *Acyl carrier protein (ACP)* gene from an aerobic bacteria-*Azospirillum brasilense* to transform *B. juncea* leading to increase in C18:1 and C18:2 levels in seed, and C18:3 in leaf, and reduced seed C22:1 (Jha et al. 2007); use of the gene, *PiD6*, encoding a homolog for *Δ6 Desaturase* from the fungi *Pythium irregulare,* under seed specific promoter to increase levels of C18:2 (6, 9), C18:3 (6, 9, 12), and C18:4 (6, 9, 12, 15)—three unsaturated fatty acids; C18:3 (6, 9, 12) was found to account for nearly 40% of total seed fatty acids in transgenic plants (Hong et al. 2002). Co-suppression based silencing of gene for *Oleate desaturase (Δ12-desaturase)* has been used to increase oleic acid by upto 73% in *B. juncea* (Stoutjesdijk et al. 2000a; b); introduction of *Δ4 Fatty Acid Desaturase* (*FAD4*) from *Thraustochytrium* sp., a marine micro-heterotroph-microkont algae) for conversion of exogenously supplied 22:5(n-3) into docosahexanoic acid (DHA; 22:6(n-3)) in *B. juncea* (Qiu et al. 2001); heterologous expression of *MlFatB* gene from *Madhuca longifolia* (*latifolia*) encoding FatB thioesterase (stearoyl-ACP thioesterase) to increase the biosynthesis of beneficial stearic acid by redirecting the carbon flux away from harmful erucic acid (Bhattacharya et al. 2015).

In order to increase the nutritional value, the cDNA for *γ-Tocopherol methyl transferase* from *A. thaliana*, which catalyzes conversion of γ-Tocopherol to α-Tocopherol was used to genetically modify *B. juncea*. The levels of α-Tocopherol, biologically active form of vitamin-E was found to be up to six-folds higher in total lipid from seeds of transgenic *B. juncea* (between 24 and 63% of total tocopherol) as compared to the non-transgenic plants (10% of total tocopherol) (Yusuf and Bhalla-Sarin 2007).

Modification and improvement of seed oil and seed coat meal quality remain a priority area of research for *Brassica* improvement. One major anti-nutritive compound that accumulates in seed is sinapine. High levels of sinapine content in seed meals of up to 15 mg/gm of dry weight are considered anti-nutritive and reduce the value of protein-rich seed meals as animal feed. Down-regulation of two genes—*SGT* (encoding *UDP-glucose: sinapate glucosyltransferase*) and *SCT* (encoding *sinapoylglucose: cholinesinapoyl transferase*) of phenylpropanoid pathway, that are involved in the final steps of sinapinebiosynthesis was undertaken, independent of each other, using antisense RNA, RNAi, and artificial miRNA technologies. Down-regulation is a seed specific manner was achieved with the help of either napin promoter, or endogenous promoter associated with *SGT* or *SCT* genes. Analysis of the seed sinapine content of transgenic plants (BjSGT-amiRNA) revealed up to a 67% reduction in total seed sinapine content (Kajla et al. 2017). The authors proposed crossing transgenic plants of BjSGT-amiRNA (with 3.8 mg/gm dry weight of seed sinapine) with those of Bj-SCTRNAi (4.95 mg/gm dry weight of seed sinapine) to further bring down the content (Kajla et al. 2017). Downregulation of *MYB28* in *B. juncea* using intron-spliced hairpin RNAi construct has also been employed to reduce glucosinolate content (Augustine and Bisht 2019).

Glucoraphanin, a glucosinolate is the precursor of an anti-cancerous compound-Sulphoraphane. However, glucoraphanin is converted into gluconapin and progoitrin by the action of the enzyme AOP (2-oxoglutarate-dependent dioxygenases) or GSL-ALK. In a recent study, the glucosinolate composition was altered by constitutive redundant silencing/downregulation of four homologs of *GSL-ALK* through RNAi leading to a significant increase in levels of Glucoraphanin and also increased tolerance against stem rot disease caused by *Sclerotinia sclerotiorum* (Augustine and Bisht 2016).

Gamma linolenic acid (GLA; omega 6 fatty acid; cis 6, 9, 12-Octadecatrienoic acid, C18:3, n-6), an unusual fatty acid, is considered nutritionally valuable but is lacking in seed oils due to absence of Δ6 desaturases in *B. juncea*. This is in spite of the two precursors, linolenic acid (C18:2) and α-linolenic acid (C18:3; omega 3 fatty acid; cis-9,12,15-octadecatrienoic acid; n-3) being present. A *delta-6 desaturase* gene from the cyanobacteria-*Synechocystis* PCC6803 was isolated and transformed into *B. juncea* under the control of 35S CaMV, and, napin promoter. The seed oil from transgenic plants was found to contain 2.76% (35S CaMV::d6D) to 0.44% in (napin::d6D) γ-linolenic acid (GLA) as compared to the complete absence of GLA in untransformed *B. juncea* var. Varuna (Das et al. 2006).

21.9.5 Imparting Tolerance/Resistance to Biotic Stresses

Phloem-sucking aphids are a major pest of *B. juncea* and cause an estimated average yield loss of upto 32% (Singh et al. 1986). Several researchers have relied upon introduction of genes from heterologous systems to introduce tolerance to aphid infestation and damage in *B. juncea*. A major strategy has been to use lectin genes to confer aphid resistance. In one such endeavor, cDNA of *wheat germ agglutinin* (*WGA*)—a chitin binding lectin was constitutively expressed in *B. juncea* var. RLM198 led to increased mortality (11%) and reduced fecundity (10 larvae as compared to 12) in aphids (Kanrar et al. 2002). In another study, the utility of *Allium sativum leaf lectin* (*ASAL*) gene was tested by transforming *B. juncea*, under both constitutive (CaMV35S) and rice phloem-specific sucrose synthase promoter (Dutta et al. 2005). Expression of *ASAL* caused significant decrease in survivability of aphids with only 20–40% of aphids surviving after 9 days when reared on transgenic plants, and the number of aphid larvae produced was also reduced by up to 60% (Dutta et al. 2005). Similarly, overexpression of *MYB28* homologs from *B. juncea* (Bju-*MYB28*) that is involved in glucosinolate biosynthesis has been shown to be a viable strategy for inducing aphid resistance (Kumar et al. 2017). Other such case studies include expression of heterologous *lectin* genes from lentils, and *protease inhibitor* genes from chickpea for introducing aphid tolerance (Rani et al. 2017a, b). Several other candidate regulatory genes such as *(E)-β-farnesene (EβF)*, a principal component of aphid alarm pheromone, from *Mentha arvensis* (Verma et al. 2015), tuber *lectin* from *Colocasia esculenta* (*CEA*; Das et al. 2018), and, *RiHSPRO2* (nematode resistance protein-like homolog) from *Rorripa indica* (Bose et al. 2019) have also been used for introducing aphid tolerance. A recent study also demonstrated tolerance against both *Alternaria brassicae* and *Erysiphe cruciferarum*, fungal pathogens that cause Alternia blight and powdery mildew, respectively when a homolog of non-expressor of pathogen-related gene 1 from *B. juncea* (BjNPR1) was over-expressed (Ali et al. 2017b).

21.9.6 Building Abiotic Stress Tolerance, and Phytoremediation

Generation of transgenic plants has been suggested as a viable option to increase the hyper-accumulation properties for phytoremediation (Fulekar et al. 2009; Pilon-Smits and LeDuc 2009; Rathore et al. 2019). Transgenic lines of *B. juncea* were obtained by introduction of *gshI* from *Escherichia coli* that encodeγ-glutamylcysteine synthetase (ECS) under transcriptional regulation of cauliflower mosaic virus CaMV35S promoter with a double enhancer sequence. Similarly, *gshII* from *E. coli* that encode glutathione synthetase (GS), and *A. thaliana APS1* encoding adenosine tri-phosphate sulfurylase (APS) were overexpressed to obtain transgenic *B. juncea* lines. The APS, ECS, and GS transgenic plants were grown under field conditions in Selenium rich soil, and were found to accumulate Se upto2.8-fold higher in ECS-transgenic, 2.3-fold higher in GS transgenics, and 4.3-fold higher in APS-transgenics demonstrating the possibility of employing genetically modified *B. juncea* for phytoremediation (Pilon-Smits et al. 1999; Van Huysen et al. 2003, 2004; Banuelos et al. 2005). Transgenic lines of *B. juncea* over-expressing *gshII* from *E. coli* encoding glutathione synthetase (GS) have been found to accumulate up to threefold higher levels of Cadmium as compared to non-transgenic plants (Zhu et al. 1999).

Several other reports on genetic modification of *B. juncea* using heterologous genes for improving phytoremediation ability have also been published. For example, *Yeast Cadmium factor*, a member of the *ATP-binding cassette (ABC) transporter* family, and, *ATM3* (*ABC transporter of themitochondrion 3*), a member of the *ATP-binding cassettetransporter* family from *A. thaliana* (*AtATM3*) for enhanced tolerance to both Cd and Pb (Bhuiyan et al. 2011a, b); γ-glutamyl cysteine synthetase (γ-ECS) and glutathione synthetase (GS) for elevated accumulation and enhanced tolerance to metals and metalloids such as As, Cd, and Cr, Pb, Zn, and mixed metals such as Zn/As (Reisinger et al. 2008); *Arabidopsis phytochelatin synthase (AtPCS1)* gene for enhanced tolerance to As, Cd, Zn (Gasic and Korban 2007a, b); overexpression of either *selenocysteine methyltransferase (SMT)* from *Astragalus bisulcatus* alone, or pyramiding of *ATP sulfurylase (APS*; from *A. thaliana*) and *selenocysteine methyltransferase (SMT*; from *Astragalus bisulcatus*) genes for ability to accumulate enhanced levels of Se (LeDuc et al. 2004, 2006); *selenocysteinelyase* (cDNA of SL derived from *Mus musculus* fused to the transit sequence of *ferredoxin* from *Silene pratensis;* cpSL) and *selenocysteine methyltransferase (SMT*; from *Astragalus bisulcatus*) also for increased Se accumulation (Banuelos et al. 2007).

21.9.7 Mineral Nutrition

Nitrogen and sulfur uptake metabolism are interlinked and are crucial to realizing the true yield potential in *B. juncea*. A *high-affinity sulfate transporter* (*HAST*) from *Lycopersicon esculentum* (*LeST1.1*) was constitutively over-expressed in *B. juncea* var. Pusa Jaikisan. The uptake rate of sulfur was twice as much as compared to untransformed plants, and, that of nitrogen was higher under both deficient (50 μM) and optimum (1000 μM) sulfate concentration; transgenic plants also accumulated higher level of biomass (Abdin et al. 2011; Akmal et al. 2014).

21.10 Gaps, Open Questions, Challenges, and Future Work

The present survey of literature shows that despite identification of a large number of regulatory genes and cis-elements, major research gaps persist which preclude translation of basic information to crop improvement programs in *B.*

juncea. The following points highlight broad areas of research that merit urgent attention:

i. The underlying evolutionary relationship shared between *B. juncea* and model plant *A. thaliana* does not ensure direct translation of vast body of knowledge available in *A. thaliana* towards *Brassica* improvement. Polyploidy induced gene and genome redundancy results in sequence, functional, and regulatory diversification of gene homologs. Evolutionary consequence and fate of homeolog diversification via neo-functionalization, sub-functionalization, pseudogenization, or functional conservation are therefore required to be independently ascertained in the context of allopolyploid *B. juncea* uses functional genomics approaches.
ii. Gene homeologs may be rewired into novel and complex gene regulatory networks plausibly presenting a case of super-modulation of regulatory cascades implied in development and adaptations in polyploid *B. juncea*. Analysis of conserved and novel gene regulatory networks and pathway analysis in the context of polyploid crop species are nascent but exciting ideas to investigate (Braynen et al. 2021).
iii. Although genome of *Brassica juncea* has been sequenced, albeit the complete benefit can only be reaped when pangenomes and super-pangenomes are generated (Golicz et al. 2016; Song et al. 2020; Mohd Saad et al. 2021). Such information, when integrated with eQTLs, and endophenotyping data are critically necessary for genomics assisted crop improvement programs in *B. juncea* (Li et al. 2018; Baker et al. 2019)
iv. Strategies such as genetic mapping, GWAS, and associative transcriptomics are being employed to understand the genetic architecture of various complex agronomic and yield quality traits such as seed weight, plant height, branches, number of pods, days to flower, days to maturity, and to identify putative candidate loci in *B. juncea* (Harper et al. 2020). However, examples of map-based cloning for discovery of regulatory genes and elements are extremely rare (Chen et al. 2018). Functional genomics of such loci that encode regulatory genes and cis-elements would provide greater insights into regulatory networks controlling morphogenesis and adaptations. There is an urgent unmet need to integrate genomics and phenomics to leverage the existing information and facilitate Brassica's improvement.
v. Investigations into regulation of developmental events of various stages of life cycle such as those of leaf, root, stamen, gynoecium, seed, and silique are almost non-existent. Similarly, studies on seedling development from a regulatory perspective are scarce. Studies on plant root-parasites to understand regulatory interactions and networks during *Brassica-Orobranche* (broomrape) infections are pending. Not much is known about photomorphogenetic regulation of developmental and adaptive events. Detailed investigations are required in line with studies undertaken by Muntha et al. (2019) describing the role of PhyA in shoot branching and flowering in *B. juncea*. Similarly, the potential of *B. juncea* as a bioenergy crop has not been investigated yet.
vi. Identification and characterization of regulatory genes and elements from cultivars of *B. juncea* are to be coupled with natural variation studies to exploit the true potential of GAB (Genomics Assisted Breeding). Cataloguing allelic diversity in gene regulators among *B. juncea* accessions is crucial and will aid in identification of novel natural alleles suitable for specific crop improvement and trait modification programs. Exotic genetic libraries and introgression lines are required to be systematically generated in *B. juncea* to identify exotic alleles conferring beneficial effects.

vii. Limited number of studies have reported splice-variants of regulatory genes in *B. juncea* (Shivaraj and Singh, 2016; Sharma et al. 2018). Large scale studies on identification and characterization of spliceforms as a means for expanding transcriptome and proteomic diversity need to be undertaken,
viii. Chromatin remodeling involving epigenetic regulation of development and adaptation in *B. juncea* is absent except a few reports, and large-scale epigenomics studies at the interface of genotype and environment are urgently required (Gupta et al. 2019b; Cao et al. 2016).
ix. Technological interventions such as deployment of high-throughput genomic tools and use of genome editing methodologies such as CRISPR-Cas9, ZFN, TALENs for gene functional analysis are also severely limited.

In our opinion, each of these gap areas holds immense potential and opportunities for future research and needs concerted efforts by research community working in *B. juncea*.

Acknowledgements We like to thank Department of Biotechnology (DBT), Science and Engineering Research Board (SERB/DST), University Grants Commission (UGC) for financial assistance for our work on *Brassica*. AS would like to acknowledge TERI-SAS, and SD would like to acknowledge Delhi University for financial, infrastructure, and administrative support. We also would like to express our apologies to the authors for a large number of excellent research articles which could not be included and cited owing to page constraints.

References

Abdin MZ, Akmal M, Ram M, Nafis T, Alam P, Nadeem M, Khan MA, Ahmad A (2011) Constitutive expression of high-affinity sulfate transporter (HAST) gene in Indian mustard showed enhanced sulfur uptake and assimilation. Protoplasma 248(3):591–600

Abraham V, Bhatia CR (1986) Development of Strains with Yellow Seedcoat in Indian Mustard (*Brassica juncea* Czern. and Coss.). Plant Breed 97:86–88

Ahmed I, Yadav D, Shukla P, Vineeth TV, Sharma PC, Kirti PB (2017) Constitutive expression of *Brassica juncea* annexin, *AnnBj2* confers salt tolerance and glucose and ABA insensitivity in mustard transgenic plants. Plant Sci 265:12–28

Akhatar J, Goyal A, Kaur N, Atri C, Mittal M, Singh MP, Kaur R, Rialch I, Banga SS (2021) Genome wide association analyses to understand genetic basis of flowering and plant height under three levels of nitrogen application in *Brassica juncea* (L.) Czern and Coss. Sci Rep 11:4278

Akmal M, Kiran U, Ali A, Abdin MZ (2014) Enhanced nitrogen assimilation in transgenic mustard (*Brassica juncea* L.) overexpressing high affinity sulfate transporter gene. Indian J Biotechnol 13(3):381–387

Ali S, Mir ZA, Tyagi A, Bhat JA, Chandrashekar N, Papolu PK, Rawat S, Grover A (2017a) Identification and comparative analysis of *Brassica juncea* pathogenesis-related genes in response to hormonal, biotic and abiotic stresses. Acta Physiol Plant 39:268

Ali S, Mir ZA, Tyagi A, Mehari H, Meena RP, Bhat JA, Yadav P, Papalou P, Rawat S, Grover A (2017b) Overexpression of *NPR1* in *Brassica juncea* confers broad spectrum resistance to fungal pathogens. Front Plant Sci 8:1693

Ali S, Mir ZA, Bhat JA, Tyagi A, Chandrashekar N, Yadav P, Rawat S, Sultana M, Grover A (2018) Isolation and characterization of systemic acquired resistance marker gene *PR1* and its promoter from *Brassica juncea*. 3 Biotechnology 8(1):10

Alvarez S, Berla BM, Sheffield J, Cahoon RE, Jez JM, Hicks LM (2009) Comprehensive analysis of the *Brassica juncea* root proteome in response to cadmium exposure by complementary proteomic approaches. Proteomics 9(9):2419–2431

Arora H, Padmaja KL, Paritosh K, Mukhi N, Tewari AK, Mukhopadhyay A, Gupta V, Pradhan AK, Pental D (2019) *BjuWRR1*, a *CC-NB-LRR* gene identified in *Brassica juncea*, confers resistance to white rust caused by *Albugo candida*. Theor Appl Genet 132 (8):2223–2236

Arora nee Joshi G, Chauhan C, Das S (2021) Sequence and functional analysis of *MIR319* promoter homologs from *Brassica juncea* reveals regulatory diversification and altered expression under stress. Mol Genet Genom 296: 731–749

Ashutosh KP, Dinesh Kumar V, Sharma PC, Prakash S, Bhat SR (2008) A novel *orf108* co-transcribed with the *atpA* gene is associated with cytoplasmic male sterility in *Brassica juncea* carrying *Moricandia arvensis* cytoplasm. Plant Cell Physiol 49(2):284–289

Atri C, Kaur B, Sharma S, Gandhi N, Verma H, Goyal A, Banga SS (2016) Substituting nuclear genome of *Brassica juncea* (L.) Czern and Coss. in cytoplasmic background of *Brassica fruticulosa* results in cytoplasmic male sterility. Euphytica 209:31–40

Atri C, Akhatar J, Gupta M, Gupta N, Goyal A, Rana K et al (2019) Molecular-genetic analysis of defensive responses of *Brassica juncea-B. fruticulose* introgression lines to *Sclerotinia* infestation. Sci Rep 9:17089

Augustine R, Bisht NC (2015) Biotic elicitors and mechanical damage modulate glucosinolate accumulation by coordinated interplay of glucosinolate biosynthesis regulators in polyploid *Brassica juncea*. Phytochemistry 117:43–50

Augustine R, Bisht NC (2016) Biofortification of oilseed *Brassica juncea* with the anti-cancer compound glucoraphanin by suppressing *GSL-ALK* gene family. Sci Rep 5:18005

Augustine R, Bisht NC (2019) Targeted silencing of genes in polyploids: lessons learned from *Brassica juncea*-glucosinolate system. Plant Cell Rep 38(1):51–57

Augustine R, Majee M, Gershenzon J, Bisht NC (2013a) Four genes encoding *MYB28*, a major transcriptional regulator of the aliphatic glucosinolate pathway, are differentially expressed in the allopolyploid *Brassica juncea*. J Exp Bot 64(16):4907–4921

Augustine R, Mukhopadhyay A, Bisht NC (2013b) Targeted silencing of *BjMYB28* transcription factor gene directs development of low glucosinolate lines in oilseed *Brassica juncea*. Plant Biotechnol J 11 (7):855–866

Baker RL, Leong WF, Brock MT, Rubin MJ, Markelz RJC, Welch S, Maloof JN, Weinig C (2019) Integrating transcriptomic network reconstruction and eQTL analyses reveals mechanistic connections between genomic architecture and *Brassica rapa* development. PLoS Genet 15(9):e1008367

Balesdent MH, Attard A, Kühn ML, Rouxel T (2002) New avirulence genes in the phytopathogenic fungus *Leptosphaeria maculans*. Phytopathol 92(10):1122–1133

Banuelos G, Terry N, LeDuc DL, Pilon-Smits EAH, Mackey B (2005) Field trial of transgenic Indian mustard plants shows enhanced phytoremediation of selenium-contaminated sediment. Environ Sci Technol 39:1771–1777

Banuelos G, LeDuc DL, Pilon-Smits EAH, Tagmount A, Terry N (2007) Transgenic Indian mustard overexpressing *selenocysteine lyase* or *selenocysteine methyltransferase* exhibit enhanced potential for selenium phytoremediation under field conditions. Environ Sci Technol 41:599–605

Bhardwaj AR, Joshi G, Pandey R, Kukreja B, Goel S, et al (2014) A genome-wide perspective of miRNAome in response to high temperature, salinity and drought stresses in *Brassica juncea* (Czern) L. PLoS One 9(3):e92456

Bhat SR, Prakash S, Kirti PB, Dineshkumar V, Chopra VL (2005) A unique introgression from *Moricandia arvensis* confers male fertility upon two different cytoplasmic male-sterile lines of *Brassica juncea*. Plant Breed 124:117–120

Bhat SR, Priya V, Ashutosh DKK, Prakash S (2006) *Diplotaxiserucoides* induced cytoplasmic male sterility in *Brassica juncea* is rescued by the *Moricandia arvensis* restorer: genetic and molecular analyses. Plant Breed 125:150–155

Bhat SR, Kumar P, Prakash S (2008) An improved cytoplasmic male sterile (*Diplotaxis berthautii*) *Brassica juncea*: identification of restorer and molecular characterization. Euphytica 159:145–152

Bhatia G, Singh A, Verma D, Sharma S, Singh K (2020) Genome-wide investigation of regulatory roles of lncRNAs in response to heat and drought stress in *Brassica juncea* (Indian mustard). Environ Exp Bot 171:103922

Bhattacharya S, Sinha S, Das N, Maiti MK (2015) Increasing the stearate content in seed oil of *Brassica juncea* by heterologous expression of *MlFatB* affects lipid content and germination frequency of transgenic seeds. Plant Physiol Biochem 96:345–355

Bhayana L, Paritosh K, Arora H, Yadava SK, Singh P, Nandan D, Mukhopadhyay A, Gupta V, Pradhan AK, Pental D (2020) A mapped locus on *LG A6* of *Brassica juncea* line Tumida conferring resistance to white rust contains a *CNL*type *R* gene. Front Plant Sci 10:1690

Bhuiyan MSU, Min SR, Jeong WJ, Sultana S, Choi KS, Lee Y, Liu JR (2011a) Overexpression of *AtATM3* in *Brassica juncea* confers enhanced heavy metal tolerance and accumulation. Plant Cell Tiss Org Cult 107:69–77

Bhuiyan MSU, Min SR, Jeong WJ, Sultana S, Choi KS, Song WY, Lee Y, Lim YP, Liu JR (2011b) Overexpression of a *yeast cadmium factor 1 (YCF1)* enhances heavy metal tolerance and accumulation in *Brassica juncea*. Plant Cell Tiss Organ Cult 105:85–91

Bisht NC, Gupta V, Ramchiary N, Sodhi YS, Mukhopadhyay A, Arumugam N, Pental D, Pradhan AK (2009) Fine mapping of loci involved with glucosinolate biosynthesis in oilseed mustard (*Brassica juncea*) using genomic information from allied species. Theor Appl Genet 118(3):413–421

Bita CE, Gerats T (2013) Plant tolerance to high temperature in a changing environment: scientific fundamentals and production of heat stress-tolerant crops. Front Plant Sci 4:273

Bogs J, Bourbouloux A, Cagnac O, Wachter A, Rausch T, Delrot S (2003) Functional characterization and expression analysis of a glutathione transporter, *BjGT1*, from *Brassica juncea*: evidence for regulation by heavy metal exposure. Plant Cell Environ 26:1703–1711

Boomiraj K, Chakrabarti B, Aggarwal PK, Choudhary R, Chander S (2010) Assessing the vulnerability of Indian mustard to climate change. Agri Ecosyst Environ 138:265–273

Borhan MH, Holub EB, Kindrachuk C, Omidi M, Bozorgmanesh-Frad G, Rimmer SR (2010) *WRR4*, a broad-spectrum *TIR-NB-LRR* gene from *Arabidopsis thaliana* that confers white rust resistance in transgenic oilseed *Brassica* crops. Mol Plant Pathol 11 (2):283–291

Bose S, Gangopadhyay G, Sikdar SR (2019) *Rorippa indica HSPRO2* expression in transgenic *Brassica juncea* induces tolerance against mustard aphid *Lipaphis erysimi*. Plant Cell Tiss Org Cult 136:431–443

Bowers JE, Chapman BA, Rong J, Paterson AH (2003) Unravelling angiosperm genome evolution by phylogenetic analysisof chromosomal duplication events. Nature 422:433–438

Braynen J, Yang Y, Yuan J, Xie Z, Cao G, Wei X, Shi G, Zhang X, Wei F, Tian B (2021) Comparative transcriptome analysis revealed differential gene expression in multiple signalling pathways at flowering in polyploid *Brassica rapa*. Cell Biosci 11(1):17

Cai Z, Chen Y, Liao J, Wang D (2020) Genome-wide identification and expression analysis of jasmonate *ZIM* domain gene family in tuber mustard (*Brassica juncea* var. tumida). PLoS One 15(6):e0234738

Cao L, Yu N, Li J, Qi Z, Wang D, Chen L (2016) Heritability and reversibility of DNA methylation induced by in vitro grafting between *Brassica juncea* and *B. oleracea*. Sci Rep 6:27233

Cartea ME, Velasco P (2008) Glucosinolates in *Brassica* foods: bioavailability in food and significance for human health. Phytochem Rev 7:213–229

Cevik V, Boutrot F, Apel W, Robert-Seilaniantz A, Furzer OJ, Redkar A et al (2019) Transgressive segregation reveals mechanisms of *Arabidopsis* immunity to *Brassica*-infecting races of white rust (*Albugo candida*). Proc Natl Acad Sci USA 116:2767–2773

Chauhan C, Joshi G, Chaudhary D, Das S (2020) Sequence and functional analysis of cis-elements associated with *MIR159* loci from *Brassica juncea* reveal functional diversification and complex transcriptional regulation. Plant Growth Regul 90:279–306

Chen C, Xiao L, Li X, Du D (2018) Comparative mapping combined with map-based cloning of the *Brassica juncea* genome reveals a candidate gene for multilocular rapeseed. Front Plant Sci 9:1744

Cheng F, Wu J, Wang X (2014) Genome triplication drove the diversification of *Brassica* plants. Hort Res 1:14024

Cheng C, Zhong Y, Cai Z, Su R, Li C (2019a) Genome-wide identification and gene expression analysis of *ABA* receptor family genes in *Brassica juncea* var. Tumida. Genes 10:470

Cheng C, Zhong Y, Wang Q, Cai Z, Wang D, Li C (2019b) Genome-wide identification and gene expression analysis of *SOS* family genes in tuber mustard (*Brassica juncea var. tumida*). PLoS One 14(11): e0224672

Clarke DB (2010) Glucosinolates, structures and analysis in food. Anal Methods 2:310–325

Cohn AS, Newton P, Gil JDB, Kuhl L, Samberg L, Ricciardi V, Manly JR, Northrop S (2017) Small holder agriculture and climate change. Annu Rev Environ Resour 42:347–375

Cong L, Chai T-Y, Zhang Y-X (2008) Characterization of the novel gene *BjDREB1B* encoding a *DRE*-binding transcription factor from *Brassica juncea* L. BiochemBiophys Res Comm 371:702–706

Dalyan E, Yuzbasıoglu E, Keskin BC, Yıldızhan Y, Memon A, Unal M, Yuksel B (2017) The identification of genes associated with Pb and Cd response mechanism in *Brassica juncea* L. by using *Arabidopsis* expression array.Environ Exp Bot 139: 105–115

Dar AA, Choudhury AR, Kancharla PK, Arumugam N (2017) The *FAD2* gene in plants: occurrence, regulation, and role. Front Plant Sci 8:1789

Das B, Goswami L, Ray S, Ghosh S, Bhattacharyya S, Das S, Majumder AL (2006) *Agrobacterium*-mediated transformation of *Brassica juncea* with a cyanobacterial (*Synechocystis* PCC6803) *delta-6 desaturase* gene leads to production of gamma-linolenic acid. Plant Cell Tiss Organ Cult 86:219–231

Das S, Sen M, Saha C, Chakraborty D, Das A, Banerjee M, Seal A (2011) Isolation and expression analysis of partial sequences of heavy metal transporters from *Brassica juncea* by coupling high throughput cloning with a molecular fingerprinting technique. Planta 234 (1):139–156

Das A, Ghosh P, Das S (2018) Expression of *Colocasia esculenta* tuber agglutinin in Indian mustard provides resistance against *Lipaphis erysimi* and the expressed protein is non-allergenic. Plant Cell Rep 37:849–863

Dhakate P, Tyagi S, Singh A, Singh A (2017) Functional characterization of a novel *Brassica LEAFY* homolog fromIndian mustard: Expression pattern and gain-of-function studies. Plant Sci 258:29–44

Duhlian L, Koramutla MK, Subramanian S, Chamola R, Bhattacharya R (2020) Comparative transcriptomics revealed differential regulation of defense related genes in *Brassica juncea* leading to successful and unsuccessful infestation by aphid species. Sci Rep 10:10583

Dutta I, Majumder P, Saha P, Ray K, Das S (2005) Constitutive and phloem specific expression of *Allium sativum* leaf agglutinin (*ASAL*) to engineer aphid (*Lipaphiserysimi*) resistance in transgenic Indian mustard (*Brassica juncea*). Plant Sci 169:996–1007

Fahey JW, Zhang Y, Talalay P (1997) Broccoli sprouts: an exceptionally rich source of inducers of enzymes that protect against chemical carcinogens. Proc Natl Acad Sci USA 94:10367–10372

Fargier E, Manceau C (2007) Pathogenicity assays restrict the species *Xanthomonas campestris* into three pathovars and reveal nine races within *X. campestris* pv. *campestris*. Plant Pathol 56:805–818

Francis A, Lujan-Toro BE, Warwick SI, Macklin JA, Martin SL (2021) Update on the Brassicaceae species checklist. Biodivers Data J 9:e58773

Fu D, Mason AS, Xiao M, Yan H (2016) Effects of genome structure variation, homeologous genes and repetitive DNA on polyploid crop research in the age of genomics. Plant Sci 242:37–46

Fulekar MH, Singh A, Bhaduri AM (2009) Genetic engineering strategies for enhancing phytoremediation of heavy metals. Afr J Biotechnol 8(4):529–535

Fusco N, Micheletto L, Dal Corso G, Borgato L, Furini A (2005) Identification of cadmium-regulated genes by cDNA-AFLP in the heavy metal accumulator *Brassica juncea* L. J Exp Bot 56(421):3017–3027

Gao Y, Jia S, Wang C, Wang F, Wang F, Zhao K (2016) *BjMYB1*, a transcription factor implicated in plant defence through activating *BjCHI1* chitinase expression by binding to a W-box-like element. J Exp Bot 67 (15):4647–4658

Garg R, Kiran MD, Santha IM, Lodha ML, Mehta SL (2000) Cloning of a gene sequence for *ω-3 desaturase* from *Brassica juncea*. J Plant Biochem Biotechnol 9:13–16

Garg R, Santha IM, Lodha ML, Mehta SL (2001) Sequence analysis of a Plastidial*Omega-3 desaturase* Gene from *Brassica juncea*. J Plant Biochem Biotechnol 10:13–17

Gasic K, Korban SS (2007a) Expression of *Arabidopsis*phytochelatin synthase in Indian mustard (*Brassica juncea*) plants enhances tolerance for Cd and Zn. Planta 225:1277–1285

Gasic K, Korban SS (2007b) Transgenic Indian mustard (*Brassica juncea*) plants expressing an *Arabidopsis phytochelatin synthase* (*AtPCS1*) exhibit enhanced As and Cd tolerance. Plant Mol Biol 64:361–369

Gautam R, Shukla P, Kirti PB (2019) Targeted expression of a *cysteine protease (AdCP)* in tapetum induces male sterility in Indian mustard *Brassica juncea*. Funct. Integr Genomics 19(5):703–714

Goel P, Bhuria M, Kaushal M, Singh AK (2016) Carbon: Nitrogen interaction regulates expression of genes involved in N-uptake and assimilation in *Brassica juncea* L. PLoS One 11(9):e0163061

Goel P, Sharma NK, Bhuria M, Sharma V, Chauhan R, Pathania S, Swarnkar MK, Chawla V, Acharya V, Shankar R, Singh AK (2018) Transcriptome and co-expression network analyses identify key genes regulatingnitrogen use efficiency in *Brassica juncea* L. Sci Rep 8(1):7451

Goel P, Singh AK (2015) Abiotic stresses downregulate key genes involved in nitrogen uptake and assimilation in *Brassica juncea* L. PLoS One 10(11):e0143645

Golicz AA, Bayer PE, Barker GC, Edger PP, Kim H, Martinez PA, Chan CK, Severn-Ellis A, McCombie WR, Parkin IA, Paterson AH, Pires JC, Sharpe AG, Tang H, Teakle GR, Town CD, Batley J, Edwards D (2016) The pangenome of an agronomically important crop plant *Brassica oleracea*. Nat Commun 7:13390

Gu Q, Ferraandiz C, Yanofsky M, Martienssen R (1998) The *FRUITFULL MADS-box* gene mediates cell differentiation during *Arabidopsis* fruit development. Development 125:1509–1517

Guo D-P, Guo Y-P, ZhaoJ-P LH, Peng Y, WangQ-M C-S, Rao G-Z (2005) Photosynthetic rate and chlorophyll fluorescence in leaves of stem mustard (*Brassica juncea* var. tsatsai) after turnip mosaic virus infection. Plant Sci 168:57–63

Gupta V, Mukhopadhyay A, Arumugam N, Sodhi YS, Pental D, Pradhan AK (2004) Molecular tagging of erucic acid trait in oilseed mustard (*Brassica juncea*) by QTL mapping and single nucleotide polymorphisms in *FAE1* gene. Theor Appl Genet 108(4):743–749

Gupta S, Akhatar J, Kaur P, Sharma A, Sharma P, Mittal M, Bharti B, Banga SS (2019a) Genetic analyses of nitrogen assimilation enzymes in *Brassica juncea* (L.) Czern and Coss. Mol Biol Rep 46 (4):4235–4244

Gupta S, Sharma N, Akhatar J, Atri C, Kaur J, Kaur G, Banga SS (2019b) Analysis of epigenetic landscape in a recombinant inbred line population developed by hybridizing natural and resynthesized *Brassica juncea* (L.) with stable C-genome introgressions. Euphytica 215:174

Hadfi K, Speth V, Neuhaus G (1998) Auxin-induced developmental patterns in *Brassica juncea* embryos. Development 125(5):879–887

Halkier BA, Gershenzon J (2006) Biology and biochemistry of glucosinolates. Annu Rev Plant Biol 57:303–333

Hanur VS, Kirti PB, Chopra VL (2004) The Seed-specific oleosin promoter of *Brassica juncea*: cloning, characterization and identification of the novel and divergent antisense open reading frame. J Plant Biochem Biotechnol 13:119–122

Harper AL, He Z, Langer S, Havlickova L, Wang L, Fellgett A, Gupta V, Kumar Pradhan A, Bancroft I (2020) Validation of an associative transcriptomics platform in the polyploid crop species *Brassica juncea* by dissection of the genetic architecture of agronomic and quality traits. Plant J 103(5):1885–1893

Hasan MJ, Rahman H (2018) Resynthesis of *Brassica juncea* for resistance to *Plasmodiophora brassicae* pathotype 3. Breed Sci 68(3):385–391

Hasan M, Strelkov SE, Howard RJ, Rahman H (2012) Screening of *Brassica* germplasm for resistance to *Plasmodiophora brassicae* pathotypes prevalent in Canada for broadening diversity in clubroot resistance. Can J Plant Sci 92:501–515

He J, He X, Chang P, Jiang H, Gong D, Sun Q (2020) Genome-wide identification and characterization of *TCP* family genes in *Brassica juncea* var. tumida. PeerJ 8:e9130

Heiss S, Schäfer HJ, Haag-Kerwer A, Rausch T (1999) Cloning sulfur assimilation genes of *Brassica juncea* L.: cadmium differentially affects the expression of a putative low-affinity sulfate transporter and isoforms of ATP sulfurylase and APS reductase. Plant Mol Biol 39:847–857

Heng S, Wei C, Jing B, Wan Z, Wen J, Yi B, Ma C, Tu J, Fu T, Shen J (2014) Comparative analysis of mitochondrial genomes between the *hau* cytoplasmic male sterility (CMS) line and its iso-nuclear maintainer line in *Brassica juncea* to reveal the origin of the CMS-associated gene *orf288*. BMC Genomics 15 (1):322

Heng S, Gao J, Wei C, Chen F, Li X, Wen J, Yi B, Ma C, Tu J, Fu T, Shen J (2018a) Transcript levels of *orf288* are associated with the *hau* cytoplasmic male sterility system and altered nuclear gene expression in *Brassica juncea*. J Exp Bot 69(3):455–466

Heng S, Liu S, Xia C, Tang H, Xie F, Fu T, Wan Z (2018b) Morphological and genetic characterization of a new cytoplasmic male sterility system (*oxa* CMS) in stem mustard (*Brassica juncea*). Theor Appl Genet 131(1):59–66

Heng S, Cheng Q, Zhang T, Liu X, Huang H, Yao P, Liu Z, Wan Z, Fu T (2020a) Fine-mapping of the *BjPur* gene for purple leaf color in *Brassica juncea*. Theor Appl Genet 133:2989–3000

Heng S, Huang H, Cui M, Liu M, Lv Q, Hu P, Ren S, Li X, Fu T, Wan Z (2020b) Rapid identification of the *BjRCO* gene associated with lobed leaves in *Brassica juncea* via bulked segregant RNA-seq. Mol Breed 40:42

Heng S, Wang L, Yang X, Huang H, Chen G, Cui M, Liu M, Lv Q, Wan Z, Shen J, Fu T (2020c) Genetic and comparative transcriptome analysis revealed DEGs involved in the purple leaf formation in *Brassica juncea.* Front Genet 11:322

Hong H, Datla N, Darwin WR, Patrick SC, Samuel LM, Xiao Q (2002) High-level production of γ-linolenic acid in *Brassica juncea* using a *Δ6-desaturase* from *Pythium irregulare.* Plant Physiol 129:354–362

http://Brassicadb.cn/#/SpeciesInfo/

http://www.nectarygenomics.org/node/558

Huang CH, Sun R, Hu Y, Zeng L, Zhang N, Cai L, Zhang Q, Koch MA, Al-Shehbaz I, Edger PP, Pires JC, Tan DY, Zhong Y, Ma H (2016) Resolution of *Brassica*ceae phylogeny using nuclear genes uncovers nested radiations and supports convergent morphological evolution. Mol Biol Evol 33(2):394–412

Inturrisi F, Bayer PE, Yang H, Tirnaz S, Edwards D, Batley J (2020) Genome-wide identification and comparative analysis of resistance genes in *Brassica juncea.* Mol Breed 40:78

Inturrisi F, Barbetti M, Tirnaz S, Patel DA, Edwards D, Batley J (2021) Molecular characterization of disease resistance in *Brassica juncea*—the current status and the way forward. Plant Pathol 70:13–34

IPCC (2014) Climate change 2014: synthesis report. Contribution of Working Groups I, II and III to the Fifth Assessment Report of the Intergovernmental Panel on Climate Change [Core Writing Team, Pachauri RK and Meyer LA (eds)]. IPCC, Geneva, Switzerland, 151pp

IPCC (2018) Global Warming of 1.5°C.An IPCC Special Report on the impacts of global warming of 1.5°C above pre-industrial levels and related global greenhouse gas emission pathways, in the context of strengthening the global response to the threat of climate change, sustainable development, and efforts to eradicate poverty [Masson-Delmotte V, Zhai P, Pörtner HO, Roberts D, Skea J, Shukla PR, Pirani A, Moufouma-Okia W, Péan C, Pidcock R, Connors S, Matthews JBR, Chen Y, Zhou X, Gomis MI, Lonnoy E, Maycock T, Tignor M, and Waterfield T (eds)], 630pp

Jain A, Anand S, Singh NK, Das S (2018) Sequence and functional characterization of *MIRNA164* promoters from *Brassica* shows copy number dependent regulatory diversification among homeologs. Funct Inter Genom 18:369–383

Jami SK, Dalal A, Divya K, Kirti PB (2009) Molecular cloning and characterization of five *annexin* genes from Indian mustard (*Brassica juncea* L. Czern and Coss). Plant Physiol Biochem 47:977–990

Jami SK, Hill RD, Kirti PB (2010) Transcriptional regulation of *annexins* in Indian Mustard, *Brassica juncea* and detoxification of ROS in transgenic tobacco plants constitutively expressing *AnnBj1*. Plant Signal Behav5:618–621

Jami SK, Clark GB, Ayele BT, Ashe P, Kirti PB (2012) Genome-wide Comparative Analysis of *Annexin* Superfamily in Plants. PLoS One 7(11):e47801

Jha JK, Sinha S, Maiti MK, Basu A, Mukhopadhyay UK, Sen SK (2007) Functional expression of an *acyl carrier protein (ACP)* from *Azospirillum brasilense* alters fatty acid profiles in *Escherichia coli* and *Brassica juncea.* Plant Physiol Biochem 45:490–500

Ji H, Pardo JM, Batelli G, Van Oosten MJ, Bressan RA, Li X (2013) The salt overly sensitive (SOS) pathway: established and emerging roles. Mol Plant 6(2):275–286

Jiang W, Wei D, Zhou W, Wang Z, Tang Q (2018) HDA9 interacts with the promoters of *SOC1* and *AGL24* involved in flowering time control in *Brassica juncea.* Biochem Biophys Res Commun 499(3):519–523

Jiang L, Sun Q, Wang Y, Chang P, Kong H, Luo C, He X (2021) Genome-wide identification and characterization of *NAC* genes in *Brassica juncea* var. tumida. PeerJ9:e11212

Jing B, Heng S, Tong D, Wan Z, Fu T, Tu J, Ma C, Yi B, Wen J, Shen J (2012) A male sterility-associated cytotoxic protein *ORF288* in *Brassica juncea* causes aborted pollen development. J Exp Bot 63(3):1285–1295

Jing-Hua Y, Ming-Fang Z, Jing-Quan Y, Shuo Z, Tao W, Zhu-Jun C (2005) Identification of alloplasmic cytoplasmic male-sterile line of leaf mustard synthesized by intra-specific hybridization. Plant Sci 168:865–871

Jouet A, Saunders DGO, McMullan M, Ward B, Furzer O, Jupe F et al (2019) *Albugo candida* race diversity, ploidy and host-associated microbes revealed using DNA sequence capture on diseased plants in the field. New Phytol 221:1529–1543

Juge N, Mithen RF, Traka M (2007) Molecular basis for chemoprevention by sulforaphane: a comprehensive review. Cell Mol Life Sci 64:1105–1127

Kajla S, Mukhopadhyay A, Pradhan AK (2017) Development of transgenic *Brassica juncea* lines for reduced seed sinapine content by perturbing phenylpropanoid pathway genes. PLoS One 12(8):e0182747

Kanrar S, Venkateswari J, Dureja P, Kirti PB, Chopra VL (2006) Modification of erucic acid content in Indian mustard (*Brassica juncea*) by up-regulation and down-regulation of the *Brassica junceaFATTY ACID ELONGATION1 (BjFAE1)* gene. Plant Cell Rep 25:148–155

Karamat U, Sun X, Li N, Zhao J (2021) Genetic regulators of leaf size in *Brassica* crops. Hort Res 8:91

Karim MM, Dakouri A, Zhang Y, Chen Q, Peng G, Strelkov SE, Gossen BD, Yu F (2020) Two clubroot-resistance genes, *Rcr3* and *Rcr9wa*, mapped in *Brassica rapa* using bulk segregant RNA sequencing. Intl J Mol Sci 21(14):5033

Kashyap P, Deswal R (2019) Two ICE isoforms showing differential transcriptional regulation by cold and hormones participate in *Brassica juncea* cold stress signalling. Gene 695:32–41

Kaur H, Banga SS (2015) Discovery and mapping of *Brassica juncea Sdt1* gene associated with determinate plant growth habit. Theor Appl Genet 128:235–245

Kaur P, Jost R, Sivasithamparam K, Barbetti MJ (2011a) Proteome analysis of the *Albugo candida–Brassica juncea*pathosystem reveals that the timing of the expression of defence-related genes is a crucial determinant of pathogenesis. J Exp Bot 62:1285–1298

Kaur P, Sivasithamparam K, Li H, Barbetti MJ (2011b) Pre-inoculation with *Hyaloperonospora parasitica* reduces incubation period and increases severity of disease caused by *Albugo candida* in a *Brassica juncea* variety resistant to downy mildew. J Gen Plant Pathol 77:101–106

Kaur C, Kumar G, Kaur S, Ansari MW, Pareek A, Sopory SK, Singla-Pareek SL (2015) Molecular cloning and characterization of salt overly sensitive gene promoter from *Brassica juncea* (*BjSOS2*). Mol Biol Rep 42(6):1139–1148

Kaur J, Akhatar J, Goyal A, Kaur N, Kaur S, Mittal M, Kumar N, Sharma H, Banga S, Banga SS (2020) Genome wide association mapping and candidate gene analysis for pod shatter resistance in *Brassica juncea* and its progenitor species. Mol Biol Rep 47(4):2963–2974

Kaushik N, Agnihotri A (2000) GLC analysis of Indian rapeseed mustard to study the variability of fatty acid composition. Biochem Soc Trans 28:581–583

Koenig D, Weigel D (2015) Beyond the thale: comparative genomics and genetics of *Arabidopsis* relatives. Nat Rev Genet 16(5):285–298

Konopka-Postupolska D, Clark G, Hofmann A (2011) Structure, function and membrane interactions of plant annexins: an update. Plant Sci 181(3):230–241

Koramutla MK, Bhatt D, Negi M, Venkatachalam P, Jain PK, Bhattacharya R (2016) Strength, stability, and cis-Motifs of in silico Identified phloem-specific promoters in *Brassica juncea* (L.). Front Plant Sci 7:457

Kumar R, Bisht NC (2020) Heterotrimeric *Gα* subunit regulates plant architecture, organ size and seed weight in the oilseed *Brassica juncea*. Plant Mol Biol 104:549–560

Kumar P, Vasupalli N, Srinivasan R, Bhat SR (2012) An evolutionarily conserved mitochondrial *orf108* is associated with cytoplasmic male sterility in different alloplasmic lines of *Brassica juncea* and induces male sterility in transgenic *Arabidopsis thaliana*. J Exp Bot 63(8):2921–2932

Kumar P, Augustine R, Singh AK, Bisht NC (2017) Feeding behaviour of generalist pests on *Brassica juncea*: implication for manipulation of glucosinolate biosynthesis pathway for enhanced resistance. Plant Cell Environ 40(10):2109–2120

Landgren M, Zetterstrand M, Sundberg E, Glimelius K (1996) Alloplasmic male-sterile *Brassica* lines containing *B. tournefortii* mitochondria express an *ORF* 3' of the *atp6* gene and a 32 kDa protein. Plant Mol Biol 32(5):879–890

Lang ML, Zhang YX, Guan ZQ, Chai TY (2005) PCR-enriched cDNA pool method for cloning of gene homologues. Plant Mol Biol Rep 23:219–226

LeDuc DL, Tarun AS, Montes-Bayo′ n M, Meija J, Malit MF, Wu CP, Abdel Samie M, Chiang C-Y, Tagmount A, de Souza MP et al (2004) Over expression of *selenocysteine methyltransferase* in *Arabidopsis* and Indian mustard increases selenium tolerance and accumulation. Plant Physiol 135:377–383

LeDuc DL, AbdelSamie M, Montes-Bayo′n M, Wu CP, Reisinger SJ, Terry N (2006) Overexpressing both *ATP sulfurylase* and *selenocysteine methyltransferase* enhances selenium phytoremediation traits in Indian mustard. Environ Pollut 144:70–76

Li H, Jiang M, Che LL, Nie L, Yang ZM (2012) *BjHO-1* is involved in the detoxification of heavy metal in India mustard (*Brassica juncea*). Biometals 25 (6):1269–1279

Li R, Jeong K, Davis JT, Kim S, Lee S, Michelmore RW, Kim S, Maloof JN (2018) Integrated QTL and eQTL mapping provides insights and candidate genes for fatty acid composition, flowering time, and growth traits in a F2 Population of a novel synthetic allopolyploid *Brassica napus*. Front Plant Sci 9:1632

Li M, Sun B, Xie F, Gong R, Luo Y, Zhang F, Yan Z, Tang H (2019a) Identification of the *GRAS* gene family in the *Brassica juncea* genome provides insight into its role in stem swelling in stem mustard. Peer J 7: e6682

Li S, Li Z, Zhang J, Wei D, Wang Z, Tang Q (2019b) Flowering signal integrator AGL24 interacts with K domain of AGL18 in *Brassica juncea*. Biochem Biophys Res Comm 518:148–153

Li M, He Q, Zhang Y, Sun B, Luo Y, Zhang Y, Chen Q, Wang Y, Zhang F, Zhang Y, Lin Y, Wang X, Tang H (2020a) New insights into the evolution of the *SBP*-box family and expression analysis of genes in the growth and development of *Brassica juncea*. Biotechnol Biotechnologic Equip 34(1):810–824

Li M, Xie F, Li Y, Gong L, Luo Y, Zhang Y, Chen Q, Wang Y, Lin Y, Zhang Y, Wang X, Tang H (2020b) Genome-wide analysis of the *Heat Shock Transcription Factor* gene family in *Brassica juncea*: structure, evolution, and expression profiles. DNA Cell Biol 39 (11):1990–2004

Li W, Chen F, Wang Y, Zheng H, Yi Q, Ren Y, Gao J (2020c) Genome-wide identification and functional analysis of *ARF* transcription factors in *Brassica juncea* var. tumida. PLoS One 15(4):e0232039

Li M, He Q, Huang Y, Luo Y, Zhang Y, Chen Q, Wang Y, Lin Y, Zhang Y, Liu Z, Wang X-R, Tang H (2021) *Sucrose synthase* gene family in *Brassica juncea*: genomic organization, evolutionary comparisons, and expression regulation. PeerJ 9:e10878

Liljegren SJ, Ditta GS, Eshed Y, Savidge B, Bowman JL, Yanofsky MF (2000) *SHATTERPROOF* MADS-box genes control seed dispersal in *Arabidopsis*. Nature 404:766–770

Liu J, Ishitani M, Halfter U, Kim CS, Zhu JK (2000) The *Arabidopsis thaliana SOS2* gene encodes a protein kinase that is required for salt tolerance. Proc Natl Acad Sci USA 97:3730–3734

Liu X, Lu Y, Yuan Y, Liu S, Guan C et al (2013) De Novo Transcriptome of *Brassica juncea* seed coat and identification of genes for the biosynthesis of flavonoids. PLoS One 8(8):e71110

Liu Y, Xu A, Liang F, Yao X, Wang Y, Liu X, Zhang Y, Dalelhan J, Zhang B, Qin M, Huang Z, Shaolin L (2018) Screening of clubroot-resistant varieties and transfer of clubroot resistance genes to *Brassica napus* using distant hybridization. Breed Sci 68(2):258–267

Lobell DB, Asner GP (2003) Climate and management contributions to recent trends in U.S. agricultural yields. Science 299 (5609):1032

Lu Y, Liu X, Liu S, Yue Y, Guan C, Liu Z (2012) A simple and rapid procedure for identification of seed coat colour at the early developmental stage of *Brassica juncea* and *Brassica* napus seeds. Plant Breed 131:176–179

Mandal MN, Santha IM, Lodha ML, Mehta SL (2000) Cloning of *acyl-acyl carrier protein (ACP) thioesterase* gene from *Brassica juncea.* Biochem Soc Trans 28(6):967–969

Mandal MNA, Santha IM, Lodha ML Mehta SL (2002) Molecular cloning and characterization of *Oleoyl-ACP Thioesterase* gene from *Brassica juncea.* J Plant Biochem Biotechnol 11:27–31

Masierowska ML (2003) Floral nectaries and nectar production in brown mustard (*Brassica juncea*) and white mustard (*Sinapis alba*) (*Brassica*ceae). Plant Syst Evol 238:97–107

Mathur V, Wagenaar R, Caissard JC, Reddy AS, Vet LE, Cortesero AM, Van Dam NM (2013) A novel indirect defence in *Brassica*ceae: structure and function of extrafloral nectaries in *Brassica juncea.* Plant Cell Environ 36(3):528–541

Meena PD, Thomas L, Singh D (2014) Assessment of yield losses in *Brassica juncea* due to downy mildew (*Hyaloperonosporabrassica*e). J Oilseed Brassica 5 (1):73–77

Meenu AR, Majee M, Pradhan AK, Bisht NC (2015) Genomic origin, expression differentiation and regulation of multiple genes encoding *CYP83A1*, a key enzyme for core glucosinolate biosynthesis, from the allotetraploid *Brassica juncea.* Planta 241(3):651–665

Meur G, Gaikwad K, Bhat SR, Prakash S, Kirti PB (2006) Homeotic-like modification of stamens to petals is associated with aberrant mitochondrial gene expression in cytoplasmic male sterile Ogura *Brassica juncea.* J Genet 85:133–139

Meur G, Shukla P, Dutta-Gupta A, Kirti PB (2015) Characterization of *Brassica juncea–Alternaria brassicicola* interaction and jasmonic acid carboxyl methyl transferase expression. Plant Gene 3:1–10

Minglin L, Yuxiu Z, Tuanyao C (2005) Identification of genes up-regulated in response to Cd exposure in *Brassica juncea* L. Gene 363:151–158

Mir ZA, Ali S, Shivaraj SM, Bhat JA, Singh A, Yadav P, Rawat S, Paplao PK, Grover A (2020) Genome-wide identification and characterization of *Chitinase* gene family in *Brassica juncea* and *Camelina sativa* in response to *Alternaria brassicae.* Genomics 112 (1):749–763

Mohammed AE, You MP, Banga SS, Barbetti MJ (2019) Resistances to downy mildew (*Hyaloperonosporabrassicae*) in diverse Brassicaceae offer new disease management opportunities for oilseed and vegetable crucifer industries. Eur J Plant Pathol 153:915–929

Mohd Saad NS, Severn-Ellis AA, Pradhan A, Edwards D, Batley J (2021) Genomics armed with diversity leads the way in Brassica improvement in a changing global environment. Front Genet 12:600789

Mondal B, Mazumder M, Mukherjee A, Ghosh S, De A, Bose R, Basu D (2020) Association of *Alternaria brassicicola* induced *NAC* transcription factors with desiccation and wound responses in Indian mustard. Physiol Mol Plant Pathol 112:101540

Mukherjee AK, Mohapatra T, Varshney A, Sharma R, Sharma RP (2001) Molecular mapping of a locus controlling resistance to *Albugo candida* in Indian mustard. Plant Breed 120:483–497

Muntha ST, Zhang L, Zhou Y, Zhao X, Hu Z, Yang J, Zhang M (2019) *Phytochrome A signal transduction 1* and *CONSTANS-LIKE 13* coordinately orchestrate shoot branching and flowering in leafy *Brassica juncea.* Plant Biotechnol J 17(7):1333–1343

Muthukumar B, Yakubov B, Salt DE (2007) Transcriptional activation and localization of expression of *Brassica juncea* putative metal transport protein *BjMTP1.* BMC Plant Biol 7:32

Naeem I, Munir I, Durrett TP, Iqbal A, Aulakh KS, Ahmad MA, Khan H, Khan IA, Hussain F, Shuaib M, Shah AA, Muhammad I, Bahadur S, Begim K, Hussain F (2020) Feasible regeneration and Agrobacterium-mediated transformation of *Brassica juncea* with *Euonymus alatus diacylglycerol acetyltransferase (EaDAcT)* gene. Saudi J Biol Sci 27:1324–1332

Nagpal K, Yadav S, Singh R (2020) Quantitative evaluation of nectar secretion and production of nectar energy from Indian mustard flowers (*Brassica juncea* L.): a study in bioenergetics. J Pharmacog Phytochem 9(2):514–517

Nakamura MT, Nara TY (2004) Structure, function, and dietary regulation of *Δ6, Δ5, and Δ9 desaturases.* Annu Rev Nutr 24:345–376

Nanjundan J, Manjunatha C, Radhamani J, Thakur AK, Yadav R, Kumar A, Meena ML, Tyagi RK, Yadava DK, Singh D (2020) Identification of new source of resistance to powdery mildew of Indian mustard and studying its inheritance. Plant PatholJ 36 (2):111–120

Nashaat NI, Heran A, Awasthi RP, Kolte SJ (2004) Differential response and genes for resistance to *Peronospora parasitica* (downy mildew) in *Brassica juncea* (mustard). Plant Breed 123:512–515

Ni WM, Chen XY, Xu ZH, Xue HW (2002a) A *pin* gene families encoding components of auxin efflux carriers in *Brassica juncea*. Cell Res 12(3–4):247–255

Ni WM, Chen XY, Xu ZH, Xue HW (2002b) Isolation and functional analysis of a *Brassica juncea* gene encoding a component of auxin efflux carrier. Cell Res 12(3–4):235–245

Nikolov LA, Shushkov P, Nevado B, Gan X, Al-Shehbaz IA, Filatov D, Bailey CD, Tsiantis M (2019) Resolving the backbone of the Brassicaceae phylogeny for investigating trait diversity. New Phytol 222(3):1638–1651

Nutan KK, Kumar G, Singla-Pareek SL, Pareek A (2018) A Salt Overly Sensitive Pathway member from *Brassica juncea BjSOS3* can functionally complement *ΔAtsos3* in *Arabidopsis*. Curr Genom 19 (1):60–69

Nyalugwe EP, Barbetti MJ, Jones RAC (2015) Studies on resistance phenotypes to Turnip mosaic virusin five species of Brassicaceae, and identification of a virus resistance gene in *Brassica juncea*. Eur J Plant Pathol 141:647–666

Nyalugwe EP, Barbetti MJ, Jones RAC (2016) Strain specificity of turnip mosaic virus resistance gene *TuRBJU01* in *Brassica juncea*. Eur J Plant Pathol 145:209–213

Østergaard L, Kempin SA, Bies D, Klee HJ, Yanofsky MF (2006) Pod shatter-resistant *Brassica* fruit produced by ectopic expression of the *FRUITFULL* gene. Plant Biotechnol J 4(1):45–51

Padmaja LK, Agarwal P, Gupta V, Mukhopadhyay A, Sodhi YS, Pental D, Pradhan AK (2014) Natural mutations in two homoeologous *TT8* genes control yellow seed coat trait in allotetraploid *Brassica juncea* (AABB). Theor Appl Genet 127(2):339–347

Palukaitis P, Kim S (2021) Resistance to turnip mosaic virus in the family Brassicaceae. Plant Pathol J 37 (1):1–23

Pandey M, Penna S (2017) Time course of physiological, biochemical, and gene expression changes under short-term salt stress in *Brassica juncea* L. Crop J 15:219–230

Panjabi-Massand P, Yadava SK, Sharma P, Kaur A, Kumar A, Arumugam N et al (2010) Molecular mapping reveals two independent loci conferring resistance to *Albugo candida* in the east European germplasm of oilseed mustard *Brassica juncea*. Theor Appl Genet 121:137–145

Pathania A, Bhat SR, Dinesh Kumar V, Ashutosh KPB, Prakash S, Chopra VL (2003) Cytoplasmic male sterility in alloplasmic*Brassica juncea* carrying *Diplotaxis catholica* cytoplasm: molecular characterization and genetics of fertility restoration. Theor Appl Genet 107(3):455–461

Pathania A, Kumar R, Kumar VD, Ashutosh DKK, Kirti PB, Prakash S, Chopra VL, Bhat SR (2007) A duplicated *coxI* gene is associated with cytoplasmic male sterility in an alloplasmic*Brassica juncea* line derived from somatic hybridization with *Diplotaxis catholica*. J Genet 86(2):93–101

Peleman JD, van der Voort JR (2003) Breeding by design. Trends Plant Sci 8(7):330–334

Phukan D, Singh B, Ravi PI, Mithra ASV, Yadava DK, Chinnusamy V, Mohapatra T (2016) Cloning and characterization of drought stress-induced *NAC* transcription factors from *Brassica juncea* and *Sinapis alba*. Indian J Genet 76(3):233–245

Pilon-Smits EAH, LeDuc DL (2009) Phytoremediation of selenium using transgenic plants. Curr Opin Biotechnol 20:207–212

Pilon-Smits EAH, Hwang SB, Lytle CM, Zhu YL, Tai JC et al (1999) Overexpression of *ATP sulfurylase* in *Brassica juncea* leads to increased selenate uptake, reduction and tolerance. Plant Physiol 119:123–132

Plissonneau C, Rouxel T, Chèvre AM, Van De Wouw AP, Balesdent MH (2018) One gene-one name: the *AvrLmJ1* avirulence gene of *Leptosphaeria maculans* is *AvrLm5*. Mol Plant Pathol 19(4):1012–1016

Prabhu KV, Somers DJ, Rakow G, Gugel RK (1998) Molecular markers linked to white rust resistance in mustard *Brassica juncea*. Theor Appl Genet 97:865–870

Pradhan AK, Pental D (2011) Genetics of *Brassica juncea*. In: Schmidt R, Bancroft I (eds) Genetics and genomics of the Brassicaceae, plant genetics and genomics: crops and models 9. Springer Science and Business Media, LLC, pp 323–346

Prakash S, Kirti PB, Bhat SR, Gaikwad K, Kumar VD, Chopra VL (1998) A *Moricandia arvensis*–based cytoplasmic male sterility and fertility restoration system in *Brassica juncea*. Theor Appl Genet 97:488–492

Qiu X, Hong H, MacKenzie SL (2001) Identification of a *Δ4 Fatty Acid Desaturase* from *Thraustochytrium* sp. involved in the biosynthesis of docosahexanoicacid by heterologous expression in *Saccharomyces cerevisiae* and *Brassica juncea*. J Biol Chem 276(34):31561–31566

Rai AN, Saini N, Yadav R, Suprasanna P (2020) A potential seedling-stage evaluation method for heat tolerance in Indian mustard (*Brassica juncea* L. Czern and Coss). 3 Biotech 10(3):114

Ram C, Annamalai M, Koramutla MK, Kansal R, Arora A, Jain PK, Bhattacharya R (2020) Characterization of *STP4* promoter in Indian mustard *Brassica juncea* for use as an aphid responsive promoter. Biotechnol Lett 42(10):2013–2033

Rana K, Atri C, Gupta M, Akhatar J, Sandhu PS, Kumar N et al (2017) Mapping resistance responses to *Sclerotinia* infestation in introgression lines of *Brassica juncea* carrying genomic segments from wild Brassicaceae *B. fruticulosa*. Sci Rep 7: 5904

Rana K, Atri C, Akhatar J, Kaur R, Goyal A, Singh MP et al (2019) Detection of first marker trait associations for resistance against *Sclerotinia sclerotiorum* in *Brassica juncea*—*Erucastrum cardaminoides* introgression lines. Front Plant Sci 10:1015

Rani S, Sharma V, Hada A, Bhattacharya RC, Koundal KR (2017a) Fusion gene construct

preparation with lectin and protease inhibitor genes against aphids and efficient genetic transformation of *Brassica juncea* using cotyledons explants. Acta Physiol Plant 39:115
Rani S, Sharma V, Hada A, Koundal KR (2017b) Efficient genetic transformation of *Brassica juncea* with *lectin* using cotyledons explants. Intl J Adv Biotechnol Res 7(1):1–12
Rao GU, Batra-Sarup V, Prakash S, Shivanna KR (1994) Development of a new cytoplasmic male-sterility system in *Brassica juncea* through wide hybridization. Plant Breed 112:171–174
Rathore SS, Shekhawat K, Dass A, Kandpal BK, Singh VK (2019) Phytoremediation mechanism in Indian mustard (*Brassica juncea*) and its enhancement through agronomic interventions. Proc Natl Acad Sci India Sect B Biol Sci 89(2):419–427
Reisinger S, Schiavon M, Terry N, Pilon-Smits EAH (2008) Heavy metal tolerance and accumulation in Indian Mustard (*Brassica juncea* L.) expressing bacterial *γ-glutamylcysteine synthetase* or *glutathione synthetase*. Intl J Phytoremed 10(5):440–454
Rout K, Yadav BG, Yadava SK, Mukhopadhyay A, Gupta V, Pental D, Pradhan AK (2018) QTL Landscape for oil content in *Brassica juncea*: analysis in multiple bi-parental populations in high and "0" erucic background. Front Plant Sci 9:1448
Saini N, Singh N, Kumar A, Vihan N, Yadav S, Vasudev S, Yadava DK (2016) Development and validation of functional CAPS markers for the *FAE* genes in *Brassica juncea* and their use in marker-assisted selection. Breed Sci 66(5):831–837
Saini N, Yashpal, Koramutla MK, Singh N, Singh S, Singh R, Yadav S, Bhattacharya R, Vasudev S, Yadava DK (2019) Promoter polymorphism in *FAE1.1* and *FAE1.2* genes associated with erucic acid content in *Brassica juncea*. Mol Breed 39:75
Saroop S, Chanda SV, Singh YD (1999) Biochemical changes associated with *Brassica junceaseed* development. IV. Acid and alkaline phosphatases. J Plant Growth Regul 18(2):69–72
Saroop S, Chanda SV, Singh YD (2002) Changes in soluble and ionically bound peroxidase activities during *Brassica juncea* seed development. Bulg J Plant Physiol 28(3–4):26–34
Savadi S, Naresh V, Kumar V, Bhat SR (2016) Seed-specific overexpression of *ArabidopsisDGAT1* in Indian mustard (*Brassica juncea*) increases seed oil content and seed weight. Botany 94:177–184
Schäfer HJ, Haag-Kerwer A, Rausch T (1998) cDNA cloning and expression analysis of genes encoding GSH synthesis in roots of the heavy-metal accumulator *Brassica juncea* L.: evidence for Cd-induction of a putative mitochondrial *gamma-glutamylcysteine synthetase* isoform. Plant Mol Biol 37(1):87–97
Sharma P, Sardana V, Banga SS (2013) Salt tolerance of Indian mustard (*Brassica juncea*) at germination and early seedling growth. Environ Exp Biol 11:39–46
Sharma M, Mukhopadhyay A, Gupta V, Pental D, Pradhan AK (2016) *BjuB.CYP79F1* regulates synthesis of propyl fraction of aliphatic glucosinolates in oilseed mustard *Brassica juncea*: functional validation through genetic and transgenic approaches. PLoS One 11(2):e0150060
Sharma P, Watts A, Kumar V, Srinivasan R, Siwach P (2018) Cloning, characterization and expression analysis of *APETALA2* genes of *Brassica juncea* (L.) Czern. Ind J Exp Biol 56:604–610
Shivaraj SM, Singh A (2016) Sequence variation in *Brassica AP2* and analysis of interaction of *AP2-miR172* regulatory module. Plant Cell Tiss Organ Cult 125:191–206
Shivaraj SM, Jain A, Singh A (2018) Highly preserved roles of *Brassica MIR172* in polyploid Brassicas: ectopic expression of variants of *Brassica MIR172* accelerates floral transition. Mol Genet Genomics 293 (5):1121–1138
Shopan J, Mou H, Zhang L, Zhang C, Ma W, Walsh JA, Hu Z, Yang J, Zhang M (2017) *Eukaryotic translation initiation factor 2B-beta (eIF2Bβ)*, a new class of plant virus resistance gene. Plant J 90(5):929–940
Shopan J, Liu C, Hu Z, Zhang M, Yang J (2020) Identification of eukaryotic initiation factors and the temperature dependent nature of Turnip mosaic virus epidemics in allopolyploid *Brassica juncea*. 3 Biotech 10:75
Singh A, Fulekar MH (2012) Phytoremediation of heavy metals by *Brassica juncea* in aquatic and terrestrial environment. In: Anjum N, Ahmad I, Pereira M, Duarte A, Umar S, Khan N (eds) The plant family Brassicaceae. Environ Pollution, vol 21. Springer, Dordrecht, pp 153–169
Singh VK, Singh D (2008) Diversity for fatty acid composition of advanced breeding lines of Indian mustard [*Brassica juncea* (L.) Czern and Coss]. Indian J Genet 68(3):330–333
Singh B, Singh R, Mahal MS, Brar HS (1986) Assessment of loss in yield of *Brassica juncea* by *Lipaphis erysimi*: influence of varying levels of aphid populations. Ind J Entomol 10:97–105
Singh BK, Nandan D, Ambawat S, Ram B, Kumar A, Singh T et al (2015) Validation of molecular markers for marker-assisted pyramiding of white rust resistance loci in Indian mustard (*Brassica juncea* L.). Can J Plant Sci 95:939–945
Singh S, Parihar P, Singh R, Singh VP, Prasad SM (2016) Heavy metal tolerance in plants: role of transcriptomics, proteomics, metabolomics, and ionomics. Front Plant Sci 6:1143
Singh I, Smita S, Mishra DC, Kumar S, Singh BK, Rai A (2017) Abiotic stress responsive miRNA-target network and related markers (SNP, SSR) in *Brassica juncea*. Front Plant Sci 8:1943
Singh S, Geeta R, Das S (2020) Comparative sequence analysis across Brassicaceae, regulatory diversity in *KCS5* and *KCS6* homologs from *Arabidopsis thaliana* and *Brassica juncea*, and intronic fragment as a negative transcriptional regulator. Gene Expr Patterns 38:119146

Sinha S, Jha JK, Maiti MK, Basu A, Mukhopadhyay UK, Sen SK (2007) Metabolic engineering of fatty acid biosynthesis in Indian mustard (*Brassica juncea*) improves nutritional quality of seed oil. Plant Biotech Rep 1:185–197

Sinha S, Raxwal VK, Joshi B, Jagannath A, Katiyar-Agarwal S, Goel S, Kumar A, Agarwal M (2015) Denovo transcriptome profiling of cold-stressed siliques during pod filling stages in Indian mustard (*Brassica juncea* L.). Front Plant Sci 6:932

Sinha-Jha S, Jha JK, Chattopadhyaya B, Basu A, Sen SK, Maiti MK (2010) Cloning and characterization of cDNAs encoding for *long-chain saturated acyl-ACP thioesterases* from the developing seeds of *Brassica juncea*. Plant PhysiolBiochem 48:476–480

Sivaraman I, Amurugam N, Sodhi SY, Gupta V, Mukhopadhyay A, Pradhan AK, Burma PK, Pental D (2004) Development of high oleic and low linoleic acid transgenics in a zero erucic acid *Brassica juncea* L. (Indian mustard) line by antisense suppression of the *fad2* gene. Mol Breed 13:365–375

Smith B, Sterner B (1994) Early ovule development, megasporogenesis and megagametogenesisin rapid-cycling *Brassica juncea* (L.) Cosson. J Penn Acad Sci 68(2):63–66

Somers D, Rakow G, Rimme S (2002) *Brassica* napus DNA markers linked to white rust resistance in *Brassica juncea*. Theor Appl Genet 104:1121–1124

Song JM, Guan Z, Hu J, Guo C, Yang Z, Wang S, Liu D, Wang B, Lu S, Zhou R, Xie WZ, Cheng Y, Zhang Y, Liu K, Yang QY, Chen LL, Guo L (2020) Eight high-quality genomes reveal pan-genome architecture and ecotype differentiation of *Brassica napus*. Nat Plants 6 (1):34–45

Sri T, Gupta B, Tyagi S, Singh A (2020) Homeologs of *Brassica* SOC1, a central regulator of flowering time, are differentially regulated due to partitioning of evolutionarily conserved transcription factor binding sites in promoters. Mol Phylogenet Evol 147:106777

Srivastava S, Srivastava AK, Suprasanna P, D'Souza SF (2013) Identification and profiling of arsenic stress-induced microRNAs in *Brassica juncea*. J Exp Bot 64 (1):303–315

Srivastava S, Srivastava AK, Sablok G, Deshpande TU, Suprasanna P (2015) Transcriptomics profiling of Indian mustard (*Brassica juncea*) under arsenate stress identifies key candidate genes and regulatory pathways. Front Plant Sci 6:646

Stoutjesdijk PA, Hurlestone C, Singh SP, Green AG (2000a) High-oleic acid Australian *Brassica napus* and *B. juncea* varieties produced by co-suppression of endogenous Δ 12-desaturases. Biochem Soc Transact 28(6):938–940

Stoutjesdijk PA, Hurlestone C, Singh SP, Green AG (2000b) High-oleic acid Australian *Brassica napus* and *B. juncea* varieties produced by co-suppression of endogenous Δ 12-desaturases. Biochem Soc Trans 28 (6):938–940

Sun T, Zhang Y, Chai T (2011) Cloning, characterization, and expression of the *BjEXPA1* gene and its promoter region from *Brassica juncea* L. Plant Growth Regul 64:39–51

Suresha GS, Santha IM (2013) Molecular cloning and in silico analysis of novel *oleate desaturase* gene homologues from *Brassica juncea* through sub-genomic library approach. Plant Omics J 6(1): 55–64

Suresha GS, Rai RD, Santha IM (2012) Molecular cloning, expression analysis and growth temperature dependent regulation of a novel *oleate desaturase* gene (*fad2*) homologue from *Brassica juncea*. Aust J Crop Sci 6(2):296–308

Suresha GS, Kumar S, Santha IM (2013) Functional expression of *Brassica juncea oleate desaturase* gene (*Bjfad2*) in *Escherichia coli*. Biologia 68(4):599–605

Thakur S, Sharma SS (2016) Characterization of seed germination, seedling growth, and associated metabolic responses of *Brassica juncea* L. cultivars to elevated nickel concentrations. Protoplasma 253 (2):571–580

Thakur S, Choudhary S, Bhardwaj P (2019) Comparative transcriptome profiling under cadmium stress reveals the uptake and tolerance mechanism in *Brassica juncea*. Plant Growth Regul 38:1141–1152

Thakur AK, Parmar N, Singh KH, Nanjundan J (2020) Current achievements and future prospects of genetic engineering in Indian mustard (*Brassica juncea* L. Czern and Coss.). Planta 252:56

Tyagi S, Mazumdar PA, Mayee P, Shivaraj SM, Anand S, Singh A, Madhurantakam C, Sharma P, Das S, Kumar A, Singh A (2018) Natural variation in *Brassica FT* homeologs influences multiple agronomic traits including flowering time, silique shape, oil profile, stomatal morphology and plant height in *B. juncea*. Plant Sci 277:251–266

Tyagi S, Sri T, Singh A, Mayee P, Shivaraj SM, Sharma P, Singh A (2019) *SUPPRESSOR OF OVEREXPRESSION OF CONSTANS1* influences flowering time, lateral branching, oil quality, and seed yield in *Brassica juncea* cv. Varuna. Funct Integr Genom 19(1):43–60

Vageeshbabu HS, Venkateswari J, Kirti PB, Chopra VL (1996) Molecular cloning of the gene encoding *stearoyl-acyl carrier protein desaturase* in *Brassica juncea*. J Plant BiochemBiotechnol 5:51–53

Van Huysen T, Abdel-Ghany S, Hale KL, LeDuc D, Terry N, Pilon-Smits EAH (2003) Overexpression of *cystathionine-γ -synthase* enhances selenium volatilization in *Brassica juncea*. Planta 218:71–78

Van Huysen T, Terry N, Pilon-Smits EAH (2004) Exploring the selenium phytoremediation potential of transgenic *Brassica juncea* overexpressing *ATP sulfurylase* or *cystathionine-γ-synthase*. Intl J Phytorem 6:111–118

Varshney A, Mohapatra T, Sharma RP (2004) Development and validation of CAPS and AFLP markers for white rust resistance gene in *Brassica juncea*. Theor Appl Genet 109:153–159

Varshney RK, Bohra A, Yu J, Graner A, Zhang Q, Sorrells ME (2021) Designing future crops: genomics-assisted breeding comes of age. Trends Plant Sci 26 (6):631–649

Venkateswari J, Kanrar S, Kirti PB, Malathi VG, Chopra VL (1999) Molecular cloning and characterization of *FATTY ACID ELONGATION1 (BjFAE1)* gene of *Brassica juncea*. J Plant Biochem Biotechnol 8:53–55

Vera CL, Woods DL (1982) Isolation of independent gene pairs at two loci for seed coat color in *Brassica juncea*. Can J Plant Sci 62:47–50

Verma SS, Sinha RK, Jajoo A (2015) *(E)-β-farnesene* gene reduces *Lipaphiserysimi* colonization in transgenic *Brassica juncea* lines. Plant Signal Behav 10(7): e1042636

Verma D, Lakhanpal N, Singh K (2019) Genome-wide identification and characterization of abiotic-stress responsive *SOD (superoxide dismutase)* gene family in *Brassica juncea* and *B. rapa*. BMC Genomics 20:227

Vicente JG, Holub EB (2013) *Xanthomonas campestris* pv. *campestris* (cause of black rot of crucifers) in the genomic era is still a worldwide threat to *Brassica* crops. Mol Plant Pathol 14:2–18

Vicente JG, Taylor JD, Sharpe AG, Parkin IAP, Lydiate DJ, King GJ (2002) Inheritance of race-specific resistance to *Xanthomonas campestris* pv. campestris in *Brassica* genomes. Phytopathol 92:1134–1141

Walsh JA, Jenner CE (2002) Turnip mosaic virus and the quest for durable resistance. Mol Plant Pathol 3 (5):289–300

Wan ZJ, Jing B, Tu JX, Ma CZ, Shen JX, Yi B, Wen J, Huang T, Wang XJ, Fu TD (2008) Genetic characterization of a new cytoplasmic male sterility system (*hau*) in *Brassica juncea* and its transfer to *B. napus*. Theor Appl Genet 116:355–362

Wan Z, Shi D, Zou R, Huang Y, Bie Z, Shia Z, Xua Y, Fu T (2014) Development and utilization of one new cytoplasmic male sterile line of Chinese leaf mustard (*Brassica juncea* var. rugosa Bailey). Sci Hort 165:211–217

Wang JW, Li Y, Zhang YX, Chai TY (2013) Molecular cloning and characterization of a *Brassica juncea* yellow stripe-like gene, *BjYSL7*, whose overexpression increases heavy metal tolerance of tobacco. Plant Cell Rep 32(5):651–662

Wang J, Liang S, Xiang, W. Dai H, Duan Y, Kang F, Chai T (2019) A repeat region from the *Brassica juncea HMA4* gene *BjHMA4R* is specifically involved in Cd^{2+} binding in the cytosol under low heavy metal concentrations. BMC Plant Biol 19:89

Wang Y, Jiang W, Dong Y, Ma X, Zhou W, Wang Z, Fan Y, Wei D, Tang Q (2020) AGL19 directly interacts with floral signal integrator AGL24 in flowering time control of *Brassica juncea*. Acta Physiol Planta 42:175

Wang P, Zhao Z, Zhang Z, Cai Z, Liao J, Tan Q, Xiang M, Chang L, Xu D, Tian Q, Wang D (2021) Genome-wide identification and analysis of *NPR* family genes in *Brassica juncea* var. tumida. Gene 769:145210

Warwick SI (2011) *Brassica*ceae in Agriculture. In: Schmidt R, Bancroft I (eds) Genetics and genomics of the *Brassica*ceae, plant genetics and genomics: crops and models 9. Springer Science and Business Media, LLC, pp 33–66

Warwick SI, Hall J (2009) Phylogeny of *Brassica* and wild relatives. In: Gupta SK (ed) Biology and breeding of crucifers. CRC Press, Boca Raton, pp 19–36

Wu XF, Wang CL, Xie EB, Gao Y, Fan YL, Liu PQ, Zhao KJ (2009) Molecular cloning and characterization of the promoter for the multiple stress-inducible gene *BjCHI1* from *Brassica juncea*. Planta 229 (6):1231–1242

Wu Z, Hu K, Yan M, Song L, Wen J, Ma C, Shen J, Fu T, Yi B, Tu J (2019) Mitochondrial genome and transcriptome analysis of five alloplasmic male-sterile lines in *Brassica juncea*. BMC Genomics 20(1):348

Xiao L, Zhao H, Zhao Z, Du D, Xu L, Yao Y, Zhao Z, Xing X, Shang G, Zhao H (2013) Genetic and physical fine mapping of a multilocular gene *Bjln1* in *Brassica juncea* to a 208-kb region. Mol Breed 32:373–383

Xiao L, Li X, Liu F, Zhao Z, Xu L, Chen C, Wang Y, Shang G, Du D (2018) Mutations in the CDS and promoter of *BjuA07.CLV1* cause a multilocular trait in *Brassica juncea*. Sci Rep 8:5339

Xu J, Zhang YX, Wei W, Han L, Guan ZQ, Wang Z, Chai TY (2008) *BjDHNs* confer heavy-metal tolerance in plants. Mol Biotechnol 38(2):91–98

Xu J, Chai T, Zhang Y, Lang M, Han L (2009) The cation-efflux transporter *BjCET2* mediates zinc and cadmium accumulation in *Brassica juncea* L. leaves. Plant Cell Rep 28(8):1235–1242

Xu P, Lv Z, Zhang X, Wang X, Pu Y, Wang H, Yi B, Wen J, Ma C, Tu J, Fu T, Shen J (2014a) Identification of molecular markers linked to trilocular gene (*mc1*) in *Brassica juncea* L. Mol Breed 33:425–434

Xu Y, Iacuone S, Li SF, Parish RW (2014b) *MYB80* homologues in *Arabidopsis*, cotton and *Brassica*: regulation and functional conservation in tapetal and pollen development. BMC Plant Biol 14:278

Xu P, Cao S, Hu K, Wang X, Huang W, Wang G, Lv Z, Liu Z, Wen J, Yi B, Ma C, Tu J, Fu T, Shen J (2017) Trilocular phenotype in *Brassica juncea* L. resulted from interruption of *CLAVATA1* gene homologue (*BjMc1*) transcription. Sci Rep 7(1):3498

Xue Y, Chai C, Chen B, Shi X, Wang B, Mei F, Jiang M, Liao X, Yang X, Yuan C, Zhang L, Chai Y (2020) Whole-genome mining and in silico analysis of *FAD* gene family in *Brassica juncea*. J Plant BiochemBiotechnol 29(1):149–154

Yadav P, Bhat SR, Chopra VL (2003) Cloning, Sequencing and characterization of *FAE1*genes of *Brassica juncea* cv Pusa Bold and the mutant *fae1* of *B. rapa* cv Tobin. J Plant Biochem Biotechnol 12:19–23

Yadav P, Mir ZA, Ali S, Papolu PK, Grover A (2020) A combined transcriptional, biochemical and histopathological study unravels the complexity of *Alternaria* resistance and susceptibility in *Brassica* coenospecies. Fungal Biol 124(1):44–53

Yamagishi H, Bhat SR (2014) Cytoplasmic male sterility in Brassicaceae crops. Breed Sci 64:38–47

Yan ML, Liu XJ, Guan CY, Liu LL, Lu Y, Liu ZS (2010) Cloning and SNP analysis of *TT1* gene in *Brassica juncea*. Acta Agron Sin 36:1634–1641

Yan K, Li CC, Wang Y, Wang XQ, Wang ZM, Wei DY, Tang QL (2018) *AGL18*-1 delays flowering time through affecting expression of flowering-related genes in *Brassica juncea*. Plant Biotechnol 35 (4):357–363

Yang JH, Zhang MF, Jing-Quan Y, Shuo Z, Tao W, Zhu-Jun C (2005) Identification of alloplasmic cytoplasmic male-sterile line of leaf mustard synthesized by intra-specific hybridization. Plant Sci 168(4):865–871

Yang JH, Huai Y, Zhang MF (2009) Mitochondrial *atpA* gene is altered in a new *orf220*-type cytoplasmic male-sterile line of stem mustard (*Brassica juncea*). Mol Biol Rep 36(2):273–280

Yang J, Liu X, Yang X, Zhang M (2010) Mitochondrially-targeted expression of a cytoplasmic male sterility-associated orf220 gene causes male sterility in *Brassica juncea*. BMC Plant Biol 10:231

Yang X, Liu X, Lv W, Li L, Shi Q, et al. (2012) Reduced expression of *BjRCE1* gene modulated by nuclear-cytoplasmic incompatibility alters auxin response in cytoplasmic male-sterile *Brassica juncea*. PLoS One 7 (6):e38821

Yang J, Liu X, Xu B, Zhao N, Yang X, Zhang M (2013) Identification of miRNAs and their targets using high-throughput sequencing and degradome analysis in cytoplasmic male-sterile and its maintainer fertile lines of *Brassica juncea*. BMC Genomics 14:9

Yang J, Song N, Zhao X, Qi X, Hu Z, Zhang M (2014) Genome survey sequencing provides clues into glucosinolate biosynthesis and flowering pathway evolution in allotetrapolyploid *Brassica juncea*. BMC Genomics 6(15):107

Yang J, Liu D, Wang X, Ji C, Cheng F, Liu B, Hu Z, Chen S, Pental D, Ju Y, Yao P, Li X, Xie K, Zhang J, Wang J, Liu F, Ma W, Shopan J, Zheng H, Mackenzie SA, Zhang M (2016) The genome sequence of allopolyploid *Brassica juncea* and analysis of differential homoeolog gene expression influencing selection. Nat Genet 48(10):1225–1232

Yang H, Bayer PE, Tirnaz S, Edwards D, Batley J (2021a) Genome-wide identification and evolution of*Receptor-Like Kinases (RLKs)* and *Receptor like Proteins (RLPs)* in *Brassica juncea*. Biology 10:17

Yang H, Mohd Saad NS, Ibrahim MI, Bayer PE, Neik TX, Severn-Ellis AA, Pradhan A, Tirnaz S, Edwards D, Batley J (2021b) Candidate *Rlm6* resistance genes against *Leptosphaeria maculans* identified through a genome-wide association study in *Brassica juncea* (L.) Czern. Theor Appl Genet 134(7):2035–2050

Yao K, Bacchetto RG, Lockhart KM, Friesen LJ, Potts DA, Covello PS, Taylor DC (2003) Expression of the *Arabidopsis ADS1* gene in *Brassica juncea* results in a decreased level of total saturated fatty acids. Plant Biotechnol J 1:221–229

Yao K, Lockhart KM, Kalanack JJ (2005) Cloning of dehydrin coding sequences from *Brassica juncea* and *Brassica napus* and their low temperature-inducible expression in germinating seeds. Plant Physiol Biochem 43:83–89

Yusuf MA, Bhalla-Sarin N (2007) Antioxidant value addition in human diets: genetic transformation of *Brassica juncea* with γ-*TMT* gene for increased α-tocopherol content. Trans Res 16:109–113

Zeng CL, Wang JB, Liu AH, Wu XM (2004) Seed coat microsculpturing changes during seed development in diploid and amphidiploid *Brassica* species. Ann Bot 93(5):555–566

Zeng CL, Wu XM, Wang JB (2006) Seed coat development and its evolutionary implications in diploid and amphidiploid *Brassica* species. Acta Biologica Cracoviensia Series Botanica 48(2):15–25

Zhao K-J, Chye ML (1999) Methyl jasmonate induces expression of a novel *Brassica juncea* chitinase with two chitin-binding domains. Plant Mol Biol 40:1009–1018

Zhao N, Xu X, Wamboldt Y, Mackenzie SA, Yang X, Hu Z, Yang J, Zhang M (2016) *MutS HOMOLOG1* silencing mediates *ORF220* substoichiometric shifting and causes male sterility in *Brassica juncea*. J Exp Bot 67(1):435–444

Zhao Z, Xiao L, Xu L, Xing X, Tang G, Du D (2017) Fine mapping the *BjPl1* gene for purple leaf color in B2 of *Brassica juncea* L. through comparative mapping and whole-genome re-sequencing. Euphytica 213:80

Zhao N, Liu C, Meng Y, Hu Z, Zhang M, Yang J (2019) Identification of flowering regulatory genes in allopolyploid *Brassica juncea*. Hort Plant J 5(3):109–119

Zhao N, Li Z, Zhang L, Yang X, Mackenzie SA, Hu Z, Zhang M, Yang J (2021) *MutS HOMOLOG1* mediates fertility reversion from cytoplasmic male sterile *Brassica juncea* in response to environment. Plant Cell Environ 44(1):234–246

Zhou W, Wei D, Jiang W, Wang Z, Wang H, Tang Q (2018) The protein J3 regulates flowering through directly interacting with the promoter of *SOC1* in *Brassica juncea*. BiochemBiophys Res Comm 496 (4):1217–1221

Zhu Y, Pilon-Smits EAH, Jouanin L, Terry N (1999) Overexpression of *glutathione synthetase* in *Brassica juncea* enhances cadmium tolerance and accumulation. Plant Physiol 119:73–79

Zhu L, Li Y, Ara N, Yang J, Zhang M (2012) Role of a newly cloned *alternative oxidase* gene (*BjAOX1a*) in Turnip mosaic virus (TuMV) resistance in mustard. Plant Mol Biol Rep 30:309–318

Zhu H, Zhai W, Li X, Zhu Y (2019) Two QTLs controlling clubroot resistance identified from bulked segregant sequencing in Pakchoi (*Brassica campestris* ssp. Chinensis Makino). Sci Rep 9:9228

22 Advanced Molecular Breeding for Yield

Satish K. Yadava and Nirala Ramchiary

Abstract

Improvement in crop yields in *Brassica juncea* has been stagnating for some time, a trend that is likely to exacerbate further due to the effects of various biotic and abiotic stresses like insects or diseases, drought, submergence, salinity and high temperature. The crucial increase in the crop production should necessarily come from higher yields, since there is little scope for expanding the agricultural land base. These factors must be addressed directly in targeted approaches for improvement in yield. Yield influencing traits with a large genetic component and direct correlation with yield can be considered suitable as selection criteria. Different morphological characters like number of pods per plant, number of seeds per pod and seed weight along with flowering time fulfill these criteria, and have therefore been the subject of genetic analysis by molecular mapping. The genetic dissection of principal yield components, documentation of genes and quantitative trait loci (QTLs) associated with various trait phenotypes in *B. juncea* have been undertaken in several studies which were based on a wide array of molecular markers including restriction fragment length polymorphism, amplified fragment length polymorphism, intron polymorphism, simple sequence repeats, etc. A few of these studies have also indicated that the A sub genome has played a major role compared to the B sub genome in the process of domestication of *B. juncea* as an oilseed crop. Moreover, it also appears that rigorous selection procedures were in practice for the selection of superior alleles during the course of domestication of lines belonging to the Indian germplasm as compared to lines of the exotic gene pool. However, genes underlying the QTLs for yield and most of the yield-related traits have still not been identified. Recent developments in the sequencing technologies are anticipated to further accelerate the discovery of the genomic regions/QTLs associated with yield and related traits. This chapter provides a detailed review on molecular mapping approaches for dissection of the genetic bases of yield in *B. juncea*.

S. K. Yadava
Centre for Genetic Manipulation of Crop Plants, University of Delhi South Campus, Benito Juarez Road, New Delhi 110021, India

N. Ramchiary (✉)
School of Life Sciences, Jawaharlal Nehru University, New Mehrauli Road, New Delhi 110067, India
e-mail: nramchiary@jnu.ac.in

22.1 Introduction

The development of *Brassica juncea* cultivars with higher seed yield and their stability are the prime objectives for all breeding programs in

C. Kole and T. Mohapatra (eds.), *The Brassica juncea Genome*, Compendium of Plant Genomes,
https://doi.org/10.1007/978-3-030-91507-0_22

Asia, whereas in Europe and Canada breeding for oil and meal quality are the major objectives (Gupta 2016). The productivity of oleiferous Brassicas and mustard in India is stagnating at an average of 1.15 tons/hectare for the last 10 years (Jat et al. 2019). This suggests that there should be a greater emphasis on developing strategies in a well-planned and targeted manner which could bring the desired growth in oilseed Brassica production. Further, to investigate factors influencing yield, it is essential to understand the yield and its attributing traits, such as the primary and secondary components, which determine the final seed yield (Diepenbrock 2000). The conventional genetic analysis, although successful to some extent, has a highly restricted role in the dissection of a majority of the yield influencing traits into their components, as they are difficult to evaluate and are highly affected by the climate in addition to their complex polygenic nature influencing the final yield. However, with the progress in the discovery and use of molecular markers, it has become easy to dissect the genetic loci/QTLs of yield and its attributing traits such as branch number, siliques per plant, seeds per silique and seed weight (Dhillon et al. 1992; Ramchairy et al. 2007). Furthermore, a few of the genetic loci/QTLs have been used in marker-assisted breeding, and fine mapping and isolation of potential candidate genes. This chapter discusses the overall development of molecular mapping of yield and its related traits, and their potential application in *B. juncea* breeding for increasing the final yield. A summary of QTL mapping studies on yield and yield-related traits in *B. juncea* is provided in Table 22.1.

22.2 Molecular Mapping of Yield and Yield-Related Traits

In as early as 1994, Sharma et al. used two parental lines, Varuna and BEC-144, which were selected based on their morphological and molecular contrasts, for the development of a F_2 mapping population consisting of 50 plants. A one-way analysis of variance (ANOVA) was performed on eight quantitative traits pertaining to yield—plant height, numbers of primary and secondary branches, siliques per secondary branch, seeds per silique, flowering time and seed oil content. The F_2 plants showed extensive variation for all of the eight quantitative traits due to the presence of a large number of segregating lines with transgressive phenotypes beyond the parental lines in either/both directions, indicating that favorable alleles were dispersed between both Varuna and BEC-144. The ANOVA for pairwise marker-trait combinations identified three restriction fragment length polymorphism (RFLP) markers, BJG 59a, BJG 426 and BJG433 significantly associated with the number of primary branches, number of secondary branches and days to flowering, respectively. The QTLs linked to these markers were believed to have major roles in trait expression because they could be detected with a limited number of RFLP loci genotyped on a small F_2 population (Sharma et al. 1994).

Upadhyay et al. (1996) used the same Varuna/BEC-144 cross to analyze significant marker-trait associations for six quantitative traits viz. days to flowering, plant height, number of primary branches, secondary branches per primary branch, siliques per secondary branch, and seeds per silique. The ANOVA revealed 17 important marker-trait linkages. Most of the individual marker-trait associations accounted for more than 12% of the trait variation suggesting that the marked genomic regions had large effects on trait expression.

The genetic dissection of principal yield components, documentation of genes and QTLs linked with trait phenotypes in *B. juncea* were also investigated by Lionneton et al. (2004). The study evaluated days to flowering, plant height, thousand seed weight, seed oil content, along with fatty acid composition, sinigrin, gluconapin and total glucosinolates (GSL) contents in a doubled haploid (DH) mapping population consisting of 131 lines. The parental lines of the DH population consisted of a yellow seeded, tall and late-flowering oriental type genotype (BJ-99), and a brown seeded, shorter in stature, early flowering Indian type (BJ-70). An amplified fragment length polymorphism (AFLP) based

Table 22.1 QTL mapping studies on yield and yield-related traits in *B. juncea*

Publication	Cross	Yield-related traits investigated	QTL information
Sharma et al. (1994)	Varuna × BEC144	Plant height at maturity, number of primary branches, number of secondary branches, siliquae per primary branch, siliquae per secondary branch, seeds per silique, days to flowering and oil content	Three RFLP markers BJG 59a, BJG 426 and BJG 433 significantly associated with number of primary branches, number of secondary branches and days to flowering
Upadhyay et al. (1996)	Varuna × BEC144	Days to flowering, plant height, number of primary branches, secondary branches per primary branch, siliquae per secondary branch, and seeds per silique	Seventeen significant marker-trait associations with phenotypic variances explained ranging between 3.0% and 33.2%
Axelsson et al. (2001)	J-o-3DH1 × Rapid cycling *B. juncea* (FTBj) 22.2.69 × Rapid cycling *B. oleracea* (FTBo) R–c-50a7 × Rapid cycling *B. rapa* (FTBr)	Flowering time	Three QTLs were detected in *B. juncea* which together explained a total of 56% of the phenotypic variation in flowering In *B. rapa*, two QTLs explaining 38% of the phenotypic variation were detected In *B. oleracea*, two QTLs were identified which explained a total of 22% of the variation in flowering time
Lionneton et al. (2004)	BJ-70 × BJ-99	Oil content, days to flower, plant height, 1000-seed weight	Thirteen QTL were identified with phenotypic variances explained ranging between 7.5% and 14.6%
Mahmood et al. (2005)	RLM-514 × canola-quality inbred	Number of siliques on the main raceme, number of seeds/siliqua, siliqua length, 1000-seed weight and seed yield	Eleven QTLs were identified with four QTLs for siliqua length, one for number seeds/siliqua, five for number siliques on the main raceme and one for 1000-seed weight
Mahmood et al. (2007)	RLM-514 × canola-quality inbred	Days to maturity, days to first flowering, days to end of flowering, flowering period and plant height	Three QTLs were found to be significant for plant height Five QTLs were found to be significant for days to first flowering, Two QTLs each were identified for days to end of flowering and days to maturity, and one QTL for flowering period
Ramchairy et al. (2007)	Varuna × Heera	Plant height, days to flowering, number of primary branches, number of secondary branches, main shoot length, number of siliques on a plant, silique density, silique length, number of seeds in a silique,1,000-seed weight, and oil content	Sixty-five QTL were detected with phenotypic variances explained ranging between 5.0% and 47.9%

(continued)

Table 22.1 (continued)

Publication	Cross	Yield-related traits investigated	QTL information
Yadava et al. (2012)	TM-4 × Donskaja-IV (TD population) Varuna × Heera (VH population)	Plant height, days to flowering, number of primary branches, number of secondary branches, main shoot length, number of siliques on a plant, silique density, silique length, number of seeds in a silique, 1,000-seed weight, and oil content	Eighty-four significant QTL over three environments detected in TD population and 94 QTL over three environments detected in VH population
Kaur and Banga (2014)	Determinate genotypes × RLC1	Days to flowering, maturity and oil content, seed yield	Genetic mapping of *Brassica juncea Sdt1* gene associated with determinate plant growth habit
Dhaka et al. (2017)	EH-2 × Pusa Jaikisan (EPJ population) Donskaja-IV × EH-2 (DE population) TM-4 × Donskaja-IV (TD population) Varuna × Heera (VH population)	Thousand seed weight	Twenty-five QTL were detected in EPJ population; 10 QTL were detected in DE population; 14 QTL detected in the TD population; 16 QTL detected in the VH population

genetic map constructed previously using 273 marker loci distributed over 18 linkage groups (LGs) spanning a genetic distance of 1,641 cM was used for QTL analysis (Lionneton et al. 2002). A total of 13 QTLs were identified in the two trials undertaken at INRA experimental field at Dijon, France in 2001 and 2002. For plant height QTLs, BJ-99 provided the trait enhancing alleles at all the three QTLs, while BJ-70 provided the positive alleles at all the three QTLs for thousand seed weight. For days to flowering and seed oil content, both the parental lines contributed the positive alleles. The study also revealed that each of these traits were independent and therefore could be bred separately. Also, no genetic interactions or QTL colocalizations were detected, except for a QTL for days to flowering which overlapped a thousand seed weight QTL on LG 6.

Mahmood et al. (2005) also carried out a QTL study on yield and component traits using a RFLP based genetic map. The experiment was undertaken with 112 recombinant DH lines derived from reciprocal crosses between a black seeded, high-yielding non-canola Indian cultivar RLM-514 and a yellow seeded, low-yielding canola breeding line (Thiagarajah and Stringam 1993; Mahmood et al. 2003). Phenotyping was performed for evaluating the number of siliques on the main shoot, number of seeds per silique, silique length, thousand seed weight and seed yield. A total of 11 QTLs dispersed over six LGs were identified and were distributed as four QTLs for silique length, one for number of seeds per silique, five for number of siliques on the main shoot and one for thousand seed weight. RLM-514 contributed positive alleles at five of these loci, the other six QTLs inherited positive alleles from the low-yielding canola breeding line. The yield component traits showed both strongly positive and negative correlations among themselves than with seed yield, while no significant correlation was observed for thousand seed weight and seed yield. The authors noted that yield compensation formed the basis of strong association of yield components which competed for limited resources (Adams, 1967). The study also estimated that the effectiveness of selection increased by only 4% for seed yield based on phenotypic selection for both seed yield and yield component traits compared to direct phenotypic selection for seed yield alone. This observation was attributed to frequent negative correlations between the yield component traits, and consequently the direct effect of a yield component trait on seed yield is reduced by the negative indirect effects through other yield components. Also, inclusion of the phenotypic data along with the genotypic data in the calculation of selection indices did not result in any improvement in the efficacy of selection. Based on the estimates of expected genetic advance (indirect) for yield by including only yield component traits in calculations for selection index for *B. juncea*, it was proposed that shorter siliques and a greater number of siliques on the main shoot will result in increased yield (Mahmood et al. 2005).

A comprehensive QTL analysis in *B. juncea* for 12 yield influencing quantitative traits was performed by Ramchairy et al. (2007) using a DH population. A total of 123 DH lines (VH population) were used, which were derived from the F_1 of a cross between two lines—Varuna and Heera, belonging to the Indian and east European gene pool, respectively. Additional RFLP and simple sequence repeat (SSR) markers were included in the previously reported AFLP based linkage map of *B. juncea* (Pradhan et al. 2003), and the resultant updated map harbored 1448 markers with a total genetic distance of 1840.1 cM. The parental lines were contrasting for majority of the 12 quantitative traits. Varuna was characterized by large seeds, longer siliques and main shoot, early flowering and short height, while Heera was tall in height, had a greater number of siliques per plant and also on the main shoot, a higher number of seeds in a silique, primary and secondary branches, and showed an extended duration to maturity. The DH lines were phenotyped in three dissimilar locations in India namely, Delhi, Gwalior (Madhya Pradesh) and Leh (Jammu and Kashmir).

A total of 65 QTLs were detected from all the three environments which were distributed over 13 LGs. Of these 65 QTLs, 24 were consistently detected in at least two environments, while the remaining 41 were specific to only one environment. The Indian parent, Varuna donated positive alleles at 34 QTLs (53%), while 31 (47%) loci received the positive alleles from the east European parental line, Heera. Whereas for most of the yield contributing traits both parents contributed the positive alleles, all the QTLs for silique length and thousand seed weight inherited the trait enhancing alleles exclusively from Varuna, and all the QTLs identified for siliques per plant acquired the trait enhancing alleles from Heera. Two LGs, J7 and J10 harbored QTL hotspots with a pronounced clustering of QTLs with 9 and 12 loci on J7 and J10, respectively. The clusters on the LGs J7, J9 and J10 of the A sub genome, and J18 of the B sub genome were characterized by QTLs of different traits having agronomically antagonistic allelic effects. An important inference obtained from these results was that the isolation of desirable recombinants with trait enhancing alleles of both the parental inbreds through conventional breeding methods would be problematic.

The study by Ramchairy et al. (2007) also revealed that the A sub genome with 66% QTLs made a larger contribution compared to the B sub genome with 34% QTLs, in the process of domestication of *B. juncea* as an oilseed crop. Further, the A sub genome also mainly contributed the trait enhancing QTL alleles (58%) in the Indian parent Varuna. Conversely, the B sub genome was the principal donor of the trait increasing alleles (59%) in the east European parent Heera. These results indicated that a rigorous selection was applied for the selection of superior alleles during the course of domestication of lines belonging to the Indian germplasm compared to lines of the exotic gene pool. The study proposed that independent processes of domestication could have given rise to the two gene pools. The study also indicated that the founder effect and divergent selection could have played a major role in the polyphyletic origin of *B. juncea* and the observed diversity in the two gene pools, as supported by the available information from RFLP and phylogenetic analyses (Song et al. 1988; Pradhan et al. 1993; Srivastava et al. 2001).

Yadava et al. (2012) performed a comparative mapping of QTLs for yield related characters in *B. juncea* using segregating DH lines of the TD population derived from the F_1 of a cross of TM-4 (a short duration cultivar adapted to Indian climate) with Donskaja-IV (an east European line with high oil content and resistant to white rust). The TD linkage map was constructed with 911 markers spanning 1629.9 cM. The previously reported VH maps of Ramchairy et al. (2007) and Panjabi et al. (2008) were also merged in this study for the development of a revised VH map which harbored a wide array of 2169 genetic markers including 1,290 AFLPs, 70 RFLPs, 708 IP (intron polymorphism markers), 69 SSRs, 28 genic markers and 4 SCAR (sequenced characterized amplified region)/CAPS (cleaved amplified polymorphic sequence) markers, spanning a total genetic distance of 1902.9 cM. The first inclusive integrated map of *B. juncea* based on the two distinct DH populations was also developed by combining the revised VH map, and the TD map constructed in this study. This high-density integrated map spanned a genetic length of 1,927 cM and included 2,662 markers, of which 418 markers were identified as common markers serving as anchors for aligning the corresponding LGs of the VH and TD maps. A similar methodology for multi-environment QTL analysis for dissection of yield influencing traits as adopted by Ramchairy et al. (2007) was followed, and also included a reanalysis of the previously published QTL data. The study included the same set of traits which were also phenotyped in three dissimilar environmental conditions in India namely, Delhi, Bharatpur (Rajasthan) and Leh (Jammu and Kashmir).

QTL analysis based on the TD population identified 84 significant QTLs in the three environments distributed over 17 of the 18 LGs. A large number of environment specific (single-environment) QTLs (78.6%) were detected, while 21.4% QTLs were detected in at least two environments. Both the parents contributed an approximately equal number of positive alleles, with 44 QTLs receiving the alleles from Donskaja-IV and the remaining 40 QTLs from TM-4. Fifty-two loci were identified as major QTLs, describing more than 10% of the phenotypic variance explained (PVE) in at least one of the three experimental locations. In comparison with the 65 QTLs identified by Ramchairy et al. (2007), the reanalysis of the multi-environment phenotypic data of the VH population mapped 94 QTLs. These results closely followed the previous study by Ramchairy et al. (2007) and also detected thirty-seven major QTLs. A comparative assessment of the QTL analyses in the TD and VH populations indicated that the A sub genome as a whole contributed more QTLs than the B sub genome, thus verifying results obtained by Ramchairy et al. (2007). The LGA10 was identified as the most noteworthy group in both the maps and harbored 14 major QTLs for 12 traits in the TD map and 12 major QTLs for 11 traits in the VH map. This LG was also characterized by the most prominent clustering of loci signifying a multifunctional QTL region on this group. However, the cluster included increasing alleles of one trait in combination with trait-decreasing alleles of the other traits, both being inherited from the same parent.

The multi-environment QTL study performed by Ramchairy et al. (2007) and Yadava et al. (2012) were totally heterogeneous as they were based on mapping populations derived from dissimilar parental lines, utilized different sample sizes and markers. Yadava et al. (2012) therefore performed QTL meta-analysis (Arcade et al. 2004) to analyze and evaluate the QTL data obtained from the two sets of experiments for synthesizing the information into a single result to establish the most likely number of 'real' or 'meta-QTL' underlying a pool of QTL (Lü et al. 2008). Meta-analysis of QTLs was accomplished on nine LGs, including A10 with a prominent clustering of QTLs. A total of 187 QTLs subjected to meta-analysis were merged into 20 meta-QTLs. The highest resolution was detected in the LG A10 in which all the 54 constituent QTLs were merged forming a single-meta-QTL, showing a six-fold decrease in the confidence interval.

Epistasis has been reported to influence the genetic control and evolution of quantitative traits in several studies (Frankel and Schork, 1996; Lynch and Walsh, 1998; Wade, 2002; Carlborg et al. 2003, 2006; Kroymann and Mitchell-Olds, 2005; Malmberg et al. 2005;

Zeng, 2005). QTL analyses for identification of additive and epistatic loci performed by Yadava et al. (2012) presented the first extensive study on the role of epistasis in influencing the variation in the yield-related traits in two segregating DH populations of *B. juncea* (TD and VH). In TD population, eight sets of interacting QTLs including 14 loci were identified for six of the 12 evaluated traits. Following the classification described in Li et al. (2001), nine loci displayed significant additive effects and participated in four interacting sets of Type 1 and two interacting sets of Type 2. The phenotypic variances explained by epistatic QTLs were, however, low in the TD population. In comparison with, a total of 60 QTLs participated in 32 epistatic pairs in the VH population for 10 of the 12 traits studied. Of these 60 QTLs contributing to epistasis, 26 QTLs showed statistically significant additive effects among 13 interacting pairs of Type 1 and four interacting pairs of Type 2. In addition, some multi-locus interactions between the QTLs identified for the yield component traits in *B. juncea* were also identified which indicated the existence of higher order epistatic interactions similar to those reported in *B. napus* by Zhao et al. (2005).

22.3 Molecular Mapping of Seed Weight QTLs

Seed weight remains a key yield-related trait in *B. juncea*, and therefore, studies undertaken to dissect the genetics of this complex trait are precisely important with respect to both the commercial and nutritional aspects. Genetic advances achieved by conventional breeding approaches based on the already available natural variation for seed size/weight in *B. juncea* have resulted in the development of bold-seeded genotypes. Currently, the challenge for the genetic improvement of this trait however remains in the identification of genomic regions directly influencing seed weight. Seed weight has been shown to be a highly heritable trait (Ramchairy et al. 2007; Fanet al. 2010; Cai et al. 2012; Ding et al. 2012; Yadava et al. 2012; Yang et al. 2012; Bouchet et al. 2014; Li et al. 2014, 2019; Qi et al. 2014; Fu et al. 2015), and is under the influence of numerous QTLs with small effects and a comparatively smaller number of QTLs with large effects.

Different studies on genetic dissection of yield influencing QTLs in *B. juncea* have also included seed weight as a constituent trait (Lionneton et al. 2004; Mahmood et al. 2005; Ramchairy et al. 2007; Yadava et al. 2012) (Table 22.1). While there are numerous reports on the genetic analysis of seed weight in *B. rapa* (Li et al. 2013; Basnet et al. 2015), *B. napus* (Cai et al. 2012; Ding et al. 2012; Yang et al. 2012; Li et al. 2014, 2019; Fu et al. 2015; Geng et al. 2016) and in *B. juncea* (Ramchairy et al. 2007; Yadava et al. 2012), none of these encapsulate the large available genetic variation in the seed weight in any of these *Brassica* species.

Dhaka et al. (2017) therefore, made extensive efforts toward genetic dissection of seed weight in *B. juncea* (Table 22.1). The study assembled the broad genetic variability in seed weight into four distinct segregating DH populations for QTL analysis. The inbred parental lines of these populations belonged to the Indian (Pusa Jaikisan, Varuna and TM-4) and the east European (Heera, Donskaja-IV and EH-2) gene pools. The inbreds showed wide contrast in seed weight phenotype which ranged from as low as 2.0 g to as high as 7.6 g in separate field experiments. Two of these four populations, namely, VH and TD were previously described in Ramchairy et al. (2007) and Yadava et al. (2012), respectively. The other two populations, DE (developed from the F_1 of a Donskaja-IV/EH-2 cross) and EPJ (developed from the F_1 of a EH-2/Pusa Jaikisan cross) were reported in this study. Field experimental design for the EPJ consisted of multi-environment trials at Delhi, Bharatpur (Rajasthan) and Alwar (Rajasthan) during the crop season of winter, while the DE population was phenotyped in Delhi only in three consecutive crop growing seasons.

The QTL analysis for thousand seed weight in the four DH populations detected 65 QTLs which were distributed in the two sub genomes of *B. juncea*, with 39 and 26 QTLs in A and B sub genomes, respectively. The study also identified 16 major QTLs with PVEs greater than 10% of the total trait variation. The EPJ population which was derived from the parental lines, EH-2 and Pusa Jaikisan with a sharp contrast for the seed weight trait, revealed the maximum number of 25 QTLs, whereas the DE population developed from the inbreds Donskaja-IV and EH-2 deficient in trait variation of the seed weight phenotype, detected a total of 10 QTLs in three-year trials.

Meta-analysis was also undertaken to investigate the hypothesis that the seed weight QTLs detected in distinct environments but mapping to similar genomic locations on a linkage group represented different estimations of the same QTL. Six consensus thousand seed weight QTLs (*C-Tsw*) were identified after meta-analysis with 65 component QTLs of the trait, which were dispersed on the LGs A3, A7, A10 and B3 (Fig. 22.1). The study also deciphered a comparative allelic status of the six parental inbred lines, and the results indicated that among the six *B. juncea* lines studied, the Indian lines—Pusa Jaikisan or Varuna, possessed the most trait enhancing alleles in all of the six consensus QTLs. The QTL analysis in the VH and EPJ populations showed that extensive breeding efforts made for the genetic improvement of Indian types (Varuna and Pusa Jaikisan) have led to higher seed weight. The study also concluded that these results were important in breeding experiments for heterosis as the seed weight of the exotic lines can be improved by utilizing the positive alleles for seed weight existing in the Indian types.

22.4 Molecular Mapping of Flowering Time QTLs

Flowering, which demarcates a shift from the vegetative to the reproductive stage and represents a vital phase in the life history of a plant, is also a major factor for crop adaptation and seed yield in Brassica crops. Early flowering genotypes with short reproductive phases are more often adapted to drought as the plant genotype and its environment regulate the duration of a crop (Dingkuhn and Asch, 1999; Araus et al. 2002). Early flowering also leads to evasion of terminal heat stress experienced by plants, and also results in early maturity of the crop (Sahni et al. 2013). Therefore, short duration early flowering *B. juncea* genotypes will be tremendously relevant for multiple cropping (Sharma and Sardana 2013).

The earliest study on QTL analysis in four Brassica genomes to analyze the evolution of genes influencing flowering time was undertaken by Axelsson et al. (2001) (Table 22.1). The study used F_2 progeniesof *B. juncea*, *B. oleracea* and *B. rapa*, which were developed from crosses between an early flowering rapid-cycling male genotype (Williams and Hill 1986) and a female genotype with delayed flowering, in each of the three *Brassica* species. Interval mapping using MAPMAKER/QTL 1.1 (Lander et al. 1987; Lander and Botstein 1989; Lincoln et al. 1992) was used to analyze the genotypic and phenotypic data of the F_2 plants of the three mapping populations. To increase the power of QTL detection, 30 most extreme flowering time phenotypes were included in all the three populations (Lander and Botstein 1989). Three QTLs were identified on LGs J2, J3 and J18 of *B. juncea* which collectively described a PVE of 56% of the flowering time variation. In *B. rapa*, two QTLs explaining 38% of the phenotypic variation were detected on LGs R2 and R3. In *B. oleracea*, two QTLs on LGs O3 and O9 were identified which together explained 22% of the flowering time variation. It was observed that the rapid-cycling parent contributed alleles for reduction of flowering time at all QTLs in the three species. Comparative mapping among the *Brassica* species showed that three and six homoeologous copies, respectively, in the diploid species (*B. nigra*, *B. oleracea* and *B. rapa*) and the amphidiploid *B. juncea*, corresponded to a genomic region of *Arabidopsis thaliana* chromosome 5. The study revealed that the various QTLs for flowering period in the three *Brassica* species resulted from genome duplication. Genetic mapping of *Brassica* homologues of *A. thaliana* genes, *CONSTANS*

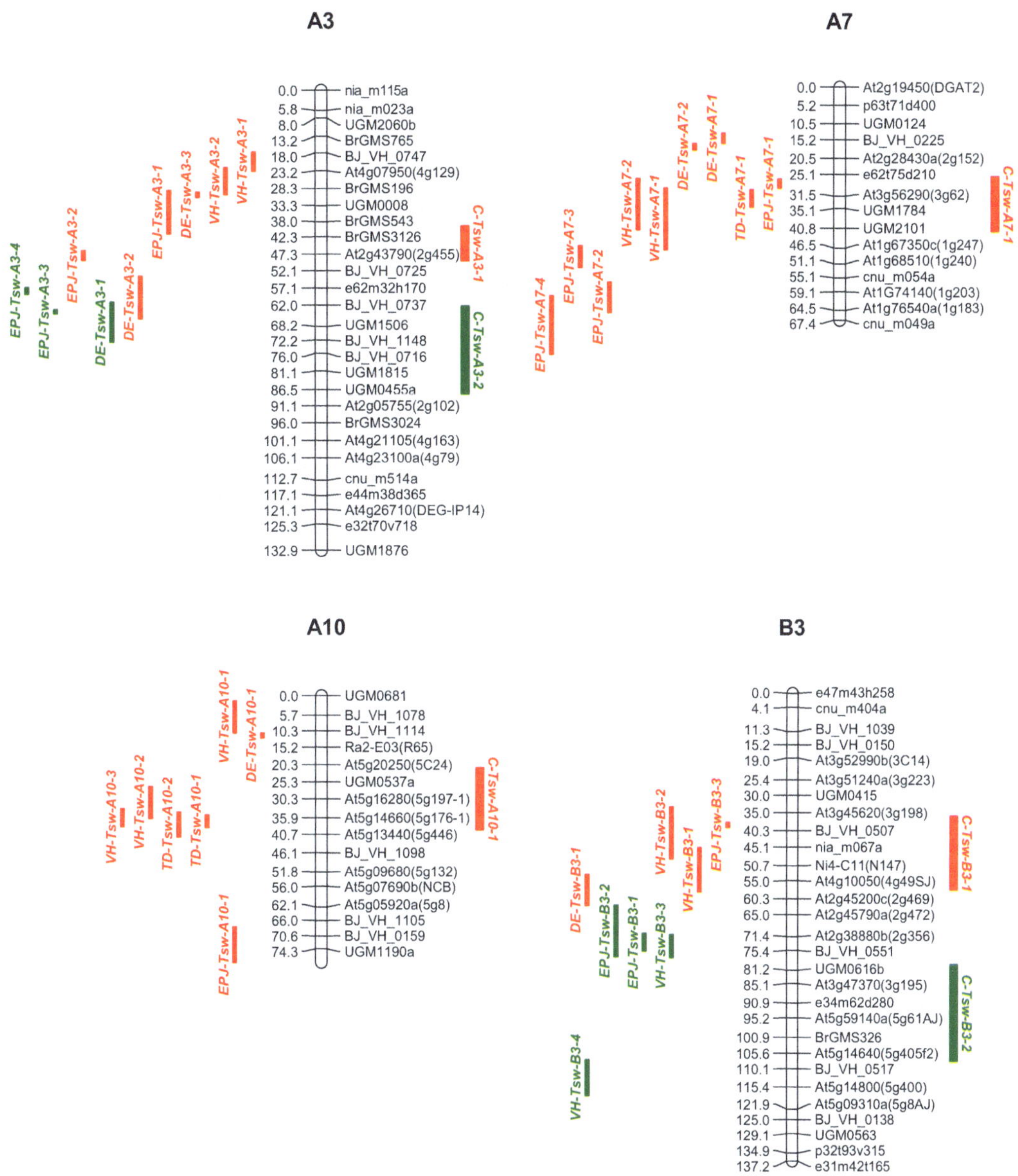

Fig. 22.1 Positions of six consensus thousand seed weight (*C-Tsw*) QTLs distributed on LGs A3, A7, A10 and B3 in *Brassica juncea* consensus map. Marker names are on the right of the LG bar and the positions in centiMorgans (cM) are on the left. The component QTLs identified in EPJ, DE, TD and VH populations integrating into the consensus QTLs are shown on the left side of the LG bar (Adapted fromFig. 2 in Dhaka et al. 2017)

(*CO*) and *Flowering Locus C* (*FLC*) (Putterill et al. 1995; Sheldon et al. 1999) revealed that duplications in an ancestral gene (*CO*) could have resulted in all the major QTLs detected in the different *Brassica* species.

Mahmood et al. (2007) analyzed QTLs for early maturity, flowering time and plant height by utilizing a DH population of *B. juncea*, derived from a previously described cross of two parental lines contrasting for days to maturity

(Thiagarajah and Stringam 1993; Mahmood et al. 2003). The efficacy of indirect selection for a primary trait (days to maturity) through selection of secondary trait(s) (days to first flowering, days to end of flowering, flowering period and plant height) was compared with direct selection for the primary trait by including the QTL information available for the two trait types. The study identified a stable major QTL on the LG B06 which simultaneously affected both the maturity and maturity-determinant traits (days to first flowering, days to end of flowering, flowering period and plant height). The results also indicated that days to first flowering, along with days to end of flowering were a potential good pointer of maturity in *B. juncea*. It was also concluded that only a restricted genetic advance for early maturity was possible through selection based on the phenotype of the maturity and/or maturity-determinant traits. However, with a combined phenotyping-based evaluation and the associated QTL data, the approximations of genetic advance for early maturity were found to be atleast threefold higher in comparison to the genetic advance based on phenotypic assessment only. Thus, marker-assisted selection was predicted to play an important role in the breeding of early maturing lines in *B. juncea*.

Genetic mapping of QTLs influencing flowering time in *B. juncea* have also been undertaken in other studies (Lionneton et al. 2004; Ramchairy et al. 2007; Yadava et al. 2012). A positive association was invariably observed in all these studies among the traits of days to flowering and plant height, which also showed colocalization of QTLs governing these traits. Yadava et al. (2012) concluded that yield improvement in the Indian types has been attained mainly through selection exercised for elongated siliques, larger seeds and higher oil content, as the positive alleles in QTLs for silique length, number of seeds in a silique, thousand seed weight and seed oil content were donated by Indian types, TM-4 and Varuna, while the east European lines, Donskaja-IV and Heera donated the positive alleles in the QTLs for plant height, number of primary branches, siliques per plant, silique density and flowering time.

An efficient plant canopy is expected to increase seed yield by an optimal mobilization of photosynthates. Breeders have been focusing on various plant architectural traits such as plant height, branches and flowering time for improving the seed yield. The plant growth in determinate genotypes stops on attaining a genetically primed vegetative biomass as their apical meristems are transformed into floral meristems, while indeterminate plants display a continued growth since the shoot apices are not transformed to floral meristems. Kaur and Banga (2015) have identified the locus governing determinacy in *B. juncea* and performed the genetic mapping of *Sdt1* gene associated with determinate plant growth habit. The authors isolated plants showing determinate growth habit from among the progeny plants of resynthesized *B. juncea* as a de novo phenotype. Indeterminacy was observed as a dominant trait since the F_1 progenies derived from crossing determinate genotypes with naturally indeterminate genotypes of *B. juncea*, were all indeterminate. Further genetic analysis in the F_2 and F_3 generations revealed that a single locus with recessive mode of inheritance controlled the trait. This locus for determinacy (Sdt_1) was mapped to LG 15 and lied between SSR loci, SJ6842 and Ni4-A10. There was substantial variation for plant height, flowering time and productivity among the determinate progenies which also exhibited high agronomic performance without any compensatory loss ascribed to a reduced silique density, productivity or oil content. The study proposed that the immense potential of the determinacy trait for major structural phenotypic modifications is not limited to *B. juncea* but could also be utilized in the genetic improvement of *B. napus*.

22.5 Future Scope of Works

Various biotic stresses like insect pests and diseases adversely affect the crop yield of *B. juncea*. Therefore, well-directed efforts are needed to breed such varieties for improving the yield, and for stabilizing the production in oilseed mustard. Approaches based on host resistance remain the most cost-effective and reliable means of disease control compared to the chemical control measures which are neither environmentally safe nor economically suitable. No released variety or hybrid of mustard (*B. juncea*) with resistance to white rust was earlier available in India. Three white rust resistant lines namely, Varuna-WRR-2, Pusa Bold-WRR2 and Rohini-WRR-2 which have been developed through a carefully designed marker-assisted backcross breeding scheme executed by the Centre for Genetic Manipulation of Crop Plants, University of Delhi (India) are currently available (Department of Biotechnology 2021). These lines have been extensively tested for resistance to the highly pathogenic strain of *Albugo candida* isolates collected from Bharatpur and Alwar (Rajasthan), Hisar (Haryana), Meerut (Uttar Pradesh), Morena (Madhya Pradesh), Pantnagar (Uttarakhand) and a site in Bihar. Prior to their release for public or commercial use, the three white rust resistant varieties were evaluated for their field performance and comparisons with their susceptible counterparts (Varuna, Pusa Bold and Rohini). The comparisons showed that the field performances of each of these three resistant varieties were at par with the normal susceptible lines. It is anticipated that more such resistance loci will be marked in the *B. juncea* genome which will allow durable resistance against white rust and other diseases of *B. juncea*.

Development of drought tolerant varieties with enhanced water use efficiency offers another promising prospect for improving crop yield under limited water availability. The genetic potential of *B. juncea*, as reproduced in terms of crop yields is exceedingly disturbed by the associated negative climatic influences which must be effectively reduced to enhance crop yields. Thus, with declining availability of water for agricultural purposes, *B. juncea* genotypes must be altered with an integral genetic facility which allows them to use water more efficiently. Moreover, for achieving the objective to meet the growing requirement for Brassica products, the cultivation of *B. juncea* will have to be extended to include areas prone to droughts, and thus, water use efficient genotypes for sustainable yields are crucial.

The growth and yield of the *B. juncea* varieties have been declining over the recent years due to various abiotic factors like frost, salinity and high temperatures at the time of sowing and maturity. It has been observed that plants briefly exposed to heat stress during the flowering and seed filling stages display accelerated senescence, diminished seed set and seed weight, and thus reduced yield (Morrison et al. 1989; Nanda et al. 1996; Siddique et al. 1999; Morrison and Stewart 2002). Development of genetically tailored genotypes can alleviate the losses in yield ascribed to these abiotic stresses.

For crossing the yield barrier and enhancing crop productivity, development of hybrids in *B. juncea* offers an opportunity for mobilizing the available genetic variability in the germplasm resources. The vast potential for heterosis breeding attributed to the two diverse gene pools has already been shown in *B. juncea* (Pradhan et al. 1993; Srivastava et al. 2001). Only a few hybrids are currently available which include DMH-1, NRCHB-506, Coral PAC 432 and Coral 437 (Directorate of Oilseeds Development 2017). Hence, in order to increase the yield potential of *B. juncea*, meticulously designed hybrid development programs are needed.

Finally, the availability of whole genome sequence information and the high throughput sequencing and genotyping platforms in conjunction with a large-scale phenotyping for various traits could be exploited to find gene(s) of interest governing yield and its attributing traits. Furthermore, the information on functionally characterized genes from related species such as *A. thaliana* could be used to isolate yield-related gene(s) in *B. juncea* through candidate gene approach. The integrated use of these approaches will be tremendously helpful in rapid and targeted breeding of *B. juncea* for yield and other economically important traits.

References

Adams MW (1967) Basis of component compensation in crop plants with special reference to the field bean, *Phaseolus vulgaris*. Crop Sci 7:505–510

Araus JL, Slafer GA, Reynolds MP, Royo C (2002) Plant breeding and drought in C3 cereals: what should we breed for? Ann Bot 89:925–940

Arcade A, Labourdette A, Falque M, Mangin B, Chardon F, Charcosset A, Joets J (2004) BioMercator: integrating genetic maps and QTL towards discovery of candidate genes. Bioinformatics 20:2324–2326

Axelsson T, Shavorskaya O, Lagercrantz U (2001) Multiple flowering time QTLs within several *Brassica* species could be the result of duplicated copies of one ancestral gene. Genome 44:856–864

Basnet RK, Duwal A, Tiwari DN, Xiao D, Monakhos S et al (2015) Quantitative trait locus analysis of seed germination and seedling vigor in *Brassica rapa* reveals QTL hotspots and epistatic interactions. Front Plant Sci 6:1032

Bouchet AS, Nesi N, Bissuel C, Bregeon M, Lariepe A, Navier H et al (2014) Genetic control of yield and yield components in winter oilseed rape (*Brassica napus* L.) grown under nitrogen limitation. Euphytica 199:183–205

Cai G, Yang Q, Yang Q, Zhao Z, Chen H, Wu J, Fan C, Zhou Y (2012) Identification of candidate genes of QTLs for seed weight in *Brassica napus* through comparative mapping among *Arabidopsis* and *Brassica* species. BMC Genet 13:105

Carlborg Ö, Kerje S, Schütz K, Jacobsson L, Jensen P, Andersson L (2003) A global search reveals epistatic interaction between QTL for early growth in the chicken. Genome Res 13:413–421

Carlborg Ö, Jacobsson L, Ahgren P, Siegel P, Andersson L (2006) Epistasis and the release of genetic variation during long-term selection. Nat Genet 38:418–420

Department of Biotechnology (2021) Annual report of the Department of Biotechnology (2020–21), Ministry of Science & Technology, Government of India. http://dbtindia.gov.in/about-us/annual-report/dbt. Accessed 10 May 2021

Dhaka N, Rout K, Yadava SK, Sodhi YS, Gupta V, Pental D, Pradhan AK (2017) Genetic dissection of seed weight by QTL analysis and detection of allelic variation in Indian and east European gene pool lines of *Brassica juncea*. Theor Appl Genet 130(2):293–307

Dhillon SS, Kumar PR, Gupta N (1992) Breeding objectives and methodologies. In: Labana KS, Banga S, Banga SK (eds) Breeding oilseed Brassicas. Springer, Berlin, Heidelberg, pp 8–20

Diepenbrock W (2000) Yield analysis of winter oilseed rape (*Brassica napus* L.): a review. Field Crops Res 67:35–49

Ding G, Zhao Z, Liao Y, Hu Y, Shi L, Long Y, Xu F (2012) Quantitative trait loci for seed yield and yield-related traits, and their responses to reduced phosphorus supply in *Brassica napus*. Ann Bot 109(4):747–759

Dingkuhn M, Asch F (1999) Phenological responses of *Oryza sativa*, *O. glaberrima* and inter-specific rice cultivars on a toposquence in West Africa. Euphytica 110:109–126

Directorate of Oilseeds Development (2017) Directorate of Oilseeds Development, Ministry of Agriculture & Farmers Welfare, Government of India. http://oilseeds.dac.gov.in/Variety/Rapeseed.aspx. Accessed 10 May 2021

Fan C, Cai G, Qin J, Li Q, Yang M, Wu J, Fu T, Liu K, Zhou Y (2010) Mapping of quantitative trait loci and development of allele specific markers for seed weight in *Brassica napus*. Theor Appl Genet 121(7):1289–1301

Frankel WN, Schork NJ (1996) Who is afraid of epistasis? Nat Genet 14:371–373

Fu Y, Wei D, Dong H, He Y, Cui Y, Mei J, Wan H, Li J, Snowdon R, Friedt W, Li X (2015) Comparative quantitative trait loci for silique length and seed weight in *Brassica napus*. Sci Rep 5:14407

Geng X, Jiang C, Yang J, Wang L, Wu X, Wei W (2016) Rapid identification of candidate genes for seed weight using the SLAF-Seq method in *Brassica napus*. PLoS One 11(1):e0147580

Gupta SK (2016) Brassicas. In: SK Gupta (ed) Breeding oilseed crops for sustainable production. Opportunities and constraints. Academic, Cambridge, pp 33–53

Jat RS, Singh VV, Sharma P, Rai PK (2019) Oilseed brassica in India: demand, supply, policy perspective and future potential. OCL 26:8

Kaur H, Banga SS (2015) Discovery and mapping of *Brassica juncea* Sdt_1 gene associated with determinate plant growth habit. Theor Appl Genet 128(2):235–245

Kroymann J, Mitchell-Olds T (2005) Epistasis and balanced polymorphism influencing complex trait variation. Nature 435:95–98

Lander ES, Green P, Abrahamson J, Barlow A, Daley M, Lincoln S, Newburg L (1987) MAPMAKER: an interactive computer package for constructing primary genetic linkage maps of experimental and natural populations. Genomics 1:174–181

Lander ES, Botstein D (1989) Mapping mendelian factors underlying quantitative traits using RFLP linkage maps. Genetics 121:185–199

Li N, Shi J, Wang X, Liu G, Wang H (2014) A combined linkage and regional association mapping validation and fine mapping of two major pleiotropic QTLs for seed weight and silique length in rapeseed (*Brassica napus* L.). BMC Plant Biol 14:114

Li N, Song D, Peng W, Zhan J, Shi J, Wang X et al (2019) Maternal control of seed weight in rapeseed (*Brassica napus* L.): the causal link between the size of pod (mother, source) and seed (offspring, sink). Plant Biotechnol J 17:736–749

Li X, Ramchiary N, Dhandapani V, Choi SR, Hur Y, Nou IS, Yoon MK, Lim YP (2013) Quantitative trait loci mapping in *Brassica rapa* revealed the structural and functional conservation of genetic loci governing morphological and yield component traits in the A, B, and C sub-genomes of Brassica species. DNA Res 20 (1):116

Li ZK, Luo LJ, Mei HW, Wang DL, Shu QY, Tabien R, Zhong DB, Ying CS, Stansel JW, Khush GS, Paterson AH (2001) Overdominant epistatic loci are the primary genetic basis of inbreeding depression and heterosis in rice. I Biomass Grain Yield. Genetics 158:1737–1753

Lincoln S, Daly M, Lander E (1992) Mapping genes controlling quantitative traits with MAPMAKER/QTL 1.1. 2nd edn. Technical Report of the Whitehead Institute, Cambridge, Massachusetts

Lionneton E, Ravera S, Sanchez L, Aubert G, Delourme R, Ochatt S (2002) Development of an AFLP-based linkage map and localization of QTLs for seed fatty acid content in condiment mustard (*Brassica juncea*). Genome 45:1203–1215

Lionneton E, Aubert G, Ochatt S, Merah O (2004) Genetic analysis of agronomic and quality traits in mustard (*Brassica juncea*). Theor Appl Genet 109:792–799

Lü X, Li X, Xie C, Hao Z, Ji H, Shi L, Zhang S (2008) Comparative QTL mapping of resistance to sugarcane mosaic virus in maize based on bioinformatics. Front Agric China 2:365–371

Lynch M, Walsh B (1998) Genetics and Analysis of quantitative traits. Sinauer Associates, Sunderland

Mahmood T, Ekuere U, Yeh F, Good AG, Stringam GR (2003) RFLP linkage analysis and mapping genes controlling the fatty acid profile of *Brassica juncea* using reciprocal DH populations. Theor Appl Genet 107:283–290

Mahmood T, Rahman MH, Stringam GR, Yeh F, Good AR (2005) Molecular markers for yield components in *Brassica juncea* – do these assist in breeding for high seed yield? Euphytica 144:157–167

Mahmood T, Rahman MH, Stringam GR, Yeh F, Good AG (2007) Quantitative trait loci forearly maturity and their potential in breeding for earliness in *Brassica juncea*. Euphytica 154:101–111

Malmberg RL, Held S, Waits A, Mauricio R (2005) Epistasis for fitness-related quantitative traits in *Arabidopsis thaliana* grown in the field and in the greenhouse. Genetics 171:2013–2027

Morrison MJ, McVetty PBE, Shaykewich CF (1989) The determination and verification of a baseline temperature for the growth of westar summer rape. Can J Plant Sci 69:455–464

Morrison MJ, Stewart DW (2002) Heat stress during flowering in summer Brassica. Crop Sci 42:797–803

Nanda R, Bhargava SC, Tomar DPS, Rawson HM (1996) Phenological development of *Brassica campestris, B. juncea*, *B. napus* and *B. carinata* grown in controlled environments and from 14 sowing dates in the field. Field Crops Res 46:93–103

Panjabi P, Jagannath A, Bisht NC, Lakshmi Padmaja K, Sharma S, Gupta V, Pradhan AK, Pental D (2008) Comparative mapping of *Brassica juncea* and *Arabidopsis thaliana* using Intron Polymorphism (IP) markers: homeologous relationships, diversification and evolution of the A, B and C Brassica genomes. BMC Genomics 9:113

Pradhan AK, Sodhi YS, Mukhopadhyay A, Pental D (1993) Heterosis breeding in Indian mustard (*Brassica juncea* L. Czern & Coss): analysis of component characters contributing to heterosis for yield. Euphytica 69:219–229

Pradhan AK, Gupta V, Mukhopadhyay A, Arumugam N, Sodhi YS, Pental D (2003) A high-density linkage map in *Brassica juncea* (Indian mustard) using AFLP and RFLP markers. Theor Appl Genet 106:607–614

Putterill J, Robson F, Lee K, Simon R, Coupland G (1995) The *CONSTANS* gene of Arabidopsis promotes flowering and encodes a protein showing similarities to zinc finger transcription factors. Cell 80:847–857

Qi L, Mao L, Sun C, Pu Y, Fu T, Ma C, Shen J, Tu J, Yi B, Wen J (2014) Interpreting the genetic basis of silique traits in *Brassica napus* using a joint QTL network. Plant Breed 133:52–60

Ramchairy N, Padmaja KL, Sharma S, Gupta V, Sodhi YS, Mukhopadhyay A, Arumugam N, Pental D, Pradhan AK (2007) Mapping of yield influencing QTL in *Brassica juncea*: implications for breeding of a major oilseed crop of dryland areas. Theor Appl Genet 115:807–817

Sahni S, Ganie SH, Narula A, Srivastava PS, Singh HB (2013) Ectopic expression of *Atleafy* in *Brassica juncea* cv. Geeta for early flowering. Physiol Mol Biol Plants 19(3):455–459

Sharma A, Mohapatra T, Sharma R (1994) Molecular mapping and character tagging *in Brassica juncea*. I. Degree, nature and linkage relationship of RFLPs and their association with quantitative traits. J Plant Biochem Biotechnol 3:85–89

Sharma P, Sardana V (2013) Screening of Indian mustard (*B. juncea*) for thermo tolerance at seedling and terminal stage. J Oilseed Brassica 4:61–67

Sheldon CC, Burn JE, Perez PP, Metzger J, Edwards JA, Peacock WJ, Dennis ES (1999) The *FLF* MADS box gene: a repressor of flowering in Arabidopsis regulated by vernalization and methylation. Plant Cell 11:445–458

Siddique KHM, Loss SP, Regan KL, Jettner RL (1999) Adaptation and seed yield of cool season gram legumes in Mediterranean environment of south-western Australia. Aust J Agric Res 50:375–387

Song K, Osborn TC, William PH (1988) Brassica taxonomy based on nuclear restriction fragment length polymorphisms (RFLP): I. Genome evolution of diploid and amphidiploid species. Theor Appl Genet 75:784–794

Srivastava A, Gupta V, Pental D, Pradhan AK (2001) AFLP-based genetic diversity assessment amongst agronomically important natural and some newly synthesized lines of *Brassica juncea*. Theor Appl Genet 102:193–199

Thiagarajah MR, Stringam GR (1993) A comparison of genetic segregation in traditional and microspore-derived populations of *Brassica juncea* L. Czern and Coss. Plant Breed 111:330–334

Upadhyay A, Mohapatra T, Pai R, Sharma R (1996) Molecular mapping and character tagging in Indian mustard (*Brassica juncea*) II. RFLP marker association with seed coat color and quantitative traits. J Plant Biochem Biotechnol 5:17–22

Wade MJ (2002) A gene's eye view of epistasis, selection and speciation. J Evol Biol 15:337–346

Williams PH, Hill CB (1986) Rapid-cycling populations of Brassica. Science (Washington, D.C.) 232:1385–1389

Yadava SK, Arumugam N, Mukhopadhyay A, Sodhi YS, Gupta V, Pental D et al (2012) QTL mapping of yield-associated traits in *Brassica juncea*: meta-analysis and epistatic interactions using two different crosses between east European and Indian gene pool lines. Theor Appl Genet 125:1553–1564

Yang P, Shu C, Chen L, Xu J, Wu J, Liu K (2012) Identification of a major QTL for silique length and seed weight in oilseed rape (*Brassica napus* L.). Theor Appl Genet 125:285–296

Zeng ZB (2005) Modeling quantitative trait loci and interpretation of models. Genetics 169:1711–1725

Zhao J, Becker HC, Zhang D, Zhang Y, Ecke W (2005) Oil content in a European Chinese rapeseed population: QTL with additive and epistatic effects and their genotype–environment interactions. Crop Sci 45:51–59 (Theor Appl Genet 123)

23 Advanced Breeding for Oil and Oil Cake Quality in *Brassica juncea*

Gurpreet Kaur, Sanjula Sharma, Sapna Langyan, Jasmeet Kaur, Pranjal Yadava, and S. S. Banga

Abstract

Mustard oil is a storehouse of essential fatty acids, linoleic acid and linolenic acid. However, its oil is still not considered good for health by many as it possesses long chain fatty acids, eicosenoic and erucic acid. These two fatty acids are associated with myocardial infractions, increase in cholesterol and cardiac lipdosis. After the extraction of oil, mustard meal is rich source of proteins with balanced aminogram, however, use of meal is restricted to only ruminants as it contains high amount of antinutritional compounds, glucosinolates (GLSs) which leads to goiter, reduce appetite and productivity of animals. Consequent upon realization of deleterious effects of erucic acid and GLSs, emphasis was given to develop rapeseed-mustard with modified seed quality. For the seed with modified fatty acids and GLSs, a new term Canola was coined. Rapeseed-mustard genotypes with low erucic acid (<2%) and GLSs (<30 umol/g of defatted meal) are referred to as Canola/'00'. Availability of zero erucic acid mutants, ZEM1 and ZEM2 and low GLSs genotype, BJ 1058 paved the way for developing Canola quality varieties in Indian mustard. Further, molecular mapping techniques have allowed identifying markers for marker-assisted breeding of these traits. Advanced sequencing technologies have enabled the sequencing of Arabidopsis and species of Brassica, including *B. rapa*, *B. nigra*, *B. oleracea*, *B. napus*, *B. juncea* and *B. carinata*. Emerging technologies, genomics, proteomics, transcriptomics and metabolomics have elucidated the function of genes involved in the biosynthetic pathways in Arabidopsis. Information available from Arabidopsis and related Brassica species has been used to dissect genes for quality improvements traits in *B. juncea*. In this chapter, success journey of quality improvement from conventional breeding to gene editing in *B. juncea* has been discussed.

G. Kaur (✉) · S. Sharma · J. Kaur · S. S. Banga
Oilseeds Section, Department of Plant Breeding and Genetics, Punjab Agricultural University, Ludhiana, Punjab 141004, India
e-mail: gurpreetkaur@pau.edu

S. Sharma
e-mail: drsanjula@pau.edu

J. Kaur
e-mail: jasmeetmann22-coapbg@pau.edu

S. S. Banga
e-mail: nppbg@pau.edu

S. Langyan
ICAR-National Bureau of Plant Genetic Resources, Pusa Campus, New Delhi 110012, India
e-mail: sapna@icar.gov.in

P. Yadava
Division of Plant Physiology, Indian Agricultural Research Institute, Pusa Campus, New Delhi 110012, India
e-mail: pranjal.yadava@icar.gov.in

C. Kole and T. Mohapatra (eds.), *The Brassica juncea Genome*, Compendium of Plant Genomes,
https://doi.org/10.1007/978-3-030-91507-0_23

23.1 Introduction

Mustard, *Brassica juncea*, is one of the most important oilseed crops of the world. Its oil is used for both edible and industrial purposes and seed cake left after oil extraction is used as cattle and poultry feed. Mustard seeds are made up of 35–45% oil, 17–25% proteins, 8–10% fibers, 10–12% other extractable substances and 6–10% moisture. Oil is made up of triacylglycerol of fatty acids (92–98%) with small quantities of tocopherols, phospholipids, glycolipids, sterols and tarpins. Brassica oil is rich in essential fatty acids (EFA) such as linoleic ($C_{18:2}$; omega 6) and linolenic ($C_{18:3}$; omega3). It also possesses high quantities of long chain fatty acid, erucic acid ($C_{22:1}$; 40–50%). Erucic acid is considered undesirable due to its link with myocardial infractions, high cholesterol and cardiac lipidosis as reported in animal models (Gopalan et al. 1974; Renard and MeGregor 1976; Ackman et al. 1977). Though similar linkages were never demonstrated in humans, many countries thought it prudent to restrict cultivation of rapeseed-mustard cultivars to <2% erucic acid for human consumption. A new term Canola was coined for rapeseed genotypes possessing 2% or less of erucic acid in the oil and <30 μmol/g of GLSs in the oil-free meal. The same term is now used for mustard genotypes with the defined seed quality attributes. The European Food Safety Authority (EFSA) has also recommended a tolerable daily intake (TDI) of erucic acid for humans to 7 mg/kg body weight per day (EFSA 2016). Meal left after the extraction of oil contains 35–40% protein, 14–15% carbohydrates, 10–12% of fiber, 6.8% moisture, 11.5% vitamins, 2–3% GLSs, tannins, sinapin and phytic acid. Antinutrients such as GLSs, tannins, sinapin, fibers and phytates decrease the quality of the seed cake. Meal value for feeding non-ruminants is lower for conventional rapeseed-mustard compared to soybean (Yoshie-Stark et al. 2006; Sadeghi and Bhagya 2009). Improving meal quality in mustard is important as it is a major export earner. India exported 11.13 lakh metric tons of rapeseed-mustard meal during 2020–2021 valued at foreign exchange equivalent of Rs. 2019 crores (SEA 2021). The mutants having genetic blocks in the biosynthetic pathway for synthesis of erucic acid and low GLSs has been used extensively utilized in hybridization programs to breed rapeseed-mustard genotypes with modified oil and meal. Genome sequence of *B. juncea* and elucidation of pathways along with functional characterization of genes have provided a better understanding of genetic mechanisms underlaying these traits. Further, the development of gene-based markers, transformation studies and gene-editing technologies has facilitated the development of mustard crop with desired traits.

23.2 Quality Considerations in Mustard

23.2.1 Oil

23.2.1.1 Fatty Acids

Fatty acid composition of oil determines the oil quality. Mustard oil contains high levels (50%) of erucic acid in the oil, which is unacceptable as per current international norms. Efforts to minimize erucic acid content involved introduction of genetic blocks in the biosynthetic pathway for fatty acid elongation. These efforts helped in the development of low erucic acid genotypes. Blocking of genetic elongation pathway is accompanied by elevated levels of oleic and linoleic acids. Though these are essential fatty acids, both linoleic and linolenic acids get easily oxidized. So, it is also imperative to breed for high oleic and low linoleic and linolenic acid varieties. Oil from these genotypes will be thermostable with better shelf-life.

23.2.1.2 Tocopherols

Tocopherols are natural lipophilic antioxidants synthesized in plastids. These exist in four forms: α-, β-, γ- and δ-tocopherols. α- and γ- are predominant tocopherols. α-tocopherols are the active form of vitamin E, whereas γ-tocopherols provide stability to oil. In Brassicas, both high levels of total tocopherols and better ratio of α- and γ-tocopherols are desirable. Oils of Brassica have a higher level of γ-tocopherol as compared to α-tocopherol.

23.2.2 Meal

Seed cake left after the extraction of oil contains protein, carbohydrates, fiber, GLSs, phytic acid and phenols etc. A good quality meal should have high digestibility with least number of antinutritional components such as GLSs, phytic acid, sinapins and fibers.

23.2.2.1 Proteins

The mustard meal contains high protein content (32–48%) with good balance of essential amino acids methionine and cystine. Mustard seed cake from traditional cultivars is mostly used for feeding ruminants. However, cake from Canola varieties can also be used to feed non-ruminants. *B. juncea* meal possesses more protein and less fiber than *B. napus*.

23.2.2.2 Glucosinolates

These are sulphonate doximethio esters of β-D-glucose derived from various amino acids. Approximately, 120 types of GLSs are known. These are classified into three main types namely, aliphatic, aromatic and indole GLSs. This classification is based on the amino acid origin of the side R chain which is derived from methionine, phenylalanine and tryptophan, respectively. The major GLSs present in Indian mustard are aliphatic types. Mustard samples from India and Pakistan include high levels of sinigrin and gluconapin (3-butenyl) whereas; mustard of Chinese and east European origin are rich in allyl isothiocyanates. Interestingly, its progenitors species, *B. rapa* and *B. nigra* possess additional types of GLS. *B. rapa* is rich in three types of GLSs; gluconapin (3-butenyl), glucobrassicin (3-indolylmethyl), gluco brassicanapin (4-pentenyl) whereas, sinigrin is the primary GLSs in *B. nigra*. The GLSs are hydrolyzed in the presence of enzyme myrosinase to yield many biologically active compounds such as oxazolidine-thiones, epithionitriles, nitriles, isothiocyanates and thiocyanatesare. The hydrolysis products are more harmful to the health of non-ruminants instead of the intact compounds. Thiocyanates bind iron from feed, whereas kidney as well as liver functions are affected by nitriles. Most mustard cultivars grown in Indian subcontinent contain high levels of GLSs (>100 μmol/g oil extracted seed meal) (Chauhan et al. 2007) and hence its seed meal is only used for feeding ruminants.

23.2.2.3 Phytic Acid

It is chemically known as myo-Inositol-1, 2, 3, 4, 5, 6-hexakis-dihydrogen phosphate. It is present in the form of mixed salts of Zn^{2+}, Ca^{2+}, Fe^{2+} and Mg^{2+} in rapeseed-mustard. It is also called phytate (Yiu et al. 1982). Phytates serve as the main form of phosphorus storage, accounting for 60–90% of the total phosphorous in the seed (Graf 1983; Raboy 2001). In Brassica oilseeds, 10% of the dry weight is contributed by the phosphorous in phytate. Defatted meal contains 2.0–5.6% phytate level, whereas protein isolates contain <1.0–9.8% of phytate (Uppström and Svensson 1980; Sadeghi et al. 2006; Sadeghi and Bhagya 2009).

23.2.2.4 Phenolic Compounds

Phenolics are secondary metabolites in rapeseed-mustard. Their content is ten to thirty folds higher than other oilseeds (Kozlowska et al. 1990). These were previously considered undesirable as these form complexes with vital enzymes, amino acids as well as other substances (Shahidi and Naczk 1992). However, phenolics are also prized for their antioxidant (Thiyam et al. 2006; Das et al. 2009; Khattab et al. 2010; Bala et al. 2011), antiviral, anti-inflammatory (Vuorela et al. 2005) as well as anticarcinogenic properties. Sinapic acid accounts for the antioxidant activity of phenolics (Brettonnet et al. 2010) and 80% of sinapic acid exists as choline esters known as sinapine (SP) (Thiyam et al. 2004) whereas only 16% of it exists as free acid (Kozlowska et al.1990).

23.2.2.5 Fiber

Crude fiber comprises non-digestible carbohydrates such as pentosans, lignin and cellulose of cell walls. Fiber in high amounts affects the digestibility of proteins as well as bioavailability of Mg and Zn minerals. High amount of indigestible fiber limits the use of rapeseed/mustard meal in the diets of monogastric animals (Slominski et al. 1994).

23.3 Biosynthetic Pathways

Improving quantity and quality of oil and meal requires a better understanding of the genes linked to their biosynthesis. A lot of efforts have already been made to modify the fatty acid profile and GLSs content in Brassica. Therefore, the biosynthesis of these two important traits has been discussed in detail.

23.3.1 Metabolic Routes for Biosynthesis of Fatty Acid and Oil Deposition in Brassica Seeds

23.3.1.1 Seed Oil Biosynthesis

Oil biosynthesis is compartmentalized into two organelles viz., plastids, and endoplasmic reticulum (ER). The process is initiated in plastids from acetyl-CoA, generated from pyruvate by the action of pyruvate dehydrogenase. Two main enzyme complexes engaged in the de novo fatty acid biosynthesis are acetyl-CoA carboxylase (ACCase) and fatty acid synthase (FAS). Plant ACCase is known to catalyze the carboxylation of acetyl-CoA (2C) to malonyl-CoA (3C) (Harwood 1988) and then to malonyl-ACP. Malonyl-CoA (3C) is an important intermediary metabolite in fatty acids synthesis. Next step of synthesis and elongation is a cyclic procedure catalyzed by four enzymes of FAS complex. These include (1) a condensation reaction catalyzed by ß-ketoacyl-ACP synthase, (2) a reduction reaction by hydroxyacyl-ACP reductase, (3) a dehydration reaction catalyzed by ß-hydroxy-ACP dehydratase followed by (4) reduction reaction catalyzed by enoyl-ACP reductase. After 7 cycles of these reactions, palmitoyl-ACP is formed as the final product, which is elongated by ß-ketoacyl-ACP synthase II (5) in combination with enzymes (2)–(4) to form stearoyl-ACP. Enzyme stearoyl-ACP desaturase (SAD) catalyzes desaturation of stearoyl-ACP to produce the unsaturated oleoyl-ACP (Harwood 1996). Palmitoyl-ACP, stearoyl-ACP and oleoyl-ACP together act as a source of all the fatty acids present in seed oils (Hernández et al. 2019). Acyl-ACP thioesterase (FatA/FatB) catalyzes the removal of ACP to generate palmitic ($C_{16:0}$), stearic ($C_{18:0}$) and oleic ($C_{18:1}$) acids. These free fatty acids can be retained in the plastid for integration inside plastid lipids via the prokaryotic pathway or these are exported to the ER for the synthesis of polyunsaturated fatty acids (PUFA), very long chain fatty acids (VLCFA) and lipid assembly via the eukaryotic pathway (Fig. 23.1). In Arabidopsis, around 62% of the de novo synthesized fatty acids are exported to ER, whereas, the remaining 38% are fused into the plastid lipids (Browse et al. 1986). Similarly, in oilseed rape major part (>90%) of the plastid synthesized fatty acids are channeled to acyl-CoA pool in ER for lipid assembly and the remaining 10% is directed to plastid lipid synthesis (Zaderimowski and Sosulski 1978).

23.3.1.2 Export of Free Fatty Acids and Biosynthesis of VLCFA in ER

The fatty acids reactivated to acyl-CoA are transported to ER through transporters, *FATTY ACID EXPORT1* (*FAX1*), *FAX2* and *FAX4* (Li et al. 2015). The biosynthesis of VLCFAs is catalyzed by a fatty acid elongase (FAE) complex, which is coded by *Fatty acid elongase 1 gene* (*FAE1*) (Millar and Kunst 1997; Denic and Weissman 2007; Haslam and Kunst 2013).

23.3.1.3 TAG Assembly in the ER

Two routes are involved in the TAG generation in ER. The first route involves the incorporation of different acyl-CoA at three positions of glycerol-3-phosphate (G3P) backbone to synthesize TAGs. The pathway occurs within the ER and encompasses four consecutive enzymatic steps. In the first two steps, successive transfer of fatty acids from acyl-CoA to sn-1 and sn-2 positions of G3P occurs. The first acylation at G3P is catalyzed by a glycerol-3-phosphate acyltransferase (GPAT; EC 2.3.1.15) to produce lysophosphatidic acid (LPA), which undergoes acylation by lysophosphatidic acid acyltransferase (LPAAT; EC 2.3.1.51) to generate phosphatidic acid (PA). Overexpression of LPAAT, increased TAG accumulation in various seeds of Arabidopsis, cotton, oilseed rape (Maisonneuve et al. 2010; Wang et al. 2017; Woodfield et al. 2019). In the third step, phosphatidic acid phosphohydrolase (PAP; EC 3.1.3.4) catalyzes the dephosphorylation of PA and releases de novo diacylglycerol (DAG). The de novo DAG generated is acylated to form TAG at sn-3 position by the action of diacylglycerol acyltransferase (DGAT; EC 2.3.1.20) (Liu et al. 2012) or could be utilized for the synthesis of membrane lipids such as phosphatidylcholine (PC) by the action of CDP-choline: diacylglycerol cholinephosphotransferase (CPT; EC 2.7.8.2) and phosphatidylcholine: diacylglycerol cholinephosphotransferase (Lu et al. 2009; Bates and Browse 2012). Another route leading to the formation of TAG in ER is acyl-CoA independent process and involves channeling of acyl groups into PC via the activity of lysophosphatidylcholine acyltransferase (LPCAT; EC 2.3.1.23), and then subsequent transfer of fatty acid from the sn-2 position of PCs (acyl donors) to generate TAG via the activity of phospholipid: diacylglycerol acyltransferase (PDAT).

23.3.1.4 The Biosynthesis of PUFAs in ER

A dominant flux of acyl moieties from acyl-CoAs pool stored in the ER are not instantly available for TAG biosynthesis, rather, these are first esterified to membrane lipid phosphatidylcholine (PC) and then undergo desaturation to form, linoleic and linolenic acids (Sasaki and Nagano 2004). The $C_{18:1}$-CoA undergo desaturation by Δ^{12} fatty acid desaturases (*FAD2*) to produce linoleic ($C_{18:2}$) subsequently converted to linolenic acid ($C_{18:3}$) by Δ^{15} fatty acid desaturases (*FAD3*) (Okuley et al. 1994).

23.3.1.5 Storage of TAGs in Oil Bodies

TAGs generated in ER are packaged with several proteins, oleosin, caleosin and steroleosins (Frandsen et al. 2001). They are stored as oil bodies in the cotyledons and the embryonic axis (Siloto et al. 2006).

23.3.2 GLSs Biosynthesis

The pathway for GLS biosynthesis is well characterized in *A. thaliana* along with regulatory genes and transcription factors (Redovniković et al. 2008; Burow et al. 2010; Sønderby et al. 2010). These can be divided into three separate steps. The first step involves the elongation of aliphatic and aromatic amino acids by the addition of methylene group to their side chains. In the second step, the core structure of GLSs is formed by reconfiguring amino acid moiety itself. In the third step, GLSs formed are modified by secondary transformations.

23.3.2.1 Amino Acid Elongation

Amino acid undergoes deamination, condensation, isomerization and transamination. Initiation of biosynthesis of methionine-derived aliphatic GLSs involves deamination of methionine to a 2-oxo acid catalyzed by branched-chain amino acid aminotransferase 4 (BCAT4), which is transported to chloroplasts via bile acid transporter 5 (BAT5). 2-oxo acid undergoes condensation with acetyl-CoA to form 2-malate derivative. The side chain of aliphatic GLSs is synthesized from methionine via successive insertion of methylene groups. This step is catalyzed by methylthioalkylmalate synthase1 (*MAM1*), *MAM2* and *MAM3* (Kroymann et al. 2001; Textor et al. 2007). In plants, amino acid can undergo as many as nine rounds of elongation. Allelic variation of the *MAM* genes determines side chain length and thus the profile of aliphatic GLSs (Kroymann et al. 2003; Heidel et al. 2006; Kumar et al. 2019).

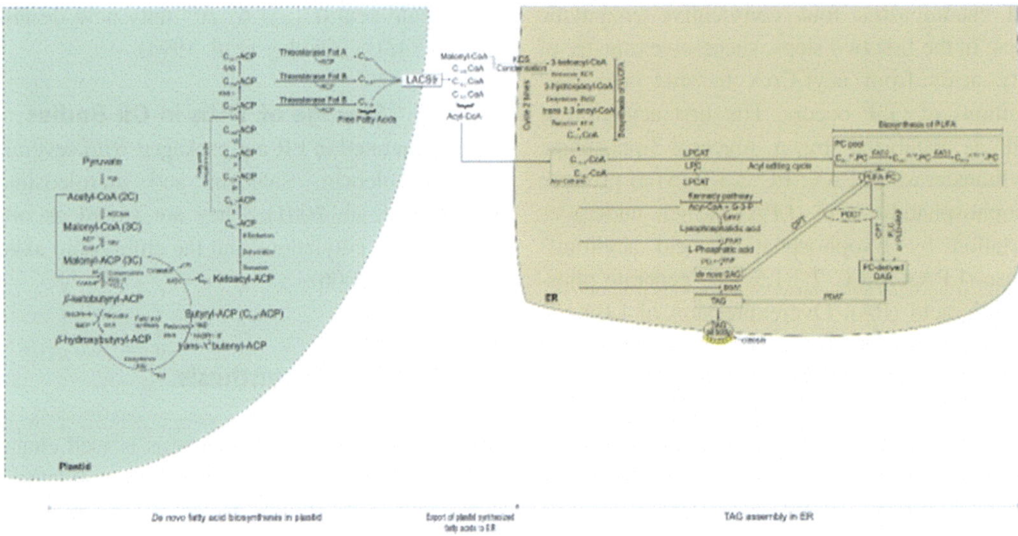

Fig. 23.1 Overview of major reactions involved in fatty acid and triacylglycerol synthesis in maturing embryos of Brassica

23.3.2.2 Core Structure Formation

Elongated amino acids are further metabolized to produce aldoximes by *CYP79* family genes. Seven genes: *CYP79A2*, *CYP79B2*, *CYP79B3*, *CYP79C1*, *CYP79C2*, *CYP79F1* and *CYP79F2* have been reported in Arabidopsis. *CYP79F1* as well as *CYP79F2* regulates the formation of aldoximes from short and long chain aliphatic amino acids respectively (Reintanz et al. 2001; Mikkelsen et al. 2002). Expression of *CYP79A1* increased total GLS (Nielsen et al. 2001). In contrast, *CYP79B2* and *CYP79B3* are involved in the biosynthesis of indole GLSs (Mikkelsen et al. 2009). *CYP79F1* mutants failed to synthesize short chain aliphatic GLSs, whereas reverse trend was observed in genotypes with overexpression of *CYP79F1*. Aldoximes are converted by *CYP83A1* into unidentified intermediates (previously assumed as aci-nitro compounds) (Grubb and Abel 2006). These intermediates undergo three steps; C-S lyase, glucosylation and sulfation to form intact GLSs. C-S lyase cleavage is catalyzed by the action of GSTF9, GSTF10, GGP1 and SUR1. Arabidopsis shows total absence of aliphatic as well as aromatic GLSs in C-S lyase mutant (Mikkelsen et al. 2004). Glucosylation catalyzed by UDP-glucosyltransferase 74C1 (UGT74) followed by sulfation catalyzed by sulphotransferase 17 (SOT17) to generate core GLS. Whereas, for the production of intact indole GLSs, glucosylation and sulfation is carried by the UGT74B1 and SOT16 (Reed et al. 1993; Sonderby et al. 2010; Hirschmann et al. 2017; Selmar 2018). Comparative analysis between Arabidopsis and *B. rapa* revealed 12 paralogs of sulfotransferases (Zang et al. 2008). The first four biosynthesis reactions take place in the chloroplast followed by last reaction in the cytosol, suggesting the role of shuttle transporters in the entire biosynthesis process (Klein et al. 2006).

23.3.2.3 Modifications

Core structure of both indole and aliphatic GLS is subjected to wide range of modifications. Flavin-monooxygenases (FMOs), alkenylhydroxalkyl-producing 2 (AOP2; At4g03060), AOP3 and 2-oxoglutarate-dependent dioxygenase (GS-OH) are involved in these modifications (Sønderby et al. 2010; Selmar 2018). 3-butenylglucosinolate and 2-propenylglucosinolate are produced by AOP2 in *B. oleracea* whereas biosynthesis of 4-(methylsulfanyl)-3-butenylglucosinolate occurs by GS-OH-related enzyme in radish (Hansen et al. 2008; Kakizaki et al 2017; Zuluaga et al. 2019).

Generally, plant metabolites, such as alkaloids, flavonoids as well as nucleic acids are biosynthesized by GS-OH family enzymes (Kawai et al. 2014). Hydroxy alkyl GLSs are formed by AOP3 (Kliebenstein et al. 2001) and it produces aliphatic GLSs with short chains with the help of MAM2 (Jensen et al. 2015). Almost all the genes involved in chain elongation of methionine as well as core structure pathways are expressed and co-ordinately regulated as revealed by the transcriptomic co-expression analysis (Hirai et al. 2007; Jensen et al. 2014). Indole GLSs undergo hydroxylation by CYP81F2, CYP86A7 and CYP71B26; leading to the formation of hydroxyindole GLSs (HIG). HIG are further metabolized into methoxyindole derivatives by indole GLSs methyltransferases 1 and 2 (IGMT1 and IGMT2) (Grubb and Abel 2006; Sonderby et al. 2010; Pfalz et al. 2011). Cytoplasmic protein phosphatase 2A regulatory subunit B′ (PP2A-B′) and IGMT5 can catalyze the methylation of 4-hydroxyindol-3-ylmethyl glucosinolate (4MI3G) and 1-hydroxyindol-3ylmethylglucosinolate respectively (Mikkelsen and Halkier 2003; Durian et al. 2016).

23.3.2.4 Breakdown of Glucosinolates

GLSs are broken down in the presence of enzyme myrosinase to many biologically active compounds. These include: oxazolidine-thiones, epithionitriles, nitriles, isothiocyanates and thiocyanatesare (Bones and Rossiter 1996, 2006; Rask et al. 2000). The hydrolysis products are considered harmful for non-ruminants as compared to the intact GLSs. The availability of iron is affected by thiocyanates whereas kidney as well as liver functions are affected by nitriles (Elfving 1980; Mithen 2001; Bellostas et al. 2007; Tripathi and Mishra 2007).

23.4 Variation and Inheritance of Seed Metabolites

23.4.1 Fatty Acids

B. juncea germplasm has been investigated for fatty acid as well as other quality traits (Iqbal et al. 2006; Kumar et al. 2013, 2018; Sharafi et al. 2015; Tahira and Saleem 2015; Ko et al. 2017). Average erucic acid content of tested genotypes ranged from 0.8 to 50%. These genotypes had oleic acid ranged between 11 and 18%, linoleic acid between 18 and 22% and linolenic acid from 11–27%. The fatty acid composition of low erucic acid genotypes differed significantly from conventional mustard germplasm. Oleic acid in these genotypes varied between 39 and 48%; linoleic acid ranged 28–38% and linolenic acid ranged between 15 and 20%. Increase of oleic acid in low erucic acid mustard was due to termination of chain elongation pathway from oleic acid to erucic acid. Germplasm variation indicated narrow germplasm base for oleic acid in mustard germplasm.

Low oleic and high linoleic acid is controlled by single gene with dominance gene action with minor additive effects in *B. juncea* (Potts et al. 1999; Woods et al. 1999). Dominance gene effect for linolenic acid has been reported by Kondra and Thomas (1975). Concentration of linolenic acid in mono galactosyl is controlled by cytoplasmic factors while nuclear cytoplasmic gene interactions are involved for concentration of linolenic acid in triacylglycerol (Diepenbrock and Wilson 1987). Involvement of only nuclear factors for linolenic acid was also reported (Pleines and Friedt 1989). Erucic acid is controlled by two genes with additive effects (Kirk and Hurlstone 1983; Tiwari 1995; Potts and Males 1999; Bhat et al. 2002; Gupta et al. 2004; Singh et al. 2015). Additive and dominance gene action for controlling erucic acid was also reported (Monpara and Jaisani 2000; Chauhan and Tyagi 2002).

23.4.2 Tocopherols

Variability for tocopherol content has been documented in rapeseed genotypes (Goffman and Becker 2002). The total tocopherol content varied from 182 to 367 mg/kg seed. Alpha-tocopherol content ranged between 63 and 157 mg/kg seed while gamma-tocopherol content varied from 114 to 211 mg/kg seed.

Whereas ratio of alpha-/gamma tocopherol contents varied from 0.36 to 1.23. Gupta et al. (2012) also observed a significant variation in tocopherol content in a recombinant inbred line (RIL) population of *B. juncea*. Tocopherol content varied between 6.43 and 91.27 mg/100 g defatted meal in *B. juncea* genotypes (Poonam et al. 2015). Gupta et al. (2012) reported that multiple alleles are responsible for controlling the tocopherol content.

23.4.3 Glucosinolates

Traditional Indian mustard genotypes possess high GLSs (90–120 u moles/g of defatted meal), whereas, Canola quality mustard genotypes possess GLSs < 30 umol/g defatted meal. Aliphatic GLSs were the major component representing 23.0–98.9% of the total GLS content in mustard whereas progoitrin (PRO), glucotropaeolin (TRO) and glucobarbarin (BAR) were present in relatively low concentrations (Assefa et al. 2019). Varied genetic control for aliphatic as well as aromatic GLSs has been reported. Indian gene pool mostly contain 2-propenyl and 3-butenyl GLSs, whereas, 2-propenyl GLSs constitute the major fraction in east European germplasm (Gland et al. 1981; Love et al. 1990b; Sodhi et al. 2002). GLSs content in Indian germplasm is controlled by 6–7 genes (Stringam and Thiagarajah 1995; Sodhi et al. 2002), whereas, digenic control was reported for east European germplasm (Love et al. 1990b). Crosses between low GLSs east European genotypes with high GLSs Indian genotype revealed the role of 6–8 loci to control GLSs content (Stringam and Thiagarajah 1995; Sodhi et al. 2002). Sporophytic inheritance along with 11 recessive alleles has been reported to control 3-butenyl, 4-pentenyl and 2-hydroxy-3-butenyl GLSs (Kondra and Stefansson 1970). Love et al. (1990b) reported multiple alleles having additive effects for 2-propenyl and 3-butenyl GLSs.

23.4.4 Phytic Acid and Crude Fiber

Very few studies have reported the variability for phytic acid and crude fiber content in *B. juncea*. Phytic acid varied between 2.92 and 3.86% in a set of tested genotypes while crude fiber ranged between 7.53 and 15.10% (Poonam et al. 2015).

23.4.5 Phenolics

Total phenols ranged between 17.11 and 21 mg/g in defatted meal of traditional genotypes of Indian mustard (Das et al. 2009; Sadeghi and Bhagya 2009; Bala et al. 2011), whereas, defatted meal from Canola varieties revealed 9.16 to 16.13 mg/g phenolics (Khattab et al. 2010).

23.5 Analytical Methods

Success at the modification of seed quality traits in rapeseed-mustard crops owe much to the availability of precise analytical methods to quantify fatty acid composition and GLSs content in the seed samples. The fatty acid composition can be determined by gas chromatography (Appelqvist 1968) or paper chromatography (Thies 1971). Fatty acid composition can be determined from whole seeds or half seeds as well. Half-seed technique has been especially useful for non-destructive estimation of fatty acids in segregating generations (Downey and Harvey 1963). Total GLSs can be easily estimated by estimation of total glucose in the crushed sample by glucose test paper (Lein 1970). This method is easy, economical and suitable for screening of large number of samples of segregating generations. However, this method has limitation of precision and subsequent confirmation can be done through HPLC and calorimetric methods. Individual fractionation of GLSs can be estimated on HPLC. Nowadays, near-infrared reflectance spectroscopy (NIRS) based equations for high throughput screening of oil, GLSs and fatty acids are available for the crop mustard (Kaur et al. 2016; Sen et al. 2018). Recently, Gohain et al. (2021) developed an NIRS equation for non-destructive estimation of fractionated components of GLSs.

23.6 Breeding Program

23.6.1 Breeding of Canola/'00' Quality Varieties in Mustard

Attempts toward quality improvement of rapeseed were initiated with identification of zero erucic acid mutants in *B. napus* (Stefansson et al. 1961) and *B. campestris* (Downey 1964) in Canada. The attempts at developing cultivars with improved quality were largely carried out in Canada, Australia and India. First really low erucic acid (<2%), ZEM1 and ZEM2 were identified in Australia (Kirk and Oram 1981). These genotypes were the pivots for breeding for low erucic acid in mustard programs first initiated in India during 1980s. ZEM1 was more widely used as donor for low erucic acid in the hybridization programs with a number of commercial varieties in India and elsewhere. Crossing followed by rigorous selection at single seed level (half seed method) led to the identification of many low erucic acid genotypes in India (Chauhan et al. 2002). Interspecific hybridization has also been made to improve quality traits (Agnihotri et al. 1995; Raney et al. 2003; Kaur et al. 2004). Efforts at breeding Canola quality mustard were initiated following availability of first low GLSs mustard genotype, BJ 1058 from Canada (Love et al. 1990a). BJ 1058 was developed by crossing *B. juncea* (non-allyl, 3-butenyl GLSs) with low GLSs *B. rapa* followed by single backcross with *B. juncea* and selfing of BC_1. Many canola quality varieties have been developed world wide utilizing these sources of low GLSs and erucic acid.

23.6.2 Breeding for Designer Fatty Acids

These '0'/'00' genotypes possess oleic acid (40–42%), linoleic acid (28–38%) and linolenic acid (15–20%). In contrast, '0'/'00' *B. napus* possesses a much higher level of oleic acid (>60%) with lesser amounts of linoleic and linolenic acid. Low oleic acid content in '0' erucic acid genotypes of *B. juncea* was attributed to strong desaturase activity of *FAD* genes located on B genome (Raney et al. 1995). Attempts have been made to introgress high oleic acid and low linolenic acid from *B. napus* in *B. juncea* using interspecific hybridization followed by one generation of backcrossing with *B. juncea* and selfing of BC_1 plants (Agnihotri and Kaushik 1998; Raney et al. 2003; Iqbal et al. 2011). Screening of single plants of BC_1 $F_{2\text{-}4}$ allowed identification of plants with low $C_{18:3}$ (5.5–6.2%) coupled with higher levels of $C_{18:1}$ (53.7%) and $C_{18:2}$ (47.2%) (Raney et al. 1995). High oleic acid trait from *B. napus* was reportedly transferred into *B. juncea* through hybridization with selected lines of *B. napus* (Agnihotri and Kaushik 1998; Kaur et al. 2004; Iqbal et al. 2006; Agnihotri et al. 2007). Increase in oleic acid in *B. juncea* type plants may be due to transfer of oleate desaturase suppressor gene from C genome of *B. napus*. However, high oleic acid level in these *B. juncea-B. napus* introgression lines was unstable and hence these stocks were not used in the breeding programs. In subsequent attempts, a low GLS *B. juncea* line J90-4253 could be improved by interspecific hybridization and plants with 60–70% oleic acid and ~11% linolenic acid could be developed (Raney et al. 2003). Mutagenesis was also used to broaden the genetic base of Brassica crops with improved fatty acid composition (Robbelen 1990). Mutants for extra high oleic acid were also reported in

B. napus (Auld et al. 1992; Rucker and Robbelen 1997; Bai et al. 2019) and *B. juncea* (Yao et al. 2005) and *B. juncea* (Bai et al. 2019). Application of doubled haploidy (DH) in combination with induced mutagenesis also provided opportunities to isolate mutants with improved oil and meal quality (Turner and Facciotti 1990; Wong and Swanson 1991; Auld et al. 1992; Huang 1992; Kott et al. 1996; Kott, 1998; Ferrie et al. 2008) with specific reference to high oleic acid. Exposure of microspore with UV resulted in mutants with reduced saturated fatty acids (Beaith et al. 2005). Variability for high erucic acid was observed in microspore derived mutants (Barro et al. 2001). Substitution of phenylalanine with a serine residue at 282 position in the polypepitide of *FAE1* through site directed mutagenesis resulted in low erucic acid *B. napus* cv. Westar (Katavic et al. 2002). However, mutant of this gene restored the elongase activity in yeast and synthesized VLCMFAs. Desaturation from oleic acid to linoleic acid is catalyzed by an enzyme delta 12 oleate desaturase coded by gene *FAD2*. Changes at single nucleotide level in the gene *FAD2* resulted in accumulation of high oleic acid. Substitution of single nucleotide resulted in synthesis of proline instead of lucine caused increase in accumulation of high oleic acid (Tanhuanpää et al. 1998). Similarly, Hu et al. (2006) also observed increase in accumulation of high $C_{18:1}$ due to substitution of single nucleotide from C to T and resulted in stop codon TAG. This led to premature termination of the peptide chain and mutated *FAD2* gene could not produce functional polypeptide and resulted in loss of desaturase activity and contributed towards high oleic acid.

23.7 Molecular Mapping and Marker Assisted Breeding

23.7.1 Fatty Acids

Different marker systems have been used to map genes associated with erucic acid biosynthesis. Restriction fragment length polymorphism (RFLPs) were first used to map two quantitative trait loci (QTLs) with unequal contribution in *B. juncea* (Cheung et al. 1998; Mahmood et al. 2003b). Whereas, only one QTL could be mapped for erucic acid by using amplified fragment length polymorphism (AFLP) markers (Lionneton et al. 2002). This was attributed to the fact that RFLP studies utilized high erucic acid parent from Indian gene pool (Cheung et al. 1998; Mahmood et al. 2003b), whereas high erucic acid parent belonging to east European gene pool was used in AFLP studies (Lionneton et al. 2002). Since, RFLP and AFLP markers are not breeder friendly as well as time consuming, attempts were also made to develop high throughput PCR-based markers. Their development was facilitated by the cloning and sequencing of two *FAE1* genes, *FAE1.1* and *FAE1.2*, representing two homologs each from low and high erucic acid mustard genotypes for the trait (Gupta et al. 2004). Sequencing of genes indicated the presence of single nucleotide polymorphisms (SNPs) for differences between high erucic acid (HEA) and low erucic acid (LEA) genotypes. *FAE1.1* and *FAE1.2* indicated four and three trait differentiating SNPs respectively. Mapping position of these SNPs coincided with earlier mapped QTLs for erucic acid on two linkage groups representing A and B genomes. These two QTLs contributed differently and each QTL mapped on A and B genome explained 60% and 40% of phenotypic variation, respectively. The identified seven SNPs could differentiate high vs low erucic acid as well as heterozygotes. These SNPs were later used to develop cleaved amplified polymorphic sequence (CAPS) markers (Saini et al 2016). CAPS are dominant markers and these cannot be used to identify recessive alleles in heterozygous state. Later on, sequence variation in the promoter regions of *FAE1* was exploited for the development of PCR-based markers (Saini et al. 2019). KASPar assays were also developed to differentiate HEA/LEA genotypes (Gill et al. 2021). These assays act as codominant markers and can identify recessive alleles in heterozygote state. Studies aimed at mapping of the fatty acids such as palmitic acid ($C_{16:0}$), stearic acid ($C_{18:0}$), oleic acid ($C_{18:1}$), linoleic acid ($C_{18:2}$) and linolenic acid ($C_{18:3}$) are summarized in Table 23.1.

Table 23.1 Summary of mapping of different fatty acids

Fatty acid	Population	Markers	QTLs	Linkage groups/ chromosomes	References
Palmitic acid ($C_{16:0}$)	F_1 derived DH population	AFLP	3 QTLs		Lionneton et al. (2002)
			E4M1_4	LG 2	
			E4M1_15	LG 3	
			E1M2_11	LG 6	
	RILs	AFLP	2QTLs		Singh et al. (2013)
			qpal-2.1	LG 2	
			qpal-5.1	LG 5	
Stearic acid ($C_{18:0}$)	F_1 derived DH population	AFLP	1 QTLs		Lionneton et al. (2002)
			E4M1_4	LG 2	
Oleic acid ($C_{18:1}$)	F_1 derived DH population	AFLP	2 QTLs		Lionneton et al. (2002)
			E4M1_4	L G 2	
			E1M2_6	LG 6	
	Reciprocal DH populations	RFLP	2 QTLs		Mahmood et al. (2003b)
			E1a	Chr A8	
			E1b	Chr B5	
Linoleic acid ($C_{18:2}$)	F_1 derived DH population	AFLP	1 QTLs E4M1_4	LG 2	Lionneton et al. (2002)
	RILs	AFLP	2QTLs		Singh et al. (2013)
			qlino-2.1	LG 2	
			qlino-3.1	LG 3	
	Reciprocal DH populations	RFLP	2 QTLs		Mahmood et al. (2003b)
			E1a	Chr A8	
			E1b	Chr B5	
Linolenic acid ($C_{18:3}$)	F_1 derived DH population	AFLP	2 QTLs		Lionneton et al. (2002)
			E4M1 4	LG 2	
			E6M2_1	LG 6	
	RILs	AFLP	3QTLs		Singh et al. (2013)
			qlin-1.1	LG 2	
			qlin-6.1	LG6	
			qlin-6.2		
	Reciprocal DH populations	RFLP	3 QTLs		Mahmood et al. (2003b)
			E1a	Chr A8	
			E1b	Chr B5	
			LN 2	Chr A4	
			LN 3	Chr A7	
			LN 4	Cht A5	

(continued)

Table 23.1 (continued)

Fatty acid	Population	Markers	QTLs	Linkage groups/ chromosomes	References
Ecosenoic acid ($C_{20:1}$)	F_1 derived DH population	AFLP	3 QTLs		Lionneton et al. (2002)
			E4M7_3	Chr 1	
			E3M6 4	Chr2	
			E4M1_4		
	RILs	AFLP	2QTLs		Singh et al. (2013)
			qeic-2.1	LG 2	
			qeic-5.1	LG 5	

23.7.2 Tocopherols

Seven QTLs for total tocopherol, α-, γ-tocopherol and α/γ ratio were mapped in winter oilseed rape (Marwede et al. 2005). These QTLs revealed additive and/or epistatic effects and explained 72% of the phenotypic variance in the rapeseed population. Gupta et al. (2015) mapped two QTLs in *B. juncea* through linkage analysis. These QTLs explained 28 and 19% phenotypic variations, respectively. Association also facilitated the identification of two genes, *BnaX.VTE3a* and *BnaA.PDS1.c* in *B. napus*. Allelic variations of these QTLs were also used to develop CAPS marker for use in marker-assisted selection (Fritsche et al. 2012).

23.7.3 Glucosinolates

Cheung et al. (1998) mapped QTLs for aliphatic GLSs using RFLP map developed following genotyping of a DH population. Two QTLs were mapped for 2-propenyl and another three for 3-butenyl GLSs. Mahmood et al. (2003a) identified four QTLs which were stable across different environments for aliphatic, 2-propenyl GLSs and 3-butenyl GLSs by using AFLP map developed on DH population. Four of these QTLs: GSL-A2a, GSL-A2b, GSL-F and GSL-B could be used for marker-assisted selection. Lionneton et al. (2004) mapped two major QTLs for both 2-propenyl and 3-butenyl GLSs by using an AFLP map. Most of these mapped QTLs were not related to any of the candidate genes involved in the GLSs biosynthetic pathways. Mapping of the genes involved in the biosynthetic pathway for aliphatic GLSs have been reported previously (Ramchiary et al. 2007; Bisht et al. 2009). Inheritance of GLSs is complex and is influenced by complexity of genetic background involved. Recurrent selection backcross (RSB) along with DH was utilized to identify QTLs which were consistent across generations. Mapping of QTLs in F_1DH and BC_4DH generations led to the identification of six 'true' QTLs for marker assisted selection (Ramchiary et al. 2007). Further, these QTLs were fine mapped by using sequence information of candidate genes from *A. thaliana* and *B. oleracea* (Bisht et al. 2009). Seventeen paralogs representing six gene families are known in *B. juncea*. However, paralogs of candidate gene belonging to three gene families GSL-ELONG, MYB 28 and GSL-ALK were linked to previously mapped QTLs. QTL intervals on chromosome J2 and J3 harbored candidate genes for GSL-ELONG family. Paralogs of transcription factor Myb 28 were mapped to QTL interval on chromosomes J3 and J17. Paralogs of *GSL-ALK* were mapped to QTL interval on chromosome J3. Three QTLs, *J2Gsl1*, *J3Gsl2* and *J9Gsl3* were found useful for marker assisted selection (Bisht et al. 2009).

23.8 Characterizing Genes Linked to Oil and Meal Quality

Several genomic techniques like high next generation sequencing (NGS) platforms, genome-wide association studies (GWAS) and candidate gene-based association mapping (CG-AM) are now used to elucidate novel genes and markers trait associations for various quality traits.

23.8.1 Fatty Acids

23.8.1.1 Erucic Acid

β-ketoacyl-CoA synthase (KCS) is a rate-limiting enzyme and it catalyzes the biosynthesis of a very long chain monounsaturated fatty acid (VLCMFA), erucic acid. Fatty acid elongase 1 (*FAE1*) encodes this enzyme (James et al. 1995). Mutation in *FAE1* genes causes dysfunctional enzyme activity (Katavic et al. 2002). Single copy of gene exists in diploid species, whereas, two copies have been identified in amphiploid species, *B. juncea* and *B. napus* (Gupta et al. 2004; Wu et al. 2008; Harper et al. 2012; Havlickova et al. 2018). *FAE1.1* and *FAE1.2* of *B. napus* co-located with positions of already mapped QTLs on A08 and C03 (Ecke et al. 1995; Qiu et al. 2006). *FAE1.1* was also mapped to chromosome A08 in *B. juncea*, but the physical position of *FAE1.2* is ambiguous. Using a yeast transgenic system, the role of missense mutations responsible for low erucic acid trait has been verified (Roscoe et al. 2001). Expression analysis of *FAE1* gene revealed that low erucic acid resulted from reduced expression level of *FAE1* gene in LEA genotypes of *B. rapa*, whereas no relationship was found between the expression level and erucic acid content in *B. napus* (Wu et al. 2008; Hu et al. 2009). Overexpression of *BjFAE1* gene increased erucic acid, whereas reverse was true for the downregulation mediated by antisense gene in *B. juncea* (Kanrar et al. 2006). Similarly, overexpression of *FAE1.1* in *B. napus* transgenic led to significant increase in the eicosenoic and erucic acid levels (Han et al. 2001).

23.8.1.2 Characterization of *FAE1*

FAE 1 gene was first isolated from *A. thaliana* using transposon tagging (James et al. 1995). Sequence information from Arabidopsis helped to clone and sequence *FAE1* gene from *B. napus* (Fourmann et al. 1998), *B. oleracea* (Das et al. 2002), *B. rapa* (Yan et al. 2015; Gill et al. 2021), *B. juncea* (Gupta et al. 2004; Gill et al. 2021), *B. nigra* (Gill et al. 2021). Sequence analysis of *FAE1* gene of 1521 bp indicated single exon encoding 507 aa in *B. napus*, *B. juncea*, *B. rapa* and *B. nigra* (Xu et al. 2010, Gill et al. 2021). These genes have been isolated from both HEA and LEA genotypes. Comparison of sequences between HEA and LEA genotypes reported several SNPs in *B. napus*, *B. rapa* and *B. juncea* (Wu et al. 2008; Wang et al. 2010; Yan et al. 2015; Gill et al. 2021). In *B. napus*, loss of activity of *FAE1* was caused by the deletion of two to four bases in C genome homologs (Fourmann et al. 1998; Wu et al. 2008), while, transition substitution (C/T) was indicated for a genome homolog (Han et al. 2001). Whereas, substitution have been reported for both the homologs of *B. juncea*. In *B. juncea*, four and three SNPs for genes *FAE1.1* and *FAE1.2* could differentiate low erucic from high erucic types (Gupta et al. 2004). However, Ghanevati and Jaworski (2001) have reported that the deletion in the *FAE1* led to the frame shift mutation which caused premature termination of translation, resulting in LEA *B. juncea*. In *B. rapa*, four substitution SNPs have been found for LEA. These were at positions 591 (G/A), 735 (C/T) and 968 (C/T) (Wang et al. 2010). Of these, only SNP at position 968 caused change in amino acid. Low erucic acid in *B. rapa* was also observed due to deletion of 24 bp in AT rich region of the promoter region (Yan et al. 2015). Single SNP resulted in substitution at position 282 bringing change in the amino acid from serine to phenylalanine which led to the development of LEA *B. napus* cv. Oro (Katavic et al. 2002). Targeting induced local lesions in genomes (TILLING), a reverse genetic strategy, identified 19 mutants of the *BnFAE1* from screening of 1344 M_2 plants. Of these, three M_3 plants showed reduced erucic acid (Wang et al. 2010).

23.8.1.3 Oleic Acid

Fatty acid desaturases (*FADs*) regulate compositions of unsaturated fatty acids. These insert double bonds into the fatty acid hydrocarbon chains (Xue et al. 2018). *FAD2* (oleoyl-PC Δ12-desaturase) and *FAD3* (delta-15 desaturase) catalyzes the desaturation of oleic acid (18:1) by forming double bond at the delta 12 position (Yang et al. 2012; Lee et al. 2016). QTLs for oleic acid have been mapped in *B. napus* using both linkage and association studies (Burns et al. 2003; Hu et al. 2006; Zhao et al. 2008; Smooker et al. 2011; Yan et al. 2011; Yang et al. 2012; Wen et al. 2015; Korber et al. 2016; Qu et al. 2017; Bao et al. 2018; Chen et al. 2018). However, similar studies are lacking in *B. juncea*. QTLs identified through linkage analysis were mapped on chromosomes A1, A5, C1 and C5 in *B. napus*. Most of the mapped QTLs for oleic acid in *B. napus* coincide with *FAD2* (Scheffler et al. 1997; Schierholt et al. 2000; Hu et al. 2006) and *B. rapa* (Tanhuanpää et al. 1996; Yang et al. 2012). Genome-wide analysis of haplotype blocks identified six haplotype regions associated with C18:1 and these regions were mapped to chromosomes A02, A07, A08, C01, C02 and C03. Further, whole-genome sequencing revealed three genes; *BnmtACP2-A02*, *BnABCI13-A02* and *BnECI1-A02* in the haplotype region of A02 and two genes, *BnFAD8-C02* and *BnSDP1-C02* in the C02 for association with oleic acid (Qu et al. 2017). Candidate gene-based association study identified three functional candidate genes on chromosomes A07, A08 and C03 for association with oleic acid. A8 region contributed to synthesis of oleic acid (Zafar et al. 2020). GWAS using 375 low erucic acid *B. napus* identified a novel QTL for *FAD2* on the A09 chromosome. This QTL explained 6.2–11.7% of phenotypic variation over three crop seasons. Introgression of this locus along with already mapped QTLs could enhance oleic acid content to 80%. Fine mapping of QTL allowed identification of three potential candidate genes and all three could be validated through gene expression assays. Closely linked marker, BnA129 has also been developed to facilitate marker assisted selection for oleic acid (Zhao et al. 2019). Expression studies in *B. napus* indicated that four homologs of *FAD2* genes, *BnFAD2-1*, *BnFAD2-2*, *BnFAD2-3* and *BnFAD2-4* showed tissue specific expression. *BnFAD2-1*(*Bna.A5.FAD2*), and *BnFAD2-2*(*Bna.C5.FAD2*) are constitutively expressed in all tissues, while expression of *BnFAD2-3*(*Bna.A1.FAD2*) and *BnFAD2-4*(*Bna.C1.FAD2*) is confined to seeds and roots (Okuzaki et al. 2018). The expected activities of fatty acid desaturases in plants, such as *FAD2* from *A. thaliana*, *Phaeodactylum tricornutum* and cotton, have been widely confirmed using yeast systems (Covello and Reed 1996; Domergue et al. 2003; Vrinten et al. 2005; Zhang et al. 2009). Functional analysis of *BnFAD2-1* and *BnFAD2-2* genes in yeast confirmed that SNPs caused loss-of-function mutations, thereby, restricting the conversion of oleic acid to linoleic acid and accumulating a significant amount of oleic acid (Long et al. 2018). Similarly, functional analysis of *BnaFAD2-A5* and *BnaFAE1-A8* genes using yeast system also validated these genes in formation of linoleic and erucic acid, respectively (Zafar et al. 2020). Expression of the four copies in yeast confirmed that *BnaFAD2.A1* is non-functional (Lee et al. 2013). Silencing of *FAD2* through antisense RNA constructs from cDNA of *B. rapa* led to a modified fatty acid profile and enhanced oleic acid content from 53 to 73% in *B. juncea*. Increase of linoleic and linolenic acid was also observed (Sivaraman et al. 2004). Simultaneously, RNA interference (RNAi) silencing of *BnaFAD2* along with *BnaFAE1* genes led to high oleic acid (80%) and low PUFA (9%) levels in *B. napus* (Peng et al. 2010). Huang et al. (2020) utilized clustered regularly interspaced short palindromic repeats/CRISPR associated protein 9 (CRISPR/Cas9) system to mutate all the orthologs of *BnaFAD2*. Resultant mutants had high oleic acid (80%) content. *BnaFAD2.A5* allele mutations generally produced more changes in fatty acid profile than the mutations on *BnaFAD2.C5* alleles. According to Stoutjesdijk et al. (2000), the oleic acid content of *B. juncea* has increased up to 73% along with the reduction in the $C_{18:2}$ and $C_{18:3}$ PUFA fractions <16% due to co-suppression of *FAD2*. RNAi induced suppression of *FAE1* and *FAD2* also led to increased oleic acid content in *B. napus* (Shi et al. 2017).

23.8.1.4 Characterization of *FAD* Genes

Four copies of *FAD2* have been identified in allotetraploid *B. napus* (Yang et al. 2012; Gacek et al. 2017). *BnaFAD2.A5* and *BnaFAD2.A1* correspond to *B. rapa* ($2n = 20$, AA), whereas, *BnaFAD2.C5* and *BnaFAD2.C1* to *B. oleracea* ($2n = 18$, CC) (Scheffler et al. 1997). Three orthologs each of *FAD3* are present in *B. rapa* and *B. oleracea* diploids while six orthologs are known in allotetraploid *B. napus* (Yang et al. 2012). *FAD2* and *FAD3* genes were characterized in *B. napus* as well as its progenitor species *B. rapa* and *B. oleracea*. A single nucleotide mutation (Hu et al. 2006) or four-base-pair (bp) insertion in *BnaA.FAD2.a* on A5 (Yang et al. 2012) and mutation in the *BnaFAD2.C5* (Wells et al. 2014) enhanced oleic acid content. Knockout mutants of three functional homologs of *FAD2* could enhance oleic acid up to 84–85% in *B. napus* (Wells et al. 2014). Single nucleotide mutation in the *FAD2* gene of *B. rapa* resulted in substitution of leucine with proline and increase in the $C_{18:1}$ content (Tanhuanpää et al. 1998). Cloning and sequence analysis of *FAD2* indicated that single nucleotide mutation, C to T in *FAD2* created a stop codon (TAG) and produced a truncated polypepitide chain causing dysfunctional *FAD2*. Sequence comparison of *FAD3* gene also revealed single nucleotide mutation. SNPs for both of these genes were converted into SNP markers using invader technology (Hu et al. 2006). Genome analysis of mutants for high oleic acid indicated 2-bp deletion in the upstream region in *FAD2-2* resulted in truncated polypeptide and led to increase in oleic acid (Lee et al. 2018).

23.8.2 Tocopherols

The content and composition of tocopherols (vitamin E) are important for oil stability and human health. Fritsche et al. (2012) utilized the established pathway for tocopherol biosynthesis in *A. thaliana* to carry out a targeted association mapping study of tocopherol content in 229 *B. napus* genotypes. Re-sequencing of 13 candidate genes facilitated the detection of nucleotide variation. Two major candidate genes, *BnaX. VTE3.a* and *BnaA.PDS1.c*, were identified in the region close to already mapped QTLs on chromosomes A7 and A10, respectively.

23.8.3 Glucosinolates

In most of the cases, QTLs mapped through linkage analysis did not show any association with candidate genes involved in the biosynthetic pathway of GLSs (Ramchiary et al. 2007). Bisht et al. (2009) used DNA sequences of genes involved in biosynthesis of aliphatic GLSs from *B. oleracea* and Arabidopsis to amplify the genes in low and high GLSs genotypes of *B. juncea*. Twenty paralogs were identified. Of these, 17 genes representing six gene families could be mapped—six paralogs of GSL-ELONG family gene (4 from A genome and 2 from B genome), four paralog belonging to GSL pro family (two each from A and B genome), four paralogs for GSL-ALK family (2 each from A and B genome). Other genes, *CYP79F1, CYP79C2, CYP83A1*, sulphotransferase and UDPgylcosyltransferase were also identified. Of these, candidate genes of gene family GSL-ELONG (*BjuA.GSL-ELONG.a BjuA.GSLELONG.c, BjuA.GSL-ELONG.d*), GSL-ALK(*BjuA.GSL-ALK.a*) and transcription factors (Myb 28) were co-mapped with known QTLs. GSL-ELONG and GSL-PRO are involved in the side chain elongation, whereas, GSL-ALK are involved in side chain modifications. Sønderby et al. (2007) also found that Myb 28 was located within a genomic region of QTL identified for aliphatic GLSs. Genome-wide association studies allowed identification of eight candidates genes (*HB16, SK1, AT2G35450*, (*CM1, JMT, LINC4, CYP81G1* and *MYB44*) on chromosomes A4, A5, A6, B3, B4 and B6 of *B. juncea* (Akhatar et al 2020). *SK1* is involved in a step of shikimate kinase (SK) synthesis and *HB16* controls carbon influxes from the central metabolism pool to a broad range of secondary metabolites. *CM 1* and *JMT* are involved in the biosynthesis of trytophan and indole GLSs. Transcription factor, *MYB44* is important for regulating the expression of transporter gene, *GTR2*. This gene is important for the transportation of GLSs from the apoplast into the phloem (Akhatar et al. 2020). Many orthologous have been identified in *B. napus* and its progenitors species through combination of syntenic and non-syntenic homology analyses with Arabidopsis (Lu et al. 2020). Genome-wide association studies identified many homologs in *B. rapa* (Zang et al. 2009). Role of HAG1 (Myb 28) was also demonstrated through associative transcriptomics analysis in *B. napus* (Hirai et al. 2007). Transcription factors regulate biosynthesis of both aliphatic as well as indole GLSs. Transcription factors, MYB28, MYB29 and MYB76 were found to be associated with aliphatic GLSs, whereas, altered tryptophan regulation 1 (ATR1/MYB34) regulate the indole GLSs. Transactivation potential of MYB28, MYB29 and MYB76 for synthesis of GLSs have been emphasized in *B. napus* (Gigolashvili et al. 2007a, 2008). Overexpression of these transcription factors led to increased expression of genes associated with aliphatic GLSs and repression of indole GLSs. This resulted in the increased accumulation of aliphatic GLS. Role of MYB28 and MYB29 was also confirmed by knockout mutants. MYB28 knockout mutants showed reduced levels of both short chain as well as long chain GLSs, whereas Myb29 mutants were only effective for short chain (Gigolashvili et al. 2007a; Sønderby et al. 2007). The double mutant for Myb28 and Myb29 lacked the biosynthesis of aliphatic GLSs, emphasizing the role of Myb28 and Myb29 as master regulators (Gigolashvili et al. 2007a; Hirai et al. 2007; Sønderby et al. 2007). Augustine et al. (2013) used RNAi-based transcription suppression of a HAG1 orthologs (BjMYB28) in *B. juncea* to replicate the low seed GLSs phenotype. A dominant overexpression allele, atr1D, altered the expression of CYP79B2, CYP79B3 and CYP83B1 involved in the production of indole GLSs. Overexpression of this allele enhanced the accumulation of indole GLSs whereas a loss of function mutant resulted in decreased levels of indole GLSs (Celenza et al. 2005). MYB51 and MYB122 have been found as positive regulators for biosynthesis of indole GLSs (Gigolashvili et al. 2007b).

23.9 Engineering Seed Quality Modifications

23.9.1 Transgenics

Genetic engineering enables precise transfer of novel genes beyond the gene pool. In Brassicas, transgenics for herbicide resistance, biotic stresses, male sterility and restorer gene for development of hybrids have been developed. Some of transgenics have also commercialized in advanced countries. Omics approaches have been successful in identification, characterization and functional validation of genes for quality-related traits in many crop plants. Genetic engineering has also facilitated the transfer of genes from other systems to improve seed and meal quality of Brassica crops (Table 23.2).

23.9.2 Gene Editing

CRISPR/Cas9 is now widely used for functional analysis and precise genome modifications for crop improvement. Brassicas are ancient polyploids and there is significant level of genome and gene redundancies. Despite these limitations, gene editing using CRISPR/Cas9 has been used successfully for seed quality modifications in oilseed Brassica species (Table 23.3).

23.10 Conclusions and Way Forward

Conventional plant breeding has been successful in developing varieties with Canola quality traits in Indian mustard. However, linkage drag associated with these characters is a key limiting factor for enhancing productivity potential of these genotypes. Integration of genetic, genomic, metabolic, molecular and molecular inferences coupled with the availability of whole genome assemblies and improved gene annotations has opened up new opportunities for engineering seed quality traits through simple plant breeding or gene editing techniques to produce Brassicas

Table 23.2 Summary of transgenic Brassicas developed for improved quality traits

Category	Targeted genes	Source	Quality modifications	References
Fatty acids: oleic acid	Antisense construct of *FAD2* gene	*B. rapa*	Increased oleic and decreased linoleic acid contents in *B. juncea*	Sivaraman et al. (2004)
Fatty acids: caprylate and caprate	Overexpression of Cn FatB2 thioesterase cDNA	*Cuphea hookeriana*	Increased levels of caprylate and caprate in *B. napus*	Dehesh et al. (1996)
Fatty acids: Stearic and erucic acid	Ml Fat B gene	*Madhuca longifolia*	Stearic acid increased upto 16 fold and erucic acid decreased upto 71% in *B. juncea*	Bhattacharya et al. (2015)
Fatty acids: Saturated fatty acids	*ADS 1 gene*	*Arabidopsis*	Decreased levels of saturated fatty acids of *B. juncea*	Yao et al. (2003)
Fatty acids: Erucic acid	Upregulation and downregulation of *BjFAE1* gene	*B. juncea*	Overexpression lead to 36% increase in erucic acid and downregulation caused 86% decrease in erucic acid	Kanrar et al. (2006)
	Sense and antisense constructs of *FAE* gene	*B. napus*	Sense construct increases the erucic acid levels and antisense construct decreases the erucic acid levels	Zebarjadi et al. (2006)
	RNAi knockdown of *FAE1*	*B. napus*	Greatly decreased levels of erucic acid, Largely increased levels of oleic acid, slightly increased PUFA	Shi et al. (2015)
GLSs	Silencing of MAM gene family	*B. napus* Canola	Reduction of aliphatic GLSs and total GLS content	Liu et al. (2011)
	Overexpression of STM gene	*B. napus*	Reduction in seed GLS levels	Elhiti et al. (2012)
	One of the seven and four of 12 GTR orthologs	*B. napus*	Decreased GLSs by 70% in *B. rapa* and *B. juncea*	Nour-Eldin et al. (2017)
	Targeted silencing of BjMYB28 transcription factor gene	*B. juncea*	Decreased GLSs	Augustine et al. (2013)
	Redirection of tryptophan into trytamine	*B. napus*	Lower levels of tryptophan derived indole GLSs	Chavadej et al. (1994)
Tocopherol	γ tocopherol methyl transferase cDNA	*Arabidopsis*	Increased seed alpha tocopherol	Yusuf and Sarin (2007)
Phyatate	Phy A and app A genes	*A. niger*	Lowers the phytate	Wang et al. (2013)
	Phy and Bar genes	–	Phytase activity 3–20 fold higher	Li et al. (2019)
Fiber	Suppression of COMT gene	*B. napus*	Decrease in ADF and NDF	Oraby et al. (2015)

for diverse uses as edible or industrial raw materials. We expect greater application of molecular markers to improve the precision of breeding programmes and rectify the yield limiting factors and deploy superior alleles of seed quality in more productive agronomic basis.

Table 23.3 Genome editing using CRISPR/Cas9 system in mustard for improvement of quality traits

Crop	Targeted gene	Trait improvement	References
B. napus	*fad2* genes	High oleic/low linoleic acid	Huang et al. (2020)
	fad2 genes	High oleic/low linoleic acid	Okuzaki et al. (2018)
	BnPAT2 BnLPAT5	High seed oil content	Zhang et al. (2019)
	BnTT8	Oil and protein content	Zhai et al. (2020)
	BnITPK	Reduced phytic acid	Sashidhar et al. (2020)
	BnSFAR4 BnSFAR5	High seed oil content	Karunarathna et al. (2020)
	BnTT2	Oil and fatty acid composition with higher linoleic acid and linolenic acid	Xie et al. (2020)
B. juncea and *B. napus*	GTR transporters	Reduced accumulation of seed GLSs	Nour-Eldin et al. (2017)
B. oleracea	*Myb28*	Reduced accumulation of glucoraphanin	Neequaye et al. (2020)

References

Ackman RG, Eaton CA, Sipos JC, Loew FM, Hancock D (1977) Comparison of fatty acids from high levels of erucic acid of RSO and partially hydrogenated fish oil in non-human primate species in a short-term exploratory study. Bibl Nutr Diet 25:170–185

Agnihotri A, Kaushik N (1998) Transgressive segregation and selection of zero erucic acid strains from intergeneric crosses of *Brassica*. Indian J Plant Genet Resour 11:251–255

Agnihotri A, Kaushik N, Singh NK, Raney JP, Downey RK (1995) Selection for better agronomical and nutritional characteristics in Indian rapeseed-mustard. In: Proceedings of the 9th international rapeseed congress, 4–7 July 1995. Cambridge, UK, GCIRC, pp 425–427

Agnihotri A, Sarkar G, Kaushik N, Gupta K, Prem D (2007) Introgression of high oleic acid in Indian mustard through interspecific hybridization. In: Proceedings of the 12th GCIRC international rapeseed congress, pp 188–190

Akhatar J, Singh MP, Sharma A, Kaur H, Kaur N, Sharma S, Bharti B, Sardana VK, Banga SS (2020) Association mapping of seed quality traits under varying conditions of nitrogen application in *Brassica juncea* L. Czern & Coss. Front Genet 11:744

Appelqvist LA (1968) Rapid methods of lipid extraction and fatty acid ester preparation for seed and leaf tissue with special remarks on preventing the accumulation of lipid contaminants. Ark Kenci 28:351–370

Assefa AD, Choi S, Lee JE, Sung JS, Hur OS, Ro NY, Lee HS, Noh J, Hwang A, Ko HC, Chung YJ (2019) Determination of glucosinolate contents in Brassica germplasm collections and inter-& intra-leaves distribution pattern using UPLC-MS/MS multiple reaction monitoring scan mode. BioRxiv 1:569889

Augustine R, Mukhopadhyay A, Bisht NC (2013) Targeted silencing of BjMYB28 transcription factor gene directs development of low glucosinolate lines in oilseed *Brassica juncea*. Plant Biotechnol J. https://doi.org/10.1111/pbi.12078

Auld DL, Heikkinen MK, Erickson DA, Sernyk JL, Romero JE (1992) Rapeseed mutants with reduced levels of polyunsaturated fatty acids and increased levels of oleic acid. Crop Sci 32:657–662

Bai S, Engelen S, Denolf P, Wallis JG, Lynch K, Bengtsson JD, Van Thournout M, Haesendonckx B, Browse J (2019) Identification, characterization and field testing of *Brassica napus* mutants producing high-oleic oils. Plant J 98:33–41

Bala M, Kumar S, Singh L (2011) Antioxidant potential of Rapeseed-Mustard seed meal extracts. Indian J Agri Biochem 24:55–59

Bao B, Chao H, Wang H, Zhao W, Zhang L, Raboanatahiry N, Li M (2018) Stable, environmental specific and novel QTL identification as well as genetic dissection of fatty acid metabolism in Brassica napus. Front Plant Sci 9:1018

Barro F, Fernandez-Escobar J, De La Vega M, Martin A (2001) Double haploid lines of *Brassica carinata* with modified erucic acid content through mutagenesis by EMS treatment of isolated microspores. Plant Breed 120:262–264

Bates PD, Browse J (2012) The significance of different diaclglycerol synthesis on plant oil composition and bioengineering. Front Plant Sci 3:147–152

Beaith ME, Fletcher RS, Kott LS (2005) Reduction of saturated fats by mutagenesis and heat selection in *Brassica napus* L. Euphytica 144:1–9

Bellostas N, Sorensen AD, Sorensen JC, Sorensen H, Gupta SK, Delseny M, Kader JC (2007) Genetic variation and metabolism of glucosinolates. Adv Bot Res 45:369–415

Bhat MA, Gupta ML, Banga SK, Raheja RK, Banga SS (2002) Erucic acid heredity in *B. juncea*—some additional information. Plant Breed 121:456–458

Bhattacharya S, Sinha S, Das N, Maiti MK (2015) Increasing the stearate content in seed oil of Brassica juncea by heterologous expression of MlFatB affects lipid content and germination frequency of transgenic seeds. Plant Physiol Biochem 96:345–355

Bisht NC, Gupta V, Ramchiary N, Sodhi YS, Mukhopadhyay A, Arumugam N, Pental D, Pradhan AK (2009) Fine mapping of loci involved with glucosinolate biosynthesis in oilseed mustard (*B. juncea*) using genomic information from allied species. Theor Appl Genet 118:413–421

Bones AM, Rossiter JT (1996) The myrosinase-glucosinolate system, its organisation and biochemistry. Physiol Plant 97:194–208

Bones AM, Rossiter JT (2006) The enzymic and chemically induced decomposition of glucosinolates. Phytochemi 67:1053–1067

Brettonnet A, Hewavitarana A, DeJong S, Lanari MC (2010) Phenolic acids composition and antioxidant activity of canola extracts in cooked beef, chicken and pork. Food Chem 121:27–933

Browse J, Warwick N, Somerville CR, Slack CR (1986) Fluxes through the prokaryotic and eukaryotic pathways of lipid synthesis in the '16:3' plant *Arabidopsis thaliana*. Biochemistry 235(1):25–31

Burns MJ, Barnes SR, Bowman JG, Clarke MHE, Werner CP, Kearsey MJ (2003) QTL analysis of an intervarietal set of substitution lines in Brassica napus: (i) Seed oil content and fatty acid composition. Heredity 90:39–48

Burow M, Halkier BA, Kliebenstein DJ (2010) Regulatory networks of glucosinolates shape Arabidopsis thaliana fitness. Curr Opin Plant Biol 13:348–353

Celenza JL, Quiel JA, Smolen GA, Merrikh H, Silvestro AR, Normanly J, Bender J (2005) The *Arabidopsis* ATR1 Myb transcription factor controls indolic glucosinolate homeostasis. Plant Physiol 137:253–262

Chauhan JS, Tyagi MK (2002) Inheritance of erucic acid content in two crosses of Indian mustard (*Brassica juncea* L.). SABRAO J Breed Genet 34:19–26

Chauhan JS, Tyagi MK, Kumar PR, Tyagi P, Singh M, Kumar S (2002) Breeding for oil and seed meal quality in rapeseed-mustard in India: a review. Agri Rev 23:71–92

Chauhan JS, Bhadauria VPS, Singh M, Singh KH, Kumar A (2007) Quality characteristics and their interrelationship in Indian rapeseed-mustard (*Brassica* sp) varieties. Indian J Agric Sci 77:616–620

Chavadej S, Brisson N, McNeil JN, De Luca V (1994) Redirection of tryptophan leads to production of low indole glucosinolate canola. Proc Natl Acad Sci USA 15(91):2166–2170

Chen F, Zhang W, Yu K, Sun L, Gao J, Zhou X., Peng Q, Fu S, Hu M, Long W, Pu H, Chen S, Wang X, Zhang J (2018) Unconditional and conditional QTL analyses of seed fatty acid composition in *Brassica napus* L. BMC Plant Biol 8:1–14

Cheung WY, Landry BS, Raney P, Rakow GFW (1998) Molecular mapping of seed quality traits in Brassica juncea L. Czern., and Coss. Acta Hort 459:139–147

Covello PS, Reed DW (1996) Functional expression of the extraplastidial Arabidopsis thaliana oleate desaturase gene (*FAD2*) in *Saccharomyces cerevisiae*. Plant Physiol 111:223–226

Das S, Roscoe TJ, Delseny M, Srivastava PS, Lakshmikumaran M (2002) Cloning and molecular characterization of the fatty acid elongase 1 (*FAE1*) gene from high and low erucic acid lines of Brassica campestris and Brassica oleracea. Plant Sci 162:245–250

Das R, Bhattacherjee C, Ghosh S (2009) Preparation of mustard (*Brassica juncea* L.) protein isolate and recovery of phenolic compounds by ultrafiltration. Indust Eng Chem Res 48:4939–4947

Dehesh K, Jones A, Knutzon DS, Voelker TA (1996) Production of high levels of 8:0 and 10:0 fatty acids in transgenic canola by overexpression of Ch FatB2, a thioesterase cDNA from *Cuphea hookeriana*. Plant J 9 (2):167–172

Denic V, Weissman JS (2007) A molecular caliper mechanism for determining very long-chain fatty acid length. Cell 130(4):663–677

Diepenbrock W, Wilson RF (1987) Genetic regulation of linolenic acid concentration in rapeseed 1. Crop Sci 27 (1):75–77

Domergue F, Spiekermann P, Lerchl J, Beckmann C, Kilian O, Kroth PG, Boland W, Zähringer U, Heinz E (2003) New insight into *Phaeodactylum tricornutum* fatty acid metabolism. Plant Physiol 131:1648–1660

Downey RK (1964) A selection of *Brassica campestris* containing no erucic acid in its seed oil. Can J Plant Sci 44:295

Downey RK, Harvey BL (1963) Methods of breeding for oil quality in rape. Can J Plant Sci 43:271–275

Durian G, Rahikainen M, Alegre S, Brosché M, Kangasjärvi S (2016) Protein phosphatase 2A in the regulatory network underlying biotic stress resistance in plants. Front Plant Sci 7:812

Ecke W, Uzunova M, Weibleder K (1995) Mapping the genome of rapeseed (*Brassica napus* L.). II. Localization of genes controlling erucic acid synthesis and seed oil content. Theor Appl Genet 91:972–977

EFSA (2016) Scientific opinion on erucic acid in feed and food. EFSA J 14:1–173

Elfving S (1980) Studies on the naturally occurring goitrogen 5-vinyl-2-thiooxazolidone. Metabolism and antithyroid effect in the rat. Ann Clin Res 28 (Suppl):1–47

Elhiti M, Yang C, Chan A, Durnin DC, Belmonte MF, Ayele BT, Tahir M, Stasolla C (2012) Altered seed oil and glucosinolate levels in transgenic plants overexpressing the *Brassica napus* shoot meristem less gene. J Exp Bot 63:4447–4461

Ferrie AMR, Taylor DC, MacKenzie SL, Rakow G, Raney JP, Keller WA (2008) Microspore mutagenesis of *Brassica* species for fatty acid modifications: a preliminary evaluation. Plant Breed 127:501–506

Fourmann M, Barret P, Renard M, Pelletier G, Delourme R, Brunel D (1998) The two genes homologous to Arabidopsis *FAE1* co-segregate with the two loci governing erucic acid content in *Brassica napus*. Theor Appl Genet 96:852–858

Frandsen GI, Mundy J, Tzen JT (2001) Oil bodies and their associated proteins, oleosin and caleosin. Plant Physiol 112:301–307

Fritsche S, Wang X, Li J, Benjamin S, Friedrich KO, Endrigkeit J, Leckband G, Dreyer F, Friedt W, Meng J, Jung C (2012) A candidate gene-based association study of tocopherol content and composition in rapeseed (*Brassica napus*) Plant Sci 3:129

Gacek K, Bayer PE, Bartkowiak-Broda I, Szala L, Bocianowski J, Edwards D, Batley J (2017) Genome-wide association study of genetic control of seed fatty acid biosynthesis in *Brassica napus*. Front Plant Sci 7:2062

Ghanevati M, Jaworski JG (2001) Active-site residues of a plant membrane-bound fatty acid elongase β-ketoacyl-CoA synthase, FAE1 KCS. Biochim Biophys Acta 1530:77–85

Gigolashvili T, Berger B, Mock HP, Müller C, Weisshaar B, Flügge UI (2007a) The transcription factor HIG1/MYB51 regulates indolic glucosinolate biosynthesis in *Arabidopsis thaliana*. Plant J 50:886–901

Gigolashvili T, Yatusevich R, Berger B, Müller C, Flügge UI (2007b) The R2R3-MYB transcription factor HAG1/MYB28 is a regulator of methionine-derived glucosinolate biosynthesis in *Arabidopsis thaliana*. Plant J 51:247–261

Gigolashvili T, Engqvist M, Yatusevich R, Müller C, Flügge UI (2008) HAG2/MYB76 and HAG3/MYB29 exert a specific and coordinated control on the regulation of aliphatic glucosinolate biosynthesis in *Arabidopsis thaliana*. New Phytol 177:627–642

Gill KS, Kaur G, Kaur G, Kaur J, Kaur Sra S, Kaur K, Gurpreet K, Sharma M, Bansal M, Chhuneja P, Banga SS (2021) Development and validation of Kompetitive Allele-Specific PCR Assays for erucic acid content in Indian Mustard [Brassica juncea (L.) Czern and Coss.]. Front Plant Sci 12:738805

Gland A, Röbbelen G, Thies W (1981) Variation of alkenyl glucosinolates in seeds of Brassica species. Z Pflanzenzüchtg 87:96–110

Goffman FD, Becker HC (2002) Genetic variation of tocopherol content in a germplasm collection of *Brassica napus* L. Euphytica 125(2):189–196

Gohain B, Kumar P, Malhotra B, Augustine R, Pradhan AK, Bisht NC (2021) A comprehensive vis-NIRS equation for rapid quantification of seed glucosinolate content and composition across diverse *Brassica* oilseed chemotypes. Food Chem 354:12957

Gopalan C, Krishnamurthi D, Shenolikar IS, Krishnamachari KAVR (1974) Myocardial changes in monkeys fed mustard oil. Ann Nutr Metab 16:352–365

Graf E (1983) Applications of phytic acid. Amer Oil Chem 60:1861–1867

Grubb CD, Abel S (2006) Glucosinolate metabolism and its control. Trends Plant Sci 11:89–100

Gupta V, Mukhopadhyay A, Arumugam N, Sodhi YS, Pental D, Pradhan AK (2004) Molecular tagging of erucic acid trait in oilseed mustard (*Brassica juncea*) by QTL mapping and single nucleotide polymorphisms in *FAE1* gene. Theor Appl Genet 108:743–749

Gupta S, Sangha MK, Kaur G, Atwal AK, Kaur P, Kumar H, Banga S, Banga SS (2012) Variation in glucosinolate and tocopherol concentrations in a germplasm collection of *Brassica juncea* (L.) Czern. + Coss. Biochem Anal Biochem 1:121

Gupta S, Sangha MK, Kaur G, Banga S, Gupta M, Kumar H, Banga SS (2015) QTL analysis for phytonutrient compounds and the antioxidant molecule in mustard (*Brassica juncea* L.). Euphytica 201:345–356

Han J, Luhs W, Sonntag K, Zahringer U, Borchardt DS, Wolter FP, Heinz E, Frentzen M (2001) Functional characterization of β-ketoacyl-CoA synthase genes from *Brassica napus* L. Plant Mol Biol 46:229–239

Hansen BG, Kerwin RE, Ober JA, Lambrix VM, Mitchell-Olds T, Gershenzon J, Halkier BA, Kliebenstein DJ (2008) A novel 2-oxoacid-dependent dioxygenase involved in the formation of the goiterogenic 2-hydroxybut-3-enyl glucosinolate and generalist insect resistance in Arabidopsis. Plant Physiol 148:2096–2108

Harper AL, Trick M, Higgins J, Fraser F, Clissold L, Wells R, Hattori C, Werner P, Bancroft I (2012) Associative transcriptomics of traits in the polyploid crop species *Brassica napus*. Nat Biotechnol 30:798–804

Harwood JL (1988) Fatty acid metabolism. Annu Rev Plant Physiol Plant Mol Biol 39:101–138

Harwood JL (1996) Recent advances in the biosynthesis of plant fatty acids. Bioch Biophys Acta (BBA) Lipids Lipid Metabolism 1301:7–56

Haslam TM, Kunst L (2013) Extending the story of very-long-chain fatty acid elongation. Plant Sci 210:93–107

Havlickova L, He Z, Wang L, Langer S, Harper AL, Kaur H, Broadley MR, Gegas V, Bancroft I (2018) Validation of an updated associative transcriptomics platform for the polyploid crop species *Brassica napus* by dissection of the genetic architecture of erucic acid and tocopherol isoform variation in seeds. Plant J 93:181–192

Heidel AJ, Clauss MJ, Kroymann J, Savolainen O, Mitchell-Olds T (2006) Natural variation in *MAM* within and between populations of *Arabidopsis lyrata* determines glucosinolates phenotype. Genetics 173:1629–1636

Hernández ML, Sicardo MD, Alfonso M, Martínez-Rivas JM (2019) Transcriptional regulation of stearoyl-acyl carrier protein desaturase genes in response to abiotic stresses leads to changes in the unsaturated fatty acids composition of olive mesocarp. Front Plant Sci 10:251

Hirai MY, Sugiyama K, Sawada Y, Tohge T, Obayashi T, Suzuki A, Araki R, Sakurai N, Suzuki H, Aoki K et al (2007) Omics-based identification of Arabidopsis Myb transcription factors regulating aliphatic glucosinolate biosynthesis. Proc Natl Acad Sci USA 104:6478–6483

Hirschmann F, Krause F, Baruch P, Chizhov I, Mueller JW, Manstein DJ, Papenbrock J, Fedorov R (2017) Structural and biochemical studies of sulphotransferase 18 from *Arabidopsis thaliana* explain its substrate specificity and reaction mechanism. Sci Rep 7:4160

Hu X, Sullivan-Gilbert M, Gupta M, Thompson SA (2006) Mapping of the loci controlling oleic and linolenic acid contents and development of *fad2* and *fad3* allele-specific markers in canola (*Brassica napus* L.). Theor Appl Genet 113:497–507

Hu Y, Wu G, Cao Y, Wu Y, Xiao L, Li X, Lu C (2009) Breeding response of transcript profiling in developing seeds of *Brassica napus*. BMC Mol Biol 10:49–65

Huang AHC (1992) Oil bodies and oleosins in seeds. Plant Physiol Plant Mol Biol 43:177–200

Huang H, Cui T, Zhang L et al (2020) Modifications of fatty acid profile through targeted mutation at *BnaFAD2* gene with CRISPR/Cas9-mediated gene editing in *Brassica napus*. Theor Appl Genet 133:2401–2411

Iqbal MCM, Weerakoon SR, Peiris PKD (2006) Variability of fatty acid composition in interspecific hybrids of mustard *Brassica juncea* and *Brassica napus*. Ceylon J Sci 35:17–23

Iqbal MCM, Weerakoon SR, Geethanjalie HDN, Peiris PKD, Weerasena OVDSJ (2011) Changes in the fatty acids in seeds of interspecific hybrids between *Brassica napus* and *Brassica juncea*. Crop Pasture Sci 62:390–395

James DW, Lim E, Keller J, Plooy I, Ralston E, Dooner HK (1995) Directed tagging of the Arabidopsis fatty acid elongation 1 (FAE1) gene with the maize transposon activator. Plant Cell 7:309–319

Jensen LM, Halkier BA, Burow M (2014) How to discover a metabolic pathway? An update on gene identification in aliphatic glucosinolate biosynthesis, regulation and transport. Biol Chem 395:529–543

Jensen LM, Kliebenstein DJ, Burow M (2015) Investigation of the multifunctional gene AOP3 expands the regulatory network fine-tuning glucosinolate production in Arabidopsis. Front Plant Sci 6:762

Kakizaki T, Kitashiba H, Zou Z, Li F, Fukino N, Ohara T, Nishio T, Ishida M (2017) A 2-xoglutarate-dependent dioxygenase mediates the biosynthesis of glucoraphasatin in radish. Plant Physiol 173:1583–1593

Kanrar S, Venkateswari J, Dureja P, Kirti PB, Chopra VL (2006) Modification of erucic acid content in Indian mustard (*Brassica juncea*) by up-regulation and down-regulation of the *Brassica juncea FAT TY ACID ELONGATION1 (BjFAE1)* gene. Plant Cell Rep 25:148–155

Karunarathna NL, Wang H, Harloff HJ, Jiang L, Jung C (2020) Elevating seed oil content in a polyploid crop by induced mutations in seed fatty acid reducer genes. Plant Biotech 18(11):2251–2266

Katavic V, Mietkiewska E, Barton DL, Giblin EM, Reed DW, Taylor DC (2002) Restoring enzyme activity in nonfunctional low erucic acid *Brassica napus* fatty acid elongase 1 by a single amino acid substitution. Eur J Biochem 269:5625–5631

Kaur G, Banga SK, Banga SS (2004) Introgression of desaturase suppressor gene(s) from *Brassica napus* L. to enhance oleic acid content in *Brassica juncea* L. Coss. In: Proceedings of the 4th international crop science congress. Brisbane, Australia

Kaur B, Sangha MK, Kaur G (2016) Calibration of NIRS for the estimation of fatty acids in *Brassica juncea*. J Amer Oil Chem Soc 93:673–680

Kawai Y, Ono E, Mizutani M (2014) Evolution and diversity of the 2-oxoglutarate-dependent dioxygenase superfamily in plants. Plant J 78:328–343

Khattab R, Eskin M, Aliani M, Thiyam U (2010) Determination of sinapic acid derivatives in canola extracts using high-performance liquid chromatography. J Am Oil Chem Soc 87: 147–155

Kirk JTO, Oram RN (1981) Isolation of erucic acid free lines of *Brassica juncea*: Indian mustard now a potential oilseed crop in Australia. J Aust Inst Agric Sci 47:51–52

Kirk JTO, Hurlstone CJ (1983) Variation and inheritance of erucic acid content in *Brassica juncea*. Z Pflanzenzuchtg 90:331–338

Klein M, Reichelt M, Gershenzon J, Papenbrock J (2006) The three desulfoglucosinolate sulfotransferase proteins in *Arabidopsis* have different substrate specificities and are differentially expressed. FEBS J 273:122–136

Kliebenstein DJ, Lambrix VM, Reichelt M, Gershenzon J, Mitchell-Olds T (2001) Gene duplication in the diversification of secondary metabolism: Tandem 2-oxoglutarate-dependent dioxygenases control glucosinolate biosynthesis in Arabidopsis. Plant Cell 13:681–693

Ko HC, Sung JS, Hur OS, Baek HJ, Lee MC, Luitel BP, Ryu KY, Rhee JH (2017) Variation in agronomic traits and fatty acid compositions of the seed oil in germplasm collection of *Brassica* spp. Kor J Plant Res 30(6):590–600

Kondra ZP, Stefansson BR (1970) Inheritance of major glucosinolates of rapeseed (*Brassica napus*) meal. Can J Plant Sci 50:643–647

Kondra ZP, Thomas PM (1975) Inheritance of oleic, linoleic and linolenic acids in seed oil of rapeseed (*Brassica napus*). Can J Plant Sci 55:205–210

Korber N, Bus A, Li JQ, Parkin IAP, Wittkop B, Snowdon RJ, Stich B (2016) Agronomic and seed quality traits dissected by genome-wide association mapping in *Brassica napus*. Plant Sci 7:386

Kott L (1998) Application of double haploid technology in breeding of oilseed *Brassica napus*. Agric Biotechnol News Info 10:69N-74N

Kott L, Wong R, Swanson E, Chen J (1996) Mutation and selection for improved oil and meal quality in

Brassica napus utilizing microspore culture. In: Jain SM, Sopory SK, Veilleux RE (eds) In vitro haploid production in higher plants, vol 2. Springer, New York, pp 151–167

Kozlowska H, Naczk M, Shahidi F, Zadernowski R (1990) In: Shahidi F (ed) Canola and rapeseed. Production, chemistry, nutrition and processing technology. Van Nostrand Reinhold, United States of America, pp 193–210

Kroymann J, Textor S, Tokuhisa JG, Falk KL, Bartram S, Gershenzon J, Mitchell-Olds T (2001) A gene controlling variation in Arabidopsis glucosinolate composition is part of the methionine chain elongation pathway. Plant Physiol 127:1077–1088

Kroymann J, Donnerhacke S, Schnabelrauch D, Mitchell-Olds T (2003) Evolutionary dynamics of an *Arabidopsis* insect resistance quantitative trait locus. Proc Natl Acad Sci USA 100:14587–14592

Kumar B, Pandey A, Singh SK (2013) Genetic diversity for agro-morphological and oil quality traits in Indian mustard (*Brassica juncea* L. Czern & Coss). Bioscan 8(3):771–775

Kumar A, Tudu VK, Pande A, Haider ZA, Rani V, Sah RP, Medha P, Yadav DK (2018) Use of biochemical indices as selection criteria for improvement in quality traits of Indian mustard (*Brassica juncea* L.). J Pharm Phytochem SP1 1831:1837

Kumar R, Lee SG, Augustine R, Reichelt M, Vassão DG, Palavalli MH, Allen A, Gershenzon J, Jez JM, Bisht NC (2019) Molecular basis of the evolution of methyl thio alkylmalate synthase and the diversity of methionine-derived glucosinolates. Plant Cell 31:1633–1647

Lee KR, Sohn SI, Jung JH, Kim SH, Roh KH, Kim JB et al (2013) Functional analysis and tissue differential expression of four FAD2 genes in amphidiploid *Brassica napus* derived from *Brassica rapa* and *Brassica oleracea*. Gene 531:253–262

Lee JM, Lee H, Kang S, Park WJ (2016) Fatty acid desaturases, polyunsaturated fatty acid regulation and biotechnological advances. Nutrients 8:23

Lee YH, Park W, Kim KS, Jang YS, Lee JE, Cha YL, Lee K (2018) EMS-induced mutation of an endoplasmic reticulum oleate desaturase gene (*FAD2-2*) results in elevated oleic acid content in rapeseed (*Brassica napus* L.). Euphytica 214:228

Lein KA (1970) Quantitative bestimmungsmethoden for samenglucosinolate in Brassica Arten and Ihre Anwendung in der Zuchtung Von Glucosinolataramen Raps. Z Pflanzenzuchtg 63:137–154

Li N, Gügel IL, Giavalisco P, Zeisler V, Schreiber L, Soll J, Philippar K (2015) FAX1, a novel membrane protein mediating plastid fatty acid export. PLoS Biol 13(2):e1002053

Li X, Zeng L, Li R, Wang C, Fan L, Huan Y, Yan H, Chen K, Fang X (2019) Marker-free lines of phytase-transgenic *Brassica napus* show enhanced ability to utilize phytate. Plant Cell Tiss Org Cult. https://doi.org/10.1007/s11240-019-01706-3

Lionneton E, Ravera S, Sanchez L, Aubert G, Delourme R, Ochatt S (2002) Development of an AFLP based linkage map and localization of QTLs for seed fatty acid content in condiment mustard (*Brassica juncea*). Genome 45:1203–1215

Lionneton E, Aubert G, Ochatt S, Merah O (2004) Genetic analysis of agronomic and quality traits in mustard (*Brassica juncea*). Theor Appl Genet 109:792–799

Liu Z, Hammerlindl J, Keller W et al (2011) *MAM* gene silencing leads to the induction of C3 and reduction of C4 and C5 side-chain aliphatic glucosinolates in *Brassica napus*. Mol Breed 27:467–478

Liu Q, Siloto RMP, Lehner R, Stone SJ, Weselake RJ (2012) Acyl-CoA:diacyl glycerol acyl transferase: molecular biology, biochemistry and biotechnology. Prog Lipid Res 51:350–377

Long W, Hu M, Gao J, Chen S, Zhang J, Cheng L, Pu H (2018) Identification and functional analysis of two new mutant BnFAD2 alleles that confer elevated oleic acid content in rapeseed. Front Genet 9:399

Love HK, Rakow G, Raney JP, Downey RK (1990a) Development of low glucosinolate mustard. Can J Plant Sci 70:419–424

Love HR, Rakow G, Raney JP, Downey RK (1990b) Genetic control of 2-propenyl and 3-butenyl glucosinolate synthesis in mustard. Can J Plant Sci 70:425–429

Lu C, Xin Z, Ren Z, Miquel M (2009) An enzyme regulating triacylglycerol composition is encoded by the ROD1 gene of arabidopsis. Proc Natl Acad Sci USA 106(44):18837–18842

Lu K, Snowdon R, Jiana Li (2020) Case study for trait-related gene evolution: Glucosinolates In: Liu S, Snowdon R, Chalhoub B (eds) The *Brassica napus* Genome 199–222. Springer, Cham

Mahmood T, Ekuere U, Yeh F, Good AG, Stringam GR (2003a) Molecular mapping of seed aliphatic glucosinolates in *Brassica juncea*. Genome 46:753–760

Mahmood T, Ekuere U, Yeh F, Good AG, Stringam GR (2003b) RFLP linkage analysis and mapping genes controlling the fatty acid profile of *Brassica juncea* using reciprocal DH populations. Theor Appl Genet 107:283–290

Maisonneuve S, Bessoule JJ, Lessire R, Delseny M, Roscoe TJ (2010) Expression of rapeseed microsomal lysophosphatidic acid acyltransferase isozymes enhances seed oil content in Arabidopsis. Plant Physiol 152(2):670–684

Marwede V, Gul MK, Becker HC, Ecke W (2005) Mapping of QTL controlling tocopherol content in winter rapeseed. Plant Breed 124:20–26

Mikkelsen MD, Halkier BA (2003) Metabolic engineering of valine-and isoleucine-derived glucosinolates in Arabidopsis expressing CYP79D2 from cassava. Plant Physiol 131:773–779

Mikkelsen MD, Petersen BL, Olsen CE, Halkier BA (2002) Biosynthesis and metabolic engineering of glucosinolates. Amino Acids 22:279–295

Mikkelsen MD, Naur P, Halkier BA (2004) Arabidopsis mutants in the C-S lyase of glucosinolates biosynthesis establish a critical role for indole-3-acetaldoxime in auxinhomeostasis. Plant J 37:770–777

Mikkelsen MD, Fuller V, Hansen BG, Nafisi M, Nielsen HB, Olsen CE, Halkier BA (2009) Controlled indole-3-acetaldoxime production through ethanol-induced expression of *CYP79B2*. Planta 229:1209–1217

Millar AA, Kunst L (1997) Very-long-chain fatty acid biosynthesis is controlled through the expression and specificity of the condensing enzyme. Plant J 12 (1):121–131

Mithen R (2001) Glucosinolates—biochemistry, genetics and biological activity. Plant Growth Regu 34:91–103

Monpara BA, Jaisani BG (2000) Inheritance of nutritional quality traits in Indian mustard. In: National seminar on oilseed and oils-research and development needs in the millennium, 2–4 Feb 2000. Hyderabad, India, p 323

Neequaye M, Stavnstrup S, Lawrenson T, Hundleby P, Troncoso-Rey P, Saha S, Harwood W, Traka M, Mithen R, Ostergaard L (2020) CRISPR-Cas9-mediated editing of myb28 impairs glucoraphanin accumulation of *Brassica oleracea* in the field. BioRxiv

Nielsen JK, Hansen ML, Agerbirk N, Petersen BL, Halkier BA (2001) Responses of the flea beetles *Phyllotreata nemorum*and *P. cruciferae* to metabolically engineered *Arabidopsis thaliana* with an altered glucosinolates profile. Chemoecology 11:75–83

Nour-Eldin HH, Madsen SR, Engelen S, Jørgensen ME, Olsen CE, Andersen S, Seynnaeve D, Verhoye T, Fulawka R, Denolf P, Halkier BA (2017) Reduction of antinutritional glucosinolates in Brassica oilseeds by mutation of genes encoding transporters. Nat Biotechnol 35(4):377–382

Okuley J, Lightner J, Feldmann K, Yadav N, Lark E (1994) Arabidopsis *FAD2* gene encodes the enzyme that is essential for polyunsaturated lipid synthesis. Plant Cell 6(1):147–158

Okuzaki A, Ogawa T, Koizuka C, Kaneko K, Inaba M, Imamura J, Koizuka N (2018) CRISPR/Cas9-mediated genome editing of the fatty acid desaturase 2 gene in *Brassica napus*. Plant Physiol Biochem 131:63–69

Oraby HF, Ramadan MF (2015) Impact of suppressing the caffeic acid *O*-methyltransferase (*COMT*) gene on lignin, fiber, and seed oil composition in *Brassica napus* transgenic plants. Eur Food Res Technol 240:931–938

Peng Q, Hu Y, Wei R, Zhang R, Guan C, Ruan Y, Liu C (2010) Simultaneous silencing of *FAD2* and *FAE1* genes affects both oleic acid and erucic acid contents in *Brassica napus* seeds. Plant Cell Rep 29:317–325

Pfalz M, Mikkelsen MD, Bednarek P, Olsen CE, Halkier BA, Kroymann J (2011) Metabolic engineering in Nicotiana benthamiana reveals key enzyme functions in Arabidopsis indole glucosinolate modification. Plant Cell 23:716–729

Pleines S, Friedt W (1989) Genetic control of linolenic acid concentration in seed oil of rapeseed (*Brassica napus* L.). Theor Appl Genet 78(6):793–797

Poonam P, Punetha H, Usha P, Pant AK, Om P (2015) Phytochemical screening and nutritional potentials of some high oil yielding exotic collections of *Brassica juncea*. Res J Chem Sci 5:66–71

Potts DA, Males DR (1999a) Inheritance of fatty acid composition in *Brassica juncea*. In: 10th International Rapeseed Congress, Canberra, Australia, p 9

Potts DA, Rakow GW, Males DR (1999b) Canola-quality *Brassica juncea*, a new oilseed crop for the Canadian prairies. In: Proceedings of the 10th international rapeseed congress, Canberra, Australia

Qiu D, Morgan C, Shi J, Long Y, Liu J, Li R, Zhuang X, Wang Y, Tan X, Dietrich E (2006) A comparative linkage map of oilseed rape and its use for QTL analysis of seed oil and erucic acid content. Theor Appl Genet 114:67–80

Qu C, Jia L, Fu F, Zhao H, Lu K, Wei L, Xu X, Liang Y, Li S, Wang R, Li J (2017) Genome-wide association mapping and Identification of candidate genes for fatty acid composition in *Brassica napus* L. BMC Genomics 18:1–17

Raboy V (2001) Seeds for a better future: 'low phytate' grains help to overcome malnutrition and reduce pollution. Trends Plant Sci 6:458–462

Ramchiary N, Bisht NC, Gupta V, Mukhopadhyay A, Arumugam N, Sodhi YS, Pental D, Pradhan AK (2007) QTL analysis reveals context-dependent loci for seed glucosinolate trait in the oilseed *Brassica juncea*: importance of recurrent selection backcross scheme for the identification of 'true' QTL. Theor Appl Genet 116:77–85

Raney JP, Rakow G, Olson TV (1995) Development of zero erucic, low linolenic *Brassica juncea* utilizing interspecific crossing. In: Murphy D (ed) Proceedings of the 9th international rapeseed congress. John Innes Centre, Norwich, pp 413–415

Raney JP, Olson TV, Rakow G, Ripley VL (2003) *Brassica juncea* with a canola fatty acid composition from an interspecific cross with *Brassica napus*. In: Proceedings of the 11th international rapeseed congress. Copenhagen, Denmark, pp 281–283

Rask L, Andréasson E, Ekbom B, Eriksson S, Pontoppidan B, Meijer J (2000) Myrosinase gene family evolution and herbivory defense in Brassicaceae. Plant Mol Biol 42:93–113

Redovniković Vana R, Glivetić T, Delonga K, Vorkapić-Furač J (2008) Glucosinolates and their potential role in plant. Period Biolog 110:297–309

Reed DW, Davin L, Jain JC, Deluca V, Nelson L, Underhill EW (1993) Purification and properties of UDP-glucose: thiohydroximate glucosyltransferase from *Brassica napus* L. seedlings. Arch Biochem Biophys 305:526–532

Reintanz B, Lehnen M, Reichelt M, Gershenzon J, Kowalczyk M et al (2001) *Bus*, a bushyrabidopsis*CYP79F1* knockout mutant with abolished synthesis of short-chain aliphatic glucosinolates. Plant Cell 13:351–367

Renard S, McGregor L (1976) Antithormbo genetic effects of erucic acid poor rapeseed oils in the rats. Rev Fr Cross 25:339–396

Robbelen G (1990) Mutation breeding for quality improvement a case study for oilseed crops. Mutat Breed Rev 6:1–43

Roscoe TJ, Lessireb R, Puyaubertb J, Renardc M, Delsenya M (2001) Mutations in the fatty acid elongation 1 gene are associated with a loss of L-ketoacyl-CoA synthase activity in low erucic acid rapeseed. FEBS Lett 492:107–111

Rucker B, Robbelen G (1997) Mutants of *Brassica napus* with altered seed lipid fatty acid composition. In: Proceedings of the 12th international symposium plant lipids. Kluwer Academic Publisher, Dordrecht, pp 316–318

Sadeghi MA, Bhagya S (2009) Effect of recovery method on different property of mustard protein. Dairy Food Sci 4:100–106

Sadeghi MA, AppuRao AG, Bhagya S (2006) Evaluation of mustard (*Brassica juncea*) protein isolate prepared by steam injection heating for reduction of antinutritional factors. Food Sci Technol 39:911

Saini N, Singh N, Kumar A, Vihan N, Yadav S, Vasudev S, Yadava DK (2016) Development and validation of functional CAPS markers for the *FAE* genes in *Brassica juncea* and their use in marker-assisted selection. Breed Sci 66:831–837

Saini N, Koramutla MK, Singh N, Singh S, Singh R, Yadav S, Bhattacharya R, Vasudev S, Yadava DK (2019). Promoter polymorphism in *FAE1.1* and *FAE1.2* genes associated with erucic acid content in *Brassica juncea*. Mol Breed 39:75

Sasaki Y, Nagano Y (2004) Plant acetyl-CoA carboxylase: structure, biosynthesis, regulation, and gene manipulation for plant breeding. Biosci Biotechnol Biochem 68(6):1175–1184

Sashidhar N, Harloff HJ, Potgieter L, Jung C (2020) Gene editing of three BnITPK genes in tetraploid oilseed rape leads to significant reduction of phytic acid in seeds. Plant Biotechnol J 18(11):2241–2250

Scheffler JA, Sharpe AG, Schmidt H, Sperling P, Parkin IAP, Lühs W et al (1997) Desaturase multigene families of *Brassica napus* arose through genome duplication. Theor Appl Genet 94:583–591

Schierholt A, Becker H, Ecke W (2000) Mapping a high oleic acid mutation in winter oilseed rape (*Brassica napus* L.). Theor Appl Genet 101:897–901

SEA (2021) https://seaofindia.com/export-of-oil-meals-april-2021/

Selmar D (2018) Biosynthesis of cyanogenic glycosides, glucosinolates and non-protein amino acids. Annu Plant Rev 15:92–181

Sen R, Sharma S, Kaur G, Banga SS (2018) Near-infrared reflectance spectroscopy calibrations for assessment of oil, phenols, glucosinolates and fatty acid content in the intact seeds of oilseed Brassica species J Sci Food Agri 98:4050–4057

Shahidi F, Naczk M (1992) An overview of the phenolics of canola and rapeseed: chemical, sensory and nutritional significance. J Am Oil Chem Soc 69:917–924

Sharafi Y, Majidi MM, Goli SAH, Rashidi F (2015) Oil content and fatty acids composition in *Brassica* species. J Food Prop 18:2145–2154

Shi J, Lang C, Wu X, Liu R, Zheng T, Hang D, Chen J, Wu G (2015) RNAi knockdown of fatty acid elongase1 alters fatty acid composition in *Brassica napus*. Biochem Biophy Res Comm 466(3):518–522

Shi J, Lang C, Wang F, Wu X, Liu R, Zheng T, Zhang D et al (2017) Depressed expression of FAE1 and FAD2 genes modifies fatty acid profiles and storage compounds accumulation in *Brassica napus* seeds. Plant Sci 263:177–182

Siloto RM, Findlay K, Lopez-Villalobos A, Yeung EC, Nykiforuk CL, Moloney MM (2006) The accumulation of oleosins determines the size of seed oil bodies in Arabidopsis. Plant Cell 18(8):1961–1974

Singh S, Mohapatra T, Singh R, Hussain Z (2013) Mapping of QTLs for oil content and fatty acid composition in Indian mustard [*Brassica juncea* (L.) Czern. and Coss.]. J Plant Biochem Biotechnol 22:80–89

Singh J, Yadava DK, Vasudev S, Singh N, Muthusamy V, Prabhu KV (2015) Inheritance of low erucic acid in Indian mustard [*Brassica juncea* (L.) Czern. and Coss.]. Indian J Genet 75:264–266

Sivaraman I, Arumugam N, Sodhi YS, Gupta V, Mukhopadhyay A, Pradhan AK, Burma PK, Pental D (2004) Development of high oleic and low linoleic acid transgenics in a zero erucic acid *Brassica juncea* L. Mol Breed 13:365–375

Slominski BA, Campbell LD, Guenter W (1994) Carbohydrates and dietary fibre components of yellow and brown seeded canola. J Agric Food Chem 42:704–707

Smooker AM, Wells R, Morgan C, Beaudoin F, Cho K, Fraser F, Bancroft I (2011) The identification and mapping of candidate genes and QTL involved in the fatty acid desaturation pathway in *Brassica napus*. Theor Appl Genet 122:1075–1090

Sodhi YS, Mukhopadhyay A, Arumugam N, Verma JK, Gupta V, Pental D (2002) Genetic analysis of total glucosinolate in crosses involving a high glucosinolate Indian variety and low glucosinolate line of *Brassica juncea*. Plant Breed 121:508–511

Sønderby IE, Hansen BG, Bjarnholt N, Ticconi C, Halkier BA, Kliebenstein DJ (2007) A systems biology approach identifies a R2R3 MYB gene subfamily with distinct and overlapping functions in regulation of aliphatic glucosinolates. PLoS ONE 19:2 (12):e1322

Sønderby IE, Geu-Flores F, Halkier BA (2010) Biosynthesis of glucosinolates–gene discovery and beyond. Trends Plant Sci 15:283–290

Stefansson BR, Hougen FW, Downey RK (1961) Note on the isolation of rape plants with seed oil free from erucic acid. Can J Plant Sci 41(1):218–219

Stoutjesdijk PA, Hurlestone C, Singh SP, Green AG (2000) High-oleic acid Australian *Brassica napus* and *B. juncea* varieties produced by co-suppression of endogenous Δ12-desaturases. Biochem Soc Trans 28:938–940

Stringam GR, Thiagarajah MR (1995) Inheritance of alkenyl glucosinolates in traditional and microspore-derived doubled haploid populations of *Brassica juncea* L. Czern and Coss. In: Proceedings of 9th international rapeseed congress, 4–7 July 1995. Cambridge, UK, pp 804–806

Tahira R, Saleem M (2015) Studies on Variability for Quality Traits, Association and Path Analysis in Raya (*Brassica juncea*) Germplasm. Intl J Agric Biol 30:17(2)

Tanhuanpää PK, Vilkki JP, Vilkki HJ (1996) Mapping of a QTL for oleic acid concentration in spring turnip rape (*Brassica rapa* ssp. *oleifera*). Theor Appl Genet 92:952–956

Tanhuanpää PK, Vilkki JP, Vihinen M (1998) Mapping and cloning of *FAD2* gene to develop allele-specific PCR for oleic acid in spring turnip rape (*Brassica rapa* ssp. *oleifera*). Mol Breed 4:543–550

Textor S, De Kraker JW, Hause B, Gershenzon J, Tokuhisa JG (2007) MAM3 catalyzes the formation of all aliphatic glucosinolate chain lengths in Arabidopsis. Plant Physiol 144:60–71

Thies W (1971) Schnelle und einfache Analysen der Fettsaurezusammen setzung in einzelnen Rape Kotyledonen I. Gaschromatographische und papier Chromatographise. Z Pflanzenzucht 65:181–202

Thiyam U, Kuhlmann A, Stöckmann H, Schwarz K (2004) Prospects of rapeseed oil byproducts with respect to antioxidative potential. CR Chimie 7:611–616

Thiyam U, Stöckmann H, Felde TZ, Schwarz K (2006) Antioxidative effect of the main sinapic acid derivatives from rapeseed and mustard oil by-products. Eur J Lipid Sci Technol 108:239–248

Tiwari AS (1995) Improved quality of oil and meal in Indian mustard. In: Rapeseed today and tomorrow, 9th international rapeseed congress. Cambridge, UK, pp 434–436

Tripathi MK, Mishra AS (2007) Glucosinolates in animal nutrition: a review. Anim Feed Sci Technol 132:1–27

Turner J, Facciotti D (1990) High oleic acid *Brassica napus* from mutagenized microspores. In: McFerson JR, Kresovich S, Dwyer SG (eds) Proceedings of the 6th crucifer genetics workshop. Geneva, New York, p 24

Uppström B, Svensson R (1980) Determination of phytic acid in rapeseed meal. J Sci Food Agric 1:651–656

Vrinten P, Hu Z, Munchinsky MA, Rowland GR, Qiu X (2005) Two *FAD3* desaturase genes control the level of linolenic acid in flax seed. Plant Physiol 79:87

Vuorela S, Kreander K, Karonen M, Nieminen R, Hamalainen M, Laitinen L, Galkin A, Salminen JP, Moilanen E, Pihalaja K, Vuorela H, Vuorela P, Heinonen M (2005) Preclinical evaluation of rapeseed, raspberry and pine bark phenolics for health related effects. J Agric Food Chem 53:5922–5931

Wang N, Shi L, Tian F, Ning H, Wu X, Long Y, Meng J (2010) Assessment of FAE1 polymorphisms in three Brassica species using EcoTILLING and their association with differences in seed erucic acid contents. BMC Plant Biol 10:137

Wang Y, Ye X, Ding G, Xu F (2013) Overexpression of phyA and appA genes improves soil organic phosphorus utilisation and seed phytase activity in *Brassica napus*. PLoS ONE 8(4):e60801

Wang N, Ma J, Pei W, WuM, Li H, Li X, Yu Shuxun, Zhang J, Yu J (2017) A genome-wide analysis of the lysophosphatidate acyltransferase (LPAAT) gene family in cotton: organization, expression, sequence variation, and association with seed oil content and fiber quality. BMC Genomics 18(1):1–18

Wells R, Trick M, Soumpourou E, Clissold L, Morgan C, Werner P, Gibbard C, Clarke M, Jennaway R, Bancroft I (2014) The control of seed oil polyunsaturate content in the polyploid crop species *Brassica napus*. Mol Breed 33:349–362

Wen J, Xu J, Long Y et al (2015) Mapping QTLs controlling beneficial fatty acids based on the embryo and maternal plant genomes in *Brassica napus* L. J Amer Oil Chem Soc 92:541–552

Wong RSC, Swanson E (1991) Genetic modification of canola oil: high oleic acid canola. Genetic modification of canola oil: high oleic acid canola. In: Haberstroh C, Morris CE (eds) Fat and cholesterol reduced food. Gulf, Houston, Texas, pp 154–164

Woodfield HK, Fenyk S, WallingtonE BRE, Brown A, Guschina IA, Marillia EF, Taylor DC, Fell D, Harwood JL, Fawcett T (2019) Increase in lysophosphatidate acyltransferase activity in oilseed rape (*Brassica napus*) increases seed triacylglycerol content despite its low intrinsic flux control coefficient. New Phytol 224(2):700–711

Woods DL, Potts DA, Males DR (1999) Genetic control of C18 fatty acid in *Brassica juncea*. In: 10th International Rapeseed Congress, Canberra, Australia, p 142

Wu G, Wu Y, Xiao L, Li X, Lu C (2008) Zero erucic acid trait of rapeseed (Brassica napus L.) results from a deletion of four base pairs in the fatty acid elongase 1 gene. Theor Appl Genet 116:491–499

Xie T, Chen X, Guo T, Rong H, Chen Z, Sun Q, Batley J, Jiang J, Wang Y (2020) Targeted knockout of BnTT2 homologues for yellow-seeded *Brassica napus* with reduced flavonoids and improved fatty acid composition. J Agric Food Chem 68:5676–5690

Xu A, Huang Z, Ma C, Xiao E, Zhang X, Tu J, Fu T (2010) FAE1 sequence characteristics and its relationship with erucic acid content in Brassica juncea. Acta Agron Sin 36:794–800

Xue Y, Chen B, Wang R, Win AN, Li J, Chai Y (2018) Genome-wide survey and characterization of fatty acid desaturase gene family in *Brassica napus* and its parental species. Appl Biochem Biotechnol 184:582–598

Yan XY, Li JN, Wang R et al (2011) Mapping of QTLs controlling content of fatty acid composition in rapeseed (*Brassica napus*). Genes Genom 33:365–371

Yan G, Li D, Cai M, Gao G, Chen B, Xu K, Li J, Li F, Wang N, Qiao J (2015) Characterization of FAE1 in the zero erucic acid germplasm of *Brassica rapa* L. Breed Sci 65:257–264

Yang Q, Fan C, Guo Z, Qin J, Wu J, Li Q, Fu T, Zhou Y (2012) Identification of *FAD2* and *FAD3* genes in *Brassica napus* genome and development of allele-specific markers for high oleic and low linolenic acid contents. Theor Appl Genet 125:715–729

Yao K, Bacchetto RG, Lockhart KM, Friesen LJ, Potts DA, Covello PS, Taylor DC (2003) Expression of the Arabidopsis ADS1 gene in *Brassica juncea* results in a decreased level of total saturated fatty acids. Plant Biotechnol J 1(3):221–229

Yao K, Potts DA, Males DR (2005) United States Patent application publication Pub. No.: US 2005/0039233 A1, Pub. Date: 17 Feb 2005

Yiu SH, Poon H, Fulcher RG, Altosaar I (1982) The Microscopic structure and chemistry of rapeseed and its products. Food Microstru 1:135–143

Yoshie-Stark Y, Wada Y, Schott M, Wäsche A (2006) Functional and bioactive properties of rapeseed protein concentrates and sensory analysis of food application with rapeseed protein concentrates. Food Sci Technnol 39:503–512

Yusuf MA, Sarin NB (2007) Antioxidant value addition in human diets: genetic transformation of *Brassica juncea* with *γ-TMT* gene for increased α-tocopherol content. Transgenic Res 16:109–113

Zaderimowski R, Sosulski F (1978) Composition of total lipids in rapeseed. J Amer Oil Chem Soc 55(12):870–872

Zafar S, Tang MQ, Wang YK, Sarwar R, Liu SY, Tan XL (2020) Candidate genes-association study to identify loci related to oleic acid in *Brassica napus* using SNP markers and their heterologous expression in yeast. Plant Physiol Biochem 146:294–302

Zang YX, Lim MH, Park BS, Hong SB, Kim DH (2008) Metabolic engineering of indole glucosinolates in Chinese cabbage plants by expression of *Arabidopsis CYP79B2*, *CYP79B3*, and *CYP83B1*. Mol Cells 25:231–241

Zang YX, Kim HU, Kim JA, Lim MH, Jin M, Lee SC, Kwon SJ, Lee SI, Hong JK, Park TH, Mun JH, Seol YJ, Hong SB, Park BS (2009) Genome-wide identification of glucosinolates synthesis genes in *Brassica rapa*. FEBS J 276:3559–3574

Zebarjadi A, Jalali JM, Karimzadeh G, Moeini A, Mousavi A, Salmanian A (2006) Transformation of rapeseed (*Brassica napus* L.) Plants with sense and antisense constructs of the fatty acid elongase gene. Iran J Biotec 4:79–87

Zhai Y, Yu K, Cai S, Hu L, Amo O, Xu L, Yang Y, Ma B, Jiao Y, Zhang C, Khan MHU (2020) Targeted mutagenesis of BnTT8 homologs controls yellow seed coat development for effective oil production in *Brassica napus* L. Plant Biotechnol J 18(5):1153–1168

Zhang D, Pirtle IL, Park SJ, Nampaisansuk M, Neogi P, Wanjie SW, Pirtle RM, Chapman KD (2009) Identification and expression of a new delta-12 fatty acid desaturase (FAD2-4) gene in upland cotton and its functional expression in yeast and *Arabidopsis thaliana* plants. Plant Physiol Biochem 47:462–471

Zhang K, Nie L, Cheng Q, Yin Y, Chen K, Qi F, Zou D, Liu H, Zhao W, Wang B, Li M (2019) Effective editing for lysophosphatidic acid acyltransferase 2/5 in allotetraploid rapeseed (*Brassica napus* L.) using CRISPR-Cas9 system. Biotechnol Biofuel 12(1):1–18

Zhao J, Dimov Z, Becker HC, Ecke W, Möllers C (2008) Mapping QTL controlling fatty acid composition in a doubled haploid rapeseed population segregating for oil content. Mol Breed 21:115–125

Zhao Q, Wu J, Cai G, Yang Q, Shahid M, Fan C, Zhang C, Zhou Y (2019) A novel quantitative trait locus on chromosome A9 controlling oleic acid content in *Brassica napus*. Plant Biotechnol J 17:2313–2324

Zuluaga DL, Graham NS, Klinder A, van Ommen KAEE, Marcotrigiano AR, Wagstaff C, Verkerk R, Sonnante G, Aarts MGM (2019) Overexpression of the MYB29 transcription factor affects aliphatic glucosinolate synthesis in *Brassica oleracea*. Plant Mol Biol 101:65–79

24 Advances in Breeding Strategies for Improving Stress Tolerance in Brassicas

Pratibha Pant, S. Hamsa, and Jagreet Kaur

Abstract

Brassica juncea, a major crop of the Indian subcontinent, is gaining greater acceptance worldwide especially in the low rainfall areas due to its better performance under semi-arid conditions, low pod shattering and tolerance to abiotic and biotic stresses. Despite several years of focussed research aimed at improving yield, the commercial growth of Indian mustard has been facing stagnation in productivity. This has been further compounded by severe yield losses caused by abiotic and biotic factors. The integrated information from various omics-based approaches can significantly contribute to increasing the genetic gain and decrease the time taken to achieve more precise improvement in agronomic traits. This chapter provides a glimpse of the core omics-based technologies that have been used in Brassicas for screening pre-breeding material, identifying stress tolerance QTLs in natural accessions, mapping populations, introgression lines and how these have improved our current understanding of the mechanisms adopted by the host in mitigating abiotic and biotic stresses. The chapter summarizes how the data generated through high throughput genotyping and phenotyping approaches, GWAS, comparative genomics, transcriptomics and metabolomics have been integrated to identify candidate genes for improving the *Brassica* response to various biotic and abiotic stresses.

24.1 Introduction

Brassica genus members are globally among the ten most economically important crops. They are used for food, fodder, industrial purposes, phytoremediation of heavy metals and biodiesel production (Thakur et al. 2019; Poveda et al. 2020). The worldwide production of oilseed *Brassicas* is second to soybean. In 2018, the production of oilseed Brassica reached 75 million tonnes and was cultivated over 38 million acres in 63 countries (Cantila et al. 2021). Rapeseed (*Brassica napus*) is cultivated more extensively in Canada, Australia, Europe and China, whereas Indian mustard (*B. juncea*) is predominantly

Pratibha Pant, S. Hamsa—Contributed equally.

P. Pant · S. Hamsa · J. Kaur (✉)
Department of Genetics, University of Delhi, South Campus, Benito Juarez Road, Dhaula Kuan, New Delhi 110021, India
e-mail: jagreet@south.du.ac.in

P. Pant
e-mail: pratibha.pant@south.du.ac.in

S. Hamsa
e-mail: shamsa@south.du.ac.in

C. Kole and T. Mohapatra (eds.), *The Brassica juncea Genome*, Compendium of Plant Genomes, https://doi.org/10.1007/978-3-030-91507-0_24

cultivated in the Indian subcontinent. Presently, *B. juncea* is grown as a major crop in Bangladesh, Central Africa, China, India, Japan, Nepal, Pakistan and some parts of Russia. The worldwide country-specific distribution of *B. juncea* has been enlisted in the Invasive Species Compendium (CABI 2021; FAOSTAT). *B. juncea* has several advantages over *B. napus* such as low water requirement, ability to grow under harsh conditions, enhanced tolerance to biotic stresses, lower pod shattering and higher oil and protein content. Similar to *Brassica carinata*, it has shown potential to be used for biofuel production and industrial usage. These advantages have led to a rapid increase in cultivation area in low rainfall countries like Canada and Australia (Burton et al. 2007).The estimated area, production and yield of rapeseed-mustard across the world were reported as 30.74 million hectares, 59.93 million tons and 1950 kg/ha, respectively, during 2009–10 which increased to 36.59 million ha, 72.37 million tons, and 1980 kg/ha, respectively, in 2018–19 (Jat et al. 2019).

Despite its economic importance, the productivity of oilseed mustard has not increased significantly in the last ten years to meet the current edible oil demands. The stagnation in production trend is partly due to several biotic and abiotic constraints that have been challenging crop yields (All India coordinated research project on Rapeseed-mustard: research highlights 2019–20, DRMR; www.drmr.res.in). Additionally, the changing climatic conditions, in the form of increased temperature and erratic rainfalls, especially in the semi-arid regions has enhanced the incidence of yield losses due to both biotic and abiotic stresses.

24.2 Various Biotic and Abiotic Stresses Faced by Brassicas

Plants routinely encounter several biotic and abiotic stresses in the environment—fungi, arthropods, aphids, insects and drought, salinity, cold/frost, high temperature and heavy metals throughout their life span. Among the various biotic constraints, Alternaria leaf blight caused by the necrotrophic fungal pathogen *Alternaria brassicae*, stem rot (SR) caused by an aggressive, generalist fungal pathogen *Sclerotinia sclerotiorum*, white rust caused by obligate biotrophic oomycete *Albugo candida*, downy mildew caused by *Peronospora brassicae*, powdery mildew caused by *Erisyphae* are the major ones affecting the *B. juncea* productivity globally. Additionally, major insect pests—aphid *L. erysimi,* mustard sawfly (*Athalia proxima*) and painted bugs (*Bagrada hilaris*) also negatively impact the oilseed Brassica productivity. Combined biotic stress comprising pests, pathogens and weeds causes approximately 45% loss of yield on annual basis (Sekhon and Brar 1997).

Besides biotic stresses, abiotic stresses like drought, soil salinity, extreme temperatures, presence of heavy metals or mineral imbalance in soil, etc. also adversely affect the Brassica crop productivity and the quality of oil produced. Drought and salinity cause hyperosmotic stress that impedes physiological and biochemical reactions (Paul et al. 2019). A combination of both drought and salt stress alters the redox homeostasis in the cells, vacuole ion sequestration, cellular energy, chloroplast formation, impacts the electron transport chain (ETC) and subsequently the photosynthesis (Alamri et al. 2020). The exposure of plants to high temperatures during the reproductive stages—'terminal heat stress' mainly due to late sowing, inhibits the photosynthesis process and increases disease incidence, leading to lower seed weight and yield. Alternatively, cold stress causes membrane rigidification, excessive Ca^{2+} accumulation, thus altering plant cellular homeostasis. The various prevalent abiotic and biotic stresses plaguing *B. juncea* productivity are enlisted in Table 24.1.

Besides cultural practices, chemicals and biological/genetic resistance form the common multi-pronged strategy to tackle the menace of pathogens and pests that severely impacts crop productivity. Unfortunately, the indiscreet use of agrochemicals has negatively impacted the environment and the diversity of crop species. Once again, the emphasis is shifting back towards breeding for resistance. The success of both classical and unconventional breeding

Table 24.1 Main stresses in Brassica production

Brassicaceae pathogen	Yield loss	Mode of action and symptom	Plant stage affected	References
White rust (*Albugo candida*)	20–60% loss in India, Australia, Canada	Airborne	Rosette leaves, pods	Arora et al. (2019a)
Blackleg (*Leptosphaeria maculans*)	10–0% globally	Airborne via spores; off-white lesions, canker	All stages	Cantila et al. (2021)
Alternaria sp.	32–50%% (India)	Airborne; necrotic lesions with concentric rings, chlorosis	Flowering stage leaves, stems, and branches	Meena et al. (2010)
Clubroot (*P. brassicae*)	1–45% globally	Galls, wilting due to nutrition loss	Roots	Dixon (2009)
Sclerotinia sclerotiorum	5–100% globally in *B. juncea*; Impacts *B. napus* in Australia, Canada, China, Europe, and India	Soil-borne (via sclerotia), airborne (spores)	Flowering stage stems, branches, and leaves	Singh et al. (2021)
Downey mildew (*Hyaloperonospora brassicae*)	50–100% globally	Chlorosis, brown lesions	Seedling and rosette stage leaves	Saharan et al. (2017)
Powdery mildew (*Erysiphe cruciferarum*)	25–30% loss globally	Airborne; grey/white patches on leaf and stem	All stages	Uloth et al. (2018)
Aphids	35–96% yield loss in mustard in India	Phloem feeding via aphids	All growth stages	Kumar et al. (2017), Dotasara et al. (2018)
Beet western yellow virus (BWYV)	35–46% in canola in Europe and Western Australia	Stunting, purple/red lower leaves; via aphids	All growth stages	Jones et al. (2007)
Turnip yellow virus	>40% loss in rapeseed crops globally; 19.26% loss in *B. juncea* var. rugosa	Stunting, mosaic pattern, necrotic spots	All growth stages	Congdon et al. (2019)
Orobanche aegyptiaca	15–49% loss in mustard in India, 80% in France, 5–100% in semi-arid regions	Wilting, nutrition loss due to competition	NA	Véronési et al. (2009), Jat and Singh (2018), Das et al. (2020, Das Laha et al. 2020)
Abiotic stress				
Drought	29.5% seed loss (reduced silique number and seeds per silique), 31.7% oil loss	Hyperosmotic stress that affects vacuole ion sequestration and cellular energy	All stages	Ahmar et al. (2019)
Heat	NA	Wilting, reduction in seed weight and seeds per silique; disturbs plant cellular homeostasis	All stages	Bhatia et al. (2020)

(continued)

Table 24.1 (continued)

Brassicaceae pathogen	Yield loss	Mode of action and symptom	Plant stage affected	References
Salinity	38% reduction in oil content in *B. juncea*; 60–80% reduction in seed yield	Hyperosmotic shock, ROS production	Seed germination and seedling stage	Alamri et al. (2020)
Cold		Anthocyanin production, Plasma membrane rigidification, Ca^{2+} accumulation, wilting	Seedling and pod filling stage	Sinha et al. (2015), Lohani et al. (2020)
Heavy metal stress Pb	Reduced growth and yield	Oxidative stress and damage inhibited plant growth and chlorophyll	All stages	Zhang et al. (2020a)
Cadmium	Reduced growth and yield loss	Growth retardation, leaf chlorosis, and necrosis; disruption of absorption and movement of essential ions, inhibition of photosynthesis, respiration, and ROS detoxification	All stages	Alessandro et al. (2013), Zhang et al. (2021)
Arsenic	Growth and biomass reduction	ROS production	All stages	Thakur et al. (2020b)

approaches mainly relies on the available variability in the crop population/diversity sets for a particular trait, followed by selection, availability of high-density genetic maps for mapping the locus, high-throughput genotyping and robust phenotyping platforms.

24.3 Classical Breeding Strategies for Improving Stress Tolerance

Significant efforts have been made in the past to improve Brassica crops for resistance against biotic and abiotic stresses through plant breeding. Most of the studies related to the identification of resistance/tolerance quantitative trait loci (QTLs) against various stresses in Brassicas have used bi-parental populations derived from a cross between parents with contrasting traits. The ease of making the doubled haploid population in Brassicas through microspore culture has shortened the time taken to develop immortal segregating populations for identification of the QTLs associated with the trait. In the past, several QTLs have been identified for resistance against white rust (Panjabi-Massand et al. 2010), blackleg (Kaur et al. 2009; Larkan et al. 2016b), Sclerotinia stem rot (Wu et al. 2013), clubroot (Suwabe et al. 2003, 2006) using the polymerase chain reaction (PCR)-based molecular markers. However, due to the limited number of markers available, the QTLs identified had very poor resolution and thus had a very limited impact on breeding programs aimed at improving stress tolerance, especially where the trait was polygenic and quantitative.

Considering the limited variability available within the natural Brassica germplasm, researchers had used a mutation breeding approach to introduce novel genetic variability (Parry et al. 2009). Somaclonal variations were also tapped into to identify biotic and abiotic stress resistance lines (Sacristan et al. 1988; Kirti et al. 1991; Liu et al. 2005). Since sources of resistance to major diseases are elusive in the primary gene pool, the secondary gene pool represented by the wild relatives has been explored as a potential source of usable genetic variation. Alien gene transfer has been exploited to broaden the gene pools and harness the available genetic variability within the closely related species. Introgression of various agronomic traits, including tolerance for biotic and abiotic stresses, from the wild-type relatives and allied species into the oilseed rape and rapeseed has been attempted for a long time (Quezada-Martinez et al. 2021). As with other breeding approaches, the introgression of a single major dominant gene has been more successful in comparison to polygenic traits influenced by environmental factors. Several single, dominant R genes conferring resistance against white rust and black rot defined within the *B. juncea* germplasm were introduced into elite varieties following conventional marker-assisted back-cross breeding approaches (Quezada-Martinez et al. 2021). Similarly, the appropriate quantitative resistance against blackleg has been introduced to the cultivated varieties (Rahman et al. 2016). Resistance to stem rot identified in the wild relatives *Erucastrum cardaminoides*, *Diplotaxis tenuisiliqua* and *E. abyssinicum* have been introgressed into *B. napus* and *B. juncea* using wide hybridizations followed by embryo rescue or ovule/ovary culture (Garg et al. 2010). Three major QTLs for SR resistance identified in *B. incana* have been introgressed into *B. napus*, and a 35% gain in resistance has been recorded (Mei et al. 2020). Similarly, introgression of SR resistance from *B. fruticulosa* to *B. juncea* has been successful as lines with around 60% restricted lesion expansion as compared to the wild type have been identified. Successful transfer of B genome residing resistance against blackleg being from *B. rapa* subspecies *silvestrys* to *B. napus* has been reported (Yu et al. 2012).

Several groups have reported their attempts to improve resistance against Alternaria leaf blight using intergenic somatic hybridization between *B juncea* and *Sinapis alba* (Kumari et al. 2017). In vitro ovary and ovule culture were attempted to rescue the embryos from the intergenic crosses. Protoplast fusion has also been used for transferring resistance against Alternaria leaf blight from *Sinapis alba* to *B. napus* and *B. juncea* (Kumari et al. 2020). Reports of successful introgressions against blackleg resistance from *B. juncea* to *B. napus* (Roy 1984; Sacristan and Gerdemann 1986) and from *B. nigra* to *B. napus* (Struss et al. 1992; Chèvre et al. 1997). More recently, the C genome residing resistance against *S. sclerotiorum* has been transferred from *B. oleracea* to *B. napus* (Mei et al. 2020).The introgression of flavonoid-associated resistance against cabbage seedpod weevil available in *S. alba* to *B. napus* was also reported (Lee et al. 2014).

However, the rate at which these genetic gains are made indicates that the reliance on these approaches alone is not enough to meet the increasing demands of edible oil. Therefore, efforts have been made in developing the high throughput omics platforms to accelerate the breeding programs.

24.4 Advancing the Brassica Genomics Resources

The six important species of the brassicas, i.e. the diploid species *B. rapa* (AA, $2n = 20$), *B. nigra* (BB, $2n = 16$) and *B. oleracea* (CC, $2n = 18$), and the cultivated allo tetraploid species *B. juncea* (AABB, $2n = 36$), *B. napus* (AACC, $2n = 38$) and *B. carinata* (BBCC, $2n = 34$) are closely related to each other and the U's triangle explains the evolutionary relationship between them (Nagaharu 1935). Advancements in next-generation sequencing (NGS) and integrating the single-molecule long-read technology (SMRT), optical mapping, Hi-C sequencing have enabled the near-complete, highly contiguous chromosome-level genome assemblies of all the six species of the U's triangle.

Paritosh et al. (2020) and Perumal et al. (2020) used Oxford Nanopore sequencing technology (ONT), Illumina sequencing and optical mapping to identify 57,249 protein-encoding genes in the B genome (522 Mb) of *B. nigra*. A comparative analysis of the B genomes of *B. nigra* and *B. juncea* showed a high gene colinearity and similarity in the gene block arrangements within them and a clear divergence from the A and C genome. Combining the PacBio SMRT sequencing with optical mapping facilitated the assembly of the 922 Mb genome of the *B. juncea* var. Varuna (AABB) and revealed a total of 79,108 annotated genes. It also reported higher Long Terminal Repeats (LTR)/Gypsy retrotransposon content in the B genome component as compared to the A genome. Both the A and B genome components showed high colinearity with their progenitors (Paritosh et al. 2021). The 442.9 Mb genome assembly of the *B. rapa* estimated 45,985 protein-encoding genes within the genome. The study also identified a new LTR-RT (long terminal repeat-retrotransposon) expansion event in the *B. rapa* genome that occurred 1.2 bn years ago (Zhang et al. 2018). The recent high-quality genome assembly of *B. napus* variety ZS11 provides valuable insights into the structural and evolutionary aspects (Chen et al. 2021). The recent addition of the high-quality genome assemblies for *B. oleracea* (Cai et al. 2020; Guo et al. 2021) and *B. carinata* (Song et al. 2021c) also provide a treasured resource to understand the evolutionary relationship between the six species of the U's triangle. The available genomic resources enable the researchers to address the possible events that might have occurred at the genomic level during the evolution and domestication of these species. The sequence information for all the six species is available at the BRAD database (http://brassicadb.cn/) and Brassica Genomics Database—BGD (http://brassicadb.bio2db.com/), Bolbase: provides a comprehensive database for *Brassica oleracea* (Yu et al. 2013).

Although the reference genomes for all the representative species of the U's triangle are now available, however, they do not adequately represent the genetic variation present within and between species. The reduced sequencing cost and improved genome assembly algorithms have stimulated the sequencing of several individual genomes of a species. Golicz et al. (2016) sequenced nine morphologically distinct varieties of *B. oleracea* and a wild relative—*Brassica macrocarpa*. Sequence comparison among these revealed several novel single nucleotide polymorphisms (SNPs), presence–absence variations (PAVs), Indels within the elite accessions as well as the wild type which were missing from the reference genome. Incorporating all the existing variation amore representative pan-genome for *B. oleracea* has been constructed and is available at http://appliedbioinformatics.com.au/gb2/gbrowse/BolePan/. Another similar study carried out a genome-wide comparative analysis of eight *B. napus* genomes to identify core, dispensable and specific gene clusters. It also helped identify PAVs responsible for silique length, seed weight and flowering time (Song et al. 2020). A **Pan-genome** thus ensures a more holistic reference genome and allows a single reference point to study the genetic variation within a species ultimately leading to the identification of novel genes that could be missing from a single reference genome. Besides SNP information, it also provides novel information/insights about the structural variations like PAVs, copy number variations (CNV), inversions, etc. Pangenomic study of resistance genes (R genes) in *B. napus* and *B. oleracea* also revealed significant structural differences (Bayer et al. 2019; Dolatabadian et al. 2020). BnPIR (*Brassica napus* Pan-genome Information Resource), a comprehensive database forming a pan-genome of 1688 *B. napus* accessions has been curated (Song et al. 2021a). A similar construction of the *B. juncea* pan-genome is the need of the hour and will provide opportunities for agronomic trait improvement. Sequencing of *B. oleracea* wild species *B. macrocarpa* also uncovered the presence of additional genes which have been lost in the cultivated elite varieties during domestication. Therefore, in the future, the construction of a **super pan-genome**, comprising of the complete pan-genomes and including wild relatives will give a complete picture of the existing genetic variation (Khan et al. 2020).

24.5 Advanced Genotyping Platforms in the Genomics Era

Molecular marker-based linkage maps for several Brassicas including *B. juncea* have been developed for trait mapping. In the pre-genomics era, researchers had mainly used low-throughput PCR-based markers—RFLP, RAPD, AFLP,

SSR and IP markers. Moreover, the density of these markers was not sufficient to develop saturated linkage maps which had a direct consequence on the resolution of the trait-associated QTL identification. Additionally, the introgression of the larger genomic region was invariably associated with linkage drag. The reduced cost of high-throughput genome sequencing and advanced comparative analysis/programs have opened up new opportunities/platforms to incorporate a multitude of genome-wide DNA markers for crop genotyping. This has resulted in accurate genotyping of populations and diversity sets generally using SNPs.

24.5.1 Genome-Dependent Genotyping Systems: SNP Infinium Array

Whole-genome resequencing (WGRS) at low depth (1-5X), called Skim-sequencing has allowed the identification of a large number of good quality intra- and inter-specific SNP and InDel markers without any bias (Bayer et al. 2015). Based on 52,157 SNPs available in *B. napus*, the International Brassica SNP Consortium developed an Illumina Infinium™ 60K SNP array which can simultaneously genotype between 3000 and 1 million polymorphic sites for a single sample (Clarke et al. 2016). This chip-based genotyping is highly preferred due to its easy sample preparation, data generation, analysis and reproducibility. It has since been upgraded to the 90K Illumina Infinium Array to include *B. carinata* genome markers as well, thus providing a comprehensive representation of all the three subgenomes (A, B, and C) (Zou et al. 2018). It is more of a closed system where existing SNPs can be tested across several accessions and germplasm but no new SNP could be identified. Infinium arrays have been used in several studies for mapping QTLs and identifying candidate genes for biotic and abiotic stress tolerance like blackleg resistance (Larkan et al. 2016a), black rot resistance in *B. oleracea* (Lee et al. 2015), stem rot resistance (Wu et al. 2016), salinity and drought tolerance (Tan et al. 2017), drought response (Fletcher et al. 2015), water stress tolerance (Zhang et al. 2015). However, the limitation of Infinium array is that the SNP information is drawn from a single or few genomes representing each subgenome so the diversity is not well represented.

24.5.2 Sequence-Independent Genotyping Approaches

In the absence of the pan-genome sequence information genotyping by sequencing (GBS), RAD seq, and DArTseq strategy enable SNP mining across multiple samples. Unlike chip-based assay, these techniques are open-ended approaches where new variations present between accessions can be identified. These approaches have been successfully adopted across crop plants due to their generic nature. These approaches involve digesting the entire genome with a methylation-specific restriction enzyme, followed by adapter ligation, PCR amplification and sequencing the pool of amplified amplicons using NGS. Further, multiplexing the sequencing using the barcodes makes the whole process cost-effective. In addition to SNP, one can also identify small insertions, deletions and microsatellite markers as well (Baird et al. 2008; Elshire et al. 2011). GBS can identify SNPs with low minor allele frequency (MAF). GBS has been especially recommended for high-throughput genotyping for polyploid species including Brassicas.

The SNPs identified can be converted to high throughput Kompetitive Allele-Specific PCR (KASP) assays that are based on competitive allele-specific PCR to detect polymorphisms at a specific locus. KASP markers were utilized for detecting 1167 polymorphisms in 70 inbred lines of *B. rapaL.* subsp. *chinensis*. A combination of GBS and KASP assays has also assisted in mapping resistance genes in *B. rapa* and *B. napus* and analyse genome-wide variations (Cheng et al. 2016; Fu et al. 2019; Karim et al. 2020). These two are generally used as complementary approaches.

24.6 Advances in High-Throughput Phenotyping Platforms: Phenomics

Huge data sets being generated by the genomics need to be correlated with a large-scale collection of phenotypic data under varying environmental conditions to understand the genotype × environment-related traits. Functional phenomics involves developing a phenotypic platform to analyse the genetic trait in a population that improves our understanding of big data analytics at different growth and development stages of the plant (Dwivedi et al. 2020). The major bottleneck in phenomics is the development of high-throughput reliable and efficient phenotyping tools for screening huge populations, especially for stress biology where manual phenotyping is laborious and requires proper training (Marsh et al. 2021). A commercial imaging device morphological analysis tool (MAT) was developed to allow accurate representation of plant geometric characters like leaf shape and leaf projected area in *Arabidopsis*. Multi-sensor imaging tool, growth-imaging approach, pulse-modulated chlorophyll fluorescence imaging and carbon isotope discrimination have been utilized to study drought stress in *Arabidopsis*. Another phenomics method of infrared thermography to screen osmotic and drought tolerance in economically important crops has gained importance in recent years (Furbank and Tester 2011). Image-based phenomics and digital biovolume were adopted to study genetic variation to salinity stress and water stress in wheat and tomato (Danzi et al. 2019). A phenotypic dataset generated by an automatic phenotyping platform along with Genome-Wide Association Studies (GWAS) and transcriptome revealed major genetic loci involved in drought resistance in cotton (Li et al. 2020a).

For biotic stress, phenomics has been applied to explore various aspects of phenotyping plant health through differences in disease severity, metabolic changes, phytohormone quantification through biosensors, quantification of pathogen mass and examining parameters related to plant biomass mainly on the field level (Mahlein et al. 2012). Imaging of chlorophyll fluorescence and fluorescence spectroscopy was developed for selecting resistant genotypes and accessing molecular changes in leaves at early infection stages in plant–fungal interactions (Scholes et al. 2009).

Phenotyping tools which are non-destructive like volume of flowers, total triose phosphate use and Vcmax for carboxylation rate of RuBisco have been applied in Brassica to understand the behaviour of various genotypes under heat or drought stress at early reproductive stages (Chen et al. 2019). It was used to study various biophysical and agronomic traits and phenotype root architecture in *Brassica napus* (Shi et al. 2013; Jaradat 2018; Arifuzzaman et al. 2019). Recently, a new low-cost and non-destructive tool electrical impedance tomography (EIT) was developed to visualize the development of roots of *B. napus* infected with *P. brassicae* (Corona-Lopez et al. 2019). Another platform, SeedGerm, was able to process multiple image series of germinating seedlings by scoring the timing of radical emergence and measuring various morphological traits in *B. napus* varieties. In combination with associative transcriptomics, SeedGerm was able to identify a gene involved in abscisic acid (ABA) signalling (Colmer et al. 2020). Phenomics platforms have helped breed for aphid resistance where traditional methods for estimating plant resistance and tracking the feeding behaviour of aphids through electrical penetration graph (EPG) are laborious and have associated drawbacks. A high-throughput video tracking platform was established for screening a large diversity set to identify lines resistant to aphid and other phloem-feeding insect pests (Kloth et al. 2015). Since aphids also transmit plant viruses, Chen et al. (2012) developed a quick and efficient method to identify mutant lines resistant against the aphid *Myruspersicae* using turnip yellow virus as a reliable indicator. Hayat et al. (2010) selected salt-tolerant varieties of *B. juncea* based on photosynthetic attributes. Owing to the limitation of digital imaging in field setup, a machine-based phenotyping tool, real-time plant phenotyping (RTPP) has been developed for identifying physical and biochemical changes in plants which can be a potential tool to explore various parameters in Brassica under field conditions when subjected to biotic stress (Arunachalam and Andreasson 2021). However, the utilization of advanced phenomics approaches is still in infancy in the *B. juncea* research.

Publicly accessible database like Plant Genomics and Phenomics Research Data Repository (PGP) has been started to submit and share the experimental phenotype data (Arend et al. 2016). A centralized repository Brassica Information Portal has been established for efficient storage and management of phenotypic data collected from various breeding and research experiments on Brassica crops (Eckes et al. 2017). Advances in phenomics through ariel phenotyping platforms, machine learning, artificial intelligence and robotics can address the challenges faced by ground-based platforms for multi-stress phenotypic data analysis in future crop breeding programs (Song et al. 2021b).

24.7 Omics-Facilitated QTL Identification and Characterization

Although the conventional linkage mapping has effectively identified several QTLs associated with stress tolerance in Brassicas, the major drawback is that it can capture the allelic diversity available between the two contrasting parents only and the genomic resolution has limitations too.

24.7.1 Association Mapping Studies

Linkage disequilibrium (LD) mapping or association mapping (AM) has opened an advanced platform for identifying polymorphisms associated with a trait in the large natural diverse gene pool. Since the approach utilizes thousands of ancestral recombination events that have accumulated over generations to draw statistically significant marker-phenotype associations, the resolution of the QTL identified is high (Fan et al. 2006). GWAS has the potential to identify the causative mutation or at least provide tightly linked markers. The resolution of the QTLs identified by AM depends upon the extent of genetic variation available in the population, size of the diversity set, the density of markers, population structure and kinship.

Several GWAS studies have been carried out to predict loci conferring resistance to Brassica crops in response to pathogen infection. A GWAS study in 167 *B. juncea* lines identified seven SNPs associated with *rlm6* mediated resistance against *L. maculens* on chromosomes A07 and B04. Sixteen candidate R genes were predicted in the region-5 on chromosome A07 (4 of the LRR-RLK type and 1 LRR-RLP type) and 11 on chromosome B04 (2 NLR genes, 3 LRR-RLK, and 6 LRR-RLP genes) (Yang et al. 2021). To study the genetic basis for *S. sclerotiorum* resistance to *B. napus*, GWAS was carried out using 448 *B. napus* accessions and 26 SNPs corresponding to 3 loci—*DSRC 4, DSRC 6* and *DSRC 8*—were identified. Of these, the *DSRC 4* and *8* were found to be novel loci, and 39 candidate genes were also predicted in these loci (Wu et al. 2016). Another study utilized 347 *B. napus* accessions and the 60K Brassica SNP array to associations for resistance to *S. sclerotiorum*. Seventeen such SNPs and candidate genes—glucosidase, PR-2 genes, LR family R genes, ZFP type transcription factor and indole glucosinolate methyltransferase (IGM) were identified (Wei et al. 2016). In *B. juncea–Erucastrum cardaminoides* introgression lines harbouring the genomic region responsible for *S. sclerotiorum* resistance, six resistance-conferring loci were identified in both the A and B genomes. A GWAS of 84 Introgression lines identified several SNPs associated with resistance—a majority of which lay on chromosomes A03 and A06. The linked genetic regions were annotated to be responsible for HR, signal transduction pathways and anti-fungal proteins and metabolites. 885 kb and 74 kb regions on chromosome A03 and B03, respectively, were found to house a clustering of SNPs conferring stem rot resistance (Rana et al. 2017).

Similarly, the genetic variation responsible for drought stress tolerance was studied using GWAS and identified 314 closely associated novel markers. These were found to contain 85 genes orthologous to drought-responsive genes in *A. thaliana* (Khanzada et al. 2020). Another study genotyped 66 accessions using 25,495 SNP markers and identified 16 loci associated with the water stress response. They identified 11 genes homologous with *A. thaliana* that were upregulated during water stress—LEA family genes, MAPKKK, Caleosin-related family genes. To study the genetics of salt stress in *B. napus*, a GWAS study was conducted in 368 accessions, which were genotyped using the 60K Array. The study yielded 75 SNPs associated with four salt-tolerance-related traits that mapped into 25 QTLs. Additionally, 38 potential candidate genes were also identified in the regions associated with salt tolerance including aquaporins, transcription factors and enzymes (Wan et al. 2017). Sandhu et al. (2019) identified 24 SNPs for seed yield traits associated with terminal heat stress. Breeding programs aimed at lead (Pb) tolerance in Brassicas have been restricted due to the limited availability of information regarding the molecular mechanisms governing Pb tolerance and detoxification. Aiming to study lead tolerance in *B. napus,* a GWAS study was conducted using 472 accessions to identify the genetic loci responsible for the same. The study identified six Pb-tolerant genotypes and four QTLs conferring this resistance. Promising candidate genes conferring Pb resistance—glutathione transferases, ubiquitin-specific proteases, aminotransferases and heavy metal-associated isoprenylated plant proteins induced in tolerant accessions upon encountering Pb stress were also identified (Zhang et al. 2020a).

To reduce the efforts involved in genotyping and phenotyping large diversity sets while identifying SNP-trait associations, the GWAS is modified to extreme phenotype GWAS (XP-GWAS) which measures allele frequencies in two bulks of individuals representing the extreme phenotypes selected from the association/diversity panel to identify SNP-trait association. XP-GWAS was first used in maize to identified 145 trait-associated variants (TAVs), which were comparable to the conventional GWAS (Yang et al. 2015). Recently, XP-GWAS and allele frequency diversity divergence (AFDD) was utilized to identify SNP variation in virulent (vcr2) pathotype and avirulent (Avcr2) *Cronartium ribibola* and identify effector candidate genes through BSA RNA-Seq (Liu et al. 2021).

24.7.2 Associative Transcriptomics (AT)

Although GWAS is efficient in identifying the causative sequence variation associated with a trait, the analysis has its limitation when it comes to polyploid crops with complex genomes. Thus to circumvent the genome complexity issue, the associative transcriptomics (AT) approach correlates the variation in the quantitative trait with the SNP markers identified by mRNA seq/Exomeseq and has been widely used in polyploid crops including Brassicas (Harper et al. 2012). There are some limitations of detecting SNPs through transcriptomics which include the inability to provide information on noncoding RNAs and missing SNPs which have low transcript expression. AT has been useful in dissecting associations with trait variation in *B.napus* in several traits like yield, oil content, erucic acid, flowering time, glucosinolate and anion homeostasis (Harper et al. 2012; Havlickova et al. 2018; Xiao et al. 2019; Jian et al. 2020). Extending this approach to stress biology, recently (Huang et al. 2020b) identified 22 candidate genes by analysing genetic variation for photosynthetic gas exchange parameters under freezing stress in 123 accessions of *B. napus*. Of the 22 candidate genes, transcriptome analysis further narrowed down to one probable candidate *BnTR1* which was later characterized to enhance freezing tolerance in *Arabidopsis*. GWAS of five *B. napus* varieties combined with transcriptome identified polymorphisms at five loci consisting of 38 genes and the major role of photosynthesis and oxidative stress mechanisms associated with freezing tolerance after deacclimation (Horvath et al. 2020). Hejna et al. (2019) used 245 accessions of *B. napus* and dissected out 86 SNPs to be significantly associated with the clubroot disease resistance. These SNPs were located in nine loci (7 minor and 2 major) indicating a quantitative nature of resistance and a further transcriptomics approach identified 82 candidate genes from these loci which may be involved in resistance to clubroot. Harper et al. (2020) established AT in *B. juncea* by creating a pan transcriptome reference from 204 accessions for association studies in important agronomic traits. A similar pan transcriptome platform for *B. napus* representing the A and C genome had provided novel insights into the genome structural variation during the evolution of the polyploidy crop (Harper et al. 2012).

24.7.3 Comparative Genomics: Aiding Gene Discovery for Improving Stress Tolerance

The whole-genome sequence information available from all the six cultivated species of Brassicas along with the model plant Arabidopsis has revealed the evolutionary relationships and structural conservation between the homologous chromosomes of the three subgenomes A, B, C (Paritosh et al. 2021). Given the close phylogenetic relationship comparative analysis offers means to identify genes with probable functional equivalence which can then be the starting point for dissecting the molecular pathway involved in several agronomic traits including stress tolerance. The molecular dissection and in-depth understanding of various processes involved in stress tolerance in the model plant *Arabidopsis* have revolutionized the field of comparative genomics in Brassicas. However, the complex genome organization of the Brassica genomes poses difficulties in identifying the orthologous relationship of genes between genomes. Nevertheless, comparative mapping of genomic regions, defined by GWAS, AT and traditional mapping associated with agronomic traits in Brassicas, with *A. thaliana* genome helps in the identification of potential candidate genes.

Most of the initial comparative genomic studies were focussed on *B. rapa* and *B. napus*; however, due to the presence of desirable traits like better tolerance to heat, drought, salinity, heavy metal accumulation and resistance to seed shattering and the recent improvements in the available genomics resources has brought *B. juncea* to the forefront as well.

Comparative analysis of *B. napus* and Arabidopsis unraveled the importance of phospholipase genes *BnaPLDα1s* and *BnaPLDδs* in drought stress (Lu et al. 2019); similarly role of transport inhibitor response1/Auxin signaling, F-box protein (*TIR1/AFB*) genes in *B. juncea* var Tumida (Cai et al. 2019), and role of Jasmonate ZIM domain proteins (*JAZ*) genes—repressors of the JA mediated pathway in the abiotic and biotic stress response—were identified in the *B. juncea* genome (Cai et al. 2020). *A. thaliana PHT* genes encoding $H_2PO_4^-/H^+$ co-transporters are well characterized for their role in the absorption and transport of inorganic phosphorus for growth, development and protection against heavy metals accumulation. A comparative study of the phosphorus transporter (*PHT*) in Brassicas identified 85 genes in *B. juncea* while 54 were found in *B. rapa* and 55 in *B. nigra*. The expression pattern of *PHT* genes also varied under As^{3+} and Cd^{2+} stress in *B. napus* indicating a role for the genes in heavy metal stress response (Wan et al. 2020). Similarly, 130 multidrug and toxic compound extrusion (*MATE*) genes responsible for the metabolism of organic acids and toxic compounds were identified in the *B. juncea* genome, while 85 and 79 were identified in *B. rapa* and *B. nigra*, respectively, and 23 genes also showed colinearity with *A. thaliana* genes. To investigate the role of Brassica *MATE* genes in heavy metal detoxification their expression patterns were studied under arsenic and cadmium stress in *B. napus*. Group 1, 2 and 4 MATE genes were found to be induced under stress, while Group 3 genes were suppressed (Qiao et al. 2020).

Similarly, 12 Salt Overly Sensitive (*SOS*) pathway genes were identified in *B. juncea* var. Tumida based on their homology with *A. thaliana SOS* genes. The different *SOS* genes encode for Na^+/H^+ anti-transport proteins, serine/threonine-protein kinase and Ca^{2+} binding proteins which help in ion sensing and efflux. All the predicted genes were responsive to salt stress at 12 h and several genes were induced under low temperature and *P. brassicae* infection (Cheng et al. 2019). Phylogenetic analysis of *B. juncea* heat shock factors (HSF), NAC transcription factors and *SOD* genes with other Brassica species have been analysed to understand their role in various stress mechanisms (Verma et al. 2019; Li et al. 2020b; Jiang et al. 2021).

24.7.3.1 R Gene Studies

The host resistance genes form the most essential component of the plants innate immune system. They detect pathogen virulence factors (Avr genes) leading to hypersensitive response and activation of various defence-related genes. However, when genetically uniform lines with a single R gene conferring resistance against a pathogen is used year after year on a large scale, there are chances that the resistance is defeated by newly emerging pathogen races. Therefore, there is a need for constantly upgrading the R gene arsenal in the breeder's kitty and deploying multiple R genes than a single R gene to reduce the selection pressure on the pathogen. Pangenomic studies have shown a large amount of variability and dispensability among host R genes indicating the need for a large repertoire to compensate for resistance breakdown and selection pressure of the arms race. However, the complex genome of polyploidy *B. juncea* presents a major challenge for the isolation of individual genes.

Bioinformatically, R genes can be identified using homology-based methods—BLAST, sequence alignment with R gene database (PRGdb) or domain sequence homology. Due to the larger allotetraploid genome and evolutionary gene duplication events, *B. juncea* and other ***Brassicas*** was found to contain a larger number of genes as compared to other plants, including its progenitors-289 NLR genes were found in the *B. juncea* genome, along with 202 genes in *B. rapa* and 282 in *B. nigra*. 24% *B. juncea* R genes contained all the typical R genes structural elements—TIR/CC, NBS, LRR, while the rest were partially deleted or truncated (Inturrisi et al. 2020). Their identification and evolutionary understanding has been further aided by the development of robust and user-friendly pipelines such as RGAugury that enables the genome-wide identification of R gene analogues (RGAs), potential R gene candidates based on conserved domains and motifs (NB-ARC, LRR, TM Domains, STTK, Lysin motifs, CC and TLR domains) (Li et al. 2016). Genome-wide analysis in *B. juncea* of the two RGAs classes receptor-like kinases (RLK) and receptor-like proteins (RLPs) identified 493 RLKs and 228 RLPs (LRR and LysM type) mainly resulting from duplication events; they were found to have a 83.62% and 41.98% conservation with their progenitors. Other R genes identified in various Brassicaceae members (genome and RGA-pan-genome) have been reviewed in (Zhang et al. 2020b). Several softwares utilizing bioinformatics pipelines like NLGenomeSweeper and AgRenSeq were developed to identify R genes. A large number of RGAs have also been identified in the wild relative *B. macrocarpa*, strengthening the observation that a large R gene base and variability is available in wild crop relatives and land races that can be exploited for crop improvement (Neik et al. 2020).

Targeted sequencing where only the region of interest or apart from the whole genome can be enriched before sequencing offers a potential method to detect R gene variants in natural populations which is done by resistance gene enrichment sequencing (Ren-Seq). This approach has been applied to identify new NBS-LRR genes in potato, wheat and to identify SNP markers associated with resistance to *Phytophthora infestans* (Jupe et al. 2013). Several modifications of traditional Ren-Seq like MutRenSeq, MutChromSeq and AgRenSeq were useful to discover new R genes independent of map-based cloning and transfer of R genes from wild relatives to domesticated crops (Sánchez-Martín et al. 2016; Steuernagel et al. 2016; Arora et al. 2019b). PacBio sequencing-based SMRT-Seq was utilized to construct a pan-NLRome using 64 accessions of *Arabidopsis* which revealed the process of evolution of NLR in terms of their diversification and selection patterns (Van de Weyer et al. 2019) and also to analyse sequence variation and presence/absence of various WRR (white rust resistance) alleles in a MAGIC population of Arabidopsis (Cevik et al. 2019).

Since Ren-Seq has been widely used in Arabidopsis, it can be a potential method for building a pan-RGAome or NLRome in Brassica to understand the complexity of R gene analogues and mechanisms of disease resistance (Zhang et al. 2020b). Such detection and discovery of orthologous genes along with a phenotype–genotype correlation can also help us in identifying lines resistant to several stresses.

The availability of high-quality reference genomes of several important pathogens of Brassicas like *L. maculans* (Dutreux et al. 2018), *S. sclerotiorum* (Derbyshire et al. 2017), *A. brassicae* (Rajarammohan et al. 2019), *P. brassicae* (Daval et al. 2019), *A. candida* (Links et al. 2011) has facilitated the prediction of a large number of effectors which may play a crucial role in pathogenesis. Several bioinformatics programs have been developed to predict effector proteins based on their structural features/domains such as EffectorP.2.0 and EffHunter (Sperschneider et al. 2018; Carreón-Anguiano et al. 2020). These predicted effectors can also be effectively deployed to screen the crop germplasm and wild species to facilitate the detection of the cognate R gene.

24.7.4 Comparative Transcriptomics: Towards Candidate Gene Identification

Extracting and decoding information on differential expression of variant and regulatory genes, major pathways involved in stress regulation, understanding gene regulatory networks through systems biology are being explored in recent years (Kole et al. 2015). Initially, *Arabidopsis* cDNA microarrays were used to examine transcriptional responses in Brassica (Dalyan et al. 2017) and later Brassica-specific oligonucleotide arrays like Br135K *Brassica rapa* oligonucleotide array was used to unravel regulatory mechanisms under cold stress and profiling of transgenic lines against *Proteobacterium carotovorum* (Jung et al. 2014). Since the array design required sequence information was involved with high cost and provided inconsistent results, they have failed to gain importance in the field of transcriptomics (Lijuan et al. 2013). On the other hand, RNA sequencing, a high-throughput technique generates complete transcriptome information at a greater depth and coverage facilitating in-depth analysis of transcriptome at the whole genome level (Mortazavi et al. 2008).

Transcriptomics studies in the stress biology field have provided insights on the role of redox signalling, SOS pathway, ABA pathway and calcium signalling in salt-tolerant *B. juncea* (Srivastava et al. 2011; Sharma et al. 2015). Sinha et al. (2015) obtained de novo assembly of transcriptome data of cold stressed siliques and revealed the role of protein kinases and various transcription factors like *CBF, ICE1, ICE2, DREB, CAMTA3, LEA* and other cold responsive genes (*COR*) in stress tolerance at early and late developmental stages. The role of these genes has also been well established in *B. napus* and *B. rapa* in freezing or cold stress (Jung et al. 2014; Pu et al. 2019; Jian et al. 2020). Similarly, comparative RNA-Seq integrated with systems biology, proteomics and metabolomics has revealed the role of several genes in heat stress (Yue et al. 2021) and heavy metal stress (Mourato et al. 2015).

Identifying potential candidates involved in tolerance to more than one stress is essential for engineering multiple abiotic and biotic stresses. Bhardwaj et al. (2015) explored genome-wide transcriptome changes in *B. juncea* seedlings against two crucial abiotic stresses high temperature and drought to elucidate common genes and pathways. Gene duplication and alternative splicing are common in polyploids like Brassica, and transcriptome combined with bioinformatic approaches has been useful in exploring new duplicated allelic variants responsive to multiple stresses like *SOS*, *PYL* and *SOD* involved in salt, drought, low temperature and high temperature in *B. juncea* (Cheng et al. 2019; Verma et al. 2019; Das Laha et al. 2020).

Transcriptomics has also played a crucial role in monitoring the molecular mechanisms involved in the complex interactions between the pathogen or pest infecting Brassica. It provides an in-depth understanding of cellular mechanisms involved in providing innate and systemic acquired immunity in response to several biotic stress responses. Transcriptomics has also expanded our knowledge on non-host disease resistance mechanisms in plant–pathogen interaction where resistant sources are not known yet (Fatima et al. 2019). **Dual RNA-Seq** to study gene expression simultaneously in both plant and pathogen and **Bulk Segregant RNA-Seq (BSR-Seq)** of varieties with contrasting phenotype are advanced approaches used to identify probable candidate genes through transcriptomics (Westermann et al. 2012; Fu et al. 2019). Chu et al. (2014) combined traditional fine mapping with transcriptomics to unravel quantitative disease resistance in *B. rapa* against clubroot disease and Qasim et al. (2020) fine mapped 36 probable candidate genes from three overlapping QTLs contributing to stem rot resistance. RNA-Seq analysis provided insights on resistance mechanisms in *B. napus* recombinant inbred lines (RILs) developed by pyramiding two clubroot-resistant genes (Shah et al. 2020). Transcriptomics along with establishment of pan-genome and identification of RGA can be a potential approach to characterize and validate R genes in *B. juncea* against pathogens and pests. Due to paucity of resistant sources in Brassica against aphid, initial transcriptome studies were performed in Arabidopsis (Bhattacharya 2019). Recently, comparative transcriptomics implicated the differential role of various transcription factors, oxidative stress genes, glucosinolate pathway, and secondary metabolites in the feeding efficacy of two aphids in *B. juncea* (Duhlian et al. 2020). Ento(o)mics have been widely used for deciphering defence-related mechanisms against insect pests (Zogli et al. 2020).

For improving our current understanding of stress-related machinery, gene expression profiling of microRNA and long noncoding RNAs (lncRNA) are also gaining importance in recent years. In recent years, several microRNAs have been identified to be involved in various abiotic and biotic stress in *Brassica* sp. through the RNA-Seq approach (Bhardwaj et al. 2015; Zhu et al. 2019; Ahmed et al. 2020) In biotic stress, lncRNA has been reported to play a crucial role in defence response against *S. sclerotiorum* in *B. napus* and also in abiotic stress like heat and drought (Joshi et al. 2016; Bhatia et al. 2020).

24.7.5 Metabolomics

For adapting to biotic and abiotic stresses, the plants modulate the production and accumulation of primary and secondary metabolites for successful adaptation to stress. Therefore, a comprehensive exploration of metabolome perturbations associated with stress sensitive and tolerant germplasm/variety/landraces can provide insights into various metabolic networks linked/associated with stress (Parida et al. 2018). Analytical methods like mass spectrometry (MS) and nuclear magnetic resonance (NMR) combined with supervised and multivariate analysis are promising tools to understand the complexity of metabolites under various stress conditions. Due to limited information in Brassica, Arabidopsis has played a central role in expanding our knowledge on key metabolic pathways regulated under abiotic and biotic stress (Razzaq et al. 2019). Differences in metabolites, especially glucosinolate which are important defence regulators in the *Brassicaceae* family have been studied in various *Brassica* species through metabolomics (Misra 2016; Arena et al. 2020; Liu et al. 2020). Integrating metabolic profiling with QTL and mGWAS has been used for predicting the association between genetic polymorphisms and concentration of metabolites (Dumas 2012). Metabolomics combined with the systems biology approach is emerging as a potential tool to explore the role of metabolites in resistance to plant–pathogen interaction (Neik et al. 2020) and is being established as an important tool for genomics assisted breeding for crop improvement (Fernie and Schauer 2009). Since metabolomics is not completely understood in *B. juncea*, we can integrate the knowledge available from related Brassica crops (Table 24.2).

24.8 Omics-Assisted Breeding: Accelerating Crop Improvement

Conventional or pedigree breeding efforts have yielded positive results in developing superior varieties in Brassica but integrating omic tools with traditional methods can speed up the process of breeding. Advances in genotyping and phenotyping platforms has provided immense scope for discovering genome-wide markers and construction of high-density linkage maps to identify QTLs of interest and applying them in marker-assisted breeding (MAB)/marker-assisted selection (MAS). Still, there are some limitations in MAS such as linkage drag which refers to introgression of undesirable traits along with the desired ones, transfer of multiple traits, and thus has not been successfully employed to improve complex traits (Hu et al. 2021). To overcome these limitations, genomic selection (GS), a spin-off from genome-wide association genetics can be employed for addressing quantitative and polygenic traits in breeding programs. GS is defined in terms of genetic value or estimated breeding values (EBVs) derived from genome-wide markers and predicts accuracies based on interactions between the genotype and the environment (G * E) in a large population (Cowling and Balázs 2010; Varshney et al. 2017). It is advantageous over MAS, especially for complex traits as it does not require identifying markers linked to QTL for a trait of interest, it is less time consuming and has higher accuracy (Gemenet et al. 2020). GS has enabled identification of superior genotypes in polyploid crops like wheat (Arruda et al. 2016), *Brassica* (Würschum et al. 2014) and legumes (Varshney et al. 2017). Since most of the abiotic stress and some biotic stress (fungal diseases) in Brassica are complex, GS can be widely implemented to enhance the yield of Brassica crops and to dissect R genes since plant resistance to biotic stress is governed by several major and minor quantitative genes.

24.8.1 Gene Editing (GEd)

As discussed in the preceding sections, functional and comparative genomics have identified several candidate genes, from the Brassicas and their wild relatives, that are involved in biotic and abiotic stress responses. The knowledge about these regulatory pathways and candidate genes drawn from the multidimensional omics approach needs to be translated to develop new varieties or improve the elite cultivars with desired traits. Thus, genome editing has emerged as a promising technology for developing cultivars with improved stress-related traits (Kumar et al. 2021). CRISPR-Cas9 technology has been based on the borrowed knowledge of bacterial type II acquired immune system and has been established for several economically important diploid and polyploid crops including Brassica (Osakabe and Osakabe 2015; Pacher and Puchta 2017). The efficacy of CRISPR-Cas9 to induce mutation at one or multiple loci has been tested in *B. napus* (Chang et al. 2021), *B. rapa* (Jeong et al. 2019), *B. oleracea* (Ma et al. 2019) and *B. campestris* (Xiong et al. 2019). Huang et al. (2020a) simultaneously targeted multiple copies of the *FADS2* gene to induce allelic variations in oleic acid and fatty acid levels in *B. napus*. This technique was effectively utilized for multiplex genome editing of *WRKY11* and *WRKY70* genes involved in biotic stress in canola (Sun et al. 2018). Similarly, since plant hormone ethylene plays an important role in abiotic stress tolerance, targeting multiple ethylene response factors through CRISPR-Cas9 is a promising strategy to develop multiple abiotic stress-tolerant crops (Debbarma et al. 2019). In addition to gene knockout, the CRISPR-Cas9-based techniques have also been developed for transcriptional regulation in the plant which can be deployed to activate or suppress the expression of stress-related genes identified through

Table 24.2 Metabolic studies to investigate abiotic and biotic responses in Brassica

Species	Stress	Method	Major findings	References
B. juncea	Cadmium	HPLC, MS-QTOF	Upregulation of phosphoglycolate: role of photorespiration	D'Alessandro et al. (2013)
	Aluminium	Biochemical analysis	Upregulation of antioxidants such as SOD, CAT, APX, GR, GST	Ahmad et al. (2018)
	Alternaria brassisicola	NMR spectroscopy	Spirobrassinin is a major phytoalexin and accumulation of *N*-acetyl-3-indolylmethanamine due to detoxification of brassinin by the pathogen	Pedras et al. (2009)
	Phosphate, phosphite	HPLC	Modulation of secondary metabolites glucosinolates and flavonoids regulation of nitrate concentrations	Trejo-Téllez et al. (2019)
	Photosynthetically Active Radiation (PAR)			
	Biotic stress	HPLC	Major role of indole glucosinolates after exogenous application of biotic elicitors	Augustine and Bisht (2017)
B. napus	Drought	Biochemical analysis	Increase in phenols, flavonoids, tocopherol in drought-resistant varieties	Rezayian et al. (2018)
	Drought	HPLC	Increase in glucosinolate content	Jensen et al. (1996)
	Low carbon-di-oxide	GC-MS, UPLC-MS, MRM HPLC-MS	Changes in fatty acid metabolism, jasmonic acid biosynthesis, starch and sucrose metabolism and redox regulation	Geng et al. (2017)
	Cadmium	LC-MS	Role of anthocyanins, lignans, cell wall saccharides, jasmonic acid, ethylene and vitamin B6 in tolerance to cadmium	Mwamba et al. (2020)
	Cold or low temperature	UHLC/MS/MS2 and GC-MS	Transgenic plants overexpressing phospholipase C2 accumulated higher levels of maltose and unsaturated fatty acids at subzero temperatures, elevation of myoinositiol, raffinose, flavonoids under acclimatizing conditions	Nokhrina et al. (2014)
	Sclerotinia sclerotiorum		High levels of linoleate and linolenate, precursors of JA pathway	
	Heat or high temperature	GC-MS	Increase in the level of sugars in heat tolerant genotypes	Koscielny et al. (2018)
	Plasmodiphoa brassicae	LC-MS	Role of citric acid and gluconasturtiin in resistance	Wagner et al. (2019)
	Fusarium oxysporum and *F. germanium*	RP-UHPLC-PDA-ESI-MS	Role of glucosinolates when infected with pathogenic and non-pathogenic fungi	Andini et al. (2019)

(continued)

Table 24.2 (continued)

Species	Stress	Method	Major findings	References
B. nigra	Herbivory attack	GC and LC TOF MS	Accumulation of glucosinolates, syngirin, TCAAs and depletion of amino acids after herbivory attack	Papazian et al. (2019)
B. rapa	*Putella xylostella* and *Spodoptera exigua*	Two-dimensional NMR	Increase in levels of sinapic acid, ferulic acid and gluconapin	Widarto et al. (2006)
	F. oxysporum	NMR spectroscopy	Infected plants accumulate more flavonoids, phenyl propanoids and fumaric acid	Abdel-Farid et al. (2009)
	Leptosphaeria maculans and *F. oxysporum*	HPLC	Role of indole and aliphatic glucosinolates in two different cultivars after infection	Abdel-Farid et al. (2010)
	Drought		Increase in the levels of amino acids alanine and aspartate	Good and Zaplachinski (1994)
B. oleracea	*Xanthomonas campestris* pv. *Campestris*	GC-MS, LC-MS, NMR	Increase in alkaloids, coumarins and sphingolipids after pathogen infection	Tortosa et al. (2018)

transcriptome analysis (Lowder et al. 2015). The advances in genome editing technology, the yield, quality and plant type-related traits that have been targeted and the bottlenecks in *B. napus* have been reviewed (Chang et al. 2021). Genome editing has not been extensively used in *B. juncea* but with the availability of genome sequence and refined gene annotations, these methods can be established to further characterize and engineer multiple abiotic and biotic stress response candidate genes in *B. juncea*.

Due to bottlenecks in stable transformation techniques and off-target mutations in the traditional CRISPR-Cas9 method, DNA-free gene editing approaches have been introduced. In this method, pre-assembled ribonucleoprotein (RNP) complexes consisting of sgRNA and Cas9 are delivered into the genome through protoplast or particle bombardment (Park and Choe 2019). Murovec et al. (2018) used this method to successfully induced high mutation frequencies in two endogenous genes of *B. oleracea* and *B. rapa*.

Besides modulating the susceptibility factors/genes to improve disease resistance, the CRISPR-Cas9 approach has been extended to mutate virulence genes in pathogens to functionally validate the effectors and predict their target in the host. It has been successfully used to create a loss of function mutant for virulent genes in *S. sclerotiorum* to implicate the role of mutated genes in development and virulence (Li et al. 2018). Recently, a deletion mutant in the *AvrLm7* of *L. maculans* caused a loss in pathogenicity in *B. napus* genotypes not carrying the cognate *Rlm7* gene (Zou et al. 2020).

The various techniques used in unravelling *B. juncea* stress biology have been diagrammatically summarized in Fig. 24.1.

24.9 Improving Stress Tolerance Through Genetic Engineering (GE)

The lack of natural resistance source in the related Brassica germplasm for fungal diseases like Alternaria blight, stem rot, powdery mildew, aphid resistance, *Orobanche* resistance and abiotic stress tolerance has driven the genetic engineering interventions for improving Indian mustard (Thakur et al. 2020a, b). Although, transgenics have a lower acceptance in society

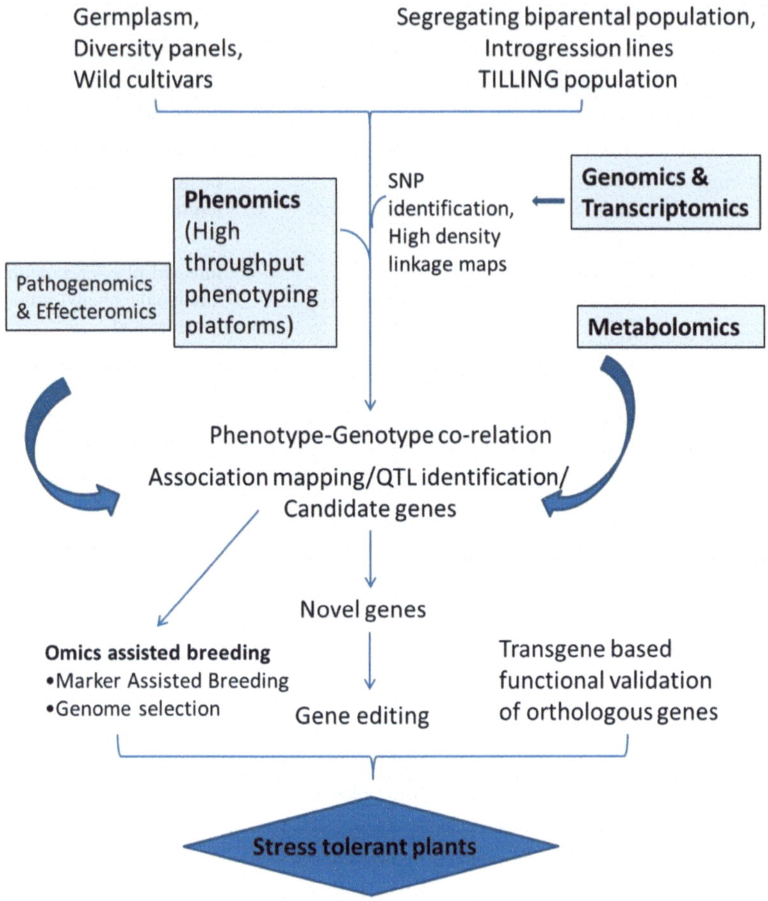

Fig. 24.1 Various omics-based techniques used in unraveling and improving *B. juncea* biotic and abiotic stress response

due to several misconceptions, it can be a beneficial tool to accelerate the development of plants resistant to abiotic and biotic stresses as it is faster in comparison to conventional breeding techniques. Additionally, it also removes the natural barriers involved in pyramiding inter-species traits. Table 24.3 showcases the glimpses of the use of genetic engineering in biotic, abiotic stress tolerance and phytoremediation in *B. juncea*.

24.10 Conclusions

The availability of reference genome sequence information for all three subgenomes (A, B and C) of Brassica along with the sequence for the model plant Arabidopsis can be extensively utilized for evolutionary and diversity analysis, construction of high-density linkage maps, QTL identification of various stress-related traits. The high-throughput omics and reliable phenotypic data have provided momentum for a sustainable crop improvement program in oilseeds by identifying candidate genes followed by functional validation through gene editing, exploring the role of various signaling and metabolic networks in stress tolerance. Efforts are now needed for capturing the complete genetic diversity within the species as well in the related wild crop species to develop pan-genome and super pan-genome. This will allow the breeders to more precisely apply genomic selection (GS), in conjugation with speed breeding (SB) for improving biotic and abiotic stress associated traits in Brassica varieties. The establishment of gene editing techniques in various *Brassica* spp. has made precision breeding a popular choice for introducing genetic variation in the genome and

Table 24.3 Table enlisting the biotic and abiotic stress tolerant transgenic *B. juncea* developed

Biotic stress tolerant B. juncea transgenics

Pest/pathogen	Source of transgene	Protein encoded	Impact	Mode of action of transgene	References
Sclerotinia sclerotiorum	*T. harzianum*	Chitinase 42 (*chit42*)	Reduced lesion formation	Chitin degrading enzyme (Hydrolyses β, 1–4 linkage)	Ojaghian et al. (2020)
	Barley	Oxalate oxidase (*Boxo*)	11–42% reduction in lesion	Degrades pathogen oxalates	Verma and Kaur (2021)
		Synthetic chimera of cecropin A and melittin cationic antimicrobial peptides (*msrA1*)	56–71% protection against *S. sclerotiorum*	Membrane antagonists	Rustagi et al. (2014)
Alternaria brassicae, Alternaria brassicicola	*T. virens*	Endochitinase (*ech42*)	Delayed onset of infection; 30–73% reduction in infection area	Degrades fungal cell wall chitinase	Kamble et al. (2016)
Alternaria brassicae Erysiphe cruciferarum	*B. juncea*	Non-repressor of pathogen-related (*NPR1*)	Delayed onset and reduced disease severity	Activation of SA mediated defence	Ali et al. (2017)
Alternaria brassicae	*msrA1*	A synthetic chimera of cecropin A and melittin cationic antimicrobial peptides	44–62% reduction in *A. brassicae* hyphal growth, 69–85% protection in transgenics	Membrane antagonist	Rustagi et al. (2014)
	Hevea brasiliensis	Lectin	Increased latency period; reduced necrosis lesions, disease intensity, and delayed senescence	Bind to carbohydrates and act as a defence tool	Kanrar et al. (2002)
	Chickpea	Lectin	30–60% reduced infection; enhanced tolerance to salinity and drought	Fungal cell wall carbohydrate immobilization	Kumar et al. (2015)

(continued)

Table 24.3 (continued)

		Class I chitinase	12–56% reduction in in vitro colony size, reduced disease onset and lesions	Fungal cell wall degradation	Mondal et al. (2003)
	Barley	Class II chitinase and ribosome inactivating protein (*RIP*)	44% reduction in fungus mycelia in vitro; reduced disease onset and lesion	RIP inactivate foreign ribosomes	Chhikara et al. (2012)
	Tomato	Class I glucanase	15–54% reduction in mycelial growth, delayed and reduced lesions	Hydrolyses fungal cell wall glucans	Mondal et al. (2007)
A. brasisicola	NA	*NIC* (Synthetic chitinase)	15–20% reduction in fungal growth following leaf extract treatment	Cell wall degradation	Munir et al. (2016)
Albugo candida	*B. juncea*	*WRR1* (CC-NB-LRR) R gene	Complete resistance against 6 isolates	R gene	Arora et al. (2019a)
Lipaphis erysimi	*Colocasia esculenta*, *Galanthus nivalis*	Tuber agglutinin (*CEA*), agglutinin (*GNA*)-related lectin	70–81% increased mortality, reduced fecundity (49.35–62.11%)		Das et al. (2018)
	Wheat	Wheat germ agglutinin (*WGA*)	65–85% mortality on feeding	Anti-metabolic and Insecticidal	Kanrar et al. (2002)
	Allium sativum	Leaf agglutinin (*ASAL*)	Reduced insect survival	blocks insect gut epithelial membrane	Dutta et al. (2005)
	Allium cepa	Agglutinin (*ACA*)	Increased insect mortality		Hossain et al. (2006)
	Lentil	Lentil lectin (*LL*)	40% inhibition of larval survival		Rani et al. (2017)

(continued)

Table 24.3 (continued)

	Chickpea	Chickpea protease inhibitor (*CPPI*)		Disrupts assimilation of dietary proteins	Rani et al. (2017)
	Mentha arvensis	((E)-β-farmesene) *Ebf*	37.5–80% repellence of aphids	Sesquiterpene compound, component of alarm pheromones	Verma et al. (2015)
	Rorripa indica	Defensin (*RiD*)	36.67% survival of aphids	Insecticidal, inhibits nutrient uptake	Sarkar et al. (2017)
	R. indica	Nematode resistance protein like homolog (*HSPRO2*)	Reduced survivability	Resistance protein. responsive to oxidative stress and SA	Bose et al. (2019)
Plutella xylostella	*Baccilus thuringensis*	*Cry1Ac* + *Cry1C*	High larval mortality, reduced eggs	Lysis of gut epithelial cells	Cao et al. (2008), Shelton et al. (2008)
Abiotic stress tolerant B. juncea transgenics					
Gene and mode of action	Source	Stress	Impact		References
Annexin (***AnnBj2***); calcium-dependant membrane-binding protein	*B. juncea cv.* Pusa Jai Kisan	Salinity	Glucose, NaCl, and ABA insensitivity, led to increased chlorophylls, water content, proline, calcium, potassium, reduced Na+ and thiobarbituric acid reactive species		Ahmed et al. (2017)
Late embryogenesis abundant (***LEA4-1***); osmoprotectant, membrane stabilizers, antioxidant and molecular chaperones	*A. thaliana*	Salinity and drought	Enhanced salt, drought tolerance		Saha et al. (2016)
Glyoxalase I, II (***glyI, II***); convert cytotoxic methylglyoxal and other 2-oxoaldehydes to their 2-hydroxyacids	*B. juncea*, *Oryza sativa*	Salinity, heavy metal, and drought	Enhanced salinity, heavy metal ($ZnCl_2$), and drought stress		Saxena (2011), Rajwanshi et al. (2016)
LEAFY; flower meristem identity gene	*A. thaliana*	Heat	Early flowering induction to escape terminal heat stress		Sahni et al. (2013)

(continued)

Table 24.3 (continued)

Transgenics developed in B. juncea for phytoremediation					
Genotype	Gene	Source	Heavy metal	Enhanced trait	References
B. juncea	Glutathione synthase (*gshII*)	*E. coli*	Cadmium	25% higher Cd accumulation in shoots	Zhu et al. (1999)
B. juncea cv. Rai-5	*Agrobacterium rhizogenes* mediated hairy roots induction		Uranium	Uranium detoxification	Eapen et al. (2003)
B. juncea	*AtPCS1* (Phytochelatin synthase)	*A. thaliana*	Cadmium	Enhanced Cd, As tolerance with increased accumulation in shoots	Gasic and Korban (2007)
B. juncea	*γ-ECS* (γ-glutamylcysteine synthase) and GS (glutathione synthase) source		Several heavy metals	Both showed enhanced tolerance to Cr, Mn, Mo, Cu in seedling stage; γ-ECS to As, Ni, and GS to Zn	Reisinger et al. (2008)
B. juncea cv. Rai-5	*AtATM3*(ATP binding cassette transporter protein)	*A. thaliana*	Cadmium and lead	1.4–1.6 and 1.3–1.7 fold increased Cd and Pb accumulation	Bhuiyan et al. (2011)

can further be exploited to fasten crop improvement. In the future, multidimensional omics along with a systems biology approach will ensure the selection of specific Brassica varieties resilient to various abiotic and biotic stresses.

References

Abdel-Farid IB, Jahangir M, van den Hondel CAMJJ, Kim HK, Choi YH et al (2009) Fungal infection-induced metabolites in *Brassica rapa*. Plant Sci 176 (5):608–615

Abdel-Farid IB, Jahangir M, Mustafa NR, van Dam NM, van den Hondel CAMJJ et al (2010) Glucosinolate profiling of *Brassica rapa* cultivars after infection by *Leptosphaeria maculans* and *Fusarium oxysporum*. Biochem Syst Ecol 38(4):612–620

Ahmad J, Baig MA, Ali AA, Al-Huqail AA, Ibrahim MM et al (2018) Differential antioxidative and biochemical responses to aluminium stress in *Brassica juncea* cultivars. Hortic Environ Biotechnol 59(5):615–627

Ahmar S, Liaqat N, Hussain M, Salim MA, Shabbir MA et al (2019) Effect of abiotic stresses on *Brassica* species and role of transgenic breeding for adaptation. Asian J Res Crop Sci 3(1):1–10

Ahmed I, Yadav D, Shukla P, Vineeth TV, Sharma PC et al (2017) Constitutive expression of *Brassica juncea* annexin, *AnnBj2* confers salt tolerance and glucose and ABA insensitivity in mustard transgenic plants. Plant Sci 265:12–28

Ahmed W, Li R, Xia Y, Bai G, Siddique KHM et al (2020) Comparative analysis of miRNA expression profiles between heat-tolerant and heat-sensitive genotypes of flowering Chinese cabbage under heat stress using high-throughput sequencing. Genes 11(3):264

Alamri S, Hu Y, Mukherjee S, Aftab T, Fahad S et al (2020) Silicon-induced postponement of leaf senescence is accompanied by modulation of antioxidative defense and ion homeostasis in mustard (*Brassica juncea)* seedlings exposed to salinity and drought stress. Plant Physiol Biochem 157:47–59

Alessandro AD, Taamalli M, Gevi F, Timperio AM, Zolla L et al (2013) Cadmium stress responses in *Brassica juncea*: hints from proteomics and metabolomics. J Proteome Res 12(11):4979–4997

Ali S, Mir ZA, Tyagi A, Mehari H, Meena RP et al (2017) Overexpression of *NPR1* in *Brassica juncea* confers broad spectrum resistance to fungal pathogens. Front Plant Sci 8:1–16

Andini S, Dekker P, Gruppen H, Araya-Cloutier C, Vincken J-P (2019) Modulation of glucosinolate composition in Brassicaceae seeds by germination and fungal elicitation. J Agric Food Chem 67 (46):12770–12779

Arena K, Cacciola F, Dugo L, Dugo P, Mondello L (2020) Determination of the metabolite content of *Brassica juncea* cultivars using comprehensive two-dimensional liquid chromatography coupled with a photodiode array and mass spectrometry detection. Molecules 25(5):1235

Arend D, Junker A, Scholz U, Schüler D, Wylie J et al (2016) PGP repository: a plant phenomics and genomics data publication infrastructure. Database (oxford) 2016:33

Arifuzzaman M, Oladzadabbasabadi A, McClean P, Rahman M (2019) Shovelomics for phenotyping root architectural traits of rapeseed/canola (*Brassica napus* L.) and genome-wide association mapping. Mol Genet Genom 294(4):985–1000

Arora H, Padmaja KL, Paritosh K, Mukhi N, Mukhopadhyay A et al (2019a) BjuWRR1, a CC-NB-LRR gene identified in *Brassica juncea*, confers resistance to white rust caused by Albugo Candida. Theor Appl Genet 132(8):2223–2236

Arora S, Steuernagel B, Gaurav K, Chandramohan S, Long Y et al (2019b) Resistance gene cloning from a wild crop relative by sequence capture and association genetics. Nat Biotechnol 37(2):139–143

Arruda MP, Lipka AE, Brown PJ, Krill AM, Thurber C et al (2016) Comparing genomic selection and marker-assisted selection for Fusarium head blight resistance in wheat (*Triticum aestivum* L.). Mol Breed 367 (36):1–11

Arunachalam A, Andreasson H (2021) Real-time plant phenomics under robotic farming setup: a vision-based platform for complex plant phenotyping tasks. Comput Electr Eng 92:107098

Augustine R, Bisht NC (2017) Regulation of glucosinolate metabolism: from model plant *Arabidopsis thaliana* to *Brassica* crops. Ref Ser Phytochem 163–199

Baird NA, Etter PD, Atwood TS, Currey MC, Shiver AL et al (2008) Rapid SNP discovery and genetic mapping using sequenced RAD markers. PLoS One 3(10):e3376

Bayer PE, Ruperao P, Mason AS, Stiller J, Chan CKK et al (2015) High-resolution skim genotyping by sequencing reveals the distribution of crossovers and gene conversions in *Cicer arietinum* and *Brassica napus*. Theor Appl Genet 128(6):1039–1047

Bayer PE, Golicz AA, Tirnaz S, Chan CKK, Edwards D et al (2019) Variation in abundance of predicted resistance genes in the *Brassica oleracea* pangenome. Plant Biotechnol J 17(4):789–800

Bhardwaj AR, Joshi G, Kukreja B, Malik V, Arora P et al (2015) Global insights into high temperature and drought stress regulated genes by RNA-Seq in economically important oilseed crop *Brassica juncea*. BMC Plant Biol 15(1):1–15

Bhatia G, Singh A, Verma D, Sharma S, Singh K (2020) Genome-wide investigation of regulatory roles of lncRNAs in response to heat and drought stress in *Brassica juncea* (Indian mustard). Environ Exp Bot 171:103922

Bhattacharya S (2019) *Brassica*-aphid interaction: challenges and prospects of genetic engineering for integrated aphid management. Physiol Mol Plant Pathol 108:101442

Bhuiyan MSU, Min SR, Jeong WJ, Sultana S, Choi KS et al (2011) Overexpression of *AtATM3* in *Brassica juncea* confers enhanced heavy metal tolerance and accumulation. Plant Cell Tiss Org Cult 107(1):69–77

Bose S, Gangopadhyay G, Sikdar SR (2019) *Rorippa indicaHSPRO2* expression in transgenic *Brassica juncea* induces tolerance against mustard aphid *Lipaphis erysimi*. Plant Cell Tiss Org Cult 136(3):431–443

Burton W, Salisbury P, Males D, Potts D (2007) The release of canola quality *Brassica juncea* for Australia. In: Proceedings of 12th international Rapeseed Congress, pp 291–293

Cai Z, Zeng DE, Liao J, Cheng C, Sahito ZA et al (2019) Genome-wide analysis of auxin receptor family genes in *Brassica juncea* var. Tumida. Genes (basel) 10 (2):1–15

Cai Z, Chen Y, Liao J, Wang D (2020) Genome-wide identification and expression analysis of jasmonate ZIM domain gene family in tuber mustard (*Brassica juncea* var. tumida). PLoS One 15(6):e0234738

CantilaAY SNSM, Amas JC, Edwards D, Batley J (2021) Recent findings unravel genes and genetic factors underlying *Leptosphaeria maculans* resistance in *Brassica napus* and its relatives. Intl J Mol Sci 22(1):1–25

Cao J, Shelton AM, Earle ED (2008) Sequential transformation to pyramid two Bt genes in vegetable Indian mustard (*Brassica juncea* L.) and its potential for control of diamondback moth larvae. Plant Cell Rep 27(3):479–487

Carreón-Anguiano KG, Islas-Flores I, Vega-Arreguín J et al (2020) Effhunter: a tool for prediction of effector protein candidates in fungal proteomic databases. Biomolecules 10(5):712

Cevik V, Boutrot F, Apel W, Robert-Seilaniantz A, Furzer OJ et al (2019) Transgressive segregation reveals mechanisms of Arabidopsis immunity to Brassica-infecting races of white rust (*Albugo candida*). Proc Natl Acad Sci 116(7):2767–2773

Chang T, Guan M, Zhou B, Peng Z, Xing M et al (2021) Progress of CRISPR/Cas9 technology in breeding of *Brassica napus*. Oil Crop Sci 6(2):53–57

Chen X, Vosman B, Visser RG, van der Vlugt RA, Broekgaarden C (2012) High throughput phenotyping for aphid resistance in large plant collections. Plant Methods 8(1):1–7

Chen S, Guo Y, Sirault X, Stefanova K, Saradadevi R et al (2019) Nondestructive phenomic tools for the prediction of heat and drought tolerance at anthesis in *Brassica* species. Plant Phenom 3264872. http://doi.org/10.34133/2019/3264872

Chen X, Tong C, Zhang X, Song A, Hu M et al (2021) A high-quality *Brassica napus* genome reveals expansion of transposable elements, subgenome evolution and disease resistance. Plant Biotechnol J 19(3):615–630

Cheng F, Wu J, Cai C, Fu L, Liang J et al (2016) Genome resequencing and comparative variome analysis in a *Brassica rapa* and *Brassica oleracea* collection. Sci Data 3(3):1–9

Cheng C, Zhong Y, Wang Q, Cai Z, Wang D et al (2019) Genome-wide identification and gene expression analysis of SOS family genes in tuber mustard (*Brassica juncea* var. Tumida). PLoS One 14(11):1–19

Chèvre AM, Eber F, Barret P, Dupuy P, Brace J (1997) Identification of the different *Brassica nigra* chromosomes from both sets of *B. oleracea-B. nigra* and *B. napus-B. nigra* addition lines with a special emphasis on chromosome transmission and self-incompatibility. Theor Appl Genet 94(5):603–611

Chhikara S, Chaudhury D, Dhankher OP, Jaiwal PK (2012) Combined expression of a barley class II chitinase and type I ribosome inactivating protein in transgenic *Brassica juncea* provides protection against *Alternaria brassicae*. Plant Cell Tiss Org Cult 108(1):83–89

Chu M, Song T, Falk KC, Zhang X, Liu X et al (2014) Fine mapping of *Rcr1* and analyses of its effect on transcriptome patterns during infection by *Plasmodiophora brassicae*. BMC Genomics 15:1166

Clarke WE, Higgins EE, Plieske J, Wieseke R, Sidebottom C et al (2016) A high-density SNP genotyping array for *Brassica napus* and its ancestral diploid species based on optimised selection of single-locus markers in the allotetraploid genome. Theor Appl Genet 129(10):1887–1899

Colmer J, O'Neill CM, Wells R, Bostrom A, Reynolds D et al (2020) SeedGerm: a cost-effective phenotyping platform for automated seed imaging and machine-learning based phenotypic analysis of crop seed germination. New Phytol 228(2):778–793

CongdonBS KMA, Filardo FF, Coutts BA (2019) In-field capable loop-mediated isothermal amplification detection of Turnip yellows virus in plants and its principal aphid vector *Myzus persicae*. J Virol Methods 265:15–21

Corona-Lopez DDJ, Sommer S, Rolfe SA, Podd F, Grieve BD (2019) Electrical impedance tomography as a tool for phenotyping plant roots. Plant Methods 15(1):1–15

Cowling WA, Balázs E (2010) Prospects and challenges for genome-wide association and genomic selection in oilseed Brassica species. Genome 53(11):1024–1028

D'Alessandro A, Taamalli M, Gevi F, Timperio AM, Zolla L et al (2013) Cadmium stress responses in *Brassica juncea*: hints from proteomics and metabolomics. J Proteome Res 12(11):4979–4997

Dalyan E, Yüzbaşıoğlu E, Keskin BC, Yıldızhan Y, Memon A et al (2017) The identification of genes associated with Pb and Cd response mechanism in *Brassica juncea* L. by using *Arabidopsis* expression array. Environ Exp Bot 139:105–115

Danzi D, Briglia N, Petrozza A, Summerer S, Povero G et al (2019) Can high throughput phenotyping help food security in the mediterranean area. Front Plant Sci 10:15. https://doi.org/10.3389/fpls.2019.00015

Das A, Ghosh P, Das S (2018) Expression of *Colocasia esculenta* tuber agglutinin in Indian mustard provides resistance against *Lipaphis erysimi* and the expressed protein is non-allergenic. Plant Cell Rep 37(6):849–863

Das Laha S, Dutta S, Schäffner AR, Das M (2020) Gene duplication and stress genomics in *Brassicas*: current understanding and future prospects. J Plant Physiol 255:153293

Das TK, Ghosh S, Gupta K, Sen S, Behera B et al (2020) The weed *Orobanche*: species distribution, diversity, biology and management. J Res Weed Sci 3(2):162–180

Daval S, Belcour A, Gazengel K, Legrand L, Gouzy J et al (2019) Computational analysis of the *Plasmodiophorabrassicae* genome: mitochondrial sequence description and metabolic pathway database design. Genomics 111(6):1629–1640

Debbarma J, Sarki YN, Saikia B, Boruah HPD, Singha DL et al (2019) Ethylene Response Factor (ERF) family proteins in abiotic stresses and CRISPR–Cas9 genome editing of ERFs for multiple abiotic stress tolerance in crop plants: a review. Mol Biotechnol 61(2):153–172

Derbyshire M, Denton-Giles M, Hegedus D, Seifbarghi S, Rollins J et al (2017) The complete genome sequence of the phytopathogenic fungus *Sclerotinia sclerotiorum* reveals insights into the genome architecture of broad host range pathogens. Genome Biol Evol 9(3):593–618

Dixon GR (2009) The occurrence and economic impact of *Plasmodiophorabrassicae* and clubroot disease. J Plant Growth Regul 28(3):194–202

Dolatabadian A, Bayer PE, Tirnaz S, Hurgobin B, Edwards D et al (2020) Characterization of disease resistance genes in the *Brassica napus* pangenome reveals significant structural variation. Plant Biotechnol J 18(4):969–982

DotasaraS SD, Kumawat K, Jat G, Choudhary H et al (2018) Assessment of crop loss due to insect pests in Indian mustard in Semi-arid region of Rajasthan. J Entomol Zool Stud 6:770–772

Duhlian L, Koramutla MK, Subramanian S, Chamola R, Bhattacharya R (2020) Comparative transcriptomics revealed differential regulation of defense related genes in *Brassica juncea* leading to successful and unsuccessful infestation by aphid species. Sci Rep 10:10583

Dumas M-E (2012) Metabolome 2.0: quantitative genetics and network biology of metabolic phenotypes. Mol Biosyst 8(10):2494–2502

Dutreux F, Da Silva C, D'agata L, Couloux A, Gay EJ et al (2018) Data descriptor: De novo assembly and annotation of three *Leptosphaeria* genomes using oxford nanopore minion sequencing. Sci Data 5(1):1–11

Dutta I, Majumder P, Saha P, Ray K, Das S (2005) Constitutive and phloem specific expression of *Allium sativum* leaf agglutinin (ASAL) to engineer aphid (*Lipaphis erysimi*) resistance in transgenic Indian mustard (*Brassica juncea*). Plant Sci 169(6):996–1007

Dwivedi SL, Goldman I, Ceccarelli S, Ortiz R (2020) Advanced analytics, phenomics and biotechnology approach to enhance genetic gains in plant breeding. In: Sparks DL (ed) Advance in agronomy. Elsevier Inc., New York

Eapen S, Suseelan KN, Tivarekar S, Kotwal SA, Mitra R (2003) Potential for rhizofiltration of uranium using hairy root cultures of *Brassica juncea* and *Chenopodium amaranticolor*. Environ Res 91(2):127–133

Eckes AH, Gubała T, Nowakowski P, Szymczyszyn T, Wells R et al (2017) Introducing the *Brassica* information portal: towards integrating genotypic and phenotypic *Brassica* crop data. F1000Research 6:465

Elshire RJ, Glaubitz JC, Sun Q, Poland JA, Kawamoto K et al (2011) A robust, simple genotyping-by-sequencing (GBS) approach for high diversity species. PLoS One 6(5):e19379

Fan R, Jung J, Jin L (2006) High-resolution association mapping of quantitative trait loci: a population-based approach. Genetics 172(1):663–686

FAOSTAT http://www.fao.org/faostat/en/#data/QCL. Retrieved on 25 Aug 2021

Fatima U, Bhorali P, Borah S, Senthil-Kumar M (2019) Perspectives on the utilization of resistance mechanisms from host and nonhost plants for durable protection of *Brassica* crops against *Alternaria* blight. PeerJ 26(7):e7486

Fernie AR, Schauer N (2009) Metabolomics-assisted breeding: a viable option for crop improvement. Trends Genet 25(1):39–48

Fletcher RS, Mullen JL, Heiliger A, McKay JK (2015) QTL analysis of root morphology, flowering time, and yield reveals trade-offs in response to drought in *Brassica napus*. J Exp Bot 66(1):245–256

Fu F, Liu X, Wang R, Zhai C, Peng G et al (2019) Fine mapping of *Brassica napus* blackleg resistance gene *Rlm1* through bulked segregant RNA sequencing. Sci Rep 9(9):1–10

Furbank RT, Tester M (2011) Phenomics—technologies to relieve the phenotyping bottleneck. Trends Plant Sci 16(12):635–644

Garg H, Atri C, Sandhu PS, Kaur B, Renton M et al (2010) High level of resistance to *Sclerotinia sclerotiorum* in introgression lines derived from hybridization between wild crucifers and the crop *Brassica* species *B. napus* and *B. juncea*. Field Crop Res 117(1):51–58

Gasic K, Korban SS (2007) Transgenic Indian mustard (*Brassica juncea*) plants expressing an *Arabidopsis* phytochelatin synthase (*AtPCS1*) exhibit enhanced As and Cd tolerance. Plant Mol Biol 64(4):361–369

Gemenet DC, Lindqvist-Kreuze H, De Boeck B, da Silva PG, Mollinari M et al (2020) Sequencing depth and genotype quality: accuracy and breeding operation considerations for genomic selection applications in autopolyploid crops. Theor Appl Genet 133 (12):3345–3363

Geng S, Yu B, Zhu N, Dufresne C, Chen S (2017) Metabolomics and proteomics of *Brassica napus* guard cells in response to low CO_2. Front Mol Biosci 4:51

Golicz AA, Bayer PE, Barker GC, Edger PP, Kim H et al (2016) The pangenome of an agronomically important crop plant *Brassica oleracea*. Nat Commun 7(1):1–8

Good AG, Zaplachinski ST (1994) The effects of drought stress on free amino acid accumulation and protein synthesis in *Brassica napus*. Physiol Plant 90(1):9–14

Guo N, Wang S, Gao L, Liu Y, Wang X et al (2021) Genome sequencing sheds light on the contribution of structural variants to *Brassica oleracea* diversification. BMC Biol 19(1):1–15

Harper AL, Trick M, Higgins J, Fraser F, Clissold L et al (2012) Associative transcriptomics of traits in the polyploid crop species *Brassica napus*. Nat Biotechnol 30(8):798–802

Harper AL, He Z, Langer S, Havlickova L, Wang L et al (2020) Validation of an associative transcriptomics platform in the polyploid crop species *Brassica juncea* by dissection of the genetic architecture of agronomic and quality traits. Plant J 103(5):1885–1893

Havlickova L, He Z, Wang L, Langer S, Harper AL et al (2018) Validation of an updated associative transcriptomics platform for the polyploid crop species *Brassica napus* by dissection of the genetic architecture of erucic acid and tocopherol isoform variation in seeds. Plant J 93(1):181–192

Hayat S, Mir BA, Wani AS, Hasan SA, Irfan M et al (2010) Screening of salt-tolerant genotypes of *Brassica juncea* based on photosynthetic attributes. J Plant Interact 6(1):53–60

Hejna O, Havlickova L, He Z, Bancroft I, Curn V (2019) Analysing the genetic architecture of clubroot resistance variation in *Brassica napus* by associative transcriptomics. Mol Breed 39(8):112

Horvath DP, Zhang J, Chao WS, Mandal A, Rahman M et al (2020) Genome-wide association studies and transcriptome changes during acclimation and deacclimation in divergent *Brassicanapus* varieties. Intl J Mol Sci 21(23):9148

Hossain MA, Maiti MK, Basu A, Sen S, Ghosh AK et al (2006) Transgenic expression of onion leaf lectin gene in Indian mustard offers protection against aphid colonization. Crop Sci 46(5):2022–2032

Hu D, Jing J, Snowdon RJ, Mason AS, Shen J et al (2021) Exploring the gene pool of *Brassica napus* by genomics-based approaches. Plant Biotechnol J. http://doi.org/10.1111/pbi.13636 (Epub ahead of print)

Huang H, Cui T, Zhang L, Yang Q, Yang Y et al (2020a) Modifications of fatty acid profile through targeted mutation at *BnaFAD2* gene with CRISPR/Cas9-mediated gene editing in *Brassicanapus*. Theor Appl Genet 133(8):2401–2411

Huang Y, Hussain MA, Luo D, Xu H, Zeng C et al (2020b) A *Brassica napus* reductase gene dissected by associative transcriptomics enhances plant adaption to freezing stress. Front Plant Sci 11:971

Inturrisi F, Bayer PE, Yang H, Tirnaz S, Edwards D et al (2020) Genome-wide identification and comparative analysis of resistance genes in *Brassica juncea*. Mol Breed 40(8)

Jaradat AA (2018) Integrating plant ontogeny and structure in *Brassica napus* L. I. forward phenomics. Euphytica 214(8):1–26

Jat RS, Singh D (2018) Broomrape intimidation to rapeseed-mustard production in semi-arid regions. Adv Plants Agric Res 8(6):560–562

Jat RS, Singh VV, Sharma P, Rai PK (2019) Oilseed brassica in India: demand, supply, policy perspective and future potential. OCL 26:8

Jensen CR, Mogensen VO, Mortensen G, Fieldsend JK, Milford GFJ et al (1996) Seed glucosinolate, oil and protein contents of field-grown rape (*Brassica napus* L.) affected by soil drying and evaporative demand. Field Crop Res 47(2–3):93–105

Jeong SY, Ahn H, Ryu J, Oh Y, Sivanandhan G et al (2019) Generation of early-flowering Chinese cabbage (*Brassica rapa* spp. pekinensis) through CRISPR/Cas9-mediated genome editing. Plant Biotechnol Rep 13(5):491–499

Jian H, Xie L, Wang Y, Cao Y, Wan M et al (2020) Characterization of cold stress responses in different rapeseed ecotypes based on metabolomics and transcriptomics analyses. PeerJ 8:e8704

Jiang L, Sun Q, Wang Y, Chang P, Kong H et al (2021) Genome-wide identification and characterization of NAC genes in *Brassica juncea* var. Tumida. PeerJ 9:1–19

Jones RAC, Coutts BA, Hawkes J, Jones RAC, Coutts BA et al (2007) Yield-limiting potential of Beet western yellows virus in *Brassica napus*. Aust J Agric Res 58:788–801

Joshi RK, Megha S, Basu U, Rahman MH, Kav NNV (2016) Genome wide identification and functional prediction of long non-coding RNAs responsive to *Sclerotinia sclerotiorum* infection in *Brassica napus*. PLoS ONE 11(7):1–19

Jung HJ, Dong X, Park JI, Thamilarasan SK, Lee SS et al (2014) Genome-wide transcriptome analysis of two contrasting *Brassica rapa* doubled haploid lines under cold-stresses using Br135K oligomeric chip. PLoS One 9(8):e106069

Jupe F, Witek K, Verweij W, Śliwka J, Pritchard L et al (2013) Resistance gene enrichment sequencing (R en S eq) enables reannotation of the NB-LRR gene family from sequenced plant genomes and rapid mapping of resistance loci in segregating populations. Plant J 76(3):530–544

Kamble S, Mukherjee PK, Eapen S (2016) Expression of an endochitinase gene from *Trichoderma virens* confers enhanced tolerance to Alternaria blight in transgenic *Brassica juncea* (L.) czern and coss lines. Physiol Mol Biol Plants 22(1):69–76

Kanrar S, Venkateswari J, Kirti P, Chopra V (2002) Transgenic Indian mustard (*Brassica juncea*) with resistance to the mustard aphid (*Lipaphis erysimi* Kalt.). Plant Cell Rep 20(10):976–981

Karim MM, Dakouri A, Zhang Y, Chen Q, Peng G et al (2020) Two clubroot-resistance genes, *Rcr3* and *Rcr9wa*, mapped in *Brassica rapa* using bulk segregant RNA sequencing. Intl J Mol Sci 21(14):5033

Kaur S, Cogan NO, Ye G, Baillie RC, Hand ML et al (2009) Genetic map construction and QTL mapping of resistance to blackleg (*Leptosphaeria maculans*) disease in Australian canola (*Brassica napus* L.) cultivars. Theor Appl Genet 120(1):71–83

Khan AW, Garg V, Roorkiwal M, Golicz AA, Edwards D et al (2020) Super-pangenome by integrating the wild side of a species for accelerated crop improvement. Trends Plant Sci 25(2):148–158

Khanzada H, Wassan GM, He H, Mason AS, Keerio AA et al (2020) Differentially evolved drought stress indices determine the genetic variation of *Brassica napus* at seedling traits by genome-wide association mapping. J Adv Res 24:447–461

Kirti PB, Prakash S, Chopra VL (1991) Interspecific hybridization between *Brassica juncea* and *B. spinescens* through protoplast fusion. Plant Cell Rep 9 (11):639–642

Kloth KJ, ten Broeke CJ, Thoen MP, den Brink MH, Wiegers GL et al (2015) High-throughput phenotyping of plant resistance to aphids by automated video tracking. Plant Methods 11(1):1–14

Kole C, Muthamilarasan M, Henry R, Edwards D, Sharma R et al (2015) Application of genomics-assisted breeding for generation of climate resilient crops: progress and prospects. Front Plant Sci 6:563

Koscielny CB, Gardner SW, Duncan RW (2018) Impact of high temperature on heterosis and general combining ability in spring canola (*Brassica napus* L.). Field Crop Res 221:61–70

Kumar M, Kumar D, Shekhar S, Bisht S, Kumar V et al (2015) Ectopic overexpression of lectin in transgenic *Brassica juncea* plants exhibit resistance to fungal phytopathogen and showed alleviation to salt and drought stress. J Bioeng Biomed Sci 5:1. https://doi.org/10.4172/2155-9538.1000147

Kumar S, Singh YP, Singh SP, Singh R (2017) Physical and biochemical aspects of host plant resistance to mustard aphid, *Lipaphis erysimi* (Kaltenbach) in rapeseed-mustard. Arthropod-Plant Interact 11 (4):551–559

Kumar R, Sharma V, Suresh S, Ramrao DP, Veershetty A et al (2021) Understanding omics driven plant improvement and de novo crop domestication: some examples. Front Genet 12:637141

Kumari P, Bisht DS, Bhat SR (2017) Stable, fertile somatic hybrids between *Sinapis alba* and *Brassica juncea* show resistance to *Alternaria brassicae* and heat stress. Plant Cell Tiss Org Cult 133(1):77–86

Kumari P, Singh KP, Kumar S, Yadava DK (2020) Development of a yellow-seeded stable allohexaploid *Brassica* through inter-generic somatic hybridization with a high degree of fertility and resistance to *Sclerotinia sclerotiorum*. Front Plant Sci 11:575591

Larkan NJ, Yu F, Lydiate DJ, Rimmer SR, Borhan MH (2016a) Single R gene introgression lines for accurate dissection of the *Brassica-Leptosphaeria* pathosystem. Front Plant Sci 7:1771

Larkan NJ, Raman H, Lydiate DJ, Robinson SJ, Yu F et al (2016b) Multi-environment QTL studies suggest a role for cysteine-rich protein kinase genes in quantitative resistance to blackleg disease in *Brassica napus*. BMC Plant Biol 16(1):1–6

Lee RWH, Malchev IT, Rajcan I, Kott LS (2014) Identification of putative quantitative trait loci associated with a flavonoid related to resistance to cabbage seedpod weevil (*Ceutorhynchus obstrictus*) in canola derived from an intergeneric cross, *Sinapis alba* × *Brassica napus*. Theor Appl Genet 127:419–428

Lee J, Izzah NK, Jayakodi M, Perumal S, Joh HJ et al (2015) Genome-wide SNP identification and QTL mapping for black rot resistance in cabbage. BMC Plant Biol 15:1–11

Li P, Quan X, Jia G, Xiao J, Cloutier S, You FM (2016) RGAugury: a pipeline for genome-wide prediction of resistance gene analogs (RGAs) in plants. BMC Genomics 17(1):1–10

Li J, Zhang Y, Zhang Y, Yu PL, Pan H, Rollins JA (2018) Introduction of large sequence inserts by CRISPR-cas9 to create pathogenicity mutants in the multinucleate filamentous pathogen *Sclerotinia sclerotiorum*. mBio 9(3):e00567-18

Li B, Chen L, Sun W, Wu D, Wang M et al (2020a) Phenomics-based GWAS analysis reveals the genetic architecture for drought resistance in cotton. Plant Biotechnol J 18(12):2533–2544

Li M, Xie F, Li Y, Gong L, Luo Y et al (2020b) Genome-wide analysis of the heat shock transcription factor gene family in *Brassica juncea*: structure, evolution, and expression profiles. DNA Cell Biol 39(11):1990–2004

Lijuan W, Meili X, Hayward A, Fu D (2013) Applications and challenges of next-generation sequencing in Brassicaspecies. Planta 238:1005–1024

Links MG, Holub E, Jiang RHY, Sharpe AG, Hegedus D et al (2011) De novo sequence assembly of *Albugo candida* reveals a small genome relative to other biotrophic oomycetes. BMC Genomics 12(1):1–12

Liu S, Wang H, Zhang J, Zhang Y, Yang Q et al (2005) In vitro mutation and selection of doubled-haploid *Brassica napus* lines with improved resistance to *Sclerotinia sclerotiorum*. Plant Cell Rep 243(24):133–144

Liu S, Huang H, Yi X, Fitt BDL, Xu Z et al (2020) Dissection of genetic architecture for glucosinolate accumulations in leaves and seeds of *Brassica napus* by genome-wide association study. Plant Biotechnol J 18(6):1472–1484

Liu JJ, Sniezko RA, Zamany A, Williams H, Omendja K et al (2021) Comparative transcriptomics and RNA-seq-based bulked segregant analysis reveals genomic basis underlying *Cronartium ribicola vcr2* virulence. Front Microbiol 22(12):602812

Lohani N, Jain D, Singh MB, Bhalla PL (2020) Engineering multiple abiotic stress tolerance in canola, *Brassica napus*. Front Plant Sci 11:1–26

Lowder LG, Zhang D, Baltes NJ, Paul JW, Tang X et al (2015) A CRISPR/Cas9 toolbox for multiplexed plant genome editing and transcriptional regulation. Plant Physiol 169(2):971–985

Lu S, Fadlalla T, Tang S, Li L, Ali U et al (2019) Genome-wide analysis of phospholipase D gene family and profiling of phospholipids under abiotic stresses in *Brassica napus*. Plant Cell Physiol 60 (7):1556–1566

Ma C, Liu M, Li Q, Si J, Ren X, Song H et al (2019) Efficient BoPDS gene editing in cabbage by the CRISPR/Cas9 system. Hortic Plant J 5(4):164–169

Mahlein A-K, Oerke E-C, Steiner U, Dehne H-W (2012) Recent advances in sensing plant diseases for precision crop protection. Eur J Plant Pathol 133(1):197–209

Marsh JI, Hu H, Gill M, Batley J, Edwards D et al (2021) Crop breeding for a changing climate: integrating phenomics and genomics with bioinformatics. Theor Appl Genet 134(6):1677–1690

Meena PD, Awasthi RP, Chattopadhyay C, Kolte SJ, Kumar A (2010) Alternaria blight: a chronic disease in rapeseed-mustard. J Oilseed Brassica 1(1):1–11

Mei J, Shao C, Yang R, Feng Y, Gao Y et al (2020) Introgression and pyramiding of genetic loci from wild *Brassica oleracea* into *B. napus* for improving Sclerotinia resistance of rapeseed. Theor Appl Genet 133(4):1313–1319

Misra BB (2016) Cataloging the *Brassica napus* seed metabolome. Cogent Food Agric 2:1

Mondal KK, Chatterjee SC, Viswakarma N, Bhattacharya RC, Grover A (2003) Chitinase-mediated inhibitory activity of Brassica transgenic on growth of *Alternariabrassicae*. Curr Microbiol 47(3):171–173

Mondal KK, Bhattacharya RC, Koundal KR, Chatterjee SC (2007) Transgenic Indian mustard (*Brassica juncea*) expressing tomato glucanase leads to arrested growth of *Alternaria brassicae*. Plant Cell Rep 26 (2):247–2520

Mortazavi A, Williams BA, McCue K, Schaeffer L, Wold B (2008) Mapping and quantifying mammalian transcriptomes by RNA-Seq. Nat Methods 5(7):621–628

Mourato MP, Moreira IN, Leitão I, Pinto FR, Sales JR, Martins LL (2015) Effect of heavy metals in plants of the genus Brassica. Intl J Mol Sci 16(8):17975–17998

Munir I, Hussan W, Kazi MSK, Farhatullah MAA, Iqbal A, Munir R (2016) Production of transgenic *Brassica juncea* with the synthetic chitinase gene (NIC) conferring resistance to *Alternaria brassicicola*. Pak J Bot 48(5):2063–2070

Murovec J, Guček K, Bohanec B, Avbelj M, Jerala R (2018) DNA-free genome editing of *Brassica oleracea* and *B. rapa* protoplasts using CRISPR-Cas9 ribonucleoprotein complexes. Front Plant Sci 9:1594. http://doi.org/10.3389/fpls.2018.01594

Mwamba TM, Islam F, Ali B, Lwalaba JLW, Gill RA et al (2020) Comparative metabolomic responses of low- and high-cadmium accumulating genotypes reveal the cadmium adaptive mechanism in *Brassica napus*. Chemosphere 250:126308

Nagaharu U (1935) Genome analysis of *Brassica* with special reference to the experimental formation of *B. napus* and peculiar mode of fertilization. Jpn J Bot 7:389–452

Neik TX, Amas J, Barbetti M, Edwards D, Batley J (2020) Understanding host–pathogen interactions in *Brassica napus* in the omics era. Plants 9(10):1–31

Nokhrina K, Ray H, Bock C, Georges F (2014) Metabolomic shifts in *Brassica napus* lines with enhanced *BnPLC2* expression impact their response to low temperature stress and plant pathogens. GM Crops Food 5(2):120–131

Ojaghian S, Wang L, Xie GL (2020) Effect of introducing chitinase gene on the resistance of tuber mustard against white mold. Plant Pathol J 36(4):378–383

Osakabe Y, Osakabe K (2015) Genome editing with engineered nucleases in plants. Plant Cell Physiol 56 (3):389–400

Pacher M, Puchta H (2017) From classical mutagenesis to nuclease-based breeding—directing natural DNA repair for a natural end-product. Plant J 90(4):819–833

Panjabi-Massand P, Yadava SK, Sharma P, Kaur A, Kumar A et al (2010) Molecular mapping reveals two independent loci conferring resistance to *Albugo candida* in the east European germplasm of oilseed mustard *Brassica juncea*. Theor Appl Genet 121 (1):137–145

Papazian S, Girdwood T, Wessels BA, Poelman EH, Dicke M et al (2019) Leaf metabolic signatures induced by real and simulated herbivory in black mustard (*Brassica nigra*). Metabolomics 15(10):1–16

Parida AK, Panda A, Rangani J (2018) Metabolomics-guided elucidation of abiotic stress tolerance mechanisms in plants. Plant Metab Regul Under Environ Stress 89–131

Paritosh K, Pradhan AK, Pental D (2020) A highly contiguous genome assembly of *Brassica* nigra (BB) and revised nomenclature for the pseudochromosomes. BMC Genom 21(1):1–12

Paritosh K, Yadava SK, Singh P, Bhayana L, Mukhopadhyay A et al (2021) A chromosome-scale assembly of allotetraploid *Brassica juncea* (AABB) elucidates comparative architecture of the A and B genomes. Plant Biotechnol J 19(3):602–614

Park J, Choe S (2019) DNA-free genome editing with preassembled CRISPR/Cas9 ribonucleoproteins in plants. Transgen Res 28(2):61–64

Parry MAJ, Madgwick PJ, Bayon C, Tearall K, Hernandez-Lopez A (2009) Mutation discovery for crop improvement. J Exp Bot 60(10):2817–2825

Paul K, Pauk J, Kondic-Spika A, Allahverdiyev T, Sass L, Vass I (2019) Co-occurrence of mild salinity and drought synergistically enhances biomass and grain retardation in wheat. Front Plant Sci 10:1–15

Pedras MSC, Chumala PB, Jin W, Islam MS, Hauck DW (2009) The phytopathogenic fungus *Alternaria brassicicola*: phytotoxin production and phytoalexin elicitation. Phytochemistry 70(3):394–402

Perumal S, Koh CS, Jin L, Buchwaldt M, Higgins EE et al (2020) A high-contiguity *Brassica nigra* genome localizes active centromeres and defines the ancestral Brassica genome. Nat Plants 6(8):929–941

PovedaJ FM, Cartea ME, Velasco P (2020) Development of transgenic *Brassica* crops against biotic stresses caused by pathogens and arthropod pests. Plants 9 (12):1–23

Pu Y, Liu L, Wu J, Zhao Y, Bai J et al (2019) Transcriptome profile analysis of winter rapeseed (*Brassica napus* L.) in response to freezing stress, reveal potentially connected events to freezing stress. Intl J Mol Sci 20(11):1–24

Qasim MU, Zhao Q, Shahid M, Samad RA, Ahmar S et al (2020) Identification of QTLs containing resistance genes for sclerotinia stem rot in *Brassica napus* using comparative transcriptomic studies. Front Plant Sci 11:776

Qiao C, Yang J, Wan Y, Xiang S, Guan M et al (2020) A genome-wide survey of MATE transporters in Brassicaceae and unveiling their expression profiles under abiotic stress in rapeseed. Plants 9(9):1–25

Quezada-Martinez D, Nyarko CPA, Schiessl SV, Mason AS (2021) Using wild relatives and related species to build climate resilience in Brassica crops. Theor Appl Genet 134(6):1711–1728

Rahman M, Mamidi S, del Rio L, Ross A, Kadir MM et al (2016) Association mapping in *Brassica napus* (L.) accessions identifies a major QTL for blackleg disease resistance on chromosome A01. Mol Breed 36 (7):1–15

Rajarammohan S, Pental D, Kaur J (2019) Near-complete genome assembly of Alternaria *Brassicae*—a necrotrophic pathogen of brassica crops. Mol Plant Microbe Interact 32(8):928–930

Rajwanshi R, Kumar D, Yusuf MA, DebRoy S, Sarin NB (2016) Stress-inducible overexpression of glyoxalase I is preferable to its constitutive overexpression for abiotic stress tolerance in transgenic *Brassica juncea*. Mol Breed 36(6):1–15

Rana K, Atri C, Gupta M, Akhatar J, Sandhu PS et al (2017) Mapping resistance responses to Sclerotinia infestation in introgression lines of *Brassica juncea* carrying genomic segments from wild Brassicaceae *B. fruticulosa*. Sci Rep 7(1):1–12

Rani S, Sharma V, Hada A, Bhattacharya RC, Koundal KR (2017) Fusion gene construct preparation with lectin and protease inhibitor genes against aphids and efficient genetic transformation of *Brassica juncea* using cotyledons explants. Acta Physiol Plant 39 (5):1–13

Razzaq A, Sadia B, Raza A, Hameed MK, Salem F et al (2019) Metabolomics: a way forward for crop improvement. Metabolites 9(12):303

Reisinger S, Schiavon M, Terry N, Pilon-Smits EAH (2008) Heavy metal tolerance and accumulation in Indian mustard (*Brassica juncea* L.) expressing bacterial γ-glutamylcysteine synthetase or glutathione synthetase. Intl J Phytoremed 10(5):440–454

Rezayian M, Niknam V, Ebrahimzadeh H (2018) Effects of drought stress on the seedling growth, development, and metabolic activity in different cultivars of canola. Soil Sci Plant Nutr 64(3):360–369

Roy NN (1984) Interspecific transfer of *Brassica juncea*-type high blackleg resistance to *Brassica napus*. Euphytica 33(2):295–303

Rustagi A, Kumar D, Shekhar S, Yusuf MA, Misra S, Sarin NB (2014) Transgenic *Brassica juncea* plants expressing MsrA1, a synthetic cationic antimicrobial peptide, exhibit resistance to fungal phytopathogens. Mol Biotechnol 56(6):535–545

Sacristan MD, Gerdemann M (1986) Different behavior of *Brassica juncea* and *B. carinata* as sources of phoma lingam resistance in experiments of interspecific transfer to B. napus. Plant Breed 97(4):304–314

Sacristan MD, Gerdemann-Knörck M, Schieder O (1988) Transformation of *Brassica nigra* through protoplast cocultivation with *Agrobacterium tumefaciens*. In: Puite KJ, Dons JJM, Huizing HJ, Kool AJ, Koorneef M et al (eds) Progress in plant protoplast research. Kluwer, Dordrecht, pp 252–253

Saha B, Mishra S, Awasthi JP, Sahoo L, Panda SK (2016) Enhanced drought and salinity tolerance in transgenic mustard [*Brassica juncea* (L.) Czern & Coss.] overexpressing *Arabidopsis* group 4 late embryogenesis abundant gene (AtLEA4-1). Environ Exp Bot 128:99–111

Saharan GS, Mehta N, Meena PD (2017) The disease: downy mildew. In: Downy mildew disease of crucifers: biology, ecology and disease management. Springer, Berlin, pp 35–65

Sahni S, Ganie SH, Narula A, Srivastava PS, Singh HB (2013) Ectopic expression of Atleafy in *Brassica juncea* cv. Geeta for early flowering. Physiol Mol Biol Plants 19(3):455–459

Sánchez-Martín J, Steuernagel B, Ghosh S, Herren G, Hurni S, Adamski N, Vrána J, Kubaláková M, Krattinger SG, Wicker T, Doležel J, Keller B, Wulff BBH (2016) Rapid gene isolation in barley and wheat by mutant chromosome sequencing. Genome Biol 17(1):1–7

Sandhu SK, Pal L, Kaur J, Bhatia D (2019) Genome wide association studies for yield and its component traits under terminal heat stress in Indian mustard (*Brassica juncea* L.). Euphytica 215(11):1–15

Sarkar P, Jana K, Sikdar SR (2017) Overexpression of biologically safe Rorippa indica defensin enhances aphid tolerance in *Brassica juncea*. Planta 246 (5):1029–1044

Saxena M (2011) Overexpression of the glyoxalase II gene leads to enhanced salinity tolerance in *Brassica juncea*. Open Plant Sci J 5(1):23–28

Scholes JD, Rolfe SA, Scholes JD, Rolfe SA (2009) Chlorophyll fluorescence imaging as tool for understanding the impact of fungal diseases on plant performance: a phenomics perspective. Funct Plant Biol 36(11):880–892

Sekhon BS, Brar KS (1997) Estimation of avoidable yield losses of *Brassica* species due to insect pests in south-western region of Punjab. J Aphidol 3(1–2):125–134

Shah N, Li Q, Xu Q, Liu, J, Huang F et al (2020) Crb and pbba8.1 synergically increases resistant genes expression upon infection of *Plasmodiophora Brassicae* in *Brassica napus*. Genes (Basel) 11(2):202

Sharma R, Mishra M, Gupta B, Parsania C, Singla-Pareek SL, Pareek A (2015) De novo assembly and characterization of stress transcriptome in a salinity-tolerant variety CS52 of *Brassica juncea*. PLoS ONE 10(5):1–16

Shelton AM, Hatch SL, Zhao JZ, Chen M, Earle ED, Cao J (2008) Suppression of diamondback moth using Bt-transgenic plants as a trap crop. Crop Protec 27(3–5):403–409

Shi L, Shi T, Broadley MR, White PJ, Long Y et al (2013) High-throughput root phenotyping screens identify genetic loci associated with root architectural traits in *Brassica napus* under contrasting phosphate availabilities. Ann Bot 112(2):381–389

Singh KP, Kumari P, Rai PK (2021) Current status of the disease-resistant gene(s)/QTLs, and strategies for improvement in *Brassica juncea*. Front Plant Sci 12:1–28

Sinha S, Raxwal VK, Joshi B, Jagannath A, Katiyar-Agarwal S et al (2015) De novo transcriptome profiling of cold-stressed siliques during pod filling stages in Indian mustard (*Brassica juncea* L.). Front Plant Sci 6:932

Song JM, Guan Z, Hu J, Guo C, Yang Z et al (2020) Eight high-quality genomes reveal pan-genome architecture and ecotype differentiation of *Brassica napus*. Nat Plants 6(1):34–45

Song JM, Liu DX, Xie WZ, Yang Z, Guo L et al (2021a) BnPIR: *Brassica napus* pan-genome information resource for 1689 accessions. Plant Biotechnol J 19 (3):412–414

Song P, Wang J, Guo X, Yang W, Zhao C et al (2021b) High-throughput phenotyping: breaking through the bottleneck in future crop breeding. Crop J 9(3):633–645

Song X, Wei Y, Xiao D, Gong K, Sun P et al (2021c) *Brassica carinata* genome characterization clarifies U's triangle model of evolution and polyploidy in Brassica. Plant Pahysiol 186(1):388–406

Sperschneider J, Dodds PN, Gardiner DM, Singh KB, Taylor JM (2018) Improved prediction of fungal effector proteins from secretomes with EffectorP 2.0. Mol Plant Pathol 19(9):2094–2110

Srivastava AK, Srivastava S, Penna S, D'Souza SF (2011) Thiourea orchestrates regulation of redox state and antioxidant responses to reduce the NaCl-induced oxidative damage in Indian mustard (*Brassica juncea* (L.) Czern.). Plant Physiol Biochem 49(6):676–686

Steuernagel B, Periyannan SK, Hernández-Pinzón I, Witek K, Rouse MN et al (2016) Rapid cloning of disease-resistance genes in plants using mutagenesis and sequence capture. Nat Biotechnol 34(6):652–655

Struss D, Quiros CF, Röbbelen G (1992) Mapping of molecular markers on *Brassica* B-genome chromosomes added to *Brassica napus*. Plant Breed 108:320–323

Sun Q, Lin L, Liu D, Wu D, Fang Y et al (2018) CRISPR/Cas9-mediated multiplex genome editing of the BnWRKY11 and BnWRKY70 genes in *Brassica napus* L. Int J Mol Sci 19(9):2716

Suwabe K, Tsukazaki H, Iketani H, Hatakeyama K, Fujimura M et al (2003) Identification of two loci for resistance to clubroot (*Plasmodiophora brassicae* Woronin) in *Brassica rapa* L. Theor Appl Genet 107(6):997–1002

Suwabe K, Tsukazaki H, Iketani H, Hatakeyama K, Kondo M et al (2006) Simple sequence repeat-based comparative genomics between *Brassica rapa* and *Arabidopsis thaliana*: the genetic origin of clubroot resistance. Genetics 173(1):309–319

Tan M, Liao F, Hou L, Wang J, Wei L et al (2017) Genome-wide association analysis of seed germination percentage and germination index in *Brassica napus* L. under salt and drought stresses. Euphytica 213(2):1–15

Thakur S, Choudhary S, Dubey P, Bhardwaj P (2019) Comparative transcriptome profiling reveals the reprogramming of gene networks under arsenic stress in Indian mustard. Genome 62(12):833–847

Thakur AK, Parmar N, Singh KH, Nanjundan J (2020a) Current achievements and future prospects of genetic engineering in Indian mustard (*Brassica juncea* L. Czern & Coss.). Planta 252(4):1–20

Thakur S, Choudhary S, Majeed A, Singh A, Bhardwaj P (2020b) Insights into the molecular mechanism of arsenic phytoremediation. J Plant Growth Regul 39 (2):532–543

Tortosa M, Cartea ME, Rodríguez VM, Velasco P (2018) Unraveling the metabolic response of *Brassica oleracea* exposed to *Xanthomonas campestris pv. campestris*. J Sci Food Agri 98(10):3675–3683

Trejo-Téllez LI, Estrada-Ortiz E, Gómez-Merino FC, Becker C, Krumbein A, Schwarz D (2019) Flavonoid, nitrate and glucosinolate concentrations in brassica species are differentially affected by photosynthetically active radiation, phosphate and phosphite. Front Plant Sci 10:371

Uloth M, You M, Barbetti MJ (2018) Plant age and ambient temperature: significant drivers for powdery mildew (*Erysiphe cruciferarum*) epidemics on oilseed rape (*Brassica napus*). Plant Pathol 67:445–456

Van de Weyer AL, Monteiro F, Furzer OJ, Nishimura MT, Cevik V et al (2019) A species-wide inventory of NLR genes and alleles in Arabidopsis thaliana. Cell 1260–1272

Varshney RK, Roorkiwal M, Sorrells ME (2017) Genomic selection for crop improvement: an introduction. In: Varshney R, Roorkiwal M, Sorrells M (eds) Genomic selection for crop improvement. Springer, Cham, pp 1–6

Vélez ACC (2017) *Brassica juncea* (mustard). In: Invasive species compendium. CABI, Wallingford, UK. http://doi.org/10.1079/ISC.91760.20203482778

Verma R, Kaur J (2021) Expression of barley oxalate oxidase confers resistance against *Sclerotinia sclerotiorum* in transgenic *Brassica juncea* cv Varuna. Transgen Res 30(2):143–154

Verma SS, Sinha RK, Jajoo A (2015) (E)-b-farnesene gene reduces *Lipaphis erysimi* colonization in transgenic *Brassica juncea* lines. Plant Signal Behav 10(7): e10

Verma D, Lakhanpal N, Singh K (2019) Genome-wide identification and characterization of abiotic-stress responsive SOD (superoxide dismutase) gene family in *Brassica juncea* and *B. rapa*. BMC Genomics 20 (1):1–18

Véronési C, Delavault P, Simier P (2009) Acibenzolar-S-methyl induces resistance in oilseed rape (*Brassica napus L.*) against branched broomrape (*Orobanche ramosa L.*). Crop Prot 28(1):104–108

Wagner G, Laperche A, Lariagon C, Marnet N, Renault D et al (2019) Resolution of quantitative resistance to clubroot into QTL-specific metabolic modules. J Exp Bot 70(19):5375–5390

Wan H, Chen L, Guo J, Li Q, Wen J et al (2017) Genome-wide association study reveals the genetic architecture underlying salt tolerance-related traits in rapeseed (*Brassica napus* L.). Front Plant Sci 8:593

Wan Y, Wang Z, Xia J, Shen S, Guan M et al (2020) Genome-wide analysis of phosphorus transporter genes in Brassica and their roles in heavy metal stress tolerance. Intl J Mol Sci 21(6):1–26

Wei L, Jian H, Lu K, Filardo F, Yin N et al (2016) Genome-wide association analysis and differential expression analysis of resistance to Sclerotinia stem rot in *Brassica napus*. Plant Biotechnol J 14(6):1368–1380

Westermann AJ, Gorski SA, Vogel J (2012) Dual RNA-seq of pathogen and host. Nat Rev Microbiol 10 (9):618–630

Widarto HT, Van Der ME, Lefeber AWM, Erkelens C, Kim HK et al (2006) Metabolomic differentiation of *Brassica rapa* following herbivory by different insect instars using two-dimensional nuclear magnetic resonance spectroscopy. J Chem Ecol 32(11):2417–2428

Wu J, Cai G, Tu J, Li L, Liu S et al (2013) Identification of QTLs for resistance to Sclerotinia stem rot and *BnaC. IGMT5*. a as a candidate gene of the major resistant QTL SRC6 in *Brassica napus*. PLoS One 8 (7):e67740

Wu J, Zhao Q, Liu S, Shahid M, Lan L et al (2016) Genome-wide association study identifies new loci for resistance to sclerotinia stem rot in *Brassica napus*. Front Plant Sci 7:1418

Würschum T, Abel S, Zhao Y (2014) Potential of genomic selection in rapeseed (*Brassica napus* L.) breeding. Plant Breed 133(1):45–51

Xiao Z, Zhang C, Tang F, Yang B, Zhang L et al (2019) Identification of candidate genes controlling oil content by combination of genome-wide association and

transcriptome analysis in the oilseed crop *Brassica napus*. Biotechnol Biofuels 12(1):1–16
Xiong X, Liu W, Jiang J, Xu L, Huang L, Cao J et al (2019) Efficient genome editing of *Brassica campestris* based on the CRISPR/Cas9 system. Mol Genet Genom 294(5):1251–1261
Yang J, Jiang H, Yeh CT, Yu J, Jeddeloh JA et al (2015) Extreme-phenotype genome-wide association study (XP-GWAS): a method for identifying trait-associated variants by sequencing pools of individuals selected from a diversity panel. Plant J 84(3):587–596
Yang H, Mohd Saad NS, Ibrahim MI, Bayer PE, Neik TX et al (2021) Candidate Rlm6 resistance genes against *Leptosphaeria maculans* identified through a genome-wide association study in *Brassica juncea* (L.) Czern. Theor Appl Genet 134(7):1513
Yu F, Lydiate DJ, Gugel RK, Sharpe AG, Rimmer SR (2012) Introgression of *Brassica rapa* subsp. sylvestris blackleg resistance into B. napus. Mol Breed 30(3):1495–1506
Yu J, Zhao M, Wang X, Tong C, Huang S et al (2013) Bolbase: a comprehensive genomics database for *Brassica oleracea*. BMC Genom 14(1):1–7
Yue L, Li G, Dai Y, Sun X, Li F et al (2021) Gene co-expression network analysis of the heat-responsive core transcriptome identifies hub genes in *Brassica rapa*. Planta 253(5):111
Zhang J, Mason AS, Wu J, Lagudah E (2015) Identification of putative candidate genes for water stress tolerance in canola (*Brassica napus*). Front Plant Sci 6:1–13
Zhang L, Cai X, Wu J, Liu M, Grob S et al (2018) Improved *Brassica rapa* reference genome by single-molecule sequencing and chromosome conformation capture technologies. Hortic Res 5(1):50
Zhang F, Xiao X, Xu K, Cheng X, Xie T et al (2020a) Genome-wide association study (GWAS) reveals genetic loci of lead (Pb) tolerance during seedling establishment in rapeseed (*Brassica napus* L.). BMC Genome 21(1):139
Zhang Y, Thomas W, Bayer PE, Edwards D, Batley J et al (2020b) Frontiers in dissecting and managing *Brassica* diseases: from reference-based RGA candidate identification to building pan-rgaomes. Int J Mol Sci 21 (23):1–14
Zhang D, Du Y, He D, Zhou D, Wu J et al (2021) Use of comparative transcriptomics combined with physiological analyses to identify key factors underlying cadmium accumulation in *Brassica juncea* L. Front Genet 12:1–16
Zhu YL, Pilon-Smits EAH, Jouanin L, Terry N (1999) Overexpression of glutathione synthetase in Indian mustard enhances cadmium accumulation and tolerance. Plant Physiol 119(1):73–79
Zhu H, Li X, Xi D, Zhai W, Zhang Z, Zhu Y et al (2019) Integrating long noncoding RNAs and mRNAs expression profiles of response to *Plasmodiophora Brassicae* infection in Pakchoi (*Brassica campestris ssp. chinensis* Makino). PLoS One 14:e0224927
Zogli P, Pingault L, Grover S, Louis J (2020) Ento(o)mics: the intersection of 'omic' approaches to decipher plant defense against sap-sucking insect pests. Curr Opin Plant Biol 56:153–161
Zou J, Hu D, Mason AS, Shen X, Wang X et al (2018) Genetic changes in a novel breeding population of *Brassica napus* synthesized from hundreds of crosses between *B. rapa* and *B. carinata*. Plant Biotechnol J 16(2):507–519
Zou Z, Liu F, Selin C, Fernando WGD (2020) Generation and characterization of a virulent *Leptosphaeria maculans* isolate carrying a mutated *AvrLm7* gene using the CRISPR/Cas9 system. Front Microbiol 11:1969

Genome Editing in Polyploid Brassica Crops

25

Mohan Varghese, Bhanu Malhotra, and Naveen C. Bisht

Abstract

Genome editing technology like the CRISPR/Cas9 has emerged as an excellent and rapid plant breeding tool for exploring basic plant biology as well as for increasing agricultural GDPs. Although CRISPR-based genome editing had shown its applicability in different plant genomes since 2013, the genus *Brassica* has come across this precise gene modification tool only since 2015. In polyploid *Brassica* species, CRISPR/Cas9 technology has been established in two major aspects; firstly, for the functional validation of novel genes and secondly, for improving various agriculturally important traits, like for example the enhancements of seed numbers and oil contents as well as downregulating the susceptibility genes responsible for various abiotic and biotic stresses. With the current chapter, we explicitly bring about all the case studies from the advent of CRISPR/Cas based mutagenesis in all the species so far, of the agriculturally significant genus *Brassica* from a period of half a decade to present. The current chapter not only combines all the CRISPR/Cas9 genome editing reports published exclusively from the *Brassica* species but also succinctly discusses the possibilities and challenges for overcoming its shortcomings in the case of polyploid crops. Additionally, we bring about the opportunities that remain open for targeting the key desired traits towards fulfilling the breeding milestones of the Indian oilseed mustard, *Brassica juncea*.

Abbreviations

Cas	CRISPR-associated protein
Cas9	CRISPR-associated protein 9
CRISPR	Clustered regularly interspaced short palindromic repeats
crRNA	CRISPR RNA
dCAPS	Derived cleaved amplified polymorphic sequence
DSB	Double-stranded break
GE	Genome editing
GM	Genetically modified
HDR	Homology-directed repair
Hi-TOM	High-throughput tracking of mutation
MN	Mega nucleases
NHEJ	Non-homologous end joining
nt	Nucleotide
PAM	Protospacer-adjacent motif
RNAi	RNA interference
RNP	Ribonucleo-protein

Mohan Varghese, Bhanu Malhotra—Contributed equally.

M. Varghese · B. Malhotra · N. C. Bisht (✉)
National Institute of Plant Genome Research, Aruna Asaf Ali Marg, New Delhi 110067, India
e-mail: ncbisht@nipgr.ac.in

C. Kole and T. Mohapatra (eds.), *The Brassica juncea Genome*, Compendium of Plant Genomes,
https://doi.org/10.1007/978-3-030-91507-0_25

sgRNA	Single guide RNA
SNP	Single-nucleotide polymorphism
SpCas9	*Streptococcus pyogenes* CRISPR-associated protein9
SSH	Suppression subtractive hybridization
SSN	Site-specific nucleases
TALEN	Transcription activator-like effector nuclease
tra crRNA	*trans*-Activating CRISPR RNA
USDA	United States Department of Agriculture
WRKY11	WRKY transcription factor 11
WRKY70	WRKY transcription factor 70
ZFN	Zinc-finger nuclease

25.1 Introduction

Genome editing (GE) is a type of gene modification technique that precisely alters the genetic codon at a specific locus using site-specific nucleases (SSN) in species of both plants and animals. Besides putting novel insights into the functional genomics of an organism and the basic biology, genome editing comes as an alternative package to conventional breeding approaches and as well as to genetically modified (GM) crops. Although conventional techniques like cross breeding, mutation breeding, and transgenic breeding are popular crop improvement approaches, they demand a lot of time, create random mutations, and possess disadvantages like bringing environmental and health issues along with them (Chen et al. 2019; Wolter et al. 2019). In the process of genome editing, the mutations generated are highly specific and precise in nature and can be detected in the first generation itself. It overcomes the pitfall of foreign gene incorporation, as in GM crops, by altering few bases in an organism's native genome. The alterations generated by genome editing technology are much like the changes that are widespread in naturally occurring populations. The whole idea of genome editing is to first create damage in the genomic sequence and then repair it with the desired nucleotides. For this, genome editing harnesses the native DNA repair system of a cell, which creates either small insertions or deletions at specific sites of a target gene (NHEJ repair pathway) or introduces specific sequences with desired mutations (HDR repair pathway) (Symington and Gautier 2011).

The toolbox of genome editing comprises various SSN dependent techniques like; engineered mega nucleases (MN), zinc-finger nucleases (ZFNs), transcription activator-like effector nucleases (TALENs), the clustered regularly interspaced short palindromic repeats/CRISPR-associated protein 9 (CRISPR/Cas9) system and the newly developed base editing and prime editing approaches. MN, ZFN, and TALENS are considered as the first-generation genome editing tools and the promising new CRISPR/Cas9 technology is referred to as the second-generation one (Shan et al. 2013; Manghwar et al. 2019). These first-generation technologies are crowned as the 2012 "Method of the Year" by *Nature Methods* and CRISPR/Cas9 as the 2015 "Breakthrough of the Year" by Science (Baker 2012; Travis 2015). Amongst all the case studies of genome editing, the CRISPR/Cas technology dominates over all the first-generation genome editing tools (MN, ZFN, and TALENS) majorly due to its simple construction along with efficient delivery of its expression cassette. Since the first use of the CRISPR/Cas system in plants, as in rice and common wheat, it has thereafter been widely adopted and exploited as a precise, low cost and easy to use genome editing tool for various species of plants (Jiang et al. 2013; Miao et al. 2013; Shan et al. 2013; Xie and Yang 2013; Upadhyay et al. 2013).

25.1.1 The CRISPR/Cas Technology and Its Principle

CRISPR/Cas technology is a newly emerged GE technique that has been added to the plant breeding toolkit for its ability to precisely edit multiple genes simultaneously in complex genomes of plants. This technology is designed based on an RNA-mediated DNA targeting the adaptive immune system, employed against the foreign invasive elements like phage and plasmids by the genomes of bacteria and archaea (Mojica et al. 2000). The principle of CRISPR/Cas technology lies, in how this prokaryotic adaptive immune system fights a viral invasion. Upon infection by a virus, the prokaryotic cell triggers a primitive immune defense response. Any bacteria or archaea system that happens to survive such infections cleaves and stores short DNA sequences (30–40 bps) of the invading pathogen with the help of endogenous Cas proteins (Cas1 and Cas2) (Barrangou et al. 2007; Sapranauskas et al. 2011). These short DNA sequences are called protospacers which act as a memory so that they can fight the virus efficiently in future exposures (Charpentier et al. 2013). Protospacers are then integrated between the leader sequence and the first repeat of the CRISPR array present in the CRISPR-*cas* locus, found in the vast majority of sequenced archaeal and bacterial genomes (Ishino et al.1987; Barrangou et al.2007). Protospacer after integration into the CRISPR array is termed as spacer sequence and the process is called spacer acquisition (Mojica et al. 2000; Bolotin et al. 2005). This stage of the mechanism is called the adaptive stage (Fig. 25.1). Followed by the adaptive stage, the immunological memory gathered by the CRISPR array in the form of integrated spacer sequence is transcribed into a short precursor RNA transcript that further undergoes endonucleolytic cleavage, referred to as the expression and maturation stage, to yield smaller units of RNA known as CRISPR RNAs (crRNAs) (Fig. 25.1). These crRNAs are combined with one or more Cas proteins to form the active Cas–crRNA complex and act as a guide for the Cas endonuclease proteins, which cleave the foreign target DNA. The last stage of this adaptive immune response called the interference stage happens when the bacteria encounter a second infection, during which this Cas–crRNA complex recognizes a foreign nucleic acid region that has a complementarity with the crRNA spacer sequence (immunological memory). Successful base pairing (Watson–Crick pairing) between the crRNA spacer and the complementary foreign target sequence (protospacer) activates sequence-specific degradation of the foreign nucleic acid. This hybridization is aided by a conventional 5′-NGG-3′ or non-conventional NAG motif (usually of 2–6 bps) adjacent to the protospacer sequence called the protospacer-adjacent motif (PAM) (Bolotin et al. 2005; Mojica et al. 2009). In simple terms, the CRISPR/Cas toolkit comprises majorly of two components: first, a guide RNA that recognizes the target DNA of interest, and second, a non-specific endonuclease, CRISPR-associated (Cas) nuclease, that cleaves the foreign nucleic acid sequence.

25.1.2 The CRISPR/Cas9 Genome Editing System

The type II CRISPR/Cas9 complex in human pathogen *Streptococcus pyogenes* is the widely accepted nuclease protein among the genome editing tools and is known for its precise editing of the genomic sequence in various organisms including the plant kingdom (Deltcheva et al. 2011). It is a simple two-component system, comprising a single DNA endonuclease (Cas9) and a chimeric guide RNA that base pairs with the target sequence. Thus, by changing only the guide RNA sequence one can virtually target any genomic sequences of interest, making it one of the most simplified forms of the available genome editing approaches.

The type II CRISPR-*cas* locus in *S. pyogenes* comprises four genes of Cas nucleases (*Cas9, Cas1, Cas2,* and *Csn1*), two non-coding RNA elements (crRNA and *trans*-activating crisprRNA (tracrRNA), and a characteristic array of direct repeats interspaced by short stretches of

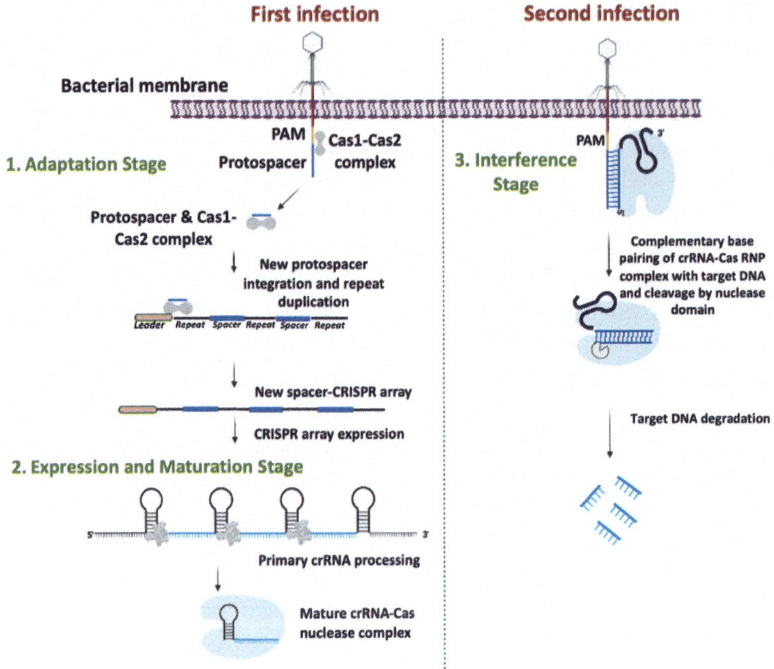

Fig. 25.1 Development of the prokaryotic CRISPR/Cas adaptive immunity. The CRISPR/Cas immunity develops through three stages; (1) *Adaptation stage*: When a prokaryotic system encounters an invader for the first time, endogenous Cas1-Cas2 complex procures a protospacer from the invader DNA and uses it as an immunological memory by integrating the protospacer into the CRISPR-array as spacer sequence. (2) *Expression and Maturation stage*: The new spacer-CRISPR array of DNA sequence is transcribed into primary crRNA and further processed into mature crRNAs, containing a transcribed spacer and part of the repeat sequence. Later this mature crRNA forms a ribonucleoprotein (RNP) complex along with the Cas proteins. (3) *Interference stage:* During a second infection, the crRNA-Cas RNP complex identifies the foreign invader carrying a protospacer sequence complementary to the transcribed spacer present in the RNP complex. Target sequence is then cleaved by the nuclease domain in the Cas protein

non-repetitive sequences (spacer) (Deltcheva et al. 2011). In the CRISPR/Cas9 system, a chimeric single guide RNA (sgRNA) is developed by combining the mature crRNA and the tracrRNA with an artificial tetraloop. The crRNA consists of a 20 nucleotide (nt) long guide sequence complementary to the target DNA and a repeat region (12 nt); while the tracrRNA comprises three tracrRNA stem loop and an anti-repeat (14 nt) region, which acts as a binding scaffold for the Cas nuclease (Nishimasu et al. 2014). Besides the scaffold, tracrRNA has a role in the maturation process of primary crRNA transcript, forming an RNA duplex with the crRNA (Deltcheva et al. 2011). The structural studies have revealed that the interaction between sgRNA and the target DNA gives a T-shaped architecture (Nishimasu et al. 2014). This dual sgRNA takes Cas9 nuclease to target any DNA sequence upstream of a very simple 5′-NGG-3′ PAM present on the non-target strand.

PAM is a short sequence of 2–6 base pairs that is recognized by the sequence-specific active site present in the Cas proteins. The presence of PAM sequence adjacent to the protospacer helps Cas nuclease to discriminate the foreign genetic material (non-self) from the sequence encoded by the CRISPR array (self), thus avoiding the CRISPR-cas autoimmunity (Mojica et al. 2009; Gleditzsch et al. 2019). However, the use of PAM for targeting a sequence limits the scope of CRISPR/Cas approach to target the sequences without PAM. To relax this requirement of PAM, studies have also discovered different variants of Cas9 nucleases for recognizing different consensus PAM sequences like; VQR-Cas9 (NGA PAM), EQR-Cas9 (NGAG PAM), VRER-Cas9 (NGCG PAM), SaKKH-Cas9 (NNNRRT PAM), xCas9 (NG, GAA, and GTA PAM), and SpCas9-NG (NG PAM). Moreover, in 2020, Walton et al. has evolved a nearly PAM-free SpCas9 nuclease variant named SpRY, through structure-guided mutagenesis of already developed VRQR variant (Kleinstiver et al. 2016). SpRY variant recognizes NRN (R = A or G) and NYN (Y = C or T), thus giving rise to the most relaxed version of PAM preference and the ability to target any sequence. However, the SpRY variant possessed a higher off-target tendency when compared to the conventional SpCas9 variant (Walton et al. 2020).

The type II signature gene *Cas9* encodes a single large multi-domain protein, Cas9 (1368 amino acids) with two DNA cleavage domains: Ruv C and His-Asn-His (HNH). Ruv C nuclease domain cleaves the DNA strand opposite to the target strand (non-target strand) and the HNH domain cleaves the DNA strand complementary to the guide RNA (target strand) (Chen et al. 2014). Cleavage by Cas9 nuclease creates double-stranded DNA snips in the gene of interest and triggers the two major repair pathways of cell: non-homologous end joining (NHEJ) pathway and homology-directed repair (HDR) pathway (Fig. 25.2). The basic difference being in these two repair systems is that the latter requires a recombining molecule, with a shared homology region, in the cell to restore the damage created by Cas nuclease, whereas the former does not (Symington and Gautier 2011; Chen et al. 2019).

Usually, a cell prefers the NHEJ pathway throughout the cell cycle without any prerequisite for a homologous repair template, thus making it a popular repair system for large-scale knockout studies in plants for creating specific deletions or insertions in the gene of interest (Chen et al. 2019). NHEJ is an error-prone DNA repair process, and hence insertions and deletions (indels) are often introduced into the gene, resulting in frameshifts and potential loss of gene functions (Fig. 25.2). However, NHEJ repair pathway unlike the HDR pathway lacks the precision required for a specific genome modification. The HDR pathway is commenced during S and G2-phase of the cell cycle and is comparatively more precise in creating a desired insertion or deletion, more specifically a point mutation (Fig. 25.2). In the HDR pathway, the repair template can be an exogenous DNA molecule, besides a sister chromatid, with desired modification to be inserted into the break site (Salsman and Dellaire 2016). Still, the efficiency of HDR-mediated gene editing in plants is not equivalent to NHEJ-mediated gene modification and its requirement for the donor molecule and delivery remains its added limitations (Chen et al. 2019). Nevertheless, the HDR-mediated gene targeting has been introduced in plants either through positive–negative selection procedure as in rice or by increasing the donor template copies by using Gemini virus-mediated delivery in tomatoes (Terada et al. 2002; Cěrmák et al. 2015).

25.1.3 The Genus *Brassica* and CRISPR Technology

Brassica is an agronomically important genus that comprises a number of species and subspecies whose different parts shoots, leaves, roots (like turnip roots), or seeds are utilized for consumption purposes across the world. Not only the vegetative parts of the plants but also their seeds are merchandised predominantly for oil production of the country accounting for approximately 25% share in total oilseed productions of India (Jat et al. 2019). Seeds from oilseed Brassicas are also used as protein-rich meals to feed cattle and are widely utilized for condiment purposes in meals. The *Brassica* genus, thus, includes crops that find use as both food and fodders possessing immense contribution in shaping the health benefits and taste characteristics for human consumptions (Punetha et al. 2020). The Brassica crops are represented via a U's triangle that contains six plant species that possess three

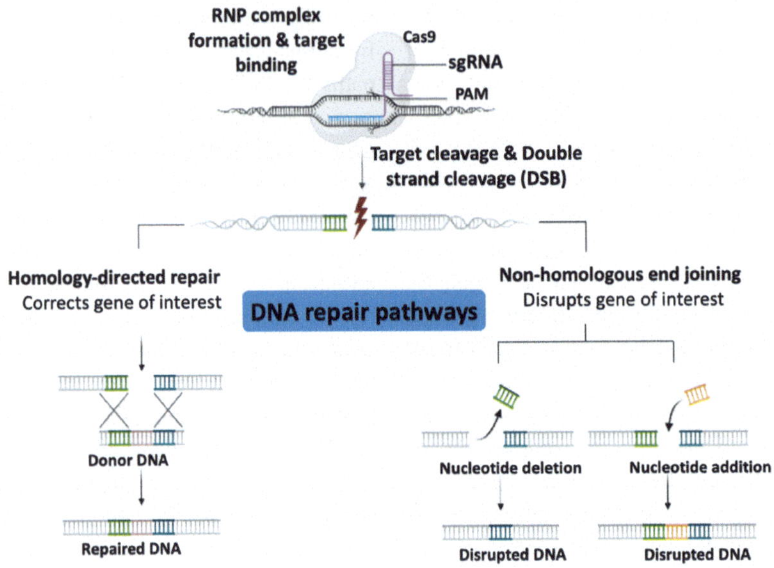

Fig. 25.2 DNA repair pathways associated with CRISPR/Cas gene editing. The double-stranded break (DSB) created by CRISPR/Cas system in the DNA causes genomic modifications by one of the two major DNA repair pathways: Homology-directed (HDR) and Non-homologous end joining (NHEJ) pathways. HDR pathway uses the homologous repair template with the desired changes to be incorporated into the DSB, while the NHEJ repair pathway is associated with the occasional insertion or deletion (called indels) at the cut site

diploids: *Brassica rapa* (AA), *Brassica nigra* (BB), and *Brassica oleracea* (CC), along with three amphidiploid species: *Brassica juncea* (AABB), *Brassica napus* (AACC) and *Brassica carinata* (BBCC), that arose after the natural interspecific hybridizations of the former three diploid species (Nagaharu 1935). Further breeding and domestication have created different agriculturally significant morphotypes across *Brassica* species, like for example *B. oleracea* which includes brussels sprouts, broccoli, cabbage, cauliflower, kale, kohlrabi, savoy cabbage; the species of *B. rapa* include morphotypes of Chinese cabbage, turnips, pak choi and oilseeds; and *B. napus* possess rapeseed (Canola), fodder rape, and rutabaga (Cheng et al. 2016). Out of which the rapeseed Canola is economically the most important oilseed crop with second-highest production levels after soybean (USDA 2021). Besides the importance of these Brassica crops in India and different parts of the world, modern biotechnological approaches still appear to be lacking for their breeding and research outputs.

In polyploid *Brassica* species, every ortholog of a gene from the model plant *A. thaliana* is represented by a minimum of two or more homologs with redundant functions. Thus, targeted genome editing of even a monogenic trait in *Brassica* species requires multiplexing to combine the mutated homologs belonging to the constituent subgenomes for obtaining a desired phenotypic effect. Conventional chemical and irradiation mutagenesis creates single mutations in either one or the other homolog, with a very low possibility of obtaining double mutations. In many such cases, the desired effect is not achieved, and this demands the use of repeated crossing and backcrossing approaches which can be further time-consuming. Here an excellent example comes from sinapine biosynthetic process, were knocking out a single homolog had no measurable effect, where the double mutant produced after backcrossing showed a reduction of sinapine in *B. napus* (Emrani et al. 2015). Therefore, the ability of a sgRNA to knock out gene families as brought about in this chapter offers superior possibilities to classical mutagenesis, without any, minimum, or no off-target effects, making CRISPR a promising approach to be used in breeding programs of Brassica crops.

The CRISPR technology has been used widely for different plant species, but its utilization for genome editing of Brassica crops came into light with the first report in 2015 (Lawrenson et al. 2015). Thus, the CRISPR/Cas 9 system has been utilized for only half a decade until now for the precise genome editing of Brassica crops. CRISPR had been used to infer gene functions, breeding of agriculturally important traits for higher yield and quality, and also recently for pathogen resistance in *Brassica* species (Lawrenson et al. 2015; Braatz et al. 2017; Okuzaki et al. 2018; Sun et al. 2018; Zhang et al. 2019; Zheng et al. 2020; Zhai et al. 2020; Pröbsting et al. 2020). As genetic analysis in *Brassica* species is desired to fulfill both the basic and applied research goals, therefore technologies like CRISPR/Cas9 offer a hopeful yet challenging future in Brassica crops. Such technologies are significantly important because of their ability of simultaneous alterations of several homologous gene copies that are needed to characterize and evolve promising agronomic traits in Brassica crops (Fig. 25.3).

25.2 CRISPR/Cas9 Based Genome Editing in Brassica Crops

CRISPR/Cas9 remains well exploited in various species of plants ranging from the simplest model plant to the polyploid crop species like wheat, cotton, etc. (Chen et al. 2019; Manghwar et al. 2019). The application of the technology for the first time in Brassica crops was published in 2015 and since then it has revolutionized the breeding objectives of *Brassica* species from genomics to the translational levels. As it began the initial studies, the CRISPR/Cas9 based genome editing in Brassica crops focused on the validation of technology using various marker genes as well as for exploitations of the screening strategies to establish the tool in these *Brassica* species. Later the use of this technology was expanded to target various important traits for yield and quality improvements and most recently for developing abiotic and biotic resistance in the *Brassica* genus. In most of the published reports since 2015, in the genus *Brassica*, the crop of choice out of all the *Brassica*'s U's triangle species for CRISPR-based genome editing was majorly the rapeseed crop, *B. napus*. The lack of genetic variation in the current genetic pool of cultivated rapeseed, and its ability to serve as the major cash crop supporting the national income make *B. napus* a hotspot for CRISPR-based applications. Based on all the published reports on CRISPR and its applications in Brassica crops so far, we here reviewed a total of 13 articles on the various aspects for which CRISPR was used in *Brassica* species from its onset to the present.

25.2.1 CRISPR/Cas9 Validation as a Genome Editing Tool in *B. oleracea*

In the year 2015, the CRISPR/Cas9 gene editing technology was initially validated in *B. oleracea*, by targeting previously well-characterized *GA4* gene homologs (Lawrenson et al. 2015). The loss-of-function mutants of *A. thaliana GA4* show a semi-dwarf phenotype and are required for efficient seed dispersal (Arnaud et al. 2010). In consistent with the CRISPR/Cas9 efficiency, the *GA4*-edited *B. oleracea* lines by CRISPR/Cas9 were found to possess modifications in the T0 generation itself, at both the genotype and the phenotype levels. For instance, in the T0 generation, dwarf phenotype lines with a series of mutant alleles at the targeted regions were obtained. Theoretically, RNA-guided Cas9 induced mutations should be stably inherited in the advanced generations from their parental generations. However, practically, the stable transmission of CRISPR-induced mutations depends upon many factors such as the extent of mutations in the germ-line cells, the Cas9 activity, and the promoters used for the expression of sgRNA-Cas9 cassette (Feng et al. 2014) and hence, variations do exist in the transmission of

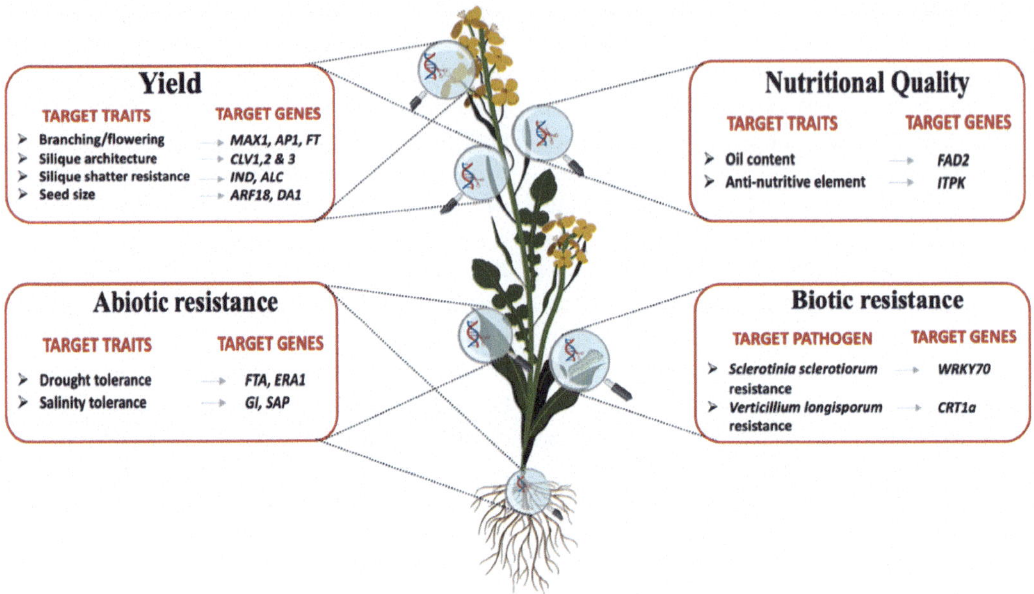

Fig. 25.3 Target genes, target organs, and traits for the *Brassica juncea* crop improvement in terms of yield, nutritional quality, abiotic and biotic resistance by CRISPR/Cas9 mediated genome editing. *MAX1, MORE AXILLARY GROWTH; AP1, APETALA; FT, FLOWERING LOCUS; CLV1*, *2* and *3*, *CLAVATA* homologs; *IND, INDEHISCENT; ALC, ALCATRAZ; ARF18, AUXIN RESPONSE FACTOR 18; DA1*, encodes ubiquitin activated peptidase; *FTA, FARNESYLTRANSFERASE A; ERA1, ENHANCED RESPONSE TO ABA1; GI, GIGANTEA; SAP, STRESS-ASSOCIATED PROTEIN; FAD2, fatty acid desaturase 2; ITPK*, encodes tetrakisphosphate kinase; *WRKY70*, plant-specific transcription factor; *CRT1a*, a plant-specific fungus susceptibility factor

Cas9 induced mutations. This was also evident in the case of *GA4*-edited T1 and T2 lines, where Lawrenson et al. (2015), observed that the CRISPR-induced mutations were not strictly transmitted to the next generation from all the independent transgenic events. Simultaneously, the CRISPR-edited *B. oleracea* lines also showed variations in the intensity of mutations between the targeted regions of *GA4* homologs. These variations are accounted for the differences in the efficiency of the guide RNAs used to target the *GA4* regions and the gRNA efficiency also depends upon the unique nucleotide sequence and its secondary structure (Uniyal et al. 2019). One of the highlights of the CRISPR/Cas9 applications in crop species is the effortless removal of the T-DNA in successive generations by either self-pollinating or crossing with the wild-type, thereby minimizing the regulatory load and facilitating commercialization. In this context, Lawrenson et al. (2015) had successfully obtained transgene-free plants with stable *GA4* mutations in the T2 generations by employing the CRISPR/Cas9 toolkit.

25.2.2 Validation of Multiplexing Feature of CRISPR/Cas9 Tool in *B. napus*

The functional studies involving gene characterizations could be ineffective in cases of polyploid crops where multiple homologs determine a trait. This can be resolved by exploiting the multiplex feature of the CRISPR/Cas9 technology, wherein a sgRNA can be used to target multiple homologs simultaneously. A case study by Yang et al. (2017), assessed the potentiality of CRISPR-based multiplex editing in *B. napus*. The study targeted 12 genes, namely four paralogs of *REPRESSOR OF GA1-3 (RGA)* gene, three paralogs of *FRUITFULL (FUL)* gene, and five paralogs of *DA2* and *DA1* genes (DA: LARGE IN CHINESE) involved in gibberellin signaling, silique dehiscence, and organ size regulation, respectively in *B. napus*. Considering the complexity of the polyploid genomes, the CRISPR/Cas9 editing could be overloaded for precise editing in *B. napus*. However, it was well surveyed that the genome size does not significantly influence the efficiency of targeted genome editing by the CRISPR/Cas9 system (Xie and Yang 2013) and this fact was also evident from the observed CRISPR-based mutation frequencies in all the paralogs targeted in *B. napus*as well (Yang et al. 2017). Another common situation faced in crop species with high ploidy levels is the existence of functional redundancy associated with paralogs in one gene family, and therefore, a single gene knockout does not lead to a phenotype in polyploid species. Such conditions were also observed in CRISPR-based mutagenesis of *B. napus,* where the phenotype of a single gene knock out mutant of *BnaRGA* was comparable to the wild-type plants, while the quadruple mutant showed longer stems than the wild-type plants (Yang et al. 2017). On that account, CRISPR/Cas9 based crop improvement comes with the remedy of using sgRNAs derived from the conserved regions of the target genes to simultaneously knockout a group of paralogs, as also observed in *B. napus*. Additionally, the RNA-guided Cas9 gene editing besides the stable transmission of mutations holds the ability to continuously modify the wild-type alleles until they get mutated, in subsequent transgenic generations. However, Yang et al. (2017), had few transgenic lines whose wild-type alleles failed to be mutated even in the advanced generations as well, probably due to the low efficiency of CRISPR/Cas9. Factors that determine the CRISPR/Cas9 efficiency are the promoters that drive the *Cas9* and *sgRNA* expressions and also the choice of such promoters varies between experiments and species (Pan et al. 2016). Yang et al. (2017) observed that in *B. napus* high mutation frequencies were generated by either using the constitutive promoters or the Arabidopsis U6 promoter. Interestingly, no off-target mutational effects were detected, suggesting the high specificity of the sgRNA in oilseed rape. The above case study suggests that CRISPR/Cas9 is a highly efficient tool for creating mutations at multiple loci as well as the fastest method for the breeding of polyploid crops.

25.2.3 CRISPR/Cas9 Based Prevention of Yield Losses in *B. napus*

CRISPR/Cas9 utility in *Brassica* species also stepped into the crop improvement objectives by targeting genes regulating the silique shattering trait in *B. napus* (Braatz et al. 2017). In the model plant *A. thaliana* two genes, namely *INDEHISCENT* (*IND*) and *ALCATRAZ (ALC)* are known to control the silique dehiscence, where the *ind* mutants are completely indehiscent since they lack both the lignified cells and the separation layer at the breaking point of siliques (Liljegren et al. 2004), while the *alc* mutants lack the separation layer only (Rajani and Sundaresan 2001). With this background, Braatz et al. (2017) enhanced the pod shatter resistance in oilseed rape, by targeting two *BnALC* homologous by CRISPR/Cas9 technology. Although they used a specific sgRNA for targeting the A genome *BnALC* homologs, a second *BnALC* homeolog in the C genome which differs from the sgRNA target by one single-nucleotide polymorphism (SNP) was also mutated, exploiting the fact that Cas9 tolerates mismatches within the target site (Lawrenson et al. 2015). At the same time, no off-target mutations were observed in potential sequences with a minimum of two SNPs. With a strong expression cassette under the control of a constitutive ubiquitin promoter, Braatz et al. (2017) could generate a 100% mutational efficiency with nonchimeric T1 plants and all the subsequent T2 off springs with mutations. Further, whole-genome sequencing of the advanced generations in *B. napus* recovered T2 plants with four mutated *BnALC* alleles that did not contain both the T-DNA and the vector sequences. Upon analysis of the seed shatter resistance, it was found that these *alc* mutants possessed an increased shatter resistance, and also longer siliques of the T2 plants were more robust than the shorter siliques, suggesting the correlation of silique lengths and shatter resistance in plants. Thus, a stronger application of the above-targeted trait and its usefulness under field conditions to prevent silique shattering can be a promising effort preventing rapeseed yield loss, which can also be thought of for other cultivated *Brassica* species.

25.2.4 PAGE-Based CRISPR/Cas9 Screening in *B. napus*

Screening strategies are considered as a rate-limiting step in the CRISPR-based breeding programs in polyploid species, where multiple genomic sites have to be evaluated for identifying simultaneous mutagenesis in the first-generation transgenic events. The identification of CRISPR/Cas9 induced mutations is based on two popular strategies of screenings namely: PCR-based sequencing and the T7 endonuclease I assay (Shan et al. 2014). In Brassica polyploid crops, both these approaches are highly time-consuming and labor-intensive when a large-scale screening is to be performed for multiple homologs. In 2018, the use of polyacrylamide gel electrophoresis (PAGE)-based screening approach allowed a large-scale identification of mutations at five designed *BnSPL3* genomic sites in *B. napus* (Li et al. 2018). PAGE-based screening demands the designing of specific primers which amplify the desired mutation region of the targeted gene homologs. The detection of heteroduplexed DNA fragments from a transgenic line corresponds to the presence of mutation/s in those screened lines and can be used as a first step for screening large-scale mutation sites. Further, the transgenic lines which showed heteroduplexed DNA fragments in PAGE in both the T0 and in advanced generations were confirmed using high-throughput sequencing and showed an editing frequency of 96.8–100%. It has also been previously shown that PAGE-based procedure although rapid, can have limitations to detect small nucleotide indels and substitutions in the mutated homozygous plants (Zhu et al. 2014). Therefore, PAGE-based identification can be applied when scanning large-scale mutations amongst a large number of transformants and after confirmation, the approach can be narrowed down to sequencing-based identifications for further confirmation of the mutation type in the subsequent generations.

25.2.5 Improvement of Oil Quality and Yield in *B. napus*

Being the second-highest oilseed producing crop species after soybean (USDA 2021), enhancement of oil quality is a major breeding objective in the rapeseed. *B. napus* seeds contain three major unsaturated fatty acids; oleic acid (C18:1), linoleic acid (C18:2), and linolenic acid (C18:3). Enhancing the level of oleic acid in seeds is known to increase the thermal stability of the oil and make it more suitable for use as edible oil. The plant-specific *stearoyl-acyl carrier protein desaturase* (*SAD*), *fatty acid desaturase 2* and *3* (*FAD2* and *FAD3*)are the genes involved in the biosynthesis of C18:1, C18:2, and C18:3 fatty acids, respectively during the seed development. Okuzaki et al. (2018) showed that premature stop codon or amino acid substitutions created by CRISPR-based mutagenesis in the FAD2 protein resulted in higher oleic acid contents without any effect on plant growth. It was previously known that, among the four homologs of *B. napus FAD2* genes (*BnaA.FAD2.a, BnaA.FAD2.b, BnaC.FAD2.a,* and *BnaC.FAD2.b*), a 4 bp insertion and a frameshift mutation in the coding region of *BnaA.FAD2.a* gene in *B. napus* varieties (SW Hickory and Cabriolet respectively) led to a high oleic acid content (Yang et al. 2012; Wells et al. 2014). Implementing the same and post-screening of transgenic lines for editing in two different *BnaA.FAD2.a* target sites, Okuzaki et al. (2018) obtained a line with a 4 bp deletion at one target site and lines with a maximum of 5 bp deletion at the second target site of the *BnaA.FAD2* genes. The mutated novel alleles caused a frame shift which led to an early stop codon in the open reading frame and created truncation in the FAD2 protein in comparison to the wild-type protein, leading to a change in the fatty acid composition. Similar to Braatz et al. (2017) and Yang et al. (2017) case studies, there were no off-target effects in the other three homologs of the *FAD2* gene, confirming the fact that precise and specific on-target editing can be generated in *B. napus* by CRISPR/Cas9. Post backcrossing and PCR analysis, Okuzaki et al., obtained lines with the mutant *fad2_Aa* allele without the T-DNA insertion. In these lines homozygous for *fad2_Aa* mutant allele, both oleic and linoleic acids were increased and decreased respectively by 7% as compared to the wild-type plants. Overall, this was the first report involving the modification of genes involved in the metabolic pathway in the species of the Brassica crops with CRISPR.

The use of CRISPR/Cas9 can also be extended to the silique-related traits, like for example targeting the regulators of silique architecture and morphology. In Brassica crops, siliques can be either bilocular or multilocular. For instance, a few genotypes like yellow sarson in *B. rapa* and *B. juncea* have been identified, which possess multilocular siliques. In contrast, *B. napus* siliques are mostly bilocular, and thus the development of varieties with increased locule numbers which translates into higher yield can be a major breeding objective of *B. napus*. A group of genes that functions in this capacity includes *CLAVATA* homologs (*CLV1, CLV2,* and *CLV3*), wherein *CLV3* encodes a small, secreted peptide which potentially interacts with its receptors encoded by *CLV1* and *CLV2* genes to repress the expression of *WUSCHEL* in the shoot apical meristems (Clark et al. 1993). In line with this, mutation of the CLV pathway genes leads to expansions of shoot apical meristems, with an increase in the number of floral organs and consequently multilocular siliques (Clark et al. 1993). The multilocular trait found in *B. rapa* and *B. juncea* has been attributed to a single-nucleotide mutation in *CLV3* and insertion in *CLV1* gene, interrupting the function of the target genes (He et al. 2003; Ze-wenet al. 2012).

Recently, Yang et al. (2018) developed *B. napus* with multilocular siliques and increased seed number by mutating the functional domains in the homologs of *BnCLV1*, *BnCLV2,* and *BnCLV3* with CRISPR/Cas9 gene editing. The study employed the multiplex ability of Cas9 based genome editing by designing the sgRNAs from the conserved regions, to target ten different sites in both the A and C sub-genome copies of *CLV1, CLV2,* and *CLV3* simultaneously. Moreover, there were no off-target mutations observed after the PCR screenings of the potential off-target loci. It is known that the GC% content and the expression levels ofCas9 and *sgRNA* are among the several crucial factors that affect the sgRNA efficacy in plants (Wang et al. 2014a). However, no such correlations were observed by Yang et al. (2018), between the mutation efficiencies and the GC% composition of these *CLAVATA* targeted sgRNAs. T0 lines with the same level of *Cas9* and *sgRNAs* expressions exhibited variable gene editing efficiencies (ranging from 0 to 46.5%) by the sgRNAs that targeted the same gene at different sites. These observations concluded that neither the GC% content nor the expression levels of *Cas9* and *sgRNA* are the major limiting factors of genome editing in *B. napus* T0 plants. Instead, the variations found in the mutation efficiencies of different sgRNAs used in Yang's case study could be attributed to the differences in their nucleotide compositions of the sgRNAs. Another important part of CRISPR-based genome editing in *B. napus* is the knowledge of the efficiency of different promoters for driving the Cas9 protein expression which affects the editing efficiencies in rapeseed. In *BnCLV3* targeted lines, Yang et al. (2018) used P_{35} and P_{ubi} promoters independently in two binary vectors to drive the *Cas9* expressions. Unlike the observations made by Ma et al. (2015), who recommended P_{ubi} rather than P_{35s} for Cas9 protein expression in dicot plants, Yang's results had a different view, where P_{35s}resulted in better Cas9 protein expression and higher efficiency of mutagenesis (14.4%) in rapeseed than those by P_{ubi} (6.1%). The above work of Cas9-based mutation of *CLAVATA* genes in rapeseed also emphasized the importance of a suitable transformation host as they had obtained a high transformation rate (66.7–92.5%) in pure *B. napus* line J9707 in comparison with that in spring cultivar Haydn (0.9%) by Braatz et al. (2017). *BnCLV3* mutated lines showed a significant increase in the number of locule, number of seed per silique, seed weight per plant than the wild-type, which suggest that *CLV3* homologs could be promising candidates for future yield improvement in other oilseed.

25.2.6 Establishing DNA-Free CRISPR/Cas9 Based Editing in *Brassica* Species

Just like any other genetic engineering tools for crop improvements like RNA interference (RNAi) and overexpression strategies, the CRISPR/Cas9 expression cassette is also carried by a T-DNA vector and gets integrated into the genome of the plant species during the targeted mutagenesis process. Therefore, it must be removed from the transgenic crops before they can be subjected to GMO-based regulations (Wolt et al. 2016). Although transgene-free crops can easily be screened from the successive genome-edited mutant generations, the transgenesis stage in the CRISPR/Cas9 based genome editing is inevitable. As an alternative for this, the use of preassembled Cas9 and ribonucleoprotein (RNP) complexes can generate DNA-free genome-edited plants, such as those previously employed in the case of maize, apple, wheat, potato, and *Petunia x hybrida* (Chen et al. 2019). Both the PEG-based transformations or biolistics have been employed to introduce these RNPs for plant regenerations.

Recently in 2018, DNA-free genome editing has been demonstrated by delivering RNPs into the protoplast of the three species of *Brassica*'s U's triangle namely, *B. oleracea*, *B. rapa,* and *B. napus* to target two endogenous reporter genes; the *phytoene desaturase* (*PDS*) gene and the vernalization determinant *FRIGIDA* (*FRI*) gene (Murovec et al. 2018). Based on an in vitro digestion assay with purified Cas9 enzyme, the protoplast of these *Brassica* species was transfected with sgRNAs showing the highest cleavage activity. In the in vitro assay, sgRNAs with the highest cleavage activity were able to cleave the PCR products from all the species and a complete cleavage was seen when a higher amount of sgRNA and Cas9 was used. Upon transfecting the protoplasts with the PEG 4000, the indel percentages observed at the cleavage site in *B. oleracea* and *B. rapa* ranged between 0.09 and 24.51%, while in the case of *B. napus* no indels could be detected in case of any of the sgRNAs used. A positive correlation could also be observed between the mutations detected and the amounts of RNPs used, and these mutations were both gene and locus-dependent. Murovec et al. in 2018, also assessed if mutational frequencies in the three species were affected with the time for which protoplasts were cultured after RNP introduction. They found that in the case of *B. rapa* the mutational frequency at 24 h after transfection was 4.89% and at 72 h was 4.30%, which were not significantly different. Although for DNA-free gene editing commercially available Cas9 enzyme can also be used, the protocol used by Murovec et al. (2018) brings high-quality RNP constituents at a decent cost. However, in the future, questions like the most suitable ratio of Cas9:sgRNA should be addressed as in different studies the results differed substantially between the loci. Also, the protoplast's viabilities after isolation and transfections can be important parameters for assessing the activities of the same RNPs in different species. Murovec's protocol for DNA-free genome editing combined with the efficient protoplast regeneration protocols will enable the development of Brassica crops with edited phenotypes without the use of transgenesis.

25.2.7 Establishing CRISPR/Cas9 Based Biotic Resistance in *B. napus*

The classic application of CRISPR/Cas9 technology for both basic science and crop improvement by targeting the pathogenesis-related genes for the first time in any *Brassica* species was conducted in *B. napus* in 2018. WRKY transcription factors that possess a DNA binding WRKY domain are implicated in plants' resistance and immune responses to diverse stressed conditions. Studies have shown the role of WRKY11 and WRKY70 in SA- and JA-induced resistance responses towards pathogens in both Arabidopsis and *B. napus* plants (Li et al. 2004; Wang et al. 2014b). According to a previous transcriptome report on *B. napus* lines, *BnWRKY11* and *BnWRKY70* genes were found to be differentially expressed in *B. napus* resistant cultivars post *Sclerotinia sclerotiorum* inoculation (Wu et al. 2016). In 2018, to develop stem rot resistance in rapeseed, NHEJ-mediated mutations in both the sub-genome copies of *WRKY11* and *WRKY70* were created using CRISPR/Cas9 technology (Sun et al. 2018). The sgRNAs used to target the above two candidate genes were driven by three different promoters: AtU6-1, AtU6-29, and AtU3b. Upon analysis of the efficiencies of gRNAs generated against the *WRKY* genes, it was found that the sgRNAs driven by AtU6-1 promoter remained non-functional. Prior screenings of the sgRNAs and changing their promoters to *B. napus* endogenous promoters could be the possible solutions to the above. Moreover, considering the chances of encountering non-functional sgRNAs in a CRISPR-based genome editing, it is highly recommended to design multiple gRNAs for a given target gene. Additionally, Sun et al. (2018) also observed that the number of mutations induced in the T0 transgenic plants was lower than those in the T1 plants. The possible explanations could be either these mutations remain undetected by sequencing experiments in the T0 due to screening of a smaller number of samples, or due to sampling of DNA for screening at the young stage rather than at the adult stage. The CRISPR/Cas9 system functions continuously as it exists in a cell and could also lead to de novo mutations of the *WRKY* genes in later stages or generations. While both the WRKY transcription factors genes showed homozygous or biallelic mutations in the T1 transgenic plants, the *BnWRKY70* mutated plants showed significantly decreased lesions, upon inoculation with mycelial plugs of *S. sclerotiorum* under detached leaf conditions than those in *BnWRKY11* mutated plants. *BnWRKY70* overexpression plants further showed an enhanced lesion area in comparison with the wild-type plants upon *S. sclerotiorum* inoculation, thus confirming both the negative regulation of *BnWRKY70* in defense against *S. sclerotiorum* as well as its suitability as a target gene to develop disease resistance in rapeseed. The case study by Sun et al. (2018) sets an example that CRISPR/Cas9 system can be useful for basic research and extrapolated for disease resistance breeding in the complex polyploid crops of the *Brassica* genus.

25.2.8 Functional Validation of Fatty Acid Pathway Genes by CRISPR/Cas9

Although *Brassica* species are one of the major sources of edible oils, the functional roles of genes involved in the production of oils such as the triacylglycerol (TAG) biosynthetic pathway genes, was unclear until 2019. A gene known as the *LPAT*, which encodes for the master enzyme called Lysophosphatidic acid acyltransferase (LPAT) involved in the production of TAGs, has been previously validated in the model plant *A. thaliana* (Angkawijaya et al. 2017) and also in *Lesquerella* seeds (Chen et al. 2016). Later in 2019, Zhang et al. used the CRISPR/Cas9 technology to elucidate the functional role of the *LPAT* genes (*LPAT2* and *LPAT5*) in *B. napus* by targeting the three exons in *BnLPAT2* and one exon in *BnLPAT5* paralogs. In allotetraploid crop *B. napus*, the *BnLPAT2* is present as seven paralogs and *BnLPAT5* as four paralogs; therefore, sgRNAs from highly conserved regions in both *BnLPAT2* and *BnLPAT5*homologs were designed to facilitate the multiplex editing. Moreover, sgRNAs with 18 nucleotide oligos were used for effective editing in *BnLPAT* genes and their paralogs simultaneously (Zhang et al. 2019), instead of the conventional sgRNAs with 20nt oligos. It is important to note that during the CRISPR-based mutations in *BnLPAT* genes, a direct correlation was observed with the GC% of sgRNAs, as the sgRNA with moderate GC contents (50%) achieved higher mutation efficiency than those with lower GC contents (33–39%). In addition, sgRNAs targeting the sense strand of *BnLPAT2* gene showed higher mutation over those targeting the antisense strand. To facilitate the mutation of three exons in *BnLPAT2* simultaneously, Zhang et al. (2019) also constructed multi-gRNAs binary construct carrying all the three sgRNAs with the potential to mutate all the three exons in all the *BnLPAT2* paralogs simultaneously. However, in such cases, multiplex editing efficiency can be affected by the position as well as the number of gRNAs, as in the case of *BnLPAT2* mutation, wherein, the cleavage activity of the gRNA targeting the third exon was inactive when it was placed at the third position in the binary vector construct, while it was active in the second position as well as individually. Cas9-based functional analysis of *BnLPAT* genes showed that the oil traits like oil contents and the morphology of oil bodies in rapeseed were affected by all the copies of both the *BnLPAT2* and *BnLPAT5* genes, as a decrease in oil content was observed in *Bnlpat2, Bnlpat5,* and *Bnlpat2/Bnlpat5* double mutant lines by 32%, 29%, and 39% respectively as well as enlarged oil bodies were observed in the double mutant seeds as compared to the wild-type under the transmission electron microscopy (Zhang et al. 2019). Overall, CRISPR/Cas9 based editing of LPAT enzymes offered a base for understanding several parameters affecting the Cas9 based mutagenesis in Brassica crops, as well as in revealing gene functions and generating agronomically important mutations in crops.

25.2.9 Exploring Multiplex Editing for Yield, Quality, and Nutritional Improvement

Among the numerous factors affecting the rapeseed yield, plant height and branch numbers are essential to plant architecture components that affect the yield of rapeseed crops directly. Strigolactones (SL), a phytohormones cum signal molecule has been functionally validated as a signal molecule in inhibiting the branching mechanism of plants (Umehara et al. 2008). Genes involved in SL biosynthesis are *MORE AXILLARY GROWTH 3* (*MAX3*), *MAX4,* and *MAX1* encoding carotenoid cleavage dioxygenases (CCD7 and CCD8) and cytochrome P450 monooxygenase (CYP711A1), respectively (Booker et al. 2005). Zheng et al. in (2020), established the *MAX1*genebased regulation of branching in rapeseed by targeting two orthologs of the Arabidopsis *MAX1* gene in *B. napus* (*BnMAX1*) each from 'A' and 'C' genomes, using the CRISPR/Cas9 technology. By employing two sgRNAs with less than 20nt oligos, high editing efficiencies in *BnMAX1* homologous were observed. Moreover, the replacement of A with G in the sgRNA targeting the *BnMAX1* gene, to suit the U6 promoter requirement, resulted in high editing levels (56.30–63.87%) in *B. napus*. In *BnMAX1* mutated lines, certain uncommon patterns were observed like the occurrence of NHEJ-mediated insertions at 4–6 bp upstream of the PAM unlike the usual occurrence of indels at 3–4 bp adjacent to the PAM (Lawrenson et al. 2015) and the insertion of Gs in most of the *BnMAX1* alleles instead of the common NHEJ-mediated insertions like As or Ts in previous studies (Ma et al. 2015). Zheng et al., believed that in *BnMAX1* mutated rapeseeds, mutations were due to both the NHEJ and HDR mechanisms, as the T0 plants observed a high rate of homozygous mutations (3.36–18.44%) in *BnMAX1* homologs simultaneously with the same mutation type. The mutations in both the *BnMAX1* homologs inherited stably and independently of the T-DNA construct in their self-pollinated T1 and T2 progenies. Interestingly, only those T2 lines with biallelic or homozygous mutations which resulted in premature stop codon in both the sub-genome copies of *BnMAX1* gene, had displayed the Arabidopsis *max1* mutant phenotype. While the lines with insertion in the A genome homolog and an amino acid deletion in the C genome homologs displayed normal phenotype similar to the wild-type. Overall, knockout of the two *BnaMAX1* homologs resulted in semi-dwarf, increased branching phenotypes, increased silique number, and yield per plant in rapeseed crop (Zheng et al.

2020). Hence, Cas9-induced mutations led to the development of a trait-oriented rapeseed ideotype as well as desirable germplasm which will facilitate further breeding for high yields in rapeseed.

Yellow seed color is the widely accepted quality trait in *B. napus* and is among the other breeding goals of rapeseed production (Hong et al. 2017). It is known from *A. thaliana* that, an MBW transcription factor complex (AtMYB123/TT2, basic helix-loop-helix/TT8, and WD40/TTG1) activates and regulates the proanthocyanidins pathway-specific genes involved in the seed coat color (Xu et al. 2014). Among these MBW TFs complex, TT8 is a central component responsible for flavonoid deposition in various crops and is also supported by the genetic studies in *Brassica* species. In *B. juncea,* the yellow seed color trait is due to the natural mutations in the two homologs of *TT8* genes (*BjuA.TT8* and *BjuB.TT8*) or due to an insertion of a transposable element in the intron of the *TT8* gene in *B. rapa* (Li et al. 2012; Padmaja et al. 2014). In *B. napus*, yellow seed genotypes so far have been developed via time-consuming and inefficient interspecific or intergeneric hybridizations (Li et al. 2012; Wang et al. 2005; Wen et al. 2012).

Zhai et al. (2020) used CRISPR/Cas9 to develop yellow-seeded mutants of rapeseed by targeting the *TT8* gene. *B. napus* contain three *TT8* homologs, one from the A genome and two tandem duplicates from the C genome. Based on the expression analysis and the presence or absence of the functional domains required for the TT8 activity, Zhai et al. targeted two of the three *TT8* homologs by designing four conserved sgRNAs against the four functional domains. As previously observed by Sun et al. in 2018, there were no transgenic plants with a mutation in the *TT8* domain targeted by the AtU6-1 driven sgRNA, suggesting that sgRNAs driven by AtU6-1 promoter are non-functional in rapeseed. However, multi-domain targeting with different gRNAs enabled the research groups to obtain yellow-seeded phenotype in the T0 generation itself, as in previous studies (Sun et al. 2018). The visible yellow-seeded knockout phenotype was only recovered after targeted mutations in both the functional copies of the *BnTT8* gene, which further emphasized that the two copies have redundant functions in rapeseed seed color formation. The P_{35} driven Cas9-based mutations in both the *TT8* copies were stably transmitted to the advanced generations (T1, T2, and T3), and their targeted mutations were verified by high-throughput tracking of mutation (Hi-TOM) sequencing analysis (Zhai et al. 2020). The oil contents in double mutant lines were increased by 9.47%, 5.89%, and 5.90% relative to the wild-type seeds from T0 to T3 generations, respectively. Simultaneously protein contents in T0, T2, and T3 double mutant lines had increased to 18.01%, 21.97%, and 20.21% respectively as compared to that of the wild-type plants (15–19%). Changes in the fatty acid profiles like increases in the palmitic acid (C16:0), linoleic acid (C18:2), and linolenic acid (C18:3) and decreases in the stearic acid (C18:0) and oleic acid (C18:1) contents were also observed in the *BnTT8* double mutant lines. Thus, simultaneously targeted mutations of *BnTT8* copies conferred a high oil yield potential with modified fatty acid compositions and improved the nutritional quality in the allotetraploid plant species.

Another research work in *B. napus* by Sashidhar et al. (2020) had demonstrated the importance of multiple paralog knockout using CRISPR/Cas9 tool. The study reduced the levels of an anti-nutritive as well as a detrimental element called Phytic acid (PA), a major phosphorous source in the seeds of cereals and oilseeds by targeting its biosynthesis enzyme, inositol tetrakisphosphate kinase (ITPK) in *B. napus*. In rapeseed plants, the content of PA ranges from 2 to 5% in Canola varieties, where its reduction so far has been carried out by either adding phytases to the seed meal or by other technical processes (El-Batal and Karem 2001; Dersjant-Li et al. 2015). Based on the previous studies and the phylogenetic analysis between the homologs of *BnITPKs* and other plants, Sashidhar et al. (2020) designed two conserved sgRNAs against two different sites in the *BnITPK1* and *BnITPK4* paralogs, out of the four homologs (*BnITPK1* to *BnITPK4*). Targeted and multiplexed editing of different paralogs of both *BnITPK1* and

BnITPK4 genes simultaneously resulted in 35% reduction of phytic acid in rapeseed. Additionally, de novo mutations were observed in the advanced generations (T2) of *BnITPKs* mutated lines that had not been found in their parents (T1), similarly as experienced in the previous study of CRISPR-based mutagenesis of *WRKY* genes in *B. napus* (Sun et al. 2018). Sashidhar et al. (2020) suggested that low Cas9 activity in T0 was due to the position of the T-DNA in the genome or due to multiple T-DNA insertions leading to gene silencing. Lines bearing a triple mutation in different *BnITPK1* alleles only displayed a significant decrease (27.2–35.3%) in the phytic acid content with a threefold increase of inorganic phosphorous in the rapeseed seeds, than the wild-type plants while the reduction in double mutants was not found to be significant. Together, it shows the indispensable role and potentials of CRISPR/Cas9 mutagenesis in crops where multiple paralogs function redundantly towards a trait.

25.2.10 Cas9-Based Modulation of Pathogenesis Factor in Rapeseed

CRISPR/Cas9 based validation of pathogenesis factors for the breeding of pathogen-resistant Brassicaceae species is very rare. Recently in 2020, by deploying suppression subtractive hybridization (SSH) library and further validation in the corresponding Arabidopsis T-DNA knockout lines, a novel *CRT1a* gene was identified in *A. thaliana* as a novel factor involved in the establishment of successful host colonization by a hemibiotrophic fungal pathogen *Verticillium longisporum* known to cause Verticillium stem stripping in rapeseed (Pröbsting et al. 2020). Database search and the expression analysis of the *CRT1a*cDNA in both infected and mock-treated rapeseed roots revealed four distinct *CRT1a* copies in the *B. napus* genome. A codon-optimized Cas9 for *B. napus* generated stable mutations in the *BnCRT1a* copies, as confirmed by the 'derived cleaved amplified polymorphic sequence' (dCAPS) assay, and mutations were stably transmitted from T0 to T2 generation (Pröbsting et al. 2020). Additionally, the Cas9 based functional analysis of *BnCRT1a* genes displayed that, the factor that attributes to the *B. napus–V. longisporum* interaction is from the A genome rather than from the C genome of rapeseed. Lines bearing a frameshift mutation in the *BnCRT1a* showed a complete knockout phenotype as decreased disease symptoms and less stunting compared to the control, even under a high infection pressure by *V. longisporum*. This suggested that the CRISPR-based modulation of *BnCRT1a* homologs could be a promising strategy to develop *V. longisporum* resistant lines in *B. napus*. Further, transgene-free lines were obtained in the T2 generation after PCR analysis, and no negative effects were observed on the agronomical traits except for slight growth suppression in these mutated lines. Thus, until now this was the most comprehensive work allowing for an efficient generation of resistance for the first time against a specialist pathogen in the rapeseed plants.

Thus far, the chapter discussed the most pressing case studies that dealt with the CRISPR/Cas9 utilities in the crops of the genus *Brassica*. Besides these, two recent studies used the multiplex ability of CRISPR/Cas9 to develop pod shattering resistance in *B. napus* by targeting multiple homologs of *INDEHISCENT* and *JAGGED* genes (Zhai et al. 2019; Zaman et al. 2019). In addition, two other research works showed the advancements made in the CRISPR/Cas9 technology with respect to transformation protocol and multiplex editing, by employing hairy root transformation protocol in *B. carinata* and endogenous tRNA-processing system in *B. oleracea* respectively (Kirchner et al. 2017; Ma et al. 2019). CRISPR/Cas9 based genome editing had successfully been utilized to modify traits such as yield, quality, and biotic and abiotic stress resistance in a few of the cultivated *Brassica* species (Fig. 25.3), however, its application is yet to be utilized in the other U's triangle species of the *Brassica*genus like *B. nigra* and *B. juncea*. Altogether if these findings are considered it can be said that the CRISPR/Cas9 technology has the potential to give an unparalleled productivity rate compared to the conventional breeding tools in economically valuable Brassica crops and thus if utilized well it can enable us to meet the future demands for oilseed productions boosting the Indian agriculture.

25.3 CRISPR and the Indian Mustard Crop, *Brassica juncea*

With 15 reports on the utilization of CRISPR/Cas9 in Brassica crops, discussed in this chapter, a proof of concept has not yet been published in the Indian mustard, *B. juncea*. Since this book deals with different concepts in *B. juncea*, like the genetic diversity and population structure, its transcriptomics, metabolomics, and proteomics studies, unfortunately, there is no single study describing the use of CRISPR/Cas9 technology for targeted gene modification in the said crop. *B. juncea* is an allotetraploid crop belonging to the U's triangle formed after the hybridization of two progenitor species *B. rapa* and *B. nigra* which constitute the A and B subgenomes respectively (Nagaharu 1935). *B. juncea* is the major oilseed crop of the Indian subcontinent which is used for both oilseed production as well as for consumption as a vegetable crop in different parts of India and across the world. It is more tolerant to low moisture conditions, and it has also been used as a replacement for *B. napus* in the western Canadian prairies (Woods et al. 1991). It is more productive than rapeseed cultivation in areas of unreliable rainfalls as it grows well under both the irrigated and rainfed conditions. And India despite being the producer of mustard still meets more than half of its requirements through imports from countries like China, Canada, and others (Jat et al. 2019). The productivity of *B. juncea* is hampered due to the requirement of large cultivation areas and the abiotic and biotic challenges this crop faces during its life cycle.

Although CRISPR/Cas9 has not been used to create mutations in *B. juncea* and few other crops of the genus *Brassica*, there still exist several open possibilities for fulfilling the breeding objectives of the Indian mustard. The genetic improvement of *B. juncea* had so far been carried out using other approaches like heterosis breeding, in vitro breeding, somaclonal variation, and others (Pradhan et al. 1993; Nehnevajova et al. 2007; Meena et al. 2015). The commercial Canola quality lines of *B. juncea* are being bred for low erucic acids (<2% of the total fatty acid pool), low seed glucosinolates (<30 μmoles/g dry weight), and fatty acid composition enhancements for increasing the oleic acid levels in seeds (>60% of the total fatty acid pool) to match the *B. napus* level. Enhancement of anti-cancerous glucosinolates (glucoraphanin), blackleg and *Pseudomonas* leaf blight tolerance, early maturing, short stature, high yielding varieties which out-perform *B. napus* in low rainfall environments are also being bred (Burton et al. 2004; Augustine et al. 2013; Pandey et al. 2013; Augustine and Bisht 2015; Inturrisi et al. 2021). Despite these, there are a number of other outlooks which are being researched for the development of superior *B. juncea* cultivars like the reduction of glucosinolates in the seeds via targeting the glucosinolate transporters for not only enhancing the seed meal quality but simultaneously retaining glucosinolates in the leaf for better tolerance against pest and pathogen-induced damages, reducing of anti-nutritive compound sinapine from seeds, and many more. This increasing effort to improve the nutritional traits of *B. juncea* seed meal to overcome the dependence on the expensive soybean imports in regions where it cannot be grown can be promising enough. In addition, a number of parameters of biotic stresses which continuously threaten and impact the productivity of *B. juncea* causing its seed quality loss should also be the concerns which are needed to be dealt. These include blackleg by *Leptosphaeria maculans*, *Sclerotinia* stem rot by *Sclerotinia sclerotiorum*, alternaria blight by *Alternaria brassicae*, powdery mildew by *Erysiphe cruciferarum*, white rust by *Albugo candida*, bacterial leaf blight by *Pseudomonas syringae*, and several others (Inturrisi et al. 2021). Thus, all these avenues remain open for the employment of CRISPR/Cas mutagenesis for targeting several traits for functional gene analysis and genetic improvements of the oilseed *B. juncea* in the near future (Fig. 25.3).

25.4 Discussion and Conclusion

The CRISPR/Cas 9 system has been utilized for only half a decade until now for the precise genome editing of Brassica crops, to infer gene functions, breeding of agriculturally important traits for higher yield and quality, and pathogen resistance. Beyond reasonable doubt, the use of CRISPR/Cas9 technology will be precise, less time-consuming, and will help in the effective generation of simultaneous mutagenesis in the allotetraploid *B. juncea* compared with other genetic modification approaches. More profound research on enhancing the ability of CRISPR vectors, precise sgRNA design, and rapid screening procedures will speed up its applications for plant species where it has not been applied yet. The use of CRISPR until now for targeting different traits across the genus *Brassica* is still in its infancy, where the future holds several opportunities like the knockout of redundant genes and parallel pathways for the improvements of these *Brassica* species.

Acknowledgements The authors acknowledge the scientific community who have contributed and published their work to establish CRISPR-based editing in agronomically valued polyploid Brassica crops. We also appreciate the funding from the Department of Biotechnology, India for CRISPR-based research at NCB Laboratory. MV and BM sincerely acknowledge the financial assistance from CSIR, India.

Author Contributions All the authors have a significant intellectual contribution to the writing and editing of the manuscript.

Conflict of Interest The authors declare that they have no conflict of interest.

References

Angkawijaya AE, Nguyen VC, Nakamura Y (2017) Enhanced root growth in phosphate-starved Arabidopsis by stimulating de novo phospholipid biosynthesis through the overexpression of lysophosphatidic acid acyltransferase 2 (LPAT2). Plant Cell Environ 40 (9):1807–1818

Arnaud N, Girin T, Sorefan K, Fuentes S, Wood TA et al (2010) Gibberellins control fruit patterning in *Arabidopsis thaliana*. Genes Dev 24(19):2127–2132

Augustine R, Bisht NC (2015) Biofortification of oilseed *Brassica juncea* with the anti-cancer compound glucoraphanin by suppressing *GSL-ALK* gene family. Sci Rep 5:18005

Augustine R, Mukhopadhyay A, Bisht NC (2013) Targeted silencing of BjMYB28 transcription factor gene directs development of low glucosinolate lines in oilseed *Brassica juncea*. Plant Biotechnol J 11 (7):855–866

Baker M (2012) Method of the year 2011. Nat Methods 9 (1)

Barrangou R, Fremaux C, Deveau H, Richards M, Boyaval P et al (2007) CRISPR provides acquired resistance against viruses in prokaryotes. Science 315 (5819):1709–1712

Bolotin A, Quinquis B, Sorokin A, Ehrlich SD (2005) Clustered regularly interspaced short palindrome repeats (CRISPRs) have spacers of extrachromosomal origin. Microbiology 151(8):2551–2561

Booker J, Sieberer T, Wright W, Williamson L, Willett B et al (2005) MAX1 encodes a cytochrome P450 family member that acts downstream of MAX3/4 to produce a carotenoid-derived branch-inhibiting hormone. Dev Cell 8:443–449

Braatz J, Harloff HJ, Mascher M, Stein N, Himmelbach A, Jung C (2017) CRISPR-Cas9 targeted mutagenesis leads to simultaneous modification of different homoeologous gene copies in polyploid oilseed rape (*Brassica napus*). Plant Physiol 174(2):935–942

Burton WA, Ripley VL, Potts DA, Salisbury PA (2004) Assessment of genetic diversity in selected breeding lines and cultivars of canola quality *Brassica juncea* and their implications for canola breeding. Euphytica 136(2):181–192

Čermák T, Baltes NJ, Čegan R, Zhang Y, Voytas DF (2015) High-frequency, precise modification of the tomato genome. Geno Biol 16:232

Charpentier E, van der Oost J, White MF (2013) crRNA biogenesis. CRISPR-Cas systems. Springer, Berlin, Heidelberg, pp 115–144

Chen H, Choi J, Bailey S (2014) Cut site selection by the two nuclease domains of the Cas9 RNA-guided endonuclease. J Biol Chem 289(19):13284–13294

Chen GQ, van Erp H, Martin-Moreno J, Johnson K, Morales E et al (2016) Expression of castor LPAT2 enhances ricinoleic acid content at the sn-2 position of Triacylglycerols in *Lesquerella* seed. Intl J Mol Sci 17 (4):507

Chen K, Wang Y, Zhang R, Zhang H, Gao C (2019) CRISPR/Cas genome editing and precision plant breeding in agriculture. Annu Rev Plant Biol 70:667–697

Cheng F, Sun R, Hou X, Zheng H, Zhang F et al (2016) Subgenome parallel selection is associated with morphotype diversification and convergent crop domestication in *Brassica rapa* and *Brassica oleracea*. Nat Genet 48(10):1218–1224

Clark SE, Running MP, Meyerowitz EM (1993) CLAVATA1, a regulator of meristem and flower development in Arabidopsis. Development 119(2):397–418

Deltcheva E, Chylinski K, Sharma CM, Gonzales K, Chao Y et al (2011) CRISPR RNA maturation by trans-encoded small RNA and host factor RNase III. Nature 471(7340):602–607

Dersjant-Li Y, Awati A, Schulze H, Partridge G (2015) Phytase in non-ruminant animal nutrition: a critical review on phytase activities in the gastrointestinal tract and influencing factors. J Sci Food Agri 95(5):878–896

El-Batal AI, Karem HA (2001) Phytase production and phytic acid reduction in rapeseed meal by *Aspergillus niger* during solid state fermentation. Food Res Inter 34(8):715–720

Emrani N, Harloff HJ, Gudi O, Kopisch-Obuch F, Jung C (2015) Reduction in sinapine content in rapeseed (*Brassica napus* L.) by induced mutations in sinapine biosynthesis genes. Mol Breed 35(1):1–11

Feng Z, Mao Y, Xu N, Zhang B, Wei P et al (2014) Multigeneration analysis reveals the inheritance, specificity, and patterns of CRISPR/Cas-induced gene modifications in Arabidopsis. Proc Nat Acad Sci USA 111(12):4632–4637

Gleditzsch D, Pausch P, Müller-Esparza H, Özcan A, Guo X, Bange G, Randau L (2019) PAM identification by CRISPR-Cas effector complexes: diversified mechanisms and structures. RNA Biol 16(4):504–517

He YT, Long WH, Hu JP, Fu TD, Li DR, Chen BY, Tu JX (2003) Anatomic and genetic studies on multicapsular character in *Brassica campestris* L. Chin J Oil Crop Sci 25:1–4

Hong M, Hu K, Tian T, Li X, Chen L et al (2017) Transcriptomic analysis of seed coats in yellow-seeded *Brassica napus* reveals novel genes that influence proanthocyanidin biosynthesis. Front Plant Sci 5(8):1674

Inturrisi FC, Barbetti MJ, Tirnaz S, Patel DA, Edwards D, Batley J (2021) Molecular characterization of disease resistance in *Brassica juncea*—the current status and the way forward. Plant Pathol 70(1):13–34

Ishino Y, Shinagawa H, Makino K, Amemura M, Nakata A (1987) Nucleotide sequence of the *IAP* gene, responsible for alkaline phosphatase isozyme conversion in *Escherichia coli*, and identification of the gene product. J Bacteriol 169(12):5429–5433

Jat RS, Singh VV, Sharma P, Rai PK (2019) Oilseed Brassica in India: demand, supply, policy perspective and future potential. OCL 26:8

Jiang W, Zhou H, Bi H, Fromm M, Yang B, Weeks DP (2013) Demonstration of CRISPR/Cas9sgRNA-mediated targeted gene modification in Arabidopsis, tobacco, sorghum, and rice. Nucl Acids Res 41(20): e188

Kirchner TW, Niehaus M, Debener T, Schenk MK, Herde M (2017) Efficient generation of mutations mediated by CRISPR/Cas9 in the hairy root transformation system of *Brassica carinata*. PLoS One 12(9): e0185429

Kleinstiver BP, Pattanayak V, Prew MS, Tsai SQ, Nguyen NT, Zheng Z, Joung JK (2016) High-fidelity CRISPR–Cas9 nucleases with no detectable genome-wide off-target effects. Nature 529(7587):490–495

Lawrenson T, Shorinola O, Stacey N, Li C, Østergaard L, Patron N et al (2015) Induction of targeted, heritable mutations in barley and *Brassica oleracea* using RNA-guided Cas9 nuclease. Geno Biol 16(1):1–13

Li J, Brader G, Palva ET (2004) The WRKY70 transcription factor: a node of convergence for jasmonate-mediated and salicylate-mediated signals in plant defense. Plant Cell 16(2):319–331

Li X, Chen L, Hong M, Zhang Y, Zu F et al. (2012) A large insertion in bHLH transcription factor BrTT8 resulting in yellow seed coat in *Brassica rapa*. PLoS One 7(9):e44145

Li C, Hao M, Wang W, Wang H, Chen F et al (2018) An efficient CRISPR/Cas9 platform for rapidly generating simultaneous mutagenesis of multiple gene homoeologs in allotetraploid oilseed rape. Front Plant Sci 9:442

Liljegren SJ, Roeder AHK, Kempin SA, Gremski K, Østergaard L et al (2004) Control of fruit patterning in Arabidopsis by indehiscent. Cell 116:843–853

Ma X, Zhang Q, Zhu Q, Liu W, Chen Y et al (2015) A robust CRISPR/Cas9 system for convenient, high-efficiency multiplex genome editing in monocot and dicot plants. Mol Plant 8(8):1274–1284

Ma C, Zhu C, Zheng M, Liu M, Zhang D et al (2019) CRISPR/Cas9-mediated multiple gene editing in *Brassica oleracea* var. capitata using the endogenous tRNA-processing system. Hortic Res 6:20

Manghwar H, Lindsey K, Zhang X, Jin S (2019) CRISPR/Cas system: recent advances and future prospects for genome editing. Trends Plant Sci 24 (12):1102–1125

Meena HS, Kumar A, Ram B, Singh VV, Meena PD, Singh BK, Singh D (2015) Combining ability and heterosis for seed yield and its components in Indian mustard (*Brassica juncea* L.).J Agric Sci Tech 17:1861–1871

Miao J, Guo D, Zhang J, Huang Q, Qin G et al (2013) Targeted mutagenesis in rice using CRISPR-Cas system. Cell Res 23(10):1233–1236

Mojica FJ, Díez-Villaseñor C, Soria E, Juez G (2000) Biological significance of a family of regularly spaced repeats in the genomes of Archaea, bacteria and mitochondria. Mol Microbiol 36(1):244–246

Mojica FJ, Díez-Villaseñor C, García-Martínez J, Almendros C (2009) Short motif sequences determine the targets of the prokaryotic CRISPR defence system. Microbiology 155(3):733–740

Murovec J, Guček K, Bohanec B, Avbelj M, Jerala R (2018) DNA-free genome editing of *Brassica oleracea* and *B. rapa* protoplasts using CRISPR-Cas9 ribonucleoprotein complexes. Front Plant Sci 9:1594

Nagaharu U (1935) Genome analysis in Brassica with special reference to the experimental formation of *B. napus* and peculiar mode of fertilization. Jpn J Bot 7 (7):389–452

Nehnevajova E, Herzig R, Erismann KH, Schwitzguébel JP (2007) *In vitro* breeding of *Brassica juncea* L. to enhance metal accumulation and extraction properties. Plant Cell Rep 26(4):429–437

Nishimasu H, Ran FA, Hsu PD, Konermann S, Shehata SI, Dohmae N et al (2014) Crystal structure of Cas9 in complex with guide RNA and target DNA. Cell 156 (5):935–949

Okuzaki A, Ogawa T, Koizuka C, Kaneko K, Inaba M, Imamura J, Koizuka N (2018) CRISPR/Cas9-mediated genome editing of the fatty acid desaturase 2 gene in *Brassica napus*. Plant Physiol Biochem 131:63–69

Padmaja LK, Agarwal P, Gupta V, Mukhopadhyay A, Sodhi YS, Pental D, Pradhan AK (2014) Natural mutations in two homoeologous *TT8* genes control yellow seed coat trait in allotetraploid *Brassica juncea* (AABB). Theor Appl Genet 127:339–347

Pan C, Ye L, Qin L, Liu X, He Y, Wang J et al (2016) CRISPR/Cas9-mediated efficient and heritable targeted mutagenesis in tomato plants in the first and later generations. Sci Rep 6(1):1–9

Pandey S, Kabdal M, Tripathi MK (2013) Study of inheritance of erucic acid in Indian mustard (*Brassica juncea* L. Czern&Coss). Octa J Biosci 1(1):77–84

Pradhan AK, Sodhi YS, Mukhopadhyay A, Pental D (1993) Heterosis breeding in Indian mustard (*Brassica juncea* L. Czern&Coss): analysis of component characters contributing to heterosis for yield. Euphytica 69 (3):219–229

Pröbsting M, Schenke D, Hossain R, Häder C, Thurau T et al (2020) Loss of function of CRT1a (calreticulin) reduces plant susceptibility to *Verticillium longisporum* in both *Arabidopsis thaliana* and oilseed rape (*Brassica napus*). Plant Biotechnol J 18:2328–2344

Punetha H, Saif NS, Dinesh P (2020) Glucosinolates in oilseed Brassica: neutraceuticals with tremendous health benefits. Biotech Today: Intern J Biol Sci 9 (2):26–32

Rajani S, Sundaresan V (2001) The Arabidopsis myc/bHLH gene ALCATRAZ enables cell separation in fruit dehiscence. Curr Biol 11:1914–1922

Salsman J, Dellaire G (2016) Precision genome editing in the CRISPR era. Biochem Cell Biol 95:187–201

Sapranauskas R, Gasiunas G, Fremaux C, Barrangou R, Horvath P, Siksnys V (2011) The *Streptococcus thermophilus* CRISPR/Cas system provides immunity in *Escherichia coli*. Nucl Acids Res 39:gkr606–gkr9282

Sashidhar N, Harloff HJ, Potgieter L, Jung C (2020) Gene editing of three *BnITPK* genes in tetraploid oilseed rape leads to significant reduction of phytic acid in seeds. Plant Biotechnol J 18(11):2241–2250

Shan Q, Wang Y, Li J, Zhang Y, Chen K et al (2013) Targeted genome modification of crop plants using a CRISPR-Cas system. Nature Biotechnol 31(8):686–688

Shan Q, Wang Y, Li J, Gao C (2014) Genome editing in rice and wheat using the CRISPR/Cas system. Nat Protoc 9(10):2395–2410

Sun Q, Lin L, Liu D, Wu D, Fang Y, Wu J, Wang Y (2018) CRISPR/Cas9-mediated multiplex genome editing of the *BnWRKY11* and *BnWRKY70* genes in *Brassica napus* L. Intl J Mol Sci 19(9):2716

Symington LS, Gautier J (2011) Double-strand break end resection and repair pathway choice. Annu Rev Genet 45:247–271

Terada R, Urawa H, Inagaki Y, Tsugane K, Iida S (2002) Efficient gene targeting by homologous recombination in rice. Nat Biotechnol 20:1030–1034

Travis J (2015) Making the cut. Science 18:350 (6267):1456–1457

Umehara M, Hanada A, Yoshida S, Akiyama K, Arite T et al (2008) Inhibition of shoot branching by new terpenoid plant hormones. Nature 455:195–200

Uniyal AP, Mansotra K, Yadav SK, Kumar V (2019) An overview of designing and selection of sgRNAs for precise genome editing by the CRISPR-Cas9 system in plants. 3 Biotechnology 9(6):223

Upadhyay SK, Kumar J, Alok A, Tuli R (2013) RNA-guided genome editing for target gene mutations in wheat. G3 (Bethesda) 3(12):2233–2238

USDA: Crop Production (2021) Summary. February 2021. ISSN: 1936-3737

Walton RT, Christie KA, Whittaker MN, Kleinstiver BP (2020) Unconstrained genome targeting with near-PAMless engineered CRISPR-Cas9 variants. Science (new York NY) 368(6488):290–296

Wang YP, Sonntag K, Rudloff E, Chen JM (2005) Intergeneric somatic hybridization between *Brassica napus* L. and *Sinapis alba* L. J Integr Plant Biol 47:84–91

Wang Y, Cheng X, Shan Q, Zhang Y, Liu J, Gao C, Qiu JL (2014a) Simultaneous editing of three homoeoalleles in hexaploid bread wheat confers heritable resistance to powdery mildew. Nat Biotechnol 32(9):947–951

Wang Z, Fang H, Chen Y, Chen K, Li G, Gu S, Tan X (2014b) Overexpression of *BnWRKY33* in oilseed rape enhances resistance to *Sclerotinia sclerotiorum*. Mol Plant Pathol 15(7):677–689

Wells R, Trick M, Soumpourou E, Clissold L, Morgan C et al (2014) The control of seed oil polyunsaturated content in the polyploid crop species *Brassica napus*. Mol Breed 33:349–362

Wen J, Zhu L, Qi L, Ke H, Yi B et al (2012) Characterization of interploid hybrids from crosses between *Brassica juncea* and *B. oleracea* and the production of yellow-seeded *B. napus*. Theor Appl Genet 125:19–32

Wolt JD, Wang K, Yang B (2016) The regulatory status of genome-edited crops. Plant Biotechnol J 14:510–518

Wolter F, Schindele P, Puchta H (2019) Plant breeding at the speed of light: the power of CRISPR/Cas to generate directed genetic diversity at multiple sites. BMC Plant Biol 19(1):1–8

Woods DL, Capcara JJ, Downey RK (1991) The potential of mustard (*Brassica juncea* (L.) Coss) as an edible oil crop on the Canadian prairies. Can J Plant Sci 71:195–198

Wu J, Zhao Q, Yang Q, Liu H, Li Q et al (2016) Comparative transcriptomic analysis uncovers the complex genetic network for resistance to *Sclerotinia sclerotiorum* in *Brassica napus*. Sci Rep 6:19007

Xie K, Yang Y (2013) RNA-guided genome editing in plants using a CRISPR-Cas system. Mol Plant 6 (6):1975–1983

Xu W, Grain D, Bobet S, Le Gourrierec J, Thévenin J et al (2014) Complexity and robustness of the flavonoid transcriptional regulatory network revealed by comprehensive analyses of MYB-bHLH-WDR complexes and their targets in Arabidopsis seed. New Phytol 202(1):132–144

Yang Q, Fan C, Gua Z, Qin J, Wu J, Li Q, Fu T, Zhou Y (2012) Identification of *FAD2* and *FAD3* genes in *Brassica napus* genome and development of allele-specific markers for high oleic and low linoleic acid contents. Theor Appl Genet 125:715–729

Yang Y, Zhu K, Li H, Han S, Meng Q, Khan S U, Fan C, Xie K, Zhou Y (2018) Precise editing of CLAVATA genes in *Brassica napus* L. regulates multilocular silique development. Plant Biotechnol J 16(7):1322–1335

Yang H, Wu JJ, Tang T, Liu KD, Dai C (2017) CRISPR/Cas9-mediated genome editing efficiently creates specific mutations at multiple loci using one sgRNA in *Brassica napus*. Sci Rep 7(1):7489

Zaman QU, Chu W, Hao M, Shi Y, Sun M et al (2019) CRISPR/Cas9-mediated multiplex genome editing of JAGGED gene in *Brassica napus* L. Biomolecules 9 (11):725

Ze-wen LV, Ping XU, Xiang-xiang Z, Bin YI, Chao-zhi MA, Ting-dong FU (2012) Primary study on anatomic and genetic characteristics of multi-loculus in *Brassica juncea*. Chin J Oil Crop Sci 34(5)

Zhai Y, Cai S, Hu L, Yang Y, Amoo O, Fan C, Zhou Y (2019) CRISPR/Cas9-mediated genome editing reveals differences in the contribution of INDEHISCENT homologues to pod shatter resistance in *Brassica napus* L. Theor App Genet 132(7):2111–2123

Zhai Y, Yu K, Cai S, Hu L, Amoo O et al (2020) Targeted mutagenesis of BnTT8 homologs controls yellow seed coat development for effective oil production in *Brassica napus* L. Plant Biotechnol J 18(5):1153–1168

Zhang K, Nie L, Cheng Q, Yin Y, Chen K et al (2019) Effective editing for lysophosphatidic acid acyltransferase 2/5 in allotetraploid rapeseed (*Brassica napus* L.) using CRISPR-Cas9 system. Biotechnol Biofuels 12(1):1–18

Zheng M, Zhang L, Tang M, Liu J, Liu H et al (2020) Knockout of two *BnaMAX1* homologs by CRISPR/Cas9-targeted mutagenesis improves plant architecture and increases yield in rapeseed (*Brassica napus* L.). Plant Biotechnol J 18(3):644–654

Zhu X, Xu Y, Yu S, Lu L, Ding M et al (2014) An efficient genotyping method for genome-modified animals and human cells generated with CRISPR/Cas9 system. Sci Rep 4:6420

The manufacturer's authorised representative in the EU is Springer Nature Customer Service Centre GmbH, Europaplatz 3, 69115 Heidelberg, Germany. If you have any concerns regarding our products, please contact ProductSafety@springernature.com

Printed and bound by CPI Group (UK) Ltd, Croydon, CR0 4YY
30/09/2026
02225496-0020